Aqueous Pretreatment of Plant Biomass for Biological and Chemical Conversion to Fuels and Chemicals

Wiley Series in Renewable Resources

Series Editor

Christian V. Stevens – Faculty of Bioscience Engineering, Ghent University, Ghent, Belgium

Titles in the Series

Forthcoming Titles

Aqueous Pretreatment of Plant Biomass for Biological and Chemical Conversion to Fuels and Chemicals

Editor

CHARLES E. WYMAN

Department of Chemical and Environmental Engineering and Center for
Environmental Research and Technology, University of California, Riverside, USA
and
BioEnergy Science Center, Oak Ridge, USA

WILEY

Contents

8 Fundamentals of Biomass Pretreatment at High pH   **145**
Rocío Sierra Ramirez, Mark Holtzapple and Natalia Piamonte

9 Primer on Ammonia Fiber Expansion Pretreatment   **169**
S.P.S. Chundawat, B. Bals, T. Campbell, L. Sousa, D. Gao, M. Jin, P. Eranki, R. Garlock, F. Teymouri, V. Balan and B.E. Dale

List of Contributors

Andy Aden URS Corporation, Denver, USA (Previously at National Renewable Energy Laboratory, Golden, USA)

Foster A. Agblevor Department of Biological Engineering, Utah State University, Logan, USA

David Martin Alonso Department of Chemical and Biological Engineering, University of Wisconsin, Madison, USA

Venkatesh Balan Department of Chemical Engineering and Materials Science and Great Lakes Bioenergy Research Center, Michigan State University, East Lansing, USA

B. Bals Department of Chemical Engineering and Materials Science and Great Lakes Bioenergy Research Center, Michigan State University, East Lansing, USA

Jesse Q. Bond Biomedical and Chemical Engineering, Syracuse University, Syracuse, USA

T. Campbell Michigan Biotechnology Institute, Lansing, USA

S.P.S. Chundawat Department of Chemical Engineering and Materials Science and Great Lakes Bioenergy Research Center, Michigan State University, East Lansing, USA

Bruce E. Dale Department of Chemical Engineering and Materials Science and Great Lakes Bioenergy Research Center, Michigan State University, East Lansing, USA

Mark F. Davis National Renewable Energy Laboratory, Golden and BioEnergy Science Center, Oak Ridge, USA

Brian H. Davison Oak Ridge National Laboratory and BioEnergy Science Center, Oak Ridge, USA

Jaclyn D. DeMartini DuPont Industrial Biosciences, Palo Alto, USA (Previously at Department of Chemical and Environmental Engineering and Center for Environmental Research and Technology, University of California, Riverside and BioEnergy Science Center, Oak Ridge, USA)

Byron S. Donohoe National Renewable Energy Laboratory, Golden and BioEnergy Science Center, Oak Ridge, USA

James A. Dumesic Department of Chemical and Biological Engineering, University of Wisconsin, Madison, USA

Richard T. Elander National Renewable Energy Laboratory, Golden, USA

P. Eranki Department of Chemical Engineering and Materials Science and Great Lakes Bioenergy Research Center, Michigan State University, East Lansing, USA

D. Gao Department of Chemical Engineering and Materials Science and Great Lakes Bioenergy Research Center, Michigan State University, East Lansing, USA

R. Garlock Department of Chemical Engineering and Materials Science and Great Lakes Bioenergy Research Center, Michigan State University, East Lansing, USA

Rajesh Gupta Chevron ETC, Houston, USA

Bassem Hallac School of Chemistry and Biochemistry, Georgia Institute of Technology, Atlanta and BioEnergy Science Center, Oak Ridge, USA

Bonnie R. Hames B Hames Consulting, Newbury Park, USA (Previously at Ceres, Inc. Thousand Oaks, USA)

Mark T. Holtzapple Department of Chemical Engineering, Texas A&M University, College Station, USA

M. Jin Department of Chemical Engineering and Materials Science and Great Lakes Bioenergy Research Center, Michigan State University, East Lansing, USA

Youngmi Kim Laboratory of Renewable Resources Engineering, Purdue University, West Lafayette, USA

Rajeev Kumar Center for Environmental Research and Technology, University of California, Riverside and BioEnergy Science Center, Oak Ridge, USA

Michael R. Ladisch Laboratory of Renewable Resource Engineering, Purdue University, West Lafayette, and Mascoma Corporation, USA

Mark Laser Thayer School of Engineering, Dartmouth College, Hanover, USA

Y. Y. Lee Department of Chemical Engineering, Auburn University, USA

Hongjia Li DuPont Industrial Biosciences, Palo Alto, USA (Previously at Center for Environmental Research and Technology and Department of Chemical and Environmental Engineering, University of California, Riverside, USA and BioEnergy Science Center, Oak Ridge, USA)

Chaogang Liu Mascoma Corporation, USA

Todd Lloyd Mascoma Corporation, USA

Lee Lynd Thayer School of Engineering, Dartmouth College, Hanover and BioEnergy Science Center, Oak Ridge, USA

Nathan S. Mosier Department of Agricultural and Biological Engineering, Laboratory of Renewable Resources Engineering, Purdue University, West Lafayette, USA

Jerry Parks Oak Ridge National Laboratory and BioEnergy Science Center, Oak Ridge, USA

Junia Pereira Department of Biological Systems Engineering, Virginia Polytechnic Institute and State University, Blacksburg, USA

Natalia Piamonte Department of Chemical Engineering, University of the Andes, Bogota, Colombia

Yunqiao Pu Georgia Institute of Technology, Atlanta and BioEnergy Science Center, Oak Ridge, USA

Qing Qing Pharmaceutical Engineering & Life Science, Changzhou University, Changzhou, China

Arthur J. Ragauskas Institute of Paper Science and Technology, and School of Chemistry and Biochemistry, Georgia Institute of Technology, Atlanta and BioEnergy Science Center, Oak Ridge, USA

Rocío Sierra Ramirez Department of Chemical Engineering, University of the Andes, Bogota, Colombia (Previously at Department of Chemical Engineering, Texas A&M University, College Station, USA)

Poulomi Sannigrahi Institute of Paper Science and Technology, Georgia Institute of Technology, Atlanta and BioEnergy Science Center, Oak Ridge, USA

Blake A. Simmons Deconstruction Division, Joint BioEnergy Institute, Emeryville and Biological and Materials Science Center, Sandia National Laboratories, Livermore, USA

Seema Singh Deconstruction Division, Joint BioEnergy Institute, Emeryville and Biological and Materials Science Center, Sandia National Laboratories, Livermore, USA

L. Sousa Department of Chemical Engineering and Materials Science, Michigan State University, East Lansing, USA

Ling Tao National Renewable Energy Laboratory, Golden, USA

F. Teymouri Michigan Biotechnology Institute, Lansing, USA

Steven R. Thomas US Department of Energy, Golden, USA (Previously at Ceres, Inc., Thousand Oaks, USA)

Heather L. Trajano Department of Chemical and Biological Engineering, University of British Colombia, Vancouver, Canada (Previously at Department of Chemical and Environmental Engineering and Center for Environmental Research and Technology, University of California, Riverside and BioEnergy Science Center, Oak Ridge, USA)

Melvin Tucker National Bioenergy Center, National Renewable Energy Laboratory, Golden, USA

Ryan Warner DuPont Industrial Biosciences, Palo Alto, USA

Charles E. Wyman Department of Chemical and Environmental Engineering and Center for Environmental Research and Technology, University of California, Riverside and BioEnergy Science Center, Oak Ridge, USA

Eduardo Ximenes Laboratory of Renewable Resources Engineering, Purdue University, West Lafayette, USA

Bin Yang Center for Bioproducts and Bioenergy, Department of Biological Systems Engineering, Washington State University, Richland, USA

Foreword

The concept of "pretreatment" arose from the observation that cocktails of glycosyl hydrolases were relatively ineffective at quantitatively depolymerizing the polysaccharides that comprise the bulk of native plant biomass. However, if biomass is first subjected to extremes of pH or temperature or various solvent extractions, enzyme cocktails were much more effective at releasing sugars from biomass. The useful effects of such pretreatments are generally understood to be due to disruptions of the native structure of plant cell walls so that cellulose and residual hemicelluloses are more exposed to enzymes than in their native condition. For instance, brief pretreatment with dilute sulfuric acid at temperatures of about 160 °C depolymerizes most of the hemicellulose that is thought to occlude cellulose microfibrils. The removal of the hemicelluloses exposes the cellulose to enzymes, increases the porosity of the cell wall, and also releases lignin that is covalently bound to hemicelluloses through linkages such as arabinose feruloly esters.

Unfortunately, each pretreatment has some inherent limitations. For example, dilute acid causes dehydration of sugars to toxic compounds such as furfural and hydroxymethyl furfural that inhibit subsequent fermentation, and neutralization of the acid leads to salt disposal issues at commercial scales. Similarly, other pretreatments have issues such as loss of sugars, high costs, safety issues, or waste disposal concerns.

The fact that no pretreatment leads cost-effectively to a clean separation of sugars and lignin increases the cost of enzymatic depolymerization; residual lignin inhibits or inactivates many glycosylhydrolases leading to a requirement for large amounts of enzyme. It usually also prevents clean separation of sugars and lignin, meaning that lignin ends up in the fermentation reactor with sugars. This in turn increases costs by preventing reuse of the cells and the glycosylhydrolases and is also inconsistent with some types of continuous fermentation that might otherwise create process efficiencies. The development of improved pretreatment methods is therefore not just about improving digestibility; for these and other related reasons, interest in new types of pretreatments has been developing.

From analysis of the past several decades of research on pretreatment I have concluded that huge importance must be placed on considering pretreatment in the context of the whole process train, from feedstock to fuel, rather than as an isolated unit operation. My impression is that significant improvements in the capital and operating costs of producing biofuels from lignocellulose appear to be economically attractive on the basis of detailed process models if the starting material is a clean sugar or polysaccharide stream, but that such improvements are precluded by the presence of insoluble materials such as lignin. In my opinion, the development of a cost-effective pretreatment technology that separates polysaccharides or sugars from all other components in feedstocks is therefore the highest priority for future research.

In line with this need, understanding pretreatment approaches and their impact on substrate–microbial interactions is important in suggesting lower-cost routes. Hopefully books such as this can provide valuable insights that will foster the development of a deeper understanding of biomass conversion to fuels and lead to low-cost pretreatments that will facilitate commercialization of biomass conversion processes with important societal benefits.

Chris Somerville
Energy Biosciences Institute
University of California
Berkeley, USA

Series Preface

Renewable resources are used and modified in a multitude of important processes having a major influence on our everyday lives. Applications can be found in the energy sector, chemistry, pharmacy, the textile industry, and paints and coatings, to name but a few.

The area connects several scientific disciplines (agriculture, biochemistry, chemistry, technology, environmental sciences, forestry, and so on), which makes it very difficult to have an expert view on the complicated interaction. The idea to create a series of scientific books focusing on specific topics concerning renewable resources has therefore been very opportune and can help to clarify some of the underlying connections in this area.

In a fast-changing world, trends are not only characteristic of fashion and political standpoints; science also has its hypes and buzzwords. The use of renewable resources is again more important nowadays; however, it is not part of a hype or a fashion. As the lively discussions among scientists continue about how many years we will still be able to use fossil fuels – opinions ranging from 50 years to 500 years – they do agree that the reserve is limited and that it is not only essential to search for new energy carriers but also for new material sources.

In this respect, renewable resources are a crucial area in the search for alternatives to fossil-based raw materials and energy. In the field of energy supply, biomass and renewable-based resources will be part of the solution alongside other alternatives such as solar energy, wind energy, hydraulic power, hydrogen technology and nuclear energy.

In the field of material sciences, the impact of renewable resources will probably be even bigger. Integral utilization of crops and the use of waste streams in certain industries will grow in importance, leading to a more sustainable way of producing materials.

Although our society was much more (almost exclusively) based on renewable resources centuries ago, this disappeared in the Western world in the nineteenth century. Now it is time to return our focus to this field of research. This does not necessarily imply a "retour à la nature," but should be a multidisciplinary effort on a highly technological level to perform research into new opportunities and to develop new crops and products from renewable resources. This will be essential to guarantee a level of comfort for a growing number of people living on our planet. *The* challenge for the coming generations of scientists is to develop more sustainable ways to create prosperity and to fight poverty and hunger in the world. A global approach is certainly favored.

This challenge can only be dealt with if scientists are attracted to this area and are recognized for their efforts in this interdisciplinary field. It is therefore also essential that consumers recognize the fate of renewable resources in a number of products.

Furthermore, scientists do need to communicate and discuss the relevance of their work. The use and modification of renewable resources may not follow the path of the genetic engineering concept in view of consumer acceptance in Europe. Related to this aspect, the series will certainly help to highlight the importance of renewable resources.

Being convinced of the value of the renewables approach for the industrial world, as well as for developing countries, I was myself delighted to collaborate on this series of books focusing on different aspects of renewable resources. I hope that readers become aware of the complexity, the interaction and interconnections, and the challenges of this field, and that they will help to communicate the importance of renewable resources.

I certainly want to thank the people of Wiley from the Chichester office, especially David Hughes, Jenny Cossham and Lyn Roberts, in seeing the need for such a series of books on renewable resources, for initiating and supporting it, and for helping to carry the project to the end.

Last but not least I would like to thank my family, especially my wife Hilde and children Paulien and Pieter-Jan, for their patience and for giving me the time to work on the series when other activities seemed to be more inviting.

Christian V. Stevens
Faculty of Bioscience Engineering
Ghent University, Belgium
Series Editor "Renewable Resources"
June 2005

Preface

The contents of this book were motivated by a long career in renewable energy research and development, the seeds of which were planted during a science fair project in junior high school and which grew through an extended opportunity to work at the Solar Energy Research Institute (SERI), now the National Renewable Energy Laboratory (NREL), blossomed in leading technology developments for start-up companies, and matured with my current career in academia. The economic and technical lessons learned at SERI while pursuing my initial interest in developing advanced technologies for extended storage of solar energy convinced me early on that the best option for long-term storage of solar energy was biomass, an unusually esthetic solar collector and storage device combined. Since that time, my passion has been to develop cost-effective processes for converting sustainable forms of solid cellulosic biomass into liquid fuels that our society so prefers. I have remained committed to this challenging but vital work for most of my long career because of the critical role biomass can play in reducing our extreme dependence on petroleum – a dependence which has tremendous economic, environmental, and security implications.

Although my work had been on chemical and catalytic processes up until that point, I chose the transition to biological conversion of biomass to fuels because of the opportunity for substantial advances through application of modern biotechnology. I initially worked primarily on enzymatic hydrolysis and fermentations to convert the carbohydrates in biomass into ethanol and other fuels. However, my experience with the start-up company BCI made it clear that the primary challenge to low-cost biological processing of biomass to liquid fuels was in effectively overcoming the recalcitrance of biomass, with biomass pretreatment playing an underappreciated but pivotal role in overcoming this barrier. It also became clear that interactions among pretreatment, plants, and their enzymatic hydrolysis to sugars are extremely complex, and enhanced knowledge of their interplay is of enormous importance. About 15 years ago, I therefore turned my attention to biomass pretreatment in support of biological conversion. A portion of my more recent aqueous pretreatment research has been directed at new opportunities in breaking biomass down into reactive intermediates, such as levulinic acid, that can be catalytically converted into liquid hydrocarbon fuels that are compatible with the existing transportation infrastructure.

In light of this background, I was extremely pleased when Sarah Hall of Wiley invited me to contribute a biomass pretreatment book, and I reached out to leaders in the field who had dedicated much of their careers to advancing biomass pretreatment and other conversion technologies to contribute chapters. Included were experts in key areas vital to understanding biomass composition in the context of pretreatment and in measuring and analyzing pretreatment streams that are important in evaluating pretreatment performance. In addition, experienced contributors were recruited to outline the societal and economic context for pretreatment and to highlight its role in biological and catalytic conversion of biomass to fuels and chemicals. It was also important to include a sense of how pretreatment affects key biomass features, comparative information on how different pretreatments perform, and the interactions among pretreatment types and downstream enzyme formulations. Chapters are also included by those knowledgeable in experimental pretreatment systems to provide insights into the equipment and procedures needed to apply and evaluate pretreatment technologies. I am extremely grateful to the authors of all of these chapters for taking time from their busy lives to contribute such insightful information. I am also grateful to Sarah Hall and Sarah Tilley and many others at Wiley for their encouragement that was so vital to making this book possible.

In closing, I would like to thank a few key people who made it possible for me to pursue a career in advancing sustainable technologies, and particularly biomass conversion. Tremendous gratitude belongs to my wife Carol and our two children Marc and Kristin for supporting this pursuit in so many ways. In all honesty, it would not have been possible without their encouragement and flexibility in taking on many challenges. I also must recognize my mother, the late Ruth A. Wyman, for instilling my interest in sustainable energy at an early age. Many other people have fostered this career path, and it would be impossible to list them all. Even then, I would run the risk of forgetting to include someone. However, I am nonetheless thankful to all who have made my career and this resulting book possible.

Charles E. Wyman
University of California at Riverside,
Department of Chemical and Environmental Engineering and
Center for Environmental Research and
Technology, Riverside, USA

BioEnergy Science Center,
Oak Ridge, USA

Acknowledgements

First, I am grateful to the authors of the chapters in this book for contributing time and effort in making their experiences and knowledge available to others. It was truly a privilege to have such a strong cast of experts in pretreatment and related fields contribute. We hope that this material will prove valuable in understanding the various aspects of pretreatment and providing references to more comprehensive information.

I would also like to thank the following experts in the field for providing thoughtful comments and suggestions on various book chapters prior to publication: Andy Aden, URS Corporation; Foster Agblevor, Utah State University; Venkatesh Balan, Michigan State University; Jesse Bond, Syracuse University; Renata Bura, University of Washington; Alain Castellan, University of Bordeaux (LCPO); Kevin Chambliss, Baylor University; Shishir Chundawat, Michigan State University; Bruce Dale, Michigan State University; Mark Davis, National Renewable Energy Laboratory; Brian Davison, Oak Ridge National Laboratory; Bruce Dien, US Department of Agriculture; Thomas Elder, US Department of Agriculture; Tom Foust, National Renewable Energy Laboratory; John Hannon, Consultant; Mark Holtzapple, Texas A&M University; George W. Huber, University of Wisconsin; David Johnson, National Renewable Energy Laboratory; Don Johnson, Grain Processing Corporation, retired; Rajeev Kumar, University of California Riverside; Michael Ladisch, Purdue University and Mascoma Corporation; Mark Laser, Dartmouth College; YY Lee, Auburn University; Hongjia Li, University of California, Riverside; Todd Lloyd, Mascoma Corporation; Nate Mosier, Purdue University; Art Ragauskas, Georgia Tech; Dan Schell, National Renewable Energy Laboratory; Xiongjun Shao, Dartmouth College; David Shonnard, Michigan Technological University; Ling Tao, National Renewable Energy Laboratory; David Templeton, National Renewable Energy Laboratory; Steve Thomas, US Department of Energy; Peter van Walsum, University of Maine; Ed Wolfrum, National Renewable Energy Laboratory; and Bin Yang, Washington State University. In addition, we are grateful to a few other reviewers who also provided valuable suggestions but chose to remain anonymous.

Finally, I would like to thank Sarah Hall of Wiley for catalyzing the development of this book and Sarah Tilley of Wiley and her colleagues for working to complete its preparation.

Charles E. Wyman
Editor

1

Introduction

Charles E. Wyman[1,2]

[1] *Department of Chemical and Environmental Engineering and Center for Environmental Research and Technology, University of California, Riverside, USA*
[2] *BioEnergy Science Center, Oak Ridge, USA*

Welcome to "Aqueous Pretreatment of Plant Biomass for Biological and Chemical Conversion to Fuels and Chemicals." This book provides insights into thermochemical preparation of cellulosic biomass such as wood, grass, and agricultural and forestry residues for aqueous conversion to fuels and chemicals as well as economic and analysis information that is broadly applicable to a wide range of aqueous biomass operations. Historically, acid catalyzed hydrolysis of biomass goes back to the early nineteenth century [1], when the emphasis was on aqueous-processing of biomass in concentrated acid or dilute acid at higher temperature to break down cellulose into glucose that could be fermented into ethanol for use as a fuel [2,3]. Because most of the hemicellulose sugars are destroyed at dilute acid conditions that realize high glucose yields from cellulose, pretreatment with dilute acid at milder conditions was employed to maximize yields of hemicellulose sugars (provided they were removed prior to treating the cellulose [4]). Then, most of the cellulose was left in the solids and could be broken down with dilute acid at more harsh conditions to fermentable glucose without sacrificing much of the hemicellulose sugars [5]. A similar approach was applied commercially to break down hemicellulose in corn cobs, sugar cane bagasse, and other hemicellulose-rich types of cellulosic biomass into xylose and arabinose sugars, and react these sugars further to marketable furfural [6]. In this case, the cellulose, lignin, and other components left in the solids were usually burned for heat and power. Application of milder conditions for hemicellulose breakdown was later found to be effective in opening up the biomass structure so enzymes could achieve high glucose yields from the recalcitrant cellulose left in the solids [5,7,8]. More recently, hemicellulose conversion to sugars or furfural has been employed followed by heterogeneous catalysis to produce hydrocarbons from biomass that are compatible with existing fossil-resource-based fuels and chemicals [9–11]. In this case, even harsher dilute acid conditions than applied to release glucose from cellulose could then be applied to the remaining cellulose-rich solids to generate 5-hydroxymethyl furfural and levulinic acid, desirable precursors for catalytic conversion into hydrocarbon fuels and chemicals. In a sense, technology for thermo-chemical breakdown of cellulosic biomass with dilute acid has come full circle from its beginnings, albeit to serve different downstream processes.

The operation to prepare biomass for downstream aqueous biological or catalytic processing is typically called pretreatment and is critical to achieving high product yields that can foster the emergence of biofuels and biochemicals industries based on biological or catalytic conversion of plants. However, the range of technologies has become broader than just the reaction of hemicellulose in dilute acid and now includes operations that also focus on lignin removal [12–14]. For biologically based processes, disruption of hemi-cellulose or lignin (and not removal) may also be adequate to realize high sugar yields from biomass in enzymatic operations. Furthermore, a wide range of combinations of reaction temperatures, pH values, and times can be effective in preparing biomass for downstream processing, depending on the technologies being applied [15–17]. Some of these aqueous pretreatments build from analogous industrial operations such as removal of lignin by reaction of biomass with caustic for the pulp and paper industry. We can there-fore now define aqueous pretreatment as the reaction of cellulosic biomass at conditions that result in the highest possible yields in subsequent biological, catalytic, or thermochemical processing.

The goal of this introductory chapter is to summarize some of the key aspects of cellulosic biomass and its aqueous pretreatment to make it compatible with downstream biological, catalytic, or thermochemical processing to provide an historical perspective for the chapters in this book and its organization. This chapter will start by providing a sense of what we mean by cellulosic biomass and why it is a vital resource for sustainable production of organic fuels and chemicals. This overview will be followed by a summary of key biomass features, including its composition. An overview will then be given of how biomass lends itself to biological and catalytic aqueous processing and the important challenges hindering commercial applica-tions. Against this background, criteria for successful pretreatment will be outlined. An overview of various pretreatment technologies will then provide a sense of options that have been investigated over the years and the rationale behind the emphasis on thermochemical pretreatments in this book. In addition, other aspects that can influence pretreatment effectiveness will be mentioned, along with limitations in our expe-rience with pretreatment. The chapter will end with an outline of the chapters that follow to help the reader utilize the information in the book.

1.1 Cellulosic Biomass: What and Why?

The word biomass encompasses any biological material derived from living or recently living organisms. The term could therefore apply to both animal and vegetable matter. However, this book focuses on cellu-losic biomass, the structural portion of plants, as a resource for the production of fuels and chemicals. Plant/cellulosic biomass contains carbon, hydrogen, and oxygen, plus typically much lower amounts of nitrogen, phosphorous, minerals, and other ingredients. The sun's energy drives the formation of plant bio-mass while releasing oxygen through the photosynthetic reaction of water with carbon dioxide. The late Dr Ray Katzen, a giant in the field of industrial biomass conversion, termed cellulosic biomass as C-water – CH_2O – in reference to the building block from which biomass sugars are made. If biomass or materials derived from biomass are burned, oxygen in the air combines with the carbon and hydrogen in biomass to release carbon dioxide and water, reversing the reactions through which plant matter was formed originally. However, as long as new biomass is planted to replace that burned or otherwise utilized, this carbon cycle results in no net change in the amount of carbon dioxide in the atmosphere. This feature of using biomass distinctly contrasts with burning fossil fuels, in which carbon from below the ground continually accumu-lates in the atmosphere. The powerful natural carbon recycle provides the potential for fuels production from cellulosic biomass to avoid contributing to the net accumulation of carbon dioxide in the atmosphere, a major driver of global climate change [18–21].

Biomass can fill a unique niche for sustainably meeting human needs. The sustainable resources are sunlight, wind, ocean/hydro, geothermal, and nuclear, and societal needs can be grouped as food, motor-driven devices, light, heat, transportation, and chemicals [22]. Electricity and thermal energy can be made from all sustainable resources as primary intermediates for human needs but only sunlight can support growth of biomass, the other primary intermediate. Biomass alone among sustainable resources can be transformed into feed for animals, human food, and organic fuels, chemicals, and materials. Plant materials could have a much greater impact if vast, low-cost untapped sources of cellulosic biomass such as agricultural and forestry residues, portions of municipal waste, and dedicated crops could be inexpensively converted into a range of fuels and commodity chemicals in large-scale biorefineries [23]. In fact, inexpensive transformation of biomass into liquid fuels and commodity chemicals will be essential if society is to sustainably and economically meet such needs [24–26].

Although the term cellulosic biomass may not be a household word, it represents the structural portion of a large group of well-known plants. Common examples include agricultural wastes such as corn stalks and corn cobs (the two together being termed corn stover) and sugar cane bagasse that are left after removal of targeted food and feed products. Forestry residues represent another familiar example of cellulosic materials as represented by sawdust, bark, and branches left after harvesting trees for commercial operations such as making paper and wood products. Large portions of municipal solid wastes, including waste paper and yard waste, are also cellulosic biomass. Paper sludge results from fines from plant biomass not captured in the final product [27,28]. Although such existing cellulosic resources can cumulatively represent a substantial resource that could provide an effective platform from which to launch a biomass-based industry, energy crops will be ultimately needed to meet the huge demand for organic fuels and chemicals. In this vein, various types of grasses can prove to be valuable feedstocks with fast-growing herbaceous plants such as switchgrass and Miscanthus being prominent examples. In addition, various trees such as poplar and eucalyptus have the high productivities desirable to maximize production potential from limited available land. Taken together, it has been estimated that the future availability of biomass for energy production in the United States could be on the order of 1.4 billion dry tons of biomass, enough to displace over 100 billion gallons of gasoline of the approximately 140 billion gallons now used in the United States [29,30]. Biomass-based fuels could make an even bigger impact if the country were to substantially reduce fuel consumption by driving more efficient vehicles and use more public transportation.

In addition to being widely available, having the potential to reduce greenhouse gas emissions, and being uniquely suited to sustainable production of liquid fuels, cellulosic biomass is inexpensive. For example, cellulosic biomass costing $60 per dry ton has about the same cost per unit mass as petroleum at about $7 per barrel. Of even more relevance for fuels production, this biomass price would be equivalent to petroleum at about $20/barrel on the basis of equivalent energy content [27,31]. The resource itself is therefore low in cost, and the challenge is how to inexpensively transform cellulosic biomass into fuels.

1.2   Aqueous Processing of Cellulosic Biomass into Organic Fuels and Chemicals

A variety of pathways can be applied to convert cellulosic biomass into fuels and chemicals [11]. For example, cellulosic biomass can be gasified to generate carbon monoxide and hydrogen. This mixture, called syngas, can in turn be catalytically converted into diesel fuel, methanol, or other products. Pyrolysis by heating biomass in the absence of air can generate oils that must be upgraded to have suitable fuel properties and be more compatible with conventional fuels. Biomass could be liquefied by application of heat and hydrogen under pressure. For such thermal routes, a proximate analysis of biomass composition may be useful to support design of a process. For example, a typical proximate analysis of switchgrass could be about 13.7% fixed carbon, 73% volatile matter, 4.9% ash, and 8.4% moisture [32]. The higher heating value could be about 17.9 MJ/kg. However, the elemental composition of biomass is likely to be more informative in that it allows development of more in-depth material and energy balances. In this case, a representative elemental analysis of switchgrass could include about 46.8% carbon, 5.1% hydrogen, 42.1% oxygen, less than 0.6% nitrogen, about 0.1% sulfur, and 5.3% minerals/ash, all being on a mass basis [32].

In reality, cellulosic biomass is more complex than simple proximate or elemental analyses suggest, with their structures evolved to support key plant functions [33]. Although the wide range of plant materials represented by cellulosic biomass are distinct in physical appearance, they all share similar structural make-ups. Generally, the most abundant portion is cellulose; about 35–50% of the weight of many plants comprises cellulose. Cellulose is a polymer of glucose sugar molecules linked together in long, straight parallel chains that are hydrogen-bonded to one another in a crystalline structure to form long fibers. Another roughly 12–25% of cellulosic biomass is a sugar polymer known as hemicellulose, which can consist of the five sugars arabinose, galactose, glucose, mannose, and xylose along with various other components such as acetyl groups and pectins [34,35]. The proportion of these components in hemicellulose varies among plants and, unlike cellulose, hemicellulose is branched and not crystalline. The other significant fraction of cellulosic biomass is lignin, a complex phenyl propene compound that is not made of sugars and whose chemical composition varies with plant type [33,36]. Cellulosic biomass also contains lesser amounts of other compounds that may include minerals/ash, soluble sugars, starch, proteins, and oils. Although often overlooked in the discussion of biomass conversion, these components are also vital to plant functions.

Aqueous processing targets processing of cellulosic biomass in water to convert the structural components in biomass into compounds dissolved in water which we call reactive intermediates (RIs) that, in turn, can be biologically, catalytically, or thermochemically converted into fuels or chemicals. Thus, biomass is broken into the basic building blocks from which it is made and not all the way down to simple molecules. For example, the arabinose and xylose in hemicellulose are five carbon sugar isomers that can be linked together in a chain n units long to form $n(C_5H_8O_4)$. As noted above, acids or enzymes can catalyze the breakdown of such chains in water to release the individual five carbon sugars from which they are made by the following hydrolysis reaction:

$$n(C_5H_8O_4) + nH_2O \rightarrow nC_5H_{10}O_5 \qquad (1.1)$$

Similarly, acids or enzymes can catalyze hydrolysis of the six carbon sugars that comprise a portion of hemicellulose and all of cellulose (glucose) into the sugar isomers glucose, galactose, or mannose as follows:

$$n(C_6H_{10}O_5) + nH_2O \rightarrow nC_6H_{12}O_6 \qquad (1.2)$$

The arabinose and xylose released from reaction (1.1) and galactose, glucose, and mannose released by reaction (1.2) can all be fermented to ethanol or other products through a choice of suitable organisms. For example, industrial yeast strains such as *Saccharomyces cerevisiae* or other yeast naturally ferment glucose and the other six carbon sugars into ethanol. Furthermore, although native yeast cannot ferment the five carbon sugars arabinose and xylose to ethanol with high yields, various bacteria such as *Escherichia coli* and yeast including *Saccharomyces cerevisiae* have been genetically engineered so they now produce ethanol from these sugars with high yields [37–40]. We can therefore view these sugars as reactive intermediates that can be biologically converted into ethanol and other final products.

A variety of acids including sulfuric, nitric, and hydrochloric have been applied to hydrolyze hemicellulose to its component sugars with yields of about 80–90% of theoretical or more, feasible in simple batch or co-current flow operations [41,42]. Dilute acids can also hydrolyze cellulose to glucose, but glucose yields are limited to about 50% of theoretical for practical operating conditions [2,43]. Enzyme catalyzed breakdown (hydrolysis) of cellulose to glucose has therefore emerged as a leading option for making commodity products because nearly theoretical glucose yields vital to economic success are possible [22,27]. Furthermore, enzyme-based processing costs have been reduced by about a factor of four [44–51], and many of the additional advances needed to make the technology competitive are achievable through application of the powerful new and evolving tools of biotechnology [48,49,52–54]. Another benefit of high-selectivity biological conversion and particularly enzymatic catalysis is minimal waste generation, reducing disposal problems. Although efforts have focused on ethanol production, a range of fuels, chemicals, and materials can be biologically derived from the same sugar intermediates [24,25,55]. However, the key obstacle to commercial use of enzymes for release of sugars from cellulosic biomass is the high doses and resulting high costs for cellulase and hemicellulase [27,31,55,56]. The most critical need to achieve low production costs is therefore the reduction of biomass recalcitrance as the major obstacle to low sugar costs [27,56].

Although sugars can be fermented into a wide range of compounds that are valuable fuels and chemicals, many are oxygenated and differ from currently employed hydrocarbons. For example, ethanol is a high octane fuel with many superior properties to gasoline, with the result that it is the fuel of choice for the Indianapolis 500 and other races for which speed and power are vital. Ethanol is also much less toxic than gasoline as evidenced by the fact we drink beer, wine, mixed drinks, and other beverages containing ethanol while no beverages contain gasoline. The fact that ethanol is different from gasoline concerns many users, however. For example, ethanol has a somewhat lower energy density, tends to separate into water when water is present, and has different solvent properties from gasoline. Thus, many desire hydrocarbon fuels that are completely fungible with the current petroleum-based infrastructure. This preference for hydrocarbons is appropriate for aviation, for example jet fuel which needs the highest possible energy density. Similarly, hydrocarbons have important advantages in compression ignition engines that are important in powering large trucks, earth-moving equipment, and other heavy-duty vehicles.

Whatever the rationale, aqueous biomass streams are now being processed into RIs including furfural, 5-hydroxymethylfurfural (5-HMF), and levulinic acid for catalytic conversion into hydrocarbon "drop-in" fuels by novel processes [9,10]. Aqueous catalysis can build off many of the same pretreatment technologies developed for biological conversions, but without enzymes or fermentations. To support catalytic processing, enzymes or acid catalyze hydrolysis of the cellulose and hemicellulose into their sugar monomers in the same way as for biological conversion. However, dilute acids also catalyze dehydration of the sugars into sugar alcohols that can be aldol condensated and hydrogenated into RIs and light alkanes by homogeneous/heterogeneous catalysts [9]. The catalysts used for these reactions include acids, bases, metals, metal oxides [10,57,58], and multifunctional catalysts. For example, ruthenium/carbon (Ru/C) and platinum/zirconium phosphate (Pt/ZrP) catalysts hydrodeoxygenate aqueous streams of xylose to xylitol at 393 K and xylitol to gasoline range products at 518 K. Bimetallic PtSn catalysts selectively hydrogenate furfural to furfural alcohol, which acids can further hydrolyze to levulinic acid (LA), a reactive building block for hydrocarbon fuels. LA can in turn be converted into gamma-valerolactone (GVL) over Ru/C catalyst. Further, GVL can be converted to equimolar amounts of butene and carbon dioxide gases through decarboxylation at elevated pressures over a silica/alumina catalyst. This stream can in turn be converted into condensable alkenes by the application of an acid catalyst (e.g., H ZSM-5, Amberlyst-70) that links butene monomers to achieve molecular weights that can be compatible with gasoline and/or jet fuel applications [59].

1.3 Attributes for Successful Pretreatment

From the above discussion, aqueous pretreatment can be applied to prepare cellulosic biomass for subsequent enzyme or acid catalyzed reactions to release sugars for fermentation to ethanol or other products.

In such cases, the primary goal for pretreatment is to work with downstream operations to achieve the highest possible product yields at the lowest costs; a variety of pretreatment approaches are promising [15–17,60]. Aqueous pretreatment is also applicable in preparing cellulosic biomass for catalytic reaction, with the goal again being to achieve the highest possible product yields and lowest costs. However, current pretreatment approaches favored for catalytic processing employ dilute acid to remove hemicellulose with high sugar or furfural yields. In addition, dilute acid can also be employed for subsequent reaction of the cellulose-enriched solids from pretreatment into HMF and/or levulinic acid. Aqueous pretreatment of biomass to support catalytic conversion can therefore avoid the high costs of enzymes that have hindered commercialization of biological routes to fuels and chemicals.

Against this background, several key attributes are vital for pretreatment to be promising for application to biological or catalytic conversion of cellulosic biomass to fuels and chemicals. Because milling of biomass to small particle sizes is energy intensive and introduces extra equipment costs [61,62], pretreatment technologies that require limited size reduction are desirable. In the case of enzymatic conversion, pretreatment must open up the biomass structure to make cellulose accessible to enzymes so they can achieve high yields from the pretreated solids and recover sugars released in pretreatment with high yields. To support catalytic processing, pretreatment must achieve high sugar or furfural yields from hemicellulose as well as serve subsequent reactions to target RIs. Regardless of the downstream operation, the concentration of RIs should be as high as possible to ensure that product concentrations are adequate to keep recovery, process equipment, and other downstream costs manageable. The requirements for chemicals in pretreatment and subsequent neutralization and conditioning for downstream operations should be minimal and inexpensive, or the chemicals should be easily recovered for reuse. Pretreatment reactors should be low in cost through minimizing their volume, requiring low pressures and temperatures, and avoiding the need for exotic materials of construction due to highly corrosive chemical environments. In addition, the pretreatment chosen must work cooperatively with other operations. For example, a pretreatment operation that separates hemicellulose sugars from glucose from cellulose may be preferred to avoid preferential glucose fermentation and associated lower yields from hemicellulose sugars due to diauxic effects. The liquid stream from pretreatment must be compatible with subsequent steps following a low-cost high-yield conditioning step. In fact, it is highly desirable to employ pretreatments that produce streams that require no conditioning to reduce costs and reduce yield losses. Any chemicals formed during hydrolyzate conditioning in preparation for subsequent steps should not present processing or disposal challenges (e.g., gypsum formed by neutralization of sulfuric acid with calcium hydroxide). An innovative pretreatment could recover lignin, protein, minerals, oils, and other materials found in biomass for use as boiler fuel, food, feed, fertilizers, and other products in a biorefinery concept that enhances revenues [63]. Such synergies would leverage biomass impact and reduce land requirements, enhancing sustainability [24,25,27,55]. Consequently, attention must be given to advancing pretreatment to make aqueous processing of biomass competitive for large-scale sustainable applications in an open market [64,65]. A number of reviews of pretreatment, enzymatic hydrolysis, and catalytic processing provide historic perspectives [e.g. 12,13,31].

In choosing a pretreatment technology, high product yields must be met to distribute total costs over as much product as possible. In addition, the capital and operating costs for pretreatment must be kept low without sacrificing product yields. We could therefore say that the best pretreatment would be free and have no costs or unwanted impacts on other operations; unfortunately however, pretreatment has been projected to be the most expensive single operation in overall biological processing in some studies [66]. Because yields suffer without pretreatment, other studies have shown that overall product unit costs are higher without pretreatment than with it, leading this author to state that "the only operation more expensive than pretreatment is no pretreatment" [31]. Ultimately, the choice of pretreatment is governed by costs of the overall process and not just the pretreatment operation [67–71].

1.4   Pretreatment Options

Over the years, a number of aqueous-based pretreatment technologies have been investigated in the search for a low-cost approach that can realize high yields of final products from both the cellulose and hemicellulose fractions [72]. Most of these have focused on supporting subsequent enzymatic hydrolysis, with only limited recent work supporting catalytic processing. Reviews have classified these pretreatment methods as (1) physical, (2) biological, and (3) chemical.

Physical pretreatments include size reduction by devices such as hammer mills, knife mills, extruders, disc refiners, and planers. Mechanical decrystallization by ball, roll, dry, and colloid mills are physical pretreatments that can increase enzymatic hydrolysis yields. Thermal pretreatment by freeze/thaw, pyrolysis, and cryomilling are also classified as physical pretreatments, as are radiation with gamma rays, microwaves, electron beams, and lasers. Many physical pretreatments are not sufficiently effective in achieving high yields, and their operating and/or capital costs are often high [73–79]. Overall, such methods are not yet considered practical to support biological processing and do not produce the RIs needed for catalytic methods. These methods are therefore not covered in depth in this book, but other sources can be checked for more information for those wishing to explore these technologies further [80].

Biological pretreatment of biomass offers some conceptually important advantages such as low chemical and energy use. Generally, organisms are sought that will preferentially attack lignin to open up biomass for subsequent attack by enzymes. Various fungi including *Fomes fomentarius*, *Phellinus igniarius*, *Ganoderma applanatum*, *Armillaria mellea*, and *Pleurotus ostreatus* are typical choices. Unfortunately, to date, biological methods tend to suffer from poor selectivity in that organisms consume cellulose and hemicellulose, hurting product yields. In addition, they require long times and are hard to control. Overall, because no biological system has been demonstrated to be effective [81–85], they are not considered further in this book and the reader should consult other sources for additional insights [86–89].

Chemical pretreatments make up the third and final class of options that employ a range of different chemicals to prepare biomass for subsequent operations [12,13]. Most also include raising the temperature to the range of 140–210 °C or so and are labeled as thermochemical pretreatments. The result is a broad range of chemical concentrations, temperatures, and times that have been applied for biomass pretreatments. Oxidizing agents such as peracetic acid, ozone, hydrogen peroxide, chlorine, sodium hypochlorite, and chlorine dioxide as well as oxygen and air have been employed for thermochemical pretreatment. Another set of options revolves around concentrated acids including sulfuric (55–75%), phosphoric (79–86%), nitric (60–88%), hydrochloric (37–42%), and perchloric (59–61%). Several solvents are effective in dissolving cellulose to improve its accessibility to enzymes, with examples being the inorganic salts lithium chloride, stannic chloride, and calcium bromide, as well as such amine salts as cadmium chloride plus ethylenediamine (cadoxen) and cobalt hydroxide plus ethylenediamine (cooxen). Biomass can also be delignified and fractionated in organosolv pretreatments that employ methanol, ethanol, butanol, or triethylene glycol. Cellulose modification to carboxymethyl cellulose, viscose, or mercerized cellulose provides another thermochemical pretreatment path. The addition of alkaline compounds such as sodium hydroxide, potassium hydroxide, calcium hydroxide, and amines has been employed to open up cellulosic biomass by removing a large portion of lignin. Kraft and soda pulping provide established routes to pretreat biomass at these higher pH levels. Ammonia provides a versatile pretreatment chemical in that it can be applied at gaseous, liquid, aqueous, or supercritical conditions at various moisture levels. Dilute sulfuric or nitric acids do a good job of removing hemicelluloses, as do gaseous hydrochloric acid and sulfur dioxide. In addition, gaseous nitrogen dioxide and carbon dioxide have been tested to reduce the pretreatment pH, although yields are not nearly as high as possible with stronger acids. Perhaps the simplest pretreatment option is to heat biomass with steam or just hot water to break down hemicellulose and dislodge lignin. This approach is sometimes classified as a physical method in that only heat is applied, but it has also been grouped with thermochemical pretreatments in light of the belief that acetic and other acids released from hemicellulose during pretreatment help catalyze hydrolysis to sugars in what is termed as autohydrolysis. Unfortunately, autohydrolysis does not achieve as high hemicellulose sugar yields as possible with stronger acids.

A number of pretreatment leaders formed a Biomass Refining Consortium for Applied Fundamentals and Innovation (CAFI) in 2000 and worked as a team for over a decade to compare results from the application of leading pretreatment technologies to biological conversion on a consistent basis. The pretreatments studied were based on dilute sulfuric acid, sulfur dioxide, neutral pH, liquid ammonia, ammonia fiber expansion (AFEX), and lime [15,17,68]. The first project focused on application of these pretreatments to corn stover through support from the US Department of Agriculture Initiative for Future Agricultural and Food Systems (IFAFS) Program, and the Office of the Biomass Program of the US Department of Energy supported two subsequent projects on pretreatment of poplar wood and switchgrass. A surprising finding of these three studies was the similarity in results between thermochemical pretreatments spanning a wide pH range from low values with dilute sulfuric acid or sulfur dioxide to high pH values with lime. Yields were particularly similar and high with corn stover for all pretreatments and nearly the same high values for switchgrass across the entire pH range. Total sugar yields from pretreatment together with enzymatic hydrolysis were more variable with poplar wood but even then were similarly high for lime and sulfur dioxide, the extremes in pH. The CAFI studies pointed out that pretreatment effectiveness could not simply be related to process conditions, but that substrate–pretreatment–enzyme interactions are complex. Thus, more detailed research is still needed to better understand how to open up the biomass structure to achieve high yields from the combined operations of pretreatment and enzymatic hydrolysis.

1.5   Possible Blind Spots in the Historic Pretreatment Paradigm

Some very important points should be kept in mind when judging and selecting pretreatment technologies. First, almost all of the past development efforts focused on pretreatment prior to enzymatic hydrolysis, with far less effort devoted to pretreating biomass for catalytic conversion. Thus, consideration of different pretreatment perspectives could be beneficial for the latter. A second vital point is that most of the pretreatment work for biological conversion has evaluated pretreatment effectiveness in terms of yields of sugars by subsequent application of fungal enzymes to the pretreated solids. Furthermore, a large portion of the evaluations of the effectiveness of pretreatment in terms of subsequent enzymatic hydrolysis have been based on high enzyme loadings that would be commercially impractical. Far more work is needed to understand how pretreatments perform at lower enzyme loadings and what features of the pretreated substrate limit high yields. In addition, very little attention has been given to determining relationships among substrate types and features, pretreatment types and conditions, and performance with other biological systems. For example, some bacteria such as the thermophile *Clostridium thermocellum* produce a complex cellulosome enzyme structure that may be more effective in hydrolyzing hemicellulose and cellulose into their component sugars with the same organism also fermenting the sugars released to final products. This simultaneous enzyme production and fermentation feature has been called consolidated bioprocessing or CBP. The close association of the enzyme-producing CBP organism with the cellulosome has also been shown to offer significant advantages [90–92]. Another important point concerns the feedstocks pretreated. Although a range of hardwoods, grasses, softwoods, forestry and agricultural residues, and municipal solid wastes have been subjected to pretreatment followed by enzymatic hydrolysis, much less effort has been devoted to determining if particular substrate features would enhance pretreatment performance. Overall, little is known about possible synergies among feedstock features, pretreatment types and conditions, and microbial systems that would greatly enhance yields while simplifying (or possibly eliminating) pretreatment and reducing enzyme loadings, therefore significantly cutting costs.

1.6   Other Distinguishing Features of Pretreatment Technologies

Pretreatment technologies can also be differentiated in ways other than whether they are biological, chemical, or physical or the type of additive used. For example, almost all laboratory experiments are conducted under batch conditions in which all contents are loaded into a reactor at the beginning where they are heated up to some target temperature, held at that temperature for a set period of time, cooled back to room temperature, and then removed for analysis and evaluation. On the other hand, many commercial ventures prefer continuous operations to obtain higher productivities by avoiding heat-up and cool-down times and non-productive periods between batches for emptying and filling reactors, as well as better heat integration. Accordingly, continuous pretreatments are often used with co-current flow of the solids and liquid; the results can be quite similar to those for batch operations if the solids and liquid move as a plug. However, high solids concentrations are also preferred to provide higher sugar concentrations from pretreatment and enzymatic hydrolysis and reduce thermal loads, and cellulosic biomass has little free liquid at such conditions [93–95]. Moving solids of this consistency presents significant challenges, particularly at high temperatures and pressures, and residence times are likely to be variable. Thus, continuous pretreatment performance may be poorer than would be expected from results with laboratory batch systems, and new tools are needed to accurately predict commercial performance.

A number of other operational features can influence performance. For example, some laboratory research has shown that flow of water through a fixed bed of biomass can remove more lignin and hemicellulose and achieve better yields from pretreatment and enzymatic hydrolysis than possible in a batch system operated at similar temperatures and times [96–99]. However, most data from such flowthrough systems has been derived from the use of finely ground biomass, and it is not known how well such systems will perform with larger-sized particles that are more commercially relevant. Bench- and pilot-scale countercurrent pretreatment systems have also shown performance advantages compared to batch operations [100], but moving solids and liquids in opposite directions at high temperatures and pressures at a large commercial scale presents challenges. Methods applied to heat up and cool down biomass can also be very influential, in that variations in temperature histories with time and space can markedly change performance. Washing pretreated biomass with hot water could also improve performance.

1.7   Book Approach

The above information presents an idea of the lay of the land for this book, and has hopefully piqued your appetite for learning more about these and other topics relevant to pretreatment. As noted at the start of this chapter, the aim of the book is to provide comprehensive information that can support research, development, and application of aqueous pretreatment technologies. Experts on biomass pretreatment, conversion, and analysis were invited to author the following 22 chapters to cover the wide range of topics appropriate to the field. These lead authors were responsible for the content of each chapter and in many cases enlisted co-authors. Their intent was to provide solid platforms from which others could understand the importance of pretreatment, developments in the field, fundamentals of the technologies, key attributes and limitations, opportunities for advances, analysis methods, and needs for additional research and development (R&D). Authors were therefore urged to focus on such things as integration into the overall process, reaction kinetics, reaction stoichiometries, reaction conditions, effects on key biomass components, component removal vs. times and temperatures, and equilibrium considerations as appropriate to the chapter topic. This could also include considerations for integration with key upstream and/or downstream operations and their interactions, such as pretreatment with enzymatic hydrolysis. It was also intended that each chapter provides a perspective on the entire topic and facts and not focus on developments in one laboratory or promote particular technologies, allowing the reader to draw their own conclusions. A particularly important goal was to provide comprehensive references to support key points and allow the reader to obtain additional insights beyond those possible in a chapter of limited length.

1.8   Overview of Book Chapters

As shown in the Table of Contents, this book provides chapters to help the reader understand the unique role of the biomass resource in sustainable fuels production, its composition and structure relevant to pretreatment, the context of aqueous biological and catalytic processing of biomass, features of prominent thermochemical pretreatment technologies, comparative data on application of leading pretreatments to a range of biomass types, economic factors to be considered in pretreatment selection, analytical methods for measuring biomass composition, and experimental systems for pretreatment and enzymatic hydrolysis.

Chapter 2 provides insights into the importance and uniqueness of cellulosic biomass as a resource to support sustainable production of organic fuels and chemicals. Chapter 3 then provides a perspective on the composition of biomass and resulting challenges its recalcitrance presents to conversion. Chapter 4 focuses on biological conversion of cellulosic biomass, with emphasis on challenges facing its incorporation with enzymes and fermentative organisms. An overview of aqueous phase catalytic processing of streams from pretreatment of cellulosic biomass, providing a perspective on the needs for this emerging application, is presented in Chapter 5. Next, fundamental insights are provided on low pH pretreatment and how it can serve both biological and catalytic processing to fuels and chemicals as well as applied to release glucose, 5-HMF, and levulinic acid from cellulose in Chapter 6. Chapters 7 and 8 provide insights into pretreatment fundamentals at nearly neutral pH and high pH to support biological conversion. Chapters are also devoted to outlining fundamental features for pretreatments by AFEX (Chapter 9), biomass fractionation (Chapter 10), and ionic liquids (Chapter 11). Armed with this background, in Chapter 12 the reader is given a summary of data developed for application of leading thermochemical pretreatment technologies to corn stover, poplar wood, and switchgrass, with Chapter 13 providing insights into how enzyme formulations must be tailored to pretreatment type to realize high yields. Chapter 14 provides fundamental insights into how physical and chemical features of pretreated biomass impact sugar release. Cost comparisons for integration of leading pretreatment technologies into biological conversion processes are offered in Chapter 15, and opportunities are defined to reduce conversion costs. Chapters 16, 17, 18 and 19 describe analytical methods that can track changes in biomass composition and other features in pretreatment and enzymatic hydrolysis. Finally, Chapters 20, 21, 22 and 23 are devoted to describing experimental systems that are applicable to pretreatment and enzymatic hydrolysis of biomass, covering scales from multiwell plates to pilot plant operations.

We sincerely hope that the reader finds this book a useful tool to better understand pretreatment of cellulosic biomass, including its importance and insights into leading thermochemical technologies as well as analytical and other supporting methods applicable to any pretreatment of cellulosic biomass.

Acknowledgements

Support by the BioEnergy Science Center (BESC), a US Department of Energy Bioenergy Research Center supported by the Office of Biological and Environmental Research in the DOE Office of Science, was vital to development of this book. Gratitude is also extended to the Ford Motor Company for funding the Chair in Environmental Engineering at the Center for Environmental Research and Technology of the Bourns College of Engineering at UCR that augments support for many projects such as this.

References

1. Sherrard, E.C. and Kressman, F.W. (1945) Review of processes in the United States prior to World War II. *Industrial Engineering Chemistry*, **37** (1), 5–8.
2. Saeman, J.F. (1945) Kinetics of wood saccharification: Hydrolysis of cellulose and decomposition of sugars in dilute acid at high temperature. *Industrial Engineering Chemistry Research*, **37**, 42–52.
3. Saeman, J.F., Bubl, J.L., and Harris, E.E. (1945) Quantitative saccharification of wood and cellulose. *Industrial and Engineering Chemistry-Analytical Edition*, **17** (1), 35–37.
4. Wright, J.D. and D'Agincourt, C.G. (1984) Evaluation of sulfuric acid hydrolysis processes for alcohol fuel production. *Biotechnology Bioengineering Symposium*, **14**, 105–123.
5. Grethlein, H.E. and Converse, A.O. (1991) Continuous acid hydrolysis of lignocelluloses for production of xylose, glucose, and furfural, in *Food, Feed, and Fuel from Biomass* (ed. D.S. Chahal), Oxford & IBH Publishing Company, New Delhi, p. 267–279.
6. Zeitsch, K.J. (2000) *The Chemistry and Technology of Furfural and Its Many By-Products*, Elsevier.
7. Grethlein, H.E., Allen, D.C., and Converse, A.O. (1984) A comparative study of the enzymatic hydrolysis of acid-pretreated white pine and mixed hardwood. *Biotechnology and Bioengineering*, **26** (2), 1498–1505.
8. Grethlein, H.E. and Converse, A.O. (1991) Common aspects of acid prehydrolysis and steam explosion for pretreating wood. *Bioresearch and Technology*, **36**, 77–82.
9. Li, N., Tompsett, G.A., Zhang, T. *et al.* (2011) Renewable gasoline from aqueous phase hydrodeoxygenation of aqueous sugar solutions prepared by hydrolysis of maple wood. *Green Chemistry*, **13** (1), 91–101.
10. Huber, G.W. and Dumesic, J.A. (2006) An overview of aqueous-phase catalytic processes for production of hydrogen and alkanes in a biorefinery. *Catalysis Today*, **111** (1–2), 119–132.
11. Huber, G.W., Iborra, S., and Corma, A. (2006) Synthesis of transportation fuels from biomass: chemistry, catalysts, and engineering. *Chemical Reviews*, **106** (9), 4044–4098.
12. McMillan, J.D. (1994) Pretreatment of lignocellulosic biomass, in *Enzymatic Conversion of Biomass for Fuels Production* (eds M.E. Himmel, J.O. Baker, and R.P. Overend), American Chemical Society, Washington, DC, p. 292–324.
13. Hsu, T-.A. (1996) Pretreatment of biomass, in *Handbook on Bioethanol, Production and Utilization* (ed. C.E. Wyman), Taylor & Francis, Washington, DC, p. 179–212.
14. Yang, B. and Wyman, C.E. (2008) Pretreatment: the key to unlocking low-cost cellulosic ethanol. *Biofuels, Bioproducts and Biorefining*, **2** (1), 26–40.
15. Wyman, C.E., Dale, B.E., Elander, R.T. *et al.* (2005) Comparative sugar recovery data from laboratory scale application of leading pretreatment technologies to corn stover. *Bioresource Technology*, **96** (18), 2026–2032.
16. Wyman, C.E., Dale, B.E., Elander, R.T. *et al.* (2009) Comparative sugar recovery and fermentation data following pretreatment of poplar wood by leading technologies. *Biotechnology Progress*, **25** (2), 333–339.
17. Wyman, C., Balan, V., Dale, B. *et al.* (2011) Comparative data on effects of leading pretreatments and enzyme loadings and formulations on sugar yields from different switchgrass sources. *Bioresource Technology*, **102** (24), 11052–11062.
18. Wyman, C.E. and Hinman, N.D. (1990) Ethanol – Fundamentals of production from renewable feedstocks and use as a transportation fuel. *Applied Biochemistry and Biotechnology*, **24–5**, 735–753.
19. Lynd, L.R., Cushman, J.H., Nichols, R.J., and Wyman, C.E. (1991) Fuel ethanol from cellulosic biomass. *Science*, **251** (4999), 1318–1323.
20. Lynd, L.R., Larson, E., Greene, N. *et al.* (2009) The role of biomass in America's energy future: framing the analysis. *Biofuels Bioprod Biorefining*, **3** (2), 113–123.
21. Wyman, C.E. (1994) Alternative fuels from biomass and their impact on carbon dioxide accumulation. *Applied Biochemistry and Biotechnology*, **45–6**, 897–915.
22. Lynd, L.R. (1996) Overview and evaluation of fuel ethanol from cellulosic biomass: Technology, economics, the environment, and policy. *Annual Review of Energy and the Environment*, **21**, 403–465.
23. Perlack, R.D., Wright, L.L., Turhollow, A. *et al.* (2005) *Biomass as a Feedstock for A Bioenergy and Bioproducts Industry: The Technical Feasibility of a Billion-Ton Annual Supply*, Oak Ridge National Laboratory, Oak Ridge, TN, April.

24. Wyman, C.E. and Goodman, B.J. (1993) Biotechnology for production of fuels, chemicals, and materials from biomass. *Applied Biochemistry and Biotechnology*, **39/40**, 41–59.
25. Wyman, C.E. (2003) Potential synergies and challenges in refining cellulosic biomass to fuels, chemicals, and power. *Biotechnology Progress*, **19**, 254–262.
26. Houghton, J., Weatherwax, S., and Ferrell, J. (June 2006) Breaking the biological barriers to cellulosic ethanol: A joint research agenda. Washington US Department of Energy, Report No. DOE/SC-0095.
27. Lynd, L.R., Wyman, C.E., and Gerngross, T.U. (1999) Biocommodity engineering. *Biotechnology Progress*, **15** (5), 777–793.
28. Sun, Y. and Cheng, J.Y. (2002) Hydrolysis of lignocellulosic materials for ethanol production: a review. *Bioresource Technology*, **83** (1), 1–11.
29. Perlack, R., Wright, L., Turhollow, A. *et al.* (April 2005) Biomass as Feedstock for a Bioenergy and Bioproducts Industry: The Technical Feasibility of a Billion-Ton Annual Supply. Oak Ridge, TN: Oak Ridge National Laboratory; 60 p.
30. U.S. Department of Enerty (2011) U.S. Billion-Ton Update: Biomass Supply for a Bioenergy and Bioproducts Industry. Oak Ridge, TN.
31. Wyman, C.E. (2007) What is (and is not) vital to advancing cellulosic ethanol. *Trends in Biotechnology*, **25** (4), 153–157.
32. Moutsoglou, A. (2012) A comparison of prairie cordgrass and switchgrass as a biomass for syngas production. *Fuel*, **95** (1), 573–577.
33. Wiselogel, A., Tyson, S., and Johnson, D. (1996) Biomass feedstock resources and composition, in *Handbook on Bioethanol: Production and Utilization*, (ed. C.E. Wyman), Taylor and Francis, Washington, DC, pp. 105–118.
34. Brigham, J.S., Adney, W.S., and Himmel, M.E. (1996) Hemicellulose: Diversity and applications, in *Handbook on Bioethanol: Production and Utilization*, (ed. C.E. Wyman), Taylor and Francis, Washington, DC, pp. 117–114.
35. Wyman, C.E., Decker, S.R., Himmel, M.E. *et al.* (2005) Hydrolysis of cellulose and hemicellulose, in *Polysaccharides: Structural Diversity and Functional Versatility*, (ed. S. Dumitriu), Marcel Dekker, Inc., New York, 995–1033.
36. Studer, M.H., DeMartini, J.D., Davis, M.F. *et al.* (2011) Lignin content in natural Populus variants affects sugar release. *Proceedings of the National Academy of Sciences*, **108** (15), 6300–6305.
37. Kuhad, R.C., Gupta, R., Khasa, Y.P. *et al.* (2011) Bioethanol production from pentose sugars: Current status and future prospects. *Renewable and Sustainable Energy Reviews*, **15** (9), 4950–4962.
38. Beall, D.S., Ohta, K., and Ingram, L.O. (1991) Parametric studies of ethanol production from xylose and other sugars by recombinant *Escherichia coli*. *Biotechnology and Bioengineering*, **38**, 296–303.
39. Ingram, L.O., Conway, T., Clark, D.P. *et al.* (1987) Genetic engineering of ethanol production in Escherichia coli. *Applied Environmental Microbiology*, **53**, 2420–2425.
40. Ho, N.W.Y., Chen, Z.D., and Brainard, A.P. (1998) Genetically engineered saccharomyces yeast capable of effective cofermentation of glucose and xylose. *Applied and Environmental Microbiology*, **64** (5), 1852–1859.
41. Knappert, D.R., Grethlein, H.E., and Converse, A.O. (1980) Partial acid hydrolysis of cellulosic materials as a pretreatment for enzymatic hydrolysis. *Biotechnology Bioengineering Symposium*, **XXI**, 1449–1463.
42. Lloyd, T.A. and Wyman, C.E. (2005) Combined sugar yields for dilute sulfuric acid pretreatment of corn stover followed by enzymatic hydrolysis of the remaining solids. *Bioresource Technology*, **96** (18), 1967–1977.
43. Brennan, A.H., Hoagland, W., and Schell, D.J. (1986) High temperature acid hydrolysis of biomass using an engineering scale plug flow reactor: results of low solids testing. *Biotechnology Bioengineering Symposium*, **17**, 53–70.
44. Hinman, N.D., Schell, D.J., Riley, C.J. *et al.* (1992) Preliminary estimate of the cost of ethanol production for SSF technology. *Applied Biochemistry Biotechnology*, **34/35**, 639–649.
45. Wyman, C.E. (1994) Ethanol from lignocellulosic biomass – Technology, economics, and opportunities. *Bioresource Technology*, **50** (1), 3–16.
46. Wyman, C.E. (1995) Economic fundamentals of ethanol production from lignocellulosic biomass, in *Enzymatic Degradation of Insoluble Carbohydrates* (eds J.N. Saddler and M.H. Penner), American Chemical Society, Washington, DC, p. 272–90.

47. Wyman, C.E. (1996) Chapter 1, Ethanol production from lignocellulosic biomass: Overview, in *Handbook on Bioethanol, Production and Utilization* (ed C.E. Wyman), Taylor & Francis, Washington, DC.

48. Wyman, C.E. (1999) Biomass ethanol: technical progress, opportunities, and commercial challenges. *Annual Review of Energy and the Environment*, **24**, 189–226.

49. US Department of Energy (1993) Evaluation of a potential wood-to-ethanol process. Washington, DC.

50. Wright, J.D., Wyman, C.E., and Grohmann, K. (1987) Simultaneous saccharification and fermentation of lignocellulose: process evaluation. *Applied Biochemistry Biotechnology*, **18**, 75–90.

51. Wyman, C.E., Decker, S.R., Himmel, M.E. *et al.* (2004) Hydrolysis of cellulose and hemicellulose, in *Polysaccharides: Structural Diversity and Functional Versatility*, 2nd edn (ed. S. Dumitriu), Marcel Dekker, Inc., New York, p. 995–1033.

52. Wyman, C.E. (2001) Twenty years of trials, tribulations, and research progress in bioethanol technology – Selected key events along the way. *Applied Biochemistry and Biotechnology*, **91–93**, 5–21.

53. Wright, J.D. (1988) Ethanol from biomass by enzymatic hydrolysis. *Chemical Engineering Progress*, **84** (8), 62–74.

54. Lynd, L.R., Elander, R.T., and Wyman, C.E. (1996) Likely features and costs of mature biomass ethanol technology. *Applied Biochemistry and Biotechnology*, **57–58**, 741–761.

55. Dale, B.E. and Artzen, C.E. (eds) (1999) *Biobased Industrial Products: Priorities for Research and Commercialization*, National Research Council, Washington, DC.

56. Lynd, L.R., Laser, M.S., Bransby, D., Dale, B.E., Davison, B., Hamilton, R., Himmel, M., Keller, M., McMillan, J.D., Sheehan, J., and Wyman, C.E. (2008) "How Biotech Can Transform Biofuels," *Nature Biotechnology*, **26** (2), 169–172.

57. Chheda, J., Huber, G., and Dumesic, J. (2007) Liquid-phase catalytic processing of biomass-derived oxygenated hydrocarbons to fuels and chemicals. *Angewandte Chemie International Edition*, **46** (38), 7164–7183.

58. Huber, G.W., Chheda, J.N., Barrett, C.J., and Dumesic, J.A. (2005) Production of liquid alkanes by aqueous-phase processing of biomass-derived carbohydrates. *Science*, **308** (5727), 1446–1450.

59. Chheda, J.N. and Dumesic, J.A. (2007) An overview of dehydration, aldol-condensation and hydrogenation processes for production of liquid alkanes from biomass-derived carbohydrates. *Catalysis Today*, **123** (1–4), 59–70.

60. da Costa Sousa, L., Chundawat, S.P.S., Balan, V., and Dale, B.E. (2009) 'Cradle-to-grave' assessment of existing lignocellulose pretreatment technologies. *Current Opinion in Biotechnology*, **20** (3), 339–347.

61. Holtzapple, M.T., Humphrey, A.E., and Taylor, J.D. (1989) Energy requirements for the size reduction of poplar and aspen wood. *Biotechnology and Bioengineering*, **33** (2), 207–210.

62. Zhu, J.Y. and Pan, X.J. (2010) Woody biomass pretreatment for cellulosic ethanol production: Technology and energy consumption evaluation. *Bioresource Technology*, **101** (13), 4992–5002.

63. Bond, J.Q., Alonso, D.M., Wang, D. *et al.* (2010) Integrated catalytic conversion of gamma-valerolactone to liquid alkenes for transportation fuels. *Science*, **327** (5969), 1110–1114.

64. Lynd, L.R., Laser, M., McBride, J. *et al.* (2007) Energy myth three - High land requirements and an unfavorable energy balance preclude biomass ethanol from playing a large role in providing energy services, in *Energy and Society: Fourteen Myths About the Environment, Electricity, Efficiency, and Energy Policy in the United States* (eds B. Sovacool and M. Brown), Springer, New York, NY.

65. Lynd, L.R. (1996) Overview and evaluation of fuel ethanol from cellulosic biomass: Technology, economics, the environment, and policy. *Annual Reviews of Energy and Environment*, **21**, 403–465.

66. Aden, A., Ruth, M., Ibsen, K. *et al.* (2002) Lignocellulosic biomass to ethanol process design and economics utilizing co-current dilute acid prehydrolysis and enzymatic hydrolysis for corn stover. National Renewable Energy Laboratory, Golden, CO, NREL/TP-510-32438.

67. Mosier, N., Wyman, C.E., Dale, B. *et al.* (2005) Features of Promising Technologies for Pretreatment of Lignocellulosic Biomass. *Bioresource Technology*, **96** (6), 673–686.

68. Wyman, C., Dale, B., Elander, R. *et al.* (2005) Coordinated development of leading biomass pretreatment technologies. *Bioresource Technology*, **96** (18), 1959–1966.

69. Wooley, R., Ruth, M., Glassner, D., and Sheehan, J. (1999) Process design and costing of bioethanol technology: A tool for determining the status and direction of research and development. *Biotechnology Progress*, **15**, 794–803.

70. Humbird, D., Mohagheghi, A., Dowe, N., and Schell, D.J. (2010) Economic impact of total solids loading on enzymatic hydrolysis of dilute acid pretreated corn stover. *Biotechnology Progress*, **26** (5), 1245–1251.

71. Humbird, D., Davis, R., Tao, L. et al. (May 2011) *Process Design and Economics for Biochemical Conversion of Lignocellulosic Biomass to Ethanol: Dilute-Acid Pretreatment and Enzymatic Hydrolysis of Corn Stover*, National Renewable Energy Laboratory, Golden, CO.

72. Mosier, N., Wyman, C., Dale, B. *et al.* (2005) Features of promising technologies for pretreatment of lignocellulosic biomass. *Bioresource Technology*, **96** (6), 673–686.

73. Rivers, D.B. and Emert, G.H. (1987) Lignocellulose pretreatment: A comparison of wet and dry ball attrition. *Biotechnology Letters*, **9** (5), 365–368.

74. Gracheck, S.J., Rivers, D.B., Woodford, L.C. *et al.* (1981) Pretreatment of lignocellulosics to support cellulase production using Trichoderma reesei QM9414. *Biotechnology and Bioengineering Symposium*, **11**, 47–65.

75. Millett, M.A., Effland, M.J., and Caulfield, D.F. (1979) Influence of fine grinding on the hydrolysis of cellulosic materials – Acid vs. enzymatic. *Advances in Chemistry Series*, **181**, 71–89.

76. Tassinari, T., Macy, C., and Spano, L. (1980) Energy-requirements and process design considerations in compression-milling pretreatment of cellulosic wastes for enzymatic-hydrolysis. *Biotechnology and Bioengineering Symposium*, **22**, 1689–1705.

77. Tassinari, T., Macy, C., and Spano, L. (1982) Technology advances for continuous compression milling pretreatment of lignocellulosics for enzymatic-hydrolysis. *Biotechnology and Bioengineering Symposium*, **24**, 1495–1505.

78. Khan, A.W., Labrie, J., and Mckeown, J. (1987) Electron-beam irradiation pretreatment and enzymatic saccharification of used newsprint and paper-mill wastes. *Radiation Physical Chemistry*, **29** (2), 117–120.

79. Horton, G.L., Rivers, D.B., and Emert, G.H. (1980) Preparation of cellulosics for enzymatic conversion. *Industrial Engineering Chemistry Product Research and Development*, **19**, 422–429.

80. Karunanithy, C., Muthukumarappan, K., and Gibbons, W.R. (2012) Extrusion pretreatment of pine wood chips. *Applied Biochemistry and Biotechnology*, **167** (1), 81–99.

81. Kumar, P., Barrett, D.M., Delwiche, M.J., and Stroeve, P. (2009) Methods for pretreatment of lignocellulosic biomass for efficient hydrolysis and biofuel production. *Industrial & Engineering Chemistry Research*, **48** (8), 3713–3729.

82. Fan, L.T., Lee, Y., and Gharpuray, M.M. (1982) The nature of lignocellulosics and their pretreatment for enzymatic hydrolysis. *Advances in Biochemical Engineering*, **23**, 157–187.

83. Detroy, R.W., Lindenfelser, L.A., St. Julian, G., and Orton, W.L. (1980) Saccharification of wheat-straw cellulose by enzymatic hydrolysis following fermentative and chemical pretreatment. *Biotechnology and Bioengineering Symposium*, **10**, 135–148.

84. Datta, R. (1981) Energy-requirements for lignocellulose pretreatment processes. *Process Biochemistry*, **16**, 16–19.

85. Kirk, T.K. and Harkin, J.M. (1973) Lignin biodegradation and the bioconversion of wood. *AIChE Symposium Series*, **133**, 124–126.

86. Lemee, L., Kpogbemabou, D., Pinard, L. *et al.* (2012) Biological pretreatment for production of lignocellulosic biofuel. *Bioresource Technology*, **117**, 234–241.

87. Zhao, L., Cao, G.L., Wang, A.J. *et al.* (2012) Fungal pretreatment of cornstalk with Phanerochaete chrysosporium for enhancing enzymatic saccharification and hydrogen production. *Bioresource Technology*, **114**, 365–369.

88. Saritha, M. Arora, A. and Lata (2012) Biological pretreatment of lignocellulosic substrates for enhanced delignification and enzymatic digestibility. *Indian Journal of Microbiology*, **52** (2), 122–130.

89. Gomez, S.Q., Arana-Cuenca, A., Flores, Y.M. *et al.* (2012) Effect of particle size and aeration on the biological delignification of corn straw using Trametes sp 44. *BioResources*, **7** (1), 327–344.

90. Lynd, L.R., van Zyl, W.H., McBride, J.E., and Laser, M. (2005) Consolidated bioprocessing of cellulosic biomass: An update. *Current Opinion in Biotechnology*, **16**, 577–583.

91. Olson, D.G., McBride, J.E., Shaw, A.J., and Lynd, L.R. (2012) Recent progress in consolidated bioprocessing. *Current Opinion in Biotechnology*, **23** (3), 396–405.

92. Shaw, A.J., Podkaminer, K.K., Desai, S.G. *et al.* (2008) Metabolic engineering of a thermophilic bacterium to produce ethanol at high yield. *Proceedings of the National Academy of Sciences of the United States of America*, **105** (37), 13769–13774.
93. Modenbach, A.A. and Nokes, S.E. (2012) The use of high-solids loadings in biomass pretreatment—a review. *Biotechnology and Bioengineering*, **109** (6), 1430–1442.
94. Wang, W., Kang, L., Wei, H. *et al.* (2011) Study on the decreased sugar yield in enzymatic hydrolysis of cellulosic substrate at high solid loading. *Applied Biochemistry and Biotechnology*, **164** (7), 1139–1149.
95. Mohagheghi, A., Tucker, M., Grohmann, K., and Wyman, C. (1992) High solids simultaneous saccharification and fermentation of pretreated wheat straw to ethanol. *Applied Biochemistry and Biotechnology*, **33** (2), 67–81.
96. Yang, B. and Wyman, C.E. (2004) Effect of xylan and lignin removal by batch and flowthrough pretreatment on the enzymatic digestibility of corn stover cellulose. *Biotechnology and Bioenginnering*, **86** (1), 88–95.
97. Liu, C. and Wyman, C.E. (2004) Effect of the flow rate of a very dilute sulfuric acid on xylan, lignin, and total mass removal from corn stover. *Industrial & Engineering Chemistry Research*, **43** (11), 2781–2788.
98. Bobleter, O. (1994) Hydrothermal degradation of polymers derived from plants. *Progress in Polymer Science*, **19**, 797–841.
99. Bobleter, O. and Concin, R. (1979) Degradation of poplar lignin by hydrothermal treatment. *Cellulose Chemistry and Technology*, **13**, 583–593.
100. Torget, R., Hatzis, C., Hayward, T.K. *et al.* (1996) Optimization of reverse-flow, two-temperature, dilute-acid pretreatment to enhance biomass conversion to ethanol. *Applied Biochemistry and Biotechnology*, **57–8**, 85–101.

2

Cellulosic Biofuels: Importance, Recalcitrance, and Pretreatment

Lee Lynd[1,2] and Mark Laser[1]
[1] *Thayer School of Engineering, Dartmouth College, Hanover, USA*
[2] *BioEnergy Science Center, Oak Ridge, USA*

2.1 Our Place in History

The two most profound societal transformations in history have been spawned by radical shifts in humankind's use of natural resources. The agricultural revolution, which spanned about two millennia beginning around 4000 BC, saw hunter-gatherer societies subsisting on wild plants and animals being largely displaced by those cultivating the land to produce crops and domesticated livestock. The industrial revolution followed, beginning around 1700 and lasting roughly two hundred years, during which time preindustrial agricultural societies gave way to those harnessing precious metals and fossil energy to develop sophisticated economies centered around machinery and factories. Now, with ever-increasing indications that resource use is exceeding the world's sustainable capacity, it is clear that a third revolution – the sustainability revolution – must begin soon and must be completed in decades, not centuries [1]. A few centuries hence, we think it is quite likely that people will look at those of us alive today, observe that "It was pretty obvious at the start of the third millennium that humanity needed to rapidly shift from resource capital to resource income," and evaluate us largely on our success at meeting this defining challenge of our time.

2.2 The Need for Energy from Biomass

As the only foreseeable sustainable source of food, organic materials, and fuels that are liquid at atmospheric pressure, plant biomass is a central and essential component of a sustainable world. Whereas biomass can be converted to high-performance liquid fuels, other large-scale sustainable energy sources are most readily converted to electricity and heat. Due to energy density considerations, it is reasonable to expect that organic fuels will meet a significant fraction of transportation energy demand for the indefinite

Aqueous Pretreatment of Plant Biomass for Biological and Chemical Conversion to Fuels and Chemicals, First Edition.
Edited by Charles E. Wyman.
© 2013 John Wiley & Sons, Ltd. Published 2013 by John Wiley & Sons, Ltd.

future. Biofuels are by far the most promising sustainable source of organic fuels and are likely to be a non-discretionary part of a sustainable transportation sector – especially for aviation and heavy-duty vehicles. I is very unlikely that anyone alive today will ride in a battery-powered jet.

In their recent analysis 'Transport Energy and CO_2', the International Energy Agency states "A revolution in technology will be needed to move toward a truly low CO_2 future. This will be built on some combination of electricity, hydrogen, and biofuels." Their BLUE Map scenario – which achieves CO_2 emissions that are 30% below 2005 levels through improvements in vehicle efficiency and introduction of advanced technologies and fuels – has biofuels responsible for about a third of total transport energy in 2050 through meeting 40% of light-duty vehicle (LDV) demand and 30% of trucking, aviation, and shipping demand. The remaining LDV energy will be met by electricity and hydrogen; petroleum fuels comprise the balance for trucking, aviation, and shipping [2]. It is notable that biomass is the largest primary energy source supporting humankind in the BLUE map scenario.

2.3   The Importance of Cellulosic Biomass

The choice of feedstock represents the most important factor impacting key bioenergy performance metrics including scale of sustainable production, productivity (i.e., yield/area/year), land availability, fossil fuel displacement, feedstock cost, conversion cost, and environmental impact. Regarding productivity, perennial cellulosic crops generally outperform annual row crops, which makes sense given that plants grow faster when their composition is optimized for photosynthesis rather than for producing components that are easy to digest or process (e.g., starch, sugar, oils). Cellulosic crops can also be grown on marginal land unsuitable for annual row crop production, reducing potential competition with food production. Biofuels production from cellulosic feedstocks offers greater potential for displacing fossil fuels, as cellulosics contain a significant fraction of energy-rich lignin that can be used to fuel the conversion process. In contrast, processes involving annual row crops typically require external fossil-energy inputs. Lignocellulosic feedstocks also appear to have a cost advantage relative to row crops and sugarcane. Corn, for example, is currently priced above $5/bushel, equivalent to $12/GJ, and soy oil at above $0.50/lb, or $30/GJ. Both commodities are likely to remain at these levels or higher for the foreseeable future. By comparison, cellulosic energy crops are likely to be valued at $60–$100/dry ton or $4–$7/GJ. Finally, cellulosic biofuels offer potentially greater environmental benefits relative to biofuels made from annual crops, including lower net greenhouse gas emissions, improved water use efficiency and water quality, reduced soil erosion, enhanced soil fertility and more positive biodiversity attributes [3]. In fact, many view the use of perennial cellulosic species as essential to achieving sustainable agriculture for reasons beyond bioenergy production. For example, in their detailed discussion, Kahn *et al.* [3] state "Perennial crops would increase soil organic matter, reduce pollution, and stabilize soils against erosion. They would help fields, forests, and rangelands retain water, thereby reducing flooding and helping aquifers recharge. Perennials would also sequester large quantities of CO_2, helping to slow climate change."

2.4   Potential Barriers

There are two primary barriers to realizing the potential of cellulosic biofuels on a large scale: (1) the recalcitrance of cellulosic biomass and (2) land-use concerns, especially those regarding competition with food production. While a detailed treatment of the latter is outside the scope of this book, we and others envision scenarios (e.g., growing crops on abandoned land; growing cool-season grasses on the same land as row crops between fall harvest and spring planting; and improving the productivity of pastureland) that gracefully reconcile large-scale bioenergy production with other priorities. The Global Sustainable Bioenergy Project (http://bioenfapesp.org/gsb/), launched in 2009, seeks to develop and evaluate such scenarios.

The recalcitrance barrier is a matter of biological function. For seeds, upon which current biofuel production in temperate climates is based, the function is to provide energy for the next generation of plants to grow and, in this capacity, to resist decay for a brief period (usually during the winter). By contrast, the biological function of lignocellulose is to hold the plant up, often including elevating the "solar collector" (leaves) of one plant above that of a competing plant. In this capacity, decay must be resisted during the summer months and quite commonly for decades. Plant cell walls therefore contain three primary polymer types – cellulose, hemicellulose, and lignin – arranged in a complex, composite matrix involving multiple layers that provide structural support and recalcitrance to attack by both microorganisms and the elements. Given these divergent functions, it is quite understandable that the carbohydrates present in cellulosic biomass are much more difficult to access than the carbohydrates present in seeds. "Recalcitrance" refers to the difficulty of accessing the carbohydrate present in lignocellulose. Overcoming this recalcitrance is the central challenge to large-scale commercial production of cellulosic biofuels.

2.5   Biological and Thermochemical Approaches to the Recalcitrance Barrier

Two broad categories of conversion technologies exist for producing cellulosic biofuels: thermochemical and biological. Thermochemical conversion involves exposing biomass feedstock to high temperatures (e.g., 300–1200 °C) under oxygen-limited conditions that serve to break the biomass polymers into light-molecule fragments. Depending on the reaction conditions, the molecular fragments can repolymerize into oily compounds, form a carbon-rich solid residue known as char, and/or remain as a gas, rich in CO and H_2. Fuels suitable for transportation (e.g., methanol, ethanol, Fischer-Tropsch diesel and gasoline, and dimethyl ether) can be made from the gas and/or liquid bio-oil by downstream processing [4]. Meanwhile, biological conversion involves the production of cellulolytic enzymes that hydrolyze the cellulose and hemicellulose fractions of biomass and fermentation of the resulting sugars to fuel products such as ethanol and butanol. These steps can be conducted separately or in varying degrees of integration, with single-step conversion (referred to as consolidated bioprocessing or CBP) representing a potential breakthrough in low-cost processing [5].

Biological conversion processes typically offer much higher product selectivity than thermochemical processes. In producing ethanol, for example, biological conversion yields ethanol and CO_2 in equal proportions on a molar basis (and approximately so on a mass basis) – a molar ethanol selectivity of 50% – while thermochemical conversion results in many additional products such as methanol, propanol, butanol, and a mixture of alkanes with ethanol selectivity generally less than 20%. Because the heat of reaction to form ethanol is small and CO_2 has no calorific value, from a fuel perspective about 98% of the energy of the sugars ends up in the ethanol, thus concentrating energy. Yields are typically higher for biological processes as well. The additional products resulting from thermochemical processing also make product recovery more challenging than for biological processes. Meanwhile, thermochemical processes are more flexible with regard to feedstock; in principle, any carbonaceous material (including e.g., manures, waste oils, food waste, and animal refuse) can be gasified and converted to fuels. They also have the advantage of being robust, well-tested processes, and commercially available today. A detailed comparative study of mature biomass conversion technology concluded that biological processing will likely prove to be the lower-cost option for processing carbohydrate, with viable process economics able to be realized at smaller scales than for thermochemical processing [6].

Biochemical and thermochemical processing however need not – and, we think, should not – be viewed as mutually exclusive. Lignin-rich residues from biological processing of carbohydrate, for example, can be converted to fuels and/or power using thermochemical processing. This configuration has the advantage that most, if not all, steam and power inputs for biological conversion can be met largely by capturing waste heat from the thermochemical process. As suggested by this example, integrated configurations involving both biological and thermochemical conversion are in general more efficient than processes that only use biological or thermochemical processing. Analysis of foreseeable mature biomass conversion technologies indicates that the most efficient and profitable configurations combine biological and thermochemical processing. Such integrated processes have the potential to realize efficiencies on a par with petroleum-based fuels and achieve production costs competitive with petroleum fuels at about $30/barrel [6].

Realizing the considerable potential of cellulosic biofuels requires that the recalcitrance of cellulosic biomass be overcome in a cost-effective manner. In the case of promising biological conversion routes, this involves the two key components of pretreatment and enzymatic hydrolysis, the central focus of this book.

2.6   Pretreatment

Like any story, the story of pretreatment of cellulosic biomass can be told beginning at many starting points. One such point is a young scientist named Elwin Reese employed at the US Army Research Lab in Natick, Massachusetts. Alarmed by the short lifetime of canvas tents in tropical climates during World War II, Dr Reese was assigned to look at microbial degradation of cellulose. Together with Dr Mary Mandels and many colleagues, Dr Reese conducted pioneering work in the field for nearly three decades; this research notably involved an aerobic fungus originally named *Trichoderma viride*, but later renamed *T. reesei* in his honor. As Drs Reese and Mandels neared retirement, much had been learned and results on hydrolyzing newsprint were promising, but the problem of how to obtain high hydrolysis yields on lignocellulosic substrates was still unresolved.

In the late 1970s, Dartmouth Professor Hans Grethlein wrote a proposal to the National Science Foundation (NSF) to study dilute acid hydrolysis as a means of making biomass accessible to enzymatic hydrolysis. The idea was that combinations of temperature, residence time, and acid concentration could be found that were sufficiently severe to remove hemicellulose and thus make cellulose accessible to enzymatic attack, but sufficiently mild to not extensively degrade solubilized hemicellulose sugars (still a key tradeoff today). Interestingly, Elwin Reese was one of the reviewers in this proposal and expressed doubt that the process could be effective since cellulose crystallinity would not be decreased. The mechanistic basis for pretreatment effectiveness (and ineffectiveness) is still a subject of active research.

Beginning in the 1980s, studies at the Solar Energy Research Institute (now the National Renewable Energy Laboratory or NREL) established a foundation for the economic evaluation of biologically based processing of cellulosic biomass. Throughout many changes in configuration and advances in performance, and analyses by many groups all over the world, pretreatment has remained among the most costly process steps [6,7]. The cost of the operation, however, extends beyond capital and operating expenses for pretreatment *per se*, due to the multiple and often pervasive impacts on downstream processing.

These impacts arise from the reactivity of pretreated solids, inhibitory compounds present in pretreatment hydrolyzates, and – depending on the process – additional compounds associated with pretreatment that require either recovery (e.g., ammonia) or can lead to operational difficulties (e.g., gypsum). We note also that the fractional cost of pretreatment generally increases as the overall process develops, that is, as the biologically mediated steps improve and as conditions become more commercially viable (e.g., increasing solids concentration).

Notwithstanding the decades-long trajectory of research on pretreatment and related topics, the field has made great strides. These have been enabled by convergent factors, including radical advances in biotechnology and analytical chemistry and, over the last five years, much higher funding from both governments and the private sector in many countries (notably including the United States). Over the next decade, it will likely become clear whether or not humanity will look to biofuels to play a key role in the historic transition to a world supported by sustainable resources addressed at the beginning of this chapter. It is therefore a particularly opportune and indeed important time to collect leading pretreatment research, and the perspectives of leading pretreatment researchers, into a volume such as this.

Acknowledgements

We thank the BioEnergy Science Center for financial support.

References

1. Lynd, L. (2010) Bioenergy: in search of clarity. *Energy & Environmental Science*, **3**, 1150–1152.
2. International Energy Agency (2009) Transport Energy and CO2: Moving Toward Sustainability. ISBN 978-92-64-07316-6.
3. Kahn, P.C., Molnar, T., Zhang, G.G., and Reed Funk, C. (2011) Investing in perennial crops to sustainably feed the world. *Issues in Science and Technology*, **27** (4), 75–81.
4. Verma, M., Godbout, S., Brar, S.K. *et al.* (2012) Biofuels Production from Biomass by Thermochemical Conversion Technologies. *International Journal of Chemical Engineering*, **2012**, 1–18.
5. Lynd, L.R., Weimer, P.J., van Zyl, W.H., and Pretorius, I.S. (2002) Microbial cellulose utilization: fundamentals and biotechnology. *Microbiology and Molecular Biology Reviews*, **66** (3), 506–577.
6. Laser, M., Larson, E., Dale, B. *et al.* (2009) Comparative analysis of efficiency, environmental impact, and process economics for mature biomass refining scenarios. *Biofuels, Bioproducts, and Biorefining*, **3**, 247–270.
7. Humbird, D., Davis, R., Tao, L. *et al.* (2011) Process Design and Economics for Biochemical Conversion of Ligno-cellulosic Biomass to Ethanol. National Renewable Energy Laboratory, Technical Report NREL/TP-5100-47764.

3

Plant Cell Walls: Basics of Structure, Chemistry, Accessibility and the Influence on Conversion

Brian H. Davison[1], Jerry Parks[1], Mark F. Davis[2] and Bryon S. Donohoe[2]

[1] *Oak Ridge National Laboratory and BioEnergy Science Center, Oak Ridge, USA*
[2] *National Renewable Energy Laboratory, Golden and BioEnergy Science Center, Oak Ridge, USA*

3.1 Introduction

This book is focused on the pretreatment of plant biomass, a necessary step for efficient conversion of plant cell-wall materials to liquid transportation fuels and other products. Pretreatment is required because it is difficult to access, separate, and hydrolyze monomeric sugars from biopolymers within biomass. Accessible sugars can be further upgraded to products through chemical processes such as aqueous phase reforming or biological routes such as fermentation of the sugars to ethanol. This resistance to degradation or difficulty to release the monomers (mostly sugars) is commonly referred to as recalcitrance [1]. Many methods can be employed to overcome recalcitrance, but the underlying cause of recalcitrance lies in the complex combination and diversity of chemical and structural features of the plant cell walls.

Recent studies by Perlack *et al.* [2,3] estimated that there is approximately 1.4 billion tons of biomass available annually in the United States, which could replace up to one-third of the petroleum-derived fuels currently used. This study determined that 1 billion tons from agricultural lands and an additional 368 million dry tons from forest lands could be sustainably harvested annually. The types and amounts of biomass available for conversion to biomass-derived fuels, whether agricultural residues, forest residues, or purpose-grown energy crops such as switchgrass, willow or poplar, is dependent on geography and climate, as shown in Figure 3.1.

Aqueous Pretreatment of Plant Biomass for Biological and Chemical Conversion to Fuels and Chemicals, First Edition.
Edited by Charles E. Wyman.
© 2013 John Wiley & Sons, Ltd. Published 2013 by John Wiley & Sons, Ltd.

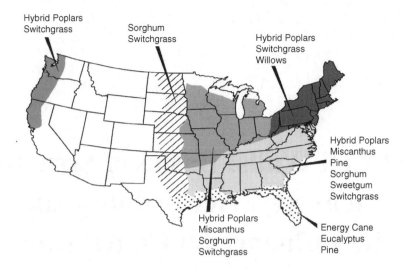

Figure 3.1   *Geographical distribution of non-agricultural residue feedstocks that can be plantation grown for biomass conversion. (Reproduced by permission of Ecological Society of America [4]).*

The wide diversity of biomass feedstocks available for conversion creates unique challenges for harvesting and conversion due to variability in both physical and chemical properties of the feedstocks. Many of the biological barriers to conversion of biomass to fuels were covered in a recent US Department of Energy Roadmap [5]. The three main classes of herbaceous, hardwood and softwood feedstocks have unique aspects of the cell-wall anatomy, macromolecular architecture, and polymer chemistry. These differences extend down to each species within these broader categories, and such differences in cell-wall composition and architecture contribute to the phenomena of biomass recalcitrance at multiple scales. Different pretreatment and bioconversion strategies have been developed to overcome biomass recalcitrance, and their success has been shown to vary with the biomass feedstock [6]. For example, dilute acid pretreatments have been shown to be very effective for the pretreatment of herbaceous materials but fail at the same thermochemical severity to be an effective pretreatment for woody feedstocks [7]. Severity is a function of temperature, duration time, and concentration (i.e., acid).

In this chapter, we will briefly survey the basics of plant cell-wall structure and composition as well as the variability of lignocellulose in plants that relate to pretreatment. We will then return to the chemical components of lignocellulose, first lignin and then cellulose. The discussion of hemicellulose will focus on its branching, decorations, and cross-linkages. We will also discuss the influences of the microscopic structure of biomass: the tissue, cellular, and macromolecular-scale architecture that contributes to recalcitrance. Finally, we will survey how molecular modeling and simulation is being used to explore biomass structure and inform the nature of recalcitrance in lignocellulosic biomass. Given the diversity of available feedstocks, the inherent variability of their plant cell walls, and the impact this variability can have on the recalcitrance of biomass, a primer on lignocellulose is appropriate to set the stage for the following chapters. The reader is referred to a modern textbook on the subject of plant cell walls for more detailed information [8].

3.2   Biomass Diversity Leads to Variability in Cell-wall Structure and Composition

The primary carbohydrate comprising the cell wall common to all plants is cellulose, a β-1,4 -linked glucose polysaccharide. Cellulose in the cell wall forms long, oriented microfibrils, which may coalesce into larger and longer fibrils [9]. The cellulose microfibrils are hydrophobic and can be highly crystalline, features that contribute greatly to the recalcitrance of biomass. For example, the 100 crystal face of a microfibril is more hydrophobic than the other faces and selectively binds the CBD (cellulose binding domain) [10]. Hemicelluloses are a class of polysaccharides that have variable compositions and structures depending on the plant source. For example, hemicelluloses isolated from herbaceous grass species, such as switchgrass, are composed of glucuronoarabinoxylans which are complex, branched polysaccharides composed mainly of pentose (five carbon) sugars. On the other hand, hemicelluloses in softwood species, such as pines, are predominantly composed of galactoglucomannans, which have a backbone of β-1,4 linked D-mannopyranose and D-glucopyranose units. Hemicelluloses are amorphous, branched, single-chain polysaccharides and are not particularly recalcitrant to conversion. Hemicellulose chains are thought to interact with more than one cellulose fibril so that they form non-covalent cross-links between cellulose bundles.

The final main structural polymer, lignin, is a complex three-dimensional polyphenolic polymer that partially encases the plant cell-wall polysaccharides and cellulose microfibrils in lignified (i.e., secondary) plant cell walls. Lignin is generally not found in the primary wall of newly formed cells. Lignin provides mechanical and elastic support, facilitates water and nutrient transport, provides a chemical barrier to microbial pathogens, and is also understood to be a key contributor to recalcitrance.

In addition to these three main polymers of lignocellulose, there are other non-structural components within the plant cell wall. These components, such as extractives, protein, ash, and pectin, vary greatly with species, tissue, plant maturity, harvest times, and storage, and are greatly influenced by environmental factors and stress. Extractives are a complex mixture of compounds which can include sugars, terpenoid compounds, and monolignols. A major theme of biomass recalcitrance that adds to the complexity is that many of the cell-wall components such as lignin, hemicellulose, and proteins can cross-link with each other to create a complex matrix that is resistant to chemical or biological attack.

Depending on the plant species, there is considerable variation in the relative amounts of each of the structural components, cellulose, hemicellulose, and lignin, within the cell walls (Table 3.1), as well as physical properties such as cell-wall thickness and porosity (Figure 3.2). The composition, structure, and interactions of the biopolymers composing the lignocellulosic matrix serve many interrelated functions for the plant, including the primary function of providing structural features that create mechanical support, allowing for internal transport of water, nutrients, and photosynthate throughout the plant. In physical terms, lignocellulose is a molecular-level structured composite material. An imperfect analogy to the macroscopic world is reinforced concrete. Here, cellulose fulfills the role of the steel rods (i.e., rebar) that provide strength over long distances. Hemicellulose represents the wire mesh or cable that wraps around the cellulose rods, providing extra strength and linkages. Lignin acts as the concrete that fills the remaining gaps and sets, holding everything in place while excluding water from the polysaccharide environment. It is important to note that terrestrial plants have co-evolved with herbivores and cell-wall degrading microbes; millions of years of evolution have therefore created living plants that are very resistant to both mechanical and biological decay.

Table 3.1   *Typical biomass composition for a variety of feedstocks, highlighting the diversity in chemical make-up (Wiselogel et al., 1996) [11].*

Feedstock	Cellulose	Hemicellulose	Lignin
Corn stover	36.4	22.6	16.6
Wheat straw	38.2	24.7	23.4
Rice straw	34.2	24.5[a]	11.9
Switchgrass	31.0	24.4	17.6
Poplar	49.9	25.1	18.1

[a] Xylan value only was reported.

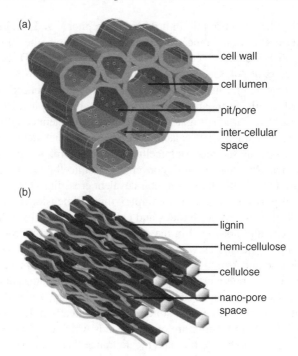

Figure 3.2 *Cartoon depiction of (a) cellular-scale and (b) macromolecular-scale structure and composition of lignocellulosic biomass. Several scales of porosity from the cell lumen to the nanopores between coated microfibrils are also indicated. (Reprinted with permission from Viamajala et al. [12] © 2010 Springer Science + Business Media B.V.).*

Previous research has contributed to a significant understanding of the biosynthesis of the various monomers from which the biopolymers found within the cell wall are built, and our knowledge of how the monomers are transported and assembled to create the plant cell wall is rapidly developing but is still incomplete. However, while monomer synthesis pathways are largely known (Figure 3.3a and c), the understanding of formation and polymerization into a specific part of the cell wall is incomplete. A plant cell wall consists of multiple layers formed during genesis and development of the cell. The composition and structure of these wall layers change depending on cell type, tissue, and location. The mature secondary cell wall contains most of the lignocellulosic biomass, but its structure and organization begins in the primary cell wall. Albersheim *et al.* [8] provide an excellent review and summary of the synthesis and nature of primary and secondary cell walls. Some of the complexity of the primary and secondary cell walls is shown in Figure 3.3.

3.3 Processing Options for Accessing the Energy in the Lignocellulosic Matrix

The challenge in pretreatment, based on the model of lignocellulose as a molecular-level structured composite material, is to disassemble this composite into its valuable constituent monomers (i.e., sugars and aromatics) without loss or damage to the sugar monomers released by pretreatment. The sugars, or polysaccharides, must be preserved in order to be a feedstock for further specific bioconversion processes into fuels or chemicals. The lignin might be used as an energy source for power but also has significant potential for higher-value uses. This is one of the great benefits of the so-called biological approaches to lignocellulose conversion: the ability to make specific products and have potentially valuable co-products.

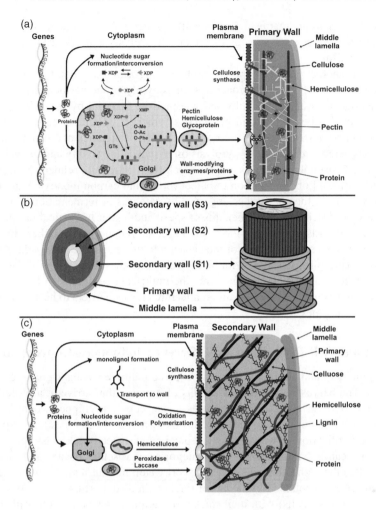

Figure 3.3   *The plant cell wall. (a) Biosynthesis of primary cell walls. (b) Plant cell-wall layers. (c) Biosynthesis of secondary plant cell walls. (Reprinted with permission from Mohnen, Bar-Peled, Sommerville, Chpt 5, page 95, in Biomass Recalcitrance: Deconstructing the Plant Cell Wall for Bioenergy. Michael E. Himmel, ed., Wiley-Blackwell publishing [1]).*

Hemicellulose is easier to remove from the cell-wall matrix than the other components during pretreatment under acidic conditions. The downside of this ease of removal from the cell wall is that the hydrolyzed free sugars and oligosaccharides will continue to react under these pretreatment conditions and degrade into undesirable compounds for biological conversion. These include furans which, in addition to being a loss of sugars and the overall process yield, are also inhibitory to subsequent biological conversion processes. While most of the hemicellulose can be easily removed by pretreatment, the covalent cross-linkages appear to create a residue of hemicellulose or lignin carbohydrate complexes, which may continue to shield or partially protect cellulose microfibrils from attack. Hemicellulose can be substituted at hydroxyl groups with different side-chain chemical groups, primarily acetyl groups, but ferulic acid and other side groups are significant as they may be more resistant to hydrolysis. The acetyl groups easily hydrolyze under thermal or acidic conditions to form dilute acetic acid, which serves to lower the pH of the pretreatment liquor and increase the rate of hydrolysis. Thus, even a hot water pretreatment has the benefit of actually being a mild autocatalyzed dilute acid pretreatment. The negative aspect is that the acetate formed must be neutralized before further processing steps such as enzymatic hydrolysis and also can be inhibitory to many fermentative microorganisms.

We note that there is another family of approaches to dealing with the difficult problem of biomass recalcitrance. These are the thermochemical approaches such as pyrolysis and gasification. These processes deal with the intimate associations of lignocellulose as a composite material by foregoing preservation of the cell-wall monomers. In pyrolysis, the solid biomass is "cracked" at temperatures of 450–650 °C in the absence of oxygen into a "biocrude," a very complex mixture of organics and water that is highly acidic and oxygenated. Considerable upgrading of pyrolysis oil to reduce the acidity and oxygen content is necessary before the material is suitable either as a blend stock for a petroleum refinery or as a finished fuel. In gasification, the goal is for all organic carbon to be converted into carbon monoxide and hydrogen, a mixture called syngas, which can then be further upgraded by catalysis or fermentation to fuels, chemicals, or other products. These approaches have the benefits of speed and higher theoretical product yields, but with the added challenges and costs of building the desired specific products from either "biocrude" or single-carbon feedstocks. They also sacrifice some of the capacity to make specific value-added by-products from lignin, sugar, or fiber that can be preserved with biological approaches. Regardless of the approach used, all biomass conversion technologies have to overcome the multiscale complexity of biomass particles. The organization of plant tissues, cells, and cell-wall architecture creates barriers to heat and mass transport and inhibits accessibility to enzymes.

3.4   Plant Tissue and Cell Types Respond Differently to Biomass Conversion

Plants are complex multicellular structures that have evolved multiple mechanisms to resist degradation by microbial enzymes. The harvested and milled chunks of biomass that arrive at a biorefinery comprise a heterogeneous mixture of tissue types including leaves, stems, cobs, and chipped wood due to the variability discussed in Section 3.2. Each of these tissues contains its own mix of cell types that respond differently to pretreatment and saccharification. If we consider a typical biomass feedstock grass species such as switchgrass, the tissue types can be simplified into the categories of vascular tissues and ground tissues [13]. Vascular tissues contain the elongated cells that transport water and nutrients through the living plant and those cells are themselves surrounded by layers of thick-walled supporting cells. These supporting cells (termed sclerenchyma or fiber cells) with their thickened, lignified secondary cell walls comprise most of the mass and therefore contain most of the carbohydrates in grass biomass particles. The other thick-walled cell types in vascular bundles of grass species are xylem cells. While the function of xylem cells in the living plant is distinct from the function of the supporting fiber cells, both cell types share the characteristic of a thickened, lignified cell wall that is especially recalcitrant during pretreatment and conversion.

In contrast to the xylem and supporting fiber cells in grass species such as switchgrass, the space between the vascular bundles and the epidermis is filled with ground tissue. Ground tissue makes up most of the stem's volume and is composed of thin-walled parenchyma cells that typically lack a secondary wall, are not lignified, and are relatively amenable to pretreatment and saccharification processes. It is not uncommon to see pretreated grass stems where the ground tissue is severely impacted and appears to have largely solubilized, while the vascular bundles and epidermis appear nearly unchanged [14]. Wood parenchymas consist of cells whose primary function is the conduction and storage of food materials. There are three types of wood parenchyma: ray parenchyma which make up the bulk of ray tissue; epithelial parenchyma which surround resin canals and therefore are only present in coniferous woods; and axial parenchyma which extend along the grain in the form of strands.

Different plant species have long been known to have different responses to pretreatment. However, evidence is increasing that there is significant variability even within a species. Studer *et al.* [15] recently presented analyses of a number of *Populus* samples for hot water pretreatment and enzymatic sugar release. These screening results showed variability from about 0.2 to almost 0.7 g carbohydrate released per gram of biomass. Surprisingly, some lower lignin content samples showed appreciable hexose release after only hot water treatment.

3.5   The Basics of Plant Cell-wall Structure

The thickness of the walls in xylem and fiber tissues is due to the formation of an extensive secondary cell wall as the cell matures. The sequence of events in the development of these cells is that after the cells have ceased growing by elongation, they begin to deposit a multilamellar secondary cell wall toward the cell lumen side of the primary cell wall [16]. At some point during and following secondary wall formation, the entire wall is infused with lignin monolignols [17]. The lignin polymerizes into the spaces in the existing wall and, consequently, significantly lowers its porosity [18]. Finally, the cell dies and its contents are absorbed, leaving a rigid water-impermeable cell-wall barrier. The biomass conversion perspective of the plant cell wall is, by necessity, heavily skewed toward the thick, lignified, secondary cell walls. These are the walls that harbor most of the mass and therefore most of the structural sugars in biomass. Unfortunately, they are also the most recalcitrant.

Chemically, the plant cell wall is understood to be composed of cellulose, hemicelluloses, pectins, proteins, and often lignin. However, it is the complex intermingling of cross-linked layers brought together largely by self-assembly and template-assembly processes that create the complexity and resilience of the plant cell wall. The complex macromolecular architecture of the cell wall is a result of the self-ordering properties of cell-wall polymers generating a high degree of structural organization and complexity [19]. Three themes that govern the architectural plan of the plant cell wall are: (1) fibrous structural units embedded in an amorphous matrix; (2) covalent and non-covalent cross-linking; and (3) a polylamellate construction [20].

The plant cell wall is sometimes envisioned as three intermingled networks that are extensively restructured during development. One is a protein network formed by the various classes of glycoproteins that contribute a scaffold to initial cell plate [21]. This network organizes cell plate membranes and the incorporation of initial cell-wall components as they are delivered to the cell plate by Golgi-derived vesicles. This original scaffold gets extensively remodeled and recycled during cell growth and maturation. While the protein network is not usually considered a significant contributor to recalcitrance, its role in establishing a template for cell-wall construction is still important. The second network is the pectic polysaccharide network. In the primary cell wall, the pectin network is credited with dictating key structural and mechanical properties of the wall including water content and porosity [22]. Again, this network is remodeled during growth and development and is usually not of critical concern for biomass conversion because the pectic polysaccharides are not especially recalcitrant to hydrolysis and extraction by pretreatment.

Finally, and arguably most importantly from the perspective of bioconversion, is the network of cellulose and hemicellulose. The main structural unit of the wall, the cellulose microfibrils thought to be composed of *c.* 36 chains of cellulose, are 5–10 nm in diameter, many micrometers in length, and spaced 20–40 nm apart [23]. They are embedded in, and non-covalently cross-linked by, a matrix created by the other major wall polysaccharides: the hemicelluloses. The cellulose/hemicellulose network is intermeshed with, but not cross-linked to, the pectic polysaccharide network. The final porosity of the cell wall is 5–10 nm [24]. This porosity is sufficient to allow some diffusion of very small proteins but too small to allow significant accessibility to cellulolytic enzymes. Mass transfer considerations suggest that pore size should be in the range of 50–100 nm to allow sufficient penetration of enzymes into cell walls. One of the primary goals of pretreatment is to increase cell-wall porosity for effective enzyme transport and penetration to the cellulose surface.

3.6   Cell-wall Surfaces and Multilamellar Architecture

Most of the biomass that is processed for bioconversion is dead plant tissue composed of chambers enclosed by cell walls referred to as lumen (see Figure 3.2). The lumen are the chambers that housed the cellular organelles and machinery that originally constructed the cell walls while the plant was alive and actively growing. Cell-wall components are typically synthesized within the living cell (by the secretory organelles or by transmembrane proteins) and then transported into the cell wall where the final steps in assembly and maturation occur. In the case of xylem and fiber cells, the cells themselves undergo programmed cell death and are recycled by the plant. The cell lumen represents the largest scale of porosity that is relevant to biomass recalcitrance. The cell lumen is in the size range of tens of micrometers in diameter and can be many hundreds of micrometers in length. In their longitudinal dimension they are often designed to transport materials; transport is more limited in the radial direction, however. This is the scale of porosity that many people imagine when they think about a biomass particle because it is the scale that is nearly visible to the naked eye. These pores do have some impact on recalcitrance because the cell lumen and intercellular spaces can trap air that may impede the bulk flow of pretreatment chemistries throughout a biomass particle.

However, the macroporosity is not the critical barrier for cellulolytic enzymes. For these enzymes, entering the cell lumen is relatively easy but only provides the enzyme access to one surface of the multilayered cell wall. Somewhat surprising to the casual observer used to viewing only 2D slices of biomass, the intercellular spaces present an additional and important access point to the cell-wall surface. The intercellular space, formed at the juncture among adjacent cells forming cell corners, is a continuous 3D space throughout the plant tissue, which provides critical access to the "back side" of the cell walls (Figure 3.4). By exploiting this route, enzymes gain access to the middle lamella and primary cell walls that have a different composition from the secondary cell wall on the cell lumen side. One of the architectural motifs of the plant cell wall is that it is constructed as multiple, concentric layers or lamellae [25]. Single lamellae are one cellulose microfibril thick but, in transmission electron microscopy (TEM) micrographs, the lamellar structure is often evident in the staining pattern of layers of varying thickness.

The structural impact of thermochemical pretreatment and enzymatic saccharification on biomass cell-wall surfaces has been studied by multiple groups [14,26–29]. These investigations have revealed the

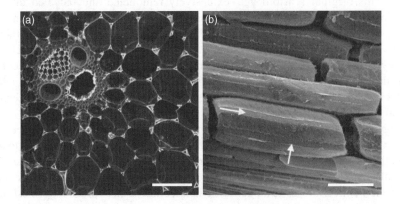

Figure 3.4   *(a) Confocal scanning laser microscopy and (b) scanning electron microscopy micrographs revealing the intercellular spaces within a biomass particle. (a) A section of untreated corn stover stem labeled with the JIM5 pectin antibody highlights the cell corners. (b) In a pretreated corn stover sample the cells have become disjoined, revealing the cell corner surfaces as regions of coalesced lignin accumulation (arrows). Scale bars = 50 μm. (Images courtesy of NRELs BSCL, unpublished).*

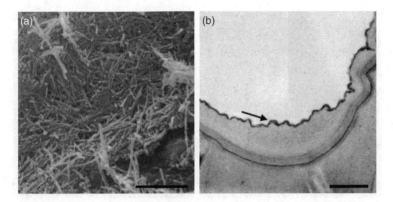

Figure 3.5   *(a) Scanning electron microscopy and (b) transmission electron microscopy micrographs revealing the impact of pretreatment on eroding the surface of switchgrass stem cell walls. (a) A lime pretreatment showing partially unsheathed layers of microfibrils. (b) An aqueous ammonia pretreated sample displays an irregularly eroded surface (arrow). Scale bars: (a) 5 μm; and (b) 1 μm. (Adapted from Donohoe et al. [29] © 2011, Royal Society of Chemistry).*

different modes of cell-wall deconstruction that directly affect the enzyme-accessible cell-wall surface area. One mode of increasing the accessible surface area of secondary cell walls is to etch away the lignin-containing matrix at the lumen surface. Etching is seen most commonly in pretreatments that employ base chemistry. This etching treatment typically does not have an impact on the wall beyond 10–20 nm into the surface, but can still provide a substantial increase in initial enzyme binding (Figure 3.5).

Another mode of increasing surface area during pretreatment is by delamination. Delamination is often seen in samples treated with acidic pretreatments, is enhanced by flowthrough reactors, and can increase the accessible surface area throughout the 3D volume of the cell wall. A final mode of increasing accessibility in pretreated biomass is through rapid pressure drops at the end of pretreatment, such as steam gun or AFEX protocols. These systems can cause delamination of the wall and also generate additional new micropore structure within the walls [29].

3.7 Cell-wall Ultrastructure and Nanoporosity

The intercellular spaces formed at the junction between cells create an interconnected network throughout plant tissues, but the cell lumen are in fact also a continuous interconnected space. In the living plant, cell-to-cell communication is mediated through plasmodesmata, the *c.* 50-nm-diameter plasma-membrane-lined tubules that connect adjacent cells through the cell-wall barrier. In senesced dried biomass, the remains of the plasmodesma are seen as pores or pits in the cell wall. Pits are regions in the cell wall where no secondary cell wall was deposited and an open pore is maintained between adjacent cell lumen (Figure 3.6). Pits may provide an escape route for the air trapped in the cell lumen of dry biomass; however, these small openings can close up due to cell-wall drying and can become occluded by relocalized lignin globules during pretreatment. Pits are an ultrastructural feature of the cell wall that can be considered part of the nanoscale porosity of a biomass particle. However, even though pits and pit fields are only 20–100 nm in diameter, they still do not represent a fundamental barrier to cellulolytic enzymes. In fact, between the interconnected intercellular space created at the cell corners and the interconnected cell lumen space created by the cell-wall pits, enzymes have good initial access to the two surfaces of the cell wall.

It is clear from multiple lines of research that the most fundamental barrier to effective enzymatic conversion of lignocellulosic biomass to sugars is enzyme accessibility to a reactive surface. At one level,

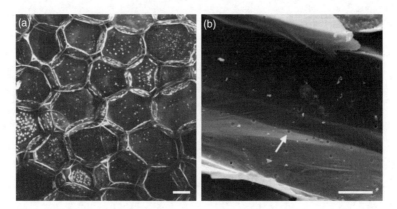

Figure 3.6 *(a) Confocal scanning laser microscopy and (b) SEM micrographs showing pits in corn stover stalk cell walls. (a) Section of dilute acid pretreated corn stover labeled with a fluorescently tagged CBM3 probe highlights the cell pits. (b) At higher resolution, an SEM micrograph reveals pits that have a pit membrane (arrow) with smaller micropores. Scale bars = 20 μm. (Unpublished images courtesy of NRELs BSCL).*

accessibility is a simple concept of physical dimensions and barriers. The architecture of the plant cell wall is a complex biomatrix with porosity on the scale of 5–10 nm. While pectins may have been the controlling factor in primary cell-wall porosity, lignin becomes the porosity gatekeeper in cells with lignified secondary cell walls. Cel7a, the main exoglucanase in a *Trichoderma reesei* enzyme mixture, is a *c.* 4 × 5 × 13 nm enzyme [30]. It would seem that the native porosity of plant cell walls would be sufficient to let these enzymes pass. In fact, the native cell wall is not nearly porous enough. Part of the reason is that the enzyme brings along with it a hydration shell that increases its effective size, requiring a much larger pore space for effective penetration. This restricted nanoporosity of the cell-wall matrix is a key aspect of biomass recalcitrance.

The primary role of pretreatment is to increase the porosity and accessibility of biomass cell walls to cellulolytic enzymes. Although enzymes can thoroughly penetrate cell walls after high severity pretreatments, incomplete cellulose conversion by cellulases suggests that additional barriers exist at the macromolecular level. One potential barrier is obstruction of the active face of cellulose microfibrils by residual lignin or hemicellulose that prevent cellulases from binding to cellulose. The critical concept therefore becomes not just physical accessibility, but accessibility to the right substrate surface.

3.8   Computer Simulation in Understanding Biomass Recalcitrance

In this section we provide an overview of selected recent applications of computer simulation applied to understand plant cell-wall chemistry and structure at the molecular level and inform the origins of recalcitrance. Various publications are highlighted to provide examples from the recent literature that show how simulation is being used to help overcome recalcitrance. Beckham *et al.* discuss the use of computation for understanding enzymatic degradation of biomass; in addition, we discuss the role of simulation in understanding lignocellulosic structure [31].

3.8.1   What Can We Learn from Molecular Simulation?

Many important biomolecular phenomena cannot be studied directly using currently available experimental techniques. Computer simulation is particularly well suited to such systems. Simulation techniques have advanced to the point that they can provide atomic-resolution insight into the structure and dynamics of macromolecular assemblies such as lignocellulose. Other techniques can be used to calculate highly accurate molecular structures and properties of lignin. For example, accurate strengths (energies and enthalpies) of key linkages in lignin are among the many useful quantities available through computational means [32].

The standard tools of molecular simulation are classical molecular dynamics (MD) simulation and quantum chemistry. In MD, a molecular system is propagated in time according to Newton's second law ($F = ma$), usually under temperature and pressure control. The potential energy of the system is governed by a force field, which includes all the energetic terms needed to describe the interactions among all atoms in the system. A typical system, consisting of solute(s), solvent, and ions, might contain tens to hundreds of thousands of atoms, but simulations of multimillion-atom systems are now possible. Timescales on the order of nanoseconds to microseconds are achievable for systems of this size. A challenge is that many of the systems of interest here are at the upper ranges of these atom numbers and timescales.

Unlike classical MD, quantum chemical calculations consider electrons explicitly. Quantum chemistry is aimed at calculating the molecular wave function (or electron density) and properties deriving from it. Such calculations can yield very accurate structures and energies of molecular systems and are particularly useful for studying chemical reactions. Numerous quantum chemical methods and approximations of varying accuracy and computational cost exist. Most commonly, density functional theory (DFT) methods are used because they can achieve high accuracy at reasonable computational costs. Computationally more demanding *ab initio* methods are also used, often as benchmarks to assess the accuracy of more approximate methods such as DFT. The following studies exemplify how the combination of theory and experiment can yield key insights into lignin structure and reactivity that might not be obtainable otherwise.

3.8.2   Simulations of Lignin

Classical and quantum chemical approaches have been used to study lignin biosynthesis and degradation. Durbeej used DFT calculations to investigate the spin density distributions of coniferyl radical and dilignol radicals [33]. Spin distributions show where an unpaired electron prefers to reside in a radical species. For coniferyl radical, the author found that the largest fraction of the unpaired electron spin was localized on the phenolic oxygen (O4), although there was also significant spin localization on the β carbon atom of the allyl alcohol group and the aromatic ring carbons (C3 and C5) at the *meta* positions relative to the allyl group [34]. As sites with the highest spin density are expected to be the most reactive, these results provide a possible explanation for the prevalence of β-O-4 linkages in natural lignin.

Durbeej and Eriksson [35] performed DFT calculations to investigate the thermodynamics and kinetics of dehydrogenative coupling of coniferyl species (radical-radical and radical-alcohol) to form guaiacyl β-O-4 dimers. They obtained reaction energies of more than *c.* 20 kcal/mol for the radical-radical coupling process, quantifying the favorability of monolignol radical coupling reactions.

Computational studies were carried out on lignin pyrolysis reactions of phenethyl phenyl ether (PPE), which is the simplest model of a β-O-4 linkage in lignin. Jarvis *et al.* [36] used accurate *ab initio* calculations and Transition State Theory to calculate rate constants for several PPE pyrolysis reactions. They observed very good agreement between computation and photo-ionization mass spectrometry and matrix-infrared (IR) experiments. Beste *et al.* [37] used DFT calculations to predict the selectivity of PPE pyrolysis reactions for the β-O-4 linkages. They calculated structures and energies of transition states for these processes, and the computed values were found to be in good agreement with experimental data.

In a recent computational study [38], bond dissociation enthalpies were computed using DFT for 69 lignin model compounds containing β-O-4, α-O-4, β-5, and biphenyl bonds. A key finding of that work was the prediction that oxidation of primary and secondary alcohol groups on the alkyl substituents of lignin macromolecules will result in lower bond dissociation energies. Exploiting this property may lead to more facile deconstruction or conversion of lignin to useful bioproducts. In a similar study, Parthasarathi *et al.* [39] quantified the homolytic bond dissociation energies of various C—C and C—O (ether) bonds in a series of 65 lignin-like model compounds. They observed a strong inverse relationship between the calculated length of a particular C—C bond and its bond strength: the longer the bond, the easier it is to break. However, no similar correlation was found for ether bonds. Instead, electronic and steric effects arising from the presence of methoxy substituents were found to have significant effects on ether bond strengths. Ether linkages were found to be weakened significantly when methoxy substituents are present at both *ortho* positions on the phenyl rings.

Sangha *et al.* [40] used DFT calculations to carry out a comprehensive study of radical-radical coupling reactions involving *p*-coumaryl, coniferyl, and sinapyl alcohol radicals. Coupling reactions with β-O-4, β-β, and β-5 linkages were computed to have the most favorable reaction enthalpies, whereas dimers with 5-O-4, 5-5, and β-1 linkages were less favorable by *c.* 5–20 kcal/mol. The authors found that the presence of *p*-coumaryl radicals has a significant effect on product regioselectivity, particularly in reactions involving coniferyl radical. In the presence of p-coumaryl radicals, coniferyl radicals enhance the favorability of C—C bond formation relative to C—O interunit bonds. The study therefore reinforced strategies involving modification of the composition of lignin in plants to improve sugar release from lignocellulosic biomass.

3.8.3   Simulations of Cellulose

Bergenstråhle *et al.* [41] used MD to study, at the molecular level, why cellulose is insoluble. They calculated the free energy required to separate solvated oligomers of glucose, cellobiose, cellotriose, and cellotetraose in different orientations. The important finding from their work was that inter-oligomer hydrophobic association and hydrogen bonding are responsible for the insolubility of cellulose. Specifically, they found a high entropic cost for hydrating glucose hydroxyl groups and no increase in configurational entropy of a short cellulose chain in going from the crystalline to the dissolved state. Taken together, these results clearly show why the crystalline state is strongly favored over the solvated state.

Matthews *et al.* [42] used MD to study models of microcrystalline cellulose Iβ. During their simulations, they observed significant conformational changes from the initial state of their models including a twist along the axis of the microcrystal, a significant tilt in the sugar rings, and significant interlayer hydrogen bonding. The most striking result of the study was the structuring of water molecules near the cellulose surface. The local water density at the interface was 1.3 times that of the bulk solvent, due to strong hydrogen bonding with hydroxyl groups of the glucose units. Based on the simulations and experimentally observed slow rates of enzymatic hydrolysis, the authors hypothesized that the local water density near the cellulose surface might create a barrier that inhibits the approach of cellulase enzymes. The detailed insight obtained from this study is difficult to obtain from experiments.

In a recent study, Matthews *et al.* [43] used available X-ray crystal structures to construct a model of a hydrated cellulose Iβ microfibril to study structural changes resulting from thermal treatment. They performed MD simulations at 500 K for 100 ns and observed a shift from a 2D to a 3D hydrogen bonding network and concomitant loss of twist in the structure. The study showed how changes in local conformations and non-bonded interactions can translate into large-scale structural metamorphoses. Their results provided atomic-level insight and helped explain several previous observations from X-ray diffraction, spectroscopy, calorimetry and thermogravimetry; they also suggested follow-up experiments based on their findings and simulations.

Beckham *et al.* [44] used MD simulations to examine the morphology dependence on enzymatic cellulose deconstruction at atomic detail. They calculated the thermodynamic work required to decrystallize a single chain from the surface of four different cellulose polymorphs. It was found that the required work decreased in the order Iβ > Iα > III$_I$ > II, consistent with experimental observations that the synthetic polymorphs cellulose II and III$_I$ are less recalcitrant to cellulolytic degradation by enzymes than cellulose I. Possible molecular-level explanations for this behavior in cellulose II and III$_I$ include increased chain fluctuations, fewer inter-chain hydrogen bonds, and greater exploration of conformational space upon decrystalization. The authors pointed out the importance of future experiments that will reveal the shapes of cellulose microfibrils at nanometer length scales, which will inform subsequent simulation studies.

Chundawat *et al.* [45] used MD simulation to show how ammonia pretreatment of cellulose Iβ induces conversion to the less recalcitrant cellulose III$_I$ polymorph. The structural rearrangement resulted in up to fivefold enhanced rates of saccharification. Analysis of MD trajectories revealed distinct differences between the two polymorphs in the torsional states of hydroxymethyl groups, as well as the total number and types of hydrogen bonds. In addition to revealing molecular-level insight into the physical origins of recalcitrance, this work may lead to improved pretreatment and enzymatic protocols for more efficient production of biofuels and bioproducts.

3.8.4   Simulation of Lignocellulosic Biomass

Besombes *et al.* [46] constructed a model of cellulose Iβ and studied the adsorption of a guaiacyl β-O-4 dimer onto various surfaces using MD. They dissected the interaction energies and found that van der Waal's interactions were the most important for adsorption of the phenolic rings of lignin onto the (200) face of the cellulose surface, whereas hydrogen bonding was also important for adsorption to other surface faces. These results were compared with Raman and IR spectroscopy and photoconductivity experiments, and the authors found them to be consistent. They concluded that lignin prefers to be oriented parallel to the surface of the secondary cell wall and that adsorption is largely dispersive in nature.

Schulz *et al.* [47] developed a method to perform accurate and efficient MD simulations of multimillion-atom biological systems on supercomputers. To test their implementation, they used a previously constructed model of cellulose Iβ, which consisted of 36 cellulose chains and was 80 glucose units in length (*c.* 40 nm). They were able to achieve 30 ns/day of MD on the 3.3 million-atom system using the Jaguar supercomputer. One of the major contributions of their work was the development of an atomistic MD method that helps bridge the gaps in size and timescales between theory and experiment.

3.8.5   Outlook for Biomass Simulations

Computer simulations enable studies of molecular detail that are inaccessible to currently available experimental techniques. Here, we have provided a brief summary of recent results from the literature demonstrating a few specific examples of how simulation can complement experimental studies. A challenge remains in how to simulate a "complete" MD model of cellulose, lignin, and hemicellulose, as well as their interactions with enzymes and chemical and thermal reactions, although progress continues to be made. Recently, models of secondary plant cell walls incorporating cellulose, xylan, water and lignin were constructed and MD simulations carried out [48]. A further challenge is how to relate these simulations to the different plant cell-wall types and structures. As simulation methods continue to improve and more studies are carried out, our understanding of the molecular basis for recalcitrance will lead to improved technologies for producing biofuels and bioproducts.

3.9   Summary

The diversity of plant anatomy, cellular structures, and cell-wall chemical composition present in herbaceous and woody crops gives rise to a broad distribution of biomass feedstocks containing unique features that can assist the conversion of the feedstocks into useful fuels, chemicals, and other products. Over time, plants have evolved cellular structures and modified their cellular composition to resist insects and microbial attacks, and this recalcitrance to decay prevents easy conversion of plant materials to products. The goal of pretreatment is to open up the cellular structures at a molecular lever to allow access by enzymes that can selectively release sugars of the carbohydrates components, often primarily cellulose. Pretreatment conditions need to be optimized to remove the less recalcitrant plant components such as the hemicelluloses while minimizing degradation reactions that reduce sugar yields and create inhibitory compounds that can interfere with further downstream biological processing. There is a large body of knowledge that has been assembled on the broad details of plant cell-wall chemistry and structure. This information shows that variability exists at the species, individual, tissue, and molecular levels; it also shows that this variability is bounded by some common features. However, we still lack a systematic understanding of how the diverse plant cell-wall features are impacted during pretreatment, and this lack of knowledge impacts our ability to develop both improved feedstocks and advanced biological processes that can efficiently and economically produce biofuels and other products. Challenges remain to measuring and detecting the specific structure and chemical changes occurring to the cell wall during pretreatment due to both plant heterogeneity and the lack of analytical methods to detect changes at the molecular level. To bridge the gap until the next generation of analytical methods are developed that can measure changes at the molecular level, computer simulations can be used to provide insight into both cell-wall construction and deconstruction. Several of the chapters that follow (Chapters 6–11) will illustrate how specific pretreatments attack different cell-wall chemistries and structures.

Acknowledgements

This work was supported and performed as part of the BioEnergy Science Center, managed by Oak Ridge National Laboratory. The BioEnergy Science Center is a US Department of Energy Bioenergy Research Center, support by the Office of Biological and Environmental Research in the US DOE Office of Science.

References

1. Himmel, M. (ed.) (June 2008) *Biomass Recalcitrance: Deconstructing the Plant Cell Wall for Bioenergy*, Wiley-Blackwell.
2. Perlack, R.D., Wright, L.L., Turhollow, A., Graham, R.L., Stokes, B., and Erbach, D.C. (2005) Biomass as Feedstock for a Bioenergy and Bioproducts Industry: the Technical Feasibility of a Billion-ton Annual Supply, DOE/GO-102005-2135, Oak Ridge National Laboratory, Oak Ridge, TN (http://feedstockreview.ornl.gov/pdf/billion ton vision.pdf).
3. Perlack, R.D. and Stokes, B.J. (2011) U.S. Department of Energy. *U.S. Billion-Ton Update: Biomass Supply for a Bioenergy and Bioproducts Industry*. R.D. Perlack and B.J. Stokes (Leads), ORNL/TM-2011/224. Oak Ridge National Laboratory, Oak Ridge, TN. 227 p. (http://www1.eere.energy.gov/biomass/pdfs/billion_ton_update.pdf).
4. Dale, V.H., Kline, K.L., Wright, L. *et al.* (2011) Interactions between bioenergy feedstock choices and landscape dynamics and land use. *Ecological Applications*, **21** (4), 1039–1054.
5. US DEO (2006) Breaking the Barriers to Cellulosic Ethanol: a joint research agenda (June, 2006) DOE/SC-0095. US Department of Energy Office of Science and Office of Energy Efficiency and Renewable Energy. http://www.doegenomestolife.org/biofuels/.
6. Elander, R.T., Dale, B.E., Holtzapple, M. *et al.* (2009) Summary of findings from the Biomass Refining Consortium for Applied Fundamentals and Innovation (CAFI): corn stover pretreatment. *Cellulose*, **16**, 649.
7. Zheng, Y., Pan, Z.L., Zhang, R.H. *et al.* (2007) Evaluation of different biomass materials as feedstock for fermentable sugar production. *Applied Biochemistry and Biotechnology*, **137**, 423.
8. Albersheim, P., Darvill, A., Roberts, K., Sederoff, R., and Staehlin, A. (eds) (2010) *Plant Cell Walls*, Garland Science, Taylor & Francis Group, LLC.

9. Somerville, C. (2006) Cellulose synthesis in higher plants. *Annual Review of Cell and Developmental Biology*, **22**, 53–78.

10. Lehtio, J., Sugiyama, J. *et al.* (2003) The binding specificity and affinity determinants of family 1 and family 3 cellulose binding modules. *Proceedings of the National Academy of Sciences of the United States of America*, **100** (2), 484–489.

11. Wiselogel, A., Tyson, S., and Johnson, D. (1996) Biomass feedstock resources and composition, in *Handbook on Bioethanol – Production and Utilization* (ed. C.E. Wyman), Taylor and Francis, Washington, DC, pp. 105–118.

12. Viamajala, S., Donohoe, B.S. *et al.* (2010) Heat and mass transport in processing of lignocellulosic biomass for fuels and chemicals, in *Sustainable Biotechnology* (eds O.V. Singh and S.P. Harvey), Springer, Netherlands, pp. 1–18.

13. Esau, K. (2006) *Esau's Plant Anatomy: Meristems, Cells, and Tissues of the Plant Body: Their Structure, Function, and Development*, John Wiley & Sons, Inc., Hoboken, New Jersey.

14. Zeng, M.J., Mosier, N.S. *et al.* (2007) Microscopic examination of changes of plant cell structure in corn stover due to hot water pretreatment and enzymatic hydrolysis. *Biotechnology and Bioengineering*, **97** (2), 265–278.

15. Studer, Michael E., DeMartini, Jaclyn D., Davis, Mark F. *et al.* (March 2011) Lignin content in natural *Populus* variants affects sugar release. *PNAS*, **108** (15), 6300–6305.

16. Lyndon, R.F. and Francis, D. (1992) Plant and organ development. *Plant Molecular Biology*, **19** (1), 51–68.

17. Bonawitz, N.B. and Chapple, C. (2010) The genetics of lignin biosynthesis: connecting Genotype to Phenotype. *Annual Review of Genetics*, **44**, 337–363.

18. Donaldson, L.A. (2001) Lignification and lignin topochemistry – an ultrastructural view. *Phytochemistry*, **57** (6), 859–873.

19. Jarvis, M.C. (1992) Self-assembly of plant-cell walls. *Plant Cell and Environment*, **15** (1), 1–5.

20. McCann, M.C., Wells, B. *et al.* (1990) Direct visualization of cross-links in the primary plant-cell wall. *Journal of Cell Science*, **96**, 323–334.

21. Cannon, M.C., Terneus, K. *et al.* (2008) Self-assembly of the plant cell wall requires an extension scaffold. *Proceedings of the National Academy of Sciences of the United States of America*, **105** (6), 2226–2231.

22. Mohnen, D. (2008) Pectin structure and biosynthesis. *Current Opinion in Plant Biology*, **11** (3), 266–277.

23. Brown, R.M. (2004) Cellulose structure and biosynthesis: What is in store for the 21st century? *Journal of Polymer Science Part A-Polymer Chemistry*, **42** (3), 487–495.

24. Carpita, N., Sabularse, D. *et al.* (1979) Determination of the pore-size of cell-walls of living plant-cells. *Science*, **205** (4411), 1144–1147.

25. Vian, B., Roland, J.C. *et al.* (1993) Primary-cell wall texture and its relation to surface expansion. *International Journal of Plant Sciences*, **154** (1), 1–9.

26. Donohoe, B.S., Decker, S.R. *et al.* (2008) Visualizing lignin coalescence and migration through maize cell walls following thermochemical pretreatment. *Biotechnology and Bioengineering*, **101** (5), 913–925.

27. Kristensen, J., Thygesen, L. *et al.* (2008) Cell wall structural changes in wheat straw pretreated for bioethanol production. *Biotechnology for Biofuels*, **1** (1), 5.

28. Kumar, R., Mago, G. *et al.* (2009) Physical and chemical characterizations of corn stover and poplar solids resulting from leading pretreatment technologies. *Bioresource Technology*, **100** (17), 3948–3962.

29. Chundawat, S.P.S., Donohoe, B.S. *et al.* (2011) Multi-scale visualization and characterization of lignocellulosic plant cell wall deconstruction during thermochemical pretreatment. *Energy & Environmental Science*, **4** (3), 973–984.

30. Abuja, P., Schmuck, M. *et al.* (1988) Structural and functional domains of cellobiohydrolase-i from trichoderma-reesei – a small-angle x-ray-scattering study of the intact enzyme and its core. *European Biophysics Journal with Biophysics Letters*, **15** (6), 339–342.

31. Beckham, G.T., Bomble, Y.J., Bayer, E.A. *et al.* (2011) Applications of computational science for understanding enzymatic deconstruction of cellulose. *Current Opinion in Biotechnology*, **22**, 231–238.

32. Parthasarathi, R., Romero, R.A., Redondo, A., and Gnanakaran, S. (2011) Theoretical study of the remarkably diverse linkages in lignin. *The Journal of Physical Chemistry Letters*, **2**, 2660–2666.

33. Durbeej, B. (2003) Spin distribution in dehydrogenated coniferyl alcohol and associated dilignol radicals. *Holzforschung*, **57**, 59–61.

34. Durbeej, B. and Eriksson, L.A. (2003) A density functional theory study of coniferyl alcohol intermonomeric cross linkages in lignin: Three-dimensional structures, stabilities and the thermodynamic control hypothesis. *Holzforschung*, **57**, 150–164.

35. Durbeej, B. and Eriksson, L.A. (2003) Formation of β-O-4 lignin models: A theoretical study. *Holzforschung*, **57**, 466–478.

36. Jarvis, M.W., Daily, J.W., Carstensen, H.-H. *et al.* (2011) Direct detection of products from the pyrolysis of 2-phenethyl phenyl ether. *Journal of Physical Chemistry A*, **115**, 428–438.

37. Beste, A., Buchanan, A.C. III, and Harrison, R.J. (2008) Computational prediction of/β selectivities in the pyrolysis of oxygen-substituted phenethyl phenyl ethers. *Journal of Physical Chemistry A*, **112**, 4982–4988.

38. Kim, S., Chmely, S.C., Nimlos, M.R. *et al.* (2011) Computational study of bond dissociation enthalpies for a large range of native and modified lignins. *Journal of Physical Chemistry Letters*, **2**, 2846–2852.

39. Parthasarathi, R., Romero, R.A., Redondo, A., and Gnanakaran, S. (2011) Theoretical study of the remarkably diverse linkages in lignin. *Journal of Physical Chemistry Letters*, **2**, 2660–2666.

40. Sangha, A.K., Parks, J.M., Standaert, R.F. *et al.* (2012) Radical coupling reactions in lignin synthesis. A density functional theory study. *The Journal of Physical Chemistry. B*, **116** (16), 4760–4768. doi: 10.1021/jp2122449

41. Bergenstråhle, M., Wohlert, J., Himmel, M.E., and Brady, J.W. (2010) Simulation studies of the insolubility of cellulose. *Carbohydrate Research*, **345**, 2060–2066.

42. Matthews, J.F., Skopec, C.E., Mason, P.E. *et al.* (2006) Computer simulation studies of microcrystalline cellulose Iβ. *Carbohydrate Research*, **341**, 138–152.

43. Matthews, J.F., Bergenstråhle, M., Beckham, G.T. *et al.* (2011) High-temperature behavior of Cellulose I. *The Journal of Physical Chemistry. B*, **115**, 2155–2166.

44. Beckham, G.T., Matthews, J.F., Peters, B. *et al.* (2011) Molecular-level origins of biomass recalcitrance: decrystallization free energies for four common cellulose polymorphs. *The Journal of Physical Chemistry. B*, **115**, 4118–4127.

45. Chundawat, S.P.S., Bellesia, G., Uppugundla, N. *et al.* (2011) Restructuring the crystalline cellulose hydrogen bond network enhances its depolymerization rate. *Journal of the American Chemical Society*, **133**, 11163–11174.

46. Besombes, S. and Mazeau, K. (2005) The cellulose/lignin assembly assessed by molecular docking. Part 1: adsorption of a threo guaiacyl β-O-4 dimer onto an Iβ cellulose whisker. *Plant Physiology and Biochemistry*, **43**, 299–308.

47. Schulz, R., Lindner, B., Petridis, L., and Smith, J.C. (2009) Scaling of multimillion-atom biological molecular dynamics simulation on a petascale supercomputer. *Journal of Chemical Theory and Computation*, **5**, 2798–2808.

48. Charlier, L. and Mazeau, K. (2012) Molecular modeling of the structural and dynamical properties of secondary plant cell walls: Influence of lignin chemistry. *The Journal of Physical Chemistry. B*, **116**, 4163–4174.

4

Biological Conversion of Plants to Fuels and Chemicals and the Effects of Inhibitors

Eduardo Ximenes[1], Youngmi Kim[1] and Michael R. Ladisch[2]

[1] *Laboratory of Renewable Resources Engineering, Purdue University, West Lafayette, USA*

[2] *Laboratory of Renewable Resources Engineering, Purdue University, West Lafayette and Mascoma Corporation, USA*

4.1 Introduction

To achieve high yields, selectivity, and scalable process conditions, the biological conversion of plants to fuels and chemicals requires a number of sequential and integrated steps. There is significant literature on the pretreatment of cellulosic materials, including woody biomass, agricultural residues, and purposely grown energy crops, particularly grasses [1–3]. All these materials have a heterogeneous chemical composition consisting of structural carbohydrates (cellulose and hemicelluloses) and lignin.

Biomass materials can be hydrolyzed to fermentable sugars that can be fermented to ethanol or other alcohols. For such conversions be cost effective, yields must be high. To achieve high hydrolysis yields, pretreatment is required to break away the lignin that is closely associated with the cellulose and disrupt the cellulose structure so enzymes can penetrate the cell wall and depolymerize cellulose and hemicellulose (Figure 4.1).

This chapter describes the chemistry and manner by which pretreatment and feedstock composition impact the efficiency of enzymatic hydrolysis and subsequent fermentation of monosaccharides to ethanol or other value-added products. The next chapter (Chapter 5) addresses chemical catalysis for obtaining sugars and converting them into high-molecular-weight alkanes. These drop-in biofuels are generally C8 through C20, whereas biomass-derived sugars are C5 or C6.

Aqueous Pretreatment of Plant Biomass for Biological and Chemical Conversion to Fuels and Chemicals, First Edition.
Edited by Charles E. Wyman.
© 2013 John Wiley & Sons, Ltd. Published 2013 by John Wiley & Sons, Ltd.

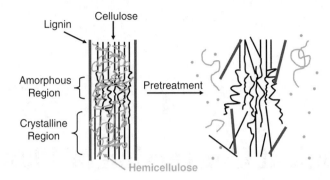

Figure 4.1   *Schematic diagram of the effect of pretreatment on cellulose structure and the lignin seal, resulting in exposure of the cellulose and access by cellulolytic enzymes that cause hydrolysis. (Reprinted with permission from Mosier et al. [1] © 2005, Elsevier). (See figure in color plate section).*

4.2   Overview of Biological Conversion

Cellulose is the major polysaccharide in plant cell walls. It is a β-1,4-glucose polymer that is hydrolyzed by the following glycoside hydrolases: *endo* (endo-β-1,4-glucananse that cleaves internally); *exo* (cellobiohydrolases that release cellobiose); or *endo*-processive (cleaves internally and progressively generates products) [4]. When these enzymes are combined in the correct ratios, the mixture greatly enhances both the rate and extent of cellulose hydrolysis (Figure 4.2).

Table 4.1 gives the compositions of various representative lignocellulosic biomass materials and also shows that glucan content (cellulose) varies from one type of feedstock to another. For agricultural residues

Table 4.1   *Compositions of different types of cellulosic biomass and the maximum ethanol yields possible for each of the compositions. These biomass materials (wood, corn stover (stalks), and switchgrass) are representative of available feedstocks for a cellulosic biorefinery. (Reprinted with permission from Ladisch et al. [7] © 2011 NSF PIRE-FHI).*

Feedstock compositions	Poplar	Red maple	Corn stover	Switchgrass	Bagasse[c]	Pinewood[d]
Cellulose	43.8	41.0	34.6	33.2	39	40
Xylan	14.9	15.0	18.3	21.0	21.8	8.9
Arabinan, mannan, galactan	5.6	0.0	2.5	3.2	2.6	19.6
Acetyl	3.6	4.7	3.5	2.5	3.3	NA
Extractives	3.6	3.0	10.8	10.2	5.7	3.5
Protein	N/A	N/A	N/A	5.7	0.5	N/A
Lignin	29.1	29.1	17.7	17.9	24.8	27.7
Ash	1.1	1.0	10.2	3.7	3.9	N/A
Total	101.7	93.8	97.6	97.4	101.6	99.7
Estimated maximum ethanol yield[a], gal/dry ton biomass	111	108	107	111	122	132
Estimated practical maximum ethanol yield[b], gal/dry ton biomass	105	103	102	105	116	125

[a] Theoretical maximum yields assuming 100% hydrolysis and 100% fermentation. Data from Laboratory of Renewable Resources Engineering, Purdue University.
[b] 95% hydrolysis/95% fermentation
[c] Templeton *et al.* [8]; Kim and Day [9]
[d] Du *et al.* [10]

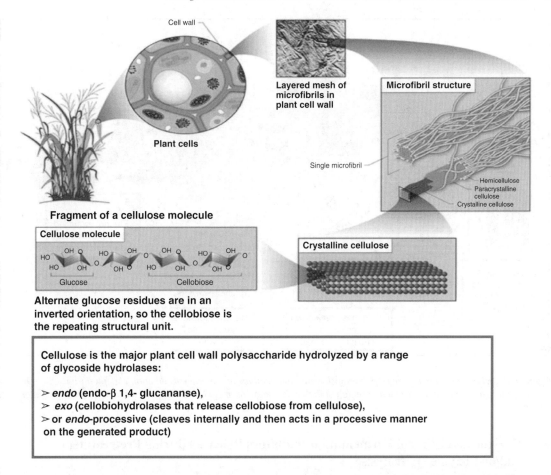

Figure 4.2 *Plant cell-wall structure and cellulolytic enzymes. (Adapted from US DOE [5,6], Office of Biological and Environmental Research of the US Department of Energy Office of Science). (See figure in color plate section).*

(e.g., corn stover, wheat straw, corn fiber, and bagasse) the compositions are similar. Hardwood is lower in hemicellulose, and pine is somewhat higher in glucan.

The hydrolysis sugars (i.e., glucose, xylose, and other 5-carbon sugars) are fermented into ethanol and CO_2 while releasing heat. Figure 4.3 shows the general schematic diagram for pretreatment, hydrolysis, and fermentation. In the consolidated bioprocessing [12,13], fermentation and hydrolysis are performed in the same vessel with the fermenting microorganism generating enzymes that can transform cellulose into sugars and then into ethanol. This approach reduces the amount of enzyme required, thereby providing another pathway for ethanol fermentation.

Pretreatment is a key step to achieving high yields and reasonable rates of cellulose hydrolysis to fermentable sugars. Pretreatments have been performed with a wide range of lignocellulosic feedstocks, which have a range of compositions and processing characteristics. When properly pretreated, all of them have the ability to yield approximately 100 gal Ethanol/dry ton of ethanol, or more.

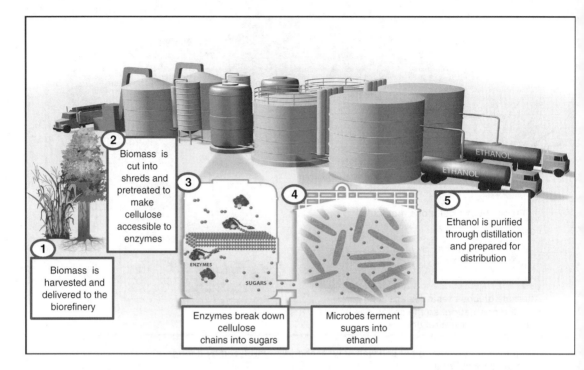

2 Biomass is cut into shreds and pretreated to make cellulose accessible to enzymes

3

4

5 Ethanol is purified through distillation and prepared for distribution

1 Biomass is harvested and delivered to the biorefinery

ENZYMES

SUGARS

Enzymes break down cellulose chains into sugars

Microbes ferment sugars into ethanol

ETHANOL

ETHANOL

Figure 4.3   *Schematic view of key steps in transforming lignocellulosic materials to ethanol. (Adapted from US DOE [11], Office of Biological and Environmental Research of the US Department of Energy Office of Science). (See figure in color plate section).*

4.3   Enzyme and Ethanol Fermentation Inhibitors Released during Pretreatment and/or Enzyme Hydrolysis

It is well known that the enzymatic hydrolysis step releases end-products (e.g., cellobiose and glucose) which are known to decrease the rate of enzymatic hydrolysis [14,15]. Cellobiose and glucose are major inhibitors of cellobiohydrolase and β-glucosidase, respectively. To overcome product inhibition, simultaneous saccharification and fermentation (SSF) removes sugars by converting them to less inhibitory products (e.g., ethanol).

Various lignocellulose pretreatments release inhibitors. An example of inhibitors resulting from pretreatment is phenolics, identified by Ximenes *et al.* [16,17]; Kim *et al.* [18] describe the impact clearly. Many years ago they were identified as plant protectants that inhibit fungal cellulase from pathogens that might attack plant tissue. For instance, Cook *et al.* [19] demonstrated that polyphenol oxidases in green apples, pears, and walnut hulls form tannins from phenols in injured tissue, which prevent entry of pathogens. These phenolics have a wide range of structures (Figure 4.4) and have different effects on cellulases. Their mode of action depends on both the type of enzyme and the microbial source from which the enzyme is derived [16,17].

Recent literature reports show that xylooligosaccharides significantly inhibit cellulases [20–22]. The following sections describe the inhibitory effects of phenolics and xylooligosaccharides on cellulase enzymes, interpret their mechanisms, and project their impact. In addition, the following sections describe ethanol fermentation inhibitors released during pretreatment with special emphasis on the effects of furans and acetic acid.

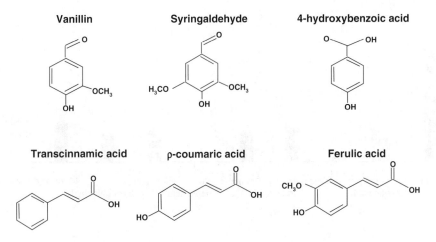

Figure 4.4 *Phenolic compounds that may act as inhibitors or deactivators of cellulases (after Ximenes et al. [16,17]).*

4.3.1 Enzyme Inhibitors Derived from Plant Cell-wall Constituents (Lignin, Soluble Phenolics, and Hemicellulose)

Although pretreatment is a significant cost-driver for industrial processes that convert lignocellulose to fermentation products, it is indispensable for enhancing biomass susceptibility to hydrolytic enzymes [23,24]. The whole slurry obtained after pretreatment will probably be used at a high solids loading, which may affect enzyme performance, because pretreatment liquid can inhibit enzymatic hydrolysis of cellulose [16,25]. Different pretreatment methods (e.g., liquid hot water, steam explosion, and dilute acid) generate soluble inhibitors that hamper enzymatic hydrolysis and fermentation of sugars to ethanol. Chapter 7 describes aqueous pretreatments that can release a range of inhibitors. In addition to sugars (e.g., oligosaccharides and monosaccharides), the pretreatment liquid may also contain aromatic compounds such as furan aldehydes, aliphatic acids, and extractives [16,18,26,27].

Recent literature has reported that phenols [16–18,24,27–29] and xylooligosaccharides [20–22] are major enzyme inhibitors formed during pretreatment. Xylans may be hydrolyzed to soluble sugars by hemicellulases [22] or removed by conversion to monosaccharides during acid pretreatment. Nonetheless, considerable amounts of lignin particles and lignin-derived soluble phenolics may remain in industrially relevant pretreated lignocellulose [1,29,30].

Lignin and Soluble Phenolics

Lignin is the most abundant renewable aromatic polymer on earth. Lignin is a component of higher plants, most of which is found within cell walls where it is intimately interspersed with hemicellulose to form a matrix that surrounds cellulose microfibrils. In wood, high lignin concentrations bind contiguous cells with the lignin forming the middle lamella [31]. Lignin is a 3D network formed of dimethoxylated (syringyl, S), monomethoxylated (guacyl, G), and non-methoxylated (p-hydroxyphenyl, H) phenylpropanoid units, all of which are derived from corresponding p-hydroxycinnamyl alcohols. These monomers are linked by ether and C—C bonds [32–34].

Lignin content, composition, distribution, cell-wall thickness, cell-wall structures, and type of tissue measurably affect enzymatic hydrolysis of cellulose in lignocellulosic feedstocks. For instance, to identify the role of structural characteristics on enzyme hydrolysis of cell walls, Zeng *et al.* [35,36] combined

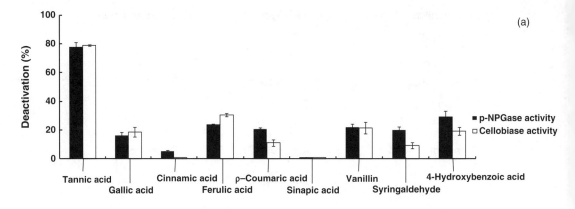

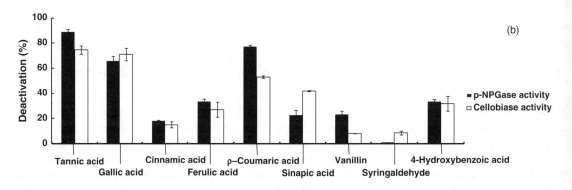

Figure 4.5 *Extents of deactivation of β-glucosidase by phenolic compounds: (a)* Aspergillus niger *β-glucosidase; and (b)* Trichoderma reesei *β-glucosidase. (Reprinted with permission from Ximenes et al. [17] © 2011, Elsevier).*

compositional analysis, pretreatment, and enzyme hydrolysis for fractionated pith, rind, and leaf tissues from a hybrid stay-green corn to identify the role of structural characteristics on enzyme hydrolysis of cell walls. The extent of enzymatic hydrolysis observed followed the sequence: rind < leaves < pith. In the best cases, 90% conversion of cellulose to glucose occurred in 24 hours.

Pretreatment often only partially degrades lignin by leaving inter-monomer bonds that result in release of mono- and oligo-aromatic compounds which increase the concentration of free phenolics [34]. Together, residual lignin and aromatic compounds inhibit both enzymatic hydrolysis and fermentation [16–18,24,37].

During biomass pretreatment, hydrolysis of lignin can release phenolic compounds that can be quite inhibitory, and indeed may deactivate cellulase [16,17]. To a greater extent, β-glucosidases are affected (Figure 4.5). Studies by Ximenes *et al.* [16,17] identified that the polymeric phenol tannic acid is a major inhibitor and deactivator for all enzyme activities tested (filter paper, endoglucanase, and β-glucosidase assays). Monomeric phenolic compounds have a less pronounced effect. Tannic, ferulic, and p-coumaric acids inactivated β-glucosidases from two different microorganisms (*Trichoderma reesei* and *Aspergillus niger*) that are commonly used to produce commercial cellulases.

Using the lignin-free cellulose Solka-Floc combined with mixtures of soluble components released during pretreatment of wood, Kim *et al.* [18] confirmed that the lower rate and extent of cellulose hydrolysis occurs due to a combination of enzyme inhibition and deactivation. They extracted the causative agents from wood pretreatment liquid using polyethylene glycol (PEG) surfactant, activated charcoal, or ethyl

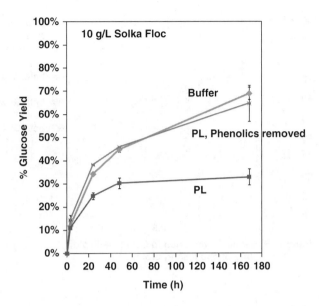

Figure 4.6   *Hydrolysis time course of Solka-Floc using 1 mg cellulase Spezyme CP per g glucan. PL: pretreatment liquid of maple. (Reprinted with permission from Kim et al. [18] © 2011, Elsevier).*

acetate. They then desorbed, recovered, and added back the causative agents to a mixture of enzyme and cellulose. The inhibitors were identified as phenolics that reduced the rate and extent of cellulose hydrolysis by half because of both inhibition and precipitation of enzymes. When the phenols were removed, full enzyme activity occurred (Figures 4.6 and 4.7). Figure 4.7 represents the activities of cellulase enzymes with varying levels of phenolics. The β-glucosidase activity of both Spezyme CP and Novozyme 188 was significantly reduced by phenolic compounds. The time-dependent loss of β-glucosidase also indicated that enzyme precipitation is the major cause of lost activity.

Tejirian and Xu [28] observed that oligomeric phenolics were more inhibitory to enzymatic hydrolysis than monomeric phenolics. They found that oligomeric phenolics could inactivate cellulases by reversibly complexing with them. Respectively, polyethylene glycol and tannase could bind and degrade the oligomeric phenolics tested and therefore mitigate their inhibition of cellulase. In another study, Olsen *et al.* [29] used tannic acid (Figure 4.8) as a general polyphenolic model compound (for lignin and soluble phenolics). Their experiments aimed to understand the mechanisms behind the acceleration of enzymatic hydrolysis of lignocellulosics by adding non-ionic surfactants (NIS). Because of its complex oligophenolic structure, tannic acid was chosen to mimic the properties of lignin and soluble phenols. This work suggested that favorable NIS–polyphenol interactions alleviate non-productive cellulase–polyphenol interactions, which may provide a mechanism for the accelerating effect of NIS.

Compared to globular proteins such as bovine serum albumin (BSA) tannic acid has a stronger interaction with random coil, proline-rich proteins such as gelatin. Less is known about the specificity of its binding to structural proteins, such as cellulases [29,38]. Taking a closer look at available crystal structures for different cellulases, Olsen *et al.* [29] concluded that enzyme-tannic acid interactions showed distinct specificity even in the same family of cellulases, which agrees with other literature reports [16–18,24,28]. For instance, cellobiohydrolase I (CBHI) and endoglycosidase I (EG1), which belong to the same GH7 family, showed different affinities and binding patterns despite their similar structures (they have comparable structures dominated by a β-sandwich composed of two anti-parallel β-sheets). The authors argued that

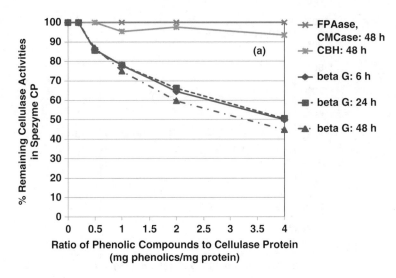

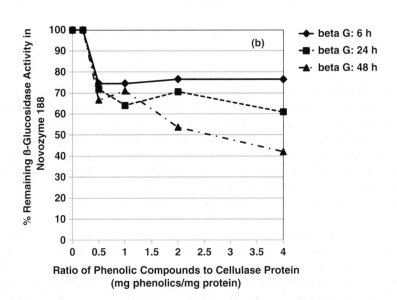

Figure 4.7   *Inhibition and deactivation of cellulase activities by phenolic compounds removed from maple pretreatment liquid after 6, 24, and 48 h incubation. (a) CMCase, CBH, and β-glucosidase activities in Spezyme CP; (b) β-glucosidase activity in Novozyme 188. Data represents an average of duplicate runs with maple wood. (Reprinted with permission from Kim et al. [18] © 2011, Elsevier).*

substantial variability in those interactions may suggest that it is feasible to find or develop cellulases with low lignin/phenol sensitivity, as also suggested by Berlin *et al.* [39].

Mechanisms

Although phenolic hydroxyl groups are key to the potency of lignin-derived aromatics in inhibiting enzymes, little information is available about the nature of the interactions or inhibitory mechanisms

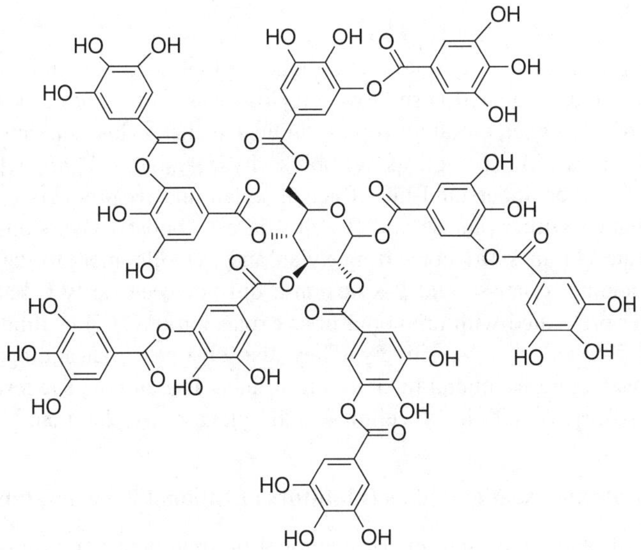

Figure 4.8   *Tannic acid structure.*

[24,40]. Kinetic analysis by Boukari *et al.* [24] demonstrated that inhibition of a GH11 endo-β-1,4-xylanase from *Thermobacillus xylanilyticus* followed a "multisite" non-competitive inhibition mechanism, indicating that more than one aromatic molecule interacts with the enzyme molecule to induce its complete inactivation. Sharma *et al.* [41] suggested that phenolic compounds in low concentrations could directly affect *in vitro* enzyme activity by forming a soluble but inactive enzyme-inhibitor complex. At high phenolic concentrations, they suggested that the solubility of enzyme proteins was reduced by forming an insoluble protein-phenolic complex.

Another interesting observation from Boukari *et al.* [24] was the cooperative effect between multiple binding sites of phenolics. They suggested that initial binding occurs at high-affinity binding sites, which then influences the binding of further inhibitors to other low-affinity sites, considered to some extent as an allosteric behavior. They also observed that the majority of interactions of phenolic compounds with xylanase likely involve residues located at the enzyme surface, which may include both hydrophobic aromatic ring stacking (between phenolic compounds and tryptophanyl side chains) and/or hydrogen interactions between their functional groups (COOH, OH) and the basic amino acid residues.

Strategies to Remove Phenols

To enhance hydrolysis and reduce enzyme usage, enzyme processes that hydrolyze cellulose to glucose must reduce inhibition and deactivation effects. Phenols inhibit and deactivate cellulolytic enzymes, particularly β-glucosidases. A key challenge is to increase enzyme access to polysaccharides and consequently increase the efficiency and economics of lignocellulose conversion processes by removing or altering lignin and other aromatic compounds. Identifying and developing β-glucosidases that resist inhibition from both sugars and phenols will also enhance cellulose hydrolysis. Other strategies include: performing enzymatic hydrolysis over shorter periods of time to decrease time-dependent deactivation; removing phenolics prior to enzymatic hydrolysis by separation methods, including washing of the solids; or using microbial, enzymatic, or chemical methods to convert phenolics to an inactive form [17,18,42].

Hemicellulose

Kumar and Wyman [20] reported that xylobiose and higher xylooligomers inhibit enzymatic hydrolysis of pure glucan, pure xylan, and pretreated corn stover. Xylose was more inhibitory than xylan but less so than xylobiose or xylotriose. Other literature reports indicated that xylose inhibits enzymatic hydrolysis of other cellulosic substrates [43], as well as cellobiose hydrolysis by *A. niger* β-glucosidase [44]. In other work designed to further understand the effect of xylan and its products on cellulose hydrolysis, cellulase was supplemented with xylanase and β-xylosidase to boost conversion of both cellulose and hemicellulose in pretreated biomass by converting xylan and xylooligomers to the less inhibitory xylose [22]. They found that adding xylanase and β-xylosidase did not necessarily enhance Avicel hydrolysis; however, for corn stover pretreated with ammonia fiber expansion (AFEX) or dilute acid, glucan conversion was increased by 27% or 8%, respectively. They also observed that adding hemicellulase several hours before cellulase was more beneficial than adding it later. The authors suggested this effect possibly resulted from a higher adsorption affinity of cellulase and xylanase to xylan than glucan.

4.3.2   Effect of Furfurals and Acetic Acid as Inhibitors of Ethanol Fermentations

To deconstruct plant cell walls, physical and/or chemical pretreatment conditions are harsh and consequently generate side products that inhibit enzymes and fermenting microorganisms [1,2,45]. These inhibitors include furan aldehydes, phenolic compounds, and aromatic and aliphatic aldehydes and acids, which are typically present in hydrolyzed biomass (Figure 4.9) [37,46–48]. Furfural and 5-hydroxy-methylfurfural

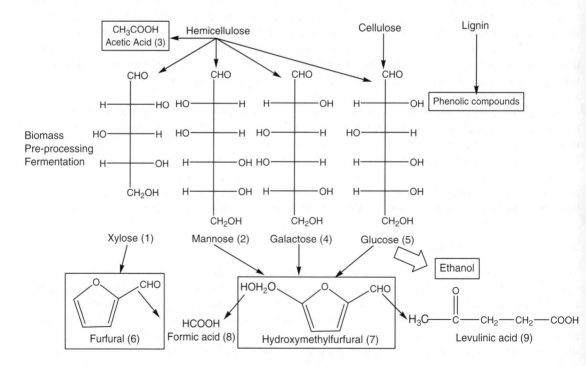

Figure 4.9   *Summary of formation of aldehydes and other degradation products from glucose and xylose. (Adapted and reprinted with permission from Palmqvist and Hann-Hägerdal [37] © 2000, Elsevier).*

(HMF) are formed by degrading pentose and hexose sugars, respectively, with formic and levulinic acids also being formed. Acetic acid is released from solubilized hemicelluloses and lignin, and hydrolysis or oxidation of the latter generates aromatic compounds (acids, aldehydes, and alcohols) [49].

Inhibitory compounds negatively impact conversion of lignocellulosic hydrolyzates because they can retard microbial growth and decrease product yield and productivity [37,46,47]. The effects are broad and affect cell membranes, synthesis of macromolecules, and glycolytic and fermentative enzymes [37,49,50].

Furans

Depending on the furan concentration and yeast strain used, HMF and furfural decrease the volumetric ethanol yield and productivity, inhibit growth, or increase the lag phase (reviewed in [47]). The inhibitory effects of furfural and HMF act synergistically with other compounds such as lignin monomers [37,51]. The concentration of furans in the lignocellulosic hydrolyzate varies according to the type of raw material and the pretreatment procedure. HMF is usually found in higher concentrations than furfural; however, the latter is often concentrated enough to be inhibitory (1 g/L) [47]. Pulse addition of 2 g/L of HMF and furfural to anaerobic batch cultures of *S. cerevisiae* converted HMF and furfural to less inhibitory alcohols. As long as both compounds remained in the culture, no growth occurred, and the specific uptake rates were lower than when 4 g/L of one component was added. On the other hand, when equimolar concentrations of each compound were compared, HMF was less inhibitory than furfural, even though it took longer to be converted by yeast [47,52].

Acetic Acid

Acetic acid is a weak acid generated by deacetylating hemicellulose during pretreatment [37,46,47,53]. The inhibitory effect of weak acids has been associated with uncoupling and intracellular anion accumulation [54]. The undissociated form of acetic acid can diffuse from the fermentation medium across the plasma membrane [55,56] and dissociate because of the higher intracellular pH, which in turn decreases the cytosolic pH [57]. This decrease in intracellular pH is then compensated by the plasma membrane adenosine-5′-triphosphate (ATPase), which pumps protons out of the cell at the expense of ATP hydrolysis, resulting in less ATP available for biomass formation [47]. Some studies indicate that low concentrations of acids stimulate the production of ATP, which would be achieved under anaerobic conditions by ethanol production. On the other hand, at higher concentrations the ATP demand would be so high that cells could not avoid acidification of the cytosol [58].

Acetic acid is present in varying concentrations in all types of biomass. For example, Lu *et al.* [59] determined that the acetyl concentration in corn stover and poplar is 5.6 and 3.6 wt%, respectively. When studying the inhibitory effect of acetic acid on microorganisms, Casey *et al.* [53] stated that process-relevant conditions for producing cellulosic ethanol at an industrial scale must be considered. Their statement assumed that the minimum ethanol concentration for economic distillation is 5%; the initial unhydrolyzed biomass concentration must therefore be approximately 20% (by weight). The result is theoretical acetic acid concentrations of 11.2 and 7.2 g/L in the hydrolyzates of corn stover and poplar, respectively, assuming no accumulation from recycling of process streams. An actual acetic acid concentration of 13 g/L has been observed in dilute-acid-pretreated corn stover hydrolyzates [59], but removing acetic acid and other inhibitors increases the cost of the overall process. A detailed study of the effects of these inhibitors on ethanol yields and production rates, especially for xylose fermentations, is important for ongoing microorganism development efforts and cellulosic ethanol commercialization.

Table 4.2   *Compositions of liquid fraction of liquid hot water pretreated biomass in units of g/L; ND: not determined.*

	Corn fiber with water[a]	Corn fiber with stillage[a]	Corn cob	Switchgrass	Poplar[b]	Maple[c]
Glucan	25.46	22.05	1.2	2.8	1.3	1.5
Glucose	0.96	6.2	2.3	0.2	0.1	0.6
Xylan/galactan	14.62	7.37	23.5	18.4	17.5	11.2
Xylose/galactose	2.82	2.79	5.1	1.3	1.1	9.2
Arabinan	5.37	1.72	0.8	1.6	0	ND
Arabinose	7.03	3	2.1	0.4	0.3	ND
Acetic acid	0.56	0.85	2	1.2	2.7	13.1
HMF (g/L)	0.01	0.3	0.32	0.04	ND	0.7
Furfural (g/L)	0.27	0.17	1.1	0.52	0.5	3.4

[a] Kim *et al.* [63]
[b] Kim *et al.* [64]
[c] Kim *et al.* [18]

4.4   Hydrolysis of Pentose Sugar Oligomers Using Solid-acid Catalysts

Hydrothermal pretreatments (steam explosion, liquid hot water) effectively remove and solubilize hemicelluloses [60–62], but the solubilized hemicellulose is mainly in the form of soluble oligosaccharides. The pretreated slurry contains high levels of pentose (xylose) sugar oligomers derived primarily from hemicellulose as well as lignin-derived soluble phenolic compounds, both of which significantly inhibit cellulases [16–18,22]. Prior to subsequent cellulose hydrolysis, it is therefore desirable to separate the dissolved sugar oligomers and other cellulase inhibitors from fibrous cellulose fraction. The separated pretreatment liquid containing pentose sugar oligomers and other enzyme inhibitors can be hydrolyzed by acids or hemicellulases into monomeric sugars, which can be further processed into fuels and other high-value chemicals. Table 4.2 presents compositions of pretreatment liquid of various feedstocks.

The constituent sugars and substitutes in hemicellulose-derived sugar oligomers are linked through different bonds that require a synergistic action of various enzyme activities for efficient hydrolysis, but commercial cellulase preparations lack the enzyme activities necessary to effectively hydrolyze these sugar oligomers [65,66]. Optimal enzyme blends for efficient hydrolysis of these sugar-oligomers greatly depend on feedstock types and pretreatment technologies. For example, hemicellulose in hardwoods is mainly methylglucuronoxylans, whereas softwoods contain galactoglucomannans and arabinomethylglucuronoxylans [67]. The structures and compositions of sugar oligomers will vary greatly depending on feedstock types and pretreatment conditions. Thus, developing broadly applicable hemicellulase systems is difficult. Although enzymes are highly selective, utilize relatively mild reaction conditions and do not degrade sugars, they require long reaction times and are costly.

4.4.1   Application of Solid-acid Catalysts for Hydrolysis of Sugar Oligomers Derived from Lignocelluloses

Acids are alternative catalysts to enzymes for hydrolysis of sugar oligomers, with sulfuric and hydrochloric acids commonly employed to hydrolyze sugar oligomers [68,69]. Although the acids themselves may be inexpensive and acid hydrolysis is faster than enzyme-catalyzed hydrolysis, liquid-acid-catalyzed hydrolysis is often expensive because of high operating temperatures, sugar degradation, corrosion, the need for neutralization after hydrolysis, and waste disposal.

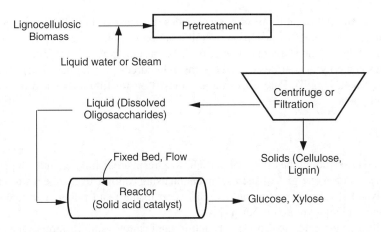

Figure 4.10   *Schematic diagram of solid-acid catalyzed hydrolysis of sugar oligomers derived from lignocellulosic biomass. (Adapted and reprinted with permission from Kim et al. [63] © 2005, American Chemical Society).*

An alternative approach is to employ solid-acid catalysts such as strong cation exchange resins, zeolites, clays, metal oxides, and sulfonated mesoporous silica [63,70–72]. Solid acids have several advantages over enzymes or acids in solution: (1) they are easily separable from a reaction mixture and can be reused; (2) they can be used in a continuous flow reactor; and (3) reaction kinetics can be controlled by adjusting catalyst properties such as pore size, acidity, functional groups, and so on. In the 1950s, sucrose was hydrolyzed to fructose and glucose using cation exchange resins [73]. Heyraud and Rinaudo [74] investigated hydrolysis of various disaccharides and malto-oligosaccharides (e.g., maltotriose, maltohexaose, and maltodextrin) by sulfonic ion exchangers in batch and continuous reactors. Recently, various forms of solid-acid catalysts were applied to directly hydrolyze cellulose into sugar alcohols, glucose, and cellooligosaccharides [63,70,75–79].

Applications of heterogeneous catalysts for hydrolysis of hemicellulose and hemicellulose-derived sugar oligomers into monomers have been limited. The catalysts studied include sulfonated strong cation exchange resins, zeolites, and mesoporous silica [63,70–73]. Although yields and efficiencies vary among these studies, solid-acid catalysts appear to be a promising alternative method to hydrolyze soluble sugar oligomers dissolved during hydrothermal pretreatment. Figure 4.10 represents a process schematic using a packed bed of solid-acid catalyst to hydrolyze sugar oligomers dissolved during hydrothermal pretreatment of lignocellulosic biomass. The pretreatment liquid, previously separated from undissolved cellulose, is hydrolyzed by solid-acid catalysts in a continuous flow reactor. The resulting fermentable sugars can be further processed to produce fuels and high-value chemicals, such as furans and sugar alcohols. As in typical acid-catalyzed hydrolysis, sugar will be lost to degradation products (e.g., hydroxymethylfurfural and furfural). It is therefore important to identify optimal conditions that minimize degradation reactions that form aldehydes and other fermentation inhibitors.

4.4.2   Factors Affecting Efficiency of Solid-acid-catalyzed Hydrolysis

Pore Size

For the hydrolysis to occur in solid-acid catalysts, sugar oligomers and polysaccharides must enter the pores and interact with internal catalytic acid sites. Unlike homogeneous acid catalysts, porous solid catalysts usually involve resistances to reaction, such as the diffusion limitation of oligosaccharides and sugars inside the pores and mass-transfer limitation of oligosaccharides diffusing from bulk solution to the catalyst surface. Consequently, the effects of diffusion on overall reaction rates have been emphasized in most studies using porous acid catalysts. The average diameter of disaccharides is *c.* 1–2 nm [80,81], while the diameter of oligosaccharides and polysaccharides can be dozens to hundreds of nanometers depending on the degree of polymerization (chain length). Pore diffusion and mass transfer limitations can therefore significantly affect hydrolysis efficiency.

Diffusion

Bodamer and Kunin [73] studied the effects of particle size, resin cross-linking, and temperature on reaction rate constants and activation energies of sucrose inversion using cationic exchange resin. The rate of sucrose inversion (hydrolysis) is limited by the sucrose diffusion in the resin pores. The rate of inversion was increased by using smaller and more porous resins.

Abbadi *et al.* [82] investigated hydrolysis of maltose and starch using different types of solid catalysts and showed that ion exchange resins are more active than other types, such as zeolites. The narrow pores of zeolites, which have a cross-section of *c.* 1 nm, appeared to be inaccessible to long-chain polymers and oligosaccharides. On the other hand, ion exchange resins, especially with low degrees of cross-linking, exhibited higher glucose yields because their active sites are more accessible than those of zeolite catalysts. Dhepe *et al.* [81] applied various types of solid-acid catalysts including sulfonated mesoporous silica, macroreticular sulfonated cationic exchange resin, and zeolites for hydrolysis of sucrose and starch. They also found that the small pore diameter of zeolite (0.5–1 nm) limits the accessibility of sugar oligomers and polysaccharides to catalytic sites, thus resulting in poor sugar yields (<5%).

Mesoporous, Gel, and Macroreticular Resins

Sulfonated mesoporous silica and porous cationic exchange resins are more effective than zeolites because of their wide pore openings. For example, typical mesoporous silica has a pore diameter greater than 2.5 nm, which is wide enough to accommodate disaccharides (*c.* 1–2 nm). Bootsma *et al.* [70] used propylsulfonic acid-functionalized mesoporous silica to hydrolyze cellobiose and oligosaccharides released from hydrothermal pretreatment of distiller's grains. The median pore size of the synthesized catalyst was *c.* 6.2 nm. They observed cellobiose was successfully hydrolyzed to yield >90% glucose with <10% glucose degradation at 175 °C within an hour. Because high-molecular-weight oligomers were inaccessible to catalytic functional sites inside the pores, hydrolysis yields of xyloarabino-oligosaccahrides were lower (60–70%) than glucose yields from cellobiose hydrolysis. Dhepe *et al.* [81] also studied hydrolysis of sucrose and starch using different types of mesoporous silica-based catalysts with pore diameter within the range 2–20 nm. Sucrose hydrolysis gave 64–90% glucose yields while starch hydrolysis resulted in 20–40% yields.

Because of their wide range of pore diameters, macroreticular-type resins and loosely cross-linked gel-type cationic exchange resins offer advantages over zeolite and mesoporous silica catalysts. Cation exchange resins are grouped into two types: gel-type resin and macro-reticular resin. Gellular resin has no distinctive pore structures; rather, it has non-permanent pores established when its polymeric matrix swells after contacting a good solvent. These porous spaces of normally cross-linked gel-type resins (amount of the cross-linking agent 2–10%) can increase significantly upon swelling of the resins [83]. Macroreticular-type resins are made by fusing gellular microporous beads in the last stages of polymerization. This type of resin has two phase structures: microspheres and macropores between clusters of microspheres. Its pore structure is permanent and rigid, and the size of macropores reaches 10–100 nm. For example, the size of the macropores of Amberlyst 15 and 35 is *c.* 20–30 nm. Unlike

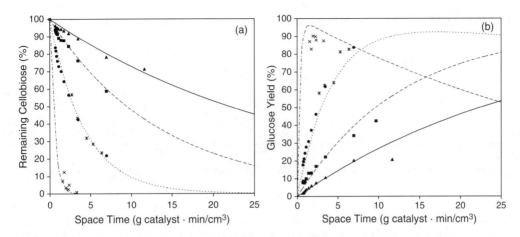

Figure 4.11 *Remaining (a) cellobiose and (b) glucose yield profiles in a packed bed of Amberlyst 35 at 110–130°C and 160°C. Symbols, experimental: circles: 130°C; squares: 120°C; triangles: 110°C; crosses: 160°C; asterisks: 130°C, 50 g/L. Lines, calculated from model: dash-dot: 160°C; dotted: 130°C; dashed: 120°C; continuous: 110°C. Data points are based on duplicate measurements. (Adapted and reprinted with permission from Kim et al. [63] © 2005, American Chemical Society).*

gel-type ion exchangers, macropores between aggregated microporous beads give reactants access to the inside of resin particles while avoiding resin swelling [84].

Kim *et al.* [63] studied a macroreticular type cationic exchanger as a catalyst for hydrolyzing a mixture of sugar oligomers derived from corn fiber, a low-lignin cellulosic biomass. Cationic exchangers functionalized with sulfonic group ($-SO_3H$) exhibit the strongest acid strength among the various functional groups. They used Amberlyst 35, a sulfonic acid functionalized macroreticular-type ion exchange resin that hydrolyzed cellobiose with 90% yield of glucose (Figure 4.11). Using a packed bed of Amberlyst 35 at 130°C, the sugar yield from hydrolyzing a mixture of sugar oligomers from corn fiber was 60% with 30% sugar loss to degradation. A packed bed of gel-type sulfonated cationic resin (Dowex 50WX2) resulted in 80% yields with 10% sugar decomposition at 150°C (Figure 4.12).

Types of Linkages in Oligosaccharides

As mentioned above, pore diffusion limits the hydrolysis efficiency of porous solid-acid catalysts (Bodamer and Kunin [73], Abbadi *et al.* [82], Kim *et al.* [63]). In addition, the difference in intrinsic hydrolysis rates of various types of oligosaccharides is also a critical factor when designing solid-acid catalysts and hydrolysis processes. Solubilized hemicellulose obtained from hydrothermal pretreatment of lignocellulosic biomass contains a complex mixture of sugar oligomers with different linkages and compositions originating from various polymers such as glucans, xylan, arabinan, and mannan. Their hydrolysis rates depend on the type of bond linkages and constituent sugar units. For example, xylooligomers have mainly β, 1–4 glycosidic bonds, whereas malto-oligosaccharides, which are derived from starch, have α, 1–4 glycosidic bonds. Table 4.3 summarizes the hydrolysis rates of various disaccharides measured for HCl catalyzed hydrolysis. Moelwin-Hughes [86] showed that α glycoside reacts 2.85 times faster than β glycoside, an effect that is independent of temperature. Wolfrom *et al.* [87] also noted that disaccharides can be grouped into three categories depending on the type of bond. They concluded that α-D linkages hydrolyze more rapidly than β-D linkages. Kim *et al.* [63] and Bootsma *et al.* [70] observed different reactivities of sugar oligomers to solid-acid hydrolysis: pentose sugars (e.g., xylose and arabinose) were released much faster than glucose from oligosaccharides.

Table 4.3   *Hydrolytic rate data for various disaccharides in 0.1-N hydrochloric acid at 80°C; k: hydrolysis rate constant; a_{H+}: activity of hydrogen ion.*

	Linkage	Unit sugar	k/a_{H+} (s^{-1})	E (cal/mole)
Cellobiose[a]	β– (1 → 4)	D-glucose	5.89×10^{-6}	30 710
Maltose[a]	α– (1 → 4)	D-glucose	1.68×10^{-5}	30 970
Sucrose[a]	(α–1) → (β–2)	D-glucose, D-fructose	1.46×10^{-2}	25 830
Lactose[a]	β–(1 → 4)	D-galactose, D-glucose	1.66×10^{-5}	26 900
Xylobiose[b]	β–(1 → 4)	D-xylose	1.58×10^{-5}	34 000

[a] Moelwin-Hughes [86]

[b] Kamiyama and Sakai [68], modifying their equation to make units consistent with Moelwyn-Hughes's data

Bond positions and length of oligosaccharides also affect hydrolysis reaction rates in solid-acid catalysts. Kamiyama and Sakai [68] showed that non-reducing ends react 1.8 times faster than reducing ends and internal bonds of xylooligosaccharides. Beltrame and Carniti [88] studied hydrolysis rates of non-reducing ends and internal bonds of malto-oligosaccharides. Their experimental results showed that the best fit was obtained by assuming that all internal bonds and reducing ends break at the same rate, whereas non-reducing ends are hydrolyzed at the same rate as maltose. The ratio of maltose hydrolysis rate to internal bond hydrolysis rate ranged from 1.79 to 1.98. Various studies have shown that internal bonds are hydrolyzed approximately 1.8 times faster than those at non-reducing ends.

Sugar Degradation

Hydrolysis yields from oligosaccharides using solid-acid catalysts also greatly depend on rates of sugar degradation. Hydrolysis of sugar oligomers is likely to occur mainly on the external surface and large pores

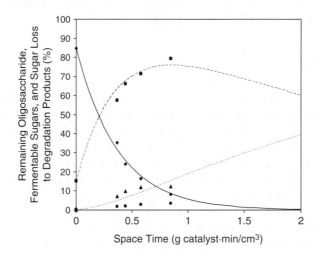

Figure 4.12   *Remaining oligosaccharides, fermentable sugar yield, and sugar loss to degradation product profiles in a packed-bed plug-flow reactor at 150°C using Dowex 50WX2. Symbols, experimental; diamonds: remaining oligosaccharides; squares: fermentable sugar yield; triangles: sugar loss to degradation products; circles: sugar loss to degradation products as measured by HPLC. Lines calculated from model. (Adapted and reprinted with permission from Kim [85] © 2005, Purdue University).*

of solid-acid catalysts. As the hydrolysis progresses, short-chain oligomers and monomeric sugars can enter smaller pores where they can further degrade into furans and aldehydes. It is well known that sugar degradation rates vary depending on the types of sugars. Saeman [89] showed that sugars degrade in the following order: xylose > arabinose, mannose > galactose, glucose. Arabinose and mannose degrade *c.* 1.5–1.7 times faster than glucose and galactose. The fastest decomposing sugar was xylose; the rate constant was *c.* 3 times higher than that of glucose. To minimize sugar degradation during solid-acid catalyzed hydrolysis of sugar oligomers, catalyst pore size and structure, as well as reaction conditions, must be carefully selected and optimized.

Catalyst Stability

For solid-acid catalyzed hydrolysis of sugar oligomers and cellulose to be an economical option, the catalysts must be stable at high temperatures and the reactants must be free of compounds that foul and deactivate resins. Operating temperatures for hydrolysis of sugar oligomers range from 100 to 200 °C. To hydrolyze sugar oligomers using porous solid-acid catalysts, high temperatures are favored because of higher selectivity to hydrolysis rather than sugar degradation. However, operating sulfonic-acid-functionalized cationic exchange resins above 120 °C causes resin desulfonation and deactivation [63]. Generally, silica-based catalysts are more heat stable than cationic exchange resins. Catalyst deactivation can also be caused by resin fouling and impurities such as proteins, lignin-derived phenolic compounds, and minerals derived from lignocellulosic biomass.

Fouling

Proteins and phenolic compounds in the pretreatment liquid may be adsorbed on polymeric particles by interactions of hydrophobic patches of proteins with hydrophobic surfaces of the adsorbent. Minerals deactivate resins by replacing hydrogen ions of sulfonic acid groups. Proteins and other impurities can be removed by pre-conditioning the pretreatment liquid using adsorbents such as hydrophobic resins and activated carbon [63,70]. Using pretreated corn fiber liquid, Kim *et al.* [63] showed that >80% of proteins and phenolic compounds were eliminated by XAD-4 and activated carbon. For woody biomass, minerals and phenolics are the major deactivators; for corn processing by-products (e.g., corn fiber and distiller's grains), proteins are the major culprit. Resin activity loss was caused mainly by minerals dissolved in the pretreatment liquid, while increased pressure drop was due to proteins. A guard bed is needed to remove components that could foul or deactivate the resin catalyst.

Costs

A preliminary economic analysis shows that if 1 lb of catalyst generates 1000 lb of glucose, the incremental cost varies between 1 and 18 g/gallon ethanol, depending on catalyst cost (Figure 4.13). Reducing processing costs of solid-acid catalyzed hydrolysis can be realized by further improving catalyst life and selectivity.

Challenges

In summary, solid-acid catalysts are promising alternative catalysts for hydrolyzing aqueous oligosaccharide streams derived from lignocellulosic feedstocks because they offer numerous advantages over homogeneous acids or enzymes. Major challenges are to increase catalyst life by removing resin foulants and deactivators, developing heat-stable catalysts, and designing advanced catalysts that are highly selective to oligosaccharides hydrolysis with minimal sugar degradation. In addition, further understanding of the

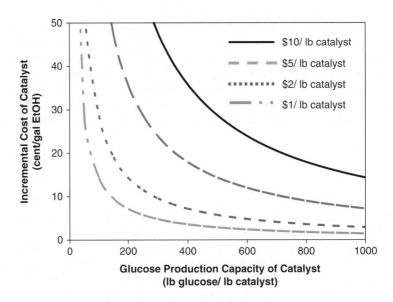

Figure 4.13   *Estimates of incremental cost of catalyst per gallon of ethanol produced from glucose as a function of glucose produced before catalyst must be replaced. (Adapted and reprinted with permission from Kim et al. [63] © 2005, American Chemical Society).*

diffusional effects on hydrolysis kinetics inside porous structures of solid-acid catalysts will benefit design of advanced catalysts with increased hydrolysis efficiency.

4.5   Conclusions

Hydrothermal pretreatment generates various soluble cellulase inhibitors, including sugar oligomers and phenolic compounds derived from lignin. If not removed, these soluble inhibitors significantly hamper enzymatic hydrolysis of cellulose in subsequent processing steps. Among the identified enzyme inhibitors, this work shows that phenols (e.g., vanillin, p-coumaric, ferulic, gallic and tannic acids) inhibit and deactivate cellulolytic enzymes and, to a greater extent, β-glucosidases. To enhance hydrolysis rates and reduce enzyme usage, processes that hydrolyze cellulose to glucose must reduce inhibition and deactivation effects. To achieve this goal, further understanding of the factors that mediate interactions between soluble phenolic compounds and/or lignin and cellulases is important. As indicated by Boukari *et al.* [24], this may include limiting the formation of hydroxyl-bearing phenolic compounds. Ximenes *et al.* [16,17] and Kim *et al.* [18] discuss other strategies such as: identifying and developing β-glucosidases that resist inhibition from both sugars and phenols; performing enzyme hydrolysis over shorter periods to decrease time-dependent deactivation; separating phenolics prior to enzyme hydrolysis; or using microbial, enzymatic, or chemical methods to convert phenolics to an inactive form.

Alternatively, to avoid enzyme inhibition/deactivation problems, pretreatment streams that contain inhibitors can be separately processed through solid-acid catalysts that efficiently hydrolyze soluble sugar oligomers into monomeric sugars. For a continuous hydrolysis process, heterogeneous solid-acid catalysts can be packed in a bed and reused, thus offering many advantages over homogeneous acids. To successfully implement technology that converts pretreated, lignocellulosic feedstocks to fermentable sugars, it is essential to design catalysts with improved diffusion characteristics and thermal stability and to develop efficient ways to remove catalyst deactivators.

Acknowledgements

This work was supported by Purdue University Agricultural Research Programs, a Mascoma Corporation Sponsored Research Grant, and Department of Energy Grant GO18103.

References

1. Mosier, N.S., Wyman, C.E., Dale, B.E. *et al.* (2005) Features of promising technologies for pretreatment of ligno-cellulosic biomass. *Bioresource Technology*, **96**, 673–686.
2. Yang, B. and Wyman, C.E. (2008) Pretreatment: the key to unlocking low-cost cellulosic ethanol. *Biofuels, Bioproducts and Biorefining*, **2**, 26–40.
3. Geddes, C.C., Nieves, I.U., and Ingram, L.O. (2011) Advances in ethanol production. *Current Opinion in Biotechnology*, **22** (3), 312–319.
4. Brás, J.L.A., Cartmell, A., Carvallho, A.L.M. *et al.* (2011) Structural insights into a unique cellulase fold and mechanism of cellulose hydrolysis. *PNAS*, **108** (13), 5237–5242.
5. US DOE (2005) *Genomics:GTL Roadmap*, DOE/SC-0090, U.S. Department of Energy Office of Science. (p. 204) (http://genomicscience.energy.gov/roadmap/#page=news).
6. US DOE (May, 2007) Biofuels Primer Placemat: From Biomass to Cellulosic Ethanol and Understanding Biomass: Plant Cell Walls, US Department of Energy Office of Science. (http://genomicscience.energy.gov/biofuels/placemat.shtml#page=news).
7. Ladisch, M., Ximenes, E., Kim, Y., and Mosier, N.S. (2011) *Biomass Chemistry and Pretreatment for Biological Processing*, NSF PIRE-FHI Summer School, Closter Secon, Germany.
8. Templeton, D.W., Scarlata, C.J., Sluiter, J.B., and Wolfrum, E.J. (2010) Compositional analysis of lignocellulosic feedstocks. 2. Method uncertainties. *Journal of Agricultural and Food Chemistry*, **58**, 9054–9062.
9. Kim, M. and Day, D.F. (2011) Composition of sugar cane, energy cane, and sweet sorghum suitable for ethanol production at Louisiana sugar mills. *Journal of Industrial Microbiology & Biotechnology*, **38** (7), 803–807.
10. Du, B., Sharma, L.N., Becker, C. *et al.* (2010) Effect of varying feedstock-pretreatment chemistry combinations on the formation and accumulation of potentially inhibitory degradation products in biomass hydrolysates. *Biotechnology and Bioengineering*, **107** (3), 430–440.
11. US DOE (June, 2007) Biofuels: Bringing Biological Solutions to Energy Challenges, US Department of Energy Office of Science (Biofuels_Flyer_2007 PDF).
12. Lynd, L.R., van Zyl, W.H., McBride, J.E., and Laser, M. (2005) Consolidated bioprocessing of cellulosic biomass: an update. *Current Opinion in Biotechnology*, **5**, 577–583.
13. Xu, Q., Singh, A., and Himmel, M.E. (2009) Perspectives and new directions for the production of bioethanol using consolidated bioprocessing of lignocelluloses. *Current Opinion in Biotechnology*, **3**, 364–371.
14. Gong, C.-S., Ladisch, M.R., and Tsao, G.T. (1977) Cellobiase from *Trichoderma viride*: purification, properties, kinetics, and mechanism. *Biotechnology and Bioengineering*, **19**, 959–981.
15. Ladisch, M.R., Gong, C.-S., and Tsao, G.T. (1977) Corn crop residues as a potential source of singlecell protein: kinetics of Trichoderma viride cellobiase action. *Developments in Industrial Microbiology Series*, **18**, 157–168.
16. Ximenes, E., Kim, Y., Mosier, N. *et al.* (2010) Inhibitors of cellulose hydrolysis in wet cake. *Enzyme and Microbial Technology*, **46**, 170–176.
17. Ximenes, E., Kim, Y., Mosier, N. *et al.* (2011) Deactivation of cellulases by phenols. *Enzyme and Microbial Technology*, **48**, 54–60.
18. Kim, Y., Ximenes, E., Mosier, N.S., and Ladisch, M.R. (2011) Soluble inhibitors/deactivators of cellulase enzymes from lignocellulosic biomass. *Enzyme and Microbial Technology*, **48**, 408–415.
19. Cook, M.T., Basset, H.F., Thompson, F., and Taubenhaus, J.J. (1911) Protective enzymes. *Science*, **33**, 624–629.
20. Kumar, R. and Wyman, C.E. (2009) Effect of additives on the digestibility of cornstover solids following pretreatment by leading technologies. *Biotechnology and Bioengineering*, **102**, 1544–1557.
21. Qing, Q., Yang, B., and Wyman, C.E. (2010) Xylooligomers are strong inhibitors of cellulose hydrolysis by enzymes. *Bioresource Technology*, **101**, 9624–9630.

22. Qing, Q. and Wyman, C.E. (2011) Supplementation with xylanase and β-xylosidase to reduce xylooligomer and xylan inhibition of enzymatic hydrolysis of cellulose and pretreated corn stover. *Biotechnology for Biofuels*, **4**, 18.

23. Wyman, C.E. (2007) What is (and is not) vital to advancing cellulosic ethanol. *Trends in Biotechnology*, **25**, 153–157.

24. Boukari, I., O'Donohue, M., Rémond, C., and Chabbert, B. (2011) Probing of a family GH11 endo-β-1,4-xylanase inhibition mechanism by phenolic compounds: Role of functional phenolic groups. *Journal of Molecular Catalysis B-Enzymatic*, **72**, 130–138.

25. Galbe, M. and Zalchi, G. (2007) Pretreatment of lignocellulosic materials for efficient bioethanol production. *Advances in Biochemical Engineering/Biotechnology*, **108**, 41–65.

26. Hahn-Hägerdal, B., Wahlbom, C.F., Gardonyi, M. *et al.* (2001) Metabolic engineering of *Saccharomyces cerevisiae* for xylose utilization. *Advances in Biochemical Engineering/Biotechnology*, **73**, 53–84.

27. Soudham, V.P., Alriksson, B., and Jönsson, L.J. (2011) Reducing agents improve enzymatic hydrolysis of cellulosic substrates in the presence of pretreatment liquid. *Journal of Biotechnology*, **155**, 244–250.

28. Terijian, A. and Xu, F. (2011) Inhibition of enzymatic cellulolysis by phenolic compounds. *Enzyme and Microbial Technology*, **48**, 239–247.

29. Olsen, S.N., Bohlin, C., Murphy, L. *et al.* (2011) Effects of non-ionic surfactants on the interactions between cellulases and tannic acid: A model system for cellulase-poly-phenol interactions. *Enzyme and Microbial Technology*, **49**, 353–359.

30. Chen, S.F., Mowery, R.A., Castleberry, V.A. *et al.* (2006) High performance liquid chromatography method for simultaneous determination of aliphatic acid, aromatic acid and neutral degradation products in biomass pretreatment hydrolysates. *Journal of Chromatography. A*, **1104**, 54–61.

31. Kirk, T.K. and Farrell, R.L. (1987) Enzyme "combustion": The microbial degradation of lignin. *Annual Review of Microbiology*, **41**, 465–505.

32. Martínez, A.T., Speranza, M., Ruiz-Dueñas, F.J. *et al.* (2005) Biodegradation of lignocellulosic: microbial, chemical, and enzymatic aspects of the fungal attack of lignin. *International Microbiology*, **8**, 195–204.

33. Del Río, J.C., Martínez, Á.T., and Gutiérrez, A. (2007) Presence of 5-hydroxyguaiacyl units as native lignin constituents in plants as seen by py-GC/MS. *Journal of Analytical and Applied Pyrolysis*, **79**, 33–38.

34. Ko, J.-J., Shimizu, Y., Ikeda, K. *et al.* (2009) Biodegradation of high molecular weight lignin under sulfate reducing conditions: Lignin degradability and degradation by-products. *Bioresource Technology*, **100**, 1622–1627.

35. Zeng, M., Ximenes, E., Ladisch, M.R. *et al.* (2012) Enzyme hydrolysis and imaging of fractionated corn stalk tissues pretreated with liquid hot water part I. *Biotechnology and Bioengineering*, **109**, 390–397.

36. Zeng, M., Ximenes, E., Ladisch, M.R. *et al.* (2012) Enzyme hydrolysis and imaging of fractionated corn stalk tissues pretreated with liquid hot water Part II. *Biotechnology and Bioengineering*, **109**, 398–404.

37. Palmqvist, E. and Hahn-Hägerdal, B. (2000) Fermentation of lignocellulosic hydrolysates. II: inhibitors and mechanisms of inhibition. *Bioresource Technology*, **74**, 25–33.

38. Deaville, E.R., Green, R.J., Mueller-Harvey, I. *et al.* (2007) Hydrolyzable tannin structures influence relative globular and random coil protein binding strengths. *Journal of Agricultural and Food Chemistry*, **55**, 4554–4561.

39. Berlin, A., Gilkes, N., Kurabi, A. *et al.* (2005) Weak lignin-binding enzymes- a novel approach to improve activity of cellulases for hydrolysis of lignocellulosics. *Applied Biochemistry and Biotechnology*, **121**, 163–170.

40. Pan, X.J. (2008) Role of functional groups in lignin inhibition of enzymatic hydrolysis of cellulose to glucose. *Biobased Materials and Bioenergy*, **2**, 25–32.

41. Sharma, A., Milstein, O., Vered, Y., and Gressel, J. (1985) Effects of aromatic compounds on hemicellulase-degrading enzymes in *Aspergillus japonicas*. *Biotechnology and Bioengineering*, **27**, 1095–1101.

42. Li, X., Ximenes, E.A., Kim, Y. *et al.* (2010) Lignin monomer composition impacts *Arabidopsis* cell wall degradability following liquid hot-water pretreatment. *Biotechnology for Biofuels*, **3** (27), 1–7.

43. Xiao, Z., Zhang, X., Gregg, D., and Saddler, J. (2004) Effects of sugar inhibition on cellulases and β-glucosidase during enzymatic hydrolysis of softwood substrates. *Applied Biochemistry and Biotechnology*, **115**, 1115–1126.

44. Dekker, RFH. (1988) Inhibitors of *Trichoderma reesei* β-glucosidase activity derived from auto-hydrolysis-exploded eucalyptous-regnans. *Applied Microbiology and Biotechnology*, **29**, 593–598.

45. Hahn-Hägerdal, B., Galbe, M., Gorwa-Grauslund, M.F. *et al.* (2006) Bio-ethanol- the fuel of tomorrow from the residues of today. *Trends in Biotechnology*, **24**, 549–556.

46. Kinkle, H.B., Thomsen, A.B., and Ahring, B.K. (2004) Inhibition of ethanol-producing yeast and bacteria by degradation products produced during pre-treatment of biomass. *Applied Microbiology and Biotechnology*, **66**, 10–26.

47. Almeida, J.R.M., Modig, T., Petersson, A. *et al.* (2007) Increased tolerance and conversion of inhibitors in ligno-cellulosic hydrolysates by *Saccharomyces cerevisiae*. *Journal of Chemical Technology and Biotechnology*, **82**, 340–349.

48. Thomsen, M.H., Thygesen, A., and Thomsen, A.B. (2009) Identification and characterization of fermentation inhibitors formed during hydrothermal treatment and following SSF of wheat straw. *Applied Microbiology and Biotechnology*, **83**, 447–455.

49. Nichols, N.N., Dien, B.S., and Cotta, M.A. (2010) Fermentation of bioenergy crops into ethanol using biological abatement for removal of inhibitors. *Bioresource Technology*, **101**, 7545–7550.

50. Gorsich, S.W., Dien, B.S., Nichols, N.N. *et al.* (2006) Tolerance of furfural-induced stress is associated with pentose phosphate pathway genes ZWF1, GND1, RPE1, and TKL1 in *Saccharomyces cerevisiae*. *Applied Microbiology and Biotechnology*, **71**, 339–349.

51. Zaldivar, J., Martinez, A., and Ingram, L.O. (1999) Effect of selected aldehydes on the growth and fermentation of ethanologenic *Escherichia coli*. *Biotechnology and Bioengineering*, **65**, 24–33.

52. Taherzadeh, M.J., Gustafsson, L., Niklasson, C., and Lidén, G. (2000) Physiologicaxl effects of 5-hydroxymethyl-furfural on *Saccharomyces cerevisiae*. *Applied Microbiology and Biotechnology*, **53**, 701–708.

53. Casey, E., Sedlak, M., Ho, N.W.Y., and Mosier, N.S. (2010) Effect of acetic acid and pH on the cofermentation of glucose and xylose to ethanol by a genetically engineered strain of *Saccharomyces cerevisiae*. *FEMS Yeast Research*, **10**, 385–393.

54. Russel, J.B. (1992) Another explanation for the toxicity of fermentation acids at low pH: anion accumulation versus uncoupling. *The Journal of Applied Bacteriology*, **73**, 363–370.

55. Verduyn, C. (1991) Physiology of yeasts in relation to biomass yields. *Antonie van Leeuwenhoek*, **60**, 325–353.

56. Verduyn, C., Postma, E., Scheffers, W.A., and Van Dijken, J.P. (1992) Effect of benzoic acid on metabolic fluxes in yeasts: a continuous-culture study on the regulation of respiration and alcoholic fermentation. *Yeast (Chichester, England)*, **8**, 501–517.

57. Pampulha, M.E. and Loureiro Dias, M.C. (1989) Combined effect of acetic acid, pH and ethanol on intracellular pH of fermenting yeast. *Applied Microbiology and Biotechnology*, **20**, 286–293.

58. Larsson, S., Palmqvist, E., Hahn-Hägerdal, B. *et al.* (1999) The generation of fermentation inhibitors during dilute acid hydrolysis of softwood. *Enzyme and Microbial Technology*, **24**, 151–159.

59. Lu, Y., Warner, R., Sedlak, M. *et al.* (2009) Comparison of glucose/xylose co-fermentation of poplar hydrolysates processed by different pretreatment technologies. *Biotechnology Progress*, **25**, 349–356.

60. Krishna, R., Kallury, M.R., Ambidge, C., and Tidwell, T.T. (1986) Rapid hydrothermolysis of cellulose and related carbohydrates. *Carbohydrate Research*, **158**, 253–261.

61. Walch, E., Zemann, A., Schinner, F. *et al.* (1992) Enzymatic saccharification of hemicellulose obtained from hydro-thermally pretreated sugar cane bagasse and beech bark. *Bioresource Technology*, **39**, 173–177.

62. Kim, Y., Mosier, N.S., and Ladisch, M.R. (2009) Enzymatic digestion of liquid hot water pretreated hybrid poplar. *Biotechnology Progress*, **25**, 340–348.

63. Kim, Y., Hendrickson, R., Mosier, N.S., and Ladisch, M.R. (2005) Plug-flow reactor for continuous hydrolysis of glucans and xylans from pretreated corn fiber. *Energy Fuels*, **19**, 2189–2200.

64. Kim, Y., Hendrickson, R., Mosier, N.S., and Ladisch, M.R. (2009) Liquid hot water pretreatment of cellulosic biomass, in *Methods in Molecular Biology: Biofuels*, **581** (ed. J.R. Mielenz), The Humana Press, Totowa, NJ, pp. 93–102.

65. Berlin, A., Maximenko, V., Gilkes, N., and Saddler, J. (2007) Optimization of enzyme complexes for lignocellulose hydrolysis. *Biotechnology and Bioengineering*, **97**, 287–296.

66. Dien, B.S., Ximenes, E.A., O'Bryan, P.J. *et al.* (2008) Enzyme characterization for hydrolysis of AFEX and liquid hot-water pretreated distillers' grains and their conversion to ethanol. *Bioresource Technology*, **99**, 5216–5225.

67. Mäki-Arvela, P., Salmi, T., Holmbom, B. *et al.* (2011) Synthesis of sugars by hydrolysis of hemicelluloses-a review. *Chemical Reviews*, **111**, 5638–5666.

68. Kamiyama, Y. and Sakai, Y. (1979) Rate of hydrolysis of xylo-oligosaccharides in dilute sulfuric acid. *Carbohydrate Research*, **73**, 151–158.

69. Ladisch, M.R. and Tsao, G.T. (1986) Engineering and economics of cellulose saccharification systems. *Enzyme and Microbial Technology*, **8**, 66–69.

70. Bootsma, J.A., Entorf, M., Eder, J., and Shanks, B.H. (2008) Hydrolysis of oligosaccharides from distillers grains using organic-inorganic hybrid mesoporous silica catalysts. *Bioresource Technology*, **99**, 5226–5231.

71. Ogaki, Y., Shinozuka, Y., Hatakeyama, M. *et al.* (2009) Selective production of xylose and xylo-oligosaccharides from bamboo biomass by sulfonated allophone solid acid catalyst. *Chemistry Letters*, **38**, 1176–1177.

72. Dhepe, P.L. and Sahu, R. (2010) A solid-acid-based process for the conversion of hemicellulose. *Green Chemistry*, **12**, 2153–2156.

73. Bodamer, G. and Kunin, R. (1951) Heterogeneous catalytic inversion of sucrose with cation exchange resins. *Industrial & Engineering Chemistry*, **43**, 1082–1085.

74. Heyraud, A. and Rinaudo, M. (1981) Hydrolysis of oligosaccharides by sulfonic ion exchangers. *European Polymer Journal*, **17**, 1167–1173.

75. Fukuoka, A. and Dhepe, P.L. (2006) Catalytic conversion of cellulose into sugar alcohols. *Angewandte Chemie-International Edition*, **45**, 5161–5163.

76. Luo, C., Wang, S., and Liu, H. (eds) (2007) Cellulose conversion into polyols catalyzed by reversibly formed acids and supported ruthenium clusters in hot water. *Angewandte Chemie-International Edition*, **46**, 7636–7639.

77. Komanoya, T., Kobayashi, H., Hara, K. *et al.* (2011) Catalysis and characterization of carbon-supported ruthenium for cellulose hydrolysis. *Applied Catalysis A-General*, **407**, 188–194.

78. Suganuma, S., Nakajima, K., Kitano, M. *et al.* (2008) Hydrolysis of cellulose by amorphous carbon bearing SO$_3$H, COOH, and OH groups. *Journal of the American Chemical Society*, **130**, 12787–12793.

79. Kobayashi, H., Ito, Y., Komanoya, T. *et al.* (2011) Synthesis of sugar alcohols by hydrolytic hydrogenation of cellulose over supported metal catalysts. *Green Chemistry*, **13**, 326–333.

80. Braun, V. (1973) Molecular organization of the rigid layer and the cell wall of *Escherichia coli*. *The Journal of Infectious Diseases*, **128**, 9–16.

81. Dhepe, P.L., Ohashi, M., Inagaki, S. *et al.* (2005) Hydrolysis of sugars catalyzed by water-tolerant sulfonated mesoporous silicas. *Catalysis Letters*, **102**, 163–169.

82. Abbadi, A., Gotlieb, K.F., and Bekkum, H.V. (1998) Study on solid acid catalyzed hydrolysis of maltose and related polysaccharides. *Starch*, **50**, 23–28.

83. Dorfner, K. (1972) *Ion Exchangers: Properties and Applications*, Ann Arbor Science, p. 34.

84. Helfferich, F.G. (1995) *Ion Exchange*, Dover Publications, Inc., New York, pp 34–35.

85. Kim, Y. (2005) Solid acid catalysts for hydrolyzing oligosaccharides derived from corn fiber. PhD thesis, Purdue University.

86. Moelwin-Hughes, E.A. (1929) The kinetics of the hydrolysis of certain glucosides, Part III. B-Methylglucoside, cellobiose, melibiose, and turanose. *Transactions of the Faraday Society*, **25**, 503–520.

87. Wolfrom, M.L., Thompson, A., and Timberlake, C.E. (1963) Comparative hydrolysis rates of the reducing disaccharides of D-glucopyranose. *Journal of the American Chemical Society*, **40**, 83–86.

88. Beltrame, P.L. and Carniti, P. (1987) Kinetics of enzymatic hydrolysis of malto-oligosaccharides: A comparison with acid hydrolysis. *Carbohydrate Research*, **166**, 71–83.

89. Saeman, J.F. (1945) Kinetics of wood saccharification; Hydrolysis of cellulose and decomposition of sugars in dilute acid at high temperature. *Industrial & Engineering Chemistry*, **37**, 43–52.

5

Catalytic Strategies for Converting Lignocellulosic Carbohydrates to Fuels and Chemicals

Jesse Q. Bond[1], David Martin Alonso[2] and James A. Dumesic[2]

[1] *Biomedical and Chemical Engineering, Syracuse University, Syracuse, USA*
[2] *Department of Chemical and Biological Engineering, University of Wisconsin, Madison, USA*

5.1 Introduction

Historically, the use of biomass as a source of energy and chemicals predates modern dependence on fossil resources; however, its use became uneconomical with the availability of low-cost fossil resources. An entire industry is dedicated to converting petroleum into consumer products – an industry that has achieved remarkable efficiency through advances in chemical reaction engineering and catalytic chemistry [1,2]. Present concerns regarding the ecological impacts and availability of fossil-derived resources (particularly petroleum) have directed increasing amounts of scientific research towards converting biomass into both fuels and chemicals to replace fossil resources with a sustainable carbon source.

Given the end-goal of common products from either renewable or fossil feedstocks, it is illustrative to contrast biorefining with petroleum refining. Major differences arise because the energy density and functionality affect the chemical conversion strategy for each feedstock. Petroleum is particularly well suited to our present infrastructure, a natural consequence of an industrial society built upon widespread availability of low-cost petroleum. It has a high energy density, which reduces the cost of transporting the feedstock from upstream sources to downstream processing facilities. Crude oil is a largely non-functionalized feedstock, which offers distinct advantages in the production of transportation fuels; however, it creates challenges in the production of many chemicals. Fossil feedstocks have minimal oxygen content and are rich in energy-dense hydrocarbons; the petroleum industry therefore converts this feedstock to bulk transportation fuels, which are generated in large volumes. In the production of petroleum-derived fuels, great effort is

Aqueous Pretreatment of Plant Biomass for Biological and Chemical Conversion to Fuels and Chemicals, First Edition.
Edited by Charles E. Wyman.
© 2013 John Wiley & Sons, Ltd. Published 2013 by John Wiley & Sons, Ltd.

expended on cracking, reforming, separating, fractionating, purifying, and refining the feedstock so different ranges of fuels may be produced. The modern petroleum refinery is so efficient that fuel prices are almost entirely governed by the cost of crude oil. In the present market, we are therefore provided with low-cost, high-volume transportation fuels.

In the production of chemicals, a non-functionalized feedstock is less easily leveraged, which requires more intensive strategies for selective transformations to tailored end-products [3]. The primary monomers that serve as building blocks to the present chemical industry are methanol, ethylene propylene, butadiene, benzene, toluene, and xylene, each of which must be obtained through energy- and resource-intensive chemical conversion of alkane feedstocks, which translates to higher costs. This system is manageable given the lower demand and smaller scale for chemical products compared to fuels [4]. The current paradigm is particularly well suited to our society, which has a high consumption of commodity products (e.g., fuels), but less extensive consumption of premium-priced specialty chemicals (e.g., pharmaceuticals). Unfortunately, widespread, low-cost availability of fossil resources is not a long-term certainty; alternative carbon sources that can supply society with both fuels and chemicals are therefore required. In this respect, lignocellulose is an attractive resource. It is reported that the United States alone can sustainably produce 1.3 billion dry tons of biomass annually without significant impact to our current infrastructure. This quantity is sufficient to offset as much as 50% of the carbon demand for US gasoline and diesel [5–7].

The challenges of lignocellulose refining are different from conventional petroleum refining. Biomass resources are extensively functionalized and have low energy densities. As such, transporting raw biomass from remote locations to central processing facilities is not energy efficient. Further, their conversion into fuels is not a matter of cracking, separating, and refining; rather, biomass requires extensive chemical transformations to improve energy density and tune physical properties (e.g., volatility) such that they match those used in the existing infrastructure [8,9]. In the production of a commodity product such as alkane fuels, this upgrading strategy is a major challenge. What has historically been a separations/purification challenge for petroleum has now become a synthetic chemistry challenge for biomass. The challenges for producing chemical products from biomass contrasts with those faced in the petroleum industry. Here, the challenge is not in difficult activation (as with alkane feeds), but in controlling selectivity in the presence of multiple functional groups.

5.2   Biomass Conversion Strategies

Lignocellulosic biomass is primarily comprised of cellulose, hemicellulose, and lignin fractions, each with unique structural properties and chemical functionalities [10]. To date, multiple avenues for converting biomass to fuels and chemicals have been developed. Each strategy uses the same approach (Figure 5.1). In an initial step, biomass undergoes deconstruction (thermal, enzymatic, chemical/catalytic, etc.) to produce functional intermediates that are used to produce a range of end-products [11]. For example, thermochemical methods allow direct processing of whole biomass and are favored by their front-end simplicity. Pyrolysis and liquefaction deconstruct raw biomass into liquid "bio-oils," which are the least expensive liquid hydrocarbons that can be obtained from lignocellulosic materials [7]. However, bio-oils are complex mixtures of small oxygenates that require substantial processing, purification, and/or fractionation to produce fuels or chemicals [12,13]. At a large scale, gasification is a viable strategy by which biomass undergoes decomposition through partial oxidation to produce synthesis gas (a H_2/CO mixture referred to as syngas) [14], which may be upgraded to methanol [15] or liquid alkanes (Fisher–Tropsch) [16]. Thermochemical strategies will certainly play an important role in future biorefining operations, particularly for producing transportation fuels, because they offer low-cost deconstruction. These methods do not require chemical pretreatments however, which is the focus of this text; thermal biomass processing strategies will

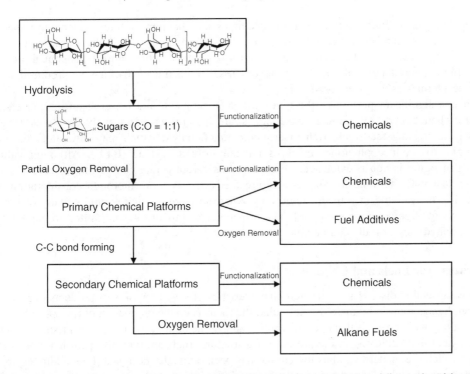

Figure 5.1 *General upgrading strategies for converting polymeric carbohydrates into fuels and chemicals. Of the possibilities outlined in this chapter, this figure is most directly applied to hydrolytic, chemical strategies for sugar isolation and upgrading. Thermochemical and biological strategies, which are not discussed at length in this chapter, are predicated on a similar general approach, although individual stages can vary considerably from those shown here.*

therefore not be further considered in this chapter. The interested reader is referred to multiple reviews of thermochemical conversion of biomass [12,13,17–26].

The target of biomass pretreatment is selective isolation of carbohydrate polymers (cellulose, hemicellulose) or sugars (glucose, mannose, xylose, arabinose) from lignin [27]. Because each biomass fraction (hemicellulose, cellulose, and lignin) has unique structure and chemistry, it is advantageous to isolate these fractions from one another. In particular, fractionation allows for selective processing strategies so that each constituent may be upgraded according to its unique chemical and structural properties, facilitating maximal carbon yield to a range of end-products [11].

There are two broad classes of biomass conversion strategies that follow pretreatment and fractionation of the feedstock. The first approach is to use biochemical methods (enzymatic hydrolysis, sugar fermentation), which is the dominant strategy for sugar conversion. Biochemical methods have multiple advantages including low environmental impact, mild process conditions, and high selectivity toward metabolic products. Chapter 4 of this text thoroughly describes available technologies for biochemical conversion.

To selectively upgrade biomass fractions, the second strategy (which is the focus of this chapter) is chemical – specifically catalytic – conversion. As with biochemical methods, such technologies are most developed for converting sugars, although promising new research is also advancing the state of lignin utilization. Relative to thermochemical methods, the primary advantages of chemical conversion are associated with comparatively mild processing conditions, which impart high selectivity toward tailored products. Additionally, chemical transformations – particularly those involving heterogeneous catalysis – allow flexible operating conditions and reduced residence times at higher temperatures, which may expedite sugar conversion. Many researchers have outlined chemical transformations of carbohydrates that provide path ways to the major fuels and chemicals used in the present infrastructure. Further, chemical processing of biomass offers the production of interesting, novel classes of chemicals that may be produced sustainably and provide alternatives to current products.

Lignin is a significant portion of the total carbon in biomass, and research on its conversion is an area where chemical or biochemical conversion strategies are less extensively developed. Lignin is a particularly interesting feedstock, rich in aromatic functionality, which may ultimately be leveraged for example, to produce phenolic resins or benzene-toluene-xylene (BTX). Although value-added processing of lignin is not as extensively considered as carbohydrate conversion, researchers are making strides that may facilitate its role as a future feedstock. A more thorough consideration of lignin and its utility is available in a recent review by Zakzeski *et al.*, which explores many of the promising pathways for lignin valorization [28]. The remainder of this chapter is limited to catalytic conversion of carbohydrates to fuels and chemicals.

5.3 Criteria for Fuels and Chemicals

The following section discusses the desired characteristics of several classes of products that may be produced from lignocellulosic biomass. In particular, the focus is on the production of transportation fuels and chemicals along with the goals that underlie each strategy. The broad classification of "fuels" includes traditional fuels and fuel additives. In the present infrastructure, fuels are typically petroleum-derived alkane blends for which the modern combustion engine has been optimally designed. Fuel additives are components that may be blended into alkane fuels, typically in limited quantities, to mitigate petroleum consumption, improve engine performance, and reduce certain classes of emissions. Although both are combusted for energy generation, they have different constraints and applications.

5.3.1 General Considerations in the Production of Fuels and Fuel Additives

Fuel additives are blended into predominately alkane fuels below a limit where they negatively affect engine performance. They may be useful in a number of ways. For example, adding oxygenates improves the combustion characteristics of fuel blends and can boost the octane number, particularly in gasoline [29]. However, oxygenated fuel additives have a somewhat lower energy density. Typically, they are subject to blending limits in conventional engines (i.e., 10–15% for ethanol, 5–20% for biodiesel) or can be used directly in modified engines. Most modern engines are designed optimally for alkane combustion; as such, oxygenates are considered as fuel additives whereas liquid alkanes (C_4–C_{20}) are considered primary transportation fuels.

Both transportation fuels and fuel additives are more energy dense than lignocellulose-derived feedstocks (cellulose, sugars), arising from a high oxygen content in biomass polymers [8]. A general goal for producing transportation fuels from biomass is to increase the energy density of the carbon source, usually by selectively removing oxygen. This removal may be achieved through a number of chemical transformations. For example, dehydration, hydrogenolysis, decarboxylation, and decarbonylation reactions all remove oxygen from parent molecules [9,11]. Dehydration and hydrogenolysis selectively cleave of C—O bonds (to release products as water), whereas decarboxylation and decarbonylation require C—C bond cleavage and the release of carbon to the gas phase. Depending on the desired outcome or product manifold, each route may play an important role. For example, reforming sugars and polyols provides green hydrogen through C—C bond cleavage. Alternatively, dehydration of xylose provides furfural, an important chemical intermediate, which retains the five carbon atoms of the parent sugar.

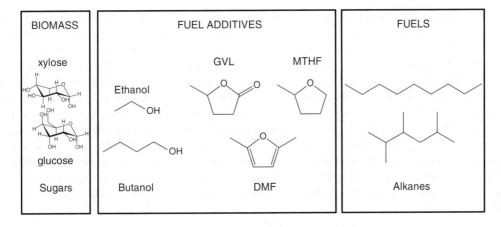

Figure 5.2 *Classes of transportation fuels and additives compared to biomass (sugars).*

A second consideration is the limitation of parent sugar molecules to five (e.g. xylose, arabinose) or six (e.g. mannose, glucose) carbon atoms. This limit is of particular interest when producing diesel or jet fuels, which may include alkanes containing on the order of 20 (diesel fuel) carbon atoms. Upgrading strategies must therefore form carbon-carbon bonds so that hydrocarbons in a desirable range of molecular weights may be built from 5- and 6-carbon monomers [30]. The extent of processing is dictated by the demand for energy density. For example, oxygenates (e.g., ethanol, butanol, γ-valerolactone (GVL), di-methylfuran (DMF), or methyltetrahydrofuran (MTHF); see Figure 5.2) have sufficient energy density to be used as fuel additives and may be blended for use in gasoline engines at levels ranging from 10 to 85% if the engine is suitably adjusted (E85, for example). In addition, they possess sufficiently low vapor pressures to be predominately transported and stored in the liquid phase and are therefore compatible with the existing infrastructure. Each of these molecules may be produced by de-oxygenating a single sugar molecule (limited to five or six carbon atoms) because the end-product itself contains less than six carbon atoms. However, oxygenated additives are subject to blending limits in most conventional engines and are not appropriate for jet fuels, which require maximal energy density. Alkanes intended as jet or diesel fuels therefore require an additional C—C coupling step followed by complete de-oxygenation. Although a more intensive strategy, the products more closely resemble current petroleum derivatives and offer greater compatibility with current engines and distribution systems.

Sugars are typically not appropriate for C—C coupling reactions because they are reactive and tend to degrade under conditions favoring C—C bond formation. Production of alkane fuels therefore proceeds in a two-stage process through an initial deoxygenation (to improve stability and energy density of the intermediate), followed by a secondary coupling step. Such intermediates are referred to as chemical platform molecules, which offer an appropriate balance of stability and chemical functionality. The most desirable characteristics of chemical platform molecules are that they may be produced selectively from biomass in a small number of stages at a low cost, and that they facilitate upgrading to a broad range of derivatives, allowing fuel and chemical production to be achieved through common intermediates [31], criteria which are presently well-met with crude oil.

Upgrading biomass-derived molecules to fuels can be illustrated as a continuum in which the energy density and physical attributes (volatility, cloud point, etc.) are continually improved to match those currently derived from petroleum. In fuel production, several factors dictate the short-term feasibility of biofuels production. First, the targeted product is a commodity resource that is intended for large-scale production to meet global energy demands. As such, biofuel production must utilize a source that can provide a sufficient volume of sustainable carbon. Lignocellulose is considered to be the only resource that can meet this criterion. Additionally, for short-term realization of lignocellulosic biofuels, it is necessary that sustainable fuel production be cost-competitive with petroleum-derived fuels. Presently, the major challenge in the production of biofuels is not a lack of platform upgrading strategies, but that many intermediates leveraged for use as fuel additives or alkane precursors are more valuable than the targeted transportation fuels. Future research should therefore strive to reduce the production cost of flexible biorefining intermediates from lignocellulosic resources. Chemical and catalytic processing of biomass can expedite this by offering a high degree of selectivity [32]. In a number of ways, catalytic conversion can also facilitate process intensification (most notably by reducing the complexity of separations and product recovery [33]). Frequently, catalytic chemistry allows multiple transformations within single reactors to yield hydrophobic or vapor-phase products, which spontaneously separate from both solid catalysts and aqueous solvents.

5.3.2   Consideration for Specialty Chemicals

The chemical industry has different objectives to that of the fuel industry, which manufactures a blend of low-cost liquid alkanes (or oxygenated additives) that are destined for combustion. In contrast, when producing chemicals, the end-products are designer molecules whose preparation requires selective processing strategies to confer specific chemical functionalities. In the petrochemical industry, the challenge in chemical production is the functionalization of a relatively inert feedstock: alkanes obtained from crude oil. On the other hand, biological feedstocks are extensively functionalized and the challenge is to selectively control the conversion of multifunctional feedstocks (sugars for example) [3]. In fuel production from biomass, the central interest is full removal of chemical functionality from parent molecules. In chemical production, functionalized feedstocks may be advantageous in a number of ways. Producing chemicals benefits from a functionalized feedstock that offers more facile avenues for upgrading, although unique challenges appear regarding control of selectivity. As with fuels production, higher-value chemicals are most readily produced not from the parent feedstock, but from intermediate platform molecules. It is therefore advantageous to identify low-cost platforms that provide pathways toward both fuels and chemicals.

5.4   Primary Feedstocks and Platforms

To logically present the myriad chemical conversion strategies possible with lignocellulosic materials, simple sugars will be considered as a primary set of platform molecules that can be produced by chemically deconstructing lignocellulose. With regard to lignin, no strategies are presently outlined to selectively isolate platform chemicals; the focus is therefore on carbohydrate conversion.

5.4.1   Cellulose

Cellulose is the most abundant component of lignocellulosic biomass, typically comprising 40–50% of the feedstock [34]. It consists of glucose units linked via β-glycoside bonds, and it is a primary structural component of cell walls. Cellulose has historically provided a raw material for the paper industry, which has been based upon the isolation of high-quality cellulose fibers from wood. Although many applications for bulk cellulose currently exist, the discussion is limited to cellulose hydrolysis for producing glucose monomers. As a feedstock for chemical processing, glucose is better suited to chemical conversion than cellulose and may be processed in solution (as opposed to recalcitrant cellulose). All of the cellulose-based chemical conversion strategies presented here proceed by depolymerizing cellulose to form glucose, which may be achieved in a number of ways. Frequently, single-pot strategies exist for producing chemical intermediates from cellulose, and they will be highlighted where relevant. Intensified processes are important in facilitating low-cost production of chemical intermediates; however, it is helpful to consider that they all rely upon the intermediate formation of glucose.

Glucose

Glucose ($C_6H_{12}O_6$) is a 6-carbon sugar containing an aldehyde moiety (an aldohexose) [35]. Although it may be sourced from a number of different biological feedstocks, in the context of this textbook glucose is obtained primarily by depolymerizing cellulose. Cellulose hydrolysis may be achieved enzymatically or chemically. Enzymatic methods are preferred for their selectivity and high glucose yields, whereas chemical methods offer low production costs and rapid conversion. The focus here is chemical hydrolysis of cellulose, which has been heavily researched in the preceding century. An excellent review of chemical methods for hydrolyzing cellulose has been prepared by Rinaldi and Schuth [36]. In general, acidic catalysts are favored for cellulose depolymerization; however, a major consideration is the crystallinity of cellulose and its limited solubility under typical reaction conditions. In practice, chemical hydrolysis of cellulose is achieved using dilute aqueous solutions of mineral acids; however, significant advantages in terms of glucose selectivity are offered by alternative approaches. Specifically, because sugars are particularly reactive in acids, glucose yields are limited when using non-selective acid hydrolysis in water. Such methods tend to produce dehydration products, such as 5-hydroxymethylfurfural and levulinic acid. Multiple alternative acid hydrolysis strategies have been proposed, and Rinaldi provides a thorough discussion. As an example, processing cellulose in concentrated HCl (6–7 mol/L) in the presence of lithium or calcium chloride allows for glucose yields of roughly 85% [37]. Alternatively, cellulose pretreatment in zinc chloride is demonstrated to improve glucose yields through acid-catalyzed hydrolysis of cellulose, with Cao *et al.* reporting nearly quantitative glucose yields through acid hydrolysis (dilute HCl) after saturation with ZnCl [38]. Finally, in ionic liquids ([EMIM]Cl), Binder and Raines have reported glucose yields of nearly 90% by gradually adding water to suppress sugar dehydration [39]. Such methods may be associated with higher processing cost and capital investment arising from expensive solvents (ionic liquids) or exotic materials of construction (required for concentrated acid processing), although they offer promising selectivities for glucose isolation. High yields of chemical intermediates such as glucose are crucial for successful biorefining; however, they must be achieved at sufficiently low cost to allow for commodity production, and it remains to be demonstrated which technology offers the appropriate balance.

Another interesting strategy that could alleviate the separations burden associated with homogeneous acids and ionic liquids is to heterogeneously hydrolyze cellulose using solid-acid catalysts. For example, sulfonated carbons are promising catalysts that hydrolyze cellulose to glucose [40] and β-1,4 glucans [41]. In particular, Onda *et al.* reported glucose yields of roughly 40% from microcrystalline cellulose [40]. Subsequent studies achieved glucose yields as high as 75% with materials synthesized through high-temperature sulfonation of activated carbon [42]. Heteropolyacids have also demonstrated moderately high yields (*c.* 50%) with good selectivity to glucose (*c.* 90%) [43]. Finally, with regard to practical application of solid acids for cellulose hydrolysis, Van de Vyver *et al.* have recently demonstrated the stability of sulfonated silica-carbon nanocomposites in multiple cellulose deconstruction cycles [44].

5.4.2 Hemicellulose

Hemicellulose is the second-most abundant fraction of lignocellulosic biomass, comprising 25–35% [34]. Similar to cellulose, it is a carbohydrate polymer and represents a potential source of feedstock sugars for producing fuels and chemicals. Hemicellulose interacts with both cellulose and lignin and is part of the protective structural matrix through which cellulose fibers are interspersed [45]. One target of biomass pretreatment is the removal of hemicellulose to facilitate access to cellulose. Ideally, hemicellulose removal is achieved by extracting and recovering simple sugars or sugar oligomers, which are facilitated by pretreating lignocellulosic biomass in dilute acids or hot water, respectively [46]. Structurally unique from the more commonly processed cellulose, hemicellulose has both advantages and disadvantages that must be addressed to completely utilize renewable feedstocks.

Hemicellulose Structure and Hydrolysis

In contrast to the crystalline structure of cellulose, hemicellulose is amorphous. As such, glycoside bonds of hemicellulose are more readily hydrolyzed than the β $1 \rightarrow 4$ linkages of cellulose. Typically, this reactivity allows hemicellulose to be extracted and depolymerized under a milder set of conditions; this may have significant implications regarding the suitability of hemicellulose as a sugar feedstock, particularly in chemical-based processing strategies [45].

Hemicellulose can be largely extracted from lignocellulosic feedstocks by contacting it with steam or hot water. Hemicellulose hydrolysis is assisted by acetic acid that is formed from acetyl groups released during pretreatment [47]. Without adding acids, this extraction product is typically rich in oligomeric sugars, which are easily hydrolyzed using dilute mineral acids [34]. Conditions for chemically hydrolyzing hemicellulose can be adjusted to reduce degradation side reactions; this makes it possible to selectively isolate sugars [48], an outcome that is more challenging when chemically deconstructing cellulose. The ability to isolate sugars that are largely undegraded makes hemicellulose attractive for catalytic upgrading.

Hemicellulose Sugars

An important consideration when processing hemicellulose is its varied composition in different biomass feedstocks. Although cellulose may be regarded as a source of glucose, hemicellulose typically contains varying proportions of 5- and 6-carbon sugars (pentoses and hexoses, respectively) in addition to smaller amounts of other compounds. In most energy crops and hardwoods, hemicellulose is a rich source of 5-carbon xylose (with lesser amounts of arabinose and glucose), leading to different upgrading strategies from those considered for glucose. In contrast, softwood hemicellulose is predominately comprised of glucomannans, offering C_6 processing strategies similar to those arising in the cellulose platform [45].

5.5   Sugar Conversion and Key Intermediates

As chemical feedstocks, sugars are remarkably versatile and may be processed in a number of different ways to yield intermediates for producing food additives, fuels, polymers, and commodity and specialty chemicals. Depending on the desired product (i.e., fuels vs. chemicals), catalytic transformation of sugars may increase or decrease functionality. In the various pathways considered, many different catalytic transformations are possible. For instance, metal catalysts offer redox activity by which sugar feedstocks may be either functionalized or defunctionalized, with material selection playing a crucial role in attaining selectivity. As another example, acid catalysts facilitate sugar dehydration, an effective strategy for creating flexible furanic intermediates. Both acidic and basic catalysts are frequently used to form carbon-carbon bonds to couple monomers for fuels production. Finally, it is often advantageous to exploit multifunctional catalysts which facilitate process intensification and provide pathways not available with monofunctional systems. The following sections describe methods and applications for sugar oxidation, sugar reduction, and sugar dehydration. Specific attention will be given to intermediates produced by the aforementioned transformations as well as applications and upgrading strategies for selected intermediates.

5.5.1 Sugar Oxidation

Sugars may be oxidized to yield aldonic, aldaric, and keto aldonic/aldaric acids, of primary interest in the production of materials, food additives, specialty chemicals, and surfactants [49,50] Although sugar oxidation may be achieved using conventional stoichiometric oxidants, in the interest of ecological impact and sustainability only catalytic transformations are considered here. Typically, these occur in the presence of air, molecular oxygen, or hydrogen peroxide and employ metal or metal oxide redox catalysts. In sugars, oxidation may occur at any carbon bearing an aldehyde, primary alcohol, or secondary alcohol to yield carboxylic acid, aldehyde, or ketone functionalities, respectively. Oxidation of a primary alcohol will frequently proceed to the carboxylic acid via oxidation of the intermediate aldehyde. Selective oxidations of carbohydrates have been well considered in the literature, and several reviews are available that discuss products and potential applications [51–53]. Figure 5.3 presents a summary of products that may be obtained by glucose oxidation. For example, gluconic acid (used in food and pharmaceutical applications) can be produced by oxidizing the anomeric (C_1) carbon in glucose. The oxidation may occur catalytically in aerobic aqueous media over supported Pd and Pt, with promoters (Bi, most notably) that maintain catalyst stability [50,54]. More recently, colloidal and supported gold catalysts have demonstrated promise in the production of gluconic acid [55–62].

A pathway to glucuronic acid can be envisioned by selective oxidation of the primary alcohol in the C_6 position. In practice, this oxidation is difficult because of the reactivity of the aldehyde in the anomeric

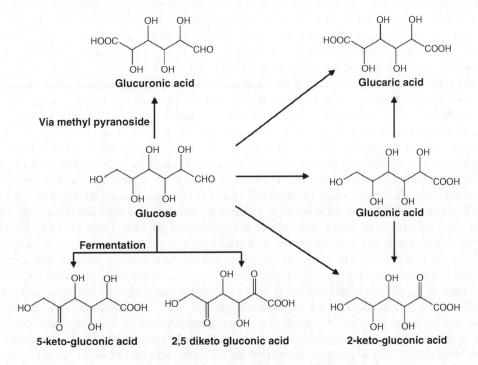

Figure 5.3 *Some products of glucose oxidation.*

position. As such, selective preparation of glucuronic acid from glucose requires protection of the anomeric carbon, which may be achieved through glucose alkylation to produce α-methyl-D-glucopyranoside. The pyranoside may be then oxidized to form methyl-α-D-glucuronic acid, which will readily hydrolyze to form α-D-glucuronic acid. Glucaric acid, bearing acid groups at each terminal carbon, may be produced by oxidizing gluconic acid (or presumably, in a single stage from glucose without protecting the anomeric carbon) using catalysts similar to those outlined above (Pt/C); however, its selective production has not been demonstrated, with limitations typically attributed to catalyst fouling [49].

Ketone functionalities may be introduced to aldonic acids (of interest in the production of Vitamin C and food additives). Generally, strategies are reported for converting gluconic acid through oxidizing either or both of the secondary alcohols in the 2 and 5 positions, yielding 2-keto-gluconic acid, 5-keto-gluconic acid, or 2,5-diketo-gluconic acid. Of these keto-acids, selective catalytic strategies have only been reported for 2-keto-gluconic acid with 5 and 2,5-diketo gluconoic acid preferentially produced by microbial fermentation [49]. 2-keto-gluconic acid may be produced by aerobic oxidation of gluconic acid (or glucose) over supported Pt with various promoters, such as Pb, Bi, or Au [63–66].

The remaining sugars typically present in lignocellulosic hemicellulose (xylose, mannose, and arabinose) are similar to glucose because they are all aldoses. Although less extensively studied than glucose oxidation, aldopentoses have been reported to have comparable trends in reactivity and conversion to aldonic and aldaric acids [67], so similar strategies should also extend to the majority of hemicellulose-derived sugars.

5.5.2   Sugar Reduction (Polyol Production)

The aldoses derived from lignocellulosic biomass (glucose, mannose, xylose, arabinose, and galactose) may be catalytically reduced with hydrogen to yield their corresponding polyols. Generally, the key transformation in polyol production is hydrogenation of the carbonyl group in the anomeric position. Hydrogenations of monomeric sugars glucose, mannose, xylose, and arabinose will yield the corresponding polyols sorbitol, mannitol, xylitol, and arabinitol (Figure 5.4). (Galactose reduction yields galactitol, an isomer of sorbitol and mannitol; however, it is not illustrated in Figure 5.4.)

Sorbitol is the most extensively studied polyol of this group, with an annual production on the order of $0.5–1.0 \times 10^6$ tons. Its predominate applications are as a non-caloric sweetener, as a precursor to Vitamin C, and in cosmetics [31,49,50,53]. Currently, sorbitol production is achieved by hydrogenating glucose obtained from starch hydrolysis (i.e., from corn starch), although the use of cellulosic sources may allow high-volume sorbitol production so that it may serve as a platform molecule to produce transportation fuels, hydrogen, or commodity chemicals. Glucose hydrogenation may be performed in batch and continuous reactors over various catalysts, which are exhaustively considered by Maki-Arvela *et al.* [50].

Historically, nickel, Raney-nickel, or promoted (Mo, Cr, Fe) nickel/Raney-nickel were preferred for glucose hydrogenation because they offer an appropriate compromise of catalytic activity and cost. However, two concerns arise when using nickel to hydrogenate glucose. These catalysts have demonstrated instability, which has been attributed to leaching of nickel or promoters and changes in particle morphology [68,69]. Additionally, nickel is susceptible to poisoning by gluconic acid formed through the Cannizaro reaction [70,71]. Because a large portion of the sorbitol demand is in the food and pharmaceutical industries, product purity is critical, and there is strict tolerance on nickel contamination. Sorbitol purification increases production costs and limits the feasibility of nickel-based systems for glucose hydrogenation. As an alternative, supported ruthenium catalysts have demonstrated promise in glucose hydrogenation. Although the cost of Ru is substantially higher than nickel, it is offset by two factors: (1) the catalytic activity of Ru is higher than that of nickel on a unit basis, and (2) Ru has demonstrated excellent stability [68,72], which has not been reported for nickel-based

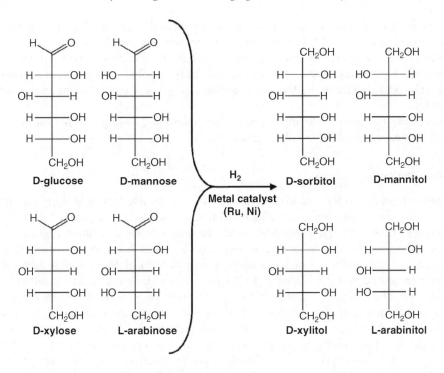

Figure 5.4   *Fischer projections of common pentoses and hexoses and corresponding polyols.*

systems. Using homogeneous Ru complexes, mannose is more susceptible to hydrogenation than glucose and has been claimed as an appropriate feedstock for polyol (specifically mannitol) production in the patent literature [73]. Similar to sorbitol, mannitol has application as a non-glycemic sweetener (in chewing gum, for example) or an inert binder in pharmaceuticals [49]. More recently, several strategies have been proposed to directly convert cellulose, which may reduce the cost associated with sorbitol production. For example, Yan *et al.* reported nearly quantitative yields of sorbitol from cellobiose using water-soluble ruthenium nanoclusters under 40 bar of H_2 [74]. Subsequently, Luo *et al.* reported sorbitol yields of roughly 45% by directly converting cellulose in hot water (>473 K) with Ru/C and 6 MPa H_2 [75]. Finally, Geboers *et al.* has demonstrated an interesting pairing of heteropolyacids and Ru/C to achieve one-pot conversion of concentrated cellulose solutions to hexitols (in yields as high as 85%) in aqueous solution under 5 MPa H_2 [76].

Regarding pentoses, xylose may be hydrogenated to yield xylitol, a sweetener. Presently, it is sourced from lignocellulosic feedstocks by purifying hardwood hydrolyzates to xylose [49]. In general, the same systems applicable for glucose hydrogenation are appropriate for xylose, and researchers have documented the use of Raney-nickel [77], promoted Raney-nickel [78], Ni/Zr alloys [79], and supported ruthenium, rhodium, and palladium [80]. Although high selectivity to xylitol is attainable in such systems, they are susceptible to catalyst deactivation in the same manner as in glucose hydrogenation; structural degradation and leaching – particularly of promoters and stabilizing agents – are primarily responsible for activity loss [81]. In contrast to glucose hydrogenation, supported ruthenium catalysts do not demonstrate on-stream stability during xylose hydrogenation [82] because of polymerization, isomerization, and degradation side reactions [83]. Research on arabinose and galactose hydrogenation is less extensively reported, although they should proceed in similar yields to arabinitol and galactitol, respectively. Because hemicellulose-derived sugar solutions are most likely mixtures of various pentoses and hexoses, hemicellulose hydrolyzates will yield, upon hydrogenation, mixed polyols in which the specific composition depends upon the biomass feedstock. Fortunately, given the functional similarity of polyols, classes of chemical transformations should be broadly applicable, enabling co-processing of mixed polyol solutions produced from hemicellulose extracts.

Polyol processing strategies, including aqueous-phase reforming, hydrogenolysis, and the combination of these methods, are described in the following sections.

Aqueous-phase Reforming

Aqueous-phase reforming (APR) is an avenue for producing renewable hydrogen from numerous oxygenated feedstocks (Figure 5.5). In contrast to alkane reforming, practiced in the petrochemical industry, oxygenate reforming offers a favorable equilibrium at low temperatures and interesting opportunities for production of hydrogen or syngas, depending on operating conditions. It is particularly effective in converting polyoxygenates containing a 1 : 1 ratio of carbon to oxygen, of which biomass-derived sugars and polyols are an appropriate example. An overview of relevant considerations has been prepared by Davda *et al.*, who have outlined a framework by which hydrogen selectivity may be maximized through appropriate selection of catalyst and operating conditions [84]. Polyols are generally preferred (over sugars) as a reforming feedstock, with higher H_2 selectivities reported for more extensively reduced oxygenates. In achieving optimal selectivity for hydrogen production, multiple factors must be considered. Hydrogen production through APR of polyoxygenated feedstocks is achieved sequentially. First, dehydrogenation and decarbonylation form H_2 and CO as primary reforming products. Second, the water-gas shift (WGS), which

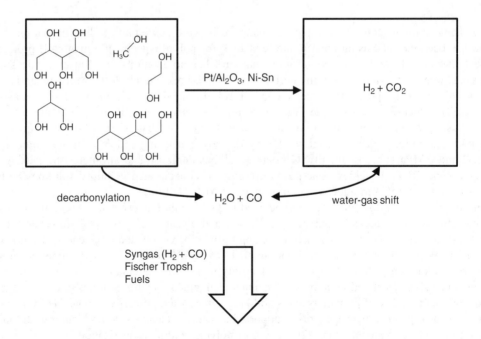

Figure 5.5   *Overview of aqueous-phase reforming (APR) of various polyols.*

establishes equilibrium according to Equation (5.1), removes surface-bound CO and produces additional H_2. Operating conditions may be varied to adjust the WGS equilibrium to favor either hydrogen [85] or syngas [86,87], depending on the intended application.

$$CO_2 + H_2 \leftrightarrow CO + H_2O \tag{5.1}$$

For H_2 production, the appropriate APR catalyst favors C—C bond cleavage (decarbonylation) instead of C—O bond cleavage (dehydration). In addition, it must also facilitate rapid equilibrium in the water-gas shift so that CO produced by decarbonylation may be further converted to CO_2 and H_2 through reduction of water. Ideal catalysts should also enable few side reactions, such as methanation or Fischer-Tropsch reactions via reduction/polymerization of carbon monoxide. By tailoring catalyst selection and experimental conditions, appropriate methods for producing clean hydrogen streams (<100 ppm CO) have been established, which may potentially be used in fuel cells [84,88].

Pt/Al_2O_3 was identified as an appropriate catalyst for producing H_2 at elevated pressures (30–60 bar) and low temperatures (498–548 K). In this system, hydrogen was most readily generated from dilute polyol solutions (<10 wt %), with the highest selectivities demonstrated from methanol and ethylene glycol. Selectivity to hydrogen by APR of glucose was lower compared to polyols, suggesting that reduced sugars (polyols) are more appropriate reforming feedstocks. In addition, polyol reforming activity correlates inversely with molecular weight [85]. Supported Pt catalysts are particularly well-suited to APR because they meet all of the criteria outlined above: they are selective toward C—C bond cleavage and decarbonylation; they promote water-gas shift; and they offer little to no methanation activity [89]. Interestingly, the catalyst support is significant in defining reforming selectivity. For example, oxide supports (such as ceria, titania, and alumina) promote the water-gas shift in metal-oxide systems through water dissociation [90,91], favoring hydrogen formation. Pt/Al_2O_3 is therefore an exceptional choice for hydrogen generation. However, carbon supports do not exhibit this characteristic and thus produce higher CO concentrations, which is useful when generating syngas. Indeed, syngas production has been demonstrated from glycerol over Pt and Pt-Re alloys supported on carbon [86,87,92]. APR for syngas production may be readily integrated with Fischer-Tropsch upgrading, over Ru/TiO_2 for example, to produce alkane fuels [93]. Finally, Pt supported on acidic silica-alumina preferentially forms light alkanes rather than syngas or hydrogen, a result which is attributed to increased C—O bond cleavage by dehydration [94].

Adding specific promoters can enhance reforming activity and hydrogen selectivity and may have important practical implications. For example, although monometallic nickel catalysts favor the formation of light alkanes, adding Sn to Raney-Ni shifts the product distribution toward hydrogen, offering comparable selectivities (*c.* 93%) to Pt/Al_2O_3 (*c.* 98%) at 498 K [95]. Bimetallic Sn-Ni systems are particularly attractive because they eliminate the need for precious metals in APR, which may reduce capital investment and improve the industrial feasibility of renewable hydrogen generation. They are unstable under APR conditions however, predominately due to sintering and leaching of active phases (Ni, Sn) and particularly when using concentrated feedstocks [96]. Other bimetallics (e.g., Pt and Pd supported on alumina and promoted by Ni, Co, and Fe) have shown promising activity in ethylene glycol reforming (compared to monometallic Pt and Pd) [97].

Although many of the initial aqueous-phase reforming studies have used ethylene glycol, the results have generalized well to sorbitol reforming [85]. Ideally, hydrogen production would be achieved directly from glucose; however, initially reported selectivities from glucose were poor, making polyols the preferred feedstock. Davda and Dumesic have demonstrated that an integrated reforming process, beginning with a glucose feed, is possible in a two-catalyst system wherein glucose undergoes an initial hydrogenation (with a hydrogen co-feed) and subsequent reforming over $Pt-Al_2O_3$ [98]. Aqueous-phase reforming continues to be a focus of research, and novel strategies are presently evolving. For example, glucose reforming for hydrogen production has been carried out over Ru/Al_2O_3 in supercritical water [99], with homogeneous Ru complexes in ionic liquids [100], and photocatalytically over La-doped tantalates [101].

Hydrogenolysis

In contrast to aqueous-phase reforming, where targeted production of hydrogen is achieved via dehydrogenation, decarbonylation, and water-gas shift, selective cleavage of C—O bonds may be achieved by C—O hydrogenolysis (Figure 5.6). Polyol hydrogenolysis may be desirable because it selectively produces monomers of potential interest in the chemical industry (e.g., ethylene glycol and 1,2-propanediol) [49]. Although initial reports regarding sugar hydrogenolysis began in the 1920s and 1930s in patent and peer-reviewed literature [102], it remains an interesting area of research with the major challenge being control of selectivity to tailored products. Typically, hydrogenolysis of polyol C—O bonds may be achieved in aqueous solution using supported metals such as Ru, Pd, Pt, Re, Ni, and Cu. Additionally, alkalinity appears to improve catalyst stability and favor production of small polyols and diols (requiring C—C cleavage) by promoting retro-aldol condensation [5,49]. Early reports from the patent literature (well documented by Corma *et al.* [49]) suggest that nickel is active for hydrogenolysis of sorbitol and xylitol to mixtures of glycerol and propylene glycol (from sorbitol) and ethylene glycol (from xylitol); however, selectivities to specific di- and triols are poor [103,104]. These results agree with those reported by Boelhouwer *et al.* for the hydrogenolysis of sucrose in methanol over nickel on kieselguhr [105]. Ru/C has demonstrated activity in the production of glycerol and propylene glycol by hydrogenolysis of sucrose and various polyols (including mannitol, sorbitol, and xylitol), with higher yields of C_3 polyols favored at higher pH [106]. Additionally, sulfur-doped Ru/C offers improved selectivity in hydrogenolysis of glycerol, xylitol, and sorbitol by limiting carbonyl hydrogenation and overall activity [107].

Copper catalysts are more active in C—O than in C—C bond cleavage. When applied to sorbitol hydrogenolysis, they can produce larger polyols (because of a smaller proportion of C—C bond cleavage), providing interesting avenues to produce polyesters, resins, and polyurethanes. Montassier *et al.* have extensively documented the catalytic properties of Cu and promoted Cu catalysts for sorbitol hydrogenolysis. In general, these systems provide pathways to hydrogenolysis products; however, they are unstable in

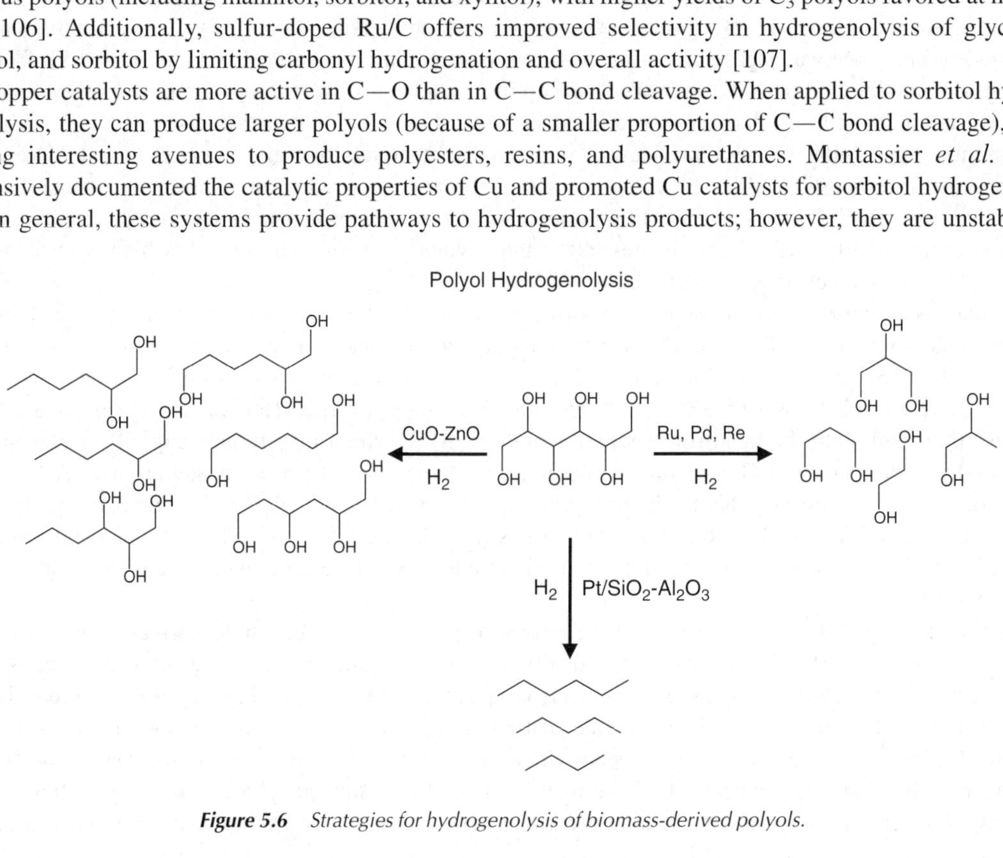

Figure 5.6   *Strategies for hydrogenolysis of biomass-derived polyols.*

aqueous-phase reactions [108,109]. Blanc *et al.* proposed an alternative system involving hydrogenolysis of sorbitol over mixed metal oxides (CuO-ZnO) to produce mixed deoxyhexitols (di-, tri-, and tetrols) in 63% yield. They were subsequently upgraded to alkyd polymers with application as a gloss paint that displayed adequate physical properties [110]. Recently, Ji *et al.* reported an interesting strategy by which cellulose may be directly converted to ethylene glycol at 61% yield by using nickel-doped tungsten carbide catalysts in aqueous solution under 6 MPa of H_2 [111]. Such strategies are particularly attractive because they do not rely upon the addition of external acids; instead, they rely on the acidity of water at high temperatures, which eases the purification burden. Additionally, the use of Ni-WC is interesting because it offers a hydrogenolysis strategy for cellulose that does not require precious metals.

Partial hydrogenolysis is achieved most commonly using metals and metal oxides and is illustrated in two of the pathways in Figure 5.6. It allows isolation of oxygenated intermediates, which may be readily upgraded for chemical production. The utility of this strategy is shown in comparison to polyol conversion over metal-acids, facilitating dehydration/hydrogenation of the polyoxygenates, and leading to total deoxygenation for alkane production.

Combined Reforming and Hydrogenolysis: Pathways to Chemical Intermediates or Transportation Fuels

To achieve partial deoxygenation of sorbitol, an interesting method involves a combination of reforming and hydrogenolysis whereby a portion of the sorbitol feedstock is converted by aqueous-phase reforming to supply (*in situ*) hydrogen required for C—O bond hydrogenolysis. In this strategy, bimetallic Pt-Re/C catalysts are favored because they offer an appropriate compromise between reforming and hydrogenolysis [112]. In contrast to traditional APR for hydrogen production, this strategy is performed at high oxygenate concentrations and relatively low partial pressures of water, thus favoring hydrogenolysis over complete reforming. To enhance production of monofunctional compounds, the addition of rhenium facilitates removal of surface-bound CO from Pt (achieved by water-gas shift in APR) under conditions of low water partial pressure. In this manner, mixtures of monofunctional intermediates (specifically carboxylic acids, ketones, secondary alcohols, and heterocyclic compounds) can be produced, in lieu of hydrogen, from the parent polyol (Figure 5.7).

Each monofunctional intermediate could have value as a solvent or in the production of various chemicals. If desired, mixtures may be fractionated and further processed for chemical applications. Alternatively, the mixture may be upgraded using various strategies for C—C bond formation which apply to general classes of functionality (Figure 5.8). For example, basic catalysts, such as ceria or BaOH [113–115], catalyze the coupling of two carboxylic acids (C_nOOH, C_mOOH) via ketonization to produce a ketone containing C_{n+m-1} carbon atoms plus CO_2 and water, a step that simultaneously removes oxygen (to improve energy density) and increases the molecular weight of the product. In this regard, ketonization is an excellent strategy for improving the quality of feedstocks rich in organic acids so that they can be used as fuels [116,117]. Secondary alcohols (via equilibrium with the corresponding ketone in the presence of metal catalysts) and ketones present in mixtures of monofunctionals may be coupled by condensation using acidic, basic, or bifunctional catalysts ($MgAlO_x$, Mg/ZrO_2, $Pd/CeZrO_x$, Pd/ZrO_2) [118–121].

In the direct application of condensation strategies that use basic catalysts, a potential challenge is catalyst poisoning in the presence of carboxylic acids. An effective solution is to couple ketonization and aldol condensation in stacked bed reactors. Basic or amphoteric solid oxides such as $CeZrO_x$ in an initial catalyst bed achieve full conversion of carboxylic acids to ketones. This catalytic step purifies the feedstock of carboxylic acids, achieves coupling of monomers to high-molecular-weight ketones (which are further condensable), and offers a product appropriate for condensation in a second bed over bifunctional materials such as $Pd/CeZrO_x$ or Pd/ZrO_2. In this manner, mixtures of linear and methyl ketones in the C_4–C_{12} carbon

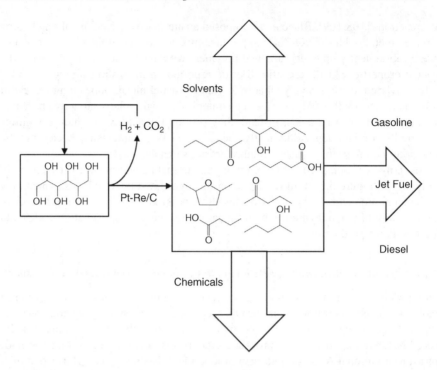

Figure 5.7   *Combined reforming and hydrogenolysis of polyols to produce mixtures of monofunctional intermediates.*

range are generated, which may undergo hydrodeoxygenation for use in gasoline, jet, or diesel fuels [122–124]. Alternatively, ketones may undergo hydrogenation to form alcohols followed by dehydration to form olefins [125], which may undergo isomerization, oligomerization, cracking, and aromatization over solid-acid catalysts [126]. Figure 5.9 gives an illustration of upgrading strategies for mixed monofunctionals.

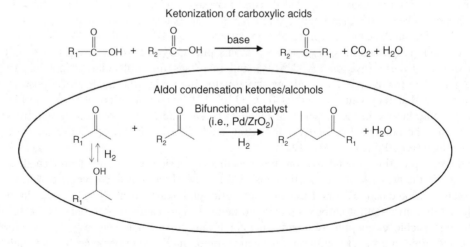

Figure 5.8   *Carbon-carbon bond-forming strategies for coupling oxygenates.*

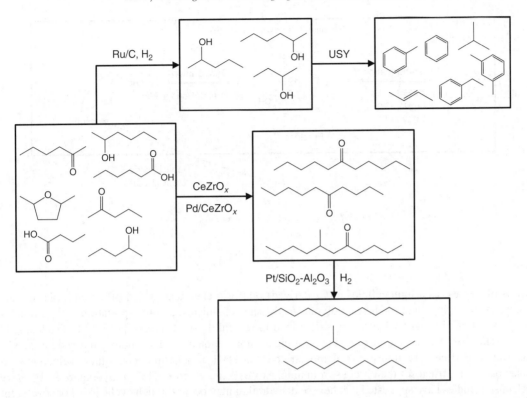

Figure 5.9 *Catalytic strategies for monofunctional upgrading to produce aromatics and alkanes for transportation fuels.*

Comparatively little has been published regarding strategies based on reforming and/or hydrogenolysis of other polyols, such as mannitol, xylitol, and arabinitol. Initial studies regarding the production of monofunctional intermediates from sorbitol suggest that xylitol may be processed in a similar manner [112,127]. Two recent studies demonstrate that xylitol undergoes catalytic deoxygenation over metal-acid catalysts (Pt supported on zirconium phosphate, specifically), with addition of externally supplied hydrogen, to yield oxygenated monofunctional compounds appropriate as gasoline additives in roughly 60% yield [29,128].

5.5.3 Sugar Dehydration (Furan Production)

Furanic species (specifically, formyl furans) may be produced by dehydrating various sugar monomers [121]. In general, dehydrating pentoses and hexoses face similar challenges because of the reactivity of both sugar molecules and dehydration products, particularly in aqueous, acidic media. Although an oversimplification, pentoses (e.g., xylose or arabinose) are feedstocks that yield furfural upon dehydration, whereas hexoses (e.g., glucose or mannose) yield 5-hydroxymethylfurfural (HMF) (Figures 5.10 and 5.11). At present, furfural production from hemicellulose-derived xylans is a reasonably mature technology and research is mainly focused upon improving the practicality of furfural production from lignocellulosic 5-carbon sugars. The more challenging pathway (and subject of the greatest amount of research) is production of HMF from 6-carbon sugars, in particular glucose derived from woody biomass.

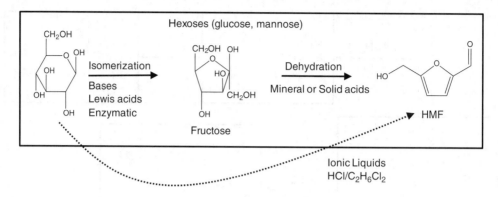

Figure 5.10   *Schematic pathway for HMF production.*

HMF

HMF is of interest as a lignocellulosic (C_6-derived) platform chemical, which offers multiple upgrading possibilities to various intermediates, polymer feedstocks, chemicals, and transportation fuels. Progress in HMF production via sugar dehydration is described in several detailed reviews [129–132]. Producing high yields of HMF from glucose is challenging, limiting its application as a lignocellulose-derived intermediate. Numerous experimental studies have demonstrated that HMF production occurs more selectively from ketohexoses (e.g., fructose) than it does from aldohexoses (e.g., glucose) [133–135] (Figure 5.10). Several mechanistic studies have suggested that hexose dehydration may occur through cyclic [136] or acyclic intermediates [129], with experimental studies by Antal *et al.* strongly supporting the role of cyclic intermediates [137]. The limited selectivity to HMF from glucose is generally attributed to the low extent of enolization (to form a necessary intermediate) arising from glucose and the tendency of glucose to undergo condensation reactions to form degradation products [49]. This limitation suggests that, on a large scale, lignocellulosic HMF production will require a glucose isomerization step (to form fructose) prior to dehydration. Typically, this step is achieved enzymatically, as in the production of high-fructose corn syrup from hydrolyzed corn starch [138].

Multiple systems for fructose dehydration have been proposed. The simplest is dehydration of fructose in aqueous media using either solid or homogeneous mineral-acid catalysts. Unfortunately, HMF is reactive under such conditions and undergoes rehydration to form levulinic and formic acids [139]. Additionally, HMF may undergo condensation reactions either with itself or other polyoxygenates to form insoluble humins, limiting HMF yields [140,141]. The problem of levulinic acid formation may be alleviated by processing fructose in non-aqueous solvents, wherein rehydration of HMF to form levulinic acid is less likely. Of the multiple non-aqueous solvents considered [49], dimethylsulfoxide (DMSO) has shown the most promise; HMF yields in excess of 90% were achieved using acidic resins. Importantly, this system demonstrated stability in continuous operation using a 0.5 mol/L solution of fructose in DMSO as the reactor feed [142]. A disadvantage of non-aqueous processing is the low solubility of sugars in organic solvents, which limits sugar loading and thus the large-scale applicability of the technology. Further, recovery of HMF from high boiling solvents such as DMSO is difficult, with elevated temperatures contributing to HMF degradation [121].

An alternative approach is to use multiphase reactors, which couple a reactive aqueous phase with an extracting solvent to simultaneously allow for relatively high fructose loadings in the aqueous phase as well as enhanced selectivity toward HMF formation. The premise of biphasic systems is that fructose will interact with acidic catalysts (mineral acids, solid acids) in the aqueous phase where it undergoes catalytic dehydration, ultimately forming HMF. The HMF product then preferentially partitions into the acid-free organic phase, minimizing the occurence of degradation and side reactions. Biphasic reactors for fructose dehydration have been exploited for a number of years, and both homogeneous and heterogeneous acid catalysts have demonstrated promise. For example, Moreau *et al.* showed that biphasic water-MIBK (methylisobutyl ketone) systems coupled with zeolites (i.e., Mordenite) of varying Si/Al ratios could achieve high HMF selectivities (>90%) at moderately high conversions (76%) of fructose [143]. Such systems are however limited by a sparse partitioning of HMF into the organic phase (MIBK), which restricts HMF concentration in the extracting phase and necessitates large quantities of solvent. To improve HMF partitioning into the aqueous phase, the use of phase modifiers was subsequently introduced to water-MIBK and other biphasic systems, allowing operation under more industrially feasible conditions (i.e., higher fructose loadings in the aqueous phase, higher HMF concentrations in the extracting phase, and lower total solvent use). For example, using systems of water/DMSO/poly(1-vinyl-2-pyrrolidinone) or PVP/MIBK/2-butanol, HMF yields approaching 75% are possible for concentrated fructose solutions (30–50% by weight) [144]. Experiments have been conducted using both homogeneous (HCl) and solid acids (ion exchange resins), with comparable selectivities demonstrated for either system. Additionally, the use of a low-boiling organic phase with favorable HMF partitioning substantially improves the possibility of HMF recovery by reducing the distillation temperature and increasing the product concentration in the organic phase [144]. Subsequent optimization of the phase-modified system produced comparable HMF yields from inulin and sucrose; however, HMF selectivity from glucose remained poor (<50%) [145]. A later study demonstrated that comparable yields of HMF could be achieved in a simpler system, which leveraged 1-butanol as an extracting solvent and the addition of NaCl to the aqueous phase, improving HMF partitioning into 1-butanol and the maximum attainable yield [146].

Although yields have been improved by using appropriate biphasic systems, HMF has been limited as a lignocellulosic platform by the necessity of glucose isomerization (historically achieved enzymatically or using basic catalysts). In the food industry, immobilized enzyme catalysts are favored for glucose isomerization, and this technology provides the dominant pathway for the production of high-fructose corn syrup (HFCS) from hydrolyzed corn starch [138]. Although the enzymatic process offers high selectivity, the materials are typically expensive and sensitive to processing conditions, limiting applicability on the commodity chemical scale. Alternatively, glucose isomerization to fructose may be achieved in basic media using homogeneous bases [147,148], anion exchange resins [149], ZrO_2 [150], and alkaline (cation or anion) exchanged zeolites, hydrotalcites, and titanosilicates [151–153]. Of particular interest, sodium-titanosilicates offered fructose yields as high as 40% via aqueous-phase isomerization of glucose [153]. More recently, comparable yields of fructose have been obtained by glucose isomerization over large-pore Lewis-acidic zeolites, specifically Sn-β, which are demonstrated to catalyze the isomerization of glucose to fructose through intramolecular hydride shift, producing fructose in rougly 30% yield from concentrated glucose solutions (*c.* 45 wt %, aqueous) [154,155]. For this reaction, an advantage of Sn-β materials is that they are sufficiently robust that they do not suffer deactivation in homogeneously acidic media. This stability offers the possibility of coupling glucose isomerization with fructose dehydration for single-pot production of HMF from glucose. Indeed, by adding Sn-β to biphasic systems (water/THF/HCl/NaCl) with an aqueous phase of pH = 1, selectivities to HMF as high as 70% have been reported from glucose and starch [156]. Application of such integrated systems may significantly improve the economics of HMF production from lignocellulosic feedstocks.

Although aqueous and biphasic-acid catalyzed dehydration of fructose are the most extensively researched methods for HMF production, they continue to suffer challenges in limited selectivity, particularly with concentrated sugar solutions. As such, several alternative processes are noteworthy. Researchers have established methods for fructose dehydration in ionic liquids (particularly imidazolium chlorides), which offer excellent HMF selectivities [157,158]. Moreau *et al.* have reported high HMF yields (92%) by fructose dehydration in 1-H-3-methyl imidazolium chloride, where the ionic liquid was demonstrated to act both as solvent and catalyst [159]. Another interesting outcome was reported by researchers at Pacific Northwest National Laboratory (PNNL), who demonstrated that combining alkyl imidazolium salts and catalytic $CrCl_2$ was an effective strategy for directly converting glucose to HMF (yields 70%), eliminating the need for glucose isomerization prior to dehydration [160]. Finally, Mascal and Nikitin have developed an interesting strategy in which dehydration of glucose, sucrose, or even microcrystalline cellulose occurs in concentrated hydrochloric acid, lithium chloride, and 1,2-dichloroethane. Under these conditions, sugar dehydration proceeds with high selectivity to chloromethylfurfural (CMF), which may be recovered in good yield (70–90%) [161]. CMF is readily converted to ethylmethylfurfural (EMF) by stirring in ethanol at ambient conditions, HMF by boiling in water for short times (86% HMF and 10% levulinic acid), methylfurfural or methyltetrahydrofuran by hydrogenation over Pd/C, levulinic acid by heating in dilute hydrochloric acid (423 K, 5 h, 94% yield of LA), or ethyl- and butyl-levulinate by refluxing in the respective alcohols at 433 K for 30 min. Importantly, chloromethylfurfural is an intermediate that may be prepared directly from cellulosic feedstocks without prior hydrolysis and isomerization. This interesting portfolio for converting CMF to a variety of useful intermediates (levulinic acid, HMF, DALA) and fuel additives (DMF, levulinate esters) has been well documented by Mascal and Nikitin [161–164] and Mascal and Dutta [165].

Furfural

Chemical conversion of pentoses (predominately xylose) to produce furfural is the most well-developed application for hemicellulose extracts (Figure 5.11). Similar to the production of HMF from hexoses, furfural production may be achieved by dehydrating pentose monomers in acidic media. As with HMF production, multiphase reaction systems facilitate high selectivities to furfural through sequestration in an organic phase with sparse partitioning of mineral acids. To date, several systems using both homogeneous and heterogeneous acid catalysts have been reported [166–171]. Moreau *et al.* have demonstrated the use of the zeolites mordenite and faujasite to dehydrate xylose in biphasic systems consisting of water and MIBK or toluene. Operating at low conversions (*c.* 30%), Moreau *et al.* reported furfural selectivities in excess of 90% [172]. Subsequently, Chheda *et al.* employed biphasic systems (using either water/HCl/MIBK/2-butanol or water/HCl/dichlormethane) in which high furfural selectivities (*c.* 75–90%) were achieved from xylose and xylans at high conversions (70–100%) [145]. To date, the biphasic reactor system for xylose dehydration has been extensively optimized and even tailored to raw hemicellulose hydrolyzates, with Xing *et al.* [173] and Weingarten *et al.* [174] reporting yields as high as 87% in a system utilizing tetrahydrofuran (THF) as an extracting solvent, NaCl as a phase modifier, and HCl as a homogeneous catalyst.

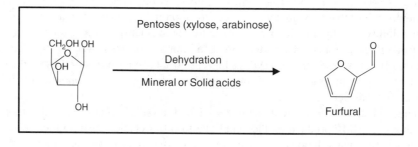

Figure 5.11   *Schematic pathway for furfural production.*

Upgrading Strategies for HMF and Furfural

Recognizing the remarkable flexibility of HMF and furfural, chemists have outlined many catalytic pathways to produce polymers, chemicals, liquid fuel additives, and liquid alkanes. Several available reviews provide extensive details regarding HMF and furfural upgrading [132,175,176]. Possibilities are highlighted in the following sections, but interested readers should refer to these reviews and original publications for a more detailed consideration.

Oxidation   Although not generally of interest to improve energy density, selective catalytic oxidation is a useful strategy by which formyl furans may be converted to a number of interesting derivatives [177]. Although the extended functionality of HMF (as compared to furfural) confers additional processing options and challenges in selectivity, oxidation of the carbonyl functionality can occur comparably in either molecule. In this manner, furfural may be oxidized catalytically to produce furoic acid, which is of interest when producing specialty chemicals such as fragrances or pharmaceuticals. Catalytically, furoic acid (Figure 5.12) may be produced using hydrogen peroxide or molecular oxygen, with quantitative yields demonstrated at low temperatures (238 K) in al kaline solution over PtPb/C catalysts [178,179].

In a similar manner, the formyl group of HMF may undergo oxidation to produce 5-hydroxymethylfuranoic acid (Figure 5.13). In addition, the 5-hydroxy group of HMF is also susceptible to catalytic oxidation and may proceed to a formyl group which, under oxidizing environments and in the presence of redox catalysts, typically oxidizes further to form a carboxylic acid. Strategies are therefore possible for producing 2,5-diformylfuran (DFF) and 2,5-furandicarboxylic acid (Figure 5.13). In practice, high yields of DFF are difficult to achieve except by using stoichiometric oxidants [180] that preferentially oxidize alcohols in the presence of aldehydes. However, catalytic systems based upon homogeneous cobalt and mixed-metal bromides [181], vanadyl phosphate in dimethyl formamide [182], Mn(III)-saleen catalysts [183], polymer-supported iodobenzoic acid [184], supported vanadyl-pyridine complexes [185], and vanadyl sulfate in ionic liquids [186] have shown promising selectivity for DFF production through HMF oxidation.

Typically, under oxidizing conditions, reactive aldehyde functionalities will preferentially form carboxylic acids (the key challenge in DFF synthesis) to yield 2,5-furandicarboxylic acid (FDCA), which is of interest as a surrogate for terephthalic acid. In general, higher selectivities are possible in producing FDCA than in isolating DFF, and various relevant systems have been demonstrated [187,188]. For instance, Pb-promoted Pt/C catalysts achieve quantitative yields of FDCA in aerobic oxidation of HMF in strongly alkaline (NaOH) aqueous media [189]. Pt/C demonstrated high selectivity and stability in the production of FDCA through aerobic oxidation of HMF in weakly basic solutions (Na_2CO_3 in water). Higher pH favors FDCA formation, whereas lower pH (acetic acid co-feed) provides increased selectivity toward DFF [177]. Recent contributions from both the laboratories of Corma and Haldor Topsoe/Technical University of Denmark have demonstrated that HMF may be oxidized under mild aerobic conditions using supported gold, specifically on TiO_2 or CeO_2, to yield either FDCA (in neutral aqueous solution) [190,191] or dimethylfuroate (in methanol) [192,193]. In addition, several intensified processes for FDCA production have been outlined. For instance, Ribeiro and Schuchardt [194] showed that silica-supported cobalt salts offer sufficient acidity and redox

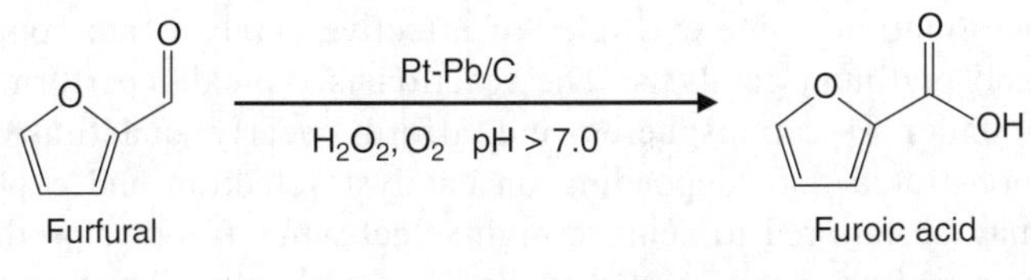

Figure 5.12   *Catalytic route for the production of furoic acid from furfural.*

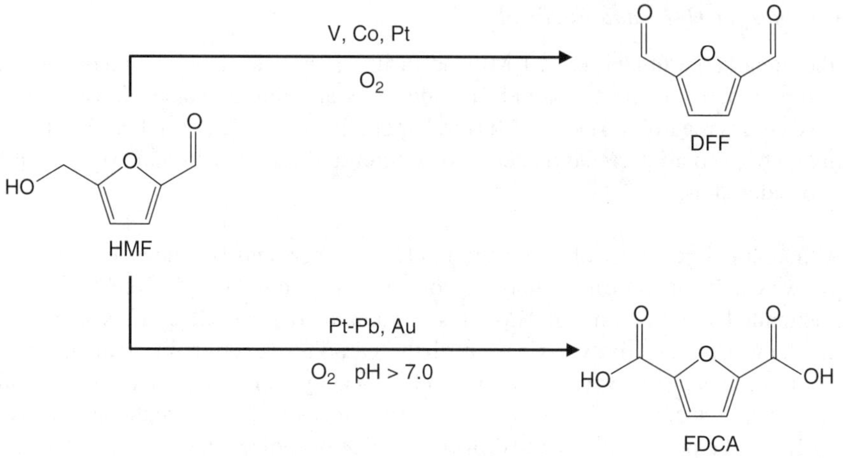

Figure 5.13   *Catalytic routes for the production of 2,5-diformylfuran (DFF) and 2,5-furandicarboxylic acid (FDCA) from HMF.*

character so that FDCA may be produced by aerobic oxidation in one-pot from fructose. Boisen *et al.* considered integrating enzymatic and chemical catalysis for integrated production of FDCA from glucose [195].

Although not considered in full detail in this text, Figure 5.14 illustrates multiple options for producing chemical intermediates or fuel additives from the HMF platform [49,132].

Fuels Production   Presently, relatively high production costs dictate that furfural and HMF are viable intermediates only for producing specialty chemicals. However, as the sciences of pretreatment, hydrolysis, and sugar dehydration advance, the cost of producing furanics should decline accordingly. In particular, developing concerted strategies for directly producing HMF from cellulose without intermediate glucose isomerization may establish their production at larger scales. At this point, the use of furans as fuel additives or fuel building blocks may become viable. As outlined in the introductory sections, when producing fuels or fuel additives, the primary objectives are improved energy density (through oxygen removal) and control of molecular weight (through C—C bond formation) so that products are compatible with current engines and transportation infrastructure. The following sections outline strategies by which the energy density of HMF or furfural may be improved by producing more extensively reduced intermediates. Additionally, to produce larger hydrocarbons appropriate as precursors to alkane fuels, furan-coupling strategies are considered.

Hydrogenation   Hydrogenating either HMF or furfural represents an interesting challenge in control of catalytic selectivity, with numerous possible derivatives illustrated in Figure 5.15. For example, selectively reducing the aldehyde functionality in HMF yields 2,5 dihydroxymethylfuran (2,5-bishydroxymethylfuran or BHMF), which is of interest when producing certain pharmaceuticals, crown ethers, and polymers [49]. However, control of selectivity is critical because more extensive hydrogenation and hydrogenolysis are possible. For this transformation, some examples of effective catalysts are copper chromite, as well as supported copper and platinum catalysts. The reaction is typically performed at relatively mild temperatures (413 K) under H_2 atmospheres and affords nearly quantitative yields of BHMF. Researchers have demonstrated that, depending on catalyst selection and experimental conditions, HMF hydrogenation may be tailored to achieve high selectivities to other products. For example, at comparable temperatures and pressures, nickel, palladium, and ruthenium favor ring saturation, thus producing 2,5-dihydroxymethyltetrahydrofuran (2,5 bishydroxymethyltetrahydrofuran), whereas Cu,

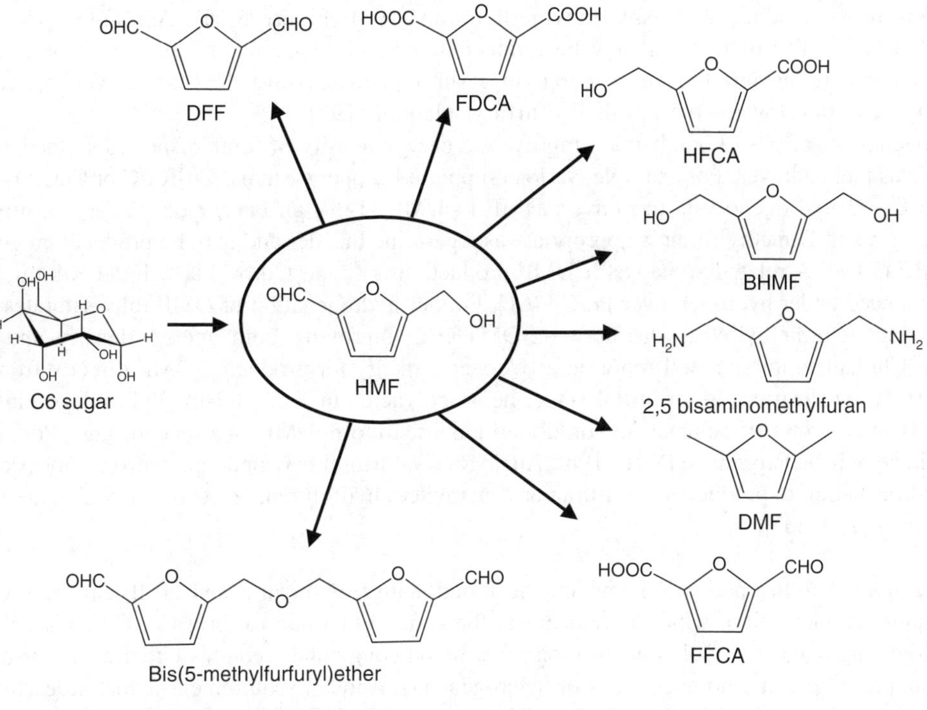

Figure 5.14 *HMF derivatives produced by catalytic upgrading.*

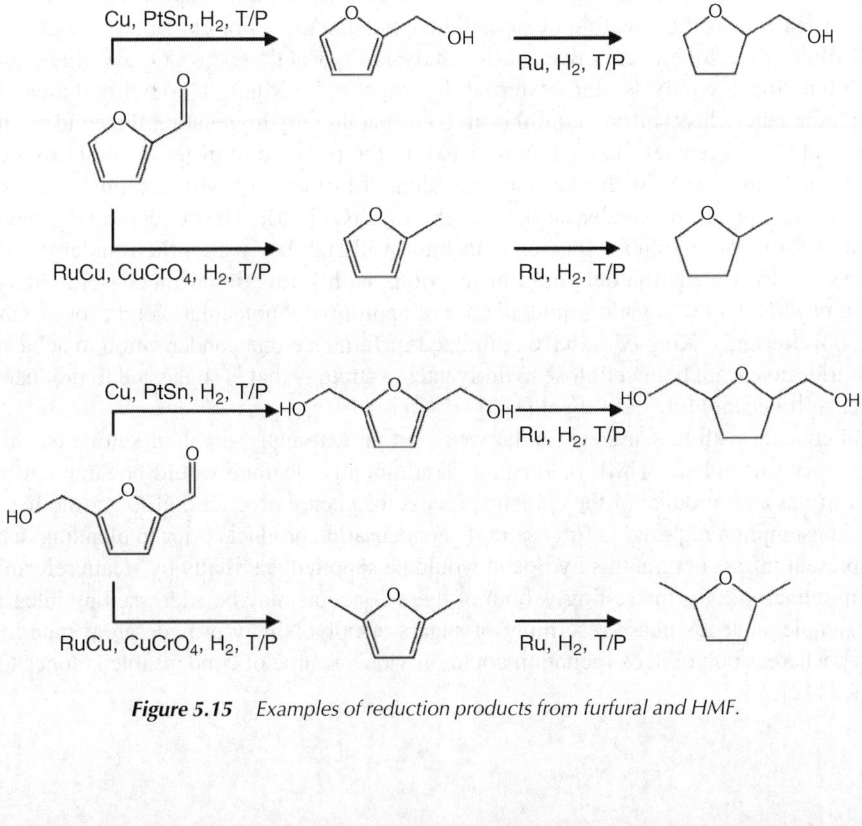

Figure 5.15 *Examples of reduction products from furfural and HMF.*

Pt and Ru alloys produce 2,5-bishydroxymethylfuran [196] (Figure 5.15). As another example, the carbonyl functionality of furfural may be reduced selectively, specifically over PtSn or Cu, to yield furfuryl alcohol (considered in more detail in a subsequent section). However, over Pd, Ru, or Ni selectivity is shifted toward the tetrahydrofurfuryl alcohol [197].

Hydrogenolysis of C—O bonds may improve the energy density of furanic species so they are readily applicable as fuel additives. For example, various supported copper systems (Cu-Ru/C or $CuCrO_4$) at 453 K facilitate C—O hydrogenolysis to convert HMF to DMF [146], an energy-dense fuel additive that is insoluble in water. Dimethylfuran is appropriate as a gasoline blender and may be produced in good yields from HMF (79%). A related strategy for DMF production is suggested by Mascal and Nikitin, by which CMF is reduced under hydrogen over Pd/C [161]. Further hydrogenation of DMF fully saturates the furan ring (producing dimethyltetrahydrofuran or DMTHF), improving both energy density and stability. Recently, Chidambaram and Bell reported a two-step strategy for producing DMF directly from glucose without HMF purification. In an initial stage, heteropolyacids in 1-ethyl-3-methylimidazolium chloride (EMIMCl) and acetonitrile achieve dehydration of glucose to form HMF. In a second stage, Pd/C is used in EMIMCl/acetonitrile to produce DMF [198]. Analogously, furfural may undergo hydrogenolysis and subsequent hydrogenation to produce methylfuran and methyltetrahydrofuran, respectively, which are promising fuel additives [199–202].

Condensation   Another pathway in which the functionality of furfural and HMF can be leveraged is through aldol condensation, which may occur at the carbonyl carbon to form C—C bonds with another carbonyl-bearing reagent. Aldol condensation of carbonyl compounds requires either acidic or basic catalysts, which may be either homogeneous or heterogeneous. Another requirement is that at least one of the two carbonyl compounds possesses an alpha-carbon having a hydrogen atom that is removed to form water. The latter requirement prevents the direct aldol condensation of two HMF or furfural monomers. However, HMF/furfural may undergo condensation with small ketones; as such, an appropriate coupling strategy is condensation of HMF or furfural with an externally supplied ketone. For example, condensation of either furfural or HMF with acetone occurs over basic catalysts (hydrotalcites, MgO, alkali and alkaline earth oxides) or bifunctional catalysts and external hydrogen (Cu-MgO, Cu-Hydrotalcites, Pd-CeZrO$_x$, Pd-ZrO$_2$), with the latter class shifting equilibrium conversion by hydrogenating the condensation product, producing C_8 and C_9 oxygenates [121] (Figure 5.16). In the production of jet or diesel fuel components, another condensation may occur with a second equivalent of furfural or HMF resulting in C_{13}–C_{15} oxygenates; high yields are reported by condensation over Pd/Mg-ZrO$_2$ [120]. The condensation products are then hydrogenated to form water-soluble species with improved stability which then undergo full saturation through successive hydrogenation/dehydration reactions with bifunctional metal-acid catalysts such as Pt/SiO$_2$-Al$_2$O$_3$ or Pt/NbOPO$_4$ to yield liquid alkanes of appropriate molecular weight for use in alkane fuel blends [203,204]. Recently, Xing *et al.* have optimized furfural/acetone condensation to achieve high-yield production of tridecane from hemicellulose hydrolyzates, a strategy that is suggested to produce diesel or jet fuel components (tridecane) for <$5.00/gal [173].

A potential concern with this strategy is the source of an externally supplied ketone (such as acetone), which is necessary to condense HMF or furfural. Traditionally, acetone would be supplied by the petrochemical industry as a by-product of the cumene process for phenol production [205]. Another limitation is the hydrogen consumption required to fully saturate condensation products prior to blending in hydrocarbon fuels. In the present infrastructure, this hydrogen would be supplied externally by steam reforming methane in a petroleum refinery [206]. Interestingly, both of these concerns may be addressed by integrated biorefineries. For example, aqueous-phase reforming of sugars or polyols may provide an avenue for renewable hydrogen [85], whereas polyol deoxygenation could provide a source of condensable ketones to replace the use of acetone [112].

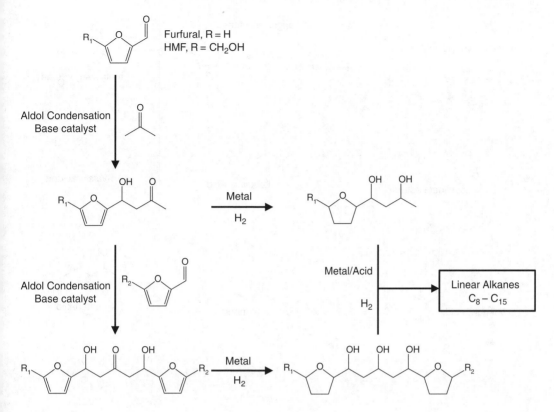

Figure 5.16 *Schematic pathway for the production of linear alkanes by HMF condensation. (Figure adapted from prior publication [203]).*

Levulinic Acid

Levulinic acid (LA) is produced by a series of acid-catalyzed reactions from either cellulose or C_6 sugars. It is widely reported that levulinic acid production proceeds through the intermediate formation of HMF, which is produced through triple dehydration of C_6 sugars, preferably ketohexoses [130]. In aqueous solution, selective production of HMF is difficult to achieve because it tends to undergo hydration to form levulinic and formic acids [140]; proposed strategies for HMF production therefore feature systems designed to sequester HMF from a reactive aqueous phase. If levulinic acid is the desired product instead, then reaction conditions may be adjusted so it is selectively produced rather than HMF and degradation products (humins) [207]. As a platform molecule, levulinic acid (4-oxopentanoic acid) is particularly interesting; its bifunctional nature enables many catalytic conversion strategies [208]. Production of LA has been demonstrated at the pilot scale by Fitzpatrick, who reports roughly 70% yields [209–211] using a two-reactor system to directly produce levulinic acid from lignocellulosic feedstocks such as wood, corn stover, and paper waste. In light of its bifunctional nature and relatively straightforward production (it can be prepared by dilute-acid hydrolysis of cellulose in a single stage), levulinic acid is presently receiving considerable attention as a platform molecule. In this section, we highlight the spectrum of levulinic-acid-upgrading strategies toward fuels, commodity and fine chemicals, and polymers. Figure 5.17 shows examples of interesting commercial products.

In the presence of homogeneous or heterogeneous acid catalysts, the acid moiety of levulinic acid can react with alcohols or olefins to yield levulinate esters [212–214] which can be used as solvents, chemical

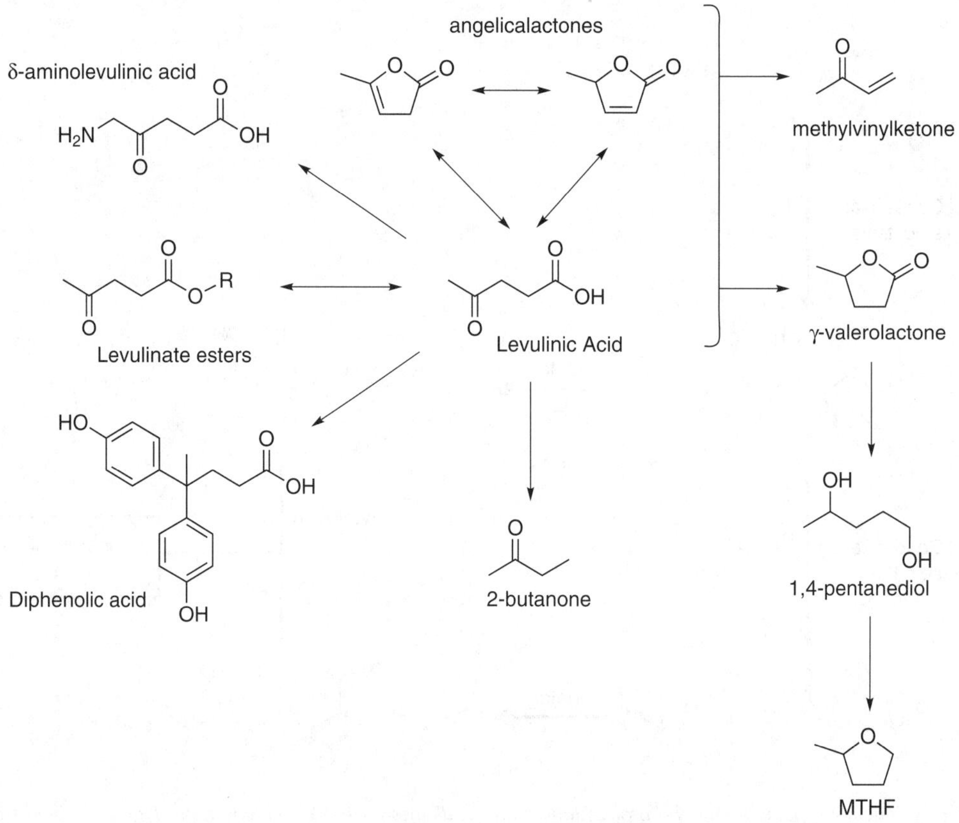

Figure 5.17   *Levulinic acid derivatives produced by catalytic conversion.*

intermediates, or diesel additives. The production of levulinate esters offers interesting opportunities in biorefining schemes. For example, a major limitation of the levulinic acid platform is its difficult recovery from dilute solutions in aqueous sulfuric acid. In acidic media, the introduction of alcohols (or alternatively, olefins) results in the formation of a separate organic phase in which the parent levulinic acid is partitioned by reactive extraction, which may ease the separation challenges facing recovery of levulinic acid from dilute solutions [214]. Further, esters of LA are demonstrated to be appropriate as blending components in diesel fuel, offering a direct lignocellulosic source of "biodiesel" that could offset consumption of petroleum resources [215].

In acidic media under mild conditions, levulinic acid undergoes ring closure and dehydration to form α- and β-angelicalactones, which exist in equilibrium with levulinic acid [208]. At low temperatures, angelicalactones can undergo photochemical decarbonylation to produce methylvinylketone [216]. Alternatively, at elevated temperatures, levulinic acid and angelicalactones undergo decarbonylation over acid catalysts. Conversion of levulinic acid may also proceed by decarboxylation, producing CO_2 and 2-butanone, over various metal catalysts. Recent reports suggest using silver persulfate or cupric oxide, with higher yields (*c.* 65%) reported for cupric oxide [208,217–219]. 2-butanone may be used as a condensation feedstock to produce high-molecular-weight fuels or in a combined hydrogenation/dehydration/oligomerization strategy to produced branched olefins, gasoline, or jet fuels.

Delta-aminolevulinic acid (DALA), a biodegradable herbicide, can be produced by aminating levulinic acid. The most successful pathway uses 5-bromolevulinic acid as an intermediate in an alcohol medium.

The LA first undergoes amination with sodium azide and is then treated with potassium phthalimide to make a second intermediate that is hydrolyzed in acidic solution to form DALA [220]. Recently, Mascal and Dutta proposed a new route through intermediate formation of 5-chloromethylfurfural (CMF). CMF undergoes amination with NaN_3 to produce 5-azidomethylfurfural, which is then irradiated with a halogen lamp in the presence of O_2 and subsequently reduced over a metal catalyst (Pd/C) to give DALA with an overall yield of 68% [165].

Another important polymer precursor, diphenolic acid, can be produced by reacting LA with phenol using acidic catalysts such as $H_3PW_{12}O_{40}$-silica composites in the absence of a solvent, a process that has demonstrated both high yield and good stability. Diphenolic acid may be used to produce polycarbonates as an alternative to petroleum-derived bisphenol A [221].

Hydrogenating the ketone moiety of levulinic acid occurs over many different metal catalysts, including Pt, Pd, Ru, and Raney-Ni. By catalytically reducing the gamma carbonyl, LA is converted to 4-hydroxypentanoic acid, which then dehydrates to form γ-valerolactone (GVL). γ-valerolactone is another interesting chemical platform molecule and, considering its wide range of applications, GVL is addressed in Section 5.5.3.6.

Formic Acid

Alongside LA, formic acid is produced in a stoichiometric quantity by dehydrating C_6 sugars [222]. The major application of formic acid is as a preservative in the food industry; however, much interest presently surrounds the use of formic acid as an *in situ* source of hydrogen, particularly for reducing levulinic acid in intensified processes for GVL production.

Formic acid is the simplest carboxylic acid, having a single carbon atom. In aqueous solution, it reacts with alcohols and alkenes to make formate esters, which may be used as solvents, chemicals, or fuel additives [215,223,224]. To reduce the separations burden encountered in cellulose hydrolysis strategies, much interest surrounds co-processing of levulinic and formic acids in the aqueous phase. One such example is the co-esterification of both acids by introducing alcohols [225] or olefins [214,215] to produce a mixture of levulinate and formate esters that may provide either a diesel additive or specialty chemicals feedstock.

An interesting possibility is the use of formic acid as an *in situ* hydrogen source to reduce levulinic acid (e.g., to produce GVL; Figure 5.18). Formic acid can undergo typical reforming reactions such as decarboxylation (forming CO_2 and H_2) and decarbonylation (forming CO and H_2O) over metal catalysts. By properly selecting catalyst (e.g., Ru, Pt, Pd, or Au) and experimental conditions, selectivity can be tuned to favor the production of H_2 and CO_2 [226]. For example, using Au/Al_2O_3 catalysts at 353 K, formic acid decomposes to a clean source of H_2 for application in fuel cells [227]. Alternatively, the decarboxylation pathway can be promoted by introducing water, which is thought to positively alter water-gas shift equilibrium, facilitating hydrogen production. This route is a particularly useful strategy over materials such as ruthenium, which favor decarbonylation [228] in the absence of water. Ruthenium catalyzes both hydrogen production (from formic acid) and levulinic acid reduction (to GVL) in aqueous solution, offering an intensified strategy for upgrading raw solutions of levulinic acid (e.g., those derived in the BioFine process) to γ-valerolactone [229].

GVL

GVL is produced by combining hydrogenation and dehydration of LA. Its production occurs over metal or metal-acid catalysts both homogeneously and heterogeneously [230,231]. There are two distinct pathways by which GVL production may occur. In the first pathway, dehydration may be performed initially (requiring acid functionality) to produce angelicalactones as intermediates. Angelicalactones may then undergo

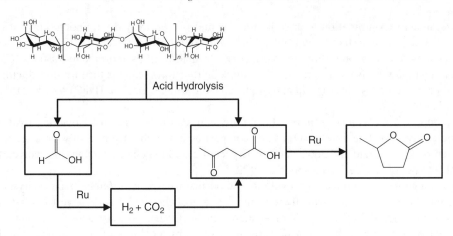

Figure 5.18   *Schematic pathway producing γ-valerolactone by reducing levulinic acid using formic acid as an internal source of H_2.*

hydrogenation – necessitating metal functionality such as that imparted by Ru, Pd, Pt, or Ni – to yield GVL by saturating carbon-carbon double bonds. Alternatively, initial hydrogenation of the gamma-ketone may be performed to yield 4-hydroxypentanoic acid, which then undergoes ring formation/dehydration to yield GVL. The highest yields of GVL have been obtained using Ru-based catalysts, with 100% conversion of the LA and >97% selectivity to GVL [232].

An interesting alternative to consuming externally supplied hydrogen is coupling formic acid decomposition with levulinic acid hydrogenation. Several studies have recently outlined such strategies to produce GVL from levulinic and formic acids. In particular, Ru-based systems are effective for both formic acid decomposition and levulinic acid hydrogenation [233] (Figure 5.18). In the absence of externally supplied hydrogen, low LA hydrogenation rates are achieved; reactions are typically performed under hydrogen atmospheres to improve the reaction rate. Importantly, if the feed contains stoichiometric quantities of levulinic and formic acid, the net hydrogen consumption is zero given that quantitative yields to H_2 and GVL are possible in each reaction [230]. A major challenge when producing GVL from raw solutions of levulinic and formic acid is the presence of residual mineral acids (typially sulfuric or hydrochloric) that remain after cellulose deconstruction and dehydration. These acids significantly impact catalyst activity and stability. Unfortunately, the cost of noble metal catalysts may be prohibitive for application in such harsh environments where they suffer marked instability. An alternative approach has been proposed by Heeres *et al.* [231] who demonstrated a single-pot strategy for the production of GVL from cellulose. In this approach, an acid catalyst (trifluoroacetic acid) and a metal catalyst (Ru/C) are combined in a single vessel to achieve both dehydration (LA production) and subsequent hydrogenation (GVL production) with optimal yields of 62% achieved from fructose. Similarly, Serrano-Ruiz *et al.* demonstrated hydrogenation activity using Ru/C for conversion of levulinic acid (to GVL) in the presence of residual sulfuric acid from cellulose deconstruction [229].

γ-valerolactone can be used directly as a fuel additive or solvent [234]. Alternatively, it may be converted into a number of different end-products (Figure 5.19). Under more severe conditions (higher temperature, higher pressure, non-selective hydrogenation catalysts), GVL formed by levulinic acid hydrogenation undergoes subsequent conversion to 1,4 pentanediol, which may be isolated for use as a precursor to synthesize polyesters, polymers, and plastics [235]. Further, α-methylen-γ-valerolactone can be produced by reacting GVL with formaldehyde, with the original synthesis described in a patent issued to DuPont. Its application is intended as a monomer for producing acrylic polymers with improved thermal stability [232].

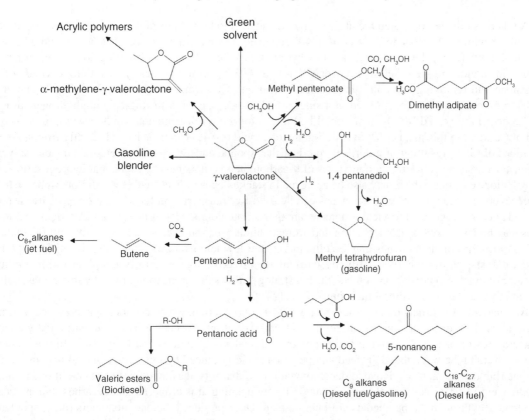

Figure 5.19　*Strategies to convert GVL into chemicals, fuels additives and fuels. (Figure adapted from prior publication [242]).*

Shell researchers have also developed a pathway to nylon intermediates by producing methyl pentenoate via ring opening GVL in methanol over solid-acid catalysts [236]. Reactive distillation has been demonstrated where GVL and the acid catalyst are retained at the bottom of the column (boiling point or b.p. 481 K) while the product, methyl pentenoate (b.p. 400 K), is continuously recovered at the top. This has successfully minimized the formation of by-products, specifically pentenoic acid and dimethylether. Once isolated, methyl pentenoate may be converted by hydroformylation, hydrocyanation, or hydroxycarbonylation to yield caprolactone, caprolactam, or adipic acid, which are typical nylon precursors [237,238].

Although GVL may be burned directly as a fuel additive, offering similar properties to ethanol [234], its unique functionality may be exploited in a number of ways to yield fuels or additives of varied energy density. For example, Lange *et al.* [239] demonstrated the use of bifunctional metal-acid catalysts (Pt/ZSM-5) to perform ring opening of GVL (to form pentenoic acid) and hydrogenation of pentenoic acid (to form pentanoic acid). In the prescence of alcohols and solid-acid catalysts (e.g., ion exchange resins), pentanoic acid undergoes esterification to form valeric esters. In this study, Lange *et al.* reported pentanoic acid selectivities of >90% from GVL over Pt/ZSM-5 at 523 K. In a second stage, they achieved selectivities of >95% to ethyl pentanoate by refluxing pentanoic acid in ethanol with acidic ion exchange resins. Because one equivalent of oxygen has been removed through dehydration/hydrogenation of LA to produce GVL, valeric esters are more energy dense and offer higher blending limits in diesel than the analogous levulinate esters. Valeric esters therefore offer a pathway for lignocellulosic biodiesel, which is currently predominately derived from triglyceride feedstocks.

A related study demonstrated the utility of pentanoic acid to build high-molecular-weight hydrocarbons. Begininng with GVL, Serrano-Ruiz *et al.* [240] achieved pentanoic acid yields of 92% over Pd/Nb$_2$O$_5$ at 35 bar and 598 K in a flow system under a hydrogen atmosphere. Instead of esterification, resultant pentanoic acid monomers were ketonized over CeZrO$_x$ at 698 K, achieving nearly quantitative yields of 5-nonanone and removing one equivalent each of water and CO$_2$. Importantly, the two-step conversion of GVL to 5-nonanone can be achieved in a single stacked-bed reactor with ring-opening/hydrogenation of GVL occurring over Pd/Nb$_2$O$_5$ in an initial bed and ketonization of pentanoic acid occurring in a second bed of CeZrO$_x$. In this manner, an overall 5-nonanone yield of 84% can be achieved directly from aqueous solutions of GVL, resulting in an organic product that spontaneously separates from the aqueous solvent. 5-nonanone may be converted into mixtures of 9-carbon olefins (nonenes) by sequential hydrogenation and dehydration reactions, which may also be achieved in stacked beds of Ru/C and H-ZSM-5 or Amberlyst-70, respectively. In a number of ways, nonenes can be used for transportation fuels. Most directly, linear nonenes may be hydrogenated to yield *n*-nonane, an appropriate component of diesel fuel. Alternatively, nonenes may undergo skeletal isomerization and aromatization over solid acids such as zeolites (ultra-stable Y or USY) [229]. Finally, nonenes may be oligomerized at low temperatures using sulfonic acid-functionalized catalysts, with Amberlyst-70 being particularly effective for selectively coupling (with minimal instance of cracking) nonene monomers. Such a strategy allows the production of C$_{18}$ hydrocarbons, which potentially may be used as diesel fuel components [241].

Alternatively, GVL may undergo acid-catalyzed decarboxylation, which occurs in the presence of water on solid-acid catalysts such as silica-alumina. Decarboxylation proceeds in nearly quantitative yields at temperatures of 623–648 K, giving stoichiometric equivalents of butene and CO$_2$ [242]. In this system, inter-conversion between the GVL feed and pentenoic acid produced by the acid-catalyzed opening of the lactone ring occurs rapidly relative to decarboxylation, so that it is quasi-equilibrated at conditions for high butene yield [242]. Given the relative ease of GVL ring opening, it is proposed that decarboxylation occurs preferentially through the pentenoic acid intermediate. Once produced, butene monomers (upon spontaneous separation from the aqueous phase) are readily oligomerized in the presence of the CO$_2$ co-product over solid acids such as Amberlyst-70 or ZSM-5, with each catalyst offering different advantages depending on the desired application. Amberlyst-70 catalyzes low-temperature oligomerization of butene (423–453 K) such that cracking side reactions are largely absent, allowing a strategy to build high-molecular-weight olefins through selective butene coupling. Aluminosilicates require elevated temperatures (473–523 K) at which cracking and aromatization reactions become significant, allowing the formation of broadly distributed (although smaller on average) olefins and aromatics [243].

Furfuryl Alcohol

Furfuryl alcohol can be prepared by selectively hydrogenating furfural, a catalytic conversion that has been extensively studied in both vapor- and liquid-phase processes. As in selective hydrogenation of HMF, furfuryl alcohol production requires selective reduction of the carbonyl functionality in the presence of unsaturated C=C bonds. Appropriate hydrogenation catalysts are generally limited to Cu-based systems (copper chromite being the most widely used commercially) [244,245]. Several supported copper catalysts have been successfully applied in vapor-phase hydrogenation of furfural, for example Cu/C [246], Cu/MgO [247], CuLa/MCM-41 [248], and Cu-Ca/SiO$_2$ [249]. Adding La and Ca improves the stability of the supported systems, with a major limitation being coke formation. Monometallic Pt systems typically catalyze formation of fully hydrogenated species such as MTHF or tetrahydrofurfuryl alcohol (THFOH) in the vapor phase, although Pt/C is selective for furfuryl alcohol production in the liquid phase [250]. Alternatively, bimetallic catalysts formed by coupling platinum with electropositive promoters such as Sn, Fe, and Ga demonstrated improved selectivity

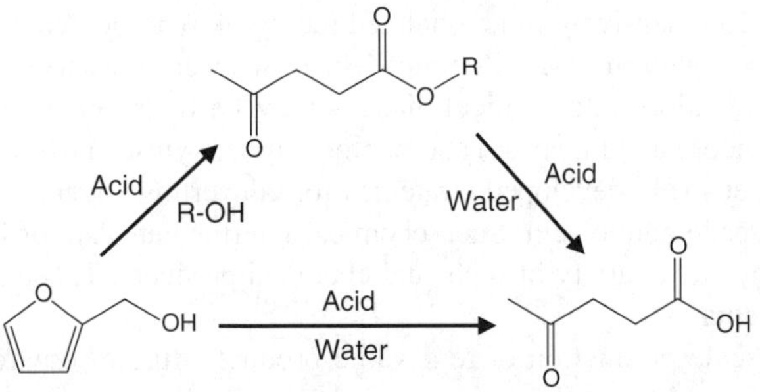

Figure 5.20 *Schematic pathway for producing levulinic acid from furfuryl alcohol.*

toward furfuryl alcohol [251]. For example, Merlo *et al.* demonstrated that PtSn catalysts offer production of furfuryl alcohol from furfural with selectivities over 96% [252].

At present, furfuryl alcohol is more highly valued as a chemical or polymer intermediate than alkane fuels or fuel additives. As such, its use as an intermediate is only feasible when producing high-value end-products. As a monomer, furfuryl alcohol may be used to prepare valuable resins. However, as the cost of furfural production (specifically from lignocellulosic feedstocks) continues to decrease, commodity-scale applications may become feasible.

An interesting option is the acid-catalyzed hydration of furfuryl alcohol to produce levulinic acid (Figure 5.20). A similar strategy may be performed in alcohols to facilitate the production of levulinate esters [253]. This strategy is attractive because levulinic acid is a chemical intermediate that enables many routes for producing both fuels and chemicals. Commonly, LA production is associated with C_6 sugars from the cellulosic fraction of biomass (as demonstrated in the technology developed by BioFine). However, a furfural pathway toward levulinic acid is interesting in a number of ways. For example, it offers another value-added application for the C_5 fraction of biomass (through intermediate furfural). Presently, commercial production of levulinic acid is achieved by the C_5 route as demonstrated by Linzi Organic Chemical, Inc. This strategy could offer synergies with biorefining strategies involving levulinic acid or GVL, which have been proposed for cellulose utilization. The major challenge associated with acid-catalyzed conversion of furfuryl alcohol is its reactivity. In aqueous solution, direct contact of concentrated furfuryl alcohol with acidic media results in uncontrolled polymerization and an associated loss of selectivity to degradation products (humins), as indicated by Lange *et al.* [253]. Several avenues have been proposed by which the conversion of furfuryl alcohol to levulinic acid may be achieved. A limitation of this technology is that the furfuryl alcohol concentration must be low levels so that conditions do not favor furfuryl alcohol polymerization.

5.6   Conclusions

Each conversion strategy outlined in this chapter is based upon selective processing of the sugars derived from lignocellulosic biomass. To establish large-scale feasibility for any of the outlined pathways, it is imperative to achieve high recovery yields for all of these sugars. Indeed, unlocking the lignocellulose matrix to selectively liberate sugar monomers is one of the most important steps to realizing the potential of lignocellulose as a commodity feedstock at appropriate scales and costs. Such an outcome will only be achieved through efficient biomass pretreatment; focused research in this area will therefore help advance the current state of biorefining.

This chapter focused specifically on catalytic valorization of carbohydrate biomass derivatives. It presented the general challenges facing the production of current industrial products from an alternative feedstock, specifically lignocellulosic biomass. A general context is provided in which the production of both fuels and chemicals from an extensively functionalized feedstock may be understood relative to the petroleum-centered strategies currently in place. Chemical processes were presented by which biomass-derived carbohydrates – namely cellulose and hemicellulose – may be hydrolyzed to yield monomeric sugars including hexoses (e.g., glucose and mannose) and pentoses (e.g., xylose and arabinose). Further, this chapter presented the most extensively developed strategies for converting sugars into chemical intermediates such as polyols, furanics, or levulinic acid. Such chemical intermediates are of interest because they offer flexible upgrading pathways to a variety of fuels and chemical products. This chapter considered the most developed of these intermediates.

To allow commercial-scale production of renewable products, future research in catalytic processing must continue to extend the spectrum of intermediate products attainable from simple sugars and improve the efficiency of the various upgrading strategies. This commercialization will likely be predicated upon future optimization of strategies for biomass pretreatment, which will enable low-cost production and recovery of intermediate platform chemicals. Catalytic chemistry also plays a key role in facilitating these outcomes, particularly by enabling rapid, highly selective transformations in systems requiring minimal energy for separations and product recovery.

Acknowledgements

Two of the authors (DMA and JQB) were supported through funding from the Defense Advanced Research Projects Agency (DARPA) and Army Research Lab (ARL) through the Defense Science Office Cooperative Agreement W911NF-09-2-0010/09-005334 B 01 (Surf-Cat: Catalysts for production of JP-8 range molecules from lignocellulosic biomass). The results from the University of Wisconsin reported here were supported in part by the US Department of Energy Office of Basic Energy Sciences, the National Science Foundation Chemical and Transport Systems Division of the Directorate for Engineering, the DOE Great Lakes Bioenergy Research Center (http://www.glbrc.org/), and by the Defense Advanced Research Project Agency. The views, opinions, and/or findings contained in this article are those of the authors and should not be interpreted as representing the official views or policies, either expressed or implied, of the Defense Advanced Research Projects Agency or the Department of Defense.

References

1. Bozell, J.J. (2008) Feedstocks for the future – biorefinery production of chemicals from renewable carbon. *CLEAN – Soil, Air, Water*, **36**, 641–647.
2. Nikolau, B.J., Perera, M.A.D.N., Brachova, L., and Shanks, B. (2008) Platform biochemicals for a biorenewable chemical industry. *The Plant Journal*, **54**, 536–545.
3. Shanks, B.H. (2010) Conversion of biorenewable feedstocks: new challenges in heterogeneous catalysis. *Industrial & Engineering Chemistry Research*, **49**, 10212–10217.
4. Vennestrom, P.N.R., Christensen, C.H., Pedersen, S. *et al.* (2010) Next-generation catalysis for renewables: combining enzymatic with inorganic heterogeneous catalysis for bulk chemical production. *Chemcatchem*, **2**, 249–258.
5. Chheda, J.N., Huber, G.W., and Dumesic, J.A. (2007) Liquid-phase catalytic processing of biomass-derived oxygenated hydrocarbons to fuels and chemicals. *Angewandte Chemie-International Edition*, **46**, 7164–7183.
6. Perlack, R.D., Wright, L.L., Turhollow, A.F. *et al.* (2005) Biomass as a feedstock for a bioenergy and bioproducts industry: the technical feasibility of a billion-ton annual supply, http://www1.eere.energy.gov/biomass/pdfs/final_billionton_vision_report2.pdf.
7. Huber, G.W. and Dale, B.E. (2009) Grassoline at the pump. *Scientific American*, **301**, 52–59.
8. Vlachos, D., Chen, J., Gorte, R. *et al.* (2010) Catalysis center for energy innovation for biomass processing: research strategies and goals. *Catalysis Letters*, **140**, 77–84.

9. Serrano-Ruiz, J.C. and Dumesic, J.A. (2011) Catalytic routes for the conversion of biomass into liquid hydrocarbon transportation fuels. *Energy & Environmental Science*, **4**, 83–99.

10. Stocker, M. (2008) Biofuels and biomass-to-liquid fuels in the biorefinery: catalytic conversion of lignocellulosic biomass using porous materials. *Angewandte Chemie-International Edition*, **47**, 9200–9211.

11. Martin Alonso, D., Bond, J.Q., and Dumesic, J.A. (2010) Catalytic conversion of biomass to biofuels. *Green Chemistry*, **12**, 1493–1513.

12. Huber, G.W., Iborra, S., and Corma, A. (2006) Synthesis of transportation fuels from biomass: Chemistry, catalysts, and engineering. *Chemical Reviews*, **106**, 4044–4098.

13. Huber, G.W. and Corma, A. (2007) Synergies between Bio- and Oil refineries for the production of fuels from biomass. *Angewandte Chemie International Edition*, **46**, 7184–7201.

14. Digman, B., Joo, H.S., and Kim, D.S. (2009) Recent progress in gasification/pyrolysis technologies for biomass conversion to energy. *Environmental Progress & Sustainable Energy*, **28**, 47–51.

15. Lee, S. and Sardesai, A. (2005) Liquid phase methanol and dimethyl ether synthesis from syngas. *Topics in Catalysis*, **32**, 197–207.

16. Iglesia, E., Reyes, S.C., Madon, R.J., and Soled, S.L. (1993) Selectivity control and catalyst design in the fischer-tropsch synthesis – sites, pellets, and reactors, in *Advances in Catalysis and Related Subjects 39* (eds D.D. Eley, P.B. Weisz, and H. Pines), Academic Press, Inc., San Diego.

17. Antal, M.J., Mok, W.S.L., Varhegyi, G., and Szekely, T. (1990) Review of methods for improving the yield of charcoal from biomass. *Energy & Fuels*, **4**, 221–225.

18. Elliott, D.C., Beckman, D., Bridgwater, A.V. *et al.* (1991) Developments in direct thermochemical liquefaction of biomass 1983–1990. *Energy & Fuels*, **5**, 399–410.

19. Bridgwater, A.V. and Cottam, M.L. (1992) Opportunities for biomass pyrolysis liquids production and upgrading. *Energy & Fuels*, **6**, 113–120.

20. Bridgwater, A.V. and Peacocke, G.V.C. (2000) Fast pyrolysis processes for biomass. *Renewable & Sustainable Energy Reviews*, **4**, 1–73.

21. Antal, M.J. and Gronli, M. (2003) The art, science, and technology of charcoal production. *Industrial & Engineering Chemistry Research*, **42**, 1619–1640.

22. Effendi, A., Gerhauser, H., and Bridgwater, A.V. (2008) Production of renewable phenolic resins by thermochemical conversion of biomass: A review. *Renewable & Sustainable Energy Reviews*, **12**, 2092–2116.

23. Kopyscinski, J., Schildhauer, T.J., and Biollaz, S.M.A. (2010) Production of synthetic natural gas (SNG) from coal and dry biomass – A technology review from 1950 to 2009. *Fuel*, **89**, 1763–1783.

24. Breault, R.W. (2010) Gasification processes old and new: a basic review of the major technologies. *Energies*, **3**, 216–240.

25. Balat, M., Balat, M., Kirtay, E., and Balat, H. (2009) Main routes for the thermo-conversion of biomass into fuels and chemicals. Part 1: Pyrolysis systems. *Energy Conversion and Management*, **50**, 3147–3157.

26. Balat, M., Balat, M., Kirtay, E., and Balat, H. (2009) Main routes for the thermo-conversion of biomass into fuels and chemicals. Part 2: Gasification systems. *Energy Conversion and Management*, **50**, 3158–3168.

27. Mosier, N., Wyman, C.E., Dale, B.E. *et al.* (2005) Features of promising technologies for pretreatment of lignocellulosic biomass. *Bioresource Technology*, **96**, 673–686.

28. Zakzeski, J., Bruijnincx, P.C.A., Jongerius, A.L., and Weckhuysen, B.M. (2010) The catalytic valorization of lignin for the production of renewable chemicals. *Chemical Reviews*, **110**, 3552–3599.

29. Li, N., Tompsett, G.A., Zhang, T. *et al.* (2011) Renewable gasoline from aqueous phase hydrodeoxygenation of aqueous sugar solutions prepared by hydrolysis of maple wood. *Green Chemistry*, **13**, 91–101.

30. Simonetti, D.A. and Dumesic, J.A. (2008) Catalytic strategies for changing the energy content and achieving C—C coupling in biomass-derived oxygenated hydrocarbons. *Chemsuschem*, **1**, 725–733.

31. Bozell, J.J. and Petersen, G.R. (2010) Technology development for the production of biobased products from biorefinery carbohydrates–the US Department of Energy's "Top 10" revisited. *Green Chemistry*, **12**, 539–554.

32. Hayes, D.J. (2009) An examination of biorefining processes, catalysts and challenges. *Catalysis Today*, **145**, 138–151.

33. Lin, Y.-C. and Huber, G.W. (2009) The critical role of heterogeneous catalysis in lignocellulosic biomass conversion. *Energy & Environmental Science*, **2**, 68–80.

34. Wyman, C.E., Dale, B.E., Elander, R.T. *et al.* (2005) Coordinated development of leading biomass pretreatment technologies. *Bioresource Technology*, **96**, 1959–1966.

35. Biermann, C.J. (1993) *Essentials of Pulping and Papermaking*, Academic Press, Inc., San Diego.

36. Rinaldi, R. and Schuth, F. (2009) Acid hydrolysis of cellulose as the entry point into biorefinery schemes. *Chemsuschem*, **2**, 1096–1107.

37. Ragg, P.L., Fields, P.R., and Tinker, P.B. (1987) The development of a process for the hydrolysis of lignocellulosic waste [and discussion]. *Philosophical Transactions of the Royal Society of London Series A, Mathematical and Physical Sciences*, **321**, 537–547.

38. Cao, N.J., Xu, Q., Chen, C.S. *et al.* (1994) Cellulose hydrolysis using zinc chloride as a solvent and catalyst. *Applied Biochemistry and Biotechnology*, **45–46**, 521–530.

39. Binder, J.B. and Raines, R.T. (2010) Fermentable sugars by chemical hydrolysis of biomass. *Proceedings of the National Academy of Sciences*, **107**, 4516–4521.

40. Onda, A., Ochi, T., and Yanagisawa, K. (2008) Selective hydrolysis of cellulose into glucose over solid acid catalysts. *Green Chemistry*, **10**, 1033–1037.

41. Suganuma, S., Nakajima, K., Kitano, M. *et al.* (2008) Hydrolysis of cellulose by amorphous carbon bearing SO3H, COOH, and OH Groups. *Journal of the American Chemical Society*, **130**, 12787–12793.

42. Pang, J., Wang, A., Zheng, M., and Zhang, T. (2010) Hydrolysis of cellulose into glucose over carbons sulfonated at elevated temperatures. *Chemical Communications*, **46**, 6935–6937.

43. Tian, J., Wang, J., Zhao, S. *et al.* (2010) Hydrolysis of cellulose by the heteropolyacid H3PW12O40. *Cellulose*, **17**, 587–594.

44. Van de Vyver, S., Peng, L., Geboers, J. *et al.* (2010) Sulfonated silica/carbon nanocomposites as novel catalysts for hydrolysis of cellulose to glucose. *Green Chemistry*, **12**, 1560–1563.

45. Mamman, A.S. (2008) Furfural: Hemicellulose/xylose derived biochemical. *Biofuels Bioproducts & Biorefining*, **2**, 438–454.

46. Wyman, C.E., Lynd, L.R., and Gerngross, T.U. (1999) Biocommodity engineering. *Biotechnology Progress*, **15**, 777–793.

47. Liu, S. (2008) A kinetic model on autocatalytic reactions in woody biomass hydrolysis. *Journal of Biobased Materials and Bioenergy*, **2**, 135–147.

48. Weiss, N., Nagle, N., Tucker, M., and Elander, R. (2009) High xylose yields from dilute acid pretreatment of corn stover under process-relevant conditions. *Applied Biochemistry and Biotechnology*, **155**, 115–125.

49. Corma, A., Iborra, S., and Velty, A. (2007) Chemical routes for the transformation of biomass into chemicals. *Chemical Reviews*, **107**, 2411–2502.

50. Maki-Arvela, P., Holmbom, B., Salmi, T., and Murzin, D.Y. (2007) Recent progress in synthesis of fine and specialty chemicals from wood and other biomass by heterogeneous catalytic processes. *Catalysis Reviews: Science and Engineering*, **49**, 197–340

51. Besson, M. and Gallezot, P. (2000) Selective oxidation of alcohols and aldehydes on metal catalysts. *Catalysis Today*, **57**, 127–141.

52. Besson, M. and Gallezot, P. (2001) Oxidation of alcohols and aldehydes on metal catalysts, in *Fine Chemicals through Heterogeneous Catalysis* (eds R.A. Sheldon and H. Van Bekkum), Wiley-VCH, Weinheim.

53. Gallezot, P. (2007) Catalytic routes from renewables to fine chemicals. *Catalysis Today*, **121**, 76–91.

54. Besson, M., Lahmer, F., Gallezot, P. *et al.* (1995) Catalytic-oxidation of glucose on bismuth-promoted palladium catalysts. *Journal of Catalysis*, **152**, 116–121.

55. Biella, S., Prati, L., and Rossi, M. (2002) Selective oxidation of D-glucose on gold catalyst. *Journal of Catalysis*, **206**, 242–247.

56. Önal, Y., Schimpf, S., and Claus, P. (2004) Structure sensitivity and kinetics of -glucose oxidation to -gluconic acid over carbon-supported gold catalysts. *Journal of Catalysis*, **223**, 122–133.

57. Beltrame, P., Comotti, M., Della Pina, C., and Rossi, M. (2006) Aerobic oxidation of glucose: II. Catalysis by colloidal gold. *Applied Catalysis A: General*, **297**, 1–7.

58. Comotti, M., Della Pina, C., Falletta, E., and Rossi, M. (2006) Is the biochemical route always advantageous? The case of glucose oxidation. *Journal of Catalysis*, **244**, 122–125.

59. Mirescu, A. and Prüße, U. (2007) A new environmental friendly method for the preparation of sugar acids via catalytic oxidation on gold catalysts. *Applied Catalysis B: Environmental*, **70**, 644–652.

60. Baatz, C. and Prüße, U. (2007) Preparation of gold catalysts for glucose oxidation by incipient wetness. *Journal of Catalysis*, **249**, 34–40.

61. Baatz, C. and Prüße, U. (2007) Preparation of gold catalysts for glucose oxidation. *Catalysis Today*, **122**, 325–329.

62. Baatz, C., Decker, N., and Prüße, U. (2008) New innovative gold catalysts prepared by an improved incipient wetness method. *Journal of Catalysis*, **258**, 165–169.

63. Besson, M., Flèche, G., Fuertes, P. *et al.* (1996) Oxidation of glucose and gluconate on Pt, Pt Bi, and Pt Au catalysts. *Recueil des Travaux Chimiques des Pays-Bas*, **115**, 217–221.

64. Smits, P.C.C., Kuster, B.F.M., van der Wiele, K., and van der Baan, S. (1987) Lead modified platinum on carbon catalyst for the selective oxidation of (2-) hydroxycarbonic acids, and especially polyhydroxycarbonic acids to their 2-keto derivatives. *Applied Catalysis*, **33**, 83–96.

65. Smits, P.C.C., Kuster, B.F.M., van der Wiele, K., and van der Baan, H.S. (1986) The selective oxidation of aldoses and aldonic acids to 2-ketoaldonic acids with lead-modified platinum-on-carbon catalysts. *Carbohydrate Research*, **153**, 227–235.

66. Abbadi, A. and van Bekkum, H. (1995) Highly selective oxidation of aldonic acids to 2-keto-aldonic acids over Pt–Bi and Pt–Pb catalysts. *Applied Catalysis A: General*, **124**, 409–417.

67. Venema, F.R., Peters, J.A., and Vanbekkum, H. (1992) Platinum-catalyzed oxidation of aldopentoses to aldaric acids. *Journal of Molecular Catalysis*, **77**, 75–85.

68. Kusserow, B., Schimpf, S., and Claus, P. (2003) Hydrogenation of glucose to sorbitol over nickel and ruthenium catalysts. *Advanced Synthesis & Catalysis*, **345**, 289–299.

69. Gallezot, P., Cerino, P.J., Blanc, B. *et al.* (1994) Glucose hydrogenation on promoted raney-nickel catalysts. *Journal of Catalysis*, **146**, 93–102.

70. Hoffer, B.W., Crezee, E., Devred, F. *et al.* (2003) The role of the active phase of Raney-type Ni catalysts in the selective hydrogenation of -glucose to -sorbitol. *Applied Catalysis A: General*, **253**, 437–452.

71. Albert, R., Strätz, A., and Vollheim, G. (1980) Die katalytische Herstellung von Zuckeralkoholen und deren Verwendung. *Chemie Ingenieur Technik*, **52**, 582–587.

72. Gallezot, P., Nicolaus, N., Flèche, G. *et al.* (1998) Glucose hydrogenation on ruthenium catalysts in a trickle-bed reactor. *Journal of Catalysis*, **180**, 51–55.

73. Kolaric, S. and Sunjic, V. (1996) Comparative study of homogeneous hydrogenation of D-glucose and D-mannose catalyzed by water soluble [Ru(tri(m-sulfophenyl)phosphine)] complex. *Journal of Molecular Catalysis A-Chemical*, **110**, 189–193.

74. Yan, N., Zhao, C., Luo, C. *et al.* (2006) One-step conversion of cellobiose to C6-Alcohols using a ruthenium nanocluster catalyst. *Journal of the American Chemical Society*, **128**, 8714–8715.

75. Luo, C., Wang, S., and Liu, H. (2007) Cellulose conversion into polyols catalyzed by reversibly formed acids and supported ruthenium clusters in hot water. *Angewandte Chemie International Edition*, **46**, 7636–7639.

76. Geboers, J., Van de Vyver, S., Carpentier, K. *et al.* (2010) Efficient catalytic conversion of concentrated cellulose feeds to hexitols with heteropoly acids and Ru on carbon. *Chemical Communications*, **46**, 3577–3579.

77. Wisniak, J., Hershkow., M., Leibowit., R., and Stein, S. (1974) Hydrogenation of Xylose to Xylitol. *Industrial & Engineering Chemistry Product Research and Development*, **13**, 75–79.

78. Karimkulova, M.P., Khakimov, Y.S., and Abidova, M.F. (1989) Activity of modified catalysts in hydrogenation of xylose to xylitol. *Chemistry of Natural Compounds*, **25**, 370–371.

79. Darsow, G. (2000) Process for the Hydrogenation of Sugars, USPTO, Patent Number 6124443.

80. Wisniak, J., Hershkowitz, M., and Stein, S. (1974) Hydrogenation of Xylose over Platinum group catalysts. *Industrial & Engineering Chemistry Product Research and Development*, **13**, 232–236.

81. Mikkola, J.-P., Vainio, H., Salmi, T. *et al.* (2000) Deactivation kinetics of Mo-supported Raney Ni catalyst in the hydrogenation of xylose to xylitol. *Applied Catalysis A: General*, **196**, 143–155.

82. Kwak, B.S., Lee, B.I., Kim, T.Y., and Kim, J.W. (2005) Method for preparing sugar alcohols by ctalytic hydrogenation of sugars, WIPO, Patent Number WO 2005/021475.

83. Shimazu, K., Tateno, Y., Magara, M. *et al.* (2002) Raney Catalyst, Process for Producing it and Process for Producing a Sugar-alcohol using the Same, USPTO, Patent Number 6414201.

84. Davda, R.R., Shabaker, J.W., Huber, G.W. *et al.* (2005) A review of catalytic issues and process conditions for renewable hydrogen and alkanes by aqueous-phase reforming of oxygenated hydrocarbons over supported metal catalysts. *Applied Catalysis B-Environmental*, **56**, 171–186.

85. Cortright, R.D., Davda, R.R., and Dumesic, J.A. (2002) Hydrogen from catalytic reforming of biomass-derived hydrocarbons in liquid water. *Nature*, **418**, 964–967.

86. Soares, R.R., Simonetti, D.A., and Dumesic, J.A. (2006) Glycerol as a source for fuels and chemicals by low-temperature catalytic processing. *Angewandte Chemie-International Edition*, **45**, 3982–3985.

87. Simonetti, D.A., Kunkes, E.L., and Dumesic, J.A. (2007) Gas-phase conversion of glycerol to synthesis gas over carbon-supported platinum and platinum-rhenium catalysts. *Journal of Catalysis*, **247**, 298–306.

88. Davda, R.R. and Dumesic, J.A. (2003) Catalytic reforming of oxygenated hydrocarbons for hydrogen with low levels of carbon monoxide. *Angewandte Chemie International Edition*, **42**, 4068–4071.

89. Shabaker, J.W., Davda, R.R., Huber, G.W. *et al.* (2003) Aqueous-phase reforming of methanol and ethylene glycol over alumina-supported platinum catalysts. *Journal of Catalysis*, **215**, 344–352.

90. Basinska, A., Kepinski, L., and Domka, F. (1999) The effect of support on WGSR activity of ruthenium catalysts. *Applied Catalysis A: General*, **183**, 143–153.

91. Gorte, R.J. and Zhao, S. (2005) Studies of the water-gas-shift reaction with ceria-supported precious metals. *Catalysis Today*, **104**, 18–24.

92. Kunkes, E.L., Simonetti, D.A., Dumesic, J.A. *et al.* (2008) The role of rhenium in the conversion of glycerol to synthesis gas over carbon supported platinum-rhenium catalysts. *Journal of Catalysis*, **260**, 164–177.

93. Simonetti, D.A., Rass-Hansen, J., Kunkes, E.L. *et al.* (2007) Coupling of glycerol processing with Fischer-Tropsch synthesis for production of liquid fuels. *Green Chemistry*, **9**, 1073–1083.

94. Shabaker, J.W., Huber, G.W., Davda, R.R. *et al.* (2003) Aqueous-phase reforming of ethylene glycol over supported platinum catalysts. *Catalysis Letters*, **88**, 1–8.

95. Huber, G.W., Shabaker, J.W., and Dumesic, J.A. (2003) Raney Ni-Sn catalyst for H-2 production from biomass-derived hydrocarbons. *Science*, **300**, 2075–2077.

96. Shabaker, J.W., Simonetti, D.A., Cortright, R.D., and Dumesic, J.A. (2005) Sn-modified Ni catalysts for aqueous-phase reforming: characterization and deactivation studies. *Journal of Catalysis*, **231**, 67–76.

97. Huber, G.W., Shabaker, J.W., Evans, S.T., and Dumesic, J.A. (2006) Aqueous-phase reforming of ethylene glycol over supported Pt and Pd bimetallic catalysts. *Applied Catalysis B: Environmental*, **62**, 226–235.

98. Davda, R.R. and Dumesic, J.A. (2004) Renewable hydrogen by aqueous-phase reforming of glucose. *Chemical Communications*, **1**, 36–37.

99. Byrd, A.J., Pant, K.K., and Gupta, R.B. (2007) Hydrogen production from glucose using Ru/Al2O3 catalyst in supercritical water. *Industrial & Engineering Chemistry Research*, **46**, 3574–3579.

100. Taccardi, N., Assenbaum, D., Berger, M.E.M. *et al.* (2010) Catalytic production of hydrogen from glucose and other carbohydrates under exceptionally mild reaction conditions. *Green Chemistry*, **12**, 1150–1156.

101. Fu, X., Wang, X., Leung, D.Y.C. *et al.* (2010) Photocatalytic reforming of glucose over La doped alkali tantalate photocatalysts for H2 production. *Catalysis Communications*, **12**, 184–187.

102. Zartman, W.H. and Adkins, H. (1933) Hydrogenolysis of sugars. *Journal of the American Chemical Society*, **55**, 4559–4563.

103. Tanikella, M.S.S.R. (1983) Hydrogenolysis of polyols to ethylene glycol in nonaqueous solvents USPTO, Patent Number 440411.

104. Caho, J.C. and Huibers, D.T.A. (1982) Catalytic hydrogenolysis of alditols to product glycerol and polyols USPTO, Patent Number 4366332.

105. Boelhouwer, C., Korf, D., and Waterman, H.I. (1960) Catalytic hydrogenation of sugars. *Journal of Applied Chemistry*, **10**, 292–296.

106. Müller, P., Rimmelin, P., Hindermann, J.P. *et al.* (1991) Transformation of sugar into glycols on a 5% Ru/C catalyst, in *Studies in Surface Science and Catalysis 59* (eds M. Guisnet, J. Barrault, C. Bouchoule *et al.*), Elsevier, Amsterdam.

107. Montassier, C., Ménézo, J.C., Hoang, L.C. *et al.* (1991) Aqueous polyol conversions on ruthenium and on sulfur-modified ruthenium. *Journal of Molecular Catalysis*, **70**, 99–110.

108. Montassier, C., Ménézo, J.C., Moukolo, J. *et al.* (1991) Polyol conversions into furanic derivatives on bimetallic catalysts: Cu–Ru, Cu–Pt and Ru–Cu. *Journal of Molecular Catalysis*, **70**, 65–84.

109. Montassier, C., Dumas, J.M., Granger, P., and Barbier, J. (1995) Deactivation of supported copper based catalysts during polyol conversion in aqueous phase. *Applied Catalysis A: General*, **121**, 231–244.

110. Blanc, B., Bourrel, A., Gallezot, P. *et al.* (2000) Starch-derived polyols for polymer technologies: preparation by hydrogenolysis on metal catalysts. *Green Chemistry*, **2**, 89–91.

111. Ji, N., Zhang, T., Zheng, M. *et al.* (2008) Direct catalytic conversion of cellulose into ethylene glycol using nickel-promoted tungsten carbide catalysts. *Angewandte Chemie International Edition*, **47**, 8510–8513.

112. Kunkes, E.L., Simonetti, D.A., West, R.M. *et al.* (2008) Catalytic conversion of biomass to monofunctional hydrocarbons and targeted liquid-fuel classes. *Science*, **322**, 417–421.

113. Renz, M. (2005) Ketonization of carboxylic acids by decarboxylation: Mechanism and scope. *European Journal of Organic Chemistry*, **6**, 979–988.

114. Corma, A., Renz, M., and Schaverien, C. (2008) Coupling fatty acids by ketonic decarboxylation using solid catalysts for the direct production of diesel, lubricants, and chemicals. *Chemsuschem*, **1**, 739–741.

115. Dooley, K.M., Bhat, A.K., Plaisance, C.P., and Roy, A.D. (2007) Ketones from acid condensation using supported CeO2 catalysts: Effect of additives. *Applied Catalysis A: General*, **320**, 122–133.

116. Gärtner, C.A., Serrano-Ruiz, J.C., Braden, D.J., and Dumesic, J.A. (2009) Catalytic upgrading of Bio-Oils by ketonization. *ChemSusChem*, **2**, 1121–1124.

117. Gaertner, C.A., Serrano-Ruiz, J.C., Braden, D.J., and Dumesic, J.A. (2009) Catalytic coupling of carboxylic acids by ketonization as a processing step in biomass conversion. *Journal of Catalysis*, **266**, 71–78.

118. Sasaki, M., Goto, K., Tajima, K. *et al.* (2002) Rapid and selective retro-aldol condensation of glucose to glycolaldehyde in supercritical water. *Green Chemistry*, **4**, 285–287.

119. Climent, M.J., Corma, A., Iborra, S. *et al.* (2004) Increasing the basicity and catalytic activity of hydrotalcites by different synthesis procedures. *Journal of Catalysis*, **225**, 316–326.

120. Barrett, C.J., Chheda, J.N., Huber, G.W., and Dumesic, J.A. (2006) Single-reactor process for sequential aldol-condensation and hydrogenation of biomass-derived compounds in water. *Applied Catalysis B: Environmental*, **66**, 111–118.

121. Chheda, J.N. and Dumesic, J.A. (2007) An overview of dehydration, aldol-condensation and hydrogenation processes for production of liquid alkanes from biomass-derived carbohydrates. *Catalysis Today*, **123**, 59–70.

122. Kunkes, E.L., Gürbüz, E.I., and Dumesic, J.A. (2009) Vapour-phase C—C coupling reactions of biomass-derived oxygenates over Pd/CeZrOx catalysts. *Journal of Catalysis*, **266**, 236–249.

123. Gürbüz, E.I., Kunkes, E.L., and Dumesic, J.A. (2010) Dual-bed catalyst system for C—C coupling of biomass-derived oxygenated hydrocarbons to fuel-grade compounds. *Green Chemistry*, **12**, 223–227.

124. Gürbüz, E.I., Kunkes, E.L., and Dumesic, J.A. (2010) Integration of C—C coupling reactions of biomass-derived oxygenates to fuel-grade compounds. *Applied Catalysis B: Environmental*, **94**, 134–141.

125. West, R.M., Braden, D.J., and Dumesic, J.A. (2009) Dehydration of butanol to butene over solid acid catalysts in high water environments. *Journal of Catalysis*, **262**, 134–143.

126. West, R.M., Kunkes, E.L., Simonetti, D.A., and Dumesic, J.A. (2009) Catalytic conversion of biomass-derived carbohydrates to fuels and chemicals by formation and upgrading of mono-functional hydrocarbon intermediates. *Catalysis Today*, **147**, 115–125.

127. Blommel, P.G., Keenan, G.R., Rozmiarek, R.T., and Cortrigth, R.D. (2008) Catalytic conversion of sugar into conventional gasoline, diesel, jet fuel, and other hydrocarbons. *International Sugar Journal*, **110**, 672–679.

128. Li, N., Tompsett, G.A., and Huber, G.W. (2010) Renewable high-octane gasoline by aqueous-phase hydrodeoxygenation of C5 and C6 carbohydrates over Pt/Zirconium phosphate catalysts. *ChemSusChem*, **3**, 1154–1157.

129. Feather, M.S. and Harris, J.F. (1973) Dehydration reactions of carbohydrates, in *Advances in Carbohydrate Chemistry and Biochemistry 28* (eds R.S. Tipson and D. Horton), Academic Press.

130. Kuster, B.F.M. (1990) 5-Hydroxymethylfurfural (Hmf) – a review focusing on its manufacture. *Starch-Starke*, **42**, 314–321.

131. Lewkowski, J. (2001) Synthesis, chemistry and applications of 5-hydroxymethyl-furfural and its derivatives. *Arkivoc*, **2**, 17–54.

132. Tong, X., Ma, Y., and Li, Y. (2010) Biomass into chemicals: Conversion of sugars to furan derivatives by catalytic processes. *Applied Catalysis A: General*, **385**, 1–13.

133. Haworth, W.N. and Jones, W.G.M. (1944) The conversion of sucrose into furan compounds. Part I. 5-Hydroxymethylfurfuraldehyde and some derivatives. *Journal of the Chemical Society*, 667–670.

134. Mednick, M.L. (1962) The acid-base-catalyzed conversion of aldohexose into 5-(hydroxymethyl)-2-furfural2 *The Journal of Organic Chemistry*, **27**, 398–403.

135. van Dam, H.E., Kieboom, A.P.G., and van Bekkum, H. (1986) The conversion of fructose and glucose in acidic media: formation of hydroxymethylfurfural. *Starch–Stärke*, **38**, 95–101.

136. Newth, F.H. (1951) The formation of furan compounds from hexoses, in *Advances in Carbohydrate Chemistry 6* (ed. C.S. Hudso and S.M. Canto), Academic Press.

137. Antal, M.J., Mok, W.S.L., and Richards, G.N. (1990) Mechanisms of formation of 5 Hydroxymethyl-2-Furaldehyde from D fructose and sucrose. *Carbohydrate Research*, **119**, 91–110.

138. Blanchard, P.H. and Geiger, E.O. (1984) Production of high fructose corn syrup in the USA. *Sugar Technology Reviews*, **11**, 1–94.

139. Kuster, B.F.M. and Temmink, H.M.G. (1977) The influence of pH and weak-acid anions on the dehydration of -fructose. *Carbohydrate Research*, **54**, 185–191.

140. Girisuta, B., Janssen, L.P.B.M., and Heeres, H.J. (2006) A kinetic study on the decomposition of 5-hydroxymethylfurfural into levulinic acid. *Green Chemistry*, **8**, 701–709.

141. Girisuta, B., Janssen, L., and Heeres, H.J. (2007) Kinetic study on the acid-catalyzed hydrolysis of cellulose to levulinic acid. *Industrial & Engineering Chemistry Research*, **46**, 1696–1708.

142. Nakamura, Y. and Morikawa, S. (1980) The dehydration of D-Fructose to 5-Hydroxymethyl-2-furaldehyde. *Bulletin of the Chemical Society of Japan*, **53**, 3705–3706.

143. Moreau, C., Durand, R., Razigade, S. *et al.* (1996) Dehydration of fructose to 5-hydroxymethylfurfural over H-mordenites. *Applied Catalysis A: General*, **145**, 211–224.

144. Roman-Leshkov, Y., Chheda, J.N., and Dumesic, J.A. (2006) Phase modifiers promote efficient production of hydroxymethylfurfural from fructose. *Science*, **312**, 1933–1937.

145. Chheda, J.N., Roman-Leshkov, Y., and Dumesic, J.A. (2007) Production of 5-hydroxymethylfurfural and furfural by dehydration of biomass-derived mono- and poly-saccharides. *Green Chemistry*, **9**, 342–350.

146. Roman-Leshkov, Y., Barrett, C.J., Liu, Z.Y., and Dumesic, J.A. (2007) Production of dimethylfuran for liquid fuels from biomass-derived carbohydrates. *Nature*, **447**, 982–985.

147. Kooyman, C., Vellenga, K., and De Wilt, H.G.J. (1977) The isomerization of D-glucose into D-fructose in aqueous alkaline solutions. *Carbohydrate Research*, **54**, 33–44.

148. Watanabe, M., Aizawa, Y., Iida, T. *et al.* (2005) Glucose reactions with acid and base catalysts in hot compressed water at 473 K. *Carbohydrate Research*, **340**, 1925–1930.

149. Rendleman, J.A. Jr. and Hodge, J.E. (1979) Complexes of carbohydrates with aluminate ion. Aldose-ketose interconversion on anion-exchange resin (aluminate and hydroxide forms). *Carbohydrate Research*, **75**, 83–99.

150. Watanabe, M., Aizawa, Y., Iida, T. *et al.* (2005) Catalytic glucose and fructose conversions with TiO_2 and ZrO_2 in water at 473 K: Relationship between reactivity and acid-base property determined by TPD measurement. *Applied Catalysis A: General*, **295**, 150–156.

151. Moreau, C., Durand, R., Roux, A., and Tichit, D. (2000) Isomerization of glucose into fructose in the presence of cation-exchanged zeolites and hydrotalcites. *Applied Catalysis A: General*, **193**, 257–264.

152. Lecomte, J., Finiels, A., and Moreau, C. (2002) Kinetic study of the isomerization of glucose into fructose in the presence of anion-modified hydrotalcites. *Starch–Stärke*, **54**, 75–79.

153. Lima, S., Dias, A.S., Lin, Z. *et al.* (2008) Isomerization of d-glucose to d-fructose over metallosilicate solid bases. *Applied Catalysis A: General*, **339**, 21–27.

154. Moliner, M., Roman-Leshkov, Y., and Davis, M.E. (2010) Tin-containing zeolites are highly active catalysts for the isomerization of glucose in water. *Proceedings of the National Academy of Sciences*, **107**, 6164–6168.

155. Román-Leshkov, Y., Moliner, M., Labinger, J.A., and Davis, M.E. (2010) Mechanism of glucose isomerization using a solid lewis acid catalyst in water. *Angewandte Chemie International Edition*, **49**, 8954–8957.

156. Nikolla, E., Roman-Leshkov, Y., Moliner, M., and Davis, M.E. (2011) One-pot synthesis of 5-(Hydroxymethyl) furfural from carbohydrates using tin-beta zeolite. *ACS Catalysis*, **1**, 408–410.

157. Qi, X.H., Watanabe, M., Aida, T.M., and Smith, R.L. (2009) Efficient process for conversion of fructose to 5-hydroxymethylfurfural with ionic liquids. *Green Chemistry*, **11**, 1327–1331.
158. Zakrzewska, M.E., Bogel-Lukasik, E., and Bogel-Lukasik, R. (2011) Ionic liquid-mediated formation of 5-Hydroxymethylfurfural-A promising biomass-derived building block. *Chemical Reviews*, **111**, 397–417.
159. Moreau, C., Finiels, A., and Vanoye, L. (2006) Dehydration of fructose and sucrose into 5-hydroxymethylfurfural in the presence of 1-H-3-methyl imidazolium chloride acting both as solvent and catalyst. *Journal of Molecular Catalysis A-Chemical*, **253**, 165–169.
160. Zhao, H.B., Holladay, J.E., Brown, H., and Zhang, Z.C. (2007) Metal chlorides in ionic liquid solvents convert sugars to 5-hydroxymethylfurfural. *Science*, **316**, 1597–1600.
161. Mascal, M. and Nikitin, E.B. (2008) Direct, high-yield conversion of cellulose into biofuel. *Angewandte Chemie-International Edition*, **47**, 7924–7926.
162. Mascal, M. and Nikitin, E.B. (2009) High yield conversion of plant biomass into the key value-added feedstocks 5-(hydroxymethyl)furfural, levulinic acid, and levulinic esters via 5-(chloromethyl)furfural. *Green Chemistry*, **12**, 370–373.
163. Mascal, M. and Nikitin, E.B. (2010) Co-processing of carbohydrates and lipids in oil crops to produce a hybrid biodiesel. *Energy & Fuels*, **24**, 2170–2171.
164. Mascal, M. and Nikitin, E.B. (2009) Dramatic advancements in the saccharide to 5-(Chloromethyl)furfural conversion reaction. *Chemsuschem*, **2**, 859–861.
165. Mascal, M. and Dutta, S. (2011) Synthesis of the natural herbicide delta-aminolevulinic acid from cellulose-derived 5-(chloromethyl)furfural. *Green Chemistry*, **13**, 40–41.
166. Lima, S., Pillinger, M., and Valente, A.A. (2008) Dehydration of D-xylose into furfural catalysed by solid acids derived from the layered zeolite Nu-6(1). *Catalysis Communications*, **9**, 2144–2148.
167. Kim, Y.C. and Lee, H.S. (2001) Selective synthesis of furfural from xylose with supercritical carbon dioxide and solid acid catalyst. *Journal of Industrial and Engineering Chemistry*, **7**, 424–429.
168. O'Neill, R., Ahmad, M.N., Vanoye, L., and Aiouache, F. (2009) Kinetics of aqueous phase dehydration of xylose into furfural catalyzed by ZSM-5 zeolite. *Industrial & Engineering Chemistry Research*, **48**, 4300–4306.
169. Herrera, A., Tellez-Luis, S.J., Ramirez, J.A., and Vazquez, M. (2003) Production of xylose from sorghum straw using hydrochloric acid. *Journal of Cereal Science*, **37**, 267–274.
170. Dias, A.S., Lima, S., Pillinger, M., and Valente, A.A. (2006) Acidic cesium salts of 12-tungstophosphoric acid as catalysts for the dehydration of xylose into furfural. *Carbohydrate Research*, **341**, 2946–2953.
171. Dias, A.S., Lima, S., Pillinger, M., and Valente, A.A. (2007) Modified versions of sulfated zirconia as catalysts for the conversion of xylose to furfural. *Catalysis Letters*, **114**, 151–160.
172. Moreau, C., Durand, R., Peyron, D. *et al.* (1998) Selective preparation of furfural from xylose over microporous solid acid catalysts. *Industrial Crops and Products*, **7**, 95–99.
173. Xing, R., Subrahmanyam, A.V., Olcay, H. *et al.* (2010) Production of jet and diesel fuel range alkanes from waste hemicellulose-derived aqueous solutions. *Green Chemistry*, **12**, 1933–1946.
174. Weingarten, R., Cho, J., Conner, W.C., and Huber, G.W. (2010) Kinetics of furfural production by dehydration of xylose in a biphasic reactor with microwave heating. *Green Chemistry*, **12**, 1423–1429.
175. James, O.O., Maity, S., Usman, L.A. *et al.* (2010) Towards the conversion of carbohydrate biomass feedstocks to biofuels via hydroxylmethylfurfural. *Energy & Environmental Science*, **3**, 1833–1850.
176. Zeitsch, K.J. (2000) *The Chemistry and Technology of Furfural and its Many By-Products*, 1st edn, Elsevier, Amsterdam.
177. Lilga, M.A., Hallen, R.T., and Gray, M. (2010) Production of oxidized derivatives of 5-hydroxymethylfurfural (HMF). *Topics in Catalysis*, **53**, 1264–1269.
178. Verdeguer, P., Merat, N., Rigal, L., and Gaset, A. (1994) Optimization of experimental conditions for the catalytic-oxidation of furfural to furoic acid. *Journal of Chemical Technology and Biotechnology*, **61**, 97–102.
179. Verdeguer, P., Merat, N., and Gaset, A. (1994) Lead platinum on charcoal as catalyst for oxidation of furfural – effect of main parameters. *Applied Catalysis A: General*, **112**, 1–11.
180. Cottier, L., Descotes, G., Lewkowski, J., and Skowronski, R. (1994) Oxidation of 5-Hydroxymethylfurfural under Sonochemical Conditions. *Polish Journal of Chemistry*, **68**, 693–698.

181. Partenheimer, W. and Grushin, V.V. (2001) Synthesis of 2,5-diformylfuran and furan-2,5-dicarboxylic acid by catalytic air-oxidation of 5-hydroxymethylfurfural. Unexpectedly selective aerobic oxidation of benzyl alcohol to benzaldehyde with metal/bromide catalysts. *Advanced Synthesis & Catalysis*, **343**, 102–111.

182. Carlini, C., Patrono, P., Galletti, A.M.R. *et al.* (2005) Selective oxidation of 5-hydroxymethyl-2-furaldehyde to furan-2,5-dicarboxaldehyde by catalytic systems based on vanadyl phosphate. *Applied Catalysis A: General*, **289**, 197–204.

183. Amarasekara, A.S., Green, D., and McMillan, E. (2008) Efficient oxidation of 5-hydroxymethylfurfural to 2,5-diformylfuran using Mn(III)-salen catalysts. *Catalysis Communications*, **9**, 286–288.

184. Yoon, H.-J., Choi, J.-W., Jang, H.-S. *et al.* (2011) Selective oxidation of 5-Hydroxymethylfurfural to 2,5-diformyl-furan by polymer-supported IBX amide. *Synlett*, **165**, 168.

185. Navarro, O.C., Canos, A.C., and Chornet, S.I. (2009) Chemicals from biomass: aerobic oxidation of 5-Hydroxy-methyl-2-furaldehyde into diformylfurane catalyzed by immobilized vanadyl-pyridine complexes on polymeric and organofunctionalized mesoporous supports. *Topics in Catalysis*, **52**, 304–314.

186. Ma, J.P., Du, Z.T., Xu, J. *et al.* (2011) Efficient aerobic oxidation of 5-Hydroxymethylfurfural to 2,5-Diformyl-furan, and synthesis of a fluorescent material. *Chemsuschem*, **4**, 51–54.

187. Skowronski, R., Grabowski, G., Lewkowski, J. *et al.* (1993) New chemical conversions of 5-Hydroxymethylfurfu-ral and the electrochemical oxidation of its derivatives. *Organic Preparations and Procedures International*, **25**, 353–355.

188. Grabowski, G., Lewkowski, J., and Skowronski, R. (1991) The electrochemical oxidation of 5-Hydroxymethylfur-fural with the nickel-oxide hydroxide electrode. *Electrochimica Acta*, **36**, 1995–1995.

189. Verdeguer, P., Merat, N., and Gaset, A. (1993) Catalytic-oxidation of Hmf to 2,5-Furandicarboxylic acid. *Journal of Molecular Catalysis*, **85**, 327–344.

190. Casanova, O., Iborra, S., and Corma, A. (2009) Biomass into chemicals: aerobic oxidation of 5-Hydroxymethyl-2-furfural into 2,5-Furandicarboxylic acid with gold nanoparticle catalysts. *ChemSusChem*, **2**, 1138–1144.

191. Gorbanev, Y.Y., Klitgaard, S.K., Woodley, J.M. *et al.* (2009) Gold-catalyzed aerobic oxidation of 5-Hydroxyme-thylfurfural in water at ambient temperature. *Chemsuschem*, **2**, 672–675.

192. Taarning, E., Nielsen, I.S., Egeblad, K. *et al.* (2008) Chemicals from renewables: Aerobic oxidation of furfural and hydroxymethylfurfural over gold catalysts. *Chemsuschem*, **1**, 75–78.

193. Casanova, O., Iborra, S., and Corma, A. (2009) Biomass into chemicals: One pot-base free oxidative esterification of 5-hydroxymethyl-2-furfural into 2,5-dimethylfuroate with gold on nanoparticulated ceria. *Journal of Catalysis*, **265**, 109–116.

194. Ribeiro, M.L. and Schuchardt, U. (2003) Cooperative effect of cobalt acetylacetonate and silica in the catalytic cyclization and oxidation of fructose to 2,5-furandicarboxylic acid. *Catalysis Communications*, **4**, 83–86.

195. Boisen, A., Christensen, T.B., Fu, W. *et al.* (2009) Process integration for the conversion of glucose to 2,5-furandi-carboxylic acid. *Chemical Engineering Research and Design*, **87**, 1318–1327.

196. Schiavo, V., Descotes, G., and Mentech, J. (1991) Catalytic-hydrogenation of 5-Hydroxymethylfurfural in aque-ous-medium. *Bulletin De La Societe Chimique De France*, **5**, 704–711.

197. Zheng, H.Y., Zhu, Y.L., Teng, B.T. *et al.* (2006) Towards understanding the reaction pathway in vapour phase hydrogenation of furfural to 2-methylfuran. *Journal of Molecular Catalysis A: Chemical*, **246**, 18–23.

198. Chidambaram, M. and Bell, A.T. (2010) A two-step approach for the catalytic conversion of glucose to 2,5-dime-thylfuran in ionic liquids. *Green Chemistry*, **12**, 1253–1262.

199. Yang, W.R. and Sen, A. (2010) One-step catalytic transformation of carbohydrates and cellulosic biomass to 2,5-dimethyltetrahydrofuran for liquid fuels. *Chemsuschem*, **3**, 597–603.

200. Wu, S.H., Wei, W., Li, B.Q. *et al.* (2003) Studies on hydrogenation of furfural to 2-methylfuran over Cu-Cr/gamma-Al₂O₃ catalyst prepared by different methods. *Chinese Journal of Catalysis*, **24**, 27–31.

201. Yuskovets, Z.G., Nekrasov, N.V., Kostyukovskii, M.M. *et al.* (1984) Investigation of the mechanism of hydrogena-tion and hydrogenolysis of 2-methylfuran on a palladium catalyst. *Kinetics and Catalysis*, **25**, 1162–1165.

202. Grizoras, E., Zolotare, N., Buimov, A. *et al.* (1969) Continuous process for preparation of 2-methylfuran by cata-lytic hydrogenation of furfural. *Pharmaceutical Chemistry Journal-USSR*, **6**, 357.

203. Huber, G.W., Chheda, J.N., Barrett, C.J., and Dumesic, J.A. (2005) Production of liquid alkanes by aqueous-phase processing of biomass-derived carbohydrates. *Science*, **308**, 1446–1450.

204. West, R.M., Liu, Z.Y., Peter, M., and Dumesic, J.A. (2008) Liquid Alkanes with targeted molecular weights from biomass-derived carbohydrates. *Chemsuschem*, **1**, 417–424.

205. Howard, W.L. (2000) Acetone, in *Kirk-Othmer Encyclopedia of Chemical Technology*, John Wiley & Sons, Inc.

206. Baade, W.F., Parekh, U.N., and Raman, V.S. (2000) Hydrogen, in *Kirk-Othmer Encyclopedia of Chemical Technology*, John Wiley & Sons, Inc.

207. Girisuta, B., Janssen, L.P.B.M., and Heeres, H.J. (2007) Kinetic study on the acid-catalyzed hydrolysis of cellulose to levulinic acid. *Industrial & Engineering Chemistry Research*, **46**, 1696–1708.

208. Serrano-Ruiz, J.C., West, R.M., and Dumesic, J.A. (2010) Catalytic conversion of renewable biomass resources to fuels and chemicals. *Annual Review of Chemical and Biomolecular Engineering*, **1**, 79–100.

209. Fitzpatrick, S.W. (1997) Production of levulinic acid from carbohydrate-containing materials, USPTO, Patent Number 5608105.

210. Fitzpatrick, S.W. (1990) Lignocellulose degradation to furfural and levulinic acid, USPTO, Patent Number 4897497.

211. Fitzpatrick, S.W. (2006) The biofine technology: A "Bio-Refinery": concept based on thermochemical conversion of cellulosic biomass, in *Feedstocks for the Future 921* (eds J.J. Bozell and M.K. Patel), American Chemical Society.

212. Bart, H.J., Reidetschlager, J., Schatka, K., and Lehmann, A. (1994) Kinetics of esterification of levulinic acid with N-butanol by homogeneous catalysis. *Industrial & Engineering Chemistry Research*, **33**, 21–25.

213. Ayoub, P.M. (2005) Process for the reactive extraction of levulinic acid, WIPO, Patent Number WO/2005/070867.

214. Gürbüz, E.I., Alonso, D.M., Bond, J.Q., and Dumesic, J.A. (2011) Reactive extraction of levulinate esters and conversion to gamma-valerolactone for production of liquid fuels. *Chemsuschem*, **4**, 357–361.

215. Fagan, P.J. and Manzer, L.E. (2006) Preparation of levulinic acid esters and formic acid esters from biomass and olefins, USPTO, Patent Number. 7153996.

216. Chapman, O.L. and McIntosh, C.L. (1971) Photochemical decarbonylation of unsaturated lactones and carbonates. *Journal of the Chemical Society D: Chemical Communications*, **8**, 383–384.

217. Timokhin, B.V., Baransky, V.A., and Eliseeva, G.D. (1999) Levulinic acid in organic synthesis. *Russian Chemical Reviews*, **68**, 73–84.

218. Gong, Y., Lin, L., Shi, J., and Liu, S. (2010) Oxidative decarboxylation of levulinic acid by cupric oxides. *Molecules*, **15**, 7946–7960.

219. Gong, Y. and Lin, L. (2011) Oxidative decarboxylation of levulinic acid by silver(i)/Persulfate. *Molecules*, **16**, 2714–2725.

220. Bozell, J.J., Moens, L., Elliott, D.C. *et al.* (2000) Production of levulinic acid and use as a platform chemical for derived products. *Resources, Conservation and Recycling*, **28**, 227–239.

221. Guo, Y.H., Li, K.X., Yu, X.D., and Clark, J.H. (2008) Mesoporous H3PW12O40-silica composite: Efficient and reusable solid acid catalyst for the synthesis of diphenolic acid from levulinic acid. *Applied Catalysis B: Environmental*, **81**, 182–191.

222. Horvat, J., Klaic, B., Metelko, B., and Sunjic, V. (1985) Mechanism of levulinic acid formation. *Tetrahedron Letters*, **26**, 2111–2114.

223. Saha, B. and Sharma, M.M. (1996) Esterification of formic acid, acrylic acid and methacrylic acid with cyclohexene in batch and distillation column reactors: Ion-exchange resins as catalysts. *Reactive & Functional Polymers*, **28**, 263–278.

224. Ballantine, J.A., Davies, M., Patel, I. *et al.* (1984) Organic-reactions catalyzed by sheet silicates - ether formation by the intermolecular dehydration of alcohols and by addition of alcohols to alkenes. *Journal of Molecular Catalysis*, **26**, 37–56.

225. Ayoub, P. and Lange, J.P. (2008) WIPO, Patent Number WO/2008/142127.

226. Fellay, C., Dyson, P.J., and Laurenczy, G. (2008) A viable hydrogen-storage system based on selective formic acid decomposition with a ruthenium catalyst. *Angewandte Chemie-International Edition*, **47**, 3966–3968.

227. Ojeda, M. and Iglesia, E. (2009) Formic acid dehydrogenation on Au-based catalysts at near-ambient temperatures. *Angewandte Chemie-International Edition*, **48**, 4800–4803.

228. Larson, L.A. and Dickinson, J.T. (1979) Decomposition of Formic-Acid on Ru(1010). *Surface Science*, **84**, 17–30.

229. Serrano-Ruiz, J.C., Braden, D.J., West, R.M., and Dumesic, J.A. (2010) Conversion of cellulose to hydrocarbon fuels by progressive removal of oxygen. *Applied Catalysis B: Environmental*, **100**, 184–189.

230. Deng, L., Li, J., Lai, D.M. *et al.* (2009) Catalytic conversion of biomass-derived carbohydrates into gamma-valerolactone without using an external H-2 supply. *Angewandte Chemie-International Edition*, **48**, 6529–6532.
231. Heeres, H., Handana, R., Chunai, D. *et al.* (2009) Combined dehydration/(transfer)-hydrogenation of C6-sugar (D-glucose and D-fructose) to gamma-valerolactone using ruthenium catalysts. *Green Chemistry*, **11**, 1247–1255.
232. Manzer, L.E. (2004) Catalytic synthesis of alpha-methylene-gamma-valerolactone: a biomass-derived acrylic monomer. *Applied Catalysis A: General*, **272**, 249–256.
233. Deng, L., Zhao, Y., Li, J.A. *et al.* (2010) Conversion of levulinic acid and formic acid into gamma-valerolactone over heterogeneous catalysts. *Chemsuschem*, **3**, 1172–1175.
234. Horvath, I.T., Mehdi, H., Fabos, V. *et al.* (2008) gamma-Valerolactone – a sustainable liquid for energy and carbon-based chemicals. *Green Chemistry*, **10**, 238–242.
235. Mehdi, H., Fabos, V., Tuba, R. *et al.* (2008) Integration of homogeneous and heterogeneous catalytic processes for a multi-step conversion of biomass: From sucrose to levulinic acid, gamma-valerolactone, 1,4-pentanediol, 2-methyl-tetrahydrofuran, and alkanes. *Topics in Catalysis*, **48**, 49–54.
236. Lange, J.P. (2007) Lignocellulose conversion: an introduction to chemistry, process and economics. *Biofuel. Bioproducts & Biorefining-Biofpr*, **1**, 39–48.
237. Beller, M. and Tafesh, A.M. (1996) Other carbonylations, in *Applied Homogeneous Catalysis with Organo metallic Compounds* (eds B. Cornils and W.A. Herrmann), Wiley-VCH, Weinheim.
238. Meessen, P., Vogt, D., and Keim, W. (1998) Highly regioselective hydroformylation of internal, functionalized olefins applying Pt/Sn complexes with large bite angle diphosphines. *Journal of Organometallic Chemistry*, **551** 165–170.
239. Lange, J.P., Price, R., Ayoub, P. *et al.* (2010) Valeric biofuels: A platform of cellulosic transportation fuels *Angewandte Chemie International Edition*, **49**, 4479–4483.
240. Serrano-Ruiz, J.C., Wang, D., and Dumesic, J.A. (2010) Catalytic upgrading of levulinic acid to 5-nonanone *Green Chemistry*, **12**, 574–577.
241. Martin Alonso, D., Bond, J.Q., Serrano-Ruiz, J.C., and Dumesic, J.A. (2010) Production of liquid hydrocarbon transportation fuels by oligomerization of biomass-derived C9 alkenes. *Green Chemistry*, **12**, 992–999.
242. Bond, J.Q., Martin Alonso, D., West, R.M., and Dumesic, J.A. (2010) gamma-Valerolactone ring-opening and decarboxylation over SiO2/Al2O3 in the presence of water. *Langmuir*, **26**, 16291–16298.
243. Bond, J.Q., Alonso, D.M., Wang, D. *et al.* (2010) Integrated catalytic conversion of gamma-valerolactone to liquid alkenes for transportation fuels. *Science*, **327**, 1110–1114.
244. Rao, R., Dandekar, A., Baker, R.T.K., and Vannice, M.A. (1997) Properties of copper chromite catalysts in hydrogenation reactions. *Journal of Catalysis*, **171**, 406–419.
245. Seo, G. and Chon, H. (1981) Hydrogenation of furfural over copper-containing catalysts. *Journal of Catalysis*, **67**, 424–429.
246. Rao, R.S., Baker, R.T.K., and Vannice, M.A. (1999) Furfural hydrogenation over carbon-supported copper. *Catalysis Letters*, **60**, 51–57.
247. Nagaraja, B.M., Kumar, V.S., Shasikala, V. *et al.* (2003) A highly efficient Cu/MgO catalyst for vapour phase hydrogenation of furfural to furfuryl alcohol. *Catalysis Communications*, **4**, 287–293.
248. Hao, X.Y., Zhou, W., Wang, J.W. *et al.* (2005) A novel catalyst for the selective hydrogenation of furfural to furfuryl alcohol. *Chemistry Letters*, **34**, 1000–1001.
249. Wu, J., Shen, Y.M., Liu, C.H. *et al.* (2005) Vapor phase hydrogenation of furfural to furfuryl alcohol over environmentally friendly Cu-Ca/SiO$_2$ catalyst. *Catalysis Communications*, **6**, 633–637.
250. Vaidya, P.D. and Mahajani, V.V. (2003) Kinetics of liquid-phase hydrogenation of furfuraldehyde to furfuryl alcohol over a Pt/C catalyst. *Industrial & Engineering Chemistry Research*, **42**, 3881–3885.
251. Marinelli, T., Ponec, L., Raab, C.G., and Lercher, J.A. (1993) Furfural – hydrogen reactions, manipulation of activity and selectivity of the catalyst, in *Heterogeneous Catalysis and Fine Chemicals Iii 78* (eds M. Guisnet, J. Barbier, J. Barrault *et al.*), Elsevier Science, Amsterdam.
252. Merlo, A.B., Vetere, V., Ruggera, J.F., and Casella, M.L. (2009) Bimetallic PtSn catalyst for the selective hydrogenation of furfural to furfuryl alcohol in liquid-phase. *Catalysis Communications*, **10**, 1665–1669.
253. Lange, J.P., van de Graaf, W.D., and Haan, R.J. (2009) Conversion of furfuryl alcohol into ethyl levulinate using solid acid catalysts. *Chemsuschem*, **2**, 437–441.

6

Fundamentals of Biomass Pretreatment at Low pH

Heather L. Trajano[1,2,*] **and Charles E. Wyman**[1,2]

[1] *Department of Chemical and Environmental Engineering and Center for Environmental Research and Technology, University of California, Riverside, USA*
[2] *BioEnergy Science Center, Oak Ridge, USA*

6.1 Introduction

A wide variety of conversion processes can be used to generate fuels and chemicals from biomass, with many including a hydrolysis and/or dehydration reaction at low pH as an initial stage. In the case of acid hydrolysis of biomass prior to enzymatic hydrolysis and fermentation, this step is called pretreatment. Due to the diversity of conversion processes, it is difficult to define a single set of objectives; the general goals however are to generate reactive intermediates in high yields for subsequent conversion to final products and minimize generation of compounds that interfere with downstream operations. For example, if lignocellulosic biomass is to be converted to ethanol through dilute acid pretreatment, enzymatic hydrolysis, and fermentation, the objectives of pretreatment are to produce high yields of hemicellulose sugars, improve the enzymatic digestibility of the remaining solids to realize high yields of glucose, and avoid generating biological inhibitors such as furfural and acetic acid. On the other hand, if lignocellulosic biomass is to be converted into jet fuel alkanes through cellulose hydrolysis to levulinic acid followed by catalytic processing of levulinic acid to alkanes, the objectives are to generate levulinic acid in high yields and avoid solid catalyst poisons such as mineral acids.

Two of the primary advantages of low-pH reactions are the ready availability of catalysts such as H_2SO_4 and SO_2 and high product yields. However, the capital costs of reactors and associated equipment used for low-pH reactions are high due to the need for expensive, corrosion-resistant materials. In addition, the solid and liquid streams resulting from low-pH pretreatment often require washing or neutralization. Both of

---

* Present address: Department of Chemical and Biological Engineering, University of British Colombia, Vancouver, Canada

Aqueous Pretreatment of Plant Biomass for Biological and Chemical Conversion to Fuels and Chemicals, First Edition.
Edited by Charles E. Wyman.
© 2013 John Wiley & Sons, Ltd. Published 2013 by John Wiley & Sons, Ltd.

these considerations create technical challenges and add cost. Finally, the conditioning process to reduce inhibitors can result in sugar losses as well as added expense. For example, Martinez *et al.* [1] found that the total sugars in pretreatment hydrolysate decreased by $8.74 \pm 4.46\%$ after overliming.

This chapter will first explore the earliest use of acid in biomass conversion that provided a foundation for extension to biological conversion of biomass to ethanol, that is, pretreatment: cellulose hydrolysis to glucose for fermentation to ethanol. Examples of the operating conditions associated with this process will be presented. Emphasis will be placed on the use of SO_2 and H_2SO_4, the most common acidifying agents, given their effectiveness and low price. Dilute acid pretreatments are now evolving to support conversion of biomass to hydrocarbon fuels with solid catalysts. Kinetic models of cellulose and hemicellulose hydrolysis at low pH conditions will also be presented. One model of pretreatment makes use of a relationship known as the combined severity (CS) factor, defined as [2,3]:

$$\log(\text{CS}) = \log\left[t \exp\left(\frac{T - 100}{14.75} \right) \right] - \text{pH} \tag{6.1}$$

where t is the time in minutes, and T is the temperature in degrees Celsius. This relationship will be discussed in further detail as a useful way of comparing the different pretreatment conditions presented.

Low-pH pretreatment is a diverse field, the references to which could easily fill a book. Therefore, the objective of this chapter is to provide a summary along with key references for the interested reader to pursue for more details. The selection of pretreatment conditions is a complex problem that depends on factors such as process objectives, biomass species, chemical costs, safety considerations, and local influences such as regulations or chemical availability. It is currently not possible to identify optimum pretreatment conditions without extensive experimental work and process engineering to arrive at the optimal overall system.

6.2   Effects of Low pH on Biomass Solids

6.2.1   Cellulose

Cellulose is a linear polymer of glucose that typically accounts for 35–50% of lignocellulosic biomass [4]. The monomer units are covalently linked by 1,4-glycosidic bonds [5]. Due to the presence of multiple hydroxyl groups, there is a high degree of intramolecular and intermolecular hydrogen bonding between the glucan chains [5]. The glucan chains form a crystalline core with a semi-crystalline shell [6,7].

Due to the crystalline nature of cellulose, very low pH, high temperatures, or extended times are required to hydrolyze significant quantities of cellulose to glucose [8]. Under conditions that favor cellulose hydrolysis, the glucose released degrades to products such as levulinic and formic acid [8]. Under less severe hydrolysis conditions, the degree of polymerization has been found to change substantially. Several researchers [9–11] hydrolyzed different cellulose substrates such as cotton linters and wood pulp with 2.45–5 M HCl or 2.5 M H_2SO_4 at 5–105 °C for 0.25–480 hours and then calculated the degree of polymerization from cuprammonium viscosity values. These treatments resulted in 2–20% loss of cellulose. They found that during hydrolysis the degree of polymerization decreased rapidly initially and then stabilized at a level-off degree of polymerization (LODP), with the time to reach the LODP typically 15–30 minutes [9–11]. It was proposed that the initial decrease in degree of polymerization (DP) was due to the rapid hydrolysis of amorphous cellulose. Nickerson and Habrle found that H_2SO_4 had similar hydrolytic effects as HCl [11]. In his study, Battista subjected wood pulp to mild hydrolysis with 5 M HCl at 18 °C for 24 hours to 44 weeks, a drastic hydrolysis with boiling 2.5 M HCl for 1–15 minutes, or mild hydrolysis followed by drastic hydrolysis [9]. In this case, the weight loss during drastic hydrolysis for the sequential process was lower than the weight loss for the single-stage drastic hydrolysis [9]. Based on these observations Battista proposed that, under mild conditions, hydrolysis occurs slowly and crystallization generates pieces that are more resistant to acid hydrolysis [9]. A study by Bouchard *et al.* revealed similar trends in the degree of polymerization following hydrolysis of α-cellulose; they also showed that after a period of slow depolymerization, the rate of depolymerization increased [12]. These three phases were attributed to endogenous attack of amorphous cellulose by acid followed by exogenous acid attack of the ends of crystalline cellulose, and finally simultaneous endogenous/exogenous hydrolysis of the remaining cellulose. Changes in the degree of polymerization of cellulose in lignocellulosic materials following acid hydrolysis have also been detected. Martínez and colleagues subjected a softwood mixture and almond shells to dilute acid pretreatment and then determined the degree of polymerization from intrinsic viscosity measurements [13]. When Martínez *et al.* plotted the degree of polymerization as a function of severity, they observed the characteristic rapid decrease in degree of polymerization for both substrates [13]. In the case of a softwood mixture, this rapid initial decrease was followed by stabilization at the level-off degree of polymerization [13]. Almond shells were not pretreated at high severity conditions, so no LODP was observed [13]. Kumar *et al.* subjected corn stover and poplar to hydrolysis with dilute H_2SO_4 and SO_2 [14]. These treatments removed 3.1–12.1% of the glucan in biomass and reduced the degree of polymerization by 65–85% compared to untreated biomass [14]. The degree of polymerization of cellulose in switchgrass has also been shown to decrease following pretreatment with 0.1 mol/m^3 H_2SO_4 at 160 °C [15].

Mild hydrolysis has also been found to affect biomass crystallinity. In work by Kumar *et al.*, the crystallinity index of the pretreated materials was measured by wide-angle X-ray diffraction and Fourier transform infrared attenuated total reflectance (FTIR-ATR) [14]. The X-ray diffraction measurements showed that the crystallinity index of corn stover and poplar increased while the FTIR spectra indicated that the ratio of amorphous to crystalline cellulose decreased [14]. The increase in crystallinity index may therefore reflect removal of amorphous components such as hemicellulose and lignin from biomass and not an increase in cellulose crystallinity.

6.2.2 Hemicellulose

The second-most plentiful carbohydrate fraction in most lignocellulosic biomass is hemicellulose. Hemicellulose typically accounts for approximately 15–35% of biomass [4]. For hardwoods, grasses, and agricultural residues, hemicellulose polymers primarily consist of pentose sugars such as xylose and arabinose. Depending on the substrate, hemicellulose typically also contains the hexose sugars glucose, mannose, and galactose, with these sugars being most prevalent in softwoods. The structure of hemicellulose is more complex than cellulose and contains many branches. Acetyl is the most common side group [4].

Due to its branched structure, hemicellulose is amorphous and much more susceptible to hydrolysis by acids than cellulose. In fact, hemicellulose can be almost completely removed with limited damage to cellulose [16]. The extent of hemicellulose removal, of course, depends upon hydrolysis conditions. For example, Öhgren *et al.* were able to recover approximately 65% of the xylan in corn stover hydrolysates following pretreatment with 2% SO_2 at 200 °C for 2 minutes, while only an 18% yield was achieved with the same corn stover pretreated at 170 °C for 2 minutes with 2% SO_2 [17].

The removal of hemicellulose appears to depend upon the acidifying agent. For example, Martín *et al.* found that pretreatment of sugarcane bagasse with H_2SO_4 resulted in complete removal and partial degradation of xylan, while SO_2 pretreatment removed less xylan but produced substantially fewer degradation products such as furfural [18]. These differences were likely due to the differences in the amount of H_2SO_4 or SO_2 absorbed by bagasse prior to pretreatment.

Hemicellulose sugars can be released into solution either as oligomers or monomers, with their ratios varying with temperature, time, and acid concentration. For example, as the temperature was increased from 201 °C to 225 °C for hydrolysis of poplar in 0.4% H_2SO_4, the fraction of monomers in the hydrolysate increased from 55 to 76% [19]. In another study, increasing the H_2SO_4 concentration increased the selective production of xylose from xylooligomers [20].

Acetyl groups are removed during acid hydrolysis as acetic acid or attached to solubilized hemicellulose [16,20,21]. Experimental results indicate that once released, acetic acid does not degrade [16,21,22]; increasing hydrolysis time therefore increases the release of acetyl monomers from hemicellulose. However, there is no consensus on the effect of temperature on the release of acetyl groups. For example, Maloney *et al.* found that the acetyl removal rate from paper birch was slightly faster than that of xylan removal at 100 and 130 °C but decreased at 150 and 170 °C [21]. In contrast, Aguilar *et al.* showed that increasing the temperature of hydrolysis of sugar cane bagasse from 100 to 122 °C with 2–6% H_2SO_4 increased the acetic acid concentration in the hydrolysate, but a second temperature increase to 128 °C reduced the acetic acid concentration slightly [16]. These differences may be a reflection of differences in the types of biomass used or the acid concentration.

Once in solution, hemicelluloses-derived oligomers can react to form products such as furfural that can react with each other or with sugars to form more complex products [23]; for example, pentose can be acetalized by furfural [24]. Furfural can further decompose to formic acid or humin char [25]. Possible reaction schemes were described by Antal *et al.* [23], Hoydonckx *et al.* [24], and Weingarten *et al.* [25]. These decomposition reactions have been observed for numerous types of biomass including corn stover [17,26], hardwoods [27], softwoods [27], and switchgrass [27]. As hydrolysis time increases, the extent of these decomposition reactions also increases. The higher the hydrolysis temperature, the sooner decomposition becomes significant. For example, the maximum xylose production from corn stover was achieved in 2 minutes at 180 °C and 5 minutes at 160 °C using 0.98% H_2SO_4; for reactions lasting longer than these times, xylose yields dropped due to degradation [26]. Similarly, as acid concentration is increased at a constant temperature, the time to maximum xylose yield or onset of significant degradation drops [26].

6.2.3 Lignin

Lignin is the third major polymer in biomass but is made up of phenol monomers, not sugars. It typically accounts for 17–33% of a plant's mass [28]. Coumaryl, coniferyl, and sinapyl alcohol are the three monomeric precursors to lignin [28]; the relative portions of each monomer vary by biomass species. Lignin's structure is highly irregular and frequently forms covalent bonds with the surrounding carbohydrates, especially hemicellulose [28].

Lignin removal during acid hydrolysis in batch reactors is typically low regardless of biomass type or acidifying agent [14,22,29]. Liu and Wyman found that pumping 0.05–0.1 w/w% H_2SO_4 through corn stover at 180 °C increased lignin removal from approximately 10% in a batch reactor to about 50% [30]. The removal of lignin is accompanied by the generation of aromatic monomers in the liquid hydrolysate, and the type and amount of phenols varies with both the biomass treated and hydrolysis conditions [31]. For example, salicylic acid was found in higher concentrations in hydrolysates produced from poplar than from corn stover and pine, and its concentration varied with H_2SO_4 concentration [31]. Other researchers who identified aromatics in hydrolysates include Martín *et al.* and Excoffier *et al.* [18,19].

The limited removal of lignin from biomass during acid hydrolysis may be somewhat deceiving as it has been shown that the carbohydrate fractions can react to form compounds that analysis procedures measure as lignin. For example, NMR analysis of loblolly pine hydrolyzed at 200 °C with sulfuric, phosphoric, or trifluoroacetic acid ($C_2HF_3O_2$) by Sievers *et al.* displayed an increase in signal intensities associated with aromatic carbon, an increase that could only result from reaction of carbohydrates to aromatics or "pseudo-lignin" [32]. Ritter and Kurth [33] provided further evidence of pseudo-lignin formation from carbohydrates by subjecting poplar-derived holocellulose to acid hydrolysis. Sannigrahi *et al.* [34] recovered holocellulose, the hemicellulose and cellulose portion of *Populus trichocarpa x deltoids*, by exposing the biomass to NaClO$_2$ and acetic acid twice at 70 °C for one hour. After acid hydrolysis of this holocellulose, a lignin-like fraction, "pseudo-lignin," was detected in the resulting solids by wet chemistry, NMR, and FTIR [33]. However, because the untreated holocellulose contained only 1.6% Klason lignin, the pseudo-lignin was primarily generated through acid-catalyzed reactions of cellulose and hemicellulose [33]. The increased production of pseudo-lignin with increasing hydrolysis severity suggests that researchers should use multiple analytical procedures when determining lignin removal following high-severity hydrolysis.

The lignin that remains in solids following acid hydrolysis is modified in several ways. At a chemical level, researchers have found evidence of acid condensation and oxidation through FTIR [14,29] and decreased bromination [29]. Microscopic examination of the solids following pretreatment revealed dramatic changes in the morphology and distribution of lignin. A number of researchers observed deposition of spherical droplets on cell walls. These droplets were observed to cluster near ultrastructural features such as cell corners [35,36] and demonstrated to contain lignin with a number of techniques including KMnO$_4$ staining [35,36], FTIR spectroscopy [36], NMR analysis [36], and antibody labeling [36]. The morphology and localization of these droplets to the natural pores of biomass led to speculation that a cycle of melting and coalescing may be responsible for lignin removal [35–37].

6.2.4   Ash

Biomass also contains inorganic material, commonly referred to as ash [38], and includes both plant structural components and inorganic materials such as in soil picked up in harvesting operations. In woody species, the structural mineral content ranges from 0.3 to 2 w/w%, while in herbaceous species and agricultural residues, the structural mineral content may account for as much as 16 w/w% [4]. The composition of this inorganic fraction varies by biomass species. Biomass cations include potassium, calcium, magnesium, sodium, manganese, and ammonium; possible anions are sulfates, phosphates, chloride and nitrate [22,39–41]. When combined with biomass, mineral acids such as H$_2$SO$_4$ are neutralized through an ion-exchange reaction between inorganic cations and hydronium ions [39]. It is difficult to determine the neutralizing capacity of biomass, but the mineral content provides an adequate estimate [39]. Due to their higher mineral content, the neutralizing capacity of herbaceous biomass and agricultural residues is generally higher than that of woody materials [42,43]. In order to simplify reaction kinetic models, it is frequently assumed that neutralization is instantaneous upon mixing of biomass and acid [22,42,43]. However, Springer and Harris demonstrated that neutralization is in fact a complex phenomenon in that the exchange of cations with hydronium ions varies with temperature and applied acid concentration, and is incomplete even under severe hydrolysis [39].

6.2.5   Ultrastructure

The ultrastructure of biomass undergoes several changes during acid hydrolysis as well. It has been shown that biomass particle sizes decrease during acid hydrolysis and that, as the severity of the treatment increases, the percentages of small particles and fines increase [44]. Additionally, it has been shown that the size of intraparticle pores changes as a result of acid hydrolysis. When Grethlein hydrolyzed birch, maple, poplar, white pine, and steam-extracted southern pine with 1% H$_2$SO$_4$ at 180–220 °C for 7.8 seconds, he found that the pore volume increased with increasing pretreatment temperature [45] and attributed these changes in hemicellulose to removal during pretreatment [45]. Excoffier *et al.* and Wong *et al.* saw a similar increase in pore volume with increasing hemicellulose removal [19,46]. However, both groups also indicated that partial removal of cellulose or lignin redistribution might also contribute to changes in pore volume [19,46].

6.2.6   Summary of Effects of Low pH on Biomass Solids

The effects of aqueous, low-pH conditions on biomass ultimately depend on the concentration of the acidifying agent, temperature, and time of the reactions. In general, low-pH aqueous treatments produce biomass solids enriched in cellulose and lignin. Some pseudo-lignin formed from hemicellulose may also be included. The acidic liquid stream or hydrolysate produced contains hemicellulose-derived sugars such as xylose and xylooligomers and associated degradation products such as furfural. Some aromatic monomers derived from lignin may also be detected in the hydrolysate. Although glucose concentrations are generally low for typically favored pretreatment conditions, the amounts of glucose and cellulose degradation products in the hydrolysate will increase as the severity of the reaction conditions is increased.

6.3   Pretreatment in Support of Biological Conversion

6.3.1   Hydrolysis of Cellulose to Fermentable Glucose

Sherrard and Kressman [47] describe how Braconnot discovered the hydrolysis of cellulose to glucose by concentrated acid in 1819. Much of the early work focused on using concentrated acids, such as 40–42% HCl [47] at atmospheric pressure, but in the late nineteenth century Simonsen erected an experimental plant in which cellulose hydrolysis was conducted with 0.5% H_2SO_4 at a pressure of 9 atm ($T_{sat} = 176\,°C$) for 15 minutes [47]. Work was also conducted using sulfurous acid, H_2SO_3, as the hydrolyzing agent. Although a number of similar facilities were constructed during the early twentieth century, these facilities did not operate for long due to numerous technical and commercial difficulties, and ultimately alcohol production from wood was abandoned.

During World War II the demand for ethanol skyrocketed. As the traditional raw material of industrial molasses was scarce, producers began using feedstocks such as wheat flour, corn, sorghum grain, and barley. However, these feedstocks became increasingly difficult to obtain by approximately 1943, so the Chemical Referee Board of the Office of Production Research and Development, War Production Board, recommended that ethanol production from wood be investigated, leading to some of the earliest pretreatment research. These early pretreatments were directed at hemicellulose removal prior to further acid hydrolysis of cellulose to recover 6-carbon sugars for fermentation to ethanol by *Sacchromocyes cerevisiae* [4]. Many of the conditions were selected for complete hydrolysis of cellulose to glucose while allowing pentose degradation, because pentoses were non-fermentable by the available organisms and considered waste [5].

One of the first commercial processes for conversion of cellulosic biomass to ethanol was the Scholler process. Sawdust and wood chips were loaded into brick-lined steel percolators and preheated to 129 °C by steam injection; batches of 0.5% H_2SO_4 were forced through at 1.14–1.24 MPa, as described by Faith [48]. Each batch reaction took approximately 45 minutes [48]. The Madison wood sugar process was a modification of the Scholler process with Douglas fir wood waste first treated with 0.5–0.6% H_2SO_4 at 150 °C for 20 minutes [49]. Additional dilute acid was added as the temperature was increased to 185 °C [49], and the reactor was maintained at this temperature until the completion of the run, typically 2.3–3.0 hours [49]. The resulting sugar solution was continuously removed, and the reducing sugar yields ranged from 35.0% to 49.0% of the theoretical possible maximum [49].

Sulfuric acid was used in both of these commercial examples, but other catalysts were investigated for the complete conversion of cellulose to glucose. Table 6.1 outlines some of the conditions tested. These catalysts were evaluated based on the rate of hydrolysis of cellulose relative to the rate of decomposition of hexose. Phosphoric acid was determined to be a poor catalyst due to the slow rate of hydrolysis and the increase in the rate of glucose degradation [50]. Sulfur dioxide was also found to be a poor catalyst for the hydrolysis of cellulose due to its relatively slow hydrolysis rate [51]. Sulfuric acid and hydrochloric acid

Table 6.1   *Selected conditions for the hydrolysis of cellulose to glucose.*

Agent	Concentration (w/w%)	Temperature (°C)	Time (min)
HCl (Harris and Kline [51])	0.2–3.2	160–190	10–320
H_3PO_4 (Harris and Lang [50])	0.2–3.2	180–195	10–180
SO_2 (Harris and Kline [51])	0.75–3.00	150–180	Various
H_2SO_4 (Harris and Kline [51])	0.04–0.16	170–190	Various
H_2SO_4 (Saeman [86])	0.4–1.6	170–190	0–90

catalysts increased the yields of reducing sugars [51]. Hydrofluoric acid, in both the liquid and vapor phase, was also considered as a catalyst for the production of glucose from lignocellulose [52,53]. Although high yields of glucose were achieved at ambient temperatures and pressures, the hazards and costs of working with hydrofluoric acid on a commercial scale limited interest and research [54].

6.3.2   Pretreatment for Improved Enzymatic Digestibility

Concerns about military equipment rotting in the South Pacific during World War II led to the discovery of cellulase enzymes [55]. Up until the late 1960s, cellulase was studied in order to avoid degradation [56]. However, the combined pressures of municipal waste disposal and the need for alternate fuel sources due to the 1970s energy crisis [57] led researchers to investigate the possibility of producing fermentable glucose from cellulose. The immediate advantages of enzymatic hydrolysis of cellulose to glucose were that only glucose was produced [58] and that its mild hydrolysis conditions required no expensive construction materials. However, it was also immediately clear that enzymatic hydrolysis of native cellulosic feedstocks was slow with low yields [58]. These significant disadvantages prompted researchers to search for pretreatments to increase cellulose accessibility.

Han and Callihan reported one of the first examples of acidic pretreatment for application of a two-stage process to sugarcane bagasse [59]. In this system, bagasse was exposed to 10–50% H_2SO_4 for 15 minutes at 121 °C in the first stage, after which the reaction mixture was diluted to 0.5–2% H_2SO_4 and reheated to 121 °C for 15 minutes to 2 hours [59]. Han and Callihan recognized that this treatment improved performance of subsequent enzymatic hydrolysis but, because the digestibility of acid-treated bagasse by *Cellulomonas* and *Alcaligenes* was much lower than that of alkali-treated material, they concluded that acid pretreatment was not feasible [59]. Nesse *et al.* showed that pretreatment of fiber from feedlot manure using 0.01–3.5% peracetic acid for 1 hour at room temperature increased its digestibility [60]. The pretreatment used by Han and Callihan [59] was relatively severe, while that by Nesse *et al.* [60] was milder. Researchers at the Lawrence Berkeley Laboratory (LBL) found that batch pretreatment with 0.9 w/w% H_2SO_4 at 100 °C lasting up to 5.5 hours significantly improved yields from enzymatic hydrolysis of several agricultural residues including wheat straw, barley straw, rice straw, sorghum straw, and corn stover [61]. Combined glucose and xylose yields (defined as grams of monomer per gram of monomer equivalent in the raw biomass) from the enzymatic hydrolysis of pretreated material were 10–41% higher than yields from the enzymatic hydrolysis of raw biomass [61]. The LBL team also showed that very little acid was consumed during the pretreatment step for most of the tested substrates, which would be commercially beneficial.

Grethlein *et al.* [58], Knappert *et al.* [62], and Grethlein and Converse [63] used a continuous plug flow reactor to pretreat a wide variety of materials including newsprint [57], corn stover [62], oak [62], white pine [58], poplar [63], and mixed hardwood [58]. The concentration of H_2SO_4 was varied from 0.4–1.2% at 160–220 °C with reactor retention times of 6.6–13.2 seconds. Some of these pretreatment conditions were very effective at increasing susceptibility of native materials such as corn stover and oak to enzymatic hydrolysis. It was found that different substrates required different pretreatments in order to achieve high enzymatic hydrolysis yields, with the improvements cautiously attributed to an increase in pore size and surface area. Many processes to increase the accessibility of cellulose to enzymes using acids were also patented; for example Foody patented a process to treat lignocellulosic material with 0.15–1 w/w% H_2SO_4 at 1.8–7.0 MPa [64].

6.3.3   Pretreatment for Improved Enzymatic Digestibility and Hemicellulose Sugar Recovery

The development of microorganisms capable of converting both pentoses and hexoses to ethanol over the past two decades has made it possible to derive value from all the sugars in hemicellulose and cellulose [65]. Consequently, the current paradigm for ethanol production from cellulosic biomass is to recover as much sugar as possible from cellulose and hemicellulose in the combined operations of pretreatment and enzymatic hydrolysis [66]. In line with this objective, successful pretreatments must not sacrifice sugars from hemicellulose, while modifying the remaining solids so that they are susceptible to enzymatic hydrolysis with high yields. It is also critical to limit formation of degradation products that inhibit enzymatic hydrolysis or fermentation [66]. Formation of degradation products also comes at the expense of fermentable sugars and thus ethanol. The goal to maximize sugar recovery is just as important for pretreatment applications to microbes now being developed to convert biomass to fuels other than ethanol such as hydrogen [67,68]. Despite the changing goals of pretreatment, many of the same chemicals such as SO_2 and H_2SO_4 are utilized because they lower pH effectively and are readily available at comparatively low prices. Additionally, SO_2 can be recovered and recycled following pretreatment; however, it is a more hazardous chemical to work with, and the recovery operations would increase capital and operating costs of a commercial operation. There is also limited literature on application of nitric, hydrochloric, or phosphoric acids to prepare biomass for biological conversion [66], and their higher costs could present economic challenges. Attempts have been made to use carbon dioxide as an acidifying agent [69–72] since it is produced during fermentation [73] and would be less corrosive than mineral acids [74]. However, yields from hydrolysis with carbon dioxide fall short of those from H_2SO_4 hydrolysis [74]. Furthermore, Jayawardhana and van Walsum [74] estimated that high-pressure carbon dioxide pretreatment reactors would be more expensive than for dilute acid pretreatment.

Since space considerations prevent inclusion of the complete body of research on low-pH pretreatments, the following sections provide a summary of thoroughly studied and commercially promising processes based on SO_2 and H_2SO_4 pretreatments. Representative works by leading investigators, both individual and institutional, are highlighted so that the reader can easily locate material for more in-depth information.

Pretreatment with Sulfur Dioxide

Sulfur dioxide has been used to treat a wide variety of biomass including softwoods such as Douglas fir [44], agricultural residues such as corn stover [75] and bagasse [18], and hardwoods such as poplar [75]. In most laboratory studies, biomass solids were impregnated with SO_2 at room temperature. After impregnation, the biomass was transferred to a pretreatment reactor, typically a steam explosion device, and injected with steam until the target reaction time was reached. At that point, a blow-down valve was opened to discharge the pretreated solid material and liquid hydrolysate to atmospheric pressure, cooling the materials almost instantly to 100 °C. A flow diagram of a sample experimental procedure and apparatus for SO_2 pretreatment was provided by Stenberg *et al.* [76]. Schell and co-workers provided a detailed process flow diagram of a potential commercial configuration for SO_2 pretreatment [77].

The length of impregnations varied. For example, Martín *et al.* [18] performed impregnations lasting 15–20 minutes while Boussaid *et al.* and Bura *et al.* [44,75] allowed impregnations to continue overnight.

Although most impregnations were performed without wetting the biomass, Öhgren *et al.* [17] and De Bari *et al.* [78] presteamed it prior to impregnation. However, as De Bari *et al.* [78] noted, when impregnated biomass is transferred from the adsorption vessel to the pretreatment reactor, some SO_2 is lost, making it difficult to compare results among researchers. Consequently, De Bari *et al.* [78] examined impregnation of aspen chips with SO_2 and the influence of moisture content using a cus-tom-designed adsorption chamber. The chamber was placed on a high-resolution industrial weighing platform and, after chips were loaded, sufficient SO_2 was added to increase the weight by 4–5% of the aspen mass. The chamber pressure was monitored to estimate adsorption from the compressibility factor equation of state. After impregnation was complete, the biomass was removed from the reactor and the decrease in mass was monitored as excess SO_2 desorbed from the biomass. Once a constant mass was reached, the biomass was pretreated. De Bari *et al.* [78] found that, even after extended impregnation times, only 50% of the available gas was adsorbed with most of it being adsorbed dur-ing the first 15 minutes. It was also found that approximately half of the adsorbed gas was lost during outgassing, leaving an adsorbed concentration of 0.6–0.9 SO_2 w/w% raw, dry biomass. Increasing the biomass moisture content slightly increased the final amount of SO_2 adsorbed. This study [78] aptly illustrated the challenges of controlling SO_2 impregnation and accurately determining the effective SO_2 concentration during pretreatment.

Two-stage pretreatment systems have been used in an attempt to increase total sugar recovery [79]. The first stage was optimized for recovery of hemicellulosic sugars, while the second step was optimized for enzymatic digestibility of biomass. Although it was possible to increase final ethanol yields and decrease enzyme usage through such two-stage pretreatments, it is unclear whether these improvements justify the additional costs and technical challenges.

Table 6.2 reveals that although SO_2 has been applied to a diverse range of biomass types, the pretreat-ment conditions were quite similar: SO_2 concentration ranged from 1.1 to 4.5%, temperatures between 170 and 220 °C, and times from 2 to 10 minutes. However, as De Bari *et al.* [78] demonstrated, the concentra-tion of SO_2 used for impregnation did not accurately reflect the amount of SO_2 adsorbed.

Pretreatment with SO_2 results in the release of hemicellulose sugars and some lignin as well as degrada-tion products at high severities. There are conflicting reports as to the effects of steam explosion with SO_2 on the degree of polymerization of the sugar products. When Boussaid *et al.* [44] and Söderström *et al.* [79] pretreated softwoods such as Douglas fir and *Picea abies* with SO_2, mannose (the primary hemicellulose

Table 6.2   *Selected conditions for pretreatment with sulfur dioxide prior to biological conversion. Yield is defined as gram carbohydrate equivalent in the liquid hydrolysate following pretreatment per gram of carbohydrate equivalent in the raw biomass.*

Substrate (Author)	SO_2 Concentration (%)	Temperature (°C)	Time (min)	Yield (w/w%)	Concentration (g/L)
Douglas fir (Boussaid *et al.* [44])	2.38–4.5	175–215	2.38–7.5	Mannose yield: 22–49%	
Corn stover, poplar (Bura *et al.* [75])	3	170–215	5–9		11.2–23.7 g xylose/L hydrolysate
Sugarcane bagasse (Martín *et al.* [18])	1.1	205	10	Xylan yield: 27%	
Corn stover (Öhgren *et al.* [17])	3	200	5		35.8 g xylose/L hydrolysate
Picea abies (Söderström *et al.* [79])	3	180–220	2–10	2-stage mannan yield: 91–96%	

sugar in softwood) was recovered as a monomer. In contrast, Bura *et al.* [75] found that SO_2 pretreatment of corn stover and poplar produced a lot of oligomers from hemicellulose, and especially xylooligomers. These differences may be due to differences in biomass or in pretreatment conditions. Biomass differences seem the more likely cause as the pretreatment conditions used by Bura *et al.* [75] overlap those used by Boussaid *et al.* [44] and Söderström *et al.* [79].

Several studies observed that little lignin was removed by SO_2 pretreatment [44,75,78]. De Bari *et al.* [78] found that the amount of lignin increased slightly for high-severity pretreatments and attributed this to formation of Klason-lignin-like by-products. A number of phenolic compounds such as p-coumaric acid, ferulic acid, and 4-hydroxybenzaldehyde were also released by SO_2 pretreatment [18].

As the severity of pretreatment was increased, sugars were degraded to 5-hydroxymethylfurfural (5-HMF-), furfural, and other non-sugar compounds [18,44,75,78,79]. Interestingly, Martín *et al.* [18] found that the concentrations of furfural, 5-HMF, levulinic acid, and formic acid in the hydrolysate from SO_2-catalyzed steam explosion were similar to concentrations found in the hydrolysate from uncatalyzed steam explosion, and the xylose yields from SO_2 pretreatment were only slightly higher than from steam explosion. This outcome may be because the effective SO_2 concentration decreased as biomass was transferred to the reactor or because SO_2 does not significantly accelerate degradation reactions.

Pretreatment with Sulfuric Acid

Dilute sulfuric acid has been by far the most common acid catalyst used for biomass pretreatment prior to biological conversion. Like SO_2, it has also been used to pretreat a wide variety of biomass types. Biomass is frequently soaked in dilute acid prior to pretreatment, and the reactors used for pretreatment with H_2SO_4 are almost as diverse as the types of biomass tested. Various possible commercial processes based on dilute sulfuric acid pretreatment have been designed over the years, with a recent design by Humbird *et al.* being one example [80]. In the case of biological conversion of the cellulose in biomass to glucose, a key research goal has been to reduce the amount of H_2SO_4 used for pretreatment.

Batch reactors are commonly applied for dilute acid pretreatments, and both unstirred reactors and stirred reactors have been used. For example, Lloyd and Wyman employed both reactor types to pretreat corn stover [26]. Small tube reactors typically have an inner diameter of about 10.8 mm to reduce temperature non-uniformity [81,82] with a length of about 100 mm, allowing reasonable quantities to be processed. A larger stirred reactor with a volume of 1.0 L and an 88.9 mm helical-impeller-provided agitation was also used by Lloyd and Wyman [26]. The reactors were loaded with a corn stover slurry at 5% solids with 0.22–0.98% H_2SO_4 by weight in the water. Pretreatment temperatures varied from 140 to 180 °C for times up to 80 minutes. After pretreatment, the solids were subjected to enzymatic hydrolysis with cellulase supplemented with beta-glucosidase. Lloyd and Wyman [26] observed significant production of xylooligomers, especially for short pretreatment times, but the fraction of xylooligomers relative to total xylose release decreased as the concentration of acid increased. This paper illustrated the classic conundrum of dilute acid pretreatment: how to maximize sugar yields from enzymatic hydrolysis favored by long reaction times, while minimizing degradation reactions during pretreatment that occur at long reaction times. In this application, Lloyd and Wyman [26] showed that the maximum yields of glucose, xylose, and glucose plus xylose together in pretreatment and enzymatic hydrolysis did not occur for the same pretreatment conditions. When selecting pretreatment conditions, it is therefore important to maximize release of all relevant sugars from pretreatment, enzymatic hydrolysis, and all subsequent operations.

Dilute sulfuric acid has also been used in steam explosion systems. Sassner *et al.* [83] used such reactors to pretreat wood chips from a *Salix* hybrid that were presoaked in 0.25–0.5% H_2SO_4 for at least 90 minutes. The solids were recovered by filtration and then transferred to a 10 L steam explosion unit, with pretreatment temperatures ranging from 180 to 210 °C and times from 4 to 12 minutes. The liquids and solids from pretreatment were subjected to fermentation and enzymatic hydrolysis, respectively. Additionally, simultaneous saccharification and fermentation (SSF) of various dilutions of the pretreated *Salix* hydrolysate slurry was performed. During SSF, biomass was combined with enzymes and a fermentative microbe, such as yeast, so that monomers produced by enzymatic saccharification were fermented to ethanol soon after release to reduce inhibition of the enzymes by sugars. One unique feature of this work was that collection and analysis of the exhaust gases and hydrolysate from pretreatment revealed high concentrations of furfural and acetic acid. Sassner *et al.* found that xylose degradation decreased when the initial moisture content of the chips was increased [83]. A high initial moisture content increased steam condensation and appeared to improve fermentations by diluting potential inhibitors [83]. The pretreated solids were easily digested during enzymatic hydrolysis despite retaining 74–95% of the lignin in the raw biomass [83].

Cahela *et al.* [84] pretreated southern red oak in a percolation reactor with 0.2% H_2SO_4. The reactor had an inner diameter of 25.4 mm and a length of 627 mm and could be loaded with 120–140 g of red oak sawdust. As described in greater detail in Section Xylose Production, they also developed a model of hemicellulose hydrolysis for a packed bed that accounted for diffusion of products from the interior of the biomass particle to the bulk liquid. In general, their experimental results and model predictions agreed within 10%, with discrepancies primarily attributed to difficulties in determining the true H_2SO_4 concentration due to the buffering effects of ash in the wood.

Mok *et al.*'s [85] work provided a second example of percolation pretreatment with H_2SO_4; however, in this work the objective was glucose recovery. The reactor consisted of two chambers: the primary chamber with a diameter of 4.6 mm and a length of 76 mm used to hold the solid substrate, and a secondary chamber with variable volume to study the influence of liquid phase reactions. Whatman no. 1 and no. 4 filter papers were used as substrates. The flow rate of 0.05% H_2SO_4 was varied from 2 to 4 mL/min, and the pretreatments lasted from 0 to 60 minutes at 190–225 °C. Although the classic Saeman model of cellulose hydrolysis [86] (see Section 6.1.1) predicted that decreasing the volume of the secondary reaction chamber would increase the yield of glucose by limiting degradation reactions, Mok *et al.* found that the glucose yield decreased; this led them to propose that cellulose hydrolysis first generates oligomers that are further hydrolyzed to glucose [85]. Mok *et al.* also suggested that the increase in glucose yield with increasing flow rate was due to the removal of soluble products prior to degradation [85].

As with the use of SO_2, the conundrum of minimizing xylose degradation during pretreatment while achieving high digestibility of cellulose in the pretreated solids can be addressed by two-stage pretreatments with dilute sulfuric acid. Nguyen *et al.* [87] tested one such configuration using a mixture of white fir (*Abies concolor*) and Ponderosa pine (*Pinus ponderosa*). Softwood chips were soaked in 0.6–2.4% H_2SO_4 for 4 hours at 60 °C and then added to a 4 L steam explosion reactor. During the first pretreatment, the temperature was varied from 180 to 215 °C for 1.6–4 minutes, giving a combined severity for the first stage pretreatment ranging from 2.18 to 3.26. At the completion of this time, the solids were washed to remove solubilized hemicellulose sugars and then soaked in 2.5% H_2SO_4 for 3 hours at ambient temperature before being treated at 210 °C for 1.6–2 minutes. The solids from this two-stage approach were employed in both enzymatic hydrolysis and SSF. The total sugar yields from two-stage pretreatment followed by enzymatic hydrolysis was slightly higher than the total sugar yields from single-stage pretreatment followed by enzymatic hydrolysis. In addition, two-stage pretreatment could potentially reduce enzyme usage as the cellulose content in the solids from the second pretreatment stage was lower than for one-stage pretreatment. However, as with the two-stage SO_2 pretreatment, it is unclear that the performance gains from two-stage pretreatment justify the additional costs.

Pretreatment conditions reviewed in this section are summarized in Table 6.3.

Table 6.3   *Selected conditions for pretreatment with sulfuric acid prior to biological conversion. Yield is defined as gram carbohydrate equivalent in the liquid hydrolysate following pretreatment per gram of carbohydrate equivalent in the raw biomass.*

Substrate (Author)	Concentration	Temperature (°C)	Time (min)	Yields
Southern red oak (Cahela *et al.* [84])	0.037–0.056 w/v%	140–160	*c.* 14–115	Xylan yield: *c.* 8.8–88%
Whatman paper no. 1 and 42 (Mok *et al.* [85])	5 mM	190–225	0–60	Glucose yield: *c.* 35–85%
Corn stover (Lloyd and Wyman [26])	0.22–0.98%	140–200	0–80	Maximum xylose yields of 71–85%
White fir and Ponderosa pine (Nguyen *et al.* [87])	0.6–2.5%	180–215	1.7–4	2 stage mannose+ galactose+ xylose+ arabinose yield: 84%
Salix hybrid (Sassner *et al.* [83])	0.25–0.5 w/w%	180–210	4–12	Xylose yield: *c.* 55–75%

6.4   Low-pH Hydrolysis of Cellulose and Hemicellulose

There has been growing interest in producing "drop-in" hydrocarbon fuels from biomass, that is to say, hydrocarbons that can be easily integrated with today's motor vehicles and airplanes. Additionally, since many of today's chemical feedstocks are derived from petroleum, there is a need to generate alternative chemical feedstocks from biomass. Although pretreatment traditionally refers to the preparation of biomass for biological conversion, researchers are now also applying low-pH reactions to produce reactive intermediates for catalytic chemical conversion to fuels and chemicals. In a sense, these reactions are also "pretreatments" and certainly share similar features to historical pretreatment technologies. However, the pretreatment conditions to support catalytic conversion tend to be much more varied due to the wider range of downstream processing objectives.

6.4.1   Furfural

One of the oldest examples of industrial chemical production from biomass is furfural manufacture from the xylan and arabinan in hemicellulose. Furfural is currently used for such applications as resins, linking foundry sand, lubrication oil extraction, and nematicides [88], but could be used to produce a wide variety of other chemicals. The first industrial production of furfural used oat hulls as a feedstock to take advantage of its high xylan content [88], and major industrial feedstocks today are corn cobs and bagasse [24].

Furfural can be produced by a one- or two-stage process [89]. In the one-stage process, biomass is combined with approximately 3% H_2SO_4 by weight in a slurry [90], and steam is introduced to bring the reactor to the desired temperature. The hemicellulose is hydrolyzed to xylose and arabinose, which in turn are dehydrated to furfural [91]. Furfural is continuously removed from the reactor in the vapor phase to reduce decomposition and recondensation reactions [89]. The reactor is typically held at 170–185 °C for 3 hours, and the process results in furfural yields of approximately 40–50% of the theoretical maximum. One of the challenges in this process is the rapid recovery of furfural, but the efficiency of recovery by steam injection is limited due to the boiling point elevation in the reactor as biomass components are solubilized. The two-stage process attempts to separate hemicellulose hydrolysis and xylose cyclodehydration reactions, with hemicellulose hydrolysis conducted at milder conditions to generate a pentose-rich liquid stream for dehydration. By separating the two reactions, it is also possible to produce a cellulose-lignin substrate that can also be converted to chemicals or fuel. A review by Mamman *et al.* [90] identified a number pretreatments resulting in high xylose yields, with some of these summarized in Table 6.4.

Table 6.4   *Selected pretreatment conditions for high xylose yields in support of furfural production [90].*

Biomass	w/w% H_2SO_4	Temperature (°C)	Time (minutes)
Oil palm empty fruit bunch	4	115	60
Corn fiber	0.75	121	30
Switchgrass	0.5	140–160	10–60

A new field of study in the production of furfural is the use of chloride salts to increase the xylose production rate [92]. Marcotullio and De Jong [92] showed that furfural yields and selectivity increased when chloride salts such as NaCl and $FeCl_3\cdot 6H_2O$ were added to the cyclodehydration of xylose with dilute hydrochloric acid; the addition of salts increased the reaction rate even for low acid concentrations. Of the salts tested, $FeCl_3\cdot 6H_2O$ seemed particularly promising due to a dramatic increase in xylose reaction rates. Although details of the mechanism for the conversion of xylose to furfural are unresolved, Marcotullio and De Jong [92] hypothesized that chloride ions promote formation of 1,2-enediol that may be an important intermediate to furfural production.

6.4.2   Levulinic Acid

Levulinic acid is a chemical that could be made from the cellulose fraction of biomass for use as a feedstock to make diesel and gasoline additives. Rackemann and Doherty recently provided a thorough review of the uses and production of levulinic acid from biomass [93]. The Biofine process was developed to produce levulinic acid from a cellulosic biomass feed [8], and a number of different substrates were tested including waste paper, waste wood, and agricultural residues. In this case, solids were combined with 2–5% H_2SO_4 at ambient temperature and then pumped into a short continuous plug flow reactor, held at 215 °C and 3.1 MPa(g) with a residence time of 15 seconds, to release glucose. The slurry was then pumped to a continuous stirred tank reactor at 193 °C and 1.4 MPa(g) with a residence time of 12 minutes. Slow conversion of sugars to levulinic acid occurred during this stage; the process generated approximately 0.5 kg levulinic acid/kg cellulose. Furfural and formic acid were also produced.

6.4.3   Drop-in Hydrocarbons

An emerging processing paradigm is to pretreat biomass in order to release reactive intermediates that can be catalytically converted to drop-in hydrocarbons. The objectives of pretreatment in support of these catalytic processes are to maximize intermediate yields and avoid use or generation of catalyst poisons [94]. One example of pretreatment for this type of process was presented by Li *et al.* [95] in which maple wood was pretreated in 0.5% H_2SO_4 in a steam gun at temperatures of 160–180 °C for 10–30 minutes [95]. They then subjected the resulting carbohydrate-rich liquid stream to low-temperature hydrogenation to produce sorbitol and xylitol, which were then converted to gasoline range hydrocarbons. Both of these steps used heterogeneous catalysts. The cellulose- and lignin-rich solids from pretreatment could then be used in other processes such as ethanol production [95].

Because mineral acids could deactivate downstream metallic catalysts [95], two possible solutions have been employed to address this challenge. In the first, the pretreatment liquor is neutralized prior to catalysis but, as discussed in Section 6.1, this approach is not ideal. The second alternative is to pretreat biomass with an organic acid such as oxalic acid ($C_2H_2O_4$) [94]. For example, Zhang *et al.* [94] applied pretreatments lasting 5–60 minutes at 160 °C using H_2SO_4, HCl, and $C_2H_2O_4$ at concentrations of 0.5–2%, and found that $C_2H_2O_4$ pretreatments resulted in slightly higher carbon recoveries than pretreatment with mineral acids

[94,95]. In general, the highest xylose monomer recovery was achieved at low acid concentrations. For example, Li *et al.* [95] reached their maximum xylose recovery using 0.5% $C_2H_2O_4$ at 180 °C for 10 minutes. At 160 °C, Zhang and Wyman's maximum total xylose recoveries, defined as xylose equivalent of monomers and oligomers in the liquid phase as a percentage of xylose equivalents available in the raw biomass, were 84.4%, 73.8%, and 87.5% with 0.5% H_2SO_4 for 30 min, 0.5% HCl for 10 min, and 0.5% $C_2H_2O_4$ for 30 min, respectively [94]. The higher cost of $C_2H_2O_4$ relative to mineral acids could however limit its use as a catalyst, but no process designs have yet been applied to estimate the tradeoffs for use of oxalic vs. mineral acids. Furthermore, the impact of pretreatment acid type on downstream operations has not been fully investigated.

Recycling $C_2H_2O_4$ will be likely key to its use on a commercial scale. Vom Stein *et al.* fractionated beechwood using a biphasic system of an aqueous phase containing wood and oxalic acid in contact with 2-methyltetrahydrofuran (2-MTHF) at temperatures of 125–150 °C [96]. Little $C_2H_2O_4$ reacted at temperatures between 125 °C and 140 °C, but only 70% of $C_2H_2O_4$ was recovered after reaction at 150 °C, suggesting side reactions may be significant at this temperature. They also found that $C_2H_2O_4$ could be recovered by crystallization after reaction and re-used [96]. However, further work is required to fully evaluate the advantages and disadvantages of using $C_2H_2O_4$ with and without recycle relative to mineral acids.

Another biomass conversion system based on dilute acid pretreatment is the production of gasoline-compatible hydrocarbons via simultaneous hydrolysis and hydrogenation. In this configuration, a metallic catalyst in an acidic aqueous solution with a hydrogen headspace promotes biomass reactions. The reactor is heated to initiate biomass hydrolysis to monomeric sugars, which are then immediately converted to sugar alcohols such as sorbitol over the metallic catalyst for conversion into hydrocarbons in downstream operations. In a study by Robinson *et al.* [97], idealized substrates and native biomasses such as switchgrass were treated in stirred batch reactors with ruthenium catalyst on a carbon support [97]. Solutions with 0.7% H_2SO_4 and 0.35–1.5% H_3PO_4 were tested at temperatures in the range 160–193 °C over a time period of 3–17 hours. A more recent example of simultaneous hydrolysis and hydrogenation was reported by Palkovits *et al.* [98] in which they tested idealized substrates such as α-cellulose and native spruce. The catalysts Pt-C, Pd-C, and Ru-C were applied in 0.5–2.5% H_3PO_4 and H_2SO_4 at 160 °C for 1–5 hours to produce 5- and 6-carbon alcohols as well as glucose and xylose. Higher conversion of cellulose was obtained using H_2SO_4 compared to phosphoric acid, likely due to the higher pKa value of H_2SO_4 [98]. Interestingly, this study showed that although overall conversion using the Ru-C catalyst was low in comparison to Pt and Pd catalysts, the Ru-C catalyst gave higher yields of desired products. Additionally, production of xylose and glucose were greater in the presence of the heterogeneous catalysts than traditional acid hydrolysis.

6.5   Models of Low-pH Biomass Reactions

For almost as long as biomass hydrolysis has been studied, there have been efforts to develop kinetic models of the system. Such models have been difficult to develop due to the complexity of the system (the solid–liquid interactions of the biomass and the aqueous phase; the challenges of determining the effective acid concentration during reaction; interactions among cellulose, hemicellulose, and lignin) and the complexity of the composition of these fractions. In addition to the complex reaction scheme, there are also the challenges in assessing and modeling mass and heat transfer within the reactor and biomass. Finally, the diversity and range of biomass and associated chemical bonds within hemicellulose and with lignin likely limit the extent to which models can be accurately applied. However, it is vital to address these challenges because of the utility of kinetic models in research and industrial production. Models provide a framework to test hypotheses in an efficient and targeted manner. Reliable kinetic models are also vital to the scaling-up of pretreatment from the lab to commercial production. Finally, models assist in determining optimum biomass feedstock and processing configurations.

6.5.1   Cellulose Hydrolysis

Glucose Production

As the early goal of low-pH biomass reactions was cellulose hydrolysis to glucose, much of the kinetic modeling literature of that time focused on acid-catalyzed hydrolysis of cellulose via the proton-catalyzed cleavage of the glycosidic bond [99]. One of the first models was based on the following series of first-order pseudo-homogeneous reactions by Saeman [86]:

$$\text{cellulose} \xrightarrow{k_1} \text{glucose} \xrightarrow{k_2} 5-\text{HMF} \qquad (6.2)$$

where

$$k_i = H_i C_a^M \exp\left(\frac{-\Delta H_{a,i}}{RT}\right) \qquad (6.3)$$

where k_i is the reaction rate constant (min^{-1}) for reaction i, H_i is a constant, C_a is the concentration of H_2SO_4 (%) and M is the reaction order, $\Delta H_{a,i}$ is the activation energy, R is the universal gas constant, and T is the absolute temperature. This model has been the basis of almost every cellulose and hemicellulose hydrolysis model since. Saeman [86] applied his model to a variety of substrates including red oak, Douglas fir, hard maple, and aspen and found that hydrolysis rates did not differ by more than 20%. These differences could be partially explained by differences in buffering capacity of the wood: substrates with the lowest ash content were found to have the highest hydrolysis rates. Saeman [86] also studied hydrolysis of Douglas fir at a variety of temperatures and acid concentrations and found that the activation energy was independent of acid concentration, while the rate constant increased by 153% for a 100% increase in acid concentration. Definition of acid concentration C_a has been challenging in part due to the neutralizing capacity of biomass. In the past, C_a has been expressed in terms of mass, molarity, and pH at room temperature [100], and variations in definition may help explain the wide variation in kinetic rate constants for hydrolysis. Lloyd and Wyman [100] demonstrated that the neutralizing capacity of biomass and the temperature have a significant influence on the pH of a system, a fact well worth considering in future modeling.

One of the first modifications to Saeman's model was adjustment of his assumption that the initial glucose concentration was zero. In particular, amorphous cellulose hydrolyzed quickly enough to be included as an initial glucose concentration [101]. Most subsequent modifications have added decomposition reactions and parallel reactions. For example, Conner *et al.* [102] added reversible formation of levoglucosan from glucose and disaccharides, and Bouchard *et al.* [103] found evidence for a parallel pathway that modifies cellulose to a structure that cannot be hydrolyzed to glucose. Abatzoglou *et al.* [104] added formation of glucoligomer intermediates as a sequential step to Saeman's model. Mok *et al.* [85] searched for additional evidence of these phenomena and, after eliminating the possibilities that chemical alteration of residual solid cellulose or glucose degradation reactions were responsible for limiting glucose yield, they concluded that unknown products that could not be hydrolyzed to glucose were produced during pretreatment.

5-HMF and Levulinic Acid Production

Levulinic acid is produced from the dehydration of hexose sugars. Production from biomass is based on cellulose first being hydrolyzed to glucose, which then undergoes dehydration to 5-HMF and its reaction to levulinic acid. However, many significant side reactions also occur, lowering the final yield of levulinic acid. Several models have been developed to describe the formation of levulinic acid from glucose [105–107]. Chang *et al.* modeled glucose conversion to 5-HMF and on to levulinic acid

(LA) as a series of first-order unimolecular reactions and incorporated parallel reactions for conversion of glucose and 5-HMF to humic solids [108]:

$$\text{biomass} \xrightarrow{k_1} \text{sugar} \xrightarrow{k_2} \text{5-HMF} \xrightarrow{k_3} \text{LA} \tag{6.4}$$

$$\text{sugar} \xrightarrow{k_4} \text{humic and unidentified products} \tag{6.5}$$

$$\text{5-HMF} \xrightarrow{k_5} \text{humic and unidentified products} \tag{6.6}$$

Girisuta *et al.* applied a similar mechanism [106]. Assary *et al.* developed quantum mechanics models for glucose conversion to levulinic acid [107]. They also showed that the first two steps, tautomerization of α-D-glucose to β-D-fructose and dehydration of β-D-fructose to an intermediate, were endothermic and indicated that initial dehydration of β-D-fructose was the rate-limiting step [107]. After developing models for glucose conversion to levulinic acid, Girisuta *et al.* developed a kinetic model for conversion of purified, crystalline cellulose to levulinic acid [109]. This model was then adapted to conversion of biomass such as water hyacinth leaves by including conversion of galactose released from hemicellulose and adding a correction factor to account for differences in cellulose properties [110]. The correction factor for conversion of hexoses to levulinic acid was found to be less than 1, indicating that the conversion of cellulose in biomass is lower than that of pure cellulose. It was also shown that the correction factor for humin formation from hyacinth leaves was approximately 2, indicating that the rate of humin production from hyacinth leaves was greater than the rate of production from pure cellulose. Chang *et al.* developed a model for levulinic acid production from wheat straw based on first-order unimolecular reactions with a power law dependence for acid concentration [108]. It was found that the reaction order of the acid concentration ranged from 0.620 to 1.434.

6.5.2   Hemicellulose Hydrolysis

Xylose Production

In general, hemicellulose hydrolysis models are less well-developed than those for cellulose; hemicellulose hydrolysis has been of interest for a shorter period of time and the composition and structure of hemicellulose is more complex than that of cellulose. Although many hemicellulose hydrolysis models were simply adaptations of Saeman's approach to describe cellulose deconstruction, Kobayashi and Sakai [111] assumed that xylose was released from fast and slow reacting fractions, as shown in Equation (6.7):

$$
\begin{array}{c}
\text{Fast hemicellulose} \searrow k_1 \\
\qquad\qquad\qquad \text{Xylose} \xrightarrow{k_3} \text{Degradation products} \\
\text{Slow hemicellulose} \nearrow k_2
\end{array}
\tag{6.7}
$$

However, each reaction was still modeled using the first-order pseudo-homogeneous system employed by Saeman. Jacobsen and Wyman [101] reported seven examples of this model. Hemicellulose hydrolysis models have also incorporated production of xylooligomer intermediates [101], and parallel reactions of other hemicellulose constituents such as acetyl groups have also been included [16].

The combined severity factor, Equation (6.1), was an important development by Abatzoglou *et al.* [2] and Chum *et al.* [3]. This parameter facilitates comparison of the combined effects of temperature, time, and acid concentration and tradeoffs among them with reaction conditions. It is important to acknowledge that although the combined severity parameter can be useful in comparing datasets collected at different temperatures and acid concentrations, it cannot reliably predict specific performance as Lloyd and Wyman [26] demonstrated with their study of dilute acid pretreatment of corn stover.

Several studies have attempted to include the effects of diffusion in biomass particles on hemicellulose hydrolysis [42,112]. Tillman and co-workers experimentally determined the diffusivity of H_2SO_4 in the longitudinal and radial directions of hardwood and showed that the diffusivity in the radial direction was much larger [112]. The diffusivity coefficient was then combined with a biphasic hemicellulose hydrolysis model to predict the spatial dependence of xylose production within hardwood particles of increasing size. The results showed that, as the reaction temperature increased, xylose yields dropped due to incomplete acid diffusion [112]. Kim and Lee expanded this work to study the transport properties of H_2SO_4 in sugar cane bagasse, corn stover, rice straw, and yellow poplar [42]. Their results showed that diffusivities in agricultural residues were significantly larger than in yellow poplar and that the diffusion of acid into biomass could significantly impact hemicellulose hydrolysis results depending on particle size, reaction temperature, and reaction time.

As stated in Section 6.3.3, Cahela *et al.* [84] developed a model to account for the diffusion of reaction products out of biomass particles. Hemicellulose was assumed to hydrolyze to xylose and xylooligomers, which then degraded to furfural via first-order homogeneous reactions. They refined their model by differentiating between the xylose concentration within the pores of the biomass particles and the xylose concentration in the bulk liquid, and the resulting coupled differential equations were solved as the reaction time became large. From this model, they predicted the maximum xylose yields and associated concentrations as well as operating conditions required to achieve these results. They also found that intraparticle diffusion of xylose oligomers could be important if the longitudinal chip dimension was greater than 4.2 mm. Because this represents a relatively small particle, mass transfer effects could well be of consequence in industrial operations with large biomass particles. Hosseini and Shah [113] took this analysis even further with a more detailed kinetic mass-transfer model that attempted to account for differences in xylooligomer reactivity by assuming that bond breakage is a function of position in the xylooligomer chain. Predictions from this approach strongly correlated with experimental data for hydrolysis of xylooligomers with a degree of polymerization less than or equal to five. Consideration of diffusion of individual oligomer products out of biomass particles showed that the concentrations of xylooligomers within the chip predicted by the model were very sensitive to the assumed value of the diffusion coefficient. This work, as well as others, demonstrated the importance of considering xylooligomer intermediates in kinetic modeling.

The effect of temperature gradients within biomass particles or reactors has also been incorporated into some models. Abasaeed *et al.* [114] and Abasaeed and Mansour [115] provided two examples of the effects of temperature gradients within biomass particles on the results of cellulose hydrolysis. The former applied a Saeman-type model to the hydrolysis of cellulose in southern red oak, determined the thermal diffusivities of southern red oak chips saturated with water, and then simultaneously solved the mass and energy balances of a wood chip. Their models showed that increasing the particle size reduced the maximum achievable glucose yield relative to that predicted based on assumed isothermal conditions, and that it took longer to reach this maximum. Increasing temperature and acid concentration exacerbated the effects of non-isothermal operation. Abasaeed and Mansour [115] modeled the effects of non-isothermal conditions in wood chips using three cellulose hydrolysis models from the literature and came to similar conclusions.

Stuhler and Wyman [82] examined the effects of temperature gradients within tubular batch reactors and applied a parameter β to represent the rate of xylan hydrolysis relative to the rate of heat conduction in the radial direction in a batch tubular reactor. They found that, when a radial temperature gradient developed within the reactor, there was a substantial reduction in xylan hydrolysis and that the erroneous assumption of isothermal conditions introduced significant errors in predicting xylan conversion. From these studies, it is clear that thermal gradients within biomass particles and reactors can have considerable impact on product yields. Heat transfer therefore must be carefully considered for experimentation, modeling, and commercialization.

Furfural Production

In the previous section, researchers were primarily focused on models that could be used to optimize xylose recovery. In this section, however, the target is primarily to maximize furfural production. However, cyclo-dehydration of xylose to furfural is accompanied by numerous side reactions that make it complex to model. As a result, greater emphasis on furfural consuming reactions is needed to predict furfural concentrations.

The simplest kinetic models of furfural destruction assumed that furfural degrades to decomposition products according to a pseudounimolecular reaction [116,117]:

$$F \rightarrow D \tag{6.8}$$

In these studies, the rate of disappearance of furfural was found to be first order with respect to furfural concentration. Williams and Dunlop [116] applied a rate law for furfural disappearance at 150–210 °C in 0.1 M H_2SO_4 to show that the rate of furfural disappearance at 160 °C in 0.05 M HCl and 0.1 M H_2SO_4 were very close, leading them to postulate that the rate of destruction of furfural was first order with respect to hydrogen ion concentration. However, because the hydrogen ion concentration was high, it remained essentially constant during a run.

Weingarten *et al.* [25] developed a slightly more detailed reaction scheme for furfural production from xylose:

$$\text{xylose} \xrightarrow{k_1} \text{furfural} + 3H_2O \tag{6.9}$$

$$\text{xylose} + \text{furfural} \xrightarrow{k_2} D_1 \tag{6.10}$$

$$\text{furfural} \xrightarrow{k_3} D_2 \tag{6.11}$$

Each reaction was modeled as first order with respect to the reactants and hydrogen ion concentration. One of the most detailed models of furfural production from xylose was developed by Antal *et al.* who tracked production and destruction of 11 different compounds [23]. Through experimentation and modeling, they determined the open-chain xylose isomer was rapidly converted to undesirable products such as formic acid while the xylopyranose ring underwent dehydration to furfural. Unfortunately xylopyranose also reacted to undesirable products.

Nimlos *et al.* [118] subsequently developed quantum mechanics models of the energetics of these reactions to estimate transition states and energy barriers associated with three different reaction schemes for xylose decomposition to furfural. Based on energy barriers, Nimlos *et al.* [118] concluded that the model of Antal *et al.* [23] for the protonation of xylopyranose followed by dehydration to furfural was the most likely furfural production mechanism.

6.5.3   Summary of Kinetic Models

Given the highly empirical nature of the kinetic models that have been developed to date, anyone planning to apply kinetic models to biomass deconstruction with dilute acids would be well advised to employ a model developed for similar biomass and hydrolysis conditions. In this regard, Table 6.5 summarizes some of the biomass types, hydrolysis conditions, and general model types that have been applied. However, because an exhaustive table is beyond the scope of this chapter, it should serve merely as a starting point for the interested reader. In addition, in light of their empirical nature, it is vital to confirm models with data collected at relevant conditions to validate their accuracy.

Table 6.5   *Kinetic models of acid hydrolysis by biomass type and hydrolysis conditions.*

Source	Biomass	Temperature (°C)	Acidifying agent	Model description
Abasaeed et al. [114]	Southern red oak	198–215	1–3% H_2SO_4	Sequential hydrolysis of cellulose with intraparticle heat transfer
Abasaeed and Mansour [115]	Cellulose	T=180	0.4–2% H_2SO_4	Sequential hydrolysis of cellulose with intraparticle heat transfer
Abatzoglou et al. [104]	Cellulose	200–240	0.2–1.0 w/w% H_2SO_4	Sequential hydrolysis of cellulose with oligomer intermediates
Abatzoglou et al. [2]	Corn stalk, alfa, *Populus tremuloides, Betula papyrifera*	100–240	0–1.8% H_2SO_4	Severity parameter to predict xylan conversion
Aguilar et al. [16]	Sugar cane bagasse	100–128	2–6% H_2SO_4	Sequential hydrolysis of glucan and xylan; models for acetic acid and furfural
Antal et al. [23]	Xylose	250	0–20 mM H_2SO_4	Detailed model of xylose dehydration
Assary et al. [107]	Glucose			Quantum mechanics modeling of glucose decomposition to levulinic acid
Brennan and Wyman [119]	Corn stover	180	0.5–1.0 w/w% H_2SO_4	Mass transfer only models
Cahela et al. [84]	Southern red oak	140–160	0.037–0.056 w/v% H_2SO_4	Sequential hydrolysis of xylan with mass transfer effects
Canettieri et al. [22]	*Eucalyptus grandis*	130–160	0.65% H_2SO_4	Biphasic, sequential hydrolysis of xylan
Carrasco and Roy [120]	Corn stover, poplar, wheat straw, bagasse, paper birch	80–260	0.5–4 w/w% H_2SO_4	Biphasic, sequential hydrolysis of xylan with oligomer intermediates
Chang et al. [105]	Glucose	170–210	1–5% H_2SO_4	Sequential destruction of glucose to levulinic acid
Chang et al. [108]	Wheat straw	190–230	1–5 w/w% H_2SO_4	Sequential destruction of cellulose to levulinic acid with parallel degradation path
Chum et al. [3]	*Populus tremuloides*	125–145	0.2–1.7 w/w% SO_2	Severity parameter to predict xylan conversion
Converse et al. [121]	90% birch, 10% maple	160–265	0.2–2.4 w/w% H_2SO_4	Sequential hydrolysis of glucan and xylan
Esteghlalian et al. [43]	Corn stover, poplar, switchgrass	140–180	0.6–1.2 w/w% H_2SO_4	Biphasic, sequential hydrolysis of xylan
Girisuta et al. [106]	Glucose	140–200	0.05–1 M H_2SO_4	Sequential decomposition of glucose to levulinic acid
Girisuta et al. [109]	Microcrysalline cellulose	150–200	0.05–1 M H_2SO_4	Sequential decomposition of cellulose to levulinic acid
Girisuta et al. [110]	Water hyacinth	150–175	0.1–1 M H_2SO_4	Sequential decomposition of cellulose to levulinic acid; correction factor for biomass matrix
Hosseini and Shah [113]	Hemicellulose	T=160C	H_2SO_4	Xylooligomer depolymerization
Jacobsen and Wyman [101]	Review of cellulose and hemicellulose hydrolysis models			
Jensen et al. [27]	Mixtures of switchgrass, balsam, red maple, aspen, basswood	175	0.5 w/w% H_2SO_4	Biphasic, sequential hydrolysis of xylan

(continued)

Table 6.5   (*Continued*)

Source	Biomass	Temperature (°C)	Acidifying agent	Model description
Kobayashi and Sakai [111]	*Fagus crenata* Blume	74–147	1–16% H_2SO_4	Biphasic, sequential hydrolysis of xylan
Lloyd and Wyman [122]	Corn stover	140	0.68–1.0% H_2SO_4	Depolymerizaton of xylan; includes oligomer intermediates
Lu and Mosier [123]	Corn stover	150–170	0.05–0.2 M $C_4H_4O_4$	Saeman sequential hydrolysis of xylan; biphasic, sequential hydrolysis of xylan
Maloney *et al.* [21]	Paper birch	100–170	0.04–0.18 M H_2SO_4	Biphasic, sequential hydrolysis of xylan; release of acetyl groups
Mok *et al.* [85]	Whatman paper no. 1 and 42	190–225	5 mM H_2SO_4	Sequential hydrolysis of cellulose
Morinelly *et al.* [124]	Aspen, balsam, switchgrass	150–175	0.25–0.75 w/w% H_2SO_4	Sequential hydrolysis of xylan; includes oligomer intermediates
Nimlos *et al.* [118]	Xylose	160	0.2 M H_2SO_4	Quantum mechanics model of xylose dehydration
Rose *et al.* [125]	Furfural	150–169	0.1 M HCl	Pseudounimolecular destruction of furfural
Saeman [86]	Cellulose	170–190	0.4–1.6% H_2SO_4	Sequential hydrolysis of cellulose and monomers
Tillman *et al.* [112]	Aspen	95–169	0.83 w/w% H_2SO_4	Biphasic, sequential hydrolysis of xylan with mass transfer of acid
Weingarten *et al.* [25]	Xylose	150–170	0.1 M HCl	Parallel dehydration of xylose
Williams and Dunlop [116]	Furfural	50–300	0.05–0.10 M H_2SO_4	Pseudounimolecular destruction of furfural

6.6   Conclusions

Biomass hydrolysis at low pH has played an important role in the long history of biomass conversion to fuels and chemicals. Low-pH reactions have been used to prepare a wide variety of biomass types including agricultural residues, grasses, hardwoods, and softwoods for subsequent conversion by biological and chemical routes as well as to make sugars, furfural, and other products directly. Acidifying agents and concentrations as well as reactor types, temperatures, and times have all been adjusted to accommodate differences in biomass types and downstream operations and objectives. Although downstream operations can vary considerably, the primary pretreatment goals are to maximize product yields, generate reactive intermediates such as enzymatically digestible solids, sugars, or furfural, and avoid generation of compounds that would negatively influence downstream operations.

As the emphasis in biomass hydrolysis evolved from the recovery of sugars from cellulose to the removal of hemicellulose to prepare biomass for subsequent acid-catalyzed cellulose hydrolysis and later enzymatic hydrolysis of cellulose, pretreatment objectives have shifted from the recovery of glucose for fermentation to the recovery of hemicellulose-based sugars and modification of cellulose for enzymatic hydrolysis. Sulfur dioxide and sulfuric acid have been the most-studied catalysts for biomass pretreatment prior to biological conversion. Most SO_2 systems used batch reactors and steam explosion. Sulfuric acid has been used in batch, percolation, steam explosion, and two-stage batch pretreatments. Other acids such as nitric, phosphoric, and hydrochloric have been occasionally studied.

Recent interest in producing chemicals or "drop-in" hydrocarbon fuels has brought renewed attention to the complete hydrolysis of hemicellulose and cellulose to sugars and their subsequent reaction to organic aldehydes and acids that can be catalytically converted to hydrocarbons. Although not traditionally referred to as pretreatment, these acidic reactions function in the same manner as traditional pretreatments, that is, production of reactive intermediates for subsequent conversion. Possible goals of these non-traditional pretreatments include maximizing production of furfural, levulinic acid, or carbohydrate-rich liquids. Another approach to fuels production is to combine hydrolysis with hydrogenation over a metallic catalyst to produce sugar alcohols.

Most models of cellulose and hemicellulose hydrolysis have been developed assuming a series of first-order homogeneous reactions in which the carbohydrate polymer is converted to monomers and then to degradation products. These models have been modified to account for parallel degradation pathways, the formation of oligomer intermediates, and differences in substrate reactivity. There are also several novel models describing hydrolysis, including the severity parameter and mass-transfer models. However, due to the empirical nature of existing models, their predictions must be validated if they are to be used with confidence.

Acknowledgements

This work was performed through the support of the BioEnergy Science Center (BESC), a US Department of Energy Bioenergy Research Center supported by the Office of Biological and Environmental Research in the DOE Office of Science. Dr Heather Trajano would also like to thank Dr Jaclyn D. DeMartini, Xiadi Gao, and Dr Hongjia Li for their inputs in organizing this chapter as well as Dr Deepti Tanjore and Dr Taiying Zhang for their valuable suggestions in initiating a literature review of acid hydrolysis in support of chemical conversion. The senior author would also like to thank the Ford Motor Company for funding the Ford Motor Company Chair in Environmental Engineering in the Center for Environmental Research and Technology within the Bourns College of Engineering (CE-CERT) of the University of California, Riverside, vital to making projects such as this possible.

References

1. Martinez, A., Rodriguez, M.E., Wells, M.L. *et al.* (2001) Detoxification of dilute acid hydrolyzates of lignocelluloses with lime. *Biotechnology Progress*, **17** (2), 287–293.
2. Abatzoglou, N.J., Chornet, E., Belkacemi, K., and Overend, R.P. (1992) Phenomenological kinetics of complex systems: the development of a generalized severity parameter and its application to lignocellulosic fractionation. *Chemical Engineering Science*, **47** (5), 1109–1122.
3. Chum, H.L., Johnson, D.K., Black, S.K., and Overend, RP. (1990) Pretreatment-catalyst effects and the combined severity parameter. *Applied Biochemistry and Biotechnology*, **24/25** (1), 1–14.
4. Wyman, C.E. (ed.) (1996) *Handbook on Bioethanol: Production and Utilization*, Taylor and Francis, United States of America.
5. Hallac, B.B. and Ragauskas, AJ. (2011) Analyzing cellulose degree of polymerization and its relevancy to cellulosic ethanol. *Biofuels, Bioproducts and Biorefining Biofpr*, **5** (2), 215–225.
6. Ding, S.Y. and Himmel, M.E. (2006) The maize primary cell wall microfibril: a new model derived from direct visualization. *Journal of Agricultural and Food Chemistry*, **54** (3), 597–606.
7. Krässig, H., Schurz, J., Steadman, R.G. *et al.* (2007) Cellulose, in *Ullmann's Encyclopedia of Industrial Chemistry* (eds M. Bohnet, G. Bellussi, J. Bus *et al.*), Wiley, New York. Also available from Wiley Online Library.
8. Fitzpatrick, S.W. (2002) Final technical report commercialization of the Biofine technology for levulinic acid production from paper sludge. MA: US Department of Energy; Report No DOE/CE/41178. 271 p.
9. Battista, O.A. (1950) Hydrolysis and crystallization of cellulose. *Industrial & Engineering Chemistry*, **42** (3), 502–507.
10. Battista, O.A. and Coppick, S. (1947) Hydrolysis of native versus regenerated cellulose structures. *Textile Research Journal*, **17** (8), 419–422.

11. Nickerson, R.F. and Habrle, J.A. (1947) Cellulose intercrystalline structure: study by hydrolytic methods. *Industrial & Engineering Chemistry*, **39** (11), 1507–1512.

12. Bouchard, J., Abatzoglou, N., Chornet, E., and Overend, R.P. (1989) Characterization of depolymerized cellulosic residues: Part 1: Residues obtained by acid hydrolysis processes. *Wood Science and Technology*, **23** (4), 343–355.

13. Martínez, J.M., Reguant, J., Montero, M.A. *et al.* (1997) Hydrolytic pretreatment of softwood and almond shells: degree of polymerization and enzymatic digestibility of the cellulose fraction. *Industrial & Engineering Chemistry Research*, **36** (3), 688–696.

14. Kumar, R., Mago, G., Balan, V., and Wyman, CE. (2009) Physical and chemical characterizations of corn stover and poplar solids resulting from leading pretreatment technologies. *Bioresource Technology*, **100** (17), 3948–3962.

15. Foston, M. and Ragauskas, A.J. (2010) Changes in lignocellulosic supramolecular and ultrastructure during dilute acid pretreatment of *Populus* and switchgrass. *Biomass & Bioenergy*, **34** (12), 1885–1895.

16. Aguilar, R., Ramírez, J.A., Garrote, G., and Vázquez, M. (2002) Kinetic study of the acid hydrolysis of sugar cane bagasse. *Journal of Food Engineering*, **55** (4), 309–318.

17. Öhgren, K., Galbe, M., and Zacchi, G. (2005) Optimization of steam pretreatment of SO_2-impregnated corn stover for fuel ethanol production. *Applied Biochemistry and Biotechnology*, **124** (1), 1055–1068.

18. Martín, C., Galbe, M., Nilvebrant, N.O., and Jönsson, L.J. (2002) Comparison of the fermentability of enzymatic hydrolysates of sugarcane bagasse pretreated by steam explosion using different impregnating agents. *Applied Biochemistry and Biotechnology*, **98–100** (1), 699–716.

19. Excoffier, G., Toussaint, B., and Vignon, M.R. (1991) Saccharification of steam-exploded poplar wood. *Biotechnology and Bioengineering*, **38** (11), 1308–1317.

20. Kumar, R. and Wyman, C.E. (2008) The impact of dilute sulfuric acid on the selectivity of xylooligomer depolymerization to monomers. *Carbohydrate Research*, **343** (2), 290–300.

21. Maloney, M.T., Chapman, T.W., and Baker, A.J. (1985) Dilute acid hydrolysis of paper birch: kinetics studies of xylan and acetyl-group hydrolysis. *Biotechnology and Bioengineering*, **27** (3), 355–361.

22. Canettieri, E.V., Rocha, G.J.M., Carvalho, J.A. Jr., and Silva, J.BA. (2007) Evaluation of the kinetics of xylose formation from dilute sulfuric acid hydrolysis of forest residues of *Eucalyptus grandis*. *Industrial & Engineering Chemistry Research*, **46** (7), 1938–1944.

23. Antal, M.J. Jr., Leesomboon, T., Mok, W.S., and Richards, G.N. (1991) Mechanism of formation of 2-furaldehyde from D-xylose. *Carbohydrate Research*, **217** (1), 71–85.

24. Hoydonckx, H.E., Van Rhijn, W.M., Van Rhijn, W. *et al.* (2007) Furfural and derivatives, in *Ullmann's Encyclopedia of Industrial Chemistry* (eds M. Bohnet, G. Bellussi, J. Bus *et al.*), Wiley, New York. Also available from Wiley Online Library.

25. Weingarten, R., Cho, J., Conner, W.C. Jr., and Huer, G.W. (2010) Kinetics of furfural production by dehydration of xylose in a biphasic reactor with microwave heating. *Green Chemistry*, **12** (8), 1423–1429.

26. Lloyd, T.A. and Wyman, C.E. (2005) Combined sugar yields for dilute sulfuric acid pretreatment of corn stover followed by enzymatic hydrolysis of the remaining solids. *Bioresource Technology*, **96** (18), 1967–1977.

27. Jensen, J., Morinelly, J., Aglan, A. *et al.* (2008) Kinetic characterization of biomass dilute sulfuric acid hydrolysis: Mixtures of hardwoods, softwood, and switchgrass. *AICHE Journal*, **54** (6), 1637–1645.

28. Saake, B. and Lehnen, R. (2007) Lignin, in *Ullmann's Encyclopedia of Industrial Chemistry* (eds M. Bohnet, G. Bellussi, J. Bus *et al.*) Wiley, New York, Also available from Wiley Online Library.

29. Shevchenko, S.M., Beatson, R.P., and Saddler, J.N. (1999) The nature of lignin from steam explosion/enzymatic hydrolysis of softwood: Structural features and possible uses. *Applied Biochemistry and Biotechnology*, **79**, 867–876.

30. Liu, C. and Wyman, C.E. (2004) The effect of flow rate of very dilute sulfuric acid on xylan, lignin, and total mass removal from corn stover. *Industrial & Engineering Chemistry Research*, **43** (11), 2781–2788.

31. Du, B., Sharma, L.N., Becker, C. *et al.* (2010) Effect of varying feedstock-pretreatment chemistry combinations on the formation and accumulation of potentially inhibitory degradation products in biomass hydrolysates. *Biotechnology and Bioengineering*, **107** (3), 430–440.

32. Sievers, C., Marzialetti, T., Hoskins, T.J.C. *et al.* (2009) Quantitative solid state NMR analysis of residues from acid hydrolysis of loblolly pine wood. *Bioresource Technology*, **100** (20), 4758–4765.

33. Ritter, G.J. and Kurth, E.F. (1933) Holocellulose, total carbohydrate fraction of extractive-free maple wood- Its isolation and properties. *Industrial & Engineering Chemistry*, **25** (11), 1250–1253.

34. Sannigrahi, P., Kim, D.H., Jung, S., and Ragauskas, A. (2011) Pseudo-lignin and pretreatment chemistry. *Energy & Environmental Science*, **4** (4), 1306–1310.

35. Donaldson, L.A., Wong, K.K.Y., and Mackie, K.L. (1988) Ultrastructure of steam-exploded wood. *Wood Science and Technology*, **22** (2), 103–114.

36. Donohoe, B.S., Decker, S.R., Tucker, M.P. *et al.* (2008) Visualizing lignin coalescence and migration through maize cell walls following thermochemical pretreatment. *Biotechnology and Bioengineering*, **101** (5), 913–925.

37. Selig, M.J., Viamajala, S., Decker, S.R. *et al.* (2007) Deposition of lignin droplets produced from dilute acid pretreatment of maize stems retards enzymatic hydrolysis of cellulose. *Biotechnology Progress*, **23** (6), 1333–1339.

38. Sluiter, A., Hames, B., Ruiz, R. *et al.* (2005) *Determination of ash in biomass: Laboratory Analytical Procedure [Internet]*, National Renewable Energy Laboratory, United States of America, Available from http://www.nrel.gov/biomass/pdfs/42622.pdf.

39. Springer, E.L. and Harris, J.F. (1985) Procedures for determining the neutralizing capacity of wood during hydrolysis with mineral acid solutions. *Industrial & Engineering Chemistry Product Research and Development*, **24** (3), 485–489.

40. Chen, S.-F., Mowery, R.A., Scarlata, C.J., and Chambliss, C.K. (2007) Compositional analysis of water-soluble materials in corn stover. *Journal of Agricultural and Food Chemistry*, **55** (15), 5912–5918.

41. Chen, S.-F., Mowery, R.A., Sevcik, R.S. *et al.* (2010) Compositional analysis of water-soluble materials in switchgrass. *Journal of Agricultural and Food Chemistry*, **58** (6), 3251–3258.

42. Kim, S.B. and Lee, Y.Y. (2002) Diffusion of sulfuric acid within lignocellulosic biomass particles and its impact on dilute-acid pretreatment. *Bioresource Technology*, **83** (2), 165–171.

43. Esteghlalian, A., Hashimoto, A.G., Fenske, J.J., and Penner, M.H. (1997) Modeling and optimization of the dilute sulfuric acid pretreatment of corn stover, poplar, and switchgrass. *Bioresource Technology*, **59** (2–3), 129–136.

44. Boussaid, A.L., Esteghlalian, A.R., Gregg, D.J. *et al.* (2000) Steam pretreatment of Douglas fir wood chips- Can conditions for optimum hemicelluloses recovery still provide adequate access for efficient enzymatic hydrolysis? *Applied Biochemistry and Biotechnology*, **84**, 693–705.

45. Grethlein, H.E. (1985) The effect of pore size distribution on the rate of enzymatic hydrolysis of cellulosic substrates. *Nature Biotechnology*, **3** (2), 155–160.

46. Wong, K.K.Y., Deverell, K.F., Mackie, K.L. *et al.* (1988) The relationship between fiber porosity and cellulose digestibility in steam-exploded *Pinus radiata*. *Biotechnology and Bioengineering*, **31** (5), 447–456.

47. Sherrard, E.C. and Kressman, F.W. (1945) Review of processes in the United States prior to World War II. *Industrial & Engineering Chemistry*, **37** (1), 5–8.

48. Faith, W.L. (1945) Development of the Scholler process in the United States. *Industrial & Engineering Chemistry*, **37** (1), 9–11.

49. Harris, E.E. and Beglinger, E. (1946) Madison wood sugar process. *Industrial & Engineering Chemistry*, **38** (9), 890–895.

50. Harris, E.E. and Lang, B.G. (1947) Hydrolysis of wood cellulose and decomposition of sugar in dilute phosphoric acid. *Journal of Physical and Colloid Chemistry*, **51** (6), 1430–1441.

51. Harris, E.E. and Kline, A.A. (1948) Hydrolysis of wood cellulose with hydrochloric acid and sulfur dioxide and the decomposition of its hydrolytic products. *Journal of Physical and Colloid Chemistry*, **53** (3), 344–351.

52. Selke, S.M., Hawley, M.C., Hardt, H. *et al.* (1982) Chemicals from wood via HF. *Industrial & Engineering Chemistry Product Research and Development*, **21** (1), 11–16.

53. Rorrer, G.L., Ashour, S.S., Hawley, M.C., and Lamport, D.T.A. (1987) Solvolysis of wood and pure cellulose by anhydrous hydrogen fluoride vapor. *Biomass*, **12** (4), 227–246.

54. Wright, J.D. and Power, A.J. (1986) Comparative technical evaluation of acid hydrolysis processes for conversion of cellulose to alcohol. United States: Solar Energy Research Institute; Report no SERI/TP-232-2957. 27 p.

55. Allen, F., Andreotti, R., Eveleigh, D.E., and Nystrom, J. (2009) Mary Elizabeth Hickox Mandels, 90, bioenergy leader. *Biotechnology for Biofuels*, **2** (1), 22–27.

56. Mandels, M. and Sternberg, D. (1976) Recent advances in cellulase technology. *Journal of Fermentation Technology*, **54** (4), 267–286.

57. Grethlein, H.E. (1978) Comparison of economics of acid and enzymatic hydrolysis of newsprint. *Biotechnology and Bioengineering*, **20** (4), 503–525.

58. Grethlein, H.E., Allen, D.C., and Converse, A.O. (1984) A comparative study of the enzymatic hydrolysis of acid-pretreated white-pine and mixed hardwood. *Biotechnology and Bioengineering*, **26** (12), 1498–1505.

59. Han, Y.W. and Callihan, C.D. (1974) Cellulose fermentation: effect of substrate pretreatment on microbial growth. *Journal of Applied Microbiology*, **27** (1), 159–165.

60. Nesse, N., Wallick, J., and Harper, J.M. (1977) Pretreatment of cellulosic wastes to increase enzyme reactivity. *Biotechnology and Bioengineering*, **19** (3), 323–336.

61. Sciamanna, A.F., Freitas, R.P., and Wilke, C.R. (1977) Composition and utilization of cellulose for chemicals from agricultural residues. United States: Lawrence Berkeley Laboratory; Report LBL-5966.

62. Knappert, D., Grethlein, H., and Converse, A. (1980) Partial acid hydrolysis of cellulosic materials as a pretreatment for enzymatic hydrolysis. *Biotechnology and Bioengineering*, **22** (7), 1449–1463.

63. Grethlein, H.E. and Converse, A.O. (1991) Common aspects of acid prehydrolysis and steam explosion for pretreating wood. *Bioresource Technology*, **36** (1), 77–82.

64. Foody, P. (1984) Method for increasing the accessibility of cellulose in lignocellulosic materials, particularly hardwoods, agricultural residues and the like. US Patent 4461648.

65. Wyman, C.E. (2001) Twenty years of trials, tribulations, and research progress in bioethanol technology. *Applied Biochemistry and Biotechnology*, **91–93** (1), 5–21.

66. Mosier, N., Wyman, C., Dale, B. *et al.* (2005) Features of promising technologies for pretreatment of lignocellulosic biomass. *Bioresource Technology*, **96** (6), 673–686.

67. Almarsdottir, A.R., Tarazewicz, A., Gunnarsson, I., and Orlygsson, J. (2010) Hydrogen production from sugars and complex biomass by *Clostridium* species AK_{14} isolated from Icelandic hot spring. *Icelandic Agricultural Sciences*, **23**, 61–71.

68. Cao, G.L., Ren, N.Q., Wang, A.J. *et al.* (2010) Effect of lignocelluloses-derived inhibitors on growth and hydrogen production by *Thermoanaerobacterium thermosaccharolyticum* W16. *International Journal of Hydrogen Energy*, **35** (24), 13475–13480.

69. Puri, V.P. and Mamers, H. (1983) Explosive pretreatment of lignocellulosic residues with high-pressure carbon dioxide for the production of fermentation substrates. *Biotechnology and Bioengineering*, **25** (12), 3149–3161.

70. McWilliams, R.C. and van Walsum, G.P. (2002) Comparison of aspen wood hydrolysates produced by pretreatment with liquid hot water and carbonic acid. *Applied Biochemistry and Biotechnology*, **98–100** (1–9), 109–121.

71. van Walsum, G.P. and Shi, H. (2004) Carbonic acid enhancement of hydrolysis in aqueous pretreatment of corn stover. *Bioresource Technology*, **93** (3), 217–225.

72. Luterbacher, J.S., Tester, J.W., and Walker, LP. (2010) High-solids biphasic CO_2-H_2O pretreatment of lignocellulosic biomass. *Biotechnology and Bioengineering*, **107** (3), 451–460.

73. Yang, B. and Wyman, C.E. (2008) Pretreatment: the key to unlocking low-cost cellulosic ethanol. *Biofuels, Bioproducts and Biorefining Biofpr*, **2** (1), 26–40.

74. Jayawardhana, K. and van Walsum, G.P. (2004) Modeling of carbonic acid pretreatment process using ASPEN-Plus®. *Applied Biochemistry and Biotechnology*, **115** (1–3), 1087–1102.

75. Bura, R., Chandra, R., and Saddler, J. (2009) Influence of xylan on enzymatic hydrolysis of steam-pretreated corn stover and hybrid poplar. *Biotechnology Progress*, **25** (2), 315–322.

76. Stenberg, K., Tengborg, C., Galbe, M., and Zacchi, G. (1998) Optimisation of steam pretreatment of SO_2-impregnated mixed softwoods for ethanol production. *Journal of Chemical Technology and Biotechnology*, **71** (4), 299–308.

77. Schell, D.J., Torget, R., Power, A. *et al.* (1991) A technical and economic analysis of acid-catalyzed steam explosion and dilute sulfuric acid pretreatments using wheat sraw or aspen wood chips. *Applied Biochemistry and Biotechnology*, **28/29** (1), 87–97.

78. De Bari, I., Nanna, F., and Braccio, G. (2007) SO_2-catalyzed steam fractionation of aspen chips for bioethanol production: optimization of the catalyst impregnation. *Industrial & Engineering Chemistry Research*, **46** (23), 7711–7720.

79. Söderström, J., Pilcher, L., Galbe, M., and Zacchi, G. (2002) Two-step steam pretreatment of softwood with SO_2 impregnation for ethanol production. *Applied Biochemistry and Biotechnology*, **98–100** (1), 5–21.

80. Humbird, D., Davis, R., Tao, L. *et al.* (2011) Process design and economics for biochemical conversion of ligno-cellulosic biomass to ethanol: dilute acid pretreatment and enzymatic hydrolysis of corn stover. CO: National Renewable Energy Laboratory; Report No. NREL/TP-5100-47764. 147. p.

81. Jacobsen, S.E. and Wyman, C.E. (2001) Heat transfer considerations in design of a batch tube reactor for biomass hydrolysis. *Applied Biochemistry and Biotechnology*, **92** (1–3), 377–386.

82. Stuhler, S.L. and Wyman, C.E. (2003) Estimation of temperature transients for biomass pretreatment in tubular batch reactors and impact on xylan hydrolysis kinetics. *Applied Biochemistry and Biotechnology*, **105** (1–3), 101–114.

83. Sassner, P., Mårtensson, C.G., Galbe, M., and Zacchi, G. (2008) Steam pretreatment of H_2SO_4-impregnated *Salix* for the production of bioethanol. *Bioresource Technology*, **99** (1), 137–145.

84. Cahela, D.R., Lee, Y.Y., and Chambers, R.P. (1983) Modeling of percolation process in hemicellulose hydrolysis. *Biotechnology and Bioengineering*, **25** (1), 3–17.

85. Mok, W.S., Antal, M.J. Jr., and Varhegyi, G. (1992) Productive and parasitic pathways in dilute acid catalyzed hydrolysis of cellulose. *Industrial & Engineering Chemistry Research*, **31** (1), 94–100.

86. Saeman, J.F. (1945) Kinetics of wood saccharification- Hydrolysis of cellulose and decomposition of sugars in dilute acid at high temperature. *Industrial & Engineering Chemistry*, **37** (1), 43–52.

87. Nguyen, Q.A., Tucker, M.P., Keller, F.A., and Eddy, FP. (2000) Two-stage dilute-acid pretreatment of softwoods. *Applied Biochemistry and Biotechnology*, **84–86** (1), 561–576.

88. De Jong, W. and Marcotullio, G. (2010) Overview of biorefineries based on co-production of furfural- existing concepts and novel developments. *International Journal of Chemical Reactor Engineering*, **8** (A69), 1–24.

89. Mansilla, H.D., Baeza, J., Urzúa, S. *et al.* (1998) Acid-catalysed hydrolysis of rice hull: evaluation of furfural production. *Bioresource Technology*, **66** (3), 189–193.

90. Mamman, A.S., Lee, J.M., Kim, Y.C. *et al.* (2008) Furfural: Hemicellulose/xylose-derived biochemical. *Biofuels, Bioproducts and Biorefining Biofpr*, **2** (5), 438–454.

91. Kottke, R.H. (2000) Furan derivatives, in *Kirk-Othmer Encyclopedia of Chemical Technology* (eds R.E. Kirk and D.F. Othmer), Wiley, New York. Also available from Wiley Online Library.

92. Marcotullio, G. and De Jong, W. (2010) Chloride ions enhance furfural formation from D-xylose in dilute aqueous acidic solutions. *Green Chemistry*, **12** (10), 1739–1746.

93. Rackemann, D.W. and Doherty, W.O.S. (2011) The conversion of lignocellulosics to levulinic acid. *Biofuels, Bioproducts and Biorefining Biofpr*, **5** (2), 198–214.

94. Zhang, T., Kumar, R., and Wyman, C.E. (2013). Sugar yields from dilute oxalic acid pretreatment of maple wood compared to those with other dilute acids and hot water. *Carbohydrate Polymers*, **92** (1), 334–344.

95. Li, N., Tompsett, G.A., Zhang, T. *et al.* (2011) Renewable gasoline from aqueous phase hydrodeoxygenation of aqueous sugar solutions prepared by hydrolysis of maple wood. *Green Chemistry*, **13**, 91–101.

96. vom Stein, T., Grande, P.M., Kayser, H. *et al.* (2011) From biomass to feedstock: one-step fractionation of ligno-cellulose components by the selective organic acid-catalyzed depolymerization of hemicellulose in a biphasic system. *Green Chemistry*, **13** (7), 1772–1777.

97. Robinson, J.M., Burgess, C.E., Bently, M.A. *et al.* (2004) The use of catalytic hydrogenation to intercept carbohydrates in a dilute acid hydrolysis of biomass to effect a clean separation from lignin. *Biomass & Bioenergy*, **26** (5), 473–483.

98. Palkovits, R., Tajvidi, K., Procelewska, J. *et al.* (2010) Hydrogenolysis of cellulose combining mineral acids and hydrogenation catalysts. *Green Chemistry*, **12** (6), 972–978.

99. Fengel, D. and Wegener, G. (1984) *Wood- Chemistry, Ultrastructure and Reactions*, Walter de Gruyter, New York, NY.

100. Lloyd, T.A. and Wyman, C.E. (2004) Predicted effects of minerals on pretreatment. *Applied Biochemistry and Biotechnology*, **113–116** (1), 1013–1022.

101. Jacobsen, S.E. and Wyman, C.E. (2000) Cellulose and hemicellulose hydrolysis models for application to current and novel pretreatment processes. *Applied Biochemistry and Biotechnology*, **84–86** (1), 81–96.

102. Conner, A.H., Wood, B.F., Hill, C.G., and Harris, J.F. (1986) Kinetic modeling of the saccharification of prehydrolyzed Southern Red Oak, in *Cellulose: Structure, Modification and Hydrolysis* (eds R.A. Young and R.M. Rowell), John Wiley and Sons, New York, NY.

103. Bouchard, J., Abatzoglou, N., Chornet, E., and Overend, R.P. (1989) Characterization of depolymerized cellulosic residues. 1. Residues obtained by acid-hydrolysis processes. *Wood Science and Technology*, **23** (4), 333–355.

104. Abatzoglou, N., Bouchard, J., and Chornet, E. (1986) Dilute acid depolymerization of cellulose in aqueous phase: experimental evidence of the significant presence of soluble oligomeric intermediates. *Canadian Journal of Chemical Engineering*, **64** (5), 781–786.

105. Chang, C., Ma, X., and Cen, P. (2006) Kinetics of levulinic acid formation from glucose decomposition at high temperature. *Chinese Journal of Chemical Engineering*, **14** (5), 708–712.

106. Girisuta, B., Janssen, P.B.M., and Heeres, H.J. (2006) Green chemicals: A kinetic study on the conversion of glucose to levulinic acid. *Chemical Engineering Research & Design*, **84** (A5), 339–349.

107. Assary, R.S., Redfern, P.C., Hammond, J.R. *et al.* (2010) Computational studies of the thermochemistry for conversion of glucose to levulinic acid. *The Journal of Physical Chemistry*, **114** (27), 9002–9009.

108. Chang, C., Ma, X., and Cen, P. (2009) Kinetic studies on wheat straw hydrolysis to levulinic acid. *Chinese Journal of Chemical Engineering*, **17** (5), 835–839.

109. Girisuta, B., Janssen, P.B.M., and Heeres, H.J. (2007) Kinetic study on the acid-catalyzed hydrolysis of cellulose to levulinic acid. *Industrial & Engineering Chemistry Research*, **46** (6), 1696–1708.

110. Girisuta, B., Danon, B., Manurung, R. *et al.* (2008) Experimental and kinetic modeling studies on the acid-catalysed hydrolysis of the water hyacinth plant to levulinic acid. *Bioresource Technology*, **99** (17), 8367–8375.

111. Kobayashi, T. and Sakai, Y. (1956) Hydrolysis rate of pentosan of hardwood in dilute sulfuric acid. *Agricultural and Biological Chemistry*, **20** (1), 1–7.

112. Tillman, L.M., Lee, Y.Y., and Torget, R. (1990) Effect of transient acid diffusion on pretreatment/hydrolysis of hardwood hemicellulose. *Applied Biochemistry and Biotechnology*, **24/25** (1), 103–113.

113. Hosseini, S.A. and Shah, N. (2009) Multiscale modeling of biomass pretreatment for biofuels production. *Chemical Engineering Research & Design*, **87** (9A), 1251–1260.

114. Abasaeed, A.E., Lee, Y.Y., and Watson, J.R. (1991) Effect of transient heat transfer and particle size on acid hydrolysis of hardwood cellulose. *Bioresource Technology*, **35** (1), 15–21.

115. Abasaeed, A.E. and Mansour, M.E. (1992) Thermal effects on acid hydrolysis of cellulose. *Bioresource Technology*, **40** (3), 221–224.

116. Williams, D.L. and Dunlop, AP. (1948) Kinetics of furfural destruction in acidic aqueous media. *Industrial & Engineering Chemistry*, **40** (2), 239–241.

117. Rose, I.C., Epstein, N., and Watkinson, A.P. (2000) Acid-catalyzed 2-furaldehyde (furfural) decomposition kinetics. *Industrial & Engineering Chemistry Research*, **39** (3), 843–845.

118. Nimlos, M.R., Qian, X., Davis, M. *et al.* (2006) Energetic of xylose decomposition as determined using quantum mechanics modeling. *The Journal of Physical Chemistry*, **110** (42), 11824–11838.

119. Brennan, M.A. and Wyman, C.E. (2004) Initial evaluation of simple mass transfer models to describe hemicellulose hydrolysis in corn stover. *Applied Biochemistry and Biotechnology*, **115** (1–3), 965–976.

120. Carrasco, F. and Roy, C. (1992) Kinetic study of dilute-acid prehydrolysis of xylan-containing biomass. *Wood Science and Technology*, **26** (3), 189–207.

121. Converse, A.O., Kwarteng, I.K., Grethlein, H.E., and Ooshima, H. (1989) Kinetics of thermochemical pretreatment of lignocellulosic materials. *Applied Biochemistry and Biotechnology*, **20/21** (1), 63–78.

122. Lloyd, T.A. and Wyman, C.E. (2003) Application of a depolymerization model for predicting thermochemical hydrolysis of hemicellulose. *Applied Biochemistry and Biotechnology*, **105** (1–3), 53–67.

123. Lu, Y. and Mosier, N.S. (2008) Kinetic modeling analysis of maleic acid-catalyzed hemicellulose hydrolysis in corn stover. *Biotechnology and Bioengineering*, **101** (6), 1170–1179.

124. Morinelly, J.E., Jensen, J.R., Browne, M. *et al.* (2009) Kinetic characterization of xylose monomer and oligomer concentrations during dilute acid pretreatment of lignocellulosic biomass from forests and switchgrass. *Industrial & Engineering Chemistry Research*, **48** (22), 9877–9884.

125. Rose, I.C., Epstein, N., and Watkinson, A.P. (2000) Acid-catalyzed 2-furaldehyde (furfural) decomposition kinetics. *Industrial and Engineering Chemistry Research*, **39** (3), 843–845.

7

Fundamentals of Aqueous Pretreatment of Biomass

Nathan S. Mosier

Department of Agricultural and Biological Engineering, Laboratory of Renewable Resources Engineering, Purdue University, West Lafayette, USA

7.1 Introduction

As described in Chapter 4, pretreatment before biological conversion of plant biomass to biofuels is necessary to improve the yield of fermentable sugars. Aqueous pretreatment resembles pretreatment at low pH in many respects, including many of the chemical reactions that occur during pretreatment. The goal of aqueous pretreatment is to balance the yield improvements delivered by acid pretreatment with lower operational and capital costs [1,2].

Aqueous pretreatment involves mixing plant biomass with water and then heating the slurry under pressure. This approach has been variously described as liquid hot water, hydrothermal, uncatalyzed steam explosion, autohydrolysis, hydrothermolysis, aqueous-controlled pH, or combinations and variations of these terms [3–6]. The technologies described by these terms share some commonalities but have different operational conditions and processing goals. One commonality between these methods is that the water is maintained as a subcritical liquid through most or all of the pretreatment process [3–6]. The specific technologies vary significantly in terms of the mechanical construction of the pretreatment vessel as well as the temperature and pressure regime during processing.

The goal of this chapter is to familiarize the reader with the basic principles of aqueous pretreatment and to highlight some of the technologies that have been developed around these principles to pretreat lignocellulosic biomass for the production of biofuels and other products through biochemical or aqueous catalytic methods. The focus is on the chemical and physical changes that occur to lignocellulosic biomass

Aqueous Pretreatment of Plant Biomass for Biological and Chemical Conversion to Fuels and Chemicals, First Edition.
Edited by Charles E. Wyman.
© 2013 John Wiley & Sons, Ltd. Published 2013 by John Wiley & Sons, Ltd.

during aqueous pretreatment and the effect these changes have on the enzymatic hydrolysis of plant cell-wall polysaccharides. Brief summaries of the results of applying these technologies to various types of biomass are also presented. The reader is provided with references for further information regarding specific details of these technologies and operational conditions for practical applications.

7.2 Self-ionization of Water Catalyzes Plant Cell-wall Depolymerization

Aqueous pretreatment utilizes the chemical and catalytic properties of liquid water at temperatures and pressures significantly above room temperature and atmospheric pressure, respectively. The ionization constant (K_w) for water is a function of both temperature and pressure [7] and is defined:

$$K_w = \frac{[H^+][OH^-]}{[H_2O]} \tag{7.1}$$

At room temperature, water has a pK_w (negative log, base ten of K_w) equivalent to 14.0. According to Equation (7.1), the corresponding pH (negative log, base ten of H$^+$ concentration) of water at room temperature is therefore 7.0. As the temperature and pressure of saturated liquid water increases, the pK_w decreases until reaching a minimum at approximately 250 °C (Figure 7.1). pK_w then increases between 250 °C and the critical point of water (375 °C). The decreasing pK_w as the temperature rises from 20 to 250 °C has the effect of increasing the concentration of protons in solution [7,8]. At 250 °C the pH of pure water is approximately 5.6. While the solution is still neutral by definition, since the pOH is also 5.6, the concentration of protons and hydroxyls available to catalyze acid and base reactions are more than 25 times higher than at room temperature [7,8]. Coupled with the higher kinetic energy available in the system at high temperatures, the rates of depolymerization reactions that break down plant cell-wall polymers are significantly increased during pretreatment with hot liquid water.

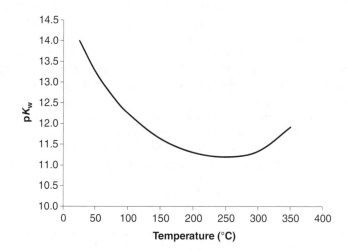

Figure 7.1   p^{K}w *of subcritical liquid water as a function of temperature at saturation pressure (plot of data from [7,8]).*

7.3 Products from the Hydrolysis of the Plant Cell Wall Contribute to Further Depolymerization

While the concentration of protons and hydroxyls both increase as the temperature of liquid water is increased, the dominant reactions during aqueous pretreatment are acid (proton) catalyzed. This is partially due to the increasingly acidic character of the aqueous medium as the pretreatment proceeds. As described in Chapter 2, plant cell walls are primarily composed of three polymers: cellulose, hemicellulose, and lignin. While cellulose is a linear polymer of glucose, hemicellulose is a branched heteropolymer of carbohydrates (e.g., xylose, arabinose, glucose, mannose) that vary depending upon plant type and plant tissue [9]. The predominant hemicellulose in most plants, especially hardwoods and grasses, is composed of a linear polymer of xylose linked by β1–4 glycosidic bonds. Short side chains of sugars and organic acids, especially acetic acid, then complete these polymers. As hemicellulose is hydrolyzed under the temperatures and pressures associated with aqueous pretreatment, these organic acids readily dissolve into the water and make the solution more acidic (more protons than hydroxyls in solution). The release of these acids from biomass during pretreatment is another reason that aqueous pretreatment has also been called autohydrolysis. Further evidence for this is that the reported final pH, measured at room temperature, for the liquid resulting from aqueous pretreatment. This ranges from 3.0 to 4.5 from a starting pH of 7.0, which is largely attributed to the release of acetyl from the hemicellulose as acetic acid although levulinic, formic, and other organic acids are also formed at low concentrations [3–5,10–16].

7.4 Mechanisms of Aqueous Pretreatment

As introduced in Section 7.2, water acts as both a solvent and a catalyst in the pretreatment of lignocellulosic biomass. In general, pretreatment significantly improves the accessibility of the cellulose present in the plant cell-wall matrix to enzyme-catalyzed hydrolysis to fermentable glucose [6]. This is achieved through several physical and chemical changes to the biomass that occur during pretreatment. Aqueous pretreatment affects hemicellulose, lignin, and cellulose in different ways to improve the enzyme accessibility and enhance the reactivity of the cell-wall polysaccharides.

7.4.1 Hemicellulose

Similar to acid pretreatment, aqueous pretreatment removes hemicellulose from the plant cell-wall matrix, improving enzyme accessibility and thus cellulose hydrolysis yields [3–5,10]. The monosaccharides that comprise hemicellulose are valuable feedstocks for conversion to biofuels or other bioproducts. Pretreatment strategies, including aqueous pretreatment, must therefore balance removal of hemicellulose from the solid matrix with degradation of the monosaccharides in the solvent to undesirable products, especially furfural [11–14,16]. Understanding of the kinetics of hemicellulose hydrolysis, solubilization, and degradation is thus critical in the design of a process to maximize overall xylose and glucose yields from lignocellulosic processing.

The kinetics of hemicellulose hydrolysis catalyzed by dilute mineral acids or water at elevated temperatures has been relatively well studied. Kinetic models have been developed to describe the series of complex reactions that occur during hemicellulose hydrolysis [17–20]. While hemicellulose is a heterogeneous branched polymer, models of hemicellulose hydrolysis usually deal with this complexity by describing the process as a set of parallel reactions that hydrolyze sugar polymers. Most models focus on the depolymerization of the xylan backbone of the hemicellulose, and the release of other sugars and organic acids are modeled as separate reactions (Figure 7.2). The common convention, therefore, is to treat each sugar and acid in hemicellulose as distinct pools of anhydro monosacchride units, referred to by replacing

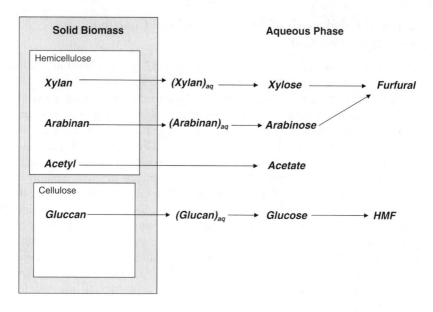

Figure 7.2   *Schematic for kinetic model of aqueous acid hydrolysis of cellulosic biomass.*

the –ose suffix of the sugar with –an (Figure 7.2). Xylan is therefore the anhydro xylose constituent of hemi-celluloses, having a molecular weight of 132 g/mol (one water mass less than xylose at 150 g/mol).

Models for acid-catalyzed hemicellulose hydrolysis generally follow the convention described by Saeman for dilute sulfuric-acid-catalyzed cellulose hydrolysis [21]:

$$\text{glucan} \rightarrow \text{glucose at rate } k \tag{7.2}$$

defined $k = k_0 \times [\text{Acid}]^n \times e^{-E_a/RT}$ where k_0 is the pre-exponential Arrhenius constant; n is the acid power factor (usually $n = 1$); E_a is the activation energy; R is the universal gas constant; and T is absolute temperature (K).

Hemicellulose hydrolysis is therefore described as consecutive homogeneous pseudo-first-order steps: xylan is first hydrolyzed to xylose, and the xylose is degraded to furfural once it is liberated from the solid matrix and exposed to the acidic solution (Figure 7.2). An improved model proposed by Kobayashi and Sakai [22] was based upon the observation of a biphasic pattern of hemicellulose hydrolysis: one part (easy-to-hydrolyze hemicellulose) of the total hemicellulose fraction tends to hydrolyze faster than the other part (hard-to-hydrolyze hemicellulose). Varying slightly among different substrates, the hard-to-hydrolyze fraction typically accounts for about 35% of the total hemicellulose [23]. Furfural is also suscep-tible to further degradation, primarily through polymerization to insoluble organic compounds [24]. Precip-itation from solution or adsorption of furfural degradation products onto the solid biomass during pretreatment has also been suggested to reduce cellulose digestibility [9,25]. Optimization of acid and aque-ous pretreatments is therefore required as it is possible to "over pretreat" cellulosic biomass and thus lose the improved digestibility of cellulose.

Recently published models are essentially built upon these two models with more in-depth considera-tions of the reaction intermediates, for example, sugar oligomers formation [26], or by including mass-transfer [27] and heat-transfer effects [28]. The use of weight-percent of acid concentration (exclusively used in papers published before 1990) does not match the non-linearity of H_3O^+ molar concentration in aqueous solutions [7,8]; more recent models therefore tend to use proton concentration [H^+] or hydronium concentration [H_3O^+] in place of the concentration of the proton donor (e.g., sulfuric acid). This change in convention allows these models to be used for the autohydrolysis of hemicellulose that occurs during aqueous pretreatment since protons, regardless of source, are able to catalyze hydrolysis reactions.

In addition, the effects of mass-transfer limitation at the solid surface become more or less important depending upon the reactor configuration (percolation versus flowthrough versus batch) and the size of biomass particles in the reactor [29,30]. Generally, oligomers were observed when reaction temperatures were low or pH was closer to neutral since the hydrolysis of soluble oligomers occurs rapidly under highly acidic conditions. Furthermore, deviation from first-order kinetics was noted in these situations where xylan oligomers are the main products of the hydrolysis reaction [31,32]. The deviations suggested that the over-simplified assumptions did not match the true complexity of the reactions occurring at the solid–liquid interface. To address these considerations, an alternate reaction scheme for xylan hydrolysis is shown in Figure 7.3. In this model, insoluble xylan is first partially hydrolyzed to soluble (aqueous) xylan. Aqueous xylan is then hydrolyzed to monomeric xylose, which can then be dehydrated to furfural. Furfural further degrades to insoluble and poorly characterized compounds. Xylose degradation is also strongly increased by the presence of ions, especially halide ions (Cl^{-1}, etc.) [33,34] which can be present in cellulosic biomass and released during pretreatment. However, models published to date do not take this factor into account.

For the design of aqueous pretreatments, the formation of longer-lived soluble xylan oligosaccharides is used as an advantage. Where reaction conditions (e.g., time, temperature, pH) can be controlled to maximize reaction 1 in Figure 7.3 (insoluble xylan to soluble xylan) and minimize reaction 2 (complete hydrolysis of soluble xylan to xylose), formation of furfural (reaction 3) is reduced. This process design approach has the advantage of improving cellulose accessibility (and thus enzymatic digestibility) by removing hemicellulose (xylan) from the cell-wall matrix while also minimizing xylose degradation [10,15]. This

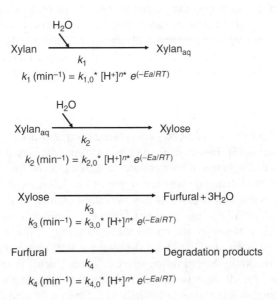

Figure 7.3 *Reactions for xylan hydrolysis where soluble (aqueous) xylan is an intermediate between insoluble xylan and monomeric xylose.*

Table 7.1 *Kinetic constants and activation energies for hemicellulose degradation during aqueous pretreatment.*

Reaction	k (at 200 °C)^a hr^{-1}	n^b	E_a (kJ/mol)	Reference(s)
1 Xylan Hydrolysis	7–25	c. 1	113–156	[4,18,20,37]
2 Xylan oligosaccharide hydrolysis	13–30	c. 1	95–120	[4,18,20,37]
3 Xylose degradation	0.5–5	c. 1	120–160	[4,18,20,37]
4 Furfural degradation	1–11	c. 1	67–73	[18,24]

^{a} Where data at 200 °C not given, constants are estimated from E_a and pre-exponential constants as provided.
^{b} Exponent associated with [H$^+$] in Figure 7.3.

approach can achieve very high yields of soluble xylan and xylose (>90%) under high-solids/low-water processing conditions [10]. Acid pretreatments that completely hydrolyze xylan to xylose usually achieve yields less than 70% under similar solids/liquid ratios [35,36]. The disadvantage of this aqueous approach is that subsequent processing step(s) must be included to complete the hydrolysis of xylan to xylose using enzymes or other catalysts before these sugars can be converted to value-added products, unless downstream operations can use oligomers effectively.

Table 7.1 summarizes published kinetic parameters for the reactions illustrated in Figure 7.3. In most models, the rate of xylose release from xylan oligosaccharides is 1–3 times that of xylan hydrolysis to soluble oligosaccharides. Where both rates are determined in a single study, the rate of soluble oligosaccharide hydrolysis is always higher than the rate of hydrolysis and dissolution of insoluble xylan. This ratio is fairly constant over a range of temperatures (150–190 °C) due to the similar activation energies (E_a) of these reactions. On the other hand, the activation energy for xylose degradation tends to be higher than the reported activation energy for xylan and xylan oligosaccharide hydrolysis. Thus, the rate of xylose degradation increases more, proportionally, than the rate of xylose release over the same temperature range. Therefore, there is a tradeoff in temperature and time during aqueous pretreatment to maximize the increase in cellulose digestibility while minimizing xylose degradation. Long pretreatment times at lower temperatures (120–140 °C) are highly effective at extracting the hemicellulose from the plant cell-wall matrix [17,38]. However, aqueous pretreatments at these temperatures do not enhance the digestibility of the remaining cellulose as much as higher temperature, but shorter time, pretreatment conditions [10,12]. This is likely due to the impact of the slightly acidic conditions on the lignin present in the plant biomass.

7.4.2 Lignin

Under acidic conditions, lignin is resistant to depolymerization and alkali conditions generally favor delignification of plant biomass [6]. However, lignin has a varying melting point between 90 and 190 °C [39,40], which corresponds to the range of temperatures commonly used for aqueous pretreatment. It has long been hypothesized that the interaction between lignin and cellulose is altered during acid and neutral aqueous pretreatment. The argument in favor of this hypothesis is that acid, and especially the slightly acidic conditions of aqueous pretreatments, does not remove a significant portion of the lignin from the biomass (<15%) [10,13,29,30]. However, the results are significantly different in pretreatment configurations where the solvent is not allowed to cool with the biomass after pretreatment, such as "flowthrough" reactors where hot solvent (neutral water or dilute aqueous sulfuric acid) is pumped through a column of stationary biomass particles [29,30]. In experiments with flowthrough reactors, the amount of lignin that can be removed from the biomass increases substantially to c. 50% under acid conditions and c. 46% under neutral conditions [29,30]. These results suggest that lignin is at least partially soluble at pretreatment conditions (high temperatures and pressure) and may precipitate from solution as the pretreated slurry cools.

More recently, advanced imaging techniques have greatly improved the understanding of the behavior of lignin during acid and neutral aqueous pretreatments. Due to its hydrophobicity, lignin tends to coalesce into droplets upon melting when in an aqueous environment. These droplets appear to migrate within the cell-wall matrix through what is hypothesized to be both Fickian diffusion as well as pressure exerted by the collapse of the cellulose's microfibrilar structure within the cell wall during pretreatment [41]. Upon cooling, these lignin droplets adsorb on the surface of the biomass or remain in coalesced pools within the cell-wall matrix instead of being more evenly distributed throughout the cell-wall matrix (Figure 7.4). This

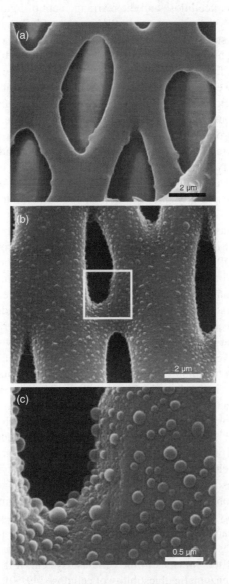

Figure 7.4 *SEM micrographs of corn stover rind xylem cell-wall surfaces (a) before and (b and c) after 0.8% H_2SO_4 dilute acid pretreatment at 150°C for 20 min. Image (c) is a higher magnification of the region boxed in (b). Numerous round droplets are distinct on the pretreated surface. These initial observations led to an extensive study of the composition and origin of these dramatic structural features. Scale bars: (a, b) 2 mm; (c) 0.5 mm. (Reproduced from reference [41] © 2005, Humana Press Inc).*

increases the pore size and volume available for cellulolytic enzymes and hydrolysis products to diffuse into and out of the cell-wall matrix [41]. In addition, this may also significantly reduce the surface area upon which cellulolytic enzymes can productively bind [42].

7.4.3   Cellulose

Aqueous pretreatment has minimal direct impacts on cellulose itself. The enhancement of reactivity is primarily due to the removal of hemicellulose and the rearrangement of lignin as described above. However, some changes to the cellulose crystal structure and cellulose microfibrils have been reported [6,43,44].

Reported changes to the crystallinity of the cellulose due to aqueous pretreatment are inconsistent. In some cases, aqueous pretreatment reduces the overall crystallinity of the biomass. The crystallinity of pretreated corn stover was reduced from a crystallinity index of 50.4 (untreated) to 44.5 through controlled pH (neutral) pretreatment [43]. However, the crystallinity of poplar pretreated by the same technique resulted in an increase in crystallinity index from 49.9 (untreated) to 54.0 [44]. Since the technique for determining cellulose crystallinity involves measuring X-ray diffraction through an entire biomass sample (including lignin and hemicellulose), interpreting these results is difficult. Since hemicellulose and lignin lack any regular crystal structure, removing these polymers while leaving the cellulose unchanged will result in an observed increase in crystallinity index because the fraction of material in the biomass that is crystalline (cellulose) has been increased. Determining the crystallinity of the cellulose itself before and after pretreatment is therefore not practical, especially since the thermal and chemical methods of isolating cellulose can alter its crystallinity [45].

A change to cellulose that has been consistently reported to occur for aqueous pretreatments is partial hydrolysis of cellulose without dissolving it. Partial hydrolysis is determined by examining the change in degree of polymerization (DP) of the cellulose chains in the microfibrils, where DP is the number of anhydro-glucose (glucan) units present in the average cellulose polymer in the microfibrils. This is usually measured by dissolving the biomass in a solvent such as 0.5 M copper ethylenediamine [46]. The chain length of the resulting cellulose solution, usually reported as DP, is then indirectly determined by measuring the viscosity of the solution and computing the viscosity degree of polymerization (DPv) [47]. However, the drawback of this approach is that incomplete dissolution of the cellulose will affect the results.

Changes in DPv for corn stover and poplar processed by various pretreatment methods were reported by Kumar *et al.* [44]. Aqueous pretreatments, such as controlled pH aqueous pretreatment, reduced DPv in corn stover by 23.5% from the initial value of *c.* 7000 while the DPv in poplar was reduced by 49.3%. This difference may be due to the fact that the pretreatment conditions for controlled pH aqueous pretreatment that resulted in the most digestible cellulose were more severe for poplar than for corn stover (210 °C for 10 minutes compared to 190 °C for 15 minutes). For both substrates, the reduction in DPv from aqueous pretreatment was less than for acid pH pretreatments, but similar to reductions measured for alkali pH pretreatments. The implication of this change is in the ability of cellulolytic enzymes to hydrolyze the pretreated cellulose. The primary enzymes involved in cellulose hydrolysis (cellobiohydrolases) act by sequentially hydrolyzing glucose pairs (cellobiose) from the ends of the cellulose chains. The cellulase system is completed with enzymes that hydrolyze cellobiose to glucose (cellobiase or β-glucosidase) or randomly hydrolyze cellulose chains (endoglucanase) to make more chain ends available for hydrolysis to cellobiose [48]. Reducing DP therefore increases the number of sites available for cellobiohydrolases to attach. However, due to the synergistic action of the enzymes in the cellulase system, it is unclear that reducing DP has a significant effect on the overall enzymatic digestibility of cellulose [49,50].

In summary, the chemical changes to biomass that occur as a result of aqueous pretreatment are removal of hemicellulose with minimized degradation of the sugars to aldehydes, rearrangement of lignin within the plant cell-wall matrix with minimal removal, and partial hydrolysis of the cellulose which results in more chain ends for cellulase attachment. Taken together, these changes increase the accessible surface area of cellulose and the number of adsorption sites for cellulases (chain ends) to hydrolyze cellulose to glucose. In addition, aqueous pretreatment partially fractionates the biomass into an aqueous phase that contains hemicellulose oligosaccharides and some hemicellulose-derived monosaccharides, and a solid phase that is mostly cellulose and lignin. This allows for flexible designs for downstream processing. Pentose sugars derived from hemicellulose can be converted with the glucose obtained from cellulose, or the pentose sugars can be separated by filtration or centrifugation from the insoluble cellulose and processed separately into biofuels or higher-value co-products. In thermochemical conversion processes, aqueous pretreatment increases the carbon to oxygen ratio by removing the hemicellulose fraction from the biomass while leaving nearly all of the lignin with the solid material [6,10,12,29,30]. This has the impact of increasing the energy density of the solid residual and thus reducing the demand for hydrogen in hydropyrolysis to deoxygenate the products into high-energy fuel molecules [51]. Sulfur also consumes hydrogen during its removal by hydrotreating [52]. Aqueous-pretreated biomass does not therefore require significant washing to remove residual sulfuric acid and avoids the thiols formed on lignin during the Kraft pulping process (1–2%) [53]. This is also an important consideration due to stringent limits on sulfur in fuels as part of the Tier 2 Vehicle and Gasoline Sulfur program (42 US Codes Service 7521; Emission standards for new motor vehicles or new motor vehicle engines, available from http://uscode.house.gov).

7.5   Impact of Aqueous Pretreatment on Cellulose Digestibility

The effect of aqueous pretreatment has been studied on numerous cellulosic feedstocks; a summary of several is presented in Table 7.2. Aqueous pretreatment has been most widely studied on herbaceous feedstocks such as straw (e.g., wheat, rice), corn stover, and switchgrass. The optimal pretreatment conditions are usually defined as the conditions that result in the highest yield of sugars from enzymatic hydrolysis following pretreatment. Table 7.2 summarizes the reported yields from hydrolysis of cellulose by enzymes following pretreatment. Comparisons between these results are difficult to make due to differences in the methods used, however. The type and amounts of enzymes used, the duration of hydrolysis, and preparation of the pretreated cellulose for hydrolysis vary significantly among these papers. For example, in some cases the whole solid/liquid slurry after pretreatment is hydrolyzed while in other cases the solids are separated from the pretreatment liquid containing the xylan and washed with water before enzymatic hydrolysis. The ratio

Table 7.2   *Hot water pretreatment conditions and cellulose enzymatic hydrolysis yields for a variety of lignocellulosic feedstocks.*

Substrate	Pretreatment conditions		Enzymatic hydrolysis of glucan (% yield)	Reference(s)
	Temperature (°C)	Time (min)		
Herbaceous				
Corn stover	190–200	15–20	90–95	[10]
Distiller's grains	160	20	>80	[54]
Wheat straw	190–200	6–12	>90	[55]
Rice straw	180	30	85–90	[56]
Sugar cane bagasse	220–230	2–10	85–90	[11,57]
Switchgrass	200–210	5–10	70–95	[58,59]
Hardwood				
Poplar	200–240	5–10	56–90	[12,15,60]
Eucalyptus	210–220	5–10[a]	>95	[61]

[a] Estimated from heat-up profile.

of pretreated solids to liquid medium in which hydrolysis is performed varies significantly from low (1–5% w/w) to high (>15% w/w). Comparison of these results is therefore difficult at best.

If the optimal conditions for aqueous pretreatment are compared to optimal conditions for dilute acid pretreatment, it is found that the temperature is significantly higher (20–50 °C higher) and the processing times longer (5–15 minutes longer) for aqueous pretreatment. As described in Chapter 6, low-pH pretreatment the combination of processing conditions (temperature, time, pH) can be expressed as the combined severity factor (CS) [62,63]:

$$\log(CS) = \log\left[t\exp\left(\frac{T-100}{14.75}\right)\right] - pH$$

where t is the time in minutes and T is the temperature in degrees Celsius. If the temperature dependence of pH for liquid water is taken into account (Figure 7.1), the combined severity factor that achieves optimal pretreatment ranges between 1 and 2 for slightly acidic pH conditions found in aqueous pretreatment. A similar combined severity factor is also optimal for low pH pretreatments, such as dilute acid. However, a longer time and a higher temperature is required to achieve a $\log(CS) = 1$–2 for aqueous pretreatment than for dilute acid pretreatment. In terms of process scale-up, neutral-pH pretreatment systems must therefore be able to attain higher operating temperatures, pressures, and retention times than low-pH pretreatment systems. The addition of acid reduces thermal energy, pressure, and volume requirements, but increases the corrosion resistance requirements for reactor design and construction. These tradeoffs should factor significantly in the design considerations and evaluation of options.

Another conclusion that can be drawn from the summary in Table 7.2 is that woody biomass requires more severe pretreatment conditions for neutral-pH pretreatment than herbaceous biomass. This is true for nearly all pretreatment approaches. Finally, very few published results can be found in the literature for neutral-pH pretreatment of softwoods. The combined severity factors required for low-pH pretreatment of softwoods are the highest of the three categories of biomass (herbaceous, hardwood, and softwood) [6,64]. The temperature/pressure and time requirements to achieve similar combined severity factors for water without the addition of mineral acids approach the critical point of water [65]. The pressures required at such high temperatures are extremely high (>200 atm) and difficult to manage in a continuous processing system where solids must be fed into a reactor at extremely high pressure and temperature.

7.6   Practical Applications of Liquid Hot Water Pretreatment

Scale-ups of liquid hot water pretreatments have been attempted [13,66]. As an example of applying the principles described above, a continuous, liquid hot water pretreatment system was designed, built, and tested in a corn wet milling facility [13]. The system made a slurry of corn fiber, a cellulosic by-product of corn wet milling, and stillage (the liquid bottoms from distillation). Stillage, rather than fresh processing water, was used to minimize the use of additional water by the entire process. The slurry contained 7.8% (w/v) dry corn fiber and 1.7% (w/v) additional suspended solids from the stillage. The stillage also contained 8.7% (w/v) of dissolved solids. This gave a total solids (dissolved and suspended) of 18.2%. Thermal energy was recovered after pretreatment by heating the incoming slurry in a spiral-wound heat exchanger. For process start-up and for trim heat during steady state, steam was directly injected into the slurry after the heat exchanger (Figure 7.5). At steady-state operation, the stillage/corn fiber slurry was processed at a rate of 165 kg/min and steam was injected at a rate of about 12 kg/min.

The pretreatment occurred as the slurry traveled through an insulated coil constructed of 8-inch (20-cm) stainless steel pipe where it was held at 160 °C. The average retention time of solids in the coil was 20 minutes. This temperature/time combination was selected based upon laboratory-scale work performed prior to

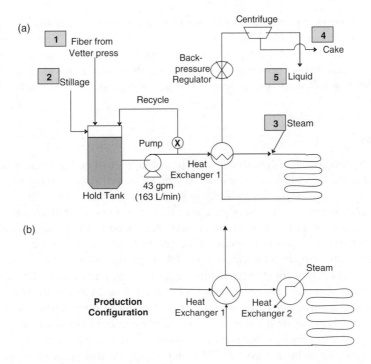

Figure 7.5 *Process flow diagram for aqueous hot water pretreatment of corn fiber. (Reproduced from [13] © 2005, Humana Press Inc).*

design of the pilot-scale equipment. The slurry leaving the pretreatment coil was cooled to 100 °C by the incoming slurry, entering at 97 °C. The pretreated slurry was then separated by a centrifuge to produce a solids cake with 74% moisture and a liquid. Nearly 50% of the fiber solids were dissolved during pretreatment [13]; however, the hydrolysis of the polysaccharides was minimal (Table 7.3). The concentration of glucose-containing oligosaccharides (mostly maltodextrins in this case) nearly tripled during pretreatment while the concentration of xylose-containing oligosaccharides increased by a factor of ten. The

Table 7.3 *Dissolved solids in stillage and pretreatment liquor. (Reproduced from [13] © 2005, Humana Press Inc).*

	Stillage (mg/mL)	Pretreatment liquor (mg/mL)
Glucan	7.1	20.2
Glucose	0.3	5.8
Xylan/galactan	0.9	8.5
Xylose/galatose	1.1	1.1
Arabinan	0.0	1.8
Arabinose	1.0	2.7
Acetic acid	0.0	0.4
Lactic acid	10.4	12.2
Glycerol	15.5	12.6
HMF	0.0	0.4
Furfural	0.0	0.3
2,3 Butanediol	3.4	not measured

concentration of monomeric xylose was largely unchanged during pretreatment and the degradation of xylose to furfural was minimal (Table 7.3). As a result, the furfural and HMF were too low to cause significant slowing of the fermentation of the sugars resulting from enzymatic hydrolysis of the pretreated solids [13]. These results suggest that the principle of limiting the hydrolysis of hemicellulose to monomeric sugar to prevent its degradation will work on a commercial scale, as suggested by the kinetic analyses performed at the laboratory scale.

7.7 Conclusions

Neutral pH aqueous pretreatment of lignocellulosic biomass capitalizes on the acidic nature of hot, pressurized liquid water and the organic acids released from biomass during processing to partially hydrolyze hemicellulose and alter the arrangement of lignin in the cell-wall matrix. While the chemical mechanism for neutral-pH aqueous pretreatment is very similar to that for acidic-pH pretreatment, no additional acid catalysts beyond what is naturally found in the biomass are added to the process. Generally, neutral-pH aqueous pretreatment results in low conversion of the hemicellulose to monosaccharides, unlike acid-pH pretreatments. The relationship between temperature and the pK_w of water allows for some degree of control over pretreatment severity without requiring the addition of an acid catalyst. This allows for tuning of pretreatment conditions to prevent complete hydrolysis of the hemicellulose in the plant biomass to monosaccharides. This has the advantage of limiting the loss of hemicellulose-derived pentoses to furfural or other products. The disadvantage in this approach is the need for enzymes or other catalysts to complete the hydrolysis to fermentable sugars in subsequent processing steps. These enzymes can be supplied exogenously or through consolidated bioprocessing (CBP) microorganisms that synthesize and secrete the needed enzymes. The best applications of neutral-pH aqueous pretreatment are where lower combined severity factors are required to achieve effective pretreatment (such as with herbaceous plant biomass); extreme temperatures and pressures are however required to attain the high severity factors necessary to achieve effective cellulose pretreatment in highly lignified feedstocks such as softwood.

References

1. Tao, L., Aden, A., Elander, R.T. *et al.* (2011) Process and technoeconomic analysis of leading pretreatment technologies for lignocellulosic ethanol production using switchgrass. *Bioresource Technology*, **102** (24), 11105–11114.
2. Eggeman, T. and Elander, R.T. (2005) Process and economic analysis of pretreatment technologies. *Bioresource Technology*, **96** (18), 2019–2025.
3. Bobleter, O., Niesner, R., and Röhr, M. (1976) The hydrothermal degradation of cellulosic matter to sugars and their fermentative conversion to protein. *Journal of Applied Polymer Science*, **20** (8), 2083–2093.
4. Garrote, G., Dominguez, H., and Parajo, J.C. (2001) Generation of xylose solutions from eucalyptus globulus wood by autohydrolysis-posthydrolysis processes: Posthydrolysis kinetics. *Bioresource Technology*, **79** (2), 155–164.
5. Garrote, G., Dominguez, H., and Parajo, J.C. (2002) Interpretation of deacetylation and hemicellulose hydrolysis during hydrothermal treatments on the basis of the severity factor. *Process Biochemistry*, **37**, 1067–1073.
6. Mosier, N., Wyman, C., Dale, B. *et al.* (2005) Features of promising technologies for pretreatment of lignocellulosic biomass. *Bioresource Technology*, **96** (6), 673–686.
7. Cooper, J.R. and Dooley, R.B. (2007) Release on the ionization constant of H_2O, The International Association for the Properties of Water and Steam. Technical report by The International Association for the Properties of Water and Steam, http://www.iapws.org/.
8. Lloyd, T. and Wyman, C. (2004) Predicted effects of mineral neutralization and bisulfate formation on hydrogen ion concentration for dilute sulfuric acid pretreatment. *Applied Biochemistry and Biotechnology*, **115** (1), 1013–1022.
9. Schädel, C., Blöchl, A., Richter, A., and Hoch, G. (2010) Quantification and monosaccharide composition of hemicelluloses from different plant functional types. *Plant Physiology and Biochemistry*, **48**, 1–8.

10. Mosier, N., Hendrickson, R., Ho, N. *et al.* (2005) Optimization of pH controlled liquid hot water pretreatment of corn stover. *Bioresource Technology*, **96**, 1986–1993.

11. Jacobsen, S.E. and Wyman, C.E. (2002) Xylose monomer and oligomer yields for uncatalyzed hydrolysis of sugar-cane bagasse hemicellulose at varying solids concentration. *Industrial & Engineering Chemistry Research*, **41** (6), 1454–1461.

12. Kim, Y., Mosier, N.S., and Ladisch, M.R. (2009) Enzymatic digestion of liquid hot water pretreated hybrid poplar. *Biotechnology Progress*, **25** (2), 340–348.

13. Mosier, N.S., Hendrickson, R., Brewer, M. *et al.* (2005) Industrial scale-up of pH controlled liquid hot water pre-treatment of corn fiber for fuel ethanol production. *Applied Biochemistry and Biotechnology*, **125** (2), 77–98.

14. Nabarlatz, D., Farriol, X., and Montane, D. (2005) Autohydrolysis of almond shells for the production of xylo-oligosaccharides: product characteristics and reaction kinetics. *Industrial & Engineering Chemistry Research*, **44** (20), 7746–7755.

15. Weil, J., Sarikaya, A., Rau, S.-L. *et al.* (1997) Pretreatment of yellow poplar sawdust by pressure cooking in water. *Applied Biochemistry and Biotechnology*, **68** (1–2), 21–40.

16. Weil, J.R., Sarikaya, A., Rau, S.-L. *et al.* (1998) Pretreatment of corn fiber by pressure cooking in water. *Applied Biochemistry and Biotechnology*, **73**, 1–17.

17. Grenman, H., Eranen, K., Krogell, J. *et al.* (2011) Kinetics of aqueous extraction of hemicelluloses from spruce in an intensified reactor system. *Industrial & Engineering Chemistry Research*, **50** (7), 3818–3828.

18. Nabarlatz, D., Farriol, X., and Montane, D. (2004) Kinetic modeling of the autohydrolysis of lignocellulosic bio-mass for the production of hemicellulose-derived oligosaccharides. *Industrial & Engineering Chemistry Research*, **43** (15), 4124–4131.

19. Shen, J. and Wyman, C.E. (2011) A novel mechanism and kinetic model to explain enhanced xylose yields from dilute sulfuric acid compared to hydrothermal pretreatment of corn stover. *Bioresource Technology*, **102** (19), 9111–9120.

20. Mittal, A., Chatterjee, S.G., Scott, G.M., and Amidon, T.E. (2009) Modeling xylan solubilization during autohy-drolysis of sugar maple and aspen wood chips: Reaction kinetics and mass transfer. *Chemical Engineering Science*, **64**, 3031–3041.

21. Saeman, J.F. (1945) Kinetics of wood saccharification hydrolysis of cellulose and decomposition of sugars in dilute acid at high temperature. *Industrial & Engineering Chemistry Research*, **37**, 42–52.

22. Kobayashi, T. and Sakai, Y. (1956) Hydrolysis rate of pentosan of hardwood in dilute sulfuric acid. *Bulletin of the Agricultural Chemical Society of Japan*, **20** (1), 1–7.

23. Wyman, C.E., Decker, S.R., Himmel, M.E. *et al.* (2005) Hydrolysis of cellulose and hemicellulose, in *Polysacchar-ides: Structural Diversity and Functional Versatility* (ed. S. Dumitriu), CRC Press, Boca Raton, pp. 995–1033.

24. Weingarten, R., Cho, J., Conner, W.C., and Huber, G.W. (2010) Kinetics of furfural production by dehydration of xylose in a biphasic reactor with microwave heating. *Green Chemistry*, **12** (8), 1423–1429.

25. Sannigrahi, P., Kim, D.H., Jung, S., and Ragauskas, A. (2011) Pseudo-lignin and pretreatment chemistry. *Energy & Environmental Science*, **4** (4), 1306–1310.

26. Abatzoglou, N., Bouchard, J., Chornet, E., and Overend, R.P. (1986) Dilute acid depolymerization of cellulose in aqueous phase – experimental-evidence of the significant presence of soluble oligomeric intermediates. *Canadian Journal of Chemical Engineering*, **64** (5), 781–786.

27. Brennan, M.A. and Wyman, C.E. (2004) Initial evaluation of simple mass transfer models to describe hemicellulose hydrolysis in corn stover. *Applied Biochemistry and Biotechnology*, **113** (16), 965–976.

28. Abasaeed, A.E., Lee, Y.Y., and Watson, J.R. (1991) Effect of transient heat-transfer and particle-size on acid-hydrolysis of hardwood cellulose. *Bioresource Technology*, **35** (1), 15–21.

29. Liu, C. and Wyman, C.E. (2003) The effect of flow rate of compressed hot water on xylan, lignin, and total mass removal from corn stover. *Industrial & Engineering Chemistry Research*, **42** (21), 5409–5416.

30. Liu, C. and Wyman, C.E. (2004) Impact of fluid velocity on hot water only pretreatment of corn stover in a flow-through reactor. *Applied Biochemistry and Biotechnology*, **113–116**, 977–987.

31. Yang, B., Gray, M., Liu, C. *et al.* (2004) Unconventional relationships for hemicellulose hydrolysis and subsequent cellulose digestion, lignocellulose biodegradation. *ACS Symposium Series* 889, pp. 100–125.

32. Wyman, C.E. (2005) Hydrolysis of cellulose and hemicellulose, polysaccharides: structural diversity and functional versatility, in *Polysaccharides: Structural Diversity and Functional Versatility* (ed. S. Dumitriu), CRC Press, Boca Raton 995–1034.

33. Liu, C. and Wyman, C.E. (2006) The enhancement of xylose monomer and xylotriose degradation by inorganic salts in aqueous solutions at 180 degrees C. *Carbohydrate Research*, **341** (15), 2550–2556.

34. Marcotullio, G. and De Jong, W. (2010) Chloride ions enhance furfural formation from D-xylose in dilute aqueous acidic solutions. *Green Chemistry*, **12** (10), 1739–1746.

35. Esteghlalian, A., Hashimoto, A.G., Fenske, J.J., and Penner, M.H. (1997) Modeling and optimization of the dilute-sulfuric-acid pretreatment of corn stover, poplar and switchgrass. *Bioresource Technology*, **59** (2–3), 129–136.

36. Lee, Y.Y., Lyer, P., and Torget, R.W. (1999) Dilute-acid hydrolysis of lignocellulosic biomass advances. *Biochemical Engineering/Biotechnology*, **65**, 93–115.

37. Pronyk, C. and Mazza, G. (2010) Kinetic modeling of hemicellulose hydrolysis from triticale straw in a pressurized low polarity water flow-through reactor. *Industrial & Engineering Chemistry Research*, **49** (14), 6367–6375.

38. Helmerius, J., von Walter, J.V., Rova, U. *et al.* (2010) Impact of hemicellulose pre-extraction for bioconversion on birch Kraft pulp properties. *Bioresource Technology*, **101** (15), 5996–6005.

39. Selig, M.J., Viamajala, S., Decker, S.R. *et al.* (2007) Deposition of lignin droplets produced during dilute acid pretreatment of maize stems retards enzymatic hydrolysis of cellulose. *Biotechnology Progress*, **23**, 1333–1339.

40. Kristensen, J.B., Thygesen, L.G., Felby, C. *et al.* (2008) Cell-wall structural changes in wheat straw pretreated for bioethanol production. *Biotechnology for Biofuels*, **1**, 5.

41. Donohoe, B.S., Decker, S.R., Tucker, M.P. *et al.* (2008) Visualizing lignin coalescence and migration through maize cell walls following thermochemical pretreatment. *Biotechnology and Bioengineering*, **101** (5), 913–925.

42. Yang, B. and Wyman, C.E. (2006) BSA treatment to enhance enzymatic hydrolysis of cellulose in lignin containing substrates. *Biotechnology and Bioengineering*, **94** (4), 611–617.

43. Laureano-Perez, L., Teymouri, F., Alizadeh, H., and Dale, B.E. (2005) Understanding factors that limit enzymatic hydrolysis of biomass: Characterization of pretreated corn stover. *Applied Biochemistry and Biotechnology*, **121–124**, 1081–1099.

44. Kumar, R., Mago, G., Balan, V., and Wyman, C.E. (2009) Physical and chemical characterizations of corn stover and poplar solids resulting from leading pretreatment technologies. *Bioresource Technology*, **100** (17), 3948–3962.

45. Weimer, P.J., Hackney, J.M., and French, A.D. (1995) Effects of chemical treatments and heating on the crystallinity of celluloses and their implications for evaluating the effect of crystallinity on cellulose biodegradation. *Biotechnology and Bioengineering*, **48** (2), 169–178.

46. ASTM (1986) *Standard test methods for intrinsic viscosity of cellulose (D 1795)*. American Society for Testing and Materials, West Conshohocken, PA.

47. Van Heiningen, A., Tunc, M.S., Gao, Y., and Perez, D.D. (2004) Relationship between alkaline pulp yield and the mass fraction and degree of polymerization of cellulose in the pulp. *Journal of Pulp and Paper Science*, **30** (8), 211–217.

48. Mosier, N.S., Hall, P., Ladisch, C.M., and Ladisch, M.R. (1999) Reaction kinetics, molecular action, and mechanisms of cellulolytic proteins. *Advances in Biochemical Engineering/Biotechnology*, **65**, 24–40.

49. Nazhad, M.M., Ramos, L.P., Paszner, L., and Saddler, J.N. (1995) Structural constraints affecting the initial enzymatic-hydrolysis of recycled paper. *Enzyme and Microbial Technology*, **17** (1), 68–74.

50. Mansfield, S.D., de Jong, E., Stephens, R.S., and Saddler, J.N. (1997) Physical characterization of enzymatically modified kraft pulp fibers. *Journal of Biotechnology*, **57** (1–3), 205–216.

51. Singh, N.R., Delgass, W.N., Ribeiro, F.H., and Agrawal, R. (2010) Estimation of liquid fuel yields from biomass, environmental. *Science & Technology*, **44** (13), 5298–5305.

52. Farrauto, R.J. and Bartholomew, C. (1997) *Fundamentals of Industrial Catalytic Processes*, Chapman & Hall, London, UK.

53. Brudin, S. and Schoenmakers, P. (2010) Analytical methodology for sulfonated lignins. *Journal of Separation Science*, **33**, 439–452.

54. Kim, Y., Hendrickson, R., Mosier, N.S. *et al.* (2008) Enzyme hydrolysis and ethanol fermentation of liquid hot water and AFEX pretreated distillers' grains at high-solids loadings. *Bioresource Technology*, **99**, 5206–5215.

55. Petersen, M.O., Larsen, J., and Thomsen, M.H. (2009) Optimization of hydrothermal pretreatment of wheat straw for production of bioethanol at low water consumption without addition of chemicals. *Biomass and Bioengineering*, **33** (5), 834–840.

56. Yu, G., Yano, S., Inoue, H. *et al.* (2010) Pretreatment of rice straw by a hot-compressed water process for enzymatic hydrolysis. *Applied Biochemistry and Biotechnology*, **160**, 539–551.

57. Laser, M., Schulman, D., Allen, S.G. *et al.* (2002) A comparison of liquid hot water and steam pretreatments of sugar cane bagasse for bioconversion to ethanol. *Bioresource Technology*, **81** (1), 33–44.

58. Hu, Z. and Ragauskas, A.J. (2011) Hydrothermal pretreatment of switchgrass. *Industrial & Engineering Chemistry Research*, **50**, 4225–4230.

59. Kim, Y., Mosier, N.S., Ladisch, M.R. *et al.* (2011) Comparative study on enzymatic digestibility of switchgrass varieties and harvests processed by leading pretreatment technologies. *Bioresource Technology*, **102** (24), 11089–11096.

60. Wyman, C.E., Dale, B.E., Elander, R.T. *et al.* (2009) Comparative sugar recovery and fermentation data following pretreatment of poplar wood by leading technologies. *Biotechnology Progress*, **25** (2), 333–339.

61. Romani, A., Garrote, G., Alonso, J.L., and Parajo, J.C. (2010) Bioethanol production from hydrothermally pre-treated *Eucalyptus globulus* wood. *Bioresource Technology*, **101**, 8706–8712.

62. Chum, H.L., Johnson, D.K., Black, S.K., and Overend, R.P. (1990) Pretreatment-catalyst effects and the combined severity parameter. *Applied Biochemistry and Biotechnology*, **24–25** (1), 1–14.

63. Abatzoglou, N.J., Chornet, E., Belkacemi, K., and Overend, R.P. (1992) Phenomenological kinetics of complex systems: The development of a generalized severity parameter and its application to lignocellulosic fractionation. *Chemical Engineering Science*, **47** (5), 1109–1122.

64. Galbe, M. and Zacchi, G. (2007) Pretreatment of lignocellulosic materials for efficient bioethanol production, in *Advances in Biochemical Engineering-Biotechnology-Biofuels*, (ed. L Olsson), Springer, **108**, pp. 41–65.

65. Kim, K.H., Eom, I.Y., Lee, S.M. *et al.* (2010) Applicability of sub- and supercritical water hydrolysis of woody biomass to produce monomeric sugars for cellulosic bioethanol fermentation. *Journal of Industrial and Engineering Chemistry*, **16** (6), 918–922.

66. Thomsen, M.H., Thygesen, A., and Thomsen, A.B. (2008) Hydrothermal treatment of wheat straw at pilot plant scale using a three-step reactor system aiming at high hemicellulose recovery, high cellulose digestibility and low lignin hydrolysis. *Bioresource Technology*, **99** (10), 4221–4228.

8

Fundamentals of Biomass Pretreatment at High pH

Rocío Sierra Ramirez[1,*], Mark Holtzapple[1] and Natalia Piamonte[2]

[1] *Department of Chemical Engineering, Texas A&M University, College Station, USA*
[2] *Department of Chemical Engineering, University of the Andes, Bogota, Colombia*

8.1 Introduction

Alkaline biomass pretreatments alter lignocellulose structure with the goal of significantly increasing enzymatic digestibility. As a general rule, alkaline biomass pretreatments and paper pulping have the same objective: remove lignin. However, there are important differences. Paper pulping aims to preserve fiber strength, but this is not important in alkaline pretreatment. Complete or nearly complete lignin removal is important for paper pulping, but is not necessary for effective biomass pretreatment. Paper pulping creates a high-value product whereas biomass pretreatment creates a low-value product. Because of these differences, technologies that are effective in paper pulping may not translate well to biomass pretreatment.

A clear understanding of the chemical alterations that occur in alkaline biomass pretreatment is fundamental for process control and optimization. This is why the chemistry of the process – with or without an oxidative agent – has been the subject of extensive work [1–3]; however, most research has been directed to applications in the paper industry.

The dominant effects of alkaline pretreatment occur on cellulose, hemicellulose, and lignin; however, it also partially removes waxes, silica, and waterproof cutins that coat plant tissue [4,5]. Most alkaline conditions are very effective at removing acetyl groups from xylan polymers [6,7]. All of these outcomes from alkaline pretreatment improve cellulase access and thereby increase digestibility [8–10].

Compared to acid pretreatments, alkaline pretreatments tend to be more gentle and therefore lose less carbohydrate. Because acid pretreatments are very aggressive, expensive vessels and careful control of

---

* Present address: Department of Chemical Engineering, University of the Andes, Bogota, Colombia

---

Aqueous Pretreatment of Plant Biomass for Biological and Chemical Conversion to Fuels and Chemicals, First Edition.
Edited by Charles E. Wyman.
© 2013 John Wiley & Sons, Ltd. Published 2013 by John Wiley & Sons, Ltd.

operating conditions are required. In contrast, alkaline pretreatments are much less aggressive, so vessel costs are lower and control of operating conditions is less critical. Furthermore, compared with acid or oxidative reagents, alkaline treatment appears to be the most effective method in breaking ester bonds between lignin and polysaccharides, and it avoids fragmenting hemicellulose polymers [11].

Regardless of the alkali employed, lignin modification is a common outcome of alkaline pretreatment. When large amounts of water are present with the alkali, lignin fragments can dissolve and be removed from the biomass [12]. When water is restricted (e.g., Ammonia Fiber Expansion or AFEX), alkali can alter the lignin structure so that it is no longer a barrier to enzymatic hydrolysis even though the lignin is not physically removed from the biomass. The AFEX process is discussed in more detail in Chapter 9. Here, attention is focused on alkaline delignification processes that have sufficient water to solubilize and remove lignin from the biomass.

In native lignocellulose, cellulose and hemicellulose are cemented together by lignin, which is responsible for integrity, structural rigidity, and prevention of fiber swelling [13]. Reducing the amount of lignin (or altering its composition) in the cell wall would increase the digestibility of the lignocellulosic material [14]. Additionally, delignification promotes biomass swelling, which increases internal surface area and median pore volume [8]. Partial delignification is sufficient to alter biomass composition so that full carbohydrate digestion is subsequently accomplished [12,15,16].

The required lignin removal depends highly on the chosen feedstock. For example, if the feedstock is woody biomass (e.g., hardwood, softwood), lignin removal is difficult because lignin is present in large quantities. In this case, extensive delignification (50–70% lignin removal) is needed to significantly increase biomass digestibility. In contrast, for herbaceous biomass characterized by a lower lignin content (e.g., grasses, agricultural residues), less extensive delignification (30–50% lignin removal) is needed to significantly increase biomass digestibility. This is probably the reason why it is often asserted that alkaline pretreatment is more effective on herbaceous than woody biomass.

This chapter reviews chemical changes that occur to lignocellulose when submitted to alkaline conditions with and without an oxidative agent. We briefly discuss and compare prominent alkaline pretreatment options: ammonia pretreatment, sodium hydroxide pretreatment, wet alkaline oxidation, and oxidative and non-oxidative lime. Finally, we review reaction kinetic modeling and other key topics such as severity and selectivity. This chapter does not discuss alkali-catalyzed pyrolysis because this procedure does not seek subsequent enzymatic digestion. It also excludes AFEX, an alkaline pretreatment involving somewhat different chemistry that is discussed in Chapter 9.

8.2   Chemical Effects of Alkaline Pretreatments on Biomass Composition

The main effect of alkaline pretreatment is delignification. Although the reactions that remove lignin have been extensively researched [17–19], they are far from completely understood. Lignin is a very complex polymer that is not readily removed from lignocellulose [20]. Research has therefore been conducted to better understand the three-dimensional structure of lignin polymer [21]; other studies have focused on using model compounds that simulate lignin to understand its chemistry. Some of these compounds include: guaiacylglycerol-β-guaiacyl ether [22]; 2-(2-methoxyphenoxy)-1-(3,4-dimethoxyphenyl)propane-1,3-diol, 2-(3,5-difluorophenoxy)-1-(4-hydroxy-3-methoxyphenyl)propane-1,3-diol, and 2-(3,5-difluorophenoxy)-1-(3,4-dimethoxyphenyl)propane-1,3-diol [23]; dihydroanisoin, veratrylglycerol-β-guaiacyl ether and veratryl alcohol [24]; 2-methoxy-4-(2-propenyl)phenol; and 4-ethyl-2-methoxyphenol [25].

Full biomass delignification is difficult because lignin is located deep within the cell wall. Lignin has additional challenges such as hydrophobicity, physical stiffness, strong poly-ring bonds (C—O—C and C—C), and its tendency to recondense [13,26]. Because of covalent bonds between lignin and hemicellulose, alkaline delignification tends to solublize hemicellulose. Additionally, alkaline conditions degrade xylan and remove acetyl groups from hemicelluloses [8,10]. Moreover, alkali also degrades some cellulose; however, it is much more resistant to attack than hemicellulose. All reactions are enhanced in the presence of an oxidative agent [2].

During alkaline pretreatment, reactions produce a large number of solubilized products in which the nature and quantity are determined by the pretreatment conditions (e.g., temperature, pressure, composition of the reacting media, and reaction time). These solubilized products may inhibit the fermentation, modify the alkalinity of the reacting media, and/or affect reaction rates; it is therefore important to understand the reactions that lead to their production.

For softwoods and hardwoods, dominant reactions in alkaline oxidative and non-oxidative media are extensively discussed in the literature for application in the paper industry [27–29]; new insights are constantly being published [30] and new feedstocks incorporated [31]. The basis for these reactions is presented in the following; however, readers are cautioned that these reactions may not be directly applicable to biomass pretreatment. Differences may result because paper pulping favors one alkali (e.g., sodium hydroxide) with the application of other bleaching enhancing agents, whereas biomass pretreatment may favor other alkalis (e.g., ammonia, lime) and seldom uses chemical enhancers. Also, paper pulping favors woody biomass, whereas biomass pretreatment may favor herbaceous biomass.

8.2.1   Non-oxidative Delignification

In alkaline media, three delignification stages are clearly distinguished: (1) initial (activation energy *c*. 61 kJ/mol, temperature <170 °C); (2) bulk (activation energy *c*. 150 kJ/mol, temperature *c*. 170 °C); and (3) terminal or residual (activation energy *c*. 120 kJ/mol, temperature *c*. 170 °C) [3]. In all stages, the primary lignin reactions depend on the phase, pH, and temperature of the reacting media. These reactions are of three types: (1) reactions that do not involve net fragmentation of the lignin macromolecule, yet yield alkali-stable structures that make additional lignin fragmentation difficult; (2) degradation reactions that form lignin fragments and eventually dissolve; and (3) condensation reactions that form alkali-stable linkages and increase the molecular size of lignin fragments, which might result in their precipitation. Although having an opposite effect, lignin fragmentation and condensation reactions are intimately connected with each other because they proceed via common intermediates [1].

The most prevalent degradation reactions are the cleavage of α-aryl ether and β-aryl ether linkages in phenolic units, which preferably occur at the initial stage of delignification [22]. This stage is characterized by low activation energy; thus, lignin fragmentation reactions must occur at the same or higher rate than diffusion. During the initial stage, lignin degradation involving only phenolic units may continue until it reaches units that are not of the α- or β-aryl ether type (peeling of lignin). The α-aryl ether linkage is easily cleaved. The reaction involves an alkali-assisted rearrangement of the phenolate structure to the corresponding quinone methide structure with elimination of the α-aryloxy substituent (Figure 8.1) [32,33].

The cleavage of β-aryl ether bonds in phenolic units also proceeds rapidly, with the initial conversion into quinone methide and constitutes the rate-determining step in the overall reaction. Phenolic β-aryl ether structures can also undergo an alkali-assisted transformation into a quinone methide, provided a suitable leaving group exists at the α-position of the propane side chain [34,35]. The quinone methide intermediate can then react in different ways to restore the aromaticity of the aromatic ring. The quinone methide could be converted into an alkali-stable enol ether structure by eliminating the terminal hydroxymethyl group as formaldehyde [34]. This reaction is more prominent in the absence of hydrosulfide ions, as occurs during alkaline pretreatment.

The cleavage of β-aryl ether linkages in non-phenolic structures is a relatively slow reaction that depends on the presence of phenylpropane units that have a hydroxyl group in the α- and γ-carbon. The result is the formation of an alkoxide anion with an oxirane intermediate and the concomitant elimination of the

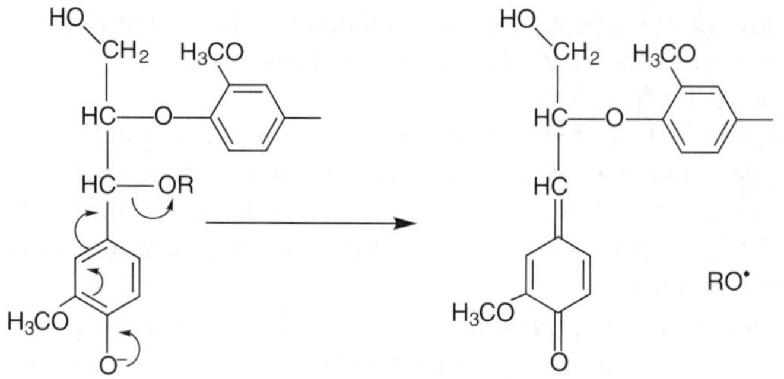

Figure 8.1   *Alkaline cleavage of the α-aryl ether linkage in phenylpropane units [3].*

β-aryloxy substituent. Finally, ions in the reacting media break the epoxide ring, producing diols or thioglycol-type structures (Figure 8.2) [3].

Lignin degradation and fragmentation reactions are believed to be counter-balanced by condensation reactions in alkaline media [36]. Primary condensation reactions occur when quinone methide intermediates are formed by eliminating an α-substituent, whereas secondary condensation reactions occur with quinone methide structures formed after an initial ether cleavage. Condensation reactions may be viewed as conjugate additions in which quinone methides, extended quinone methides, or side-chain enone structures function as acceptors, and carbanions from phenolic or enolic units serve as adding nucleophiles [3].

8.2.2   Non-oxidative Sugar Degradation

In alkaline media, carbohydrate polymers in lignocellulose are degraded according to mechanisms that depend on pretreatment conditions (i.e., pH, temperature, and time). The loss is particularly high for glucomannan in pine wood [37] and xylans in other lignocellulosic sources [16]; however, cellulose losses occur as well. All carbohydrate losses are particularly high at the beginning of pretreatment; after this unfavorable period, the selectivity for lignin degradation improves [38].

Carbohydrate degradation starts from the reducing end in the polysaccharide chains (primary peeling). From these reactions, new reducing ends emerge resulting in secondary peeling. Competing stopping reactions form alkali-stable acid end-groups, some of which are 3-deoxyhexonic acid (metasaccharinic), 2-C-methylglyceril acid, and 3-deoxy-2-C-hydroxymethylpentanoic acid (glucoisosaccharinic) [28]. Figure 8.3 depicts the Nef-Isbell mechanism for alkaline degradation of glucose.

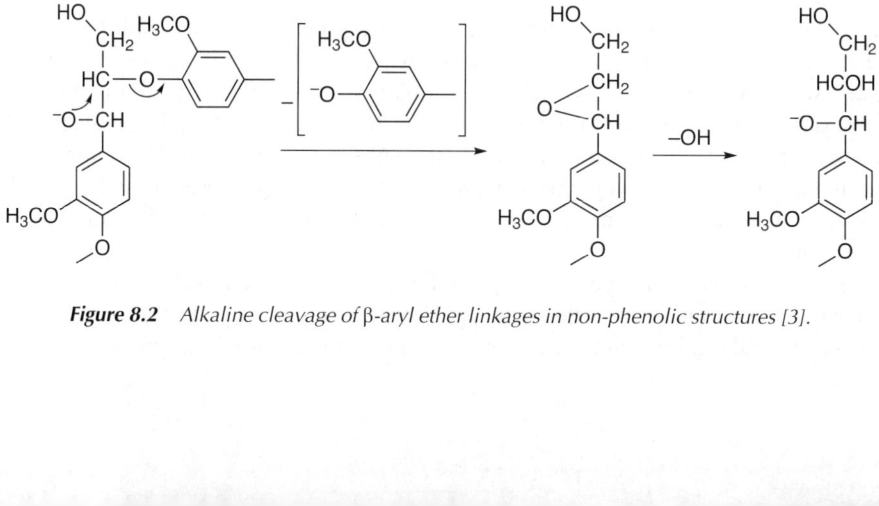

Figure 8.2   *Alkaline cleavage of β-aryl ether linkages in non-phenolic structures [3].*

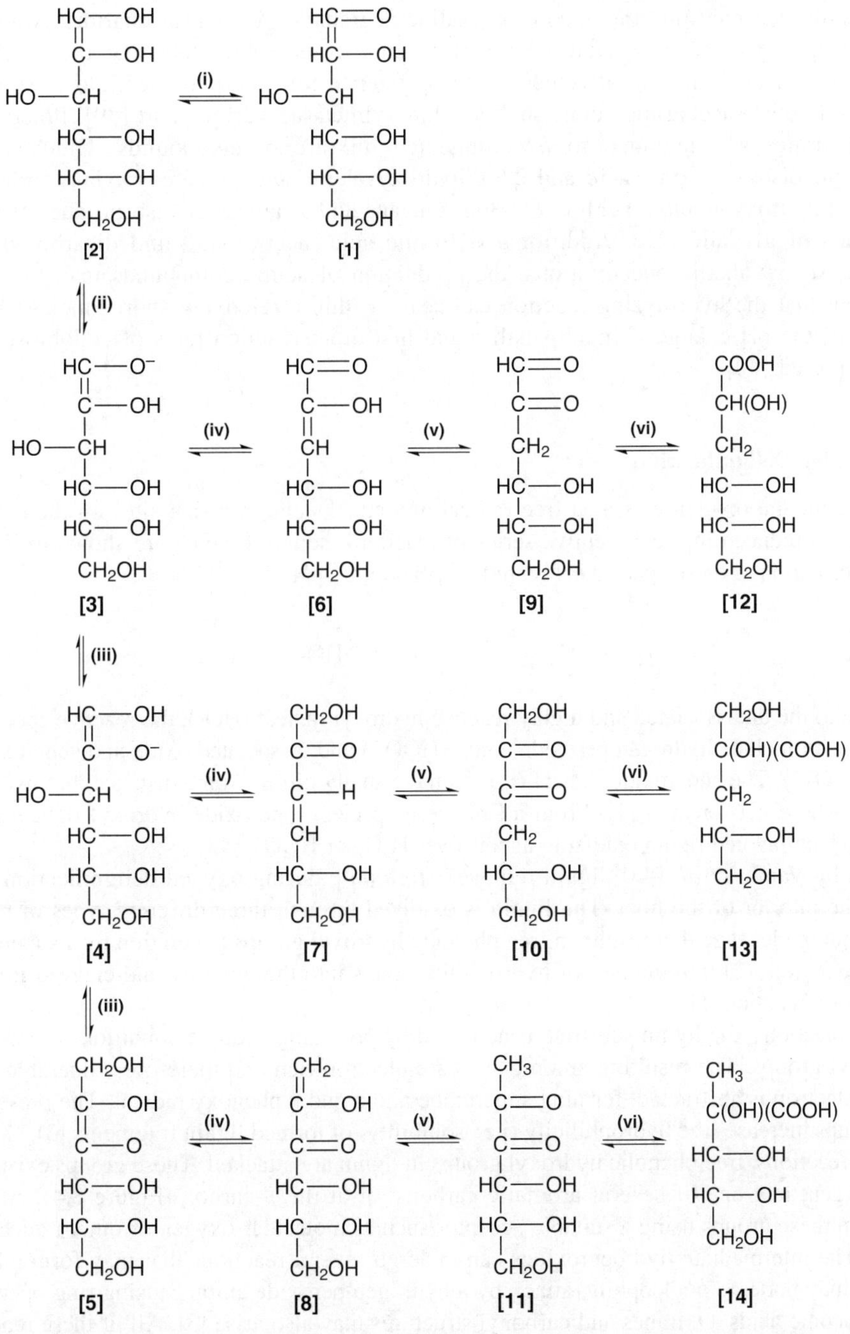

Figure 8.3 *Nef-Isbell mechanism for the formation of saccharinic acids from D-glucose in alkaline media [28]. (i) Keto-enol tautomerism; (ii) enediol deprotonation; (iii) anion isomerization; (iv) β-hydrocarbonyl elimination; (v) keto-enol tautomerism; (vi) benzilic acid rearrangement; (1) D-glucose; (2) enediol; (3–5) intermediate anions; (6–8) enols; (9–11) dicarbonyl compounds; and (12–14) deoxyaldonic (saccharinic) acids.*

It is generally accepted that the extent of alkaline degradation of polysaccharides is determined by the peeling-stopping reaction sequence. However, some researchers state that the mechanisms are much more complicated and are affected by the preference for amorphous cellulose degradation and the presence of some saccharinic acids, such as glucosylmetasaccharinic acid [39]. Practically all isolated carbohydrates are degraded to a complicated mixture of compounds. Glucans are mainly degraded to glucoisosaccharinic acid and 2,5-dihydroxyvaleric acids, whereas xylans form xyloisosaccharinic and 2-hydroxybutanoic acids [28]. Both glucans and xylans may also produce lactic acid and small amounts of glycolic acid. Additionally, formic acid, acetic acid, and dicarboxylic acids are formed [37]. At low alkali concentrations, the production of acidic compounds may lower the pH to such an extent that the hydrolyzing reaction can cease within a relatively short time. At higher alkali concentrations, the pH change is much smaller and first-order reaction rates of cellobiose degradation have been reported [40].

8.2.3 Oxidative Delignification

In an alkaline media, oxygen acts as a free radical and attacks electron-rich sites on the substrate. As a result, oxygen is reduced in a consecutive series of reactions, some of which are shown in Equation (8.1) [1,41]. The formation of these species is favored at a maximum pH of 11–11.5.

$$O_2 \xrightarrow{e^-,H^+} HO_2 \cdot \xrightarrow{e^-,H^+} H_2O_2 \xrightarrow{e^-,H^+} HO \cdot \xrightarrow{e^-,H^+} H_2O \qquad (8.1)$$

In addition to the undissociated and highly reactive hydroxyl radical ($HO\cdot$), the reactive species generated from oxygen may include hydrogen peroxide anion (HOO^-) and dissociated oxide and superoxide hydroxyl radical ($\cdot O^-$, $\cdot O_2^-$). The superoxide radical ($\cdot O_2^-$) may usually diffuse some distance, but hydroxyl radical ($HO\cdot$) reacts where it is created [1]. From all of these species, superoxide hydroxyl radical and hydroxyl radical may be the favored lignin oxidizing agents over H_2O_2 or HOO^- [42].

As shown by Vainio *et al.* [43], lignin removed from pulp during oxygen delignification is primarily taken from the interior of the fiber. The lignin is oxidized through three different types of reactions: (1) inter-unit bonding cleavage that results in free phenolic hydroxyl groups generation by a superoxide-driven Fenton-type reaction; (2) introduction of hydrophilic groups into the polymer and cleaved fragments; and (3) condensation reactions [1].

In Type 1 reactions, the lignin substrate is activated by providing alkaline conditions to ionize free phenolic hydroxyl groups. The resulting anionic sites are electron-rich and therefore vulnerable to attack by oxygen. An electron is abstracted, forming superoxide anion and a phenoxy radical. The presence of these hydroxyl groups increases the hydrophilicity (i.e., solubility) of formed lignin fragments [3].

In Type 2 reactions, free phenolic hydroxyl groups in lignin are attacked. These groups exist at either the phenolic oxygen, any one of several aromatic carbons, or at the β-carbon (Figure 8.4). An electron is removed from these groups using a suitable acceptor such as molecular oxygen or other radicals present in the system. The intermediate hydroperoxides can undergo several reactions that may form quinone structures, which may undergo nucleophilic attack by a hydrogen peroxide anion causing ring opening and formation of muconic acids. Oxiranes and carbonyl structures may also arise [3]. All of these reactions imply breakage of lignin inter-linkages, thereby increasing lignin solubility (Figure 8.5).

In Type 3 reactions, condensation products are formed. The phenoxy radicals that are formed in the initial reactions may undergo coupling reactions resulting in carbon-carbon bonds between lignin units [44]. These condensation products will increase the molecular weight of the lignin, decrease its solubility, and make the lignin unreactive to the attack of oxygen [41].

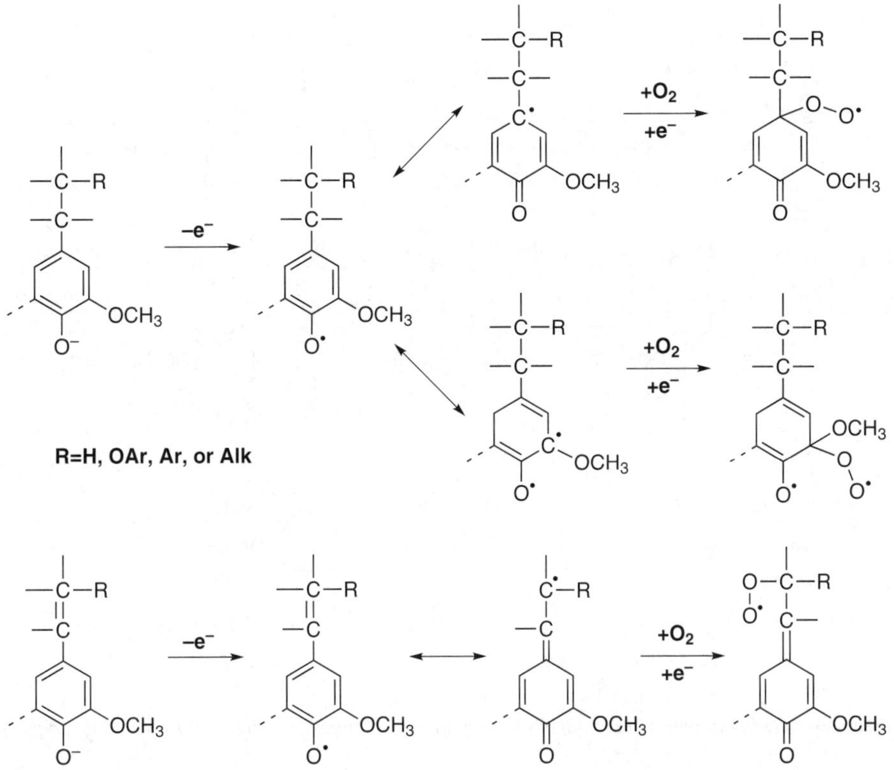

Figure 8.4 *Initial reactions in oxygen delignification: free phenolic hydroxyl groups in lignin are attacked [3].*

8.2.4 Oxidative Sugar Degradation

Some studies assert that, during oxygen delignification, up to 35–50% of lignin must be removed before significant cellulose degradation occurs [45]. Other studies affirm that carbohydrate degradation occurs concurrently with lignin degradation and may be fast initially and then slows as pretreatment proceeds [38]. During oxidative-alkaline treatment, processes involved in cellulose degradation have not been clearly elucidated; however, knowledge of the mechanisms of this degradation is needed to control this undesired event.

During oxidative alkaline treatment of lignocellulose, most studies report that carbohydrate degradation results from the presence of several highly oxidative species present in reacting media to which oxygen has been added [46]. These species include – but are not limited to – dioxygen, hydroxyl radical, and superoxide anion radical (Equation (8.1)). Literature reviews [47] assert that hydroxyl radicals are the chief species responsible for carbohydrate degradation. These radicals have been shown to cleave directly glycosidic linkages in methyl β-D-glucoside and methyl β-cellobioside. They are also responsible for degrading glycosidic linkages in 1,5-anhydrocellobitol and 2-methoxytetrahydropyran by substitution reactions displacing 1-deoxyglucose, D-glucose, tetrahydropyran-2-ol, and methanol. Once the glycosidic linkages are broken, reducing carbohydrates undergo a series of reactions forming aldonic acids and lower-order aldoses. Figure 8.6 illustrates the alkaline peeling oxidation reaction [47].

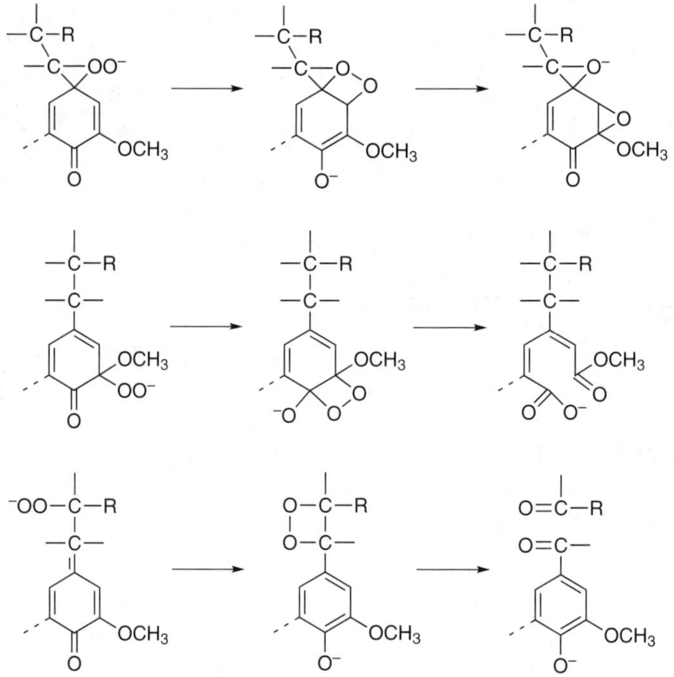

Figure 8.5   *Hydroperoxides intermediates form quinone structures, which may form muconic acids [3].*

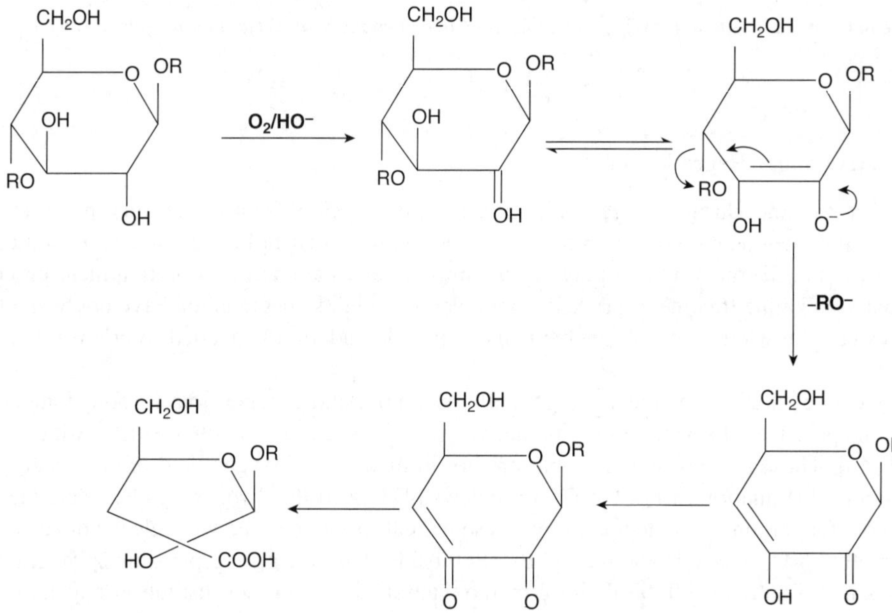

Figure 8.6   *An alkaline peeling reaction: the oxidative formation of a C-2 carbonyl group (last molecule) from the cellulose chain (first molecule) [3,46].*

Hydroxyl radical production is catalyzed by the presence of transition metals such as copper (II), iron (II), and manganese (II). Most lignocellulosic materials contain appreciable amounts of such metals which, as other compounds, are detrimental for both lignin and carbohydrates [48,49]. The literature extensively reviews methods to protect cellulose from these harmful radicals [50].

8.3 Ammonia Pretreatments

Ammonium hydroxide causes compositional changes and biomass swelling. Delignification is achieved by cleaving bonds between lignin and hemicellulose, C—C, C=O, and C—O bonds of lignin macromolecules. This means that in ammonia pretreatments, the guaiacyl type of lignin is more readily hydrolyzable than the syringyl type [51]. Ammonia causes significant changes in the biomass other than delignification. For example, ammonia increases the accessibility to carbohydrates by hydrolyzing glucuronic acid ester cross-links. Also, ammonia causes ammonolysis of uronic acid ester groups in hemicellulose, which leads to amides, changes of the cellulose fiber structure from cellulose I to cellulose III [52] and, in lignin-carbohydrate ester linkages, hydrolysis reactions that result in carboxylic acids and Maillard type reactions that produce pyrazines [53]. For biomass pretreatment, ammonia is advantageous because it is easily recoverable due to its high volatility.

Pretreatment with ammonia is successful in several modes. Some of these are Ammonia Recycled Percolation (ARP), Soaking in Aqueous Ammonia (SAA), and AFEX, which is the subject of Chapter 9.

In ARP, the feedstock is packed inside a flowthrough (percolation) reactor that is back-pressurized using nitrogen (174–425 psi). Aqueous ammonia is continuously pumped through the reactor, regenerated, and recycled (Figure 8.7). The primary factors influencing the reactions in ARP pretreatment are reaction time, temperature, ammonia concentration, and the amount of liquid throughput. Pretreatment can last as long as 90 min (not counting heating time of about 30 min), but most delignification occurs during the first 20 min of pretreatment [51,54]. Temperature may range from 150 to 175 °C. The ammonia concentration may vary from 2.5 to 20% w/w, but is preferred at 10–15%. The amount of liquid throughput is a major cost factor in ARP. Ideally, liquid input and residence times are 2.0–4.7 mL/g biomass and 10–12 min, respectively [51,54].

After pretreatment, solids are separated by filtration and extensively washed with water. Liquid samples are boiled until all the free ammonia evaporates and most of the dissolved lignin precipitates. Sugars (mainly xylooligomers) may be found in the pretreatment liquid [55].

Compositional changes from ARP pretreatment are affected by the feedstock and pretreatment conditions. Lignin removal has been reported as high as 70–85%, with hemicellulose removal of 40–60%, and cellulose retention of 85–99% [51]. The crystalline structure of the cellulose in the biomass is not modified by ARP treatment. Enzymatic digestibilities of pretreated biomass are strongly affected by feedstock and enzyme loading, and may vary from 75 to 95% at 7.5–60 filter paper units/g of glucan, respectively.

To enhance pretreatment results, successful ARP variations include:

1. Biomass presoaking in the ammonia solution (15% w/w ammonia) at room temperature overnight before pretreatment [56].
2. Two-stage pretreatment in which the ammonia percolation pretreatment explained above is preceded by a hot-water percolation stage at 190 °C. For this pretreatment, 15 g of biomass is packed in the reactor, through which either hot water or a 15% ammonia solution is continuously pumped at 5.0 mL/min for 30 min. The first stage is intended to remove hemicellulose; hydrolyzed xylan is 92–95% of which 83–86% has been recovered [51]. The second stage seeks to remove lignin; lignin removal of 75–81% has been reported. After this two-stage treatment, the cellulose recovered is 78–85%. The proposed

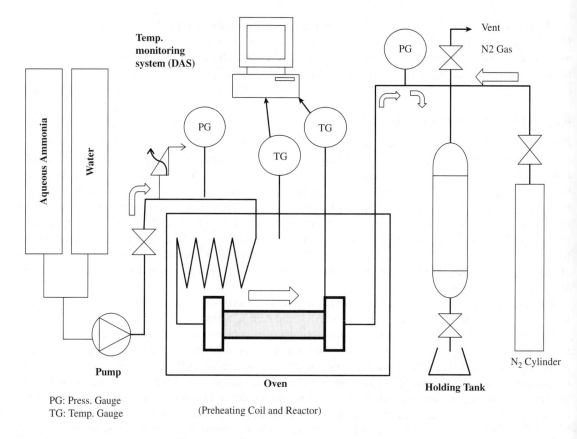

Figure 8.7   *ARP pretreatment reactor apparatus [54].*

two-stage processing is more meaningful as a fractionation than as a pretreatment method, because other one-stage pretreatment methods are equally effective [57].

3. Two-stage pretreatment process using aqueous ammonia (first stage, for delignification) followed by dilute sulfuric acid (second stage for hemicellulose removal and crystallinity alteration), or the other way around [58,59]. This strategy results in the fragmentation of biomass into its main components, which allows further usage of all fractions. The resulting cellulose is highly digestible and a very small amount of degradation products are formed, which is desirable for better fermentation.

In SAA pretreatment, batch reactors filled with the desired amounts of ammonium hydroxide (15% w/w) and biomass (1:10 of solid:liquid ratio) are placed in a convection oven set at the desired temperature (60–120 °C). No agitation is performed during pretreatment. Reaction may be allowed for up to 24 h of total reaction time [60,61]. Optionally, 2–5% H_2O_2 may be added through a syringe valve located in the reactor cap, intermittently during pretreatment, or at the time of reactor removal from the oven without cooling it [51]. The results of SAA pretreatment are not as good as for ARP pretreatment, and H_2O_2 is highly sensitive to temperature. To enhance delignification through SAA, modifications include step changes of temperature (60 °C for the first 4 h for effective oxidative reaction and 120 °C for the remaining 20 h for delignification).

This adjustment significantly improves delignification, but decomposition of carbohydrates may considerably lower the process selectivity [51].

8.4 Sodium Hydroxide Pretreatments

Among alkaline pretreatments, sodium hydroxide is perhaps the most widely used base. This may result because of the well-known success of alkaline delignification of wood with NaOH in industrial pulping and bleaching [17]. Recent studies show significant improvement in biomass digestibility as a result of pretreatments with NaOH concentrations of 1–10% (w/w) and temperatures of 25–190 °C. Depending on these factors and the lignocellulosic feedstock, the pretreatment may take from 30 min to 48 h (Table 8.1).

To enhance delignification, an oxidative agent (e.g., oxygen or hydrogen peroxide) may be applied at the beginning, or during pretreatment. Unfortunately, oxidative agents are not selective for lignin and pretreatments that use them may result in significant carbohydrate loss, particularly for feedstocks with high metals content (Fe, Co, Cu, and Mn). If this is the case, adding inorganic salts (e.g., magnesium sulfate or polyoxolates) may increase pretreatment lignin selectitivity.

Sodium hydroxide pretreatment is mostly performed in batch reactors with one or several stages.

In some NaOH pretreatments with two or more steps, a fairly long initial stage (up to 12 h) at room temperature is followed by one or several consecutive shorter stages (up to 2 h) at higher temperatures (up to 120 °C) with or without addition of oxygen, other oxidative agents (e.g., hydrogen peroxide), alkaline reagents (e.g., calcium hydroxide), or acid reagents (e.g., peracetic acid). The exact pretreatment conditions (i.e., time, reactants involved, reactants concentrations, temperature) and the number of stages substantially depend on the nature of the biomass feedstock. Factors that affect the obtained delignification during pretreatment include the amount of lignin and the structural composition of the lignin polymer, which is characterized by the proportion of the constituent monomers, the linkages within lignin monomeric units, and linkages between lignin and carbohydrates [1].

In addition to the aforementioned variations of sodium hydroxide pretreatment, Zhao *et al.* [69] reported a unique process that uses freezing temperatures (up to −15 °C) with or without addition of urea. For spruce, up to 70% glucose yield could be obtained using 7% NaOH in a 12% urea solution, but only 20% and 24% glucose yields were obtained at temperatures of 23 °C and 60 °C, respectively, when other conditions remained the same. Li *et al.* [72] used similar pretreatment conditions. In this study, pretreatment of bamboo was started with ball milling followed by an ethanol sonicated extraction. After this, alkaline (7% NaOH/12% urea) pretreatment at −12 °C was performed and was followed by successive extractions with dioxane, ethanol, and dimethyl sulfoxide. The results showed that the successive pretreatments resulted in partial removal of carbohydrates, which effectively disrupted the recalcitrance of bamboo, generating highly reactive cellulosic materials for enzymatic hydrolysis.

8.5 Alkaline Wet Oxidation

For over half a century, wet oxidation (WO) has been used to hydrothermally treat wastewater. In the process, the wastewater is mixed with an oxidative agent (O_2 is preferred) at temperatures above the normal boiling point of water (100 °C), but below its critical point (374 °C) [73]. Some wet oxidations use supercritical water. In such case, the process is termed supercritical water oxidation (SCWO) or hydrothermal oxidation (HTO). If air is used instead of oxygen, the process is called wet air oxidation (WAO). In WO, WAO, and SCWO, some reactions only occur in liquid media; thus, some water is kept liquid by pressurizing the system [73].

Table 8.1 Sugar recovery yields reported for sodium hydroxide pretreatment.

Source[a]	Biomass	Pretreatment conditions	Special pretreatment features	Cellulase loading, time[a]	Sugar recovery
[62]	Cotton stalks	2% NaOH, 90 min, 121 °C	—	40 FPU/g biomass, 72 h	68%
[63]	Switchgrass	0.1 g NaOH/g biomass, 50 g solids/L, 30 min, 190 °C	2 h presoak in 0.1% NaOH and microwave-heated	12 FPU/g biomass, 72 h	99%
[64]	Coastal bermudagrass	1% NaOH, 30 min, 121 °C	—	40 FPU/g biomass, 72 h	86%
[65]	Sugarcane bagasse	10% NaOH, 3:1 liquid:solid, 1.5 h, 90 °C	followed by 10% PAA,[c] 2.5 h, 75 °C	15 FPU/g solid, 120 h	92%
[66]	Wheat straw	pH adjusted with NaOH to 11.5, 24 h, 35 °C	pretreatment started in 2.15% v/v H_2O_2	0.16 mL/g biomass, 120 h[d]	97%
[67]	Switchgrass	0.1 g NaOH/g biomass, 6 h, room temperature	0.02 g $Ca(OH)_2$/g biomass added at the beginning	35 FPU/g biomass, 72 h	60%[b]
[68]	Prairie cord grass	1.70% NaOH, 30 min, room temperature followed by extrusion	extrusion pretreatment at 114 °C, 122 rpm, 8 mm[e]	15 FPU/g biomass, 72 h	82%
[69]	Spruce	3% NaOH soaking, 20:1 liquid: wood, −15 °C, 24 h	12% urea soaking simultaneous with NaOH soaking	20 FPU/g biomass, 72 h	60%[b]
[70]	Wheat straw	2% NaOH, 30 min, 121 °C	—	2.5 FPU/g of cellulase, 3.75 CBU/g of β-glucosidase and 1.5 FXU/g of xylanase, 48 h	6.3-fold increased digestibility
[71]	Corn stover	Titrate H_2O_2 solution with NaOH to pH 11.5, 23 °C, 24 h	10% (w/v) biomass loading, NaOH added to 0.25 g H_2O_2/g biomass to adjust to desired pH	Mixture of enzymes; total protein loading: 15 mg/g glucan	95%[b]

[a] β-glucosidase was added in different loadings.
[b] Glucose only
[c] Peracetic acid
[d] Xylanase was also added
[e] Barrel temperature, screw speed, and particle size.

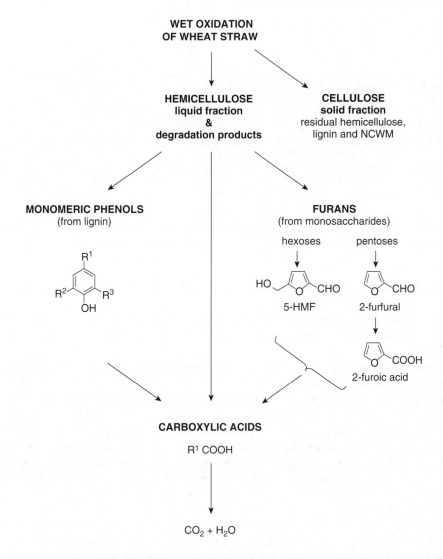

Figure 8.8 *Simplified scheme of reaction paths during WOP [74].*

Wet oxidation pretreatment (WOP) treats lignocellulose with water above 120 °C while adding either air or oxygen pressurized at 50 psi (345 kPa) or higher. This process is an effective method for separating cellulose from lignin and hemicellulose. The three feedstock constituents are affected as follows: (1) hemicellulose is extensively cleaved such that wet oxidation of this fraction results in a balance between solubilization and degradation to carboxylic acids; (2) lignin undergoes substantial cleavage and oxidation; and (3) cellulose remains mostly in the solid phase and is degraded to a highly digestible form (see Figure 8.8) [74].

WOP is exothermic, and therefore it becomes self-supporting with respect to heat after the reaction is initiated. The main reactions in wet oxidation pretreatment are the formation of acids from hydrolytic

Table 8.2   *Sugar recovery yields reported for WOP.*

Source	Biomass	Pretreatment conditions	Cellulase loading, hydrolysis time[a]	Sugar recovery[a]
[75]	Sugarcane bagasse	0.067 g Na_2CO_3/(g bagasse), 195 °C, 15 min, 12 O_2 bar	25 FPU/g biomass, 48 h	69%
[76]	Sugarcane bagasse	0.033 g Na_2CO_3/(g bagasse), 195 °C, 15 min, 12 O_2 bar	25 FPU/g biomass, 24 h	57%
[77]	Corn stover	0.067 g Na_2CO_3/(g corn stover), 195 °C, 15 min, 12 O_2 bar	25 FPU/g biomass, 24 h	85%
[78]	Rice husk	0.067 g Na_2CO_3/(g rice husk), 185 °C, 15 min, 0.5 O_2 MPa	—	70%[b]
[74]	Wheat straw	0.067 g Na_2CO_3/(g wheat straw), 185 °C, 10 min, 12 O_2 bar	—	99%[b]

[a] Reported yields of cellulose or holocellulose after pretreatment and enzymatic hydrolysis
[b] Recovered cellulose after pretreatment only (enzymatic hydrolysis yields were not reported)

processes as well as oxidative reactions; however, the control of reactor temperature is critical because of the fast rates of reaction and heat generation.

If WOP is supplemented with an alkali (the most widely employed is sodium carbonate), the pretreatment is termed alkaline wet oxidation (AWOP), and the degradation reactions are as described before (see Sections 8.2.3 and 8.2.4).

For each feedstock, pretreatment conditions are varied appropriately; Table 8.2 shows some examples of temperature, oxygen pressure, and time.

Degradation products of this pretreatment include high amounts of aliphatic acids and low quantities of furan aldehydes. For example, AWOP of wheat straw for 10 min, at 195 °C, with 12 bar oxygen, and 6.5 g/L $NaCO_3$ gave a hydrolyzate (liquid phase) in which the main constituents were soluble hemicellulose (16% w/w), low-molecular-weight carboxylic acids (11%), monomeric phenols (0.48%), and 2-furoic acid (0.01%). Formic acid and acetic acid constituted the majority of the produced acids. The main phenol monomers were 4-hydroxybenzaldehyde, vanillin, syringaldehyde, acetosyringone (4-hydroxy-3,5-dimethoxy-acetophenone), vanillic acid, and syringic acid [74]. High lignin removal from the solid fraction (62%) did not provide a corresponding increase in the phenol monomer content, but was correlated to high carboxylic acid concentrations. In the principal component analysis, the degradation products in the hemicellulose fractions co-varied with the pretreatment conditions according to their chemical structure, for example, diacids (oxalic and succinic acids), furan aldehydes, phenol aldehydes, phenol ketones, and phenol acids. These products have been reported as not inhibiting ethanol production [75,76]. AWOP is performed in stirred reactors provided with feed valves for liquids and gaseous phases; they are designed to withstand the temperature and pressure at which pretreatment occurs.

8.6   Lime Pretreatment

In lime pretreatment, the biomass is pretreated with calcium hydroxide (>0.1 g $Ca(OH)_2$/g dry biomass) and 10–15 g H_2O/g dry biomass. Depending on specific needs, lime pretreatment conditions may widely vary as follows.

1. *Long-term lime pretreatment* can be applied at very mild temperatures (25–60 °C) and atmospheric pressure. In this case, it lasts from days to weeks and air may be used as an inexpensive oxidative agent to enhance pretreatment effects [79].

Table 8.3 Highest yields of glucan[a] and xylan[a] observed after short-term lime pretreatment of poplar wood and/or enzymatic hydrolysis of remaining carbohydrates[b] [16].

Pretreatment conditions	Mode		EGY[c]	OGY[d]	EXY[e]	OXY[f]
	CP	VP				
2 h; 140 °C; 21.7 bar (abs) O_2	×		96	91	89	65
2 h; 160 °C; 14.8 bar absolute	×		96	92	90	66
2 h; 160 °C; 21.7 bar absolute	×		96	76	88	51
4 h; 180 °C; 14.8 bar absolute	×		94	65	90	32
10 h; 140 °C; 14.8 bar absolute		×	92	85	90	76
6 h; 140 °C; 7.9 bar absolute		×	92	70	78	71
6 h; 160 °C; 14.8 bar absolute		×	93	84	94	78
6 h; 160 °C; 21.7 bar absolute		×	93	88	94	75
6 h; 160 °C; 28.6 bar absolute		×	92	86	96	70

[a] Mass of glucose and xylose expressed as mass of equivalent glucan and xylan to calculate the yield based on glucan and xylan in the feedstock
[b] Enzyme hydrolysis obtained after 72 h hydrolysis at cellulase loading of 15 FPU/g glucan in raw biomass
[c] Enzymatic glucan yield (EGY) (g glucan hydrolyzed/100 g glucan in treated biomass)
[d] Overall glucan yield (OGY) (g glucan hydrolyzed/100 g glucan in raw biomass)
[e] Enzymatic xylan yield (EXY) (g xylan hydrolyzed/100 g xylan in treated biomass)
[f] Overall xylan yield (OXY) (g xylan hydrolyzed/100 g xylan in raw biomass)

2. *Short-term lime pretreatment* employs higher temperatures (100–180 °C) and may include oxygen at partial pressures from 3 to 30 bar or another oxidative agent, such as dilute hydrogen peroxide. In this case, pretreament lasts minutes to hours. In the oxidative mode with oxygen, if the gas is applied constantly as it is depleted from the reacting media, the pretreatment is termed constant pressure (CP). If O_2 is applied at the desired partial pressure just once at the beginning, the pretreatment is termed varying pressure (VP). Table 8.3 presents comparative results from these two options [16].

Compared to other alkaline agents, the use of lime is advantageous because it is inexpensive, easy to recover, safe to handle, and compatible with oxidants [80]. Because lime is not toxic, lime pretreatment has also been widely studied as a method of enhancing digestibility of crop residues for animal feed [81]. For this application, the pretreatment conditions can be very mild (ambient temperature with pretreatment times ranging from 24 h to several months); a sophisticated or expensive reactor is therefore not needed. This methodology has resulted in a moderate to good increase of *in vitro* digestibility [82]. Further research has led to the conclusion that lime pretreatment is effective in hydrolyzing protein from animal waste (such as chicken feathers, animal hair, or shrimp heads) [81], which provides a highly digestible protein component for animal feed.

Table 8.4 summarizes selected results from lime pretreatment applied to improve enzymatic digestibility. Overall, total sugar yields were >64%, and were often 70–88%. These yields were 2–9 times greater than those from raw biomass. Additionally, glucan recovery after pretreatment is high (>90% in all reported cases).

As shown by SEM analysis, lime pretreatment more strongly modifies the biomass surface than NaOH pretreatment [89]. Because lime is only partially soluble in water, it can form deposits on biomass. These deposits may prevent access of oxidative agents, thereby improving carbohydrate protection [89].

Table 8.4　Literature review assessing lime pretreatment on the basis of enzyme digestibility. (NR: not reported; NO: not optimized, exploratory studies only).

Feedstock	Lignin (%)	Pretreatment time	Oxidative agent	Temperature (°C)	Lime (g/g)	Digestibility increase	Total sugar yield (%)	Pretreatment glucan recovery (%)	Cellulase loading	Hydrolysis time (h)	Study
Sugarcane	25.8	36 h	None	70	0.40^a	NR	70.7^d	NR	3.5^i	Max^n	[83]
	21.9	4 wk	Air	57	0.12^b	NR	64.3^e	NR	5.0^i	72	[84]
Switchgrass	22.0	1 h	None	120	0.10^b	4.7×	70.0^f	93.6	5.0^i	72	[85]
	21.7	2 h	None	100	0.10^a	7×	58.1^f	90.3	5.0^i	72	[82]
	20.8	4 wk	Air	55	0.07^b	4×	88.1^f	97.8	15.0^i	72	[15]
Corn stover	21.5	5 h	None	120	0.08^b	NR	53.3^f	93.3	10.0^i	72	[86]
	21.5	5 h	None	120	0.10^b	NR	75.0^f	93.3	5.0^k	72	[87]
Newspaper N.O.	NR	3 h	7.1 bar O_2	140	0.30^a	2.2×	62.7^g	NR	5.0^l	72	[88]
	NR	3 h	None	120	0.30^b	1.7×	49.0^g	NR	5.0^l	72	[88]
Poplar wood	28.0	6 h	14.8 bar O_2	150	0.10^a	9.1×	77.0^f	98.2	5.0^l	72	[88]
	28.0	30 min	None	240	0.10^a	7.3×	43.8^g	NR	5.0^l	72	[88]
Wheatstraw	NR	24 h	None	50	0.10^a	8.7×	83.0^g	NR	5.0^l	72	[85]
	NR	1 h	None	121	0.10^a	2.4×	82.0^f	NR	0.15^k	72	[66]

[a]Lime loaded; [b]Lime consumed; [c]Lignin content measured through Van Soest Method. All other lignin contents were measured through H_2SO_4 two-stage hydrolysis method. Van Soest Method tends to give lower results; [d]Measured through DNS method (Total Reducing Sugars, TRS); Klason and soluble lignin were determined through H_2SO_4 hydrolysis. Ash was determined gravimetrically. Hollocellulose (glucose + xylose) was considered as total mass minus (lignin + ash). [e]Sugars measured by HPLC; [f]Estimated using total reducing sugars data reported in the paper; [g]Enzymatic hydrolysis time in hours; [h]FPU/g dry treated biomass; [i]FPU/g glucan. For this particular study, this translates to 5.4 FPU/g dry raw biomass; [j]FPU/g dry treated biomass with addition of Tween 20 during enzymatic hydrolysis; [k]FPU/g dry raw biomass; [l]mL/g dry treated biomass for each of three different enzyme cocktails with varying cellulase activity (Celluclast, Novozyme and Viscostar 150); [m]Hydrolysis time necessary to obtain no further changes in carbohydrates concentration.

8.7 Pretreatment Severity

Pretreatment conditions are specified to maximize yield and to increase the enzymatic digestibility of the substrate. Additional considerations include costs of capital, chemicals, and energy. A factor intended to quantify the energy intensity or *severity* of a pretreatment strategy is termed the "severity factor", R_o [90]. It is quantified by a simple calculation that uses the conditions at which pretreatment occur (i.e., temperature T, pH, and holding time t). The defining equation for severity factor follows [91,92]:

$$R_o = \int_a^b \exp\left(\frac{T(t) - 100}{14.75}\right) dt = t \exp\left(\frac{T(t) - 100}{14.75}\right) \tag{8.2}$$

$$\log(R_o{}^*) = \log(R_o) + |pH - 7| \tag{8.3}$$

High severity factors indicate extreme, energy-intensive pretreatment conditions; the severity factor has therefore been used to compare different pretreatment strategies. In fact, most of the latest pretreatment research is accompanied by calculations of the severity factor; however, the range over which glucose and xylose yields correlate with severity factor is limited. This is expected because of the complex reactions that occur during pretreatment. For example, Pedersen *et al.* [93] reported that pretreatments at the same final pH and similar severity factors resulted in two distinctly different yields. This study reveals that the monosaccharide yields obtained after subsequent enzymatic hydrolysis cannot be reliably predicted by using a one-dimensional severity factor calculation, even when this factor encompasses the pretreatment pH. It seems clear that the general use of the same fitted value (14.75) for the derived activation energy constant on different substrates is an inherent limitation. Understanding the effects of pretreatment factors on different types of lignocellulosic biomass can only be improved if more specific methodologies and benchmarks are used. In view of the widely diverging results currently available, more valid models are needed to quantitatively describe influences of the multiple factors that affect the enzymatic digestibility of pretreated lignocellulose. Therefore, in opposition to the current trend, the severity factor is not used as a means for comparison in this chapter. In contrast, the use of kinetic models for pretreatment is encouraged.

8.8 Pretreatment Selectivity

In the pretreatment literature, the definition of *selectivity* is not uniform. Although many researchers define selectivity as the ratio of lignin degradation rate to carbohydrate degradation rate, others define it as the amount of lignin remaining per unit of carbohydrates remaining, and others prefer to use the multiplicative inverse of these numbers [94]. Thus, depending on how it is defined, selectivity should be high or low. In any case, to make useful comparisons between different pretreatment options, pretreatment selectivity is a much more accurate choice than severity factor. The main features of a chosen pretreatment option that affect selectivity are (1) the selected process conditions; and (2) for alkaline pretreatments, the presence of pulp contaminants such as transition metals (e.g., copper (II), iron (II), and manganese (II)).

To accurately compare different pretreatment options, it is advisable to calculate selectivity rather than severity. If selectivity is defined as the ratio of lignin degradation rate to carbohydrate degradation rate, it may be applied to pretreatments in which both lignin and carbohydrates degradation is observed. With this, pretreatment conditions that optimize selectivity can be identified. Moreover, pretreatment conditions that can obtain specific target compositional changes in biomass can also be determined.

As applied to bleaching and pulping, one widely used model for alkaline delignification avoids descriptions of specific reactions that may occur. Instead, the following simplified model is used:

Lignin + Carbohydrates + Alkali + Oxidative agent
$$\rightarrow \text{Undegraded lignin} + \text{Undegraded carbohydrates} + \text{Degradation products}$$

For lime pretreatment, a lumped power law model is used to follow the yield Y of species i (i = lignin, glucan, or xylan). Multiple reactions j (j = slow, medium, or fast) are assumed to occur in parallel. The following rate equations are used in the model [79,95]:

$$-\frac{dY_i}{dt} = \sum_j k_{ij} P_{O_2}^{\beta_{ij}} Y_{ij} \tag{8.4}$$

where k_{ij} is the rate constant $((\text{min·bar}^{\beta_{ij}})^{-1})$, defined

$$k_{ij} = a_{ij} \exp\left(-\frac{E_{ij}}{RT}\right); \tag{8.5}$$

a_{ij} is the frequency factor $((\text{min·bar}^{\beta_{ij}})^{-1})$; E_{ij} is activation energy (kJ/mol); R is the ideal gas constant (8.314×10^{-3} kJ/(mol·K)); T is absolute temperature (K); and P_{O_2} is oxidative agent concentration, which is pressure for O_2 (bar, absolute); and β_{ij} is a dimensionless exponent ($\beta_{ij} = 0$ for non-oxidative pretreatment).

The integrated form of Equation (8.4) is

$$Y_i = \sum_j Y_{ij0} \exp\left(-k_{ij} P_{O_2}^{\beta_{ij}} t\right) \tag{8.6}$$

where Y_{ij0} is the yield of component ij at time zero (kg residual component ij/kg initial component i).

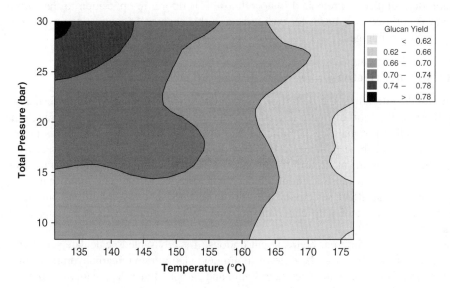

Figure 8.9 *Surface plot for oxidative lime pretreatment of poplar wood to define pretreatment conditions that optimize glucan yield for a target lignin of 0.2 g lignin/g lignin in raw biomass.*

By measuring the lignin and carbohydrate content of lignocellulose as a function of time, the lignin yields (Y_L), glucan yields (Y_G), and xylan yields (Y_X) can be used to obtain kinetic parameters in a pretreatment model. These models allow calculation of selectivity for glucan (S_G) and xylan (S_X). These models can be used in a constrained optimization problem that maximizes Y_G subject to a target Y_L, which can vary between 0.2 and 0.8 g lignin/g lignin in raw biomass. This procedure results in a surface plot (e.g., Figure 8.9), which can be used to compare different pretreatment options.

8.9   Concluding Remarks

Alkaline pretreatment of lignocellulose is employed to remove lignin while selectively retaining as much of the carbohydrate as possible. This type of pretreatment is extremely flexible and has been successfully applied to a wide variety of lignocellulose feedstocks using a multiplicity of alkaline reagents, equipment, and reaction conditions.

Although great advances have been reached through decades of research, typical experimental approaches involve the systematic investigation of a few process variables (e.g., temperature, time, reagent concentration) to identify promising reaction conditions that enhance enzymatic digestibility. Because it is very costly to investigate every possible combination of process variables, this approach is rarely able to identify a true optimum. In an attempt to more rationally design a pretreatment process, the literature widely employs the severity factor; however, this has not proven to reliably predict pretreatment effectiveness. To overcome these problems, a preferred approach is to use kinetic models that describe the degradation of each biomass component (lignin, glucan, xylan) as a function of process variables. Once the kinetic parameters are fit to the experimental data, it is possible to perform a computer optimization that identifies the process variables that selectively minimize the loss of carbohydrates while maximizing the removal of lignin. Such a rational approach improves process economics and therefore improves the likelihood of successfully replacing fossil fuels.

References

1. Sjöström, E. (1981) The chemistry of oxygen delignification. *Paperi ja puu*, **63** (6–7), 438–442.
2. Gierer, J. and Stockholm, S. (1985) Chemistry of delignification. Part 1: General concept and reactions during pulping. *Wood Science and Technology*, **19**, 289–312.
3. Ragauskas, A. (2002) *High Selectivity Oxygen Delignification*, United States of America: Institute of Paper Science and, Technology, Atlanta.
4. Freire, C., Silvestre, A., Pascoal Neto, C., and Evtuguin, D. (2006) Effect of oxygen, ozone and hydrogen peroxide bleaching stages on the contents and composition of extractives of Eucalyptus globulus kraft pulps. *Bioresource Technology*, **97** (3), 420–428.
5. Taherzadeh, M. and Karimi, K. (2008) Pretreatment of lignocellulosic wastes to improve ethanol and biogas production: A review. *International Journal of Molecular Sciences*, **9** (9), 1621–1651.
6. Baptista, C., Belgacem, N., and Duarte, A.P. (2006) The effect of wood extractives on pulp properties of maritime pine kraft pulp. *Appita Journal*, **59** (4), 311–316.
7. Garlock, R.J., Balan, V., Dale, B.E. *et al.* (2011) Comparative material balances around pretreatment technologies for the conversion of switchgrass to soluble sugars. *Bioresource Technology*, **102** (24), 11063–11071.
8. Zhu, L., O'Dwyer, J.P., Chang, V.S. *et al.* (2008) Structural features affecting biomass enzymatic digestibility. *Bioresource Technology*, **99** (9), 3817–3828.
9. Sierra, R., Holtzapple, M., and Granda, C. (2011) Long-term lime pretreatment of poplar wood. *AIChE Journal*, **57** (5), 1320–1328.
10. Chang, V. and Holtzapple, M. (2000) Fundamental factors affecting biomass enzymatic reactivity. *Applied Biochemistry and Biotechnology*, **84–86** (1), 5–37.

11. Gáspár, M., Kálmán, G., and Réczey, K. (2007) Corn fiber as a raw material for hemicellulose and ethanol production. *Process Biochemistry*, **42** (7), 1135–1139.
12. Mosier, N., Wyman, C., Dale, B. *et al.* (2005) Features of promising technologies for pretreatment of lignocellulosic biomass. *Bioresource Technology*, **96** (6), 673–686.
13. Holtzapple, M. (1993) *Lignin: Structure and Analysis. Encyclopedia of Food Science, Food Technology, and Nutrition*, Academic Press, London, p. 2731–2738.
14. Mattea, R.N. (2007) Future of cellulosic ethanol production. *Basic Biotechnology eJournal*, **3** (1), 28–33.
15. Kim, S. and Holtzapple, M. (2005) Lime pretreatment and enzymatic hydrolysis of corn stover. *Bioresource Technology*, **96** (18), 1994–2006.
16. Sierra, R., Granda, C., and Holtzapple, M. (2009) Short-term lime pretreatment of poplar wood. *Biotechnology Progress*, **25** (2), 323–332.
17. McDonough, T. (1983) Oxygen bleaching processes: an overview. IPC Technical paper series. 10. http://en.scientificcommons.org/57089089.
18. Lai, Y. and Funaoka, M. (1994) Oxygen bleaching of kraft pulp. 1. Role of condensed units. *Tappi Journal*, **48** (4), 355–359.
19. Senior, D., Hamilton, J., Ragauskas, A. *et al.* (1998) Interaction of hydrogen peroxide and chlorine dioxide in ECF bleaching. *Tappi Journal*, **81** (6), 170.
20. Bjorkman, A. (1957) Lignin and lignin-carbohydrate complexes. *Industrial & Engineering Chemistry*, **49** (9), 1395–1398.
21. Terashima, N. (2001) Possible Approaches for Studying Three Dimensional Structure of Lignin, in *Progress in Biotechnology* (eds M. Noriyuki and K. Atsushi), Elsevier, p. 257–262.
22. Imai, A., Tomoda, I., Yokoyama, T. *et al.* (2008) Application of the amount of oxygen consumption to the investigation of the oxidation mechanism of lignin during oxygen-alkali treatment. *Journal of Wood Science*, **54** (1), 62–67.
23. Ohmura, S., Yokoyama, T., and Matsumoto, Y. (2012) Progress of oxidation of non-phenolic lignin moiety in an oxygen bleaching process via the conversion of non-phenolic into phenolic lignin moiety. *Journal of Wood Science*, **58** (3), 243–250.
24. Huynh, V.-B. (1986) Biomimetic oxidation of lignin model compounds by simple inorganic complexes. *Biochemical and Biophysical Research Communications*, **139** (3), 1104–1110.
25. Binder, J.B., Gray, M.J., White, J.F. *et al.* (2009) Reactions of lignin model compounds in ionic liquids. *Biomass and Bioenergy*, **33** (9), 1122–1130.
26. Glasser, W.G. and Sarkanen, S. (1989) Lignin, properties and materials (545 p). http://www.osti.gov/energycitations/product.biblio.jsp?osti_id=6495223.
27. Chakar, F.S. and Ragauskas, A.J. (2004) Review of current and future softwood kraft lignin process chemistry. *Industrial Crops and Products*, **20** (2), 131–141.
28. Knill, C. and Kennedy, J. (2003) Degradation of cellulose under alkaline conditions. *Carbohydrate Polymers*, **51** (3), 281–300.
29. Sjöström, E. (1991) Carbohydrate degradation products from alkaline treatment of biomass. *Biomass and Bioenergy*, **1** (1), 61–64.
30. Gellerstedt, G., Majtnerova, A., and Zhang, L. (2004) Towards a new concept of lignin condensation in kraft pulping. Initial results. *Comptes Rendus Biologies*, **327** (9–10), 817–826.
31. Huang, G., Shi, J.X., and Langrish, T.A.G. (2007) NH$_4$OH–KOH pulping mechanisms and kinetics of rice straw. *Bioresource Technology*, **98** (6), 1218–1223.
32. Gierer, J. and Wannstrom, S. (1970) Reactions of lignin during pulping. Description and comparison of conventional pulping processes. *Svensk Papperstidning*, **73** (18), 571–596.
33. Iversen, T. and Wannastrom, S. (1986) Lignin-carbohydrate bonds in a residual lignin isolated from kraft pulp. *Holzforschung*, **40** (1), 19–22.
34. Donald, D. and Göran, G. (2010) *Chemistry of Alkaline Pulping, Lignin and Lignans,* CRC Press, p. 349–391.
35. Sjöström, E. (1981) *Wood Chemistry: Fundamentals and Applications.* Academic Press, New York.
36. Gierer, J. and Pettersson, I. (1977) Studies on the condensation of lignins in alkaline media. Part II. The formation of stilbene and arylcoumaran structures through neighbouring group participation reactions. *Canadian Journal of Chemistry*, **55** (4), 593–599.

37. Sjöström, E. (1977) The behavior of wood polysaccharides during alkaline pulping processes. *Tappi Journal*, **60** (9), 151–154.

38. Guay, D., Cole, B., Fort, R. *et al.* (2001) Mechanisms of oxidative degradation of carbohydrates during oxygen delignification. I. Reaction of photochemically generated hydroxyl radicals with methyl B-cellobioside. *Journal of Wood Chemistry and Technology*, **21** (1), 67–79.

39. Malinen, R. and Sjöström, E. (1972) Studies on the reactions of carbohydrates during oxygen bleaching. Part l. Oxidative alkaline degradation of cellobiose. *Papperi ja Puu*, **54**, 451.

40. Bonn, G., Binder, H., Leonhard, H., and Bobleter, O. (1985) The alkaline degradation of cellobiose to glucose and fructose. *Monatshefte für Chemie/Chemical Monthly*, **116** (8), 961–971.

41. Gierer, J. and Jansbo, K. (1993) Formation of hydroxyl radicals from hydrogen peroxide and their effect on bleaching of mechanical pulps. *Journal of Wood Chemistry and Technology*, **13** (4), 561–581.

42. Gould, J.M. (1985) Studies on the mechanism of alkaline peroxide delignification of agricultural residues. *Biotechnology and Bioengineering*, **27** (3), 225–231.

43. Vainio, U., Maximova, N., Hortling, B. *et al.* (2004) Morphology of dry lignins and size and shape of dissolved kraft lignin particles by X-ray scattering. *Langmuir*, **20** (22), 9736–9744.

44. Sjöström, E. (2003) Carbohydrate degradation products from alkaline treatment of biomass. *Biomass and Bioenergy*, **1** (1), 61–64.

45. Froass, P., Ragauskas, A., McDonough, T., and Jiang, J. (1996) Relationship between residual lignin structure and pulp bleachability. International Pulp Bleaching Conference Proceeedings, Washington, DC. United States of America.

46. Thompson, N.S. and Green, J.W. (1977) Study of the carbohydrate peeling and stopping reactions under the conditions of oxygen-alkali pulping, Project 3265, report three: a progress report to members of the Institute of Paper Chemistry. Appleton, Wisconsin: the Institute.

47. Golova, O. and Nosova, N. (1973) Degradation of cellulose by alkaline oxidation. *Russian Chemical Reviews*, **42** (4), 327–338.

48. Suchy, M. and Argyropoulos Dimitris, S. (2001) *Catalysis and Activation of Oxygen and Peroxide Delignification of Chemical Pulps: A Review. Oxidative Delignification Chemistry*, American Chemical Society, p. 2–43.

49. Sippola, V. (2006) Transition metal-catalysed oxidation of lignin model compounds for oxygen delignification of pulp. Industrial Chemistry Publication Series, 21. https://aaltodoc.aalto.fi/handle/123456789/2673.

50. Gierer, J., Torbjörn, R., Yang, E., and Byung-Ho, Y. (2001) Formation and involvement of radicals in oxygen delignification studied by the autoxidation of lignin and carbohydrate model compounds. *Journal of Wood Chemistry and Technology*, **21** (4), 313–341.

51. Gupta, R. and Lee, Y.Y. (2009) Pretreatment of hybrid poplar by aqueous ammonia. *Biotechnology Progress*, **25** (2), 357–364.

52. Lewin, M. and Roldan, L.G. (1971) The effect of liquid anhydrous ammonia in the structure and morphology of cotton cellulose. *Journal of Polymer Science Part C: Polymer Symposia*, **36** (1), 213–229.

53. Chundawat, S.P.S., Vismeh, R., Sharma, L.N. *et al.* (2010) Multifaceted characterization of cell wall decomposition products formed during ammonia fiber expansion (AFEX) and dilute acid based pretreatments. *Bioresource Technology*, **101** (21), 8429–8438.

54. Kim, T.H., Kim, J.S., Sunwoo, C., and Lee, Y.Y. (2003) Pretreatment of corn stover by aqueous ammonia. *Bioresource Technology*, **90** (1), 39–47.

55. Gupta, R. and Lee, Y. (2010) Investigation of biomass degradation mechanism in pretreatment of switchgrass by aqueous ammonia and sodium hydroxide. *Bioresource Technology*, **101** (21), 8185–8191.

56. Kim, J.-S., Kim, H., Lee, J.-S. *et al.* (2008) Pretreatment characteristics of waste oak wood by ammonia percolation. *Applied Biochemistry and Biotechnology*, **148** (1), 15–22.

57. Kim, T.H. (2004) *Bioconversion of Lignocellulosic Material into Ethanol: Pretreatment, Enzymatic Hydrolysis, and Ethanol Fermentation*, Auburn University, Auburn, Alabama, US.

58. Zhangwen, W. and Lee, Y. (1997) Ammonia recycled percolation as a complementary pretreatment to the dilute-acid process. *Applied Biochemistry and Biotechnology*, **63–64** (1), 21–34.

59. Oh, K., Kim, Y., Yoon, H., and Tae, B. (2001) Pretreatment of lignocellulosic biomass using combination of ammonia recycled percolation and dilute acid process. *Journal of Industrial and Engineering Chemistry*, **8** (1), 64–70.

60. Kim, T. and Lee, Y. (2005) Pretreatment and fractionation of corn stover by ammonia recycle percolation process. *Bioresource Technology*, **96** (18), 2007–2013.

61. Kang, K., Jeong, G., Sunwoo, C., and Park, D. (2012) Pretreatment of rapeseed straw by soaking in aqueous ammonia. *Bioprocess and Biosystems Engineering*, **35** (1), 77–84.

62. Silverstein, R.A., Chen, Y., Sharma-Shivappa, R.R. *et al.* (2007) A comparison of chemical pretreatment methods for improving saccharification of cotton stalks. *Bioresource Technology*, **98** (16), 3000–3011.

63. Hu, Z., Wang, Y., and Wen, Z. (2008) Alkali (NaOH) pretreatment of switchgrass by radio frequency-based dielectric heating, in *Biotechnology for Fuels and Chemicals* (eds W.S. Adney, J.D. McMillan, J. Mielenz, and K.T. Klasson), Humana Press, p. 589–599.

64. Wang, Z., Keshwani, D.R., Redding, P. *et al.* (2010) *Sodium Hydroxide Pretreatment and Enzymatic Hydrolysis of Coastal Bermuda Grass*, Elsevier, Kidlington.

65. Zhao, X., Peng, F., Cheng, K., and Liu, D. (2009) Enhancement of the enzymatic digestibility of sugarcane bagasse by alkali–peracetic acid pretreatment. *Enzyme and Microbial Technology*, **44** (1), 17–23.

66. Saha, B. and Cotta, M. (2006) Ethanol production from alkaline peroxide pretreated enzymatically saccharified wheat straw. *Biotechnology Progress*, **22** (2), 449–453.

67. Xu, J., Cheng, J., Sharma-Shivappa, R., and Burns, J. (2010) Sodium hydroxide pretreatment of switchgrass for ethanol production. *Energy & Fuels*, **24** (3), 2113–2119.

68. Karunanithy, C. and Muthukumarappan, K. (2011) Optimization of alkali soaking and extrusion pretreatment of prairie cord grass for maximum sugar recovery by enzymatic hydrolysis. *Biochemical Engineering Journal*, **54** (2), 71–82.

69. Zhao, Y., Wang, Y., Zhu, J. *et al.* (2008) Enhanced enzymatic hydrolysis of spruce by alkaline pretreatment at low temperature. *Biotechnology and Bioengineering*, **99** (6), 1320–1328.

70. McIntosh, S. and Vancov, T. (2011) Optimisation of dilute alkaline pretreatment for enzymatic saccharification of wheat straw. *Biomass and Bioenergy*, **35** (7), 3094–3103.

71. Banerjee, U., Chisti, Y., and Moo-Young, M. (1995) Effects of substrate particle size and alkaline pretreatment on protein enrichment by *Neurospora sitophila*. *Resources, Conservation and Recycling*, **13** (2), 139–146.

72. Li, M., Fan, Y., Xu, F., and Sun, R. (2010) Characterization of extracted lignin of bamboo (*Neosinocalamus affinis*) pretreated with sodium hydroxide/urea solution at low temperature. *Bioresources*, **5** (3), 1762–1778.

73. Luck, F. (1999) Wet air oxidation: past, present and future. *Catalysis Today*, **53** (1), 81–91.

74. Klinke, H., Ahring, B., Schmidt, A., and Thomsen, A. (2002) Characterization of degradation products from alkaline wet oxidation of wheat straw. *Bioresource Technology*, **82** (1), 15–26.

75. Martín, C., Klinke, H., and Thomsen, A. (2007) Wet oxidation as a pretreatment method for enhancing the enzymatic convertibility of sugarcane bagasse. *Enzyme and Microbial Technology*, **40** (3), 426–432.

76. Martín, C., Marcet, M., and Thomsen, A. (2008) Comparison between wet oxidation and steam explosion as pretreatment methods for enzymatic hydrolisis of sugarcane bagasse. *BioResources*, **3** (3), 670–683.

77. Varga, E., Schmidt, A., Réczey, K., and Thomsen, A. (2003) Pretreatment of corn stover using wet oxidation to enhance enzymatic digestibility. *Applied Biochemistry and Biotechnology*, **104** (1), 37–50.

78. Banerjee, S., Sen, R., Pandey, R. *et al.* (2009) Evaluation of wet air oxidation as a pretreatment strategy for bioethanol production from rice husk and process optimization. *Biomass and Bioenergy*, **33**, 1680–1686.

79. Sierra, R., Garcia, L., and Holtzapple, M. (2010) Selectivity and delignification kinetics for oxidative and nonoxidative lime pretreatment of poplar wood, part III: Long-term. *Biotechnology Progress*, **26** (6), 1685–1694.

80. Holtzapple, M. and Davidson, R. (1999) *Methods of Biomass Pretreatment*, United States of America. Patent no. 08/096,972

81. Coward-Kelly, G., Chang, V.S., Agbogbo, F., and Holtzapple, M. (2006) Lime treatment of keratinous materials for the generation of highly digestible animal feed: 1. Chicken feathers. *Bioresource Technology*, **97** (11), 1337–1343.

82. Chang, V., Burr, B., and Holtzapple, M. (1997) Lime pretreatment of switchgrass. *Applied Biochemistry and Biotechnology*, **63–65** (1), 3–19.

83. Rabelo, S., Filho, R., and Costa, A. (2008) A comparison between lime and alkaline hydrogen peroxide pretreatments of sugarcane bagasse for ethanol production. *Applied Biochemistry and Biotechnology*, **144** (1), 87–100.

84. Granda, C. (2005) *Sugarcane Juice Extraction and Preservation, and Long-Term Lime Pretreatment of Bagasse*, Texas A&M University.

85. Chang, V., Nagwani, M., and Holtzapple, M. (1998) Lime pretreatment of crop residues bagasse and wheat straw. *Applied Biochemistry and Biotechnology*, **74** (3), 135–159.
86. Kaar, W. and Holtzapple, M. (2000) Using lime pretreatment to facilitate the enzymic hydrolysis of corn stover. *Biomass and Bioenergy*, **18** (3), 189–199.
87. Kaar, W. and Holtzapple, M. (1998) Benefits from Tween during enzymic hydrolysis of corn stover. *Biotechnology and Bioengineering*, **59** (4), 419–427.
88. Chang, V., Nagwani, M., Kim, C., and Holtzapple, M. (2001) Oxidative lime pretreatment of high-lignin biomass. *Applied Biochemistry and Biotechnology*, **94** (1), 1–28.
89. López, R., Poblano, V., Licea-Claverie, A. *et al.* (2000) Alkaline surface modification of sugarcane bagasse. *Advanced Composite Materials*, **9** (2), 99–108.
90. Heitz, M., Carrasco, F., Rubio, M. *et al.* (1987) Physico-chemical characterization of lignocellulosic substrates pretreated via autohydrolysis: an application to tropical woods. *Biomass*, **13** (4), 255–273.
91. Abatzoglou, N., Chornet, E., Belkacemi, K., and Overend, R. (1992) Phenomenological kinetics of complex systems: the development of a generalized severity parameter and its application to lignocellulosics fractionation. *Chemical Engineering Science*, **47** (5), 1109–1122.
92. Chum, H., Johnson, D., Black, S., and Overend, R. (1990) Pretreatment-catalyst effects and the combined severity parameter. *Applied Biochemistry and Biotechnology*, **24–25** (1), 1–14.
93. Pedersen, M. and Meyer, A. (2010) Lignocellulose pretreatment severity – relating pH to biomatrix opening. *New Biotechnology*, **27** (6), 739–750.
94. Dang, Z. (1995) *Pulp Pretreatments for Improved Selectivity And Extended Oxygen Delignification*, Nanjung University of Science and Technology.
95. Sierra, R., Garcia, L., and Holtzapple, M. (2011) Selectivity and delignification kinetics for oxidative short-term lime pretreatment of poplar wood, part I: Constant-pressure. *Biotechnology Progress*, **27** (4), 976–985.

9

Primer on Ammonia Fiber Expansion Pretreatment

S.P.S. Chundawat[1,2], **B. Bals**[1,2], **T. Campbell**[3], **L. Sousa**[1], **D. Gao**[1,2], **M. Jin**[1,2], **P. Eranki**[1,2], **R. Garlock**[1,2], **F. Teymouri**[3], **Venkatesh Balan**[1,2] and **Bruce E. Dale**[1,2]

[1] *Department of Chemical Engineering and Materials Science, Michigan State University, East Lansing, USA*
[2] *Great Lakes Bioenergy Research Center, Michigan State University, East Lansing, USA*
[3] *Michigan Biotechnology Institute, Lansing, USA*

9.1 Historical Perspective of Ammonia-based Pretreatments

Plant biomass is a plentiful resource which can be used to co-produce fuels, chemicals and nutritional feed, driving forward a nation's bio-based economy [1]. However, the costs of chemical/biological processes that facilitate the conversion of plant polymers into monomers and finally into desired end-products have stymied the development of cellulosic biorefineries [2]. The choice of thermochemical pretreatment has a substantial impact on the economic and environmental viability of biorefineries within the gates of the refinery and beyond [3]. Ammonia fiber expansion, or AFEX (a trademark of MBI International, Lansing), is an ammonia-based pretreatment which has shown tremendous promise at cost-effectively reducing the recalcitrance of lignocellulosics towards biologically catalyzed deconstruction into fermentable sugars. Unlike other aqueous pretreatments, AFEX is a dry-to-dry process (i.e., AFEX-treated biomass composition is unchanged) that reduces plant cell-wall recalcitrance through a unique physicochemical mechanism. The aim of this chapter is to explore the recent advances in AFEX and provide a general overview of this technology from a microscopic (physicochemical mechanisms) and macroscopic (distributed regional processing and life-cycle analysis) perspective.

Ammonia was first employed by farmers in the 1970s to increase the digestibility of forages for ruminant animals using rudimentary on-farm ammoniation processes [4]. At least some minimum amount of forage (i.e., feed high in fiber such as hay or silage) is generally required in the diets of ruminant animals such as beef and dairy cattle. The digestibility of forage material is one of its most important qualities, and thus a major factor in its price. Digestible forages such as alfalfa or orchard grass hay

Aqueous Pretreatment of Plant Biomass for Biological and Chemical Conversion to Fuels and Chemicals, First Edition.
Edited by Charles E. Wyman.
© 2013 John Wiley & Sons, Ltd. Published 2013 by John Wiley & Sons, Ltd.

tend to have a high leaf to stem ratio, low fiber concentration, and very little lignin. Because these forages are expensive or may not be available year-round, ammoniation has been used to improve the digestibility of recalcitrant material such as late-harvest grasses or straw [5–7]. The ammoniation process involves pumping anhydrous ammonia under a tarp containing bales of hay at a rate of 3–4 g ammonia per 100 g dry biomass, and allowing it to stand at ambient temperatures for 6–8 weeks. This reaction slightly increases fiber digestibility (approximately 10–20% improvement in digestibility compared to untreated hay), improving ruminant uptake and reducing spoilage. An added benefit of ammoniation is that the crude protein content of the hay increases by 5–10% due to ammonia reacting with the biomass [8]. This pretreatment is limited in its application due to only a modest increase in digestibility as well as the relatively high costs associated with the process. The ammonia is not recoverable, and the increase in digestibility is not likely to make the treated forages comparable to traditional high-quality feeds. Thus, ammoniation is used primarily during the winter when fresh forages are not available, or in countries such as China where agricultural land may be limited.

Ammonia was first employed under conditions of increased temperature (90–100 °C) and pressure (300–700 psi) to defiberize/plasticize [9] wood with the aim of developing a superior chemi-mechanical pulping process called ammonia explosion pulping [10]. It was not until 1982 that pressurized, high-temperature, concentrated ammonia (>30% NH_4OH) was employed to enhance the enzymatic digestibility and fermentability of lignocellulosic biomass [11]. This process was originally called ammonia freeze explosion but was later referred to as ammonia fiber explosion and ammonia fiber expansion, or simply AFEX. Researchers at Du Pont have also successfully explored the utility of supercritical and near-critical ammonia as a pretreatment reagent to increase digestibility of both agricultural residues and hardwoods [12,13]. In recent decades there has been an emphasis on utilizing low-concentration aqueous ammonia (<15% NH_4OH) as a pretreatment reagent, either with (ammonia recycle percolation or ARP[14]) or without (Du Pont process [15]) hemicellulose/lignin fractionation. However, we shall explore the scientific/engineering advancements and related commercialization efforts made in the field of AFEX with limited focus on other aqueous-ammonia-based pretreatments, which have features similar to those discussed in Chapter 8 on high-pH pretreatments.

9.2 Overview of AFEX and its Physicochemical Impacts

One of the most attractive features of AFEX is the reversible nature of the interaction of ammonia with moist biomass. This reversibility allows most of the ammonia catalyst to be recovered and re-used and also allows the treated biomass to be recovered as a relatively dry solid with little residual ammonia from the pretreatment process. The challenge in the design of any commercial-scale AFEX pretreatment process will be to provide adequate ammonia contact with the moist biomass followed by adequate ammonia removal from the biomass, with minimal capital and operating costs. The conventional methodology for carrying out AFEX is contacting liquid ammonia (0.3–2 g NH_3/g dry biomass) with moist biomass (0.1–2 g H_2O/g dry biomass) and heating (40–180 °C) the biomass-water-ammonia mixture for the necessary time period (5–60 min) before rapidly releasing the pressure [16]. However, there are several variations to the conventional AFEX process depending on how ammonia comes in contact with the biomass and is removed and recycled after pretreatment, as briefly described in the following (Figure 9.1).

- **Conventional laboratory batch AFEX:** Most of the previously reported work on AFEX has utilized a conventional batch reactor with/without stirring to pretreat the biomass on a laboratory scale (1–750 g) [17,18]. The ammonia may be delivered in liquid or gaseous form, which reacts with the water to cause a rapid increase in temperature. The temperature is maintained by heating indirectly through a jacketed vessel or heating mantle. The ammonia is removed from the vessel by explosively releasing the pressure

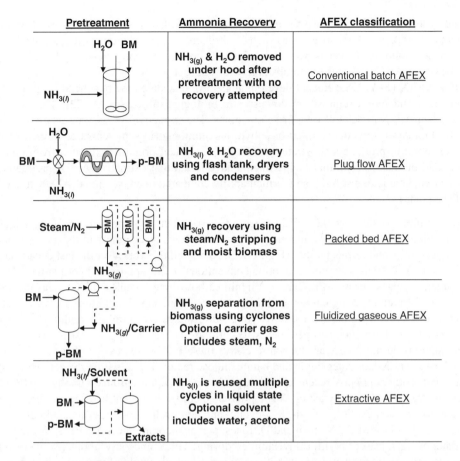

Pretreatment	Ammonia Recovery	AFEX classification
H_2O BM / $NH_{3(l)}$	$NH_{3(g)}$ & H_2O removed under hood after pretreatment with no recovery attempted	Conventional batch AFEX
H_2O / BM → p-BM / $NH_{3(l)}$	$NH_{3(l)}$ & H_2O recovery using flash tank, dryers and condensers	Plug flow AFEX
Steam/N_2 → BM BM BM / $NH_{3(g)}$	$NH_{3(g)}$ recovery using steam/N_2 stripping and moist biomass	Packed bed AFEX
BM → / $NH_{3(g)}$/Carrier / p-BM	$NH_{3(g)}$ separation from biomass using cyclones Optional carrier gas includes steam, N_2	Fluidized gaseous AFEX
$NH_{3(l)}$/Solvent / BM → p-BM ← / Extracts	$NH_{3(l)}$ is reused multiple cycles in liquid state Optional solvent includes water, acetone	Extractive AFEX

Figure 9.1 *Process flow diagrams for batch, plug flow, packed-bed, extractive and gaseous AFEX; BM and p-BM are untreated and pretreated biomass, $NH_{3(l)}$ and $NH_{3(g)}$ are liquid and gaseous ammonia.*

and allowing the biomass to dry in a fume hood overnight to remove residual ammonia. No attempt is currently made to recover the ammonia on the lab-scale AFEX [19]. The composition of the biomass is nearly identical before and after treatment, as no extractives are removed from the biomass.

- **Plug flow AFEX (PF-AFEX):** This pretreatment employs a tubular reactor (with or without internal conveyor screws) for carrying out continuous AFEX at the biorefinery scale. In principle, this is typically a continuous screw-type reactor that allows extensive contact between concentrated NH_4OH and biomass in co- or counter-current mode followed by ammonia separation using a flash tank/dryer and ammonia-water recovery using a suitable recycle system. MBI International (http://www.mbi.org) has developed a prototype plug flow tubular-type AFEX reactor (without screw-type conveyors) that has been demonstrated to work well on pumpable biomass slurries (e.g., distiller's grains and solubles).

- **Packed-bed AFEX (PB-AFEX):** This pretreatment employs a series of packed-bed reactors for contacting gaseous ammonia with moist biomass and with ammonia recycle coupled directly into the pretreatment reactor assembly. MBI International has developed a prototype packed-bed AFEX that loads ammonia in its gaseous state and uses moist biomass to recover ammonia in a cyclical fashion. This approach has been demonstrated to work well on non-flowable biomass (e.g., corn stover, oat hulls).

- **Fluidized gaseous AFEX (FG-AFEX):** Fluidized gaseous AFEX involves rapid contact of hot gaseous ammonia (with/without other carrier gases such as nitrogen/steam) and biomass in a fluidized bed reactor followed by gas-solid separation using cyclones [20]. Efforts are currently underway to develop prototype fluidized gaseous AFEX systems to establish proof-of-concept.
- **Extractive AFEX (E-AFEX):** Recent developments have elucidated the beneficial impacts of cellulose III$_I$ production and lignin removal on downstream biological processes [21,22]. There are on-going efforts to adapt AFEX to achieve both these outcomes. The extractive AFEX process involves fractionation of dry biomass using concentrated or anhydrous ammonia (in some cases combined with suitable organic solvents) to produce a substrate with physicochemical properties distinct from conventional-AFEX-treated biomass. Preliminary experiments have found that extractive-AFEX-pretreated biomass has superior enzymatic digestibility and fermentability compared to conventional-AFEX-treated biomass (unpublished data).

Despite the differences in how the AFEX process is carried out, there are certain fundamental physical and chemical effects of concentrated ammonia on the plant cell walls regardless of the method that enhance enzymatic digestibility and fermentability (Figure 9.2). Nearly all the work over the last decade characterizing the effects of AFEX has used non-extractive conventional-AFEX-pretreated corn stover. The primary physicochemical changes during AFEX are: (1) lignin carbohydrate complex (LCC) cleavage and decomposition product formation; (2) lignin/hemicellulose redistribution; and (3) cellulose decrystallization [2,3].

LCC linkages between lignin-hemicellulose (e.g., diferulates cross-linking arabinose side-chains of xylan) are thought to be primary impediments to cellulase accessibility in grasses. Ammonolysis and hydrolysis reactions during AFEX are known to cleave these LCC-based ester linkages [23]. Cleavage of diferulates during AFEX facilitates lignin and hemicellulose removal from within the cell wall to outer wall surfaces and cell corners [17]. A recent ultrahigh-performance liquid chromatography (UHPLC) tandem mass spectrometry (MS-MS) study using AFEX-pretreated corn stover identified several diferulate isomers (e.g., 8–8-cyclic- and non-cyclic-, 8–O–4-, 8–5-cyclic- and non-cyclic-, 5–5-diferulates) including diamide, acid-amide, and diacid forms (R. Vismeh and D. Jones, personal communication, 2012). More than 90% of the diferulates were released as diferuloyl amides during AFEX and only about 5% were released as diferulic acids, reflecting the relative contributions of ammonolysis (amide-forming) and hydrolysis (acid-forming) reactions during AFEX. Similarly, near-theoretical ammonolysis of acetyl ester linkages has also been observed during AFEX [23]. Ammonolysis of acetyl and uronic esters during supercritical ammonia treatment or milder subcritical ammonia treatments has been reported for hardwoods [13,25]. Detailed kinetics of ferulate and diferulate ester links cleavage during AFEX is currently being performed using model compounds (e.g., arabinose ferulate ester) and plant biomass.

A detailed characterization and quantification of the major plant cell wall decomposition products formed/released during AFEX has only recently become available [23]. More importantly, a detailed mass balance for nitrogen lost (from ammonia available for subsequent pretreatments) during AFEX due to the reaction of ammonia with various cell-wall components has also been recently determined. There was a 300% increase in the total nitrogen content of AFEX-pretreated corn stover (AFCS; 36 mg NH$_3$ equiv./g substrate) compared to untreated corn stover (UTCS; 9 mg NH$_3$ equiv./g substrate). The reacted ammonia and soluble proteins/amino acids can provide an important nutrient source during fermentation [18,23,26]. These studies suggest that 98% of ammonia should be theoretically recoverable after AFEX pretreatment of corn stover under the conditions tested, with the remaining 2% reacted with various cell-wall components. Close to 80–90% of the reacted ammonia was found to form soluble nitrogenous products during AFEX, which included acetamide, phenolic amides, and Maillard reaction products (e.g., methyl imidazoles and pyrazine derivatives). However, only 50% of the total reacted ammonia can be accounted for based on the soluble nitrogenous products identified, half of which were ammonolysis by-products. This suggests that

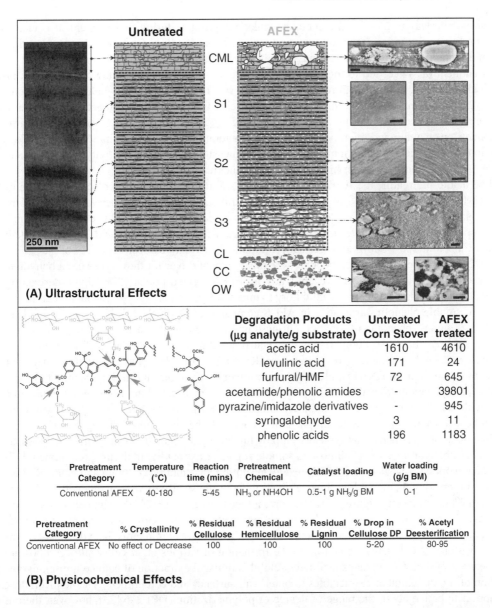

Degradation Products (μg analyte/g substrate)	Untreated Corn Stover	AFEX treated
acetic acid	1610	4610
levulinic acid	171	24
furfural/HMF	72	645
acetamide/phenolic amides	-	39801
pyrazine/imidazole derivatives	-	945
syringaldehyde	3	11
phenolic acids	196	1183

Pretreatment Category	Temperature (°C)	Reaction time (mins)	Pretreatment Chemical	Catalyst loading	Water loading (g/g BM)
Conventional AFEX	40-180	5-45	NH_3 or NH4OH	0.5-1 g NH_3/g BM	0-1

Pretreatment Category	% Crystallinity	% Residual Cellulose	% Residual Hemicellulose	% Residual Lignin	% Drop in Cellulose DP	% Acetyl Deesterification
Conventional AFEX	No effect or Decrease	100	100	100	5-20	80-95

(B) Physicochemical Effects

Figure 9.2   *Ultrastructural and physicochemical impacts of non-extractive AFEX on grass cell walls. (a) Schematic model depicting cell wall structural changes due to AFEX within compound middle lamella (CML), secondary walls (S1-3), cell lumen (CL), cell corners (CC) and outer walls (OW). (b) Figure depicts common chemical linkages cleaved and major cell-wall decomposition products formed during AFEX. Other physicochemical impacts of conventional AFEX are also shown: cell-wall composition, cellulose crystallinity and DP, acetylation, enzyme and microbial digestibility. (Adapted from [2] © 2011, Annual Reviews, Inc., [17] © 2011, Royal Society of Chemistry, [23] © 2011, Elsevier, and [24] © 2009, Elsevier). (See figure in color plate section).*

there are several other Maillard reaction products (e.g., α-amino carbonyl-type intermediates, melanoidins) and insoluble adduct products (e.g., insoluble nitrogenous phenolics) that likely account for the remaining reacted ammonia. Detailed structural characterization of AFEX-treated cell walls (and lignin in particular) is currently underway to identify the chemical sinks for nitrogen during pretreatment. It is clear that the AFEX conditions employed (reaction temperature, ammonia loading, etc.) have a major bearing on the extent of ammonia reacted with the biomass. High-severity (temperature > 130 °C, reaction time > 45 min) AFEX treatment of Avicel led to the formation of various soluble and insoluble Maillard-type products. These studies suggest that although removal of soluble sugars prior to AFEX would prevent extensive formation of Maillard end-products, ammonia may still react with insoluble polysaccharides under extreme AFEX-pretreatment conditions (high temperature, long reaction times). The impact of these nitrogenous end-products on enzymes and microbes is under investigation (briefly explored in Section 9.3). If any of the nitrogenous products are found to inhibit or stimulate downstream biological activity, it will be necessary to modulate AFEX conditions to alter kinetics of end-product reactions.

Unlike acidic pretreatments, AFEX does not yield significant degradation of carbohydrates to furans (e.g., 5-HMF, furfural). Close to 30-fold higher concentrations of furan-based derivatives were formed during dilute acid pretreatment of corn stover (DATCS) compared to AFEX [23]. However, AFEX can catalyze the decomposition of carbohydrates to organic acids (e.g., lactic acid, succinic acid) like many alkaline pretreatments, albeit at 100–1000-fold lower concentrations than what has been reported for NaOH pretreated biomass [23]. High pretreatment temperatures are the dominant driver for these types of carbohydrate degradation reactions, as highlighted in a recent study on AFEX-treated poplar [27]. Lignin-derived phenolics released during pretreatments are very inhibitory to microbial fermentations. We found three–four-fold lower amounts of soluble phenolics/aromatics in AFCS compared to DATCS [23]. However, since degradation product (e.g., furans, phenolics) formation during dilute acid pretreatment is highly dependent on the pretreatment severity, definitive comparisons between pretreatment technologies is ill-advised. Preliminary whole-cell-wall nuclear magnetic resonance (NMR) analysis of AFCS has revealed no major structural changes to lignin after AFEX [17], although minor changes to lignin during AFEX may have significant downstream implications. Sewalt *et al.* have shown that ammoniation of lignin could reduce inhibitory interactions with cellulases compared to its underivatized form [28]. Similar changes to lignin (introduction of hydrophilic side-chain moieties to lignin-phenolics) could also be taking place during AFEX.

Unlike other pretreatments that typically employ higher liquid to solids loading, the conventional AFEX process does not hydrolyze hemicellulose to soluble sugars or extract lignin during pretreatment [17]. However, recent work has revealed that AFEX results in subtle redistribution of lignin and hemicellulose within the cell wall, creating an enzyme-porous cell wall without physically extracting any cell-wall components in a separate liquid stream (Figure 9.2). The current mechanism of AFEX on plant cell walls (monocots in particular) is thought to be as follows: ammonia penetrates into the cell wall through the cell lumen or middle lamella regions where, in the presence of water, it catalyzes ammonolysis/hydrolysis-type reactions to cleave various LCC linkages, resulting in the formation of various other cell-wall decomposition products. Cleavage of the LCC linkages facilitates solubilization and extraction of hemicellulose oligomers, and migration of lignin and other extractables to outer wall surfaces and cell corners. The hemicellulose oligomers appear to be largely in the range of degree of polymerization (DP) 2–6 [23], however, there is also a significant proportion of high-DP gluco- and xylooligomers detected (R. Vismeh and D. Jones, personal communication, 2012). The rapid decompression of ammonia at the end of pretreatment results in the convective transport of ammonia-water and various cell-wall extracts towards the cell lumen and corners, followed by vaporization of ammonia that results in the formation of large pores (>10 nm in diameter) in the outer secondary cell walls [17]. The formation of pores and removal of lignin/hemicellulose to outer wall surfaces enhances accessibility of cellulases. The shape, size and spatial distribution of the pores depend on their location within the cell wall and the exact AFEX conditions employed. Electron tomography has revealed the pore network to be extensive and highly inter-connected. However, the impact of cell-wall pore tortuosity on enzyme diffusion and non-productive binding to exposed lignin/hemicellulose needs to be explored further. Interestingly, both AFEX and dilute acid pretreatments were found to significantly alter the ultrastructure of outer secondary cell walls and the middle lamella. This suggests that mass transfer is a key limiting factor for lignin and hemicellulose removal from cell walls during pretreatment. Solubility of cell-wall components in the pretreatment liquor also impacts removal of lignin and hemicellulose. Anhydrous ammonia-based pretreatments resulted in significant coalescence of lignin (i.e., lignin globules with associated polysaccharides) during AFEX compared to conventional concentrated ammonia pretreatment [17]. These results suggest that anhydrous ammonia may be an excellent solvent, facilitating lignin removal during extractive AFEX-type pretreatments. Close to 50% of the lignin from AFCS is readily extractable using acetone-water solvent mixtures, suggesting that lignin could be readily removed during extractive AFEX-type pretreatments.

Most acidic pretreatments have been found to increase cellulose crystallinity marginally [29,30]. However, by altering the ammonia pretreatment conditions, it is possible to decrystallize cellulose and in some cases produce a non-native cellulose allomorph (e.g., cellulose III$_I$) [21]. Under conventional AFEX conditions there was no significant transformation of cellulose I$_\beta$ to III$_I$ allomorph as high concentrations of water likely prevents the intercalation of ammonia into the cellulose crystalline network to disrupt its structure [17]. However, in the absence of water, cellulose III$_I$ can be readily formed using various sources of biomass (e.g., Avicel, cotton, corn stover, Cladophora) under milder reaction conditions than what has been reported by Igarashi *et al.* [31]. These results suggest that conventional AFEX can improve the digestibility of plant biomass by simply enhancing cellulose accessibility by removal of lignin/hemicellulose without any alteration of cellulose crystal structure. However, we have recently found that modifying the cellulose crystal structure using anhydrous ammonia can increase the synergistic action of exo- and endocellulases by several-fold [21]. Developing a suitable AFEX process that can also alter cellulose crystallinity would therefore be a significant improvement to the existing process.

Interestingly, the rearrangement of the hydrogen-bond network within cellulose III$_I$ increased the number of solvent-exposed glucan-chain hydrogen-bonds with water by 50% (based on extensive molecular simulations), and enhanced enzymatic saccharification rates by up to five-fold (closest to amorphous cellulose hydrolysis rates) [21]. At the same time, the maximum surface-bound cellulase capacity was reduced by 60–70% for cellulose III$_I$ compared to native cellulose. The enhancement in the activity of cellulases was attributed to the amorphous-like nature of crystalline cellulose III$_I$. Unrestricted glucan chain accessibility to active-site clefts of certain endocellulase families (e.g., Cel7B vs. Cel5A) was found to further accelerate the deconstruction of cellulose III$_I$. These findings point to how subtle alterations within the cellulose III$_I$ hydrogen-bonding network can provide an attractive solution to enhancing biomass deconstruction, while offering an insight into the nature of cellulose recalcitrance. This might eventually lead to unconventional pathways for development of novel ammonia-based pretreatments and cellulases for cost-effective biofuel production.

9.3 Enzymatic and Microbial Activity on AFEX-treated Biomass

9.3.1 Impact of AFEX Pretreatment on Cellulase Binding to Biomass

Cellulase binding to cellulose is the preliminary step for enzymatic activity. Most cellulases contain a catalytic domain (CD) and carbohydrate-binding domain (CBD). The extent of binding and processive hydrolytic action of cellobiohydrolases (CBHs) on crystalline cellulose depends on both the CBD and CD. CBDs are thought to enhance CD hydrolysis efficiency onto insoluble substrates by increasing local surface-bound enzyme concentrations. Most *Trichoderma* cellulase-derived CBDs belonging to family 1 are known to have high sequence homology, suggesting similar binding capacities for crystalline cellulose [2]. However, thermochemical pretreatments can modify the ultra-structural morphology and chemical composition of lignocellulosic substrates, strongly impacting their interaction with hydrolytic enzymes.

A high-throughput fast performance liquid chromatography (FPLC) based method has been used for separation and quantification of individual cellulases [32]. This method is based on Medve *et al.*'s FPLC method for CBH I/CBH II [33] and CBH I/EG (endoglucanases) II [34] binary mixtures hydrolyzing purified microcrystalline cellulose (but never used for pretreated lignocellulosic biomass). Using this method, it is possible to accurately quantify unbound CBH I, CBH II and EG I within untreated and pretreated corn stover hydrolyzates without interference from UV-absorbing biomass derivatives [32]. Both AFEX and dilute-acid pretreatment were found to increase cellulase binding to embedded cellulose microfibrils within cell walls. In the presence of EG I, enhanced exocellulase cooperative binding to AFEX pretreated cell walls was observed [32], likely due to the presence of residual hemicellulose that sheathed cellulose fibrils. Competitive binding among enzymes was also observed for certain substrates, cellulase combinations, and protein loadings employed. This technique can help gain a better understanding of enzyme synergism correlated to pretreatment efficacy. These studies could assist in enzyme engineering efforts to minimize nonproductive binding and help design improved pretreatments that facilitate productive binding, hence lower the necessary enzyme dosage.

9.3.2   Enzymatic Digestibility of AFEX-treated Biomass

Most acidic and alkaline pretreatments (e.g., dilute acid, steam explosion, ammonia recycle percolation) extract a significant fraction of hemicellulose and/or lignin to enhance enzyme accessibility [2,3,35]. While conventional AFEX modifies the cell wall ultra-structure to enhance enzyme accessibility, it does not physically extract any of the hemicellulose or lignin as separate fractions [17]. Thus, the inclusion of suitable hemicellulases and accessory enzymes in suitable amounts is necessary to obtain high yields of both glucose and xylose.

The most important glycosyl hydrolases (GH) necessary to digest pretreated biomass (based on protein abundances in typical fungal enzyme extracts predicted using proteomics) include mainly endoglucanases (EG), endoxylanases (EX), cellobiohydrolases (CBH I and II), β-glucosidases (βG), endoxylanases (EX) and β-xylosidases (βX) [36]. The available commercial enzymes cocktail for lignocellulosic biomass hydrolysis can normally be categorized into: cellulases such as Celluclast, Spezyme CP, Accellerase, and CTec; β-glucosidases such as Novozyme 188, Accellerase BG; hemicellulases such as Multifect Xylanase, HTec; and pectinases/accessory enzymes such as Multifect Pectinase. Most commercial cocktails have cross-activities on cellulose, hemicellulose, and other intermediate hydrolysis products such as cello- or xylooligomers. Figure 9.3 depicts the overall enzymatic digestibility of untreated and conventional-AFEX-treated biomass from various sources (monocots and dicots) using Spezyme CP and Novo 188. Supplementation with hemicellulases and other accessory enzymes permitted higher glucose and xylose yields at up to four-fold lower total protein loading for AFCS.

Adding different crude enzyme blends can improve both glucose and xylose yield for AFEX-pretreated biomass [46]. Multifect Xylanase increased both glucose and xylose yields by around 10–20% for AFCS [46]. Commercial Multifect Pectinase and Depol 740L were effective at releasing xylose and arabinose from AFEX-treated dried distiller's grains with solubles (DDGS) [47]. However, since commercial enzyme cocktails contain numerous different enzymes in varying amounts, it is impossible to understand each individual enzyme's role during hydrolysis using only commercial cellulases. Also, the impact of varying ratios for individual enzymes on pretreated biomass saccharification is not clear.

To overcome the limitation of using crude enzyme cocktails, purified enzymes can provide a straightforward way to study the role of enzyme synergy on AFEX-treated biomass saccharification. In our recent work [48], six core cellulases and hemicellulases were isolated using various purification and heterologous expression strategies. Thirty-one unique combinations of purified fungal glycosyl hydrolase mixtures were tested on AFEX-treated corn stover to determine their impact on glucose and xylose yields, at three different

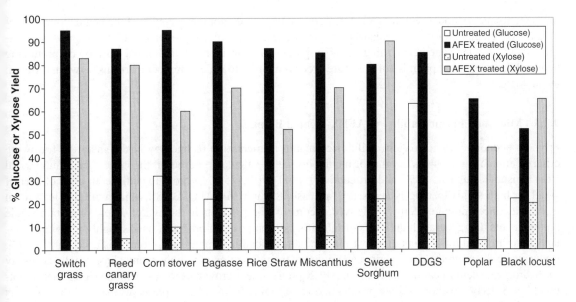

Figure 9.3   *Enzymatic digestibility of untreated and conventional AFEX-treated biomass. All hydrolysis assays were carried out using approximately 15 FPU Spezyme CP/g glucan, 64 p-NPGU Novo 188/g glucan for 168 h at 50°C and 250 rpm. (Adapted from [27] © 2009, American Chemical Society, [37–45] © 2010, Elsevier, © 2006, American Chemical Society, © 2010 Wiley Periodicals, Inc., © 2009, Springer Verlag).*

total protein loadings (8.25, 16.5 and 33 mg/g glucan; inclusive of all six enzymes) using a suitable experimental design. The optimal enzyme ratios that gave maximal hydrolysis yields were closely dependent on the total enzyme loading employed, with endoglucanase I (EG I; Cel7B) requiring the greatest amount. Increasing the proportion of hemicellulases (endoxylanase, β-xylosidase) also significantly enhanced xylose yields with no decrease in glucose yield. The optimal mixture of six core enzymes that maximized glucan and xylan hydrolysis yields for AFEX-treated corn stover was 27–30% CBH I, 17–20% CBH II, 29–35% EG I, 14–15% EX, 2–6% βX, and 1–5% βG (% protein mass composition based on fixed total protein mass loading) at a total protein mass loading of 15 mg enzyme/g glucan [48]. The protein assays used for purified and crude enzyme broths utilized the bicinchoninic (BSA protein as standard) and Kjeldahl assays, respectively. These results demonstrate the potential to rationally design enzyme mixtures targeted towards particular feedstocks or pretreatments that can help maximize hydrolysis yields and minimize enzyme usage. Further experiments have found that both α-arabinofuranosidase and α-glucuronidase can increase xylose yield by an additional 20% [49]. Similar studies on AFEX-treated corn stover have been carried out using an extended set of purified *Trichoderma reesei* cellulases (EG II, EG III, Cel61), hemicellulases (GH 10 EX) and non-hydrolytic proteins (swollenin) [50]. A relatively wide range of protein compositions can give comparable conversions. At high enzyme loadings, the optimum enzyme loading range is broad compared to lower enzyme loadings where non-productive binding to lignin and other effects reduce the range of protein compositions that maximize saccharification. For example, a ratio of cellulase/xylanase ranging from 0.18 : 1 to 3 : 1 for a relatively high enzyme loading (20 mg/g glucan) gave comparable glucose (>70%) and xylose yields (>60%) [49].

In order to achieve an unbiased comparison between pretreated samples, enzymes used for evaluation should be able to reflect the true efficacy of the pretreatment. When comparing AFEX to a pretreatment that selectively removes hemicelluloses, such as dilute acid, the necessary hemicellulases and accessory enzyme activities should be included in the enzymatic cocktail. Otherwise, the lack of specific activities could result in underestimating the true digestibility of AFEX-pretreated biomass. In addition, for other low-severity pretreatments (such as alkaline peroxide) which do not remove most of the hemicellulose fraction, the development of similar balanced enzyme cocktails to evaluate pretreatment effectiveness is necessary. In the bigger scheme of things it is important to consider the additional cost of co-producing hemicellulases on the economics of the entire process in order to make fairer comparisons between different technologies.

9.3.3 Microbial Fermentability of AFEX-treated Biomass

From an economic point of view, cellulosic ethanol fermentation technology has to meet the following criteria: ethanol titer >40 g/L, ethanol metabolic yield (calculated based on theoretical maximum yield from consumed sugar) >90%, and productivity >1.0 g/(L h) [51]. In order to achieve an ethanol titer of 40 g/L, a high solids loading is necessary (at least 18% w/w for AFCS) [18], which will also lead to an increased concentration of cell-wall-derived decomposition products in the hydrolyzate. For certain acidic pretreatments that generate high levels of inhibitory decomposition products, unless highly inhibitor-tolerant strains are used for fermentation, washing and detoxification of pretreated biomass is necessary [23,52]. Reducing acidic pretreatment severity can minimize the production of inhibitors and eliminate the necessity for hydrolyzate detoxification. In contrast, AFEX pretreatment provides biomass with no significant inhibitors (AFCS decomposition product profiles shown in Figure 9.2 and [23]), preserves many of the nutrients present in the biomass, and incorporates ammonia which can be used as a nitrogen source during fermentation [17,23]. In total, 750 ± 50 mg/L ammonia and 1231 ± 44 mg/L total amino acids with excess trace elements for yeast fermentation were found in 6% (w/w) glucan-loading AFCS hydrolyzate (Table 9.1). Since different feedstocks have slightly different nutrient compositions, cell growth patterns and fermentation kinetics vary on different AFEX-treated biomass (Table 9.2). For example, on 6% (w/w) glucan-loading AFCS, *S. cerevisiae* 424A(LNH-ST) showed the maximum cell density of around 6.0 g dry cell weight per liter under microaerobic conditions [18], while this value on AFEX bagasse and cane leaf was 4.4 and 6.9, respectively [39]. Moreover, even for the same feedstock harvested in different seasons, cell growth and fermentation patterns could be different, probably due to the differences in nutrient levels and by-products formed during AFEX [23,57].

Almost all of the tested ethanologens have been able to grow well on AFEX-treated biomass without external nutrient supplementation and detoxification. Those strains include genetic engineered xylose-fermenting ethanologens *S. cerevisiae* 424A (LNH-ST), *E. coli* KO11, *Z. mobilis* AX101, and *P. stipitis* FPL-061 [37,55,56] as well as CBP (consolidated bioprocessing) microbes such as *C. phytofermentans* [58]. *Saccharomyces cerevisiae* 424A (LNH-ST) fermentation of AFEX-treated corn stover without washing and detoxification produced a final ethanol concentration of 40 g/L or 191.5 g ethanol/kg corn stover (Figure 9.4) [18]. Moreover, AFEX decomposition products increased ethanol metabolic yield from 83.6% in yeast extract peptone (YEP) medium to 92.9% in AFCS hydrolyzate [18]. As shown in Table 9.2, similar high ethanol metabolic yields were also obtained on various AFEX-pretreated biomass hydrolyzates using *Saccharomyces cerevisiae* 424A (LNH-ST), and the maximum ethanol productivities on AFEX-treated biomass were all above 1.0 g/(L h). However, the average values did not reach even half of the maximum (all below 1.0 g/(L h)). Further improvement of pretreatment, strains, and fermentation process are necessary to reach the standard industrial requirement of an ethanol productivity >1.0 g/(L h).

A major concern with lignocellulosic biomass fermentations is achieving high conversions of xylose, the second-most abundant sugar in most lignocellulosics. Lau *et al.* compared three xylose-fermenting ethanologenic strains (*S. cerevisiae* 424A(LNH-ST), *E. coli* KO11, and *Z. mobilis* AX101) [55] in AFCS hydrolyzate. All three were able to co-ferment glucose and xylose quickly on corn steep liquor medium achieving >90% xylose consumption in 168 h. *S. cerevisiae* 424A (LNH-ST) could even completely consume 40 g/L xylose in 48 h on YEP medium [18]. In AFCS hydrolyzate, however, xylose fermentation performance was

Table 9.1 Nutrients available in AFCS hydrolyzate. (a) Total nitrogen and amino acids (short form notations are given), and (b) trace elements and vitamins. (Adapted from [26] © 2012, Royal Society of Chemistry).

(a)

Nitrogenous nutrient components	AFCS hydrolyzate (mg/L)	
	Free	Total
Ammonia	—	750
Asp	8.4	75.9
Glu	0	133.8
Ser	16.8	104.2
Gly	5.2	127.2
His	4.5	34.3
Thr	17.6	98.9
Arg	17.1	55.0
Ala	11.6	110.2
Pro	30.4	108.7
Tyr	30.0	28.6
Val	9.9	68.8
Met	2.6	19.4
Ile	7.6	55.4
Leu	0	93.6
Lys	18.4	25.7
Phe	15.7	91.6
Total amino acids	195.8	1231

(b)

Trace micronutrient components			Units	AFCS hydrolyzate
Trace elements	Mg	Magnesium	mg/L	168.42
	Ca	Calcium		242.87
	Mn	Manganase		2.32
	Co	Cobalt	μg/L	11.3
	Ni	Nickel		13.5
	Cu	Copper		116.2
	Zn	Zinc		505.7
	Mo	Molybdenum		15.9
	Fe	Iron		296.4
Vitamins		Panthothenic acid	μM	1.50
		Pyridoxine		1.26
		Nicotinic Acid		10.87
		Biotin		c. 0.05
		Thiamine		c. 0.66

poorer [18,55]. *E. coli* KO11 and *Z. mobilis* AX101 consumed less than 20% of xylose in 144 h, and *S. cerevisiae* 424A(LNH-ST) consumed 82.2% in 168 h (Table 9.2). Even in synthetic media such as YEP, xylose fermentation was much slower compared to glucose. Maximum specific xylose consumption rate of *S. cerevisiae* 424A (LNH-ST) was 0.41 g/(h g cell) [59]. However, for glucose this rate reached as high as 1.9 g/(h g cell) [60]. Because of this difference in rates, there is typically a very high ethanol productivity at the early stage of AFEX hydrolyzate fermentation (glucose fermentation period) and very low ethanol productivity at later stages (xylose fermentation period).

Table 9.2   *Summary of fermentation kinetics on AFEX-treated biomass hydrolyzates.*

AFEX pretreated biomass	Max. ethanol productivity (g/(L·h))[a]	Average ethanol productivity (g/(L·h))[b]	Ethanol metabolic yield	Ethanol titer (g/L)	% Xylose consumption[c]	Final xylose conc. (g/L)	Fermentation conditions	Ref.
Saccharomyces cerevisiae 424A(LNH-ST)								
Corn stover	1.6	0.24	92.9	40.0	82.2	5.0	Solid loading: 6%(w/w) glucan loading; Ferm. time 168 h; Initial cell density: 1.1 g/L; Temp: 30 °C; pH 5.5	[18]
Switchgrass (cave-in-rock, cut in Oct.)	1.7	0.36	89.7	34.6	94	1.9	Solid loading: 6%(w/w) glucan loading; Ferm. time 96 h; Initial cell density: 0.96 g/L; Temp: 30 °C; pH 5.5	[53]
Cane leaf	2.4	0.51	91.6	36.4	87.0	3.8	Solid loading: 6% (w/w) glucan loading; Ferm. time 72 h; Initial cell density: 0.96 g/L; Temp: 30 °C; pH 5.5	[39]
Sugarcane bagasse	3.0	0.28	91.6	33.7	62.7	12.3	Solid loading: 6% (w/w) glucan loading; Ferm. time 120 h; Initial cell density: 0.96 g/L; Temp: 30 °C; pH 5.5	[39]
Rice straw	1.7	0.26	95.3	37.0	69.2	7.4	Solid loading: 6% (w/w) glucan loading; Ferm. time 144 h; Initial cell density: 0.28 g/L; Temp: 30 °C; pH 5.5	[40]

Table 9.2 (*Continued*)

AFEX pretreated biomass	Max. ethanol productivity (g/(L·h))[a]	Average ethanol productivity (g/(L·h))[b]	Ethanol metabolic yield	Ethanol titer (g/L)	% Xylose consumption[c]	Final xylose conc. (g/L)	Fermentation conditions	Ref.
Poplar	4.8	0.74	93.0	35.5	78.7	3.5	Solid loading: 200 g/L; Ferm. time 48 h; Initial cell density: 5 g/L; Temp: 28.5 °C; pH 5.5–6.0	[54]
Forage sorghum	N/A	0.32	86.2	30.9	57.0	13.0	Solid loading: 6% (w/w) glucan loading Ferm. time 96 h Initial cell density: 0.28 g/L Temp: 30 °C; pH 6.0	[37]
Corn silage	1.1	0.39	93.2	28.4	72.0	2.1	Solid loading: 6% (w/w) glucan loading; Ferm. time 72 h; Initial cell density: 0.05 g/L; Temp: 30 °C; pH N/A	[56]
Whole corn plant	1.2	0.41	89.2	29.8	79.5	1.6	Solid loading: 6% (w/w) glucan loading; Ferm. time 72 h; Initial cell density: 0.05 g/L; Temp: 30 °C; pH N/A	[56]
P. stipitis FPL-061								
Rice straw	0.8	0.25	71.7	29.7	92.4	1.8	Solid loading: 6% (w/w) glucan loading; Ferm. time 144 h; Initial cell density: 0.28 g/L; Temp: 30 °C; pH 5.5	[40]

(*continued*)

Table 9.2   (*Continued*)

AFEX pretreated biomass	Max. ethanol productivity (g/(L·h))[a]	Average ethanol productivity (g/(L·h))[b]	Ethanol metabolic yield	Ethanol titer (g/L)	% Xylose consumption[c]	Final xylose conc. (g/L)	Fermentation conditions	Ref.
***P. stipitis* DX-26**								
Rice straw	0.52	0.23	67.7	27.6	88.4	2.8	Solid loading: 6% (w/w) glucan loading; Ferm. time 144 h; Initial cell density: 0.28 g/L; Temp: 30 °C; pH 5.5	[40]
***Z. mobilis* AX101**								
Corn stover	0.42	0.22	96.5	32.0	31.0	20.0	Solid loading: 6% (w/w) glucan loading; Ferm. time 144 h; Initial cell density: OD_{600} of 0.5; Temp: 30 °C; pH 5.5	[55]
***E. Coli* KO11**								
Corn stover	0.46	0.22	N/A	31.0	10.3	26.0	Solid loading: 6% (w/w) glucan loading; Ferm. time 144 h; Initial cell density: OD_{600} of 0.5; Temp: 37 °C; pH 6.8	[55]

[a] The maximum ethanol productivity during the fermentation process, based on the fastest ethanol production period.
[b] The average ethanol productivity during the fermentation process, based on the whole fermentation period.
[c] Xylose consumption was calculated based on the initial xylose concentration and the xylose concentration consumed during fermentation.

Jin *et al.* investigated the factors causing slow xylose fermentation in hydrolyzate [59]. Decomposition products in 6% (w/w) glucan-loading AFCS hydrolyzate resulted in 22% reduction of cell biomass production during glucose fermentation by *S. cerevisiae* 424A (LNH-ST) [59], among which nitrogenous compounds showed the most inhibitory effect followed by aliphatic acids and aromatic compounds (X. Tang, personal communication, 2012). Xylose fermentation always began after glucose was depleted. During the xylose fermentation, there were not only decomposition products but also ethanol and fermentation metabolites in the broth, which inhibited xylose consumption. According to one study [59], ethanol, fermentation

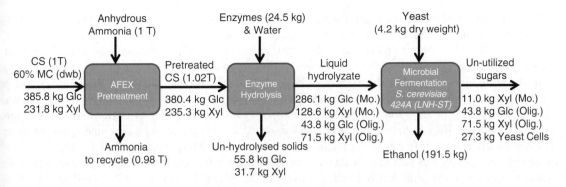

Figure 9.4 *Mass-balance for conventional AFEX, hydrolysis and fermentation for bioethanol production. The mass-balance was based on the AFEX process. All material flows are based on 1 metric ton of untreated corn stover (CS) on dry weight basis (dwb). Conventional AFEX was carried out on CS at 1 : 1 ammonia to biomass loading, 0.6 : 1 water to biomass loading, 15 min reaction time and 130°C. Enzymatic hydrolysis (15 FPU Spezyme CP and 32 p-NPGU Novo 188 added per g glucan; Multifect Xylanase and Multifect pectinase also added and reported as total enzymes) was conducted at 6% glucan loading which is equivalent to 17.6% w/w solids loading. The output streams of hydrolysis and fermentation are monomeric (Mo) sugars, glucose (Glc) and xylose (Xyl) and their respective oligomeric (Olig) sugars. (Adapted from [18] © 2009 PNAS).*

metabolites and decomposition products reduced maximum specific xylose consumption rate by 31%, 42%, and 13%, respectively. Overall, decomposition products were responsible for about 36% of the reduction in total xylose consumption in the hydrolyzate fermentation by *S. cerevisiae* 424A (LNH-ST); the other 64% of the reduction was attributed to ethanol and fermentation metabolites [59]. Among the decomposition products, furans did not show a substantial inhibitory effect but nitrogenous compounds, aliphatic acids and oligomeric sugars were major inhibitors for xylose consumption (X. Tang, personal communication, 2012). For *E. coli* KO11, robustness seemed to be the major problem affecting its tolerance to both decomposition products and the relatively high-concentration of ethanol, thus affecting xylose consumption [59]. Xylose fermentation in AFEX hydrolyzate could also be affected by nutrient levels (affecting cell density) and nutrient type (affecting xylose metabolism). Xylose fermentation performance also differed with choice of feedstock. For example, on AFEX cane leaf, 87% xylose consumption could be achieved in 72 h by *S. cerevisiae* 424A (LNH-ST) [39]. However, on AFEX sugarcane bagasse only 62.7% of xylose was consumed in 120 h (Table 9.2) [39].

CBP performances have also been tested on AFCS. At low solid loading (0.5% (w/w) glucan loading), both glucan and xylan conversions were able to reach above 80% in four days by an anaerobic thermophilic strain *Clostridium thermocellum* [61]. Mesophilic strain *Clostridium phytofermentans* also converted 76% of glucan and 88.6% of xylan in 10 days [62]. At high solid loading (4% (w/w) glucan loading), *C. phytofermentans* achieved 78% of xylan conversion and 49% of glucan conversion in 11 days on AFCS without detoxification and nutrient supplementation [63]. The end-product acetate seems to be the biggest inhibitor limiting high solid loading CBP process by *C. phytofermentans* [63].

9.4   Transgenic Plants and AFEX Pretreatment

There are two main arenas of research and development of transgenic plants for biofuel production. The first modifies cell-wall components with the hope of improving sugar yields from biomass, and the other incorporates genes for the production of cellulolytic enzymes *in planta*. To date, all of the published research on

AFEX pretreatment of transgenic plant materials has been related to enzyme production. One way to theoretically reduce the processing costs of the biorefinery is to produce the enzymes which are used during enzymatic hydrolysis within the plants themselves. One question is whether it would be necessary to extract these enzymes prior to the pretreatment step or whether sufficient activity is retained following low-severity pretreatment to allow the pretreated biomass to be used directly and hence minimize additional costs for enzyme extraction. To examine this question, AFEX pretreatment was applied to transgenic tobacco leaves which expressed a gene for a thermostable *Acidothermus cellulolyticus* endo-1,4-β-glucanase (E1) within the tobacco apoplast [64]. Both heating the biomass above 70 °C and, in a separate experiment, adding room temperature ammonia at 1.0 g NH_3:g dry biomass effectively destroyed most enzyme activity. Only 30–40% of the enzyme activity was retained for the enzymes which were extracted from tobacco leaves following pretreatment under fairly mild AFEX conditions (60 °C, 0.5 g NH_3:g dry biomass, 0.2–0.4 g H_2O:g dry biomass and 5 min residence time following heat-up).

Because the AFEX conditions required for effective biomass conversion destroys enzyme activity it seems that, compared to pretreatment and direct use of the transgenic biomass, extraction of the proteins prior to pretreatment could be a feasible option. To examine the ability to extract the enzymes from the biomass and use the extract for hydrolysis of pretreated biomass, the E1 endoglucanase was also expressed in the apoplasts of corn [65] and rice [66] and the leaves were extracted to obtain a solution of extracted proteins. When used in conjunction with β-glucosidase, the total leaf protein extract from transgenic corn was able to hydrolyze roughly 20% of the glucan from AFEX-treated corn stover and the extract from transgenic rice achieved glucose yields from AFEX-treated corn stover and rice straw of 30% and 22%, respectively [66]. While the sugar yields were not as high as those obtained with commercial enzyme mixtures, the yields could be increased by taking advantage of enzyme synergism (i.e., adding other enzyme activities such as exoglucanases and xylanases [48]).

Experiments are currently underway to examine the interactions of AFEX pretreatment with plants that have undergone modifications to the lignin synthesis pathway, either by reducing the lignin content or altering subunit composition. Reductions in total lignin content are known to improve enzymatic saccharification [67–69]; however, it might also be possible to alter the lignin composition in such a way that plant materials have improved processing characteristics. One promising material that is being examined is a poplar transgenic where the ferulate-5-hydroxylase (F5H) enzyme has been up-regulated using a cinnamate-4-hydroxylase (C4H) promoter [70–72]. The lignin in this material is composed of >97% syringyl units, and the resulting polymer chains are highly linear and lower in molecular weight compared to the wild type [72]. Initial results suggest that these structural changes increase the ability to extract the lignin from plant cell walls [70,72,73] (A. Azarpira and J. Ralph, personal communication, 2012), which can benefit increasing enzyme accessibility to the substrate following an extractive AFEX-type pretreatment.

There are many types of linkages which are possible between lignin subunits, the majority of which are highly resistant (e.g., C—O—C, C—C linkages) to chemical degradation. One idea that has been proposed is the possibility of engineering plants to produce new monomers which would be automatically incorporated into the cell wall and would produce new types of linkages (e.g., ester linkages) within the lignin polymer that are more readily cleaved under alkaline conditions [74]. This is a desirable possibility because it would avoid the negative issues sometimes associated with reductions in total lignin content, including reduced biomass yields [67,75], reduced winter hardiness [75,76], and increased susceptibility to pests and pathogens [77]. One of the monomers being considered is coniferyl ferulate which, when incorporated, would form alkali-cleavable ester linkages (or "zipped lignin") within the lignin polymer [74]. When 25% coniferyl ferulate was incorporated into model maize primary cell walls (dehydration polymer cell walls; DHP-CW), the same amount of delignification was achieved using NaOH at 100 °C compared to the material that had only coniferyl alcohol at 160 °C [78].

9.5   Recent Research Developments on AFEX Strategies and Reactor Configurations

9.5.1   Non-extractive AFEX Systems

The simplicity of the conventional, non-extractive AFEX pretreatment allows considerable latitude in reactor design. AFEX reactors may be designed to operate either batch-wise or continuously, may use either direct or indirect heating, and may or may not be mechanically agitated. In all cases, the reactor design objectives are to provide adequate ammonia and temperature distributions throughout the biomass for a reasonable residence time. The nature of ammonia-water vapor/liquid equilibrium dictates that, at the ammonia concentrations and temperatures required for AFEX treatment, reactors must operate at elevated pressure. For example, at conditions of 50 wt% ammonia concentration and 90 °C temperature, the saturation pressure will be about 19 bar (260 psig) [79]. Consequently, AFEX reactors must be designed as pressure vessels. In addition, it is vital to be able to feed biomass to the reactors at these pressures. Furthermore, all wetted reactor materials must be compatible with ammonia-water liquid mixtures. Typically, AFEX reactor vessels are fabricated from 300-series stainless steels, with ethylene propylene diene monomer (EPDM) or tetrafluoroethylene (TFE) seals. Since ammonia-air mixtures are flammable in the concentration range 16–25 vol%, AFEX reactors must be designed to exclude air from the headspace vapor. While AFEX pretreatment has yet to be practiced at a commercial scale, several appropriate reactor designs may be considered.

- **Pandia- and Kamyr-type reactors:** In the pulp and paper industry, continuous reactors have been used for decades to contact fibrous biomass with various delignification catalysts under conditions of elevated temperature and pressure. These reactor designs should be readily adaptable to AFEX pretreatment. Because these continuous reactors typically operate at high-solids (low-moisture) content, the biomass must be transported through the reactor by means of an internal screw. Horizontal screw (aka Pandia-type) reactors typically operate at low fill factors, otherwise spillover between screw flights causes broad residence time distributions and results in poorer performance. Vertical screw (aka Kamyr-type) reactors can operate at higher fill factors. While both horizontal and vertical screw reactors have proven to be economical for pulp bleaching at large scales, both may be scaled down only at very high capital cost per ton of treated biomass.
- **Tubular reactors:** Some biomass materials, such as DDGS, form highly fluid slurries when mixed with ammonia and water under typical AFEX conditions. In these cases, an internal screw is not needed to transport the slurry through the reactor vessel. Once the fluid slurry has been formed by mixing with ammonia at the appropriate temperature, the slurry may be simply pumped through a tube to provide adequate residence time. A tubular AFEX reactor system designed to treat 300 lb/hr of biomass has been demonstrated in the MBI pilot plant. This reactor system uses a progressive cavity pump to feed the biomass at 60–70 wt% moisture. Liquid anhydrous ammonia is then injected into the biomass stream. Introduction of high-pressure steam provides both turbulent mixing and heating to the target temperature. The mixed and heated biomass slurry is then pushed under pressure through a length of stainless steel tube; at the discharge end of the tube, reactor pressure is controlled by means of a lobe pump. Capital costs associated with this tubular reactor design are significantly lower than for screw reactors, but the biomass must form a fluid, pumpable slurry. Agricultural residues such as corn stover and wheat straw, and energy crops such as miscanthus and switchgrass, do not form pumpable slurries and therefore cannot be AFEX-treated in tubular reactors.
- **Packed bed reactors:** A potentially scalable and cost-effective alternative to continuous AFEX reactors is the use of ammonia absorption in packed beds of moist biomass, which is discussed in Section 9.6 (Figure 9.5).

9.5.2   Extractive AFEX Systems

The potential of utilizing ammonia as a solvent to extract lignin and/or hemicellulose from biomass has been demonstrated by several researchers [80–82]. Most studies consider this aqueous ammonia extraction process as being particularly relevant to the pulping industry. Recently, ammonia-based extraction of lignin and hemicellulose is gaining industrial relevance as a possible pretreatment process to be used in cellulosic biorefineries. The benefits of lignin extraction in the enzymatic hydrolysis performance have been widely demonstrated in the literature [82,83]. The presence of lignin increases non-productive binding of cellulases that causes enzyme inhibition and decreased sugar yields [84,85]. Moreover, lignin acts as an antimicrobial agent, reducing cell viability and ethanol yields [53]. Research on methodologies to remove lignin is therefore of special interest to the cellulosic biofuels industry.

One major advantage of using anhydrous ammonia (or concentrated ammonia with little or no water associated with the biomass) in an extractive pretreatment process is that ammonia can potentially alter the crystal structure of native cellulose I_β to III_I [21,86]. The latter cellulose allomorph has been shown to be considerably less recalcitrant to enzymatic degradation [21,29]. However, since it can only be made under specific pretreatment conditions, no cellulose III_I is formed at conventional AFEX conditions [17]. The most critical parameters that interfere with this transformation include the concentration of ammonia, the liquid to solid ratio (which must be relatively high), and the absence of water. A novel extractive AFEX pretreatment employing liquid ammonia (added to the biomass at a higher liquid-to-solid ratio than conventional AFEX), with or without other solvents such as water or organic solvents that can simultaneously produce cellulose III_I and extract lignin during pretreatment [21,22], is currently being developed. The reactions between the biomass and ammonia at temperatures of 40–120 °C facilitate the release of cell-wall extractives (e.g., lignin) that tend to be more soluble in ammonia and can be selectively extracted during this process. To complete the extraction, the liquid is filtered from the biomass at high pressure and reused for subsequent pretreatment cycles prior to collection of the extractive-rich ammonia liquor for recycle. These lignin-rich extractives can be used as raw material to produce value-added chemicals [87], contributing to the economic viability of this process.

9.5.3   Fluidized Gaseous AFEX Systems

In this approach, hot ammonia gas is used to pretreat biomass in a fluidized reactor rather than using liquid ammonia (unpublished data). The hot ammonia gas condenses on the biomass and reacts with water (used to pre-wet the biomass prior to adding the ammonia gas), causing a spontaneous increase in the temperature. During this process, biomass is more uniformly pretreated by ammonia and requires a shorter pretreatment residence time (1–15 min). This also helps reduce the formation of potentially inhibitory degradation products that might affect downstream biological processing. There is also no decompression of ammonia during pretreatment, hence allowing significant energy savings by avoiding ammonia-water mixture separation/recovery as in conventional AFEX. The process can be easily made continuous using a recycled stream of ammonia/water/inert-carrier hot gas in a fluidized/semi-fluidized bed reactor. All ammonia that did not react with the biomass could be recycled in its gaseous state directly, allowing substantial savings in ammonia recovery costs compared to flashing concentrated ammonium hydroxide and recompressing it back to the anhydrous, liquid state.

9.6   Perspectives on AFEX Commercialization

9.6.1   AFEX Pretreatment Commercialization in Cellulosic Biorefineries

AFEX pretreatment has yet to be commercialized, although large reactors have been built and process designs have been developed. Economical, scalable AFEX process designs must meet specific requirements. Ammonia, water, and biomass must be fed into the reactor and adequately mixed. Ammonia must then be removed from the treated biomass, recovered, and re-used economically. A number of different process designs have been proposed to meet these requirements.

Ammonia Recycle

In order to be economical, virtually all of the non-reacted ammonia must be recovered after the conventional AFEX process, reducing catalyst costs and eliminating the need for later neutralization. Ammonia removal from AFEX-treated biomass may be achieved in either continuous or batch operations. The high volatility of ammonia relative to water allows a high degree of ammonia removal with considerable moisture still remaining in the treated biomass. As the pressure is reduced, much of the ammonia will immediately vaporize, and a compressor can repressurize and condense the ammonia into a liquid for recycling. The remaining ammonia must be stripped via a dryer or other similar approach before being recompressed. Mechanical compressors are sensitive to water vapor, which tends to accumulate as emulsified condensate in the compressor oil. The compressor must therefore be protected by flashing off a vapor stream with low moisture content on the compressor suction side. Vapor exhaust from an ammonia removal dryer will contain significant amounts of water, which must be separated out by condensation. This condensate may be pumped back into the reactor if the biomass feed moisture is low enough, otherwise the condensate must be rectified to separate the ammonia and water. Mechanical compression equipment offers a compact footprint and moderate capital cost, but high electrical power load.

This generalized approach is the method most commonly modeled and was used in a comparative pretreatment study on the economics of ethanol from corn stover [88]. While this approach was effective in completely recovering all ammonia, it was also fairly energy intensive, particularly for the compressors. According to a comparative study performed by the National Renewable Energy Laboratory (NREL), a biorefinery with AFEX pretreatment would require nearly 25% more in electricity demand than a similar refinery with dilute-acid pretreatment [88]. Continuous ammonia removal from slurries of DDGS with ammonia and water has been demonstrated using a horizontal thin film dryer (unpublished results). Greater than 98% ammonia removal was achieved at a feed rate of 30 lb/hr of slurry per square foot of heat transfer area using a steam-jacketed horizontal thin film dryer with nitrogen sweep gas. Moisture content of the de-ammoniated biomass product was 25–30 wt%. Thin film dryers use rotating blades to spread the feed slurry in a thin turbulent film over the heated surface. Consequently, the biomass slurry must be sufficiently fluid to be de-ammoniated in this type of dryer.

Wang *et al.* considered a water extraction unit to remove ammonia remaining in the biomass rather than strip the ammonia with a dryer [89]. This extractor strips most of the ammonia out of the biomass, although some remains as a fermentation nutrient. The ammonia/water mixture is then sent to a distillation column to purify the ammonia, while the water is recycled. The vaporized ammonia is condensed to a liquid prior to being repressurized. This approach was also energy intensive, particularly in cooling the ammonia for condensation.

More recent studies indicate that ammonia for AFEX does not need to be pure [90]; Laser *et al.* proposed a quench system to reduce energy requirements [91]. Most of the ammonia is flashed off the biomass while the remaining is stripped with steam. The ammonia vapor is condensed by direct contact with water, creating a concentrated ammonia (*c.* 75% NH3) vapor-liquid mixture. The remaining vapor is condensed via chilled water before repressurizing via a pump. This process eliminates the need for both a costly dryer and the compression system. Similarly, ammonia removed from treated biomass may be absorbed in water in a spray tower absorber, and the strong solution from the absorber may be pumped up to pressure and stripped to recover ammonia. Ammonia absorption and stripping are well-known processes, as for example in sour gas scrubbing and sour water stripping operations that are widely practiced in petroleum refining.

Ammonia-water absorption eliminates the need for a compressor and the problems associated with oil contamination; however, the equipment cost and footprint may be greater.

Capital and operating costs associated with ammonia removal and recovery are anticipated to be significant [92]. Consequently, a principle objective in the design of these systems is to minimize cost while maximizing biomass throughput. Economically-viable AFEX pretreatment of biomass has yet to be demonstrated at any scale, so there is at present no particular design that can be referenced as an industry standard. However, standards and costs for ammonia and ammonia-water refrigeration equipment are well known [93]. Adaptation of standard ammonia refrigeration equipment for service in commercial-scale AFEX processes may provide moderate capital costs with well-characterized performance, reliability, safety, and service requirements. Alternatively, ammonia may be stripped from one batch of biomass to another, as discussed in detail in the following section.

AFEX Pretreatment Scale-up Designs

Initial process designs for a scaled-up AFEX reactor were based on NREL's initial development of dilute acid pretreatment using a Pandia-type reactor system. This type of reactor has been used for many years in the pulp and paper industry and can be adopted for a variety of pretreatment processes including AFEX, dilute acid, and steam explosion. The Pandia reactor system was not specifically designed for ammonia-catalyzed pretreatment and therefore has several drawbacks. The reactors would require highly skilled operators due to the complexity of operation (to perform AFEX pretreatment), require moving biomass against a pressure gradient, and tend to have a relatively short lifespan. All of these factors can greatly increase the cost of the system even before considering ammonia recovery. The capital investment in particular is quite high when considering both the reactor and ammonia recovery, and thus this system is viable only in large refineries and not economically scalable to smaller, regional operations.

Solids feed and discharge to and from a continuous AFEX reactor may use plug-screw devices, just as for a conventional dilute acid reactor. However, in a continuous Pandia-type AFEX reactor, steam and liquid ammonia are injected as the biomass is conveyed through the reactor by the screw rotation. Both the feed and discharge plug screws are designed with close-pitch screws which compact the biomass solids to form a plug, effectively sealing the reactor vessel at the feed and discharge ends and allowing the reactor vessel to be operated at elevated pressure. After discharge from the reactor into a blowdown vessel, the ammonia can then be removed from the biomass by vaporization in a biomass dryer.

To address the primary cost drivers of an AFEX system using a Pandia-type reactor for pretreatment and a dryer for ammonia removal, the MBI team recently developed a novel, simple, and unique reactor system for the AFEX process that exploits the chemical and physical characteristics of ammonia [19]. The high volatility of ammonia relative to water, combined with its ability to react reversibly and exothermically with water, can be used to treat biomass in packed beds using very simple equipment and systems with very low capital cost. A packed bed of moist biomass can be charged with NH_3 vapor under pressure, which will absorb into the moisture and saturate the biomass with NH_3. As the bed becomes saturated, the exothermic NH_3-H_2O interaction will raise the temperature of the bed. Allowing the bed to remain saturated for an adequate period of time will cause the biomass in the bed to become AFEX-treated. The bed pressure may then be released, allowing some of the absorbed NH_3 to flash off, and then the remaining NH_3 can be removed by stripping the bed with steam. The AFEX-treated biomass will then be free of NH_3 and can be removed.

A convenient way to recover the NH_3 removed from a packed bed is to absorb it onto a second bed of moist biomass. Figure 9.5 shows a concept for an arrangement of three packed beds that can be used in a cyclic bed process to treat biomass. Three vessels containing packed beds of moist biomass are connected in series (a). In Figure 9.5, the first bed in series (Bed 1) is charged to saturation with NH_3 (b), while some NH_3 breaks through Bed 1 to Bed 2. Bed 1 can be allowed to remain saturated for an adequate period of

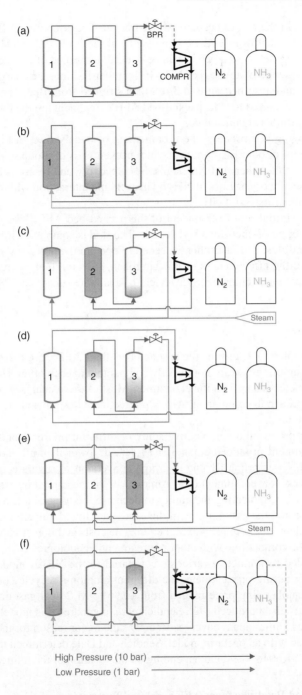

Figure 9.5 Concept for hybrid AFEX pretreatment and ammonia recovery in a system of three packed beds (1–3) in a step-wise manner (a–f). (See figure in color plate section).

time to cause the biomass in the bed to become AFEX-treated. The ammonia in Bed 1 is then removed by steam stripping, with the effluent from Bed 1 compressed and sent to Bed 2, which becomes saturated with NH_3 (c). When Bed 1 is completely stripped, it can be opened, unpacked of treated biomass, and re-packed with untreated biomass (d), without any loss of NH_3. In a similar fashion, the charge of NH_3 can be stripped from Bed 2, compressed, and absorbed on Bed 3 (e), then stripped from Bed 3 and absorbed on Bed 1 (f). The only major equipment required for this packed-bed AFEX treatment are the three bed vessel(s), an NH_3 compressor, and a boiler or other steam source.

As in a simulated moving bed process, the arrangement of beds shown in Figure 9.5 allows a single charge of ammonia to be used to AFEX-treat sequential packed beds of biomass. The treatment cycle may be repeated indefinitely. Unlike other AFEX designs, NH_3 recovery and re-use is accomplished in packed-bed AFEX without the use of a dryer. Loss of NH_3 by irreversible interaction with biomass may be compensated as needed by addition of make-up NH_3.

MBI has demonstrated initial proof of concept of the packed-bed AFEX design using a test skid with three 4-inch-outer-diameter by 48-inch-long bed vessels. The skid is equipped with a small NH_3 compressor and all of the valves, manifolds, and instruments needed to operate the process shown in Figure 9.5. Initial results have demonstrated the effectiveness of this approach, showing that pretreatment efficacy matches that of lab-scale batch reactions, achieving >75% yield of available sugars and 98% recovery of ammonia from the biomass.

Techno-economic Analyses

Several studies have attempted to determine the cost and benefit of AFEX pretreatment. All of these studies are based on different assumptions, so are not directly comparable to each other. However, they can provide insight into the key parameters that affect the economics. Most of these analyses consider either the NREL approach or the Laser approach to commercial AFEX pretreatment design, and are generally considered for a biorefinery producing ethanol.

Eggeman and Elander produced a comparative study of different pretreatments on corn stover [88]. In this study, AFEX pretreatment produced ethanol at $1.41/gal, lower than all other pretreatments studied except for dilute acid. This was primarily due to a high conversion of sugars compared to other pretreatments, as well as a relatively low equipment cost. Almost 75% of the capital investment in the pretreatment was due to ammonia recovery, indicating the ability to decrease the total cost if a packed-bed type system were used. Kazi *et al.* updated the assumptions in this study and found a much higher cost, approximately $5.14/gal gasoline equivalent, due primarily to higher enzyme costs and higher capital costs [94]. However, AFEX was still found to be competitive with other pretreatment technologies.

The Laser model tended to predict lower costs compared to the NREL model. This is partly due to reduced ammonia loading during pretreatment, the different ammonia recycle configuration, and reduced enzyme loading. The ethanol selling price ranged from $0.96 to $1.24/gal gasoline equivalent, depending on the multiple co-products also produced [92]. Sendich *et al.* compared this model to the NREL model and confirmed that the change in ammonia recovery system was the major reason for the reduction in price [90]. Using this model combined with an on-farm model, Sendich and Dale determined that the minimum ethanol selling price ranged from $1.60 to $1.90/gal depending on the location of the refinery [95].

9.6.2   Novel Value-added Products from AFEX-related Processes

Animal Feed

As stated previously, ammoniation of forages is a common practice for improving the digestibility of animal feeds. AFEX, as a more extreme treatment than forage ammoniation, has the potential to greatly improve on this existing technology. A recent study compared the difference in fiber digestibility by rumen microbes between untreated and AFEX-treated samples for 11 different forages and potential energy crops [96]. All crops showed increased digestibility after AFEX treatment, with the greatest improvement coming from material that was very difficult to digest without ammonia treatment. AFEX, for example, improved switch-grass digestibility by 128% over untreated and 78% over traditionally ammoniated samples. One animal feeding trial was performed using AFEX-treated rice straw as 7% of the total diet in lactating dairy cows [97]. Total dry matter intake did not change significantly, but milk production increased slightly by 3% ($p = 0.02$). One concern, however, is that hyperexcitability of cows has occasionally occurred when fed diets containing forages treated with ammonia at high temperatures, possibly due to the production of 4-methylimidazole from soluble reducing sugars [98]. Although the exact mechanism of hyperexcitability due to ammonia-treated feeds (including the role of AFEX treatment) is currently unknown, removal of free soluble sugars from the biomass and/or reducing pretreatment temperature can minimize formation of imidazole-derived compounds that are potential neuro-toxins [23]. If AFEX-treated feeds could displace high-value forages such as alfalfa, then they could be sold for $50–100/ton and significantly improve the economics of biofuel production [95,99].

Leaf Protein Concentrate

Leaf protein concentrates (LPC) are considered a possible alternative to soybean meal in a bioenergy economy due to the fact that protein and fuel can be co-produced on the same land [100,101]. One method of obtaining LPCs is to extract the protein using an alkaline solution such as ammonia or sodium hydroxide. Given the need to obtain an alkaline solution, efforts have been made to use AFEX treatment as a method of enhancing protein solubility. Urribarrí *et al.* compared the extraction of protein from AFEX versus untreated cassava [102] and dwarf elephant grass [103], and found protein solubilization improved after AFEX pretreatment by 45% and 350%, respectively. A later study with switchgrass did not see improved yields after AFEX treatment, however, and also discovered more polyphenolics in the protein extract [104,105]. These polyphenolics could be bound to the protein, lowering their nutritional value. An economic analysis of AFEX integrated with protein extraction was performed by Laser *et al.*, and it was determined that LPC production had the potential to lower the selling price of ethanol by $0.01–0.05/L [92]. Creating animal feed co-products (or, in the case of AFEX for forage feed, alternative products) from biofuel processing can largely eliminate the "food vs. fuel" controversy surrounding bioenergy. A recent study considered adding these two co-products (as well as double cropping) to United States animal feeding operations and determined that 400 billion liters of ethanol could be produced while maintaining food production and exports at their current levels [106]. This result is partly due to displacing relatively unproductive hay and soybean land with more productive switchgrass and protein-rich forages.

Distiller's-grain-derived Proteins and Amino Acids

Distiller's grains and solubles (DGS) are a by-product of corn ethanol production generally sold as an animal feed, but have drawn interest as a potential source for cellulosic biofuel [107,108]. DGS contains approximately 30% protein, and this protein must preserve its value if the fiber fraction is to be converted to biofuel. As with leaf proteins, the DGS protein can be extracted with the help of AFEX pretreatment. Bals *et al.* also tested protein extracted from DGS, and AFEX improved the extractability to 32% of the total protein when using 0.1 M sodium hydroxide as the solvent [109]. In addition, AFEX improved both the rate and extent of fiber hydrolysis in both dry DGS and wet distiller's grains without solubles (WDG) [38,110]. A unique opportunity for value addition was investigated by Brehmer *et al.*, who considered using proteases to solubilize the protein for later use as value-added precursors to bioplastics [111]. An analysis of this fractionation method to produce both ethanol and bioplastics from WDG suggested that 175 g of amino acids and 140 g of ethanol could be produced per kilogram WDG on a dry basis, resulting in a net energy yield of 3.6 GJ/ton WDG.

Natural Enzyme Inducers and Microbial Nutrients

AFEX pretreatment partially breaks down hemicellulose, thus increasing the amount of water-soluble xylooligomers in the biomass [17,23]. After extracting these xylooligomers and other microbial nutrients using hot water [23], these extractives can be used to induce cellulase/hemicellulase production from native *T. reesei* strain [26]. We found that the enzymes induced by the water extract of AFEX-treated corn stover compared to those induced by lactose alone gave three-fold higher sugar hydrolysis yields from pretreated biomass. The induced enzyme mixture was nearly as effective as commercially available enzymes, potentially reducing the need for an exogenous enzyme source by 90% (w/w of total enzyme protein needed for saccharification) for the same hydrolysis yields. A preliminary economic analysis of this approach suggests that the profit to the biorefinery could increase by $15/Mg biomass by including on-site enzyme induction using AFEX-created xylooligomers and nutrients [26].

Lignin and Amidated Products

With the development of extractive AFEX-based pretreatments, it would be possible to selectively remove lignin during pretreatment for use as precursors for the chemical/polymer industry. This would be further facilitated by the use of transgenic plants containing easily extractable, highly homogenous lignin (see Section 9.4 for details). Other chemical products that could be isolated from AFEX-pretreated biomass in high yields include amides (e.g., acetamide, phenolic amides). Oxidized, ammoniated lignin from lignin extractives or from residual solids following enzymatic hydrolysis could also be used as a slow-release nitrogen fertilizer in place of chemical fertilizers [112].

9.6.3   AFEX-centric Regional Biomass Processing Depot

While most of the focus on developing a cellulosic biofuel economy has been on creating large, centralized refineries, there is a growing consensus that such facilities present logistical challenges. Such large refineries will be valuable for reducing the capital cost per gallon of fuel produced, but require contracting with potentially thousands of farmers and transporting biomass upwards of 50 miles to the refinery. An alternative is to create smaller processing centers that can pre-process and densify biomass prior to transport to the refinery [113]. In the most basic form, these centers would include grinding, densification into pellets or briquettes, and potentially storage, although further processing is possible. Packed-bed AFEX pretreatment, requiring relatively low capital investment and leaving a dry, stable intermediate after pretreatment, is a uniquely suitable pretreatment that can be decoupled from the biorefinery and included in these regional biomass processing depots (RBPDs). In addition, the lignin extracted to the biomass surface after AFEX pretreatment acts as a natural binder [17,114], allowing for easier and potentially cheaper densification. Likewise, the potential of using AFEX-treated biomass as an animal feed allows for a secondary market for these RBPDs.

In addition to improved logistics, RBPDs offer several advantages over a centralized refinery. One advantage is that the rural community could own these regional centers, possibly as a farmer cooperative ("co-op"). This would increase local economic development, a key concern for the emerging bioeconomy. It would also encourage the development of biofuels, as farmers would have an incentive to grow new cellulosic feedstocks for the RBPDs. The emergence of alternative markets (utilizing AFEX-treated animal feeds) would promote the development of the biofuel economy by allowing the infrastructure/logistics for cellulosic feedstocks and RBPDs to be developed prior to the emergence of costly centralized biorefineries. Furthermore, it is possible to adopt these RBPDs for multiple co-products, increasing the efficiency of land use and reducing the "food vs. fuel" issue [106].

Distributed biomass processing plants will have to operate on narrow margins, which will make them highly sensitive to capital costs. The capital investment of a plant based on packed-bed AFEX technology will probably be less than one-third that of a plant based on conventional plug flow AFEX technology (unpublished data). The lower mechanical complexity of packed-bed AFEX also means that the plant can be built in less time and will be operable for a longer period without maintenance. Because of these advantages, the RBPDs can break even in less time, with a greatly reduced risk for the owner and a significantly greater ultimate return on investment. Improved return on investment with lower risk means that packed-bed AFEX reactors will be much more likely to be built for RBPDs than other types of AFEX reactor.

Currently, few studies have considered the impact of RBPDs on the commercialization of cellulosic biofuels. Carolan *et al.* used the Laser AFEX model to compare the economics of RBPDs at various sizes and with or without the possibility of animal feeds [99]. This study required large RBPDs (at least 666 tons/day of biomass) due to a high capital cost, but the packed-bed AFEX approach should decrease the cost and allow for smaller RBPDs (100–500 tons/day). This study also showed that AFEX-treated animal feed was an essential element to the economics of the process, reducing the price of AFEX-treated feedstock for biofuels production by up to $9/ton if 25% of the biomass were used as animal feed at $98/ton. While no other studies have been published on RBPD economics, the current approach is to enhance these RBPDs with multiple technologies to produce a multitude of co-products. These RBPDs will be flexible and suited for the local landscape and crops grown in the area.

Low-temperature and long-residence-time pretreatments have been proposed as an alternative to conventional pretreatments within a centralized biorefinery, allowing for a decentralized pretreatment without high energy costs [115]. AFEX pretreatment at 40 °C and 8 hours produced comparable sugar and ethanol yields as conventional AFEX pretreatment at high temperatures and short residence time during subsequent hydrolysis and fermentation. This study suggests a greater flexibility in AFEX pretreatment conditions than previously thought, allowing for an alternative approach for decentralized facilities if the economic conditions are appropriate [116].

9.7   Environmental and Life-cycle Analyses for AFEX-*centric* Processes

Currently, environmental life-cycle analyses (LCAs) on lignocellulosic feedstocks are in their infancy. Most of these analyses consider only one technology and are focused primarily on differences in crop management and land use. Those that consider technology changes generally use currently established models such as the NREL or Laser models. Wu *et al.* combined the Laser model with the greenhouse gases, regulated emissions, and energy use in transportation (GREET) model to obtain a wells-to-wheel LCA [117]. This study found that, regardless of the configuration used to produce biofuels and various co-products (electricity, protein, Fischer–Tropsch fuels), total fossil energy and petroleum use were reduced by 88–93% compared to conventional gasoline, while greenhouse gas (GHG) emissions were reduced by 82–87%. Of the various biorefinery configurations considered, ethanol paired with a Rankine cycle was the worst performer, whereas ethanol paired with Fischer–Tropsch diesel and gas turbine combined cycle power production performed the best in terms of GHG emissions. Bai *et al.* also considered switchgrass-based ethanol with AFEX pretreatment, and compared ethanol and gasoline against numerous environmental indicators [118]. Ethanol greatly reduced global warming potential, abiotic depletion, and ozone layer depletion, but increased toxicity, eutrophication (due to fertilizer runoff), and photochemical oxidation potential (due to acetaldehydes formed during fermentation).

Some studies have also compared pretreatments. Chouinard-Dussault *et al.* applied the NREL model to explicitly compare the environmental performance of refining corn stover into ethanol using AFEX, hot water, or dilute-acid pretreatments, as well as considering multiple processing options associated with dilute-acid pretreatment [119]. This study relied primarily on process integration to reduce energy requirements in the refinery. In this study, AFEX performed the worst out of the three pretreatments, reducing GHGs by 12 g CO_2/MJ ethanol produced while dilute-acid pretreatment reduced GHGs by 20 g CO_2/MJ ethanol. This appears to be primarily caused by high energy costs in conventional AFEX pretreatment. Spatari *et al.* also found that dilute-acid pretreatment produced less GHGs than AFEX for both corn stover and switchgrass, although AFEX required less petroleum [120]. This result is primarily due to increased energy use for ammonia recovery, which reduces the excess electricity available from burning lignin. However, the study also analyzed the Laser model, which reduced fossil fuel inputs by 3 MJ/L and GHG emissions by 350 g CO_2/L. While no studies have been performed on the packed-bed AFEX approach, the results of these studies strongly suggest that reducing the energy costs for ammonia recycle is a key opportunity to improve the environmental performance of biofuel production based on AFEX pretreatment.

Previous LCAs were based on separating farming from biomass processing and calculating the environmental impacts of each operation. While this is a valid scenario, it does not consider the unique aspects offered by better integration of AFEX treatment with rural operations. To that end, two studies have been performed on a more holistic approach that considers changes to both the farmland and refinery. Sendich and Dale [95] developed a holistic model combining crop management, animal feeding operations, and the biorefinery, modeling the performance of numerous sites and possible farmland configurations throughout the United States. Both economic and environmental performance of these landscapes varied significantly, primarily in terms of the location and whether or not animal feed operations were involved. However, a clear trend emerged that adding biorefinery operations to the landscape significantly improved the economics and environmental performance in the area. Eranki *et al.* [113] considered the differences between regional processing and centralized processing combined with a variety of landscape scenarios. Performing regional processing slightly increased the net energy yield (NEY) for perennial grass-dominated landscapes compared to centralized processing, but slightly decreased the NEY for corn (including grain, stover, and double-crop rye) dominated landscapes. This study confirmed that perennial grasses combined with AFEX processing may reduce carbon emissions, primarily due to the carbon sequestration effect of one of the perennial grasses in the landscape (switchgrass). The study also established that combining high-yielding perennial grasses with regional processing that includes densification of AFEX-treated biomass prior to transportation to the biorefinery may generate greater energy yields compared to centralized processing.

9.8 Conclusions

Ammonia fiber expansion (AFEX) pretreatment is a unique, non-extractive alkaline pretreatment method that can greatly improve the digestibility and fermentability of lignocellulosic biomass in a cost-effective and environmentally sustainable manner. AFEX is a dry-to-dry process that requires no conditioning of the pretreated biomass or nutrient supplementation to achieve high enzyme/microbial activity. Several novel configurations are being developed (e.g., extractive/non-extractive AFEX systems, cellulose III$_I$ formation during AFEX, ammonolysis kinetics) in parallel with a focus on commercialization to realize cheap, hybrid AFEX reactors with in-built low-cost ammonia recovery systems. On-going techno-economic and life-cycle analysis efforts have shown that integration of the AFEX process into a regional biomass processing depot can significantly benefit bioeconomies centered on cellulosic biomass.

Acknowledgements

This work was funded by Great Lakes Bioenergy Research Center (http://www.greatlakes-bioenergy.org/) supported by the U.S. Department of Energy, Office of Science, Office of Biological and Environmental Research, through Cooperative Agreement DEFC02-07ER64494 between the Board of Regents of the University of Wisconsin System and the U.S. Department of Energy.

References

1. Dale, B.E. (1999) Biobased industrial products: bioprocess engineering when cost really counts (editorial). *Biotechnology Progress*, **15** (5), 775–776.
2. Chundawat, S.P.S., Beckham, G.T., Himmel, M., and Dale, B.E. (2011) Deconstruction of lignocellulosic biomass to fuels and chemicals. *Annual Review of Chemical and Biomolecular Engineering*, **2**, 121–145.
3. da Costa Sousa, L., Chundawat, S.P.S., Balan, V., and Dale, B.E. (2009) 'Cradle-to-grave' assessment of existing lignocellulose pretreatment technologies. *Current Opinion in Biotechnology*, **20** (3), 339–347.
4. Solaiman, S.G., Horn, G.W., and Owens, F.N. (1979) Ammonium Hydroxide Treatment on Wheat Straw. *Journal of Animal Science*, **49** (3), 802–808.
5. Dean, D.B., Adesogan, A.T., Krueger, N.A., and Littell, R.C. (2008) Effects of treatment with ammonia or fibrolytic enzymes on chemical composition and ruminal degradability of hays produced from tropical grasses. *Animal Feed Science and Technology*, **145**, 68–83.
6. Solaiman, S.G., Horn, G.W., and Owens, F.N. (1979) Ammonium hydroxide treatment on wheat straw. *Journal of Animal Science*, **49** (3), 802–809.
7. Van Soest, P.J. (2006) Rice straw, the role of silica and treatments to improve quality. *Animal Feed Science and Technology*, **130**, 137–171.
8. Belasco, I.J. (1954) New Nitrogen Feed Compounds for Ruminants–A Laboratory Evaluation. *Journal of Animal Science*, **13** (3), 601–610.
9. Schuerch, C., Burdick, M.P., and Mahdalik, M. (1966) Liquid ammonia-solvent combinations in wood plasticization: chemical treatments. *Industrial & Engineering Chemistry Product Research and Development*, **5** (2), 101–105.
10. O'Connor, J. (1972) Ammonia explosion pulping: A new fiber separation process. *Tappi*, **55** (3), 353–358.
11. Dale, B.E. and Moreira, M.J. (1982) A freeze-explosion technique for increasing cellulose hydrolysis. *Biotechnology and Bioengineering*, **12**, 31–43.
12. Chou, Y. (1986) Supercritical ammonia pretreatment of lignocellulosic materials. *Biotechnology & Bioengineering Symposium*, **17**, 19–32.
13. Weimer, P., Chou, Y., Weston, W., and Chase, D. (1986) Effect of supercritical ammonia on the physical and chemical structure of ground wood. *Biotechnology & Bioengineering Symposium*, **17**, 5–18.
14. Yoon, H.H., Wu, Z.W., and Lee, Y.Y. (1995) Ammonia-recycled percolation process for pretreatment of biomass feedstock. *Applied Biochemistry and Biotechnology*, **51–52** (1), 5–19.
15. Dunson, J.R., Elander, R.T., Tucker, M., and Hennessey, S.M. (2007) Treatment of biomass to obtain fermentable sugars (USPTO 2007/0031918). USA.
16. Balan, V., Bals, B., Chundawat, S.P. *et al.* (2009) *Lignocellulosic Biomass Pretreatment Using AFEX. Biofuels: Methods and Protocols*, Humana Press, p. 61–77.
17. Chundawat, S.P.S., Donohoe, B.S., Sousa, L. *et al.* (2011) Multi-scale visualization and characterization of plant cell wall deconstruction during thermochemical pretreatment. *Energy & Environmental Science*, **4** (3), 973–984.
18. Lau, M.W. and Dale, B.E. (2009) Cellulosic ethanol production from AFEX-treated corn stover using Saccharomyces cerevisiae 424A(LNH-ST). *Proceedings of the National Academy of Sciences of the United States of America*, **106** (5), 1368–1373.
19. Campbell, T.J., Teymouri, F., Bals, B., Glassbrook, J., Nielson, C.D., and Videto, J. (2013) A packed bed AFEX reactor system for pretreatment of agricultural residues at regional depots. *Biofuels*, **4** (1), 23–34.

20. Balan, V., Chundawat, S.P.S., Sousa, L., and Dale, B.E. (2009) Methods for pretreating biomass (Provisional patent: TEC2010-0068). USA.

21. Chundawat, S.P.S., Bellesia, G., Uppugundla, N. *et al.* (2011) Restructuring the crystalline cellulose hydrogen bond network enhances its depolymerization rate. *Journal of American Chemical Society*, **133** (29), 11163–11174.

22. Chundawat, S.P.S., Sousa, L., Cheh, A. *et al.* (2010) Digestible lignocellulosic biomass and extractives and methods for producing same (Provisional patent application: TEC2010-0090). USA.

23. Chundawat, S.P.S., Vismeh, R., Sharma, L. *et al.* (2010) Multifaceted characterization of cell wall decomposition products formed during ammonia fiber expansion (AFEX) and dilute-acid based pretreatments. *Bioresource Technology*, **101**, 8429–8438.

24. Kumar, R., Mago, G., Balan, V., and Wyman, C.E. (2009) Physical and chemical characterizations of corn stover and poplar solids resulting from leading pretreatment technologies. *Bioresource Technology*, **100** (17), 3948–3962.

25. Wang, P., Bolker, H., and Purves, C. (1964) Ammonolysis of uronic ester groups in birch xylan. *Canadian Journal of Chemistry*, **42**, 2434–2439.

26. Lau, M.W., Bals, B., Chundawat, S.P.S. *et al.* (2012) An integrated paradigm for cellulosic biorefineries: Utilization of lignocellulosic biomass as self-sufficient feedstocks for fuel, food precursors and saccharolytic enzyme production. *Energy and Environmental Science*, **5**, 7100–7110.

27. Balan, V., Sousa, L.d.C., Chundawat, S.P.S. *et al.* (2009) Enzymatic digestibility and pretreatment degradation products of AFEX-treated hardwoods (*Populus nigra*). *Biotechnology Progress*, **25** (2), 365–375.

28. Sewalt, V., Glasser, W., and Beauchemin, K. (1997) Lignin impact on fiber degradation. 3. Reversal of inhibition of enzymatic hydrolysis by chemical modification of lignin and additives. *Journal of Agricultural and Food Chemistry*, **45**, 1823–1828.

29. Väljamäe, P., Sild, V., Nutt, A. *et al.* (1999) Acid hydrolosis of bacterial cellulose reveals different modes of synergistic action between cellobiohydrolase I and endoglucanase I. *European Journal of Biochemistry*, **266** (2), 327–334.

30. Samuel, R., Pu, Y., Foston, M., and Ragauskas, A.J. (2009) Solid-state NMR characterization of switchgrass cellulose after dilute acid pretreatment. *Biofuels*, **1** (1), 85–90.

31. Igarashi, K., Wada, M., and Samejima, M. (2007) Activation of crystalline cellulose to cellulose III results in efficient hydrolysis by cellobiohydrolase. *FEBS Journal*, **274** (7), 1785–1792.

32. Gao, D., Chundawat, S.P.S., Uppugundla, N. *et al.* (2011) Binding characteristics of trichoderma reesei cellulases on untreated, ammonia fiber expansion and dilute-acid pretreated lignocellulosic biomass. *Biotechnology and Bioengineering*, **108** (8), 1788–1800.

33. Medve, J., Stahlberg, J., and Tjerneld, F. (1994) Adsorption and synergism of cellobiohydrolase I and II of *Trichoderma reesei* during hydrolysis of microcrystalline cellulose. *Biotechnology and Bioengineering*, **44** (9), 1064–1073.

34. Medve, J., Karlsson, J., Lee, D., and Tjerneld, F. (1998) Hydrolysis of microcrystalline cellulose by cellobiohydrolase I and endoglucanase II from *Trichoderma reesei*: Adsorption, sugar production pattern, and synergism of the enzymes. *Biotechnology and Bioengineering*, **59** (5), 621–634.

35. Chundawat, S.P.S., Balan, V., Sousa, L., and Dale, B.E. (2010) Thermochemical pretreatment of lignocellulosic biomass, in *Bioalcohol Production: Biochemical Conversion of Lignocellulosic Biomass* (ed. K. Waldron), Woodhead Publishing, Cambridge, UK.

36. Chundawat, S.P.S., Lipton, M.S., Purvine, S.O. *et al.* (2011) Proteomics based compositional analysis of complex cellulase-hemicellulase mixtures. *Journal of Proteome Research*, **10** (10), 4365–4372.

37. Li, B.Z., Balan, V., Yuan, Y.J., and Dale, B.E. (2010) Process optimization to convert forage and sweet sorghum bagasse to ethanol based on ammonia fiber expansion (AFEX) pretreatment. *Bioresource Technology*, **101** (4), 1285–1292.

38. Bals, B., Dale, B.E., and Balan, V. (2006) Enzymatic hydrolysis of distiller's dry grain and solubles (DDGS) using ammonia fiber expansion pretreatment. *Energy & Fuels*, **20** (6), 2732–2736.

39. Krishnan, C., Sousa, L.D., Jin, M.J. *et al.* (2010) Alkali-based AFEX pretreatment for the conversion of sugarcane bagasse and cane leaf residues to ethanol. *Biotechnology and Bioengineering*, **107** (3), 441–450.

40. Zhong, C., Lau, M.W., Balan, V. *et al.* (2009) Optimization of enzymatic hydrolysis and ethanol fermentation from AFEX-treated rice straw. *Applied Microbiology and Biotechnology*, **84** (4), 667–676.

41. Alizadeh, H., Teymouri, F., Gilbert, T.I., and Dale, B.E. (2005) Pretreatment of switchgrass by ammonia fiber explosion (AFEX). *Applied Biochemistry and Biotechnology*, **124** (1–3), 1133–1141.

42. Bradshaw, T.C., Alizadeh, H., Teymouri, F. *et al.* (2007) Ammonia fiber expansion pretreatment and enzymatic hydrolysis on two different growth stages of reed canarygrass. *Applied Biochemistry and Biotechnology*, **137–140** (1–12), 395–405.

43. Teymouri, F., Laureano-Perez, L., Alizadeh, H., and Dale, B.E. (2005) Optimization of the ammonia fiber explosion (AFEX) treatment parameters for enzymatic hydrolysis of corn stover. *Bioresource Technology*, **96** (18), 2014–2018.

44. Murnen, H.K., Balan, V., Chundawat, S.P.S. *et al.* (2007) Optimization of ammonia fiber expansion (AFEX) pretreatment and enzymatic hydrolysis of Miscanthus x giganteus to fermentable sugars. *Biotechnology Progress*, **23** (4), 846–850.

45. Garlock, R.J., Wong, Y.C., Balan, V., and Dale, B.E. (2011) AFEX pretreatment and enzymatic conversion of black locust (Robinia pseudoacacia L.) to soluble sugars. *BioEnergy Research*. doi: 10.1007/s12155-011-9134-6

46. Garlock, R., Chundawat, S., Balan, V., and Dale, B. (2009) Optimizing harvest of corn stover fractions based on overall sugar yields following ammonia fiber expansion pretreatment and enzymatic hydrolysis. *Biotechnology for Biofuels*, **2** (1), 29.

47. Dien, B.S., Ximenes, E.A., O'Bryan, P.J. *et al.* (2008) Enzyme characterization for hydrolysis of AFEX and liquid hot-water pretreated distillers' grains and their conversion to ethanol. *Bioresource Technology*, **99** (12), 5216–5225.

48. Gao, D., Chundawat, S.P.S., Krishnan, C. *et al.* (2010) Mixture optimization of six core glycosyl hydrolases for maximizing saccharification of ammonia fiber expansion (AFEX) pretreated corn stover. *Bioresource Technology*, **101** (8), 2770–2781.

49. Gao, D., Uppugundla, N., Chundawat, S. *et al.* (2011) Hemicellulases and auxiliary enzymes for improved conversion of lignocellulosic biomass to monosaccharides. *Biotechnology for Biofuels*, **4** (1), 5.

50. Banerjee, G., Car, S., Scott-Craig, J.S. *et al.* (2010) Synthetic multi-component enzyme mixtures for deconstruction of lignocellulosic biomass. *Bioresource Technology*, **101** (23), 9097–9105.

51. Dien, B., Cotta, M., and Jeffries, T. (2003) Bacteria engineered for fuel ethanol production: current status. *Applied Microbiology and Biotechnology*, **63** (3), 258–266.

52. Lau, M., Gunawan, C., and Dale, B. (2009) The impacts of pretreatment on the fermentability of pretreated lignocellulosic biomass: a comparative evaluation between ammonia fiber expansion and dilute acid pretreatment. *Biotechnology for Biofuels*, **2** (1), 30.

53. Jin, M., Lau, M.W., Balan, V., and Dale, B.E. (2010) Two-step SSCF to convert AFEX-treated switchgrass to ethanol using commercial enzymes and Saccharomyces cerevisiae 424A(LNH-ST). *Bioresource Technology*, **101** (21), 8171–8178.

54. Lu, Y.L., Warner, R., Sedlak, M. *et al.* (2009) Comparison of glucose/xylose cofermentation of poplar hydrolysates processed by different pretreatment technologies. *Biotechnology Progress*, **25** (2), 349–356.

55. Lau, M.W., Gunawan, C., Balan, V., and Dale, B.E. (2010) Comparing the fermentation performance of Escherichia coli KO11, Saccharomyces cerevisiae 424A(LNH-ST) and Zymomonas mobilis AX101 for cellulosic ethanol production. *Biotechnology for Biofuels*, **3**, http://www.biotechnologyforbiofuels.com/content/3/1/11.

56. Shao, Q., Chundawat, S.P.S., Krishnan, C. *et al.* (2010) Enzymatic digestibility and ethanol fermentability of AFEX-treated starch-rich lignocellulosics such as corn silage and whole corn plant. *Biotechnology for Biofuels*, **3**, http://www.biotechnologyforbiofuels.com/content/3/1/12.

57. Bals, B., Rogers, C., Jin, M.J. *et al.* (2010) Evaluation of ammonia fibre expansion (AFEX) pretreatment for enzymatic hydrolysis of switchgrass harvested in different seasons and locations. *Biotechnology for Biofuels*, **3**, http://www.biotechnologyforbiofuels.com/content/3/1/1.

58. Jin, M., Gunawan, C., Balan, V., and Dale, B.E. (2012) Consolidated bioprocessing (CBP) of AFEX-pretreated corn stover for ethanol production using Clostridium phytofermentans at a high solids loading. *Biotechnology and Bioengineering*, **109** (8), 1929–1936.

59. Jin, M., Balan, V., Gunawan, C., and Dale, B.E. (2012) Quantitatively understanding of reduced xylose fermentation performance in AFEX treated corn stover hydrolysate using Saccharomyces cerevisiae 424A (LNH-ST) and Escherichia coli KO11. *Bioresource Technology*, **111**, 294–300.

60. Casey, E., Sedlak, M., Ho, N.W.Y., and Mosier, N.S. (2010) Effect of acetic acid and pH on the cofermentation of glucose and xylose to ethanol by a genetically engineered strain of Saccharomyces cerevisiae. *Fems Yeast Research*, **10** (4), 385–393.

61. Shao, X., Jin, M., Guseva, A. *et al.* (2011) Conversion for Avicel and AFEX pretreated corn stover by Clostridium thermocellum and simultaneous saccharification and fermentation: Insights into microbial conversion of pretreated cellulosic biomass. *Bioresource Technology*, **102** (17), 8040–8045.

62. Jin, M., Balan, V., Gunawan, C., and Dale, B.E. (2011) Consolidated bioprocessing (CBP) performance of Clostridium phytofermentans on AFEX treated corn stover for ethanol production. *Biotechnology and Bioengineering*, **108** (6), 1290–1297.

63. Jin, M., Balan, V., Lau, M.W., and Dale, B.E. (eds) (2010) Consolidated bioprocessing (CBP) of AFEX treated corn stover by Clostridium phytofermentans. The 32nd Symposium on Biotechnology for Fuels and Chemicals, Clearwater Beach, Florida.

64. Teymouri, F., Alizadeh, H., Laureano-Perez, L. *et al.* (2004) Effects of ammonia fiber explosion treatment on activity of endoglucanase from *Acidothermus cellulolyticus* in transgenic plant. *Applied Biochemistry and Biotechnology*, **116** (1–3), 1183–1192.

65. Ransom, C., Balan, V., Biswas, G. *et al.* (2007) Heterologous *Acidothermus cellulolyticus* 1, 4-β-endoglucanase E1 produced within the corn biomass converts corn stover into glucose. *Applied Biochemistry and Biotechnology*, **137–140** (1–12), 207–219.

66. Oraby, H., Venkatesh, B., Dale, B.E. *et al.* (2007) Enhanced conversion of plant biomass into glucose using transgenic rice-produced endoglucanase for cellulosic ethanol. *Transgenic Research*, **16** (6), 739–749.

67. Chen, F. and Dixon, R. (2007) Lignin modification improves fermentable sugar yields for biofuel production. *Nature Biotechnology*, **25**, 759–761.

68. Jackson, L., Shadle, G., Zhou, R. *et al.* (2008) Improving saccharification efficiency of alfalfa stems through modification of the terminal stages of monolignol biosynthesis. *BioEnergy Research*, **1** (3), 180–192.

69. Shen, H., Fu, C., Xiao, X. *et al.* (2009) Developmental control of lignification in stems of lowland switchgrass variety alamo and the effects on saccharification efficiency. *BioEnergy Research*, **2** (4), 233–245.

70. Franke, R., McMichael, C.M., Meyer, K. *et al.* (2000) Modified lignin in tobacco and poplar plants over-expressing the Arabidopsis gene encoding ferulate 5-hydroxylase. *The Plant Journal*, **22** (3), 223–234.

71. Marita, J., Ralph, J., Hatfield, R., and Chapple, C. (1999) NMR characterization of lignins in Arabidopsis altered in the activity of ferulate 5-hydroxylase. *Proceedings of the National Academy of Sciences of the United States of America*, **96**, 12328–12332.

72. Stewart, J., Akiyama, T., Chapple, C. *et al.* (2009) The effects on lignin structure of overexpression of ferulate 5-hydroxylase in hybrid poplar. *Plant Physiology*, **150**, 621–635.

73. Ong, R.G. (2011) Interactions between biomass feedstock characteristics and bioenergy production: From the landscape to the molecular scale. Ph.D. thesis, Michigan State University.

74. Simmons, B.A., Loqué, D., and Ralph, J. (2010) Advances in modifying lignin for enhanced biofuel production. *Current Opinion in Plant Biology*, **13** (3), 312–319.

75. Pedersen, J.F., Vogel, K.P., and Funnell, D.L. (2005) Impact of reduced lignin on plant fitness. *Crop Science*, **45**, 812–819.

76. Casler, M.D., Buxton, D.R., and Vogel, K.P. (2002) Genetic modification of lignin concentration affects fitness of perennial herbaceous plants. *Theoretical and Applied Genetics*, **104** (1), 127–131.

77. Delgado, N.J., Casler, M.D., Grau, C.R., and Jung, H.G. (2002) Reactions of smooth bromegrass clones with divergent lignin or etherified ferulic acid concentration to three fungal pathogens. *Crop Science*, **42**, 1824–1831.

78. Grabber, J.H., Hatfield, R.D., Lu, F., and Ralph, J. (2008) Coniferyl Ferulate Incorporation into Lignin Enhances the Alkaline Delignification and Enzymatic Degradation of Cell Walls. *Biomacromolecules*, **9** (9), 2510–2516.

79. Tillner-Roth, R. and Friend, D.G. (1998) A helmholtz free energy formulation of the thermodynamic properties of the mixture {Water + Ammonia}. *Journal of Physical and Chemical Reference Data*, **27** (1), 63–96.

80. Huang, G., Zhang, C., and Chen, Z. (2006) Pulping of wheat straw with caustic potash-ammonia aqueous solutions and its kinetics. *Chinese Journal of Chemical Engineering*, **14** (6), 729–733.

81. Zhang, Y.-J., Li, Z.-Q., Su, Y.-F., and Cao, C.-Y. (2011) Effects of supercritical ammonia on bamboo pulping. *Forestry Studies in China*, **13** (1), 80–84.

82. Kim, T.H. and Lee, Y.Y. (2006) Fractionation of corn stover by hot-water and aqueous ammonia treatment. *Bioresource Technology*, **97** (2), 224–232.

83. Kim, S. and Holtzapple, M.T. (2006) Effect of structural features on enzyme digestibility of corn stover. *Bioresource Technology*, **97** (4), 583–591.

84. Nakagame, S., Chandra, R.P., and Saddler, J.N. (2009) The effect of isolated lignins, obtained from a range of pretreated lignocellulosic substrates, on enzymatic hydrolysis. *Biotechnology and Bioengineering*, **105**, 871–879.

85. Pan, X.J. (2008) Role of functional groups in lignin inhibition of enzymatic hydrolysis of cellulose to glucose. *Journal of Biobased Materials and Bioenergy*, **2** (1), 25–32.

86. Wada, M., Chanzy, H., Nishiyama, Y., and Langan, P. (2004) Cellulose III$_I$ crystal structure and hydrogen bonding by synchrotron X-ray and neutron fiber diffraction. *Macromolecules*, **37** (23), 8548–8555.

87. Pandey, M.P. and Kim, C.S. (2010) Lignin depolymerization and conversion: a review of thermochemical methods. *Chemical Engineering & Technology*, **34** (1), 29–41.

88. Eggeman, T. and Elander, R.T. (2005) Process and economic analysis of pretreatment technologies. *Bioresource Technology*, **96** (18), 2019–2025.

89. Wang, L., Dale, B.E., Yurttas, L., and Goldwasser, I. (1998) Cost estimates and sensitivity analyses for the ammonia fiber explosion process. *Applied Biochemistry and Biotechnology*, **70–72** (1), 51–66.

90. Sendich, E., Laser, M., Kim, S. *et al.* (2008) Recent process improvements for the ammonia fiber expansion (AFEX) process and resulting reductions in minimum ethanol selling price. *Bioresource Technology*, **99** (17), 8429–8435.

91. Laser, M., Jin, H., Jayawardhana, K., and Lynd, L.R. (2009) Coproduction of ethanol and power from switchgrass. *Biofuels, Bioproducts and Biorefining*, **3** (2), 195–218.

92. Laser, M., Larson, E., Dale, B. *et al.* (2009) Comparative analysis of efficiency, environmental impact, and process economics for mature biomass refining scenarios. *Biofuels Bioprod Biorefining*, **3** (2), 247–270.

93. ASHRAE (2010) *Chapter 2, Ammonia refrigeration systems*, in *Handbook - Refrigeration*, American Society of Heating, Refrigeration, and Air,-Conditioning Engineers, Atlanta GA.

94. Kazi, F.K., Fortman, J.A., Anex, R.P. *et al.* (2010) Techno-economic comparison of process technologies for biochemical ethanol production from corn stover. *Fuel*, **89** (Supplement 1), S20–S30.

95. Sendich, E. and Dale, B.E. (2009) Environmental and economic analysis of the fully integrated biorefinery. *GCB Bioenergy*, **1** (5), 331–345.

96. Bals, B., Murnen, H., Allen, M., and Dale, B. (2010) Ammonia fiber expansion (AFEX) treatment of eleven different forages: Improvements to fiber digestibility *in vitro*. *Animal Feed Science and Technology*, **155** (2–4), 147–155.

97. Weimer, P.J., Mertens, D.R., Ponnampalam, E. *et al.* (2003) FIBEX-treated rice straw as a feed ingredient for lactating dairy cows. *Animal Feed Science and Technology*, **103** (1–4), 41–50.

98. Perdok, H.B. and Leng, R.A. (1987) Hyperexcitability in cattle fed ammoniated roughages. *Animal Feed Science and Technology*, **17**, 121–143.

99. Carolan, J.E., Joshi, S.V., and Dale, B.E. (2007) Technical and financial feasibility analysis of distributed bioprocessing using regional biomass pre-processing centers. *Journal of Agricultural & Food Industrial Organization*, **5** (10), 1–29.

100. Betschart, A. and Kinsella, J.E. (1973) Extractability and solubility of leaf protein. *Journal of Agricultural and Food Chemistry*, **21** (1), 60–65.

101. Fiorentini, R. and Galoppini, C. (1981) Pilot plant production of an edible alfalfa protein concentrate. *Journal of Food Science*, **46**, 1514–1520.

102. Urribarrí, L., Chacón, D., González, O., and Ferrer, A. (2009) Protein Extraction and Enzymatic hydrolysis of ammonia-treated cassava leaves (Manihot esculenta Crantz). *Applied Biochemistry and Biotechnology*, **153** (1), 94–102.

103. Urribarrí, L., Ferrer, A., and Colina, A. (2005) Leaf protein from ammonia-treated dwarf elephant grass (Pennisetum purpureum Schum cv. Mott). *Applied Biochemistry and Biotechnology*, **121–124**, 721–730.

104. Bals, B., Teachworth, L., Dale, B.E., and Balan, V. (2007) Extraction of proteins from switchgrass using aqueous ammonia within an integrated biorefinery. *Applied Biochemistry and Biotechnology*, **143** (2), 187–198.

105. Bals, B. and Dale, B.E. (2011) Economic comparison of multiple techniques for recovering leaf protein in biomass processing. *Biotechnology and Bioengineering*, **108**, 530–537.

106. Dale, B.E., Bals, B.D., Kim, S., and Eranki, P. (2010) Biofuels done right: land efficient animal feeds enable large environmental and energy benefits. *Environmental Science & Technology*, **44**, 8385–8389.

107. Tucker, M.P., Nagle, N.J., Jennings, E.W. *et al.* (2004) Conversion of distiller's grain into fuel alcohol and a higher-value animal feed by dilute-acid pretreatment. *Applied Biochemistry and Biotechnology*, **115** (1–3), 1139–1159.

108. Ladisch, M. and Dale, B. (2008) Distillers grains: On the pathway to cellulose conversion. *Bioresource Technology*, **99** (12), 5155–5156.

109. Bals, B., Balan, V., and Dale, B. (2009) Integrating alkaline extraction of proteins with enzymatic hydrolysis of cellulose from wet distiller's grains and solubles. *Bioresource Technology*, **100** (23), 5876–5883.

110. Kim, Y., Hendrickson, R., Mosier, N.S. *et al.* (2008) Enzyme hydrolysis and ethanol fermentation of liquid hot water and AFEX pretreated distillers' grains at high-solids loadings. *Bioresource Technology*, **99** (12), 5206–5215.

111. Brehmer, B., Bals, B., Sanders, J., and Dale, B. (2008) Improving the corn-ethanol industry: Studying protein separation techniques to obtain higher value-added product options for distillers grains. *Biotechnology and Bioengineering*, **101** (1), 767–777.

112. Meier, D., Zúñiga-Partida, V., Ramírez-Cano, F. *et al.* (1994) Conversion of technical lignins into slow-release nitrogenous fertilizers by ammoxidation in liquid phase. *Bioresource Technology*, **49** (2), 121–128.

113. Eranki, P., Bals, B., and Dale, B.E. (2011) Advanced regional biomass processing depots - A key to the logistical challenges of the cellulosic biofuel industry. *Biofuels, Bioproducts, and Biorefining*, **5** (6), 621–630.

114. Huda, M.S., Balan, V., Drzal, L.T. *et al.* (2007) Effect of ammonia fiber expansion (AFEX) and silane treatments of corncob granules on the properties of renewable resource based biocomposites. *Journal of Biobased Materials and Bioenergy*, **1** (1), 127–136.

115. Digman, M.F., Shinners, K.J., Casler, M.D. *et al.* (2010) Optimizing on-farm pretreatment of perennial grasses for fuel ethanol production. *Bioresource Technology*, **101** (14), 5305–5314.

116. Bals, B., Teymouri, F., Campbell, T. *et al.* (2011) Low temperature and long residence time AFEX pretreatment of corn stover. *Bioenergy Research*, **5** (2), 372–379.

117. Wu, M., Wu, Y., and Wang, M. (2006) Energy and emission benefits of alternative transportation liquid fuels derived from switchgrass: a fuel life cycle assessment. *Biotechnology Progress*, **22** (4), 1012–1024.

118. Bai, Y., Luo, L., and van der Voet, E. (2010) Life cycle assessment of switchgrass-derived ethanol as transport fuel. *The International Journal of Life Cycle Assessment*, **15** (5), 468–477.

119. Chouinard-Dussault, P., Bradt, L., Ponce-Ortega, J., and El-Halwagi, M. (2011) Incorporation of process integration into life cycle analysis for the production of biofuels. *Clean Technologies and Environmental Policy*, **13** (5), 673–685.

120. Spatari, S., Bagley, D.M., and MacLean, H.L. (2010) Life cycle evaluation of emerging lignocellulosic ethanol conversion technologies. *Bioresource Technology*, **101** (2), 654–667.

10

Fundamentals of Biomass Pretreatment by Fractionation

Poulomi Sannigrahi[1,2] and Arthur J. Ragauskas[1,2,3]

[1] *BioEnergy Science Center, Oak Ridge, USA*

[2] *Institute of Paper Science and Technology, Georgia Institute of Technology, Atlanta, USA*

[3] *School of Chemistry and Biochemistry, Georgia Institute of Technology, Atlanta, USA*

10.1 Introduction

With the rise in global energy demand and environmental concerns about the use of fossil fuels, the need for rapid development of alternative fuels from sustainable, non-food sources is now well acknowledged. The effective utilization of low-cost high-volume agricultural and forest biomass for the production of transportation fuels and bio-based materials will play a vital role in addressing this concern [1]. The processing of lignocellulosic biomass, especially from mixed agricultural and forest sources with varying composition, is currently significantly more challenging than the bioconversion of corn starch or cane sugar to ethanol [1,2]. This is due to the inherent recalcitrance of lignocellulosic biomass to enzymatic and microbial deconstruction, imparted by the partly crystalline nature of cellulose and its close association with hemicellulose and lignin in the plant cell wall [2,3]. Pretreatments that convert raw lignocellulosic biomass to a form amenable to enzymatic degradation are therefore an integral step in the production of bioethanol from this material [4]. Chemical or thermochemical pretreatments act to reduce biomass recalcitrance in various ways. These include hemicellulose removal or degradation, lignin modification and/or delignification, reduction in crystallinity and degree of polymerization of cellulose, and increasing pore volume. Biomass pretreatments are an active focus of industrial and academic research efforts, and various strategies have been developed.

Among commonly studied pretreatments, organosolv pretreatment, in which an aqueous organic solvent mixture is used as the pretreatment medium, results in the fractionation of the major biomass components, cellulose, lignin, and hemicellulose into three process streams [5,6]. Cellulose and lignin are recovered as separate solid streams, while hemicelluloses and sugar degradation products such as furfural and hydroxymethylfurfural (HMF) are released as a water-soluble fraction. The combination of ethanol as the solvent and

Aqueous Pretreatment of Plant Biomass for Biological and Chemical Conversion to Fuels and Chemicals, First Edition.
Edited by Charles E. Wyman.
© 2013 John Wiley & Sons, Ltd. Published 2013 by John Wiley & Sons, Ltd.

sulfuric acid as the delignification catalyst has been applied to several common biomass feedstocks [7–11]. This approach allows for an efficient utilization of all of the major biomass components, making it potentially attractive from an economic perspective if good value can be obtained from each in a true biorefinery concept. Many other pretreatments such as steam explosion, dilute acid, and AFEX produce cellulose-rich solids that can be hydrolyzed with enzymes to produce glucose but do not result in a clean fractionation into separate lignin, cellulose, and hemicellulose streams [4,5,12]. While hemicellulose can be potentially utilized by fermentation with pentose fermenting microbes, lignin is usually too degraded to be useful as a co-product. In these pretreatments, the lignin is usually expected to be burned as an energy source.

In this chapter, we focus on organosolv pretreatment as a method for biomass fractionation to recover high-quality streams of each of the major biomass components. We provide a short historical perspective on organosolv pulping and pretreatment, followed by a more detailed overview of the latter. Different solvents and catalysts used in organosolv pretreatment and results from biomass fractionation are discussed with an emphasis on ethanol, as it is an excellent solvent when the production of bio-ethanol is desired. The chemistry of organosolv delignification, a key component of biomass fractionation by this method, and the nature of organosolv lignin are described. Other aspects, including modifications in the structure and crystallinity of cellulose after organosolv pretreatment and the co-products of biomass fractionation by ethanol organosolv pretreatment, are also discussed.

10.2   Organosolv Pretreatment

10.2.1   Organosolv Pulping

Kleinert and Tayenthal proposed the use of aqueous ethanol for the delignification of wood in 1931, as described by Johansson *et al.* [13]. Along with addressing some of the environmental concerns associated with traditional Kraft and sulfite pulping, the use of organic solvents enabled the recovery of lignin and other dissolved components such as extractives in an un-degraded form, thus enabling a more efficient use of the lignocellulosic feedstock [13]. The most common organosolv pulping systems apply ethanol or methanol with mineral acids as catalysts. Alkaline organosolv pulping with methanol-water-NaOH has been investigated, but the requirement of an additional chemical recovery system for the alkali is a major drawback. Comparison of results from organosolv pulping to conventional Kraft and sulfite pulping shows that yields of organosolv softwood pulps are higher than that of conventional pulps at equivalent values of the Kappa number [13]. The Kappa number is a parameter representing the residual lignin content of a wood pulp. The strength properties of these pulps are comparable to Kraft and sulfite pulps and do not show much variation based on solvents and cooking methods. Organosolv pulp mills can be operated on a smaller scale (300 tons of pulp/day) than Kraft pulp mills (1000 tons of pulp/day) and still remain economically attractive [14]. The pulp produced can be bleached easily without chlorine, adding to the environmental benefits of this process. Further details on organosolv pulping have been reviewed by Aziz and Sarkanen [15], Johansson *et al.* [13] and Muurinen [16].

10.2.2   Overview of Organosolv Pretreatment

Organosolv pretreatment is similar to organosolv pulping, but does not require the equivalent degree of delignification as the latter. Further, while slight degradation of the cellulose structure can aid in enzymatic hydrolysis during the production of biofuels, preservation of fiber quality is important during pulping. Different cooking conditions may therefore be optimal for organosolv pulping versus pretreatment of the same lignocellulosic feedstock. In order to lower costs, both processes require that most of the solvent be recycled. This is usually achieved by flashing to atmospheric pressure, precipitation of lignin by dilution with water, and distillation of the precipitation liquor. Up to 98% of methanol and 99% ethanol can be recovered during organosolv pulping [17].

The ethanol organosolv process was originally designed to produce a clean biofuel for turbine generators by researchers at General Electric and University of Pennsylvania in the 1970s [18]. It was subsequently developed by the Canadian pulp and paper industry in to the Alcell pulping process for hardwoods. The Lignol Corporation adapted this technology as part of a commercial lignocellulosic biorefinery platform [19]. While the pretreatment conditions in the Lignol process vary with the feedstock being processed, the general ranges are: cooking temperature of 180–195 °C; cooking time of 30–90 min; ethanol concentration of 35–70% (w/w) and a liquor to solids ratio from 4:1 to 10:1 (w/w). Lignin is recovered as a precipitate by flashing the pulping liquor to atmospheric pressure, followed by dilution with water. Hemicellulose sugars and furfural are recovered as co-products from the water-soluble stream.

10.2.3 Solvents and Catalysts for Organosolv Pretreatment

A wide range of solvents and catalysts has been studied for their suitability to organosolv pulping and pretreatment. The following section describes the application of different solvent-catalyst systems and their application to lignocellulosic biomass. Ethanol and methanol are commonly used as solvents for organosolv pretreatment primarily due to their low cost, low boiling point, and ease of recovery. The Hildebrand solubility parameter or δ-value of a solvent can be used to estimate the solubility of lignin or other polymers. In general, the Hildebrand solubility parameter provides a numerical estimate of the interaction between different materials, with similar values indicating good solubility. Solvents which display good lignin solubility have δ values close to 11 [20], with acetic acid (δ = 10.1), formic acid (δ = 12.1), ethanol (δ = 12.9), and acetone (δ = 9.7) being good examples. 75% of a mixture of dioxane, ethanol, and acetone with 25% water were also found to have δ-values close to lignin and exhibited the ability to dissolve both high- and low-molecular-weight lignin fractions [21].

Ethanol and Methanol

From the viewpoint of bioethanol production, ethanol is a good solvent as losses can be made up by feeding back some of the ethanol produced by fermentation and the ethanol lost in pretreatment could end up in the product. This choice also reduces process complexity by eliminating an additional solvent stream. A general schematic of the ethanol organosolv pretreatment is given in Figure 10.1. Ethanol organosolv pretreatment can be performed with or without a catalyst with auto-catalyzed pretreatments being performed at higher temperatures (185–210 °C). The severity of the pretreatment conditions can be represented by the combined severity factor CS, which is a function of pH, cooking time, and temperature [22] and is calculated:

$$CS = \log\left[t\exp\left(\frac{T - T_{ref}}{14.7}\right)\right] - pH$$

where t is pretreatment time (min); T is pretreatment temperature (°C); and T_{ref} is 100 °C.

A wide range of catalysts have been explored for ethanol and methanol organosolv pretreatment including mineral acids, magnesium sulfate, magnesium, calcium or barium chloride or nitrate, sodium bisulfate, and sodium hydroxide [23]. While organosolv delignification benefits by addition of mineral acids as catalysts, the lignin recovered is also more degraded, potentially making it less useful as a co-product. The factors affecting organosolv delignification, mechanisms of delignification, and results from characterization of organosolv lignin and the residual lignin on the biomass are discussed in Section 10.3.

Four main reactions or processes occur during ethanol or methanol organosolv pretreatments: (1) hydrolysis of lignin hemicellulose linkages and internal lignin bonds results in hemicellulose and lignin

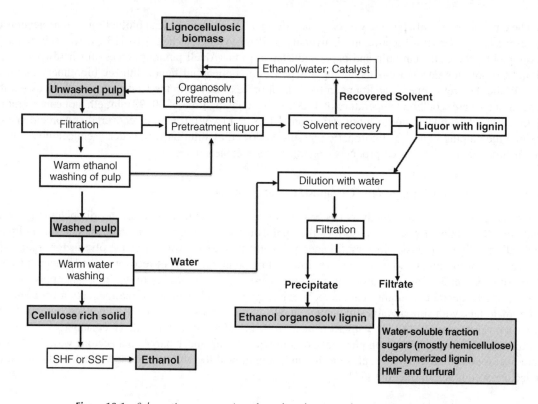

Figure 10.1   *Schematic representation of an ethanol organosolv pretreatment approach.*

solubilization by cleavage of 4-*O*-methylglucuronic acid ester bonds to the α-carbons of lignin and cleavage of α and β-*O*-aryl ether linkages respectively; (2) glycosidic bonds are cleaved in hemicelluloses and less frequently in cellulose, with the extent of cellulose degradation a function of the pretreatment severity as discussed further in Section 10.4; (3) acid-catalyzed degradation of monosaccharides to furfural, HMF, and further degradation products such as levulinic acid and formic acid; and (4) the lignin condensation reactions also discussed in Section 10.3 can occur (especially in acid-catalyzed organosolv pretreatment).

Ethanol organosolv pretreatment has been applied to a wide range of lignocellulosic feedstocks including softwoods, hardwoods, agro-energy crops, and agricultural residues [7,8,10,11,24]. Pretreatment conditions used for different types of biomass are compiled in Table 10.1. Sulfuric acid is the most frequently used catalyst for ethanol organosolv pretreatment and has been applied to several commonly used feedstocks including pine, hybrid poplar, Miscanthus, and switchgrass. Sulfuric acid concentrations are based on percentage dry weight of the biomass and are usually between 0.5 and 1.75%. Higher acid concentrations lead to greater delignification, but greater hemicellulose degradation may be an undesired side effect. In case of Miscanthus, an additional dilute sulfuric acid (0.15 M) presoaking step in which the biomass is extracted overnight with acid under reflux, was beneficial in recovering hemicellulose sugars prior to organosolv pretreatment [7]. Sulfur dioxide (SO_2) behaved similarly to sulfuric acid in terms of lignin removal and cellulose recovery when used as a catalyst for the organosolv pretreatment of Lodgepole pine [28]. When subjected to enzyme hydrolysis, the acid-catalyzed organosolv substrate showed faster and greater (100% vs. 70%) cellulose to glucose conversion than the SO_2 catalyzed substrate. The lower weight-average degree of polymerization (DP_w) of cellulose after sulfuric-acid-catalyzed organosolv pretreatment may be an important factor in determining its enzymatic digestibility and is discussed further in Section 10.4.

Table 10.1 *Solvent concentrations, pretreatment conditions, and catalysts used for ethanol organosolv pretreatment of different biomass feedstocks. The conditions given are optimal for high cellulose recovery and cellulose to glucose conversion after enzymatic hydrolysis.*

Biomass	Ethanol concentration (% by volume)	Catalyst (% dry wt. biomass)	Temperature, time (°C, min)	Reference
Eucalyptus	75	1% acetic acid	200, 60	[25]
Baggase	75	1% acetic acid	200, 60	[25]
Hybrid poplar	60	1.3% sulfuric acid	180, 60	[8]
Lodgepole pine	65	1.1% sulfuric acid	170, 60	[9]
Loblolly pine	65	1.1% sulfuric acid	170, 60	[11]
Miscanthus	80	0.5% sulfuric acid	170, 60	[7]
Kanlow switchgrass	75	0.9% sulfuric acid	180, 60	[26]
Buddleja davidii	50	1.8% sulfuric acid	180, 40	[24]
Lodgepole pine	78	0.03M $MgCl_2$	200, 60	[28]
Lodgepole pine	65	20% NaOH	170, 69	[28]
Lodgepole pine	65	1% sulfur dioxide	170, 60	[28]
Barley straw	50	0.1M $FeCl_3$	170, 60	[27]

For most feedstocks, the cellulose-rich substrate produced during ethanol organosolv pretreatment exhibits high glucose yields after enzymatic hydrolysis. One of the exceptions is base- (NaOH) catalyzed pretreatment of Lodgepole pine, which resulted in high lignin removal but low glucose recovery and xylan removal and showed low cellulose to glucose conversion of the resulting substrate [28]. Results from enzymatic digestibility of cellulose obtained after organosolv pretreatment are included in Table 10.2, which compiles results from biomass fractionation. In most cases, 98% cellulose to glucose conversion is achieved in 48 hours of enzyme hydrolysis with an enzyme loading on 20 FPU cellulase and 40 IU beta-glucosidase per gram of cellulose. These studies were mostly carried out at a solids loading of 2 g cellulose in 100 mL solution. Several characteristics of organosolv-pretreated biomass such as low hemicellulose and lignin content, decreased cellulose chain length and molecular weight [24,28], and increased pore volume contribute to this improved digestibility.

Fermentation of organosolv-pretreated substrates has been performed with simultaneous saccharification and fermentation (SSF) and separate hydrolysis and fermentation (SHF). At an industrial scale, SSF is likely to be the preferred approach as it leads to overall lower processing times, less enzymatic inhibition by hydrolysis products (e.g., cellulase inhibition by cellobiose), and hence lower capital and operating costs [29]. Pan *et al.* [18] compared the ethanol yields upon SSF and SHF of ethanol-organosolv-pretreated mixed softwoods. SHF on a substrate with 6.4% residual lignin resulted in 90% of the theoretical ethanol conversion (assuming 1 g glucose yields 0.51 g ethanol) in 8 hours. A substrate with 8.5% residual lignin produced 84% of its theoretical yield in 24 h with SSF [18]. At a low concentration of fermentation inhibitors, SSF has been seen to yield 99.5% theoretical ethanol yield from acetone-organosolv-pretreated *Pinus radiata* [30]. Thus SSF and SHF both appear suitable for the production of ethanol after biomass pretreatment and fractionation using the organosolv method. It should be mentioned that all the above studies employed the hexose-fermenting yeast *Saccharomyces cerevisiae*. Fermentation of the pentose sugars with specialized microorganisms or other high-value uses of these sugars is vital to good process economics.

Butanol

Butanol is an excellent delignification agent due to its hydrophobicity [28]. It can also be produced from lignocellulosic biomass by a fermentation of sugars released after pretreatment and enzymatic hydrolysis,

Table 10.2 *Results from biomass fractionation by ethanol organosolv pretreatment according to conditions presented in Table 10.1. All results are given as % dry weight of untreated biomass.*

Biomass feedstock	Cellulose			Hemicellulose			Lignin			
	Untreated	Liquid fraction[a]	Solid fraction[b]	Untreated	Liquid fraction	Solid fraction	Untreated	EOL[c]	Liquid fraction	Solid fraction
Loblolly pine [11]	42.0	3.0	33.3 (55)	21.6	0.5	15.3	29.9	5.4	12.4	11.6
Lodgepole pine [10]	50.5	4.2	37.6 (100)	23.9	1.2	11.1	25.1	19.6	4.8	4.2
Hybrid poplar [8]	48.9	0.6	43.2 (98)	22.4	4.9	11.2	23.3	15.5	5.2	6.2
Miscanthus [7]	37.7	1.7	35.5 (98)	37.3	3.3	25.1	26.3	18.1	0.2	7.8
B. davidii [24]	38.9	1.4	32.3 (98)	26.1	5.7	11.0	30.2	8.9	3.4	19.0

[a] Effluent and wash solution after pretreatment
[b] Solid substrate recovered after pretreatment. Numbers in parentheses represent the % cellulose to glucose conversion after 48 h enzymatic hydrolysis with 20 FPU cellulase and 40 IU beta-glucosidase per gram cellulose. 8 FPU cellulase and 16 IU beta-glucosidase per gram of cellulose were used in this study.
[c] Ethanol organosolv lignin

using the microorganism *Clostridium acetobutylicium* [31]. Butanol has value as a fuel additive and as a platform chemical for producing materials and value-added chemicals. When an aqueous butanol mixture is used as a solvent for organosolv pretreatment, given the limited miscibility of butanol in water it is possible to concentrate hemicelluloses in the aqueous layer, lignin in the butanol layer, and cellulose in the solid fraction. Thus, pretreatment with butanol may also be an efficient means of biomass fractionation. Del Rio *et al.* [28] studied the effects of different catalysts on ethanol and butanol organosolv treatment of Lodgepole pine. Pretreatment with butanol/water consistently yielded substrates that were more readily hydrolysable with enzymes. The limited miscibility of butanol and water was suggested to lead to higher pretreatment severity, which in turn led to the formation of substrates with lower hemicellulose content, lower cellulose DP_w, and increased pore size. Among the solvent-catalyst systems studied, butanol-SO_2 resulted in the fastest cellulose to glucose conversion with 82% conversion in 12 hours [28].

Polyhydroxy Alcohols

Ethylene glycol and glycerol are the most commonly employed higher-boiling-point alcohols for organosolv pretreatments. One of the main advantages of using such solvents is that the pretreatment can be performed at atmospheric pressure, which reduces energy costs and the need for pressure vessels. Aqueous glycerol was found to be effective for delignification of wood chips [32]. Auto-catalyzed aqueous organosolv pretreatment at 240 °C for 4 h resulted in 95% cellulose recovery and 70% lignin removal from wheat straw [33]. Results from recovery of lignin and hemicellulose in glycerol pretreatment are not available. Similar to steam explosion, aqueous glycerol pretreatment dissociates the guaiacyl lignin subunits and is seen to have a smaller effect on syringyl lignin [23]. In lieu of using high-grade glycerol, crude glycerol produced as a by-product of biodiesel generation can be used in these pretreatments, leading to further cost reduction. The high-energy costs of solvent recovery, which is considered a disadvantage of glycerol pretreatment, can also be partially offset by using crude glycerol.

Organic Acids and Peracids

As mentioned in Section 10.2.3, acetic acid and formic acid have solubility parameter values similar to lignin and are good lignin solvents. However, their use in biomass pretreatment has been somewhat limited due to their corrosive nature. Vazquez *et al.* [34] pretreated Eucalyptus wood chips with an HCl-catalyzed 70% acetic acid solution. Xylan removal and delignification were observed as the main effects of this pretreatment. However, high rates or extent of xylan and lignin removal did not translate into faster or greater enzymatic cellulose hydrolysis. Acetylation of cellulose, in which hydroxyl groups of cellulose are substituted by acetyl groups, inhibited productive binding of cellulase to cellulose via hydrogen bonds. Pretreatment of beech hardwood with 80% formic acid under different conditions showed that it was capable of extensive delignification (up to 90% at 130 °C and 150 min) and high cellulose recovery (average of 98%; [35]). Further, xylose was reported as the main hemicellulose hydrolysis product and did not undergo significant conversion to furfural. The enzymatic digestibility of fractionated cellulose was not examined but, in a manner analogous to acetic acid pretreatment, formylation of cellulose during formic acid pretreatment can hinder enzymatic digestibility of the pretreated biomass.

Peracetic and performic acid have also been investigated as organosolv pretreatment reagents. Performic acid is used in the Milox pulping process; however, there are major concerns about its stability and the associated safety issues [36]. Peracetic acid is produced by reacting acetic acid with hydrogen peroxide in the presence of sulfuric acid. While peracetic acid pretreatments can be performed at ambient temperatures for longer time periods to save on energy costs, increasing the temperature to 80–90 °C can significantly decrease the pretreatment time. Treatment with sodium hydroxide to remove part of the lignin and swell the biomass prior to peracetic acid pretreatment was found to be very effective in producing material from hybrid poplar and sugarcane baggase that was very amenable to enzymatic deconstruction [37]. This two-step method had high carbohydrate yields together with negligible formation of furfural and HMF due to low carbohydrate dehydration at low pretreatment temperatures. The corrosive nature of peracetic acid, reagent costs in producing it, and operational concerns are some of the drawbacks to its large-scale application.

Phosphoric Acid

Zhang *et al.* [38] exploited the differential solubilities of cellulose, hemicelluloses, and lignin in different solvents to develop a novel pretreatment strategy by which lignocellulosic biomass could be fractionated into amorphous cellulose, hemicellulose, lignin, and acetic acid. In this method, the biomass is brought in contact with concentrated phosphoric acid (>82%) at 50 °C for 30–60 min, which acts to disrupt the lignin-carbohydrate complex bonds, break up hydrogen bonding in carbohydrate chains, weakly hydrolyze cellulose and hemicellulose to low degree of polymerization (DP) fragments, and remove acetyl groups from hemicellulose to produce acetic acid [38]. The cellulose and hemicellulose dissolved in acetic acid are precipitated by the addition of acetone, which also results in the partial dissolution of lignin. Phosphoric acid is washed from the precipitated solids with the addition of more acetone and the liquor comprising phosphoric acid, acetone, acetone-soluble lignin, and acetic acid derived from the hemicellulose fraction is distilled to separate the acetone and acetic acid based on their different volatility. This pretreatment was shown to effectively produce enzyme hydrolysable cellulose from diverse feedstocks including corn stover, switchgrass, hybrid poplar, and Douglas fir [38]. With the exception of Douglas fir cellulose, which showed a 75% conversion to glucose in 24 h, cellulose from the other biomass materials showed 97% conversion to glucose in the same time period. While this process has promise as a method for biomass fractionation, the corrosive nature of phosphoric acid stream may impede its implementation on an industrial scale.

Acetone and Methyl Isobutyl Ketone

Acetone is an excellent lignin solvent, and both auto-catalyzed and catalyzed (usually with mineral acids) acetone organosolv pretreatments have been successfully applied to the fractionation of a variety of biomass feedstocks. Araque *et al.* [30] optimized the conditions for organosolv pretreatment of *Pinus radiata* with aqueous acetone (50%) and 0.9% sulfuric acid. After pretreatment at an H factor (a parameter representing pretreatment severity which is a function of heat-up time, cooking temperature, and cooking time) of 939, up to 70.9% of the glucan could be recovered in the solid fraction. Almost all the hemicelluloses were solubilized or degraded, and about 47% of the lignin was recovered as organosolv lignin, leading to a relatively efficient fractionation of the biomass. While modest (72%) enzymatic glucose yields were obtained, the extent of delignification was not found to have a clearly discernible effect on the extent of cellulose hydrolysis. Huijgen *et al.* [39] fractionated a key agricultural residue, wheat straw, using auto-catalyzed-acetone-organosolv pretreatment. A 50% acetone-water mixture gave the highest lignin recovery (61%), but it decreased at higher acetone concentrations due to higher solution pH. Slowing lignin bond cleavage at higher pH lowered the extent of delignification. Huijgen *et al.* [39] conducted a series of pretreatments with 50% acetone-water at different reaction times and temperature. Good biomass fractionation was obtained after acetone pretreatment at 205 °C for 1 h, which resulted in 82% hemicellulose hydrolysis, 79% delignification, and 93% cellulose recovery. A glucose yield of 87% was obtained after enzymatic hydrolysis. Delignification increased with process severity up to a temperature of 205 °C. At 220 °C, lignin condensation reactions were suggested to increase the residual lignin content in the pulp. However, the amount of lignin which could be recovered from the organosolv liquor increased with process severity, and 100% could be recovered after pretreatment at 220 °C for 120 minutes.

Hemicellulose hydrolysis is auto-catalyzed by the pH drop resulting from the formation of acetic acid by dehydration of hemicellulose acetyl side groups. Increased pH resulting from higher acetone concentrations in the cooling liquor also results in reduced hemicellulose hydrolysis. At high temperatures, xylose was found to be increasingly dehydrated to furfural [39]. Separate reaction conditions were found to be optimal for the recovery of cellulose, hemicellulose, and lignin. Pretreatment conditions should therefore be chosen after taking into account the net revenues of the different process streams.

Acetone-based fractionation of biomass has also been carried out in a two-step process in which biomass is first treated with water at 180 °C to recover hemicellulose. This is followed by extraction with a flowing water-acetone mixture at 230 °C and 10 MPa pressure to separate the lignin and the cellulose [40]. These results suggest that acetone is a promising solvent for biomass fractionation.

In a recent study, Bozell *et al.* [41] developed a novel organosolv biomass fractionation process which they termed "Clean Fractionation." In this method, lignocellulosic material is separated with a ternary mixture of methyl isobutyl ketone, ethanol and water in the presence of sulfuric acid, which selectively dissolves lignin and hemicellulose, leaving a cellulose residue. Treatment of the single-phase liquor with water results in the separation of a lignin-rich organic phase and a hemicelluloses-rich aqueous phase. For woody feedstocks, the yield of the cellulose fraction across all separations averaged 47.7 wt% (±1.1). The authors reported that while the cellulose and lignin fractions obtained using this method were quite pure, the impurities in the hemicellulose stream could be removed by ion exchange chromatography.

10.2.4 Fractionation of Biomass during Organosolv Pretreatment

Organosolv pretreatment, especially with ethanol as a solvent, has been shown to be very effective in fractionating lignocellulosic biomass into a cellulose-rich solid, a water-soluble hemicellulose stream, and a solid organosolv lignin fraction. Information on the mass balance of the major biomass components available for several promising feedstocks is compiled in Table 10.2. As the results in Table 10.2 indicate, the cellulose recovered in the solid fraction can range from 74% [9] to 94% [7]; when combined with that dissolved in the pretreatment liquor, up to 99% of the cellulose in the untreated biomass can be accounted for. Results from glucose released after 48 hours of enzymatic hydrolysis of this cellulose-rich solid substrate are also included in Table 10.2, providing further evidence of the effectiveness of organosolv pretreatment. The hemicellulose content of the pretreated solids was usually low, and a greater proportion of these sugars are measured in the liquid stream. A much greater fraction of hemicelluloses are degraded during organosolv pretreatment compared to cellulose, with hemicellulose recoveries ranging from 51% [9] to 76% in Table 10.2 [7]. Hemicelluloses are easily degraded in pretreatments carried out at low pH to form furfural and HMF and acetic acid. Organosolv pretreatment conditions can be optimized further to improve hemicellulose yields, but the net revenue of the hemicellulose stream or from its conversion to ethanol by pentose-fermenting organisms compared to that from greater cellulose recovery and corresponding ethanol yield should be considered.

The generation of a relatively pure, undegraded, and sulfur-free lignin stream is one of the main advantages of organosolv pretreatment; at appropriate processing conditions, most of the lignin in the biomass can be recovered. However, conditions that result in the highest degree of delignification usually involve higher acid concentrations and cooking temperatures [8], which may not be optimal for producing substrates amenable to enzymatic hydrolysis. Under conditions which generated pretreated biomass with good enzymatic sugar release (≥70%), 20–78% of the total lignin (acid insoluble + acid soluble) in the untreated biomass was recovered as ethanol organosolv lignin (Table 10.2). Lignin depolymerization during pretreatment leads to its partial solubilization in the cooking and wash solutions, and this lignin is not recovered in the organosolv lignin stream. However, it still leads to delignification of biomass and aids in enzymatic hydrolysis. As seen in the results compiled in Table 10.2, up to 42% of the lignin may be dissolved in the combined ethanol and water wash solutions. Overall, the mass closure for lignin was much higher than for hemicellulose and cellulose and, in some cases, exceeded 100% due to the presence of extractives, lignin fragments, tannins, and related phenol compounds in the water-soluble fraction, resulting in inflated measurements of the acid-soluble lignin content [8].

10.3 Nature of Organosolv Lignin and Chemistry of Organosolv Delignification

Lignin is an amorphous, cross-linked phenolic polymer that is biosynthesized from three mono-lignols: coniferyl alcohol, sinapyl alcohol, and *p*-coumaryl alcohol [42]. The proportions of different mono-lignols involved depend on the plant species and undergo radical polymerization to form lignin inter-unit linkages. Some common lignin inter-unit linkages are β-*O*-aryl ether (β-*O*-4), resinol (β-β′), phenylcoumaran (β-5′), biphenyl (5-5′) and 1,2-diarylpropane (β-1′) [42].

10.3.1 Composition and Structure of Organosolv Lignin

Organosolv pretreatment results in lignin solubilization into the pretreatment solvent and washing liquor, from which it is precipitated by lowering the solution pH for recovery as a separate fraction. The lignin recovered in this process is generally termed organosolv lignin. To understand pathways for organosolv delignification, native (milled wood) lignin from biomass and organosolv lignin can be characterized using elemental analysis, quantitative ^{1}H, ^{13}C and ^{31}P nuclear magnetic resonance (NMR), and molecular weight analysis. Wet chemistry techniques, such as thioacidolysis and nitrobenzene oxidation, coupled with gas chromatography can also be applied for precise determination of specific structural moieties in lignin. However, unlike spectroscopic techniques, these methods do not have the ability to analyze the whole lignin structure. Recently, results from detailed characterization of these two lignin fractions from Loblolly pine, Miscanthus, and *Buddleja davidii* have been obtained [43–45]. Table 10.3 summarizes quantitative ^{13}C NMR results, which shed information on the proportions of different lignin functional groups and compositional parameters such as degree of condensation that can be calculated from these results. For all three biomass feedstocks, the largest change after organosolv pretreatment was a decrease (~50%) in Cγ in β-*O*-4. This is consistent with acid-catalyzed scission of these linkages, which has been proposed as the major mechanism for lignin dissolution in organosolv systems. Such distinct changes are not seen for the β-β′ and β-5′ linkages. In the aromatic region, ^{13}C NMR can provide information on the protonated (aromatic C—H), oxygenated (aromatic C—O), and condensed (aromatic C—C) functional groups. In lignin isolated after organosolv pretreatment, the oxygenated aromatic region does not change compared to milled wood

Table 10.3 *Lignin functional group content and degree of condensation for milled wood lignin (MWL) and ethanol organosolv lignin (EOL).*

Equivalences per aromatic ring	Pine MWL [43]	Pine EOL[a] [43]	Miscanthus MWL [44]	Miscanthus EOL[b] [44]	*B. davidii* MWL [45]	*B. davidii* EOL[a] [45]
Methoxyl	1.0	0.9	1.0	1.1	1.2	1.1
β-*O*-4	0.6	0.3	0.5	0.3	0.6	0.3
Cβ in β-β and β-5	0.2	0.1	—	—	0.2	0.1
Aromatic C-O	2.0	2.1	1.9	1.5	2.2	2.2
Aromatic C-C	1.5	2.1	1.7	2.0	1.4	1.8
Aromatic C-H	2.6	2.0	2.5	2.3	2.4	2.1
Condensation	0.4	1.1	—	—	0.6	1.0

[a] The organosolv pretreatment conditions are given in Table 10.1.
[b] Organosolv pretreatment of Miscanthus was performed at 65% ethanol, 1.2% sulfuric acid, 190 °C, and 60 min.

Table 10.4   *Hydroxyl group contents of MWL and EOL determined by [31] P NMR spectroscopy.*

Lignin	Hydroxyl (OH) content (mmol/g lignin)			Carboxylic OH (δ 136.6–133.6)
	Aliphatic OH (δ 150.0–145.5)	Condensed phenolic OH (δ 144.7–141.0)	Guaiacyl OH (δ 141.0–138.0)	
Pine MWL [43]	7.3	0.1	0.5	0.0
Pine EOL[a] [43]	4.2	0.6	1.4	0.3
B. davidii MWL [45]	4.5	0.3	0.4	0.0
B. davidii EOL[a] [45]	2.5	1.0	1.5	0.2
Miscanthus MWL [44]	4.0	0.2	0.7	0.1
Miscanthus EOL[b] [44]	1.2	0.1	1.3	0.2

[a] The organosolv pretreatment conditions are given in Table 10.1.
[b] Organosolv pretreatment of Miscanthus was performed at 65% ethanol, 1.2% sulfuric acid, 190 °C, and 60 min.

lignin. However, for all three biomasses, an increase in the condensed aromatic C content and a decrease in protonated aromatic C is seen in the organosolv lignin (Table 10.3). This implies an increase in the degree of condensation due to the formation of C—C linkages. The methoxyl (OCH_3) content of lignin does not change appreciably, except after severe pretreatments [44,45], indicating no demethylation.

Following phosphitylation, the different free hydroxyl (OH) functionalities in lignin can be quantitatively estimated with ^{31}P NMR spectroscopy [46–48]. Quantification is performed based on the known OH content of an internal standard such as cyclohexanol. ^{31}P NMR experiments require less instrument time than quantitative ^{13}C NMR, and the spectra display nicely resolved peaks which can be ascribed to aliphatic, condensed and uncondensed phenolic, and carboxylic OH groups. ^{31}P NMR data from organosolv lignin and their comparison to milled wood lignin showed a decrease in aliphatic OH content after organosolv pretreatment (Table 10.4). This was accompanied by an increase in the phenolic OH and an increase in degree of condensation as shown by ^{13}C NMR results and also evident from the ^{31}P results. Carboxylic OH content of organosolv lignin fractions was higher than the corresponding milled wood lignin (MWL), indicating the hydrolysis of ester bonds during pretreatment.

The molecular weight distribution of lignin can be determined with gel permeation chromatography (GPC), with acetylated lignin dissolved in tetrahydrofuran and results quantified against polystyrene standards. Number average $(\overline{M_n})$ and weight average $(\overline{M_w})$ molecular weights and the polydispersity index $(D = \overline{M_w}/\overline{M_n})$ of lignin can be estimated using this method. The polydispersity index provides an estimate of the distribution of molecular weights in a polymeric material. Results listed in Table 10.5 show that the molecular weight of organosolv lignin is lower than milled wood lignin, even at low pretreatment severity. This leads to better solubility of organosolv lignin in various solvents and allows for a variety of practical applications that are discussed later. The polydispersity does not vary significantly but is often lower in

Table 10.5   *Molecular weight distributions and polydispersities of MWL and EOL.*

Lignin	$\overline{M_n}$ (g/mol)	$\overline{M_w}$ (g/mol)	D $(\overline{M_n}/\overline{M_w})$
Pine MWL [43]	7590	13 500	1.77
Pine EOL[a] [43]	3070	5410	1.77
B. davidii MWL [45]	7260	16 800	2.31
B. davidii EOL[a] [45]	661	2350	3.56
Miscanthus MWL [44]	8300	13 700	1.65
Miscanthus EOL[b] [44]	4690	7060	1.51

[a] The organosolv pretreatment conditions are given in Table 10.1.
[b] Organosolv pretreatment of Miscanthus was performed at 65% ethanol, 1.2% sulfuric acid, 190 °C, and 60 min.

organosolv lignin. Low molecular weights coupled with low polydispersity (indicating a narrow distribution of molecular weights) are ideal for further valorization of organosolv lignin. With an increase in pretreatment severity, the molecular weight of organosolv lignin falls to a certain extent, after which it does not change significantly [24]. Thus, ethanol organosolv pretreatment appears to attack certain lignin linkages and solubilize a fraction of the lignin. As severity increases, more of the same linkages are hydrolyzed leading to greater delignification. A drop in molecular weight with increasing severity has also been seen for formic acid and acetic acid organosolv pretreatment of wheat straw [49].

Effects of Pretreatment Severity on Composition of Organosolv Lignin

With increasing pretreatment severity, a greater proportion of lignin in a biomass feedstock is recovered as organosolv lignin [7–9,24,50,51]. Under conditions of optimum glucose recovery, up to 70% of the lignin could be removed from Miscanthus [7]. When the effects of the different pretreatment parameters, temperature, ethanol concentrations, and acid concentrations are considered separately, temperatures above 195 °C result in decreased lignin yield from hybrid poplar due to excessive depolymerization which reduces the lignin recovery [50]. The maximum lignin recovery was obtained at 65% ethanol concentration. While low ethanol concentrations lead to lower solution pH and promote the cleavage of lignin aryl ether linkages, the degraded lignin has limited solubility due to the low solvent concentration. Lignin solubility in ethanol-water mixtures is highest at *c.* 70% ethanol concentration [52]. Organosolv pretreatment at higher ethanol concentrations therefore leads to lower lignin recovery due to lower solution pH and limited lignin solubility. Longer reaction times and higher acid concentrations both lead to greater delignification [50].

With increasing pretreatment severity, the organosolv lignin recovered has higher phenolic content due to the enhanced cleavage of α- and β-aryl ether linkages between lignin units. This leads to the formation of new phenolic units [53]. A decrease in the number of these linkages observed in ^{13}C and ^{1}H NMR results and in aliphatic OH groups from ^{31}P NMR provides evidence for this mechanism (Table 10.6). An increase in lignin condensation is also seen to occur at high severity, and the reaction responsible is discussed in the following section. Other structural trends with increased severity include increased carboxylic content in organosolv lignin as a result of enhanced ester hydrolysis. The molecular weight distribution and polydispersity of organosolv lignin decreased with increasing temperature and acid concentration (Table 10.6). Increasing ethanol concentration produced organosolv lignin with higher molecular weights and polydispersity, due to the reduced cleavage of aryl ether linkages. Reaction time did not have a significant effect on molecular weight [49].

Table 10.6   *Variations in ethanol organosolv lignin compositions and molecular weights as a function of CS.*

Lignin Parameter	Miscanthus [53]			B. davidii [43]			Hybrid Poplar [51]	
	CS 1.75	CS 2.39	CS 2.93	CS 2.40	CS 2.92	CS 1.68	CS 2.47	CS 2.87
% biomass recovered as EOL	3.7	11.0	13.1	4.3	19.0	5.7	17.2	4.4
β-O-4 (#/Ar group)	0.9	0.2	0.1	0.4	0.2	—	—	—
Aliphatic OH[a] (mmol/g lignin)	3.1	1.1	1.3	2.7	1.9	5.0	3.2	3.3
Phenolic OH[a] (mmol/g lignin)	2.3	4.0	3.9	2.6	2.7	2.2	4.1	4.6
$\overline{M_n}$ (g/mol)	2500	3100	2300	578	645	1515	1123	783
$\overline{M_w}$ (g/mol)	6500	4300	3200	2740	2490	3877	1890	1105
D	2.6	1.4	1.4	4.7	3.9	2.6	1.7	1.4

[a] Aliphatic and phenolic OH contents for Miscanthus and *B. davidii* were determined with ^{31}P NMR spectroscopy, while those for Hybrid poplar were determined with ^{1}H NMR spectroscopy.

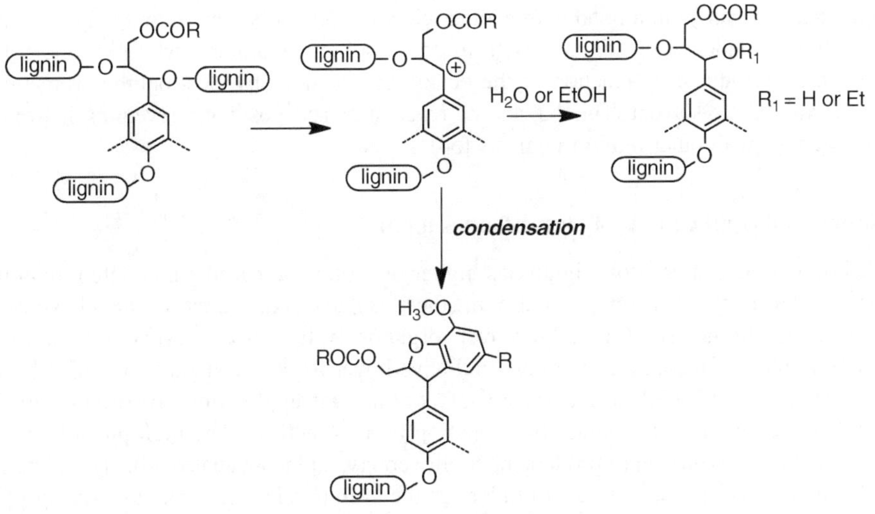

Figure 10.2   *Solvolytic splitting of α-O-aryl ether linkages in lignin [53].*

10.3.2   Mechanisms of Organosolv Delignification

It is widely acknowledged that the cleavage of aryl ether linkages is primarily responsible for lignin breakdown during the organosolv pretreatment. Of these, α-O-aryl ether bonds are cleaved more easily, whereas β-O-aryl ether bonds are broken under more severe conditions, especially at higher acid concentrations [54]. The cleavage of α-O-aryl ether bonds is the rate-controlling step in organosolv delignification [55], and several pathways have been proposed for this reaction [54] including: (1) solvolytic splitting of α-O-aryl ether linkages via the quinone methide intermediate; (2) solvolytic cleavage by nucleophilic substitution benzylic position by an S_N2 mechanism; and (3) the reaction via formation of a benzyl carbocation under acidic conditions (Figure 10.2). β-O-aryl ether linkages can be broken by homolytic cleavage with the loss of terminal γ-methylol groups as formaldehyde. This mechanism has been shown to give rise to stilbenes [45]. Formation of Hibbert's ketones (evidenced by the presence of carbonyl groups) can also occur via cleavage of β-O-aryl ether bonds [53]. Mechanisms of solvolytic splitting of β- and γ-O-aryl ether linkages are shown in Figure 10.3. Under more acidic conditions, lignin condensation reactions occur which are counter-productive to organosolv delignification, as observed by several researchers [43,54,56]. Condensation reactions lead to the formation of higher-molecular-weight lignin fractions which are not readily soluble in the organosolv pretreatment solvent and hence cannot be recovered. The benzyl carbocation intermediate of α-O-aryl and

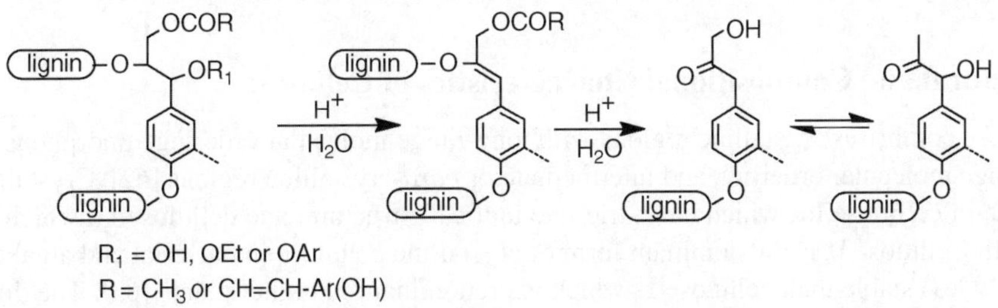

R_1 = OH, OEt or OAr
R = CH_3 or CH=CH-Ar(OH)

Figure 10.3   *Solvolytic splitting of β- and γ-O-aryl ether linkages in lignin [53].*

β-O-aryl ether cleavage can form a bond with another electron-rich carbon atom in a neighboring lignin unit, thus increasing lignin condensation (Figure 10.2). In organosolv pretreatments with phenolic solvents, lignin condensation can be prevented by reaction of the benzyl carbocations by electrophilic aromatic substitution on the aromatic ring of the solvent. This blocks the reactive benzyl position, preventing it from undergoing condensation reactions with other lignin fragments [54].

10.3.3   Commercial Applications of Organosolv Lignin

As seen in Section 10.3.1, organosolv lignin has higher phenolic and carboxylic content than native lignin from biomass. Under most pretreatment conditions, these isolated lignin fractions also have lower molecular weight and hence higher solubility. Moreover, organosolv lignins have very low sulfur content (as opposed to lignin obtained from Kraft pulping) and have lower oxygen content than native biomass lignin. These characteristics could allow for a variety of commercial applications of organosolv lignin, thus enhancing process economics of biomass fractionation by this method. The high phenolic OH content of organosolv lignin can be exploited to produce phenolic, epoxy, and isocyanate resins [57]. Phenolic powder resins have been successfully applied as a binder for the commercial-scale manufacturing of automotive brake pads and molding. Polyurethane and polyisocyanate foams, with 17% and 26% lignin substitution for polyols, have high density [58]. Investigation of the antioxidant activity of different organosolv lignin preparations has shown that free phenolic groups contribute to higher antioxidant activity [51]. Stabilization of phenoxyl radicals by conjugated double bonds or substituents such as methoxyl groups at the *ortho* position can further enhance antioxidant activity. Lower molecular weight and polydispersity of organosolv lignin also correlates with higher antioxidant activity [51]. Polyphenols can be beneficial for human health in various ways. These compounds can inhibit the oxidation of low-density proteins and decrease the risk of heart disease. Polyphenols also have anti-inflammatory and anti-carcinogenic properties.

High-purity lignin such as organosolv lignin, which has low carbohydrate content, can be used as a precursor for the production of chemicals such as vanilla, phenol, and ethylene and can be converted to carbon fibers which are of high value [58]. Other large-scale applications of organosolv lignin include dispersants, soil-conditioning agent, adsorbents, and adhesives. As an energy source, organosolv lignin has a higher heating value of 26 MJ/kg and, upon combustion, can provide energy in excess of that required for the pretreatment and ethanol distillation [59]. For added value, organosolv lignin can be valorized by catalytic hydrogenation to low-molecular-weight molecules, which are more amenable to liquefaction and production of fuels and fuel additives. Depolymerization and deoxygenation of organosolv lignin is essential for its successful conversion to a form usable as a fuel or fuel additive. Organosolv lignin can also be thermally depolymerized by pyrolysis to produce pyrolytic lignin oil [60]. These oils have some limitations such as low pH, high corrosiveness, and instability, which necessitate upgrading of the crude bio-oil to impart favorable fuel properties. Advances in catalyst research have made important contributions to this field [61].

10.4   Structural and Compositional Characteristics of Cellulose

Native cellulose comprises crystalline regions with long-range molecular ordering, amorphous regions with only short-range molecular ordering, and intermediate or *para*-crystalline regions [62]. Crystalline cellulose is a composite of cellulose Iα, which has a triclinic unit cell structure, and cellulose Iβ, which has a monoclinic unit cell. Cellulose I$_\alpha$ is the dominant form of crystalline cellulose in bacterial and algal cellulose and is energetically less stable than cellulose Iβ which is predominant in higher plants [63]. The different forms of crystalline cellulose, *para*-crystalline cellulose, amorphous cellulose at accessible fibril surfaces (i.e., in contact with water), and amorphous cellulose at inaccessible fibril surfaces (fibril-fibril contact surfaces and surfaces resulting from distortions in the fibril interior) can be characterized by solid-state ^{13}C NMR spectroscopy. The parameter crystallinity index (CrI) is a measure of the relative proportion of crystalline cellulose and can be estimated by various analytical techniques including solid-state ^{13}C NMR, Fourier transform infrared (FTIR) spectroscopy, and X-ray diffraction [64]. Due to their greater level of molecular ordering, crystalline cellulose is thought to be more resistant to chemical and enzymatic degradation. The effect of cellulose crystallinity on enzymatic action is however a subject of considerable debate and is beyond the scope of this chapter.

The molecular weight distribution of cellulose can be determined by GPC after derivatization by tricarbanilation and can in turn provide information on the DP. The weight-average degree of polymerization (DP$_w$) of cellulose, which is most commonly reported, can be calculated from its molecular weight data [65]. Cellulose DP values can also be rapidly determined from viscometry measurements and denoted DP$_v$. If a pretreatment reduces cellulose DP without causing extensive structural degradation, it produces more reducing ends available for exoglucanase enzymes to act on, thus enhancing sugar release. In case of organosolv pretreatment, cellulose DP is seen to vary with solvent, catalyst and, to some extent, pretreatment severity [24,28,66]. Values of cellulose DP$_w$ compiled from the literature are given in Table 10.7. Organosolv pretreatment of Lodgepole pine with butanol as the solvent and MgCl$_2$ and SO$_2$ as catalysts resulted in much lower cellulose DP$_w$ than the equivalent pretreatment with ethanol [28]. However, sulfuric-acid-catalyzed organosolv pretreatment produced cellulose with similar DP$_w$ for both solvents. In general, for the different solvent-catalyst systems investigated, lower cellulose DP$_w$ translated to greater cellulose hydrolysis after 12 hours. Organosolv pretreatment of wheat straw with ethanol-produced cellulose with the lowest DP, followed by similar values for methanol and acetic acid and the lowest cellulose depolymerization (highest DP) resulted from combined acetic acid/formic acid pretreatment [66]. For the same solvent, an increase in the proportion of solvent helped preserve cellulose DP$_w$ (Table 10.7). Overall, an increase in pretreatment severity results in lower DP of the cellulose to some extent [10,24], which can help in enzymatic cellulose hydrolysis; however, care should be taken to avoid extremely severe conditions that cause cellulose loss.

The effect of organosolv pretreatment on cellulose CrI has not been extensively investigated but, from the limited amount of data available, it appears to be a function of the initial CrI of biomass and pretreatment severity. The cellulose CrI values discussed in this section were determined with solid-state ^{13}C NMR of cellulose isolated from biomass. Ethanol organosolv pretreatment of Switchgrass [26] and Buddleja [24],

Table 10.7 *Degree of polymerization (DP$_w$) of cellulose before and after organosolv pretreatment of various biomass feedstocks. (EtOH: ethanol; SA: sulfuric acid; AcOH: acetic acid; FA: formic acid; MeOH: methanol; BuOH: butanol.).*

Biomass	Pretreatment conditions	Cellulose DP$_w$	
		Untreated	Pretreated
Kanlow Switchgrass [26]	75% EtOH, 0.9% SA, 180 °C, 1 h	2900	2412
B. davidii [24]	50% EtOH, 1.8% SA, 180 °C, 40 min	1000	530
Wheat straw [66]	75% AcOH, 0.1% HCl, 85 °C, 4 h	2600	1594
	90% AcOH, 0.1% HCl, 85 °C, 4 h	2600	1952
	20% AcOH, 60% FA, 0.1% HCl, 85 °C, 4 h	2600	2182
	60% MeOH, 0.1% HCl, 85 °C, 4 h	2600	1519
	60% EtOH, 0.1% HCl, 85 °C, 4 h	2600	1356
Lodgepole pine [28]	78% EtOH, 0.025M MgCl$_2$, 200 °C, 1 h	—	1400
	78% BuOH, 0.025M MgCl$_2$, 200 °C, 1 h	—	848
	65% EtOH, 1.1% SA, 170 °C, 1 h	—	1062
	65% BuOH, 1.1% SA, 170 °C, 1 h	—	1060
	65% EtOH, 1.1% SO$_2$, 170 °C, 1 h	—	1200
	65% BuOH, 1.1% SO$_2$, 170 °C, 1 h	—	769

both of which have cellulose CrI values of around 0.50, resulted in an insignificant change in CrI after pretreatment. However, in the case of Loblolly pine [11] with cellulose CrI of 0.63, ethanol organosolv pretreatment produced cellulose with CrI of 0.53, suggesting that pretreatment is capable of altering the crystalline cellulose component. The effect of pretreatment severity on cellulose crystallinity is not well defined as contradictory results have been reported in the literature. While ethanol organosolv pretreatment of Lodgepole pine produced cellulose with higher CrI at higher severities, the opposite effect was seen for Buddleja which exhibited an 11% reduction in CrI at higher pretreatment severity. Cellulose isolated from organosolv-pretreated Lobolly pine after 72 hours of enzymatic hydrolysis in which 70% of the cellulose was hydrolyzed had a CrI of 0.81, indicating preferential hydrolysis of amorphous cellulose [11]. No dominant trends were seen in switchgrass cellulose CrI after 2, 4 and 8 hours of enzymatic hydrolysis [26].

Organosolv pretreatment of biomass can therefore result in good cellulose recovery coupled with high cellulose to glucose conversion by enzymes. This is likely a combined effect of reducing cellulose DP and altering crystallinity. However, with the limited information available in the literature, clear trends in variations in these parameters with pretreatment conditions cannot be discerned, and additional studies are necessary to fully understand the effects of organosolv pretreatment on biomass cellulose.

10.5 Co-products of Biomass Fractionation by Organosolv Pretreatment

In order to maximize revenues from a lignocellulosic biorefinery, the three major biomass components should be utilized to the maximum extent possible. In the simplest biorefinery scenario for ethanol production, all sugars are converted to ethanol and the residual lignin is burned to provide a heat source for pretreatment, distillation, and other operations. One idealized scenario would convert glucose to ethanol; while xylose and the other minor sugars and their degradation products are recovered and sold separately for conversion to value-added products, lignin is used as a high-value polymeric material and acetic acid is sold separately [58]. Such an ideal scenario could potentially increase revenues from $150/ton (for the simple case) to $640/ton when most of the biomass is valorized. Along with ethanol, which is produced from the cellulosic fraction, biomass fractionation by organosolv pretreatment produces several valuable co-products such as organosolv lignin, hemicellulose sugars, acetic acid, furfural, HMF, and levulinic acid. Under conditions optimized for cellulose recovery and good enzymatic digestibility from hardwoods, furfural is formed at slightly higher concentrations (0.5 g/100 g biomass) than HMF (0.1 g/100 g biomass), as there is more xylose than in softwoods [8]. During organosolv pretreatment of the softwood Lodgepole pine [10] however, the hexose degradation products HMF (2.1 g/100 g biomass), levulinic acid (13.5 g/100 g biomass), and formic acid (1.5 g/100 g biomass) are formed at much higher concentrations than furfural (2.1 g/100 g biomass). While most of these degradation products are fermentation inhibitors, they have a variety of potential uses which increases the importance of extracting them prior to SSF and utilizing them as co-products. They could also be used as reactants to make alkanes and other products by catalytic processes now being developed [67–69].

10.5.1 Hemicellulose

The composition of the hemicellulose stream obtained after biomass fractionation is determined by the biomass feedstock. Softwoods generally have lower hemicellulose contents (25–35%) and mannose is the main component of the hemicellulose fraction [70,71]. Hardwoods and herbaceous biomass have higher hemicellulose contents (20–40%), and xylose is the primary constituent of the hemicellulose stream [70,71]. Arabinose and galactose are the least-abundant sugars in most biomass hemicelluloses. During biomass pretreatment and fractionation, most of the hemicelluloses are fractionated into the water-soluble stream; about 50% of the sugars within this fraction are oligomers, with the rest being monomers.

Pentose sugars can be fermented to ethanol with specialized microorganisms but also have several other potential practical applications. Hemicelluloses have been used as a plant gum for thickeners, adhesives, protective colloids, emulsifiers, and stabilizers [72]. They have recently been applied to produce biodegradable oxygen-barrier films [73]. Animal feed additives are also a common use for hemicellulose sugars. Xylose, which is one of the major monomeric sugars in the hemicellulose from a variety of feedstocks, can be fermented to produce xylitol which is used as a sucrose replacement for diabetics and in dental hygiene products. Quesada-Medina *et al.* [20] explored the use of different organosolv systems for lignin extraction from almond shells which had been previously utilized for xylose production by acid hydrolysis. In this manner, two value-added streams could be extracted from almond shells, an agricultural waste.

As mentioned earlier, hemicellulose sugars are typically the fraction with lowest recovery yields as they are easily degraded even under moderately severe pretreatment conditions. In order to utilize these sugars, pre-extraction of the hemicelluloses before organosolv fractionation is an option that should be considered. In case of Miscanthus, presoaking with dilute sulfuric acid (0.15 M) followed by extraction overnight with acid under reflux was beneficial in recovering 57% of the hemicellulose sugars prior to organosolv pretreatment [7].

10.5.2 Furfural

Furfural, a key co-product of biomass pretreatment, is formed by the acid-catalyzed dehydration of pentose sugars (Figure 10.4). It has a wide range of industrial applications as a solvent in lubricants, coatings, adhesives, and furan resins [74] and for production of polytetramethylene ether glycol, which is used to manufacture Lycra and Spandex. Earlier, furfural was used in the production of Nylon but was replaced by petrochemicals [58]. With the advent of lignocellulosic biorefineries, furfural can now return as a green replacement for petrochemicals in Nylon manufacture.

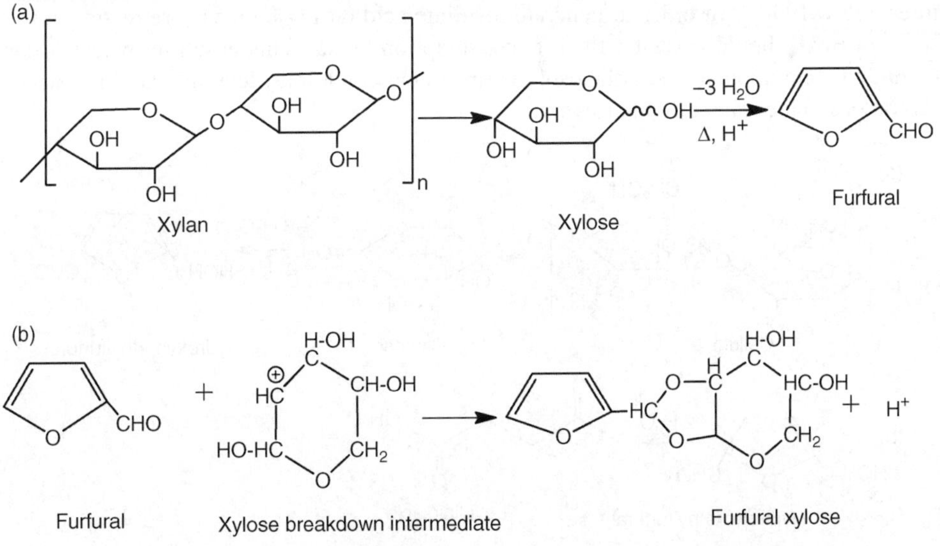

Figure 10.4   *Reactions showing (a) the hydrolysis of xylan to xylose and its dehydration to furfural and (b) condensation reaction of furfural and xylose, which can occur at high temperatures [74].*

Results from organosolv biomass fractionation have shown that xylose dehydration to furfural mostly occurs at temperatures of 190–205 °C [39,50]. Below 190 °C, xylose formation by the hemicellulose hydrolysis is the predominant reaction. At higher temperatures (>205 °C), furfural concentration in the water-soluble fraction decreases due to self-condensation or reaction of xylose with furfural to produce insoluble polymeric material (Figure 10.4), which have been referred to as pseudo-lignin or humins [50,75]. This material may behave similarly to lignin and account for the >100% lignin mass closures for organosolv pretreatments at high severities [39]. The effect of the organosolv cooking parameters of ethanol concentration, acid concentration, and temperature on the generation of furfural has been recently studied [50]. At low temperatures, xylose to furfural conversion is favored only at low ethanol concentrations as it results in lower pH at the same acid loading. Increasing temperature enhanced furfural condensation and lowered its yield. For softwoods, the highest xylose to furfural conversion (70% molar yield) was obtained at 200 °C, 25 min, 60% ethanol and 3% sulfuric acid [50]. Higher ethanol concentrations retarded furfural condensation at high temperatures.

10.5.3 Hydroxymethylfurfural (HMF)

Acid-catalyzed dehydration of hexose sugars results in the formation of HMF, which can further degrade to levulinic acid and formic acid or form condensation products similar to those from furfural (Figure 10.5). HMF is also an inhibitor of fermentations, but can be converted to liquid transportation fuels through the pathways of aldol condensation and hydrogenation [76]. The molar conversion yield of HMF is highest at high temperatures (up to 200 °C), low pH, and longer pretreatment times [50]. Beyond 200 °C, formation of levulinic acid and condensation starts to occur. Ethanol concentration in organosolv has been reported as not having a significant effect on HMF production.

10.5.4 Levulinic Acid

Levulinic acid can be used for several value-added products such as dyes, polymers, flavoring agents, plasticizers, solvents, and fuel additives [50]. Levulinic acid is produced from acid-catalyzed carbohydrate dehydration and is formed by the degradation of HMF, as seen above. Its formation is therefore favored at temperatures above 190 °C. In order to generate levulinic acid during biomass pretreatment, the reaction forming it from HMF should be faster than its consumption by self-condensation, which occurs at high temperature. Under commonly used organosolv pretreatment conditions, levulinic acid is detected only in low concentrations in the water-soluble fraction.

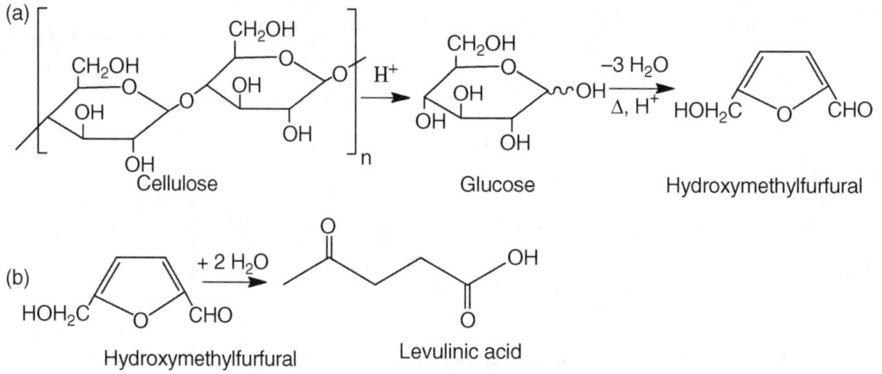

Figure 10.5 *Reactions showing (a) cellulose hydrolysis to glucose and its dehydration to HMF and (b) levulinic acid formation from HMF.*

10.5.5   Acetic Acid

The acetyl groups in hemicelluloses are hydrolyzed during acid-catalyzed pretreatments to form acetic acid. Acetic acid further lowers the pH of the cooking liquor and promotes acid-catalyzed hydrolysis of the bio-molecules in biomass. Acetic acid is one of the highest-volume chemical commodities traded throughout the world [19]. It is used to manufacture acetic anhydride for production of cellulose acetate fibers and membranes. Acetic acid is also converted to vinyl acetate, which is used in making paint and paper coatings. In the Lignol process, acetic acid is one of the major co-products of biomass fractionation and ethanol production [19]. In order to avoid problems associated with high acetic acid concentrations in the cooking liquor after solvent reuse (as the acetic acid is also volatilized during flashing to recover the solvent), soda is added to the base of the distillation tower, and acetic acid is recovered as sodium acetate.

10.6   Conclusions and Recommendations

Organosolv pretreatment is capable of fractionating lignocellulosic biomass into separate streams rich in lignin, hemicellulose, and cellulose. The cellulosic fraction is very amenable to enzymatic deconstruction and subsequent fermentation to ethanol. Of the three major biocomponents, the hemicelluloses are the most degraded during fractionation, but once recovered can be fermented to ethanol or utilized as a separate value-added stream. Hemicellulose and glucose degradation products such as furfural, acetic acid, hydrox-ymethyl furfural, levulinic acid, and formic acid also have a wide range of industrial applications. One of most economically attractive advantages of biomass fractionation by organosolv pretreatment is the ability to recover a large proportion of biomass lignin as a pure sulfur-free low-molecular-weight material. Compositional characteristics of organosolv lignin, such as its high phenolic content and low molecular weight, enable its valorization to a wide range of products including antioxidants, phenolic resins, and chemicals. Other notable advantages of the organosolv pretreatment include increased cellulose enzymatic digestibility and pore volume and reduced DP.

There are tremendous opportunities for future research in this area, including development of efficient solvent and co-product recovery systems and catalysts for lignin conversion to fuels and fuel additives. The effect of pre-extraction of hemicelluloses and extractive compounds and their contribution to improving process economics should be investigated. In terms of fundamental research, the changes in cellulose structure and crystallinity during organosolv pretreatment and their effect on the enzymatic digestibility of the substrate are not fully understood and should be further explored.

Acknowledgements

This work was supported and performed as part of the BioEnergy Science Center, managed by Oak Ridge National Laboratory (ORNL, the manager partner and home facility for the BioEnergy Science Center). The BioEnergy Science Center is a US Department of Energy Bioenergy Research Center supported by the Office of Biological and Environmental Research in the DOE Office of Science. ORNL is managed by UT-Battelle, LLC, under contract DE-AC05-00OR 22725 for the US Department of Energy.

References

1. Lal, R. (2005) World crop residues production and implications of its use as a biofuel. *Environment International*, **31** (4), 575–584.
2. Ragauskas, A., Williams, C., Davison, B. *et al.* (2006) The path forward for biofuels and biomaterials. *Science*, **311**, 484–489.
3. Himmel, M.E., Ding, S., Johnson, D.K. *et al.* (2007) Biomass recalcitrance: Engineering plants and enzymes for biofuels production. *Science*, **315**, 804–807.

4. Yang, B. and Wyman, C. (2008) Pretreatment: the key to unlocking low-cost cellulosic ethanol. *Biofuels, Bioproducts, Biorefining*, **2**, 26–40.

5. Kumar, P., Barrett, D., Delwiche, M., and Stroeve, P. (2009) Methods for pretreatment of lignocellulosic biomass for efficient hydrolysis and biofuel production. *Industrial and Engineering Chemical Research*, **48** (8), 3713–3729.

6. Pu, Y., Zhang, D., Singh, P., and Ragauskas, A. (2008) The new forestry biofuels sector. *Biofuels, Bioproducts, Biorefining*, **2**, 58–73.

7. Brosse, N., Sannigrahi, P., and Ragauskas, A. (2009) Pretreatment of miscanthus x giganteus using the ethanol organosolv process for ethanol production. *Industrial & Engineering Chemistry Research*, **48** (18), 8328–8334.

8. Pan, X., Gilkes, N., Kadla, J. *et al.* (2006) Bioconversion of hybrid poplar to ethanol and co-products using an organosolv fractionation process: Optimization of process yields. *Biotechnology and Bioengineering*, **94** (5), 851–861.

9. Pan, X., Xie, D., Yu, R. *et al.* (2007) Biorefining of Lodgepole pine killed by mountain pine beetle using ethanol organo-solv process: Fractionation and process optimization. *Industrial Engineering Chemical Research*, **46** (8), 2609–2617.

10. Pan, X., Xie, D., Yu, R., and Saddler, J. (2008) The bioconversion of mountain pine beetle-killed Lodgepole pine to fuel ethanol using the organosolv process. *Biotechnology and Bioengineering*, **101** (1), 39–48.

11. Sannigrahi, P., Miller, S.J., and Ragauskas, A.J. (2010) Effects of organosolv pretreatment and enzymatic hydrolysis on cellulose structure and crystallinity in Loblolly pine. *Carbohydrate Research*, **345** (7), 965–970.

12. Mosier, N., Wyman, C., Dale, B. *et al.* (2005) Features of promising technologies for pretreatment of lignocellulosic biomass. *Bioresource Technology*, **96**, 673–686.

13. Johansson, A., Aaltonen, O., and Ylinen, P. (1987) Organosolv pulping – methods and pulp properties. *Biomass*, **13** (1), 45–65.

14. Pye, E. and Lora, J. (1991) The Alcell process, a proven alternative to kraft pulping. *TAPPI Journal*, **74** (3), 113–118.

15. Aziz, S. and Sarkanen, K.V. (1989) Organosolv pulping – a review. *TAPPI Journal*, **72** (3), 169–175.

16. Muurinen, E. (2000) Organosolv pulping—a review and distillation study related to peroxyacid pulping. Ph.D. thesis, University of Oulu, Finland

17. Botello, J., Gilarranz, M.A., Rodriguez, F., and Oliet, M. (1999) Recovery of solvents and by-products from organosolv black liquor. *Separation Science and Technology*, **34** (12), 2431–2445.

18. Pan, X., Arato, C., Gilkes, N. *et al.* (2005) Biorefining of softwoods using ethanol organosolv pulping: Preliminary processing of process streams for manufacture of fuel-grade ethanol and co-products. *Biotechnology and Bioengineering*, **90** (4), 473–481.

19. Arato, C., Pye, E., and Gjennestad, G. (2005) The Lignol approach to biorefining of woody biomass to produce ethanol and chemicals. *Applied Biochemistry and Biotechnology*, **121–124**, 871–882.

20. Quesada-Medina, J., Lopez-Cremades, F.J., and Olivares-Carrillo, P. (2010) Organosolv extraction of lignin from hydrolyzed almond shells and application of the [delta]-value theory. *Bioresource Technology*, **101** (21), 8252–8260.

21. Pan, X. and Sano, Y. (1999) Atmospheric acetic acid pulping of rice straw IV: Physico-chemical characterization of acetic acid lignins from rice straw and woods. Part 1. physical characteristics. *Holzforschung*, **53** (5), 511–518.

22. Chum, H.L., Black, S.K., Johnson, D.K. *et al.* (1988) Organosolv pretreatment for enzymatic hydrolysis of poplars: isolation and quantitative structural studies of lignins. *Clean Technologies and Environmental Policy*, **1** (3), 187–198.

23. Zhao, X., Cheng, K., and Liu, D. (2009) Organosolv pretreatment of lignocellulosic biomass for enzymatic hydrolysis. *Applied Microbiology and Biotechnology*, **82** (5), 815–827.

24. Hallac, B.B., Sannigrahi, P., Pu, Y. *et al.* (2010) Effect of ethanol organosolv pretreatment on enzymatic hydrolysis of buddleja davidii stem biomass. *Industrial & Engineering Chemistry Research*, **49** (4), 1467–1472.

25. Teramoto, Y., Lee, S., and Endo, T. (2008) Pretreatment of woody and herbaceous biomass for enzymatic saccharification using sulfuric acid-free ethanol cooking. *Bioresource Technology*, **99** (18), 8856–8863.

26. Cateto, C., Hu, G., and Ragauskas, A. (2011) Enzymatic hydrolysis of organosolv Kanlow switchgrass and its impact on cellulose crystallinity and degree of polymerization. *Energy & Environmental Science*, **4** (4), 1516–1521.

27. Kim, Y., Yu, A., Han, M. *et al.* (2010) Ethanosolv pretreatment of barley straw with iron(III) chloride for enzymatic saccharification. *Journal of Chemical Technology and Biotechnology*, **85** (11), 1494–1498.

28. Del Rio, L., Chandra, R., and Saddler, J. (2010) The effect of varying organosolv pretreatment chemicals on the physicochemical properties and cellulolytic hydrolysis of mountain pine beetle-killed lodgepole pine. *Applied Biochemistry and Biotechnology*, **161** (1), 1–21.
29. Munoz, C., Mendonca, R., Baeza, J. *et al.* (2007) Bioethanol production from bio-organosolv pulps of Pinus radiata and Acacia dealbata. *Journal of Chemical Technology and Biotechnology*, **82**, 767–774.
30. Araque, E., Parra, C., Freer, J. *et al.* (2008) Evaluation of organosolv pretreatment for the conversion of Pinus radiata D. Don to ethanol. *Enzyme and Microbial Technology*, **43**, 214–219.
31. Wackett, L.P. (2008) Biomass to fuels via microbial transformations. *Current Opinion in Chemical Biology*, **12** (2), 187–193.
32. Demirbas, A. (1998) Aqueous glycerol delignification of wood chips and ground wood. *Bioresource Technology*, **63** (2), 179–185.
33. Sun, F. and Chen, H. (2008) Organosolv pretreatment by crude glycerol from oleochemicals industry for enzymatic hydrolysis of wheat straw. *Bioresource Technology*, **99** (13), 5474–5479.
34. Vazquez, G., Antorrena, G., Gonzalez, J. *et al.* (2000) The influence of acetosolv pulping conditions on the enzymatic hydrolysis of Eucalyptus pulps. *Wood Science and Technology*, **34**, 345–354.
35. Dapia, S., Santos, V., and Parajo, J. Study of formic acid as an agent for biomass fractionation. *Biomass & Bioenergy*, **22** (3), 213–221.
36. Zhao, X., Peng, F., Cheng, K., and Liu, D. (2009) Enhancement of the enzymatic digestibility of sugarcane bagasse by alkali-peracetic acid pretreatment. *Enzyme and Microbial Technology*, **44** (1), 17–23.
37. Teixeira, L., Linden, J., and Schroeder, H. (1999) Alkaline and peracetic acid pretreatments of biomass for ethanol production. *Applied Biochemistry and Biotechnology*, **77** (1), 19–34.
38. Zhang, Y.P., Ding, S., Mielenz, J.R. *et al.* (2007) Fractionating recalcitrant lignocellulose at modest reaction conditions. *Biotechnology and Bioengineering*, **97** (2), 214–223.
39. Huijgen, W., Reith, J., and den Uil, H. (2010) Pretreatment and fractionation of wheat straw by an acetone-based organolv process. *Industrial & Engineering Chemistry Research*, **49** (20), 10132–10140.
40. Paszner, L., Quinde, A., and Meshgini, M. (1985) ACOS- accelerated hydrolysis of wood by acid catalyzed organosolv means, Vancouver, B.C., p. 235–240.
41. Bozell, J., Black, S.K., Myers, M. *et al.* (2011) Solvent fractionation of renewable woody feedstocks: Organosolv generation of biorefinery process streams for the production of biobased chemicals. *Biomass & Bioenergy*, **35**, 4197–4208.
42. Boerjan, W., Ralph, J., and Baucher, M. (2003) Lignin biosynthesis. *Annual Reviews in Plant Biology*, **54**, 519–546.
43. Sannigrahi, P., Ragauskas, A.J., and Miller, S.J. (2010) Lignin structural modifications resulting from ethanol organosolv treatment of loblolly pine. *Energy & Fuels*, **24** (1), 683–689.
44. El Hage, R., Brosse, N., Chrusciel, L. *et al.* (2009) Characterization of milled wood lignin and ethanol organosolv lignin from Miscanthus. *Polymer Degradation and Stability*, **94**, 1632–1638.
45. Hallac, B.B., Pu, Y., and Ragauskas, A.J. (2010) Chemical transformations of buddleja davidii lignin during ethanol organosolv pretreatment. *Energy & Fuels*, **24** (4), 2723–2732.
46. Granata, A. and Argyropoulos, D. (1995) 2-chloro-4,4,5,5-tetramethyl-1,3,2-dioxaphospholane, a reagent for the accurate determination of the uncondensed and condensed phenolic moieties in lignins. *Journal of Agricultural and Food Chemistry*, **43**, 1538–1544.
47. Wu, S. and Argyropoulous, D. (2003) An improved method of isolating lignin in high yield and purity. *Journal of Pulp and Paper Science*, **29**, 235–240.
48. Zawadski, M. and Ragauskas, A. (2001) N-hydroxyl compounds as new internal standards for the ^{31}P NMR determination of lignin hydroxyl functional groups. *Holzforschung*, **55**, 283–285.
49. Xu, F., Sun, J., Sun, R. *et al.* (2006) Comparative study of organosolv lignins from wheat straw. *Industrial Crops and Products*, **23**, 180–193.
50. Kim, D. and Pan, X. (2010) Preliminary study on converting hybrid poplar to high-value chemicals and lignin using organosolv ethanol process. *Industrial & Engineering Chemistry Research*, **49** (23), 12156–12163.
51. Pan, X., Kadla, J.F., Ehara, K. *et al.* (2006) Organosolv ethanol lignin from hybrid Poplar as a radical scavenger: Relationship between lignin structure, extraction conditions and antioxidant activity. *Journal of Agricultural and Food Chemistry*, **54** (16), 5806–5813.

52. Ni, Y. and Hu, Q. (1995) Alcell® lignin solubility in ethanol–water mixtures. *Journal of Applied Polymer Science*, **57** (12), 1441–1446.

53. El Hage, R., Brosse, N., Sannigrahi, P., and Ragauskas, A. (2010) Effects of process severity on the chemical structure of Miscanthus ethanol organosolv lignin. *Polymer Degradation and Stability*, **95** (6), 997–1003.

54. McDonough, T. (1993) The chemistry of organosolv delignification. *TAPPI Journal*, **76** (8), 186–193.

55. Meshgini, M. and Sarkanen, K.V. (1989) Synthesis and kinetics of acid-catalyzed hydrolysis of some alpha-aryl ether lignin model compounds. *Holzforschung*, **43** (4), 239–243.

56. Li, J., Henriksson, G., and Gellerstedt, G. (2007) Lignin depolymerization/repolymerization and its critical role for delignification of aspen wood by steam explosion. *Bioresource Technology*, **98**, 3061–3068.

57. Lora, J.H. and Glasser, W.G. (2002) Recent industrial applications of lignin: a sustainable alternative to non-renewable materials. *Journal of Polymers and the Environment*, **10** (1), 39–48.

58. Zhang, P.Y. (2008) Reviving the carbohydrate economy via multi-product lignocellulose biorefineries. *Journal Industrial Microbiology and Biotechnology*, **35**, 367–375.

59. White, R. (1987) Effect of lignin content and extractives on the higher heating value of wood. *Wood and Fiber Science*, **19**, 446–452.

60. de Wild, P., Van der Laan, R., Kloekhorst, A., and Heeres, E. (2009) Lignin valorisation for chemicals and (transportation) fuels via (catalytic) pyrolysis and hydrodeoxygenation. *Environmental Progress & Sustainable Energy*, **28** (3), 461–469.

61. Zakzeski, J., Bruijnincx, P., Jongerius, A.L., and Weckhuysen, B. (2010) The catalytic valorization of lignin for the production of renewable chemicals. *Chemical Reviews*, **110** (6), 3552–3599.

62. Larsson, P., Wickholm, K., and Iverson, T. (1997) A CP/MAS ^{13}C NMR investigation of molecular ordering in celluloses. *Carbohydrate Research*, **302**, 19–25.

63. Hult, E., Larsson, P., and Iverson, T. (2000) A comparative CP/MAS ^{13}C NMR study of cellulose structure in spruce wood and kraft pulp. *Cellulose*, **7**, 35–55.

64. Park, S., Baker, J., Himmel, M.E. *et al.* (2010) Cellulose crystallinity index: measurement techniques and their impact on interpreting cellulase performance. *Biotechnology for Biofuels*, **3** (10), http://www.biotechnologyforbiofuels.com/content/3/1/10.

65. Hallac, B. and Ragauskas, A. (2011) Analyzing cellulose degree of polymerization and its relevancy to cellulosic ethanol. *Biofuels, Bioproducts, Biorefining*, **5** (2), 215–225.

66. Sun, X.F., Sun, R.C., Fowler, P., and Baird, M.S. (2004) Isolation and characterisation of cellulose obtained by a two-stage treatment with organosolv and cyanamide activated hydrogen peroxide from wheat straw. *Carbohydrate Polymers*, **55** (4), 379–391.

67. Patel, A.D., Serrano-Ruiz, J.C., Dumesic, J.A., and Anex, R.P. (2010) Techno-economic analysis of 5-nonanone production from levulinic acid. *Chemical Engineering Journal*, **160** (1), 311–321.

68. Serrano-Ruiz, J.C. and Dumesic, J.A. (2011) Catalytic routes for the conversion of biomass into liquid hydrocarbon transportation fuels. *Energy & Environmental Science*, **4** (1), 83–99.

69. Xing, R., Subrahmanyam, A.V., Olcay, H. *et al.* (2010) Production of jet and diesel fuel range alkanes from waste hemicellulose-derived aqueous solutions. *Green Chemistry*, **12** (11), 1933–1946.

70. Ragauskas, A., Nagy, M., Kim, D. *et al.* (2006) From wood to fuels, integrating biofuels and pulp production. *Industrial Biotechnology*, **2** (1), 55–65.

71. Carroll, A. and Somerville, C. (2009) Cellulosic biofuels. *Annual Reviews in Plant Biology*, **60**, 165–182.

72. Kamm, B. and Kamm, M. (2004) Principles of biorefineries. *Applied Microbiology and Biotechnology*, **64** (2), 137–145.

73. Hartman, J., Albertsson, A., Lindblad, M.S., and Sjöberg, J. (2006) Oxygen barrier materials from renewable sources: Material properties of softwood hemicellulose-based films. *Journal of Applied Polymer Science*, **100** (4), 2985–2991.

74. Zeitsch, K. (2000) *The Chemistry and Technology of Furfural and its Many Byproducts*, Elsevier, Amsterdam.

75. Sannigrahi, P., Kim, D.H., Jung, S., and Ragauskas, A. (2011) Pseudo-lignin and pretreatment chemistry. *Energy & Environmental Science*, **4** (4), 1306–1310.

76. Roman-Leshkov, Y., Barrett, C.J., Liu, Z.Y., and Dumesic, J.A. (2007) Production of dimethylfuran for liquid fuels from biomass-derived carbohydrates. *Nature*, **447** (7147), 982–985.

11

Ionic Liquid Pretreatment: Mechanism, Performance, and Challenges

Seema Singh[1,2] and Blake A. Simmons[1,2]

[1] *Deconstruction Division, Joint BioEnergy Institute, Emeryville, USA*
[2] *Biological and Materials Science Center, Sandia National Laboratories, Livermore, USA*

11.1 Introduction

The development of an effective, scalable, and economically viable biomass pretreatment technology that can handle a wide range of non-food feedstocks (e.g., agricultural residues, woody biomass, dedicated energy crops) and produce high fermentable sugar yields (>90–95%) remains elusive but is intensely pursued within the biofuels community. Several structural and compositional attributes of biomass are thought to influence the conversion of lignocellulose to fermentable sugars including cellulose crystallinity [1,2], the presence and composition of lignin [3], and the extent and type of functional groups present on hemicelluloses [4]. An efficient biomass pretreatment strategy should therefore be capable of effectively disrupting and removing the cross-linked matrix of lignin and hemicelluloses present in the plant cell walls, and increasing the porosity and surface area of cellulose for subsequent enzymatic hydrolysis.

One relatively recent addition to the pretreatment technologies being developed is based on ionic liquids (ILs), typically defined as salts composed of anions and cations that are poorly coordinated relative to other salts such as NaCl, with melting points typically under 100 °C. ILs were originally developed as electrolytes for batteries and as alternatives to organic solvents, and have been under development since the early 1980s. There are over a thousand ILs listed in curated databases (e.g., International Union of Pure and Applied Chemistry or IUPAC, Dortmund Data Bank or DDBST, and the Institute of Process Engineering at the Chinese Academy of Sciences in Beijing or IPE/Beijing) whose properties are defined by a wide range of anions and cations, including miscibility with water and organic solvents. The anion has a primary effect on water miscibility and the cation a secondary effect. The specific combination of the anion and cation determines the thermodynamic and physicochemical properties of the IL and enables the selection of an IL for a particular task.

Aqueous Pretreatment of Plant Biomass for Biological and Chemical Conversion to Fuels and Chemicals, First Edition.
Edited by Charles E. Wyman.
© 2013 John Wiley & Sons, Ltd. Published 2013 by John Wiley & Sons, Ltd.

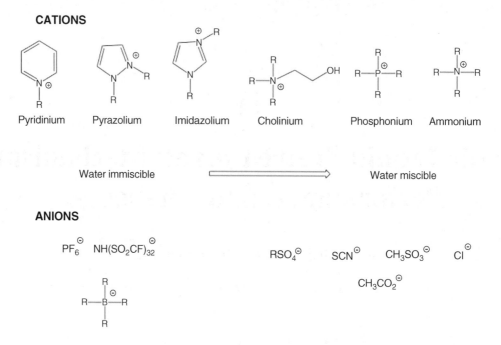

Figure 11.1   *Examples of common anions and cations used for ionic liquid pretreatment of cellulose and lignocellulose. (Adapted from [23] © 2007, Elsevier).*

The release of fermentable sugars for biofuel production from lignocellulosic biomass using ILs as a pretreatment technology is a growing area of global research and development [5,6]. The first report of using pyridine-based ILs to dissolve cellulose occurred in 1934 [7], but there was a long gap before the next report of using other hydrophilic ILs, such as 1-butyl-3-methylimidazolium chloride (abbreviated here as [C$_4$mim]Cl) appeared in 2002 detailing how this IL could be used to readily solubilize microcrystalline cellulose (MCC) [8]. The cellulose can be recovered using an anti-solvent such as water or ethanol, and it was subsequently reported that the recovered cellulose using this approach was much easier to hydrolyze using cellulases as compared to the initial MCC due to a loss in crystallinity induced by the IL pretreatment and recovery process [9–11]. Although the majority of the initial work focused on the processing of MCC and various methods to functionalize the solubilized cellulose, subsequent reports of IL biomass pretreatment have demonstrated their great promise as efficient solvents for biomass dissolution with easy recovery of cellulose and hemicelluloses upon anti-solvent addition [12,13]. A summary of the common anions and cations used in IL biomass pretreatment is presented in Figure 11.1.

There are two general routes that use IL-pretreated materials to produce fermentable sugars from lignocellulose, as shown in Figure 11.2, that will be discussed in this chapter. In the enzymatic route (Figure 11.2a), hydrolysis is accomplished by the addition of glycoside hydrolases whereas in the catalytic route (Figure 11.2b), acid and metal catalysts are added in place of enzymes to break down the polysaccharides into fermentable sugars and/or directly convert polysaccharides into fuels. In general, these ionic liquid process routes generate a substrate that is more efficiently converted into monomeric sugar than other pretreatment technologies using either route. This is a direct result of the reduced cellulose crystallinity and decreased lignin content after IL pretreatment [12]. The total process time (biomass → sugar) of the ionic liquid pretreatment combined with hydrolysis is significantly

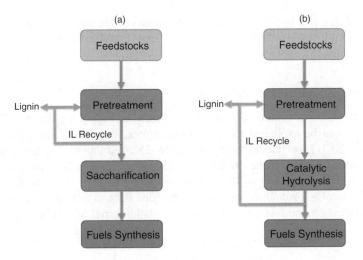

Figure 11.2 *The two general routes of IL-based sugar production and fuel production from lignocellulosic feedstocks: (a) enzymatic and (b) catalytic hydrolysis routes. Lignin is typically a separate process output from the generation of fermentable sugars and/or fuels that can be recovered as a function of IL recycle, dependent on IL pretreatment process conditions used (See figure in color plate section).*

shorter than conventional technologies as a result [14,15]. ILs offer a unique combination in terms of impacts on biomass and the resultant ease of liberating fermentable sugars from biomass: (1) the ability to fully solubilize some or all components of the plant cell wall; and (2) the ability to perform specific chemical reactions. This combination of effects is something that most of the other leading pretreatment technologies are incapable of achieving, but there remain several challenges that must be resolved before IL pretreatment becomes commercially viable.

Relative to other leading pretreatment technologies, the IL pretreatment process can be carried out under relatively mild thermal (100–160 °C) conditions and therefore does not generate a significant amount of inhibitors from the spontaneous degradation of cellulose and hemicelluloses that occurs at elevated temperatures. However, it has been shown that the ionic liquid itself, such as 1-ethyl-3-methylimidazolium acetate ($[C_2mim][OAc]$), can be a strong inhibitor for certain microbes and therefore efficient means of recycling the IL and/or engineering IL-tolerant strains must be developed [16]. ILs have been demonstrated for the production of sugars and biofuels at the laboratory and pre-pilot scale and hold significant promise [17], but several challenges must be addressed before this technology is fully realized at the commercial biorefinery scale [18,19]. This chapter gives a general overview of the mechanisms of IL pretreatment, followed by detailed discussions of both routes of sugar production shown in Figure 11.2. The advantages and challenges of each route will be presented, as well as major obstacles that must be overcome before these technologies can be viewed as commercially viable and scalable.

11.2 Ionic Liquid Pretreatment: Mechanism

In order to take full advantage of the unique properties of ILs and their application to biomass, a fundamental understanding of how they interact with biomass is required. The eventual realization of a predictive tool to identify novel combinations of anions and cations could provide significant breakthroughs in all aspects of the IL pretreatment/sugar production process. The mechanisms by which ILs interact with biomass depend on the particular combination of anion and cation, as well as the particular biomass component (cellulose, hemicelluloses, and lignin) of interest. The majority of the work in establishing these fundamental interactions has been focused on those that occur between IL and cellulose; far less is understood about the mechanisms by which ILs interact with lignin and hemicelluloses.

11.2.1   IL Polarity and Kamlet–Taft Parameters

One solvent parameter that has been used to define the characteristics of a given IL is that of polarity. The polarity of a solvent can be evaluated using many different approaches and methodologies including dielectric constants [20,21], solvatochromic probes [22–25], and chromatographic techniques [26–28]. While these single-parameter approaches to defining solvent polarity have been used successfully in certain applications, the inherent complexity of ILs sometimes precludes their use. The Kamlet–Taft system of quantifying the polarity of a solvent has been used to describe ILs and their interactions with biomass components using a modified general linear solvation energy relationship (LSER) [29–31]:

$$(XYZ) = (XYZ)_o + a\,\alpha + b\,\beta + s\,\pi^* \tag{11.1}$$

where XYZ represents the solvent-dependent property to be measured, $(XYZ)_o$ is the regression point, a, b, and s are solvent-independent coefficients characteristic of the process and indicate the sensitivity of the property studied, α is a measure of the IL to act as a hydrogen bond donor, β is a measure of the IL to act as a hydrogen bond acceptor, and π^* is a measure of the dipolarity/polarizability ratio of the IL. The combination of these three parameters, each of which can be determined experimentally using solvatochromic dyes such as 4-nitroaniline and N,N-diethyl-4-nitroaniline and UV-Vis spectroscopy [32] for each IL, can describe the polarity of an IL with a much higher degree of resolution than those that use a single-parameter scale. These measurements are typically carried out in simple solutions or mixtures with no biomass present, but the parameters determined are then correlated to the observed impact on the substrate of interest. For example, a variant of the original Kamlet–Taft system has been successfully demonstrated as a means to predict the cellulose/biomass solubilization properties of several ILs [32–35]. Furthermore, the Kamlet–Taft system has been used to demonstrate that the ability of an IL as a hydrogen bond acceptor can be directly linked to cellulose solubility, whereas increases in the hydrogen-bond donating ability of the IL will decrease the solubility of the cellulose [36].

11.2.2   Interactions between ILs and Cellulose

MCC is a very challenging material in terms of solubility, and the solvents that are most commonly used to solubilize significant amounts of MCC include the organic zwitterionic solvent N-morpholine-N-oxide [37,38], aqueous mixtures with inorganic salts, organic solvent mixtures with inorganic salts, liquid ammonia [39–41], concentrated phosphoric acid [42,43], and ILs [9,10,44–46]. The interactions between ILs and cellulose have been studied using a wide range of techniques, including experimental studies using nuclear magnetic resonance (NMR) and computational studies using molecular dynamics (MD) simulations.

NMR studies have reported that the hydroxyl groups present in glucan oligomers are strong hydrogen bond donors to certain anions, such as chloride [47,48] and acetate [49], and that this interaction is the basis for how these ILs solubilize MCC. In addition to the hydrogen bonds formed with these anions, NMR data indicate that hydrogen bonding also occurs between hydroxyls of cellobiose and 1-ethyl-3-methylimidazolium ($[C_2mim]^+$). In particular, the aromatic H2 acidic protons of $[C_2mim]^+$ were found to associate with the O_2 atoms of cellobiose hydroxyls, and it was estimated that the stoichiometric ratio of $[C_2mim][OAc]$:[OH] is between 3:4 and 1:1 in the primary solvation shell. This indicates that one anion or cation forms hydrogen bonds with two cellulose hydroxyl groups simultaneously [49].

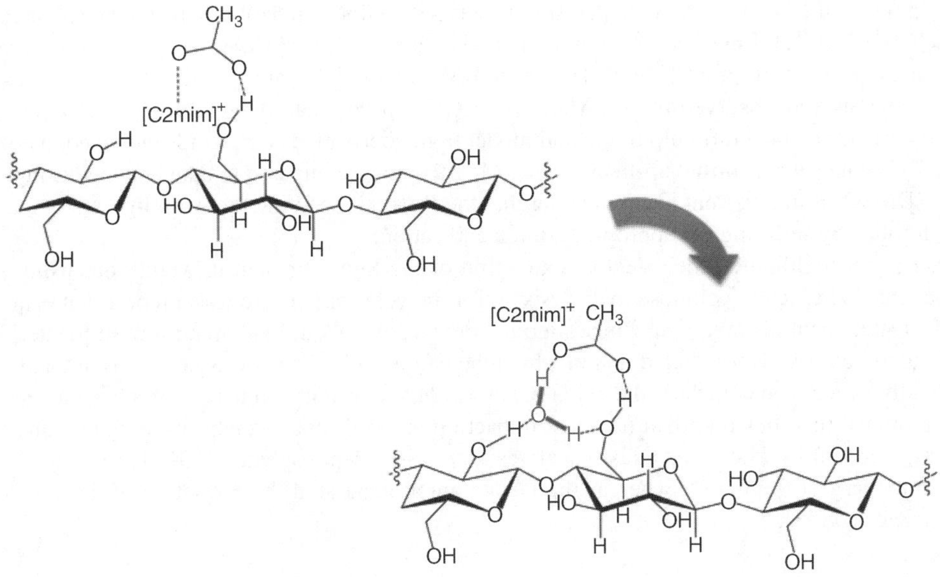

Figure 11.3 *Proposed mechanism of cellulose precipitation from [C₂mim][OAc] upon addition of water. (Adapted from [69] © 2009, Elsevier). (See figure in color plate section).*

In addition to experimental studies of the interactions between ILs and cellulose, MD has been used to provide molecular-level insights into the solubilization process and the forces that govern performance. One recent study used MD to study two distinct states of cellulose in the presence of [C₄mim]Cl: (1) as an intact crystalline microfiber; and (2) in a dissociated state in which all of the glucan chains are separated by at least four solvation shells [48]. The results of this study reveal that the perturbation of solvent structures by solubilized glucan chains can be a crucial element in determining overall solubility, and that both the Cl⁻ and the [C₄mim]⁺ ions interact with the glucan residues that form the intersheet fibrillar contacts between glucan oligomers.

Another MD study of cellulose and [C₄mim]Cl proposed that ions interact with glucan chains along the axial direction and disrupt the intersheet contacts of cellulose [50]. MD simulations of cellulose oligomers in water-[C₂mim][OAc] mixtures have been used to develop an understanding of the roles of the anion and cation in the solvated state and to provide more insights into the mechanism of subsequent step of cellulose precipitation as a function of anti-solvent addition. The results obtained for [C₂mim][OAc] are in general agreement with the results obtained for [C₄mim]Cl in that anions and cations interact with cellulose and drive solubility, and that there is a proposed intermediate transition state (Figure 11.3) in the presence of water that disrupts the IL hydrogen bonding with cellulose and initiates cellulose precipitation [51].

11.2.3   Interactions between ILs and Lignin

Very few studies have been carried out to determine how ILs solubilize and interact with lignin, primarily due to the inherent complexity of lignin and the lack of a "model lignin" that would enable a robust comparative analysis. The published studies to date have been experimental studies on how ILs impacted the structure and composition of lignin that has been isolated from lignocellulose. One experimental study evaluated the solubilization properties of several ILs on lignin isolated from pine kraft pulp [3]. Up to 20 wt% of this type of lignin could be dissolved in the ILs 1-hexyl-3-methylimidazolium trifluoromethanesulfonate ([C$_6$mim][CF$_3$SO$_3$]), 1,3-dimethylimidazolium methylsulfate ([C$_1$mim][MeSO$_4$]) and 1-butyl-3-methylimidazolium methylsulfate ([C$_4$mim][MeSO$_4$]). For the ILs containing [C$_4$mim]$^+$, the order of lignin solubility for varying anions was observed to be: [MeSO$_4$]$^-$ > Cl$^-$, approximately equal to Br$^-$ ≫ [PF$_6$]$^-$ [3]. ^{13}C NMR analyses of the pine kraft pulp lignin and model lignin derivative compounds indicated that the corresponding ^{13}C signals were shifted up-field by δ = 0.1–1.9 ppm as compared to data obtained using dimethyl sulfoxide (DMSO) as the solvent [3], indicating that the ILs do indeed interact with lignin in a manner that can be controlled by selecting the appropriate anion and cation.

Another paper highlighted the selective extraction of the lignin present in wood flour using [C$_2$mim][OAc] that enabled efficient cellulose hydrolysis and IL recycle [52]. There was a report that examined the impact of IL treatment on several technical lignins: organosolv, alkali, and alkali low sulfonate [53]. Two types of experiments were conducted: (1) in which the cation ([C$_2$mim]$^+$) was held constant and the anion systematically varied, and (2) where the anion (Cl$^-$) was held constant and the cation systematically varied. It was determined that the anion had the most impact on the molecular weight of the lignin, and followed the general trend of sulfates > lactates > acetates > chlorides > phosphates [53]. The sulfate-based ILs resulted in the largest fragments, whereas the lactate anion generated the formation of the smallest-sized fragments detected [53].

11.3 Ionic Liquid Biomass Pretreatment: Enzymatic Route

The IL process scheme that liberates fermentable sugars using enzymes (Figure 11.2a) after pretreatment has received significant attention in the scientific literature; it is the most analogous conventional biomass conversion using pretreatments derived from the pulp and paper industry. The ability of certain ILs, such as those based on imidazolium cations, to significantly increase substrate accessibility to hydrolytic enzymes enables more efficient use at lower loading levels as compared to other pretreatment technologies. At the same time, IL pretreatment using this general route poses other challenges in terms of lignin recovery, IL recycle, and IL contamination downstream that can inhibit enzymes and microbes. Indeed, the extent to which full plant cell-wall solubilization in the presence of ILs is desired and/or required to overcome biomass recalcitrance and generate high yields of fermentable sugars remains unknown, although there are several reaction chemistries that can be carried out more efficiently in the solubilized state. The overall performance of the IL pretreatment process depends on several factors such as biomass loading, water content, process temperature, process time, viscosity, mixing, reactor type, recovery mechanism, and the type of lignocellulose that is targeted for conversion. The following sections highlight the progress made to date via the IL pretreatment for the enzymatic conversion route as a function of three major types of lignocellulosic bioenergy feedstocks: grasses, agricultural residues, and woody biomass. While certainly not comprehensive in terms of the potential types of bioenergy feedstocks available globally, these three general classes do have very different chemical and structural compositions that highlight the dynamic interplay between the structural and chemical changes that occur as a function of IL pretreatment type and severity.

11.3.1 Grasses

Herbaceous feedstocks, such as *Panicum virgatum* (switchgrass) and *Miscanthus giganteus* (miscanthus), have quickly risen to the top of the list of potential "dedicated energy crops" that could be grown on marginal lands not suitable for food production in the United States, potentially require little or no irrigation, reduce demand for synthetic fertilizer, enhance carbon and nitrogen soil sequestration, and grow to reasonably high energy densities per acre [54,55]. The chemical composition of these materials relative to woody biomass, typically of higher hemicellulose and lower lignin content, may make them more suitable for

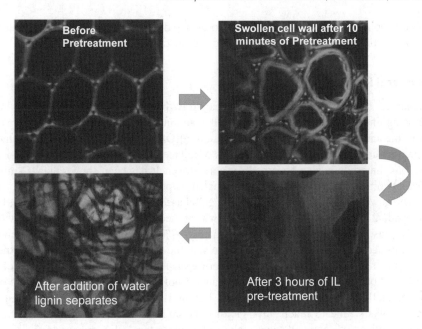

Figure 11.4 *Auto-fluorescence images of plant cell walls of switchgrass before, during, and after IL pretreatment with [C₂mim] [OAc]. (Adapted from [74] © 2010, Royal Society of Chemistry). (See figure in color plate section).*

efficient conversion. Certain hydrophilic ILs, such as [C₂mim][OAc], have been shown to be capable of solubilizing major plant cell-wall components of these herbaceous feedstocks at relatively mild processing temperatures ($T = 120–160\,°C$) and can efficiently fractionate the biomass into polysaccharide-rich and lignin-rich streams upon anti-solvent addition [56,57]. Visualization studies (Figure 11.4) indicated rapid swelling of the switchgrass plant cell walls in the presence of [C₂mim][OAc] attributed to disruption of inter- and intra-molecular hydrogen bonding between cellulose fibrils and lignin. Addition of the anti-solvent induced precipitation of the majority of the cellulose and a significant fraction of hemicelluloses, but the majority of the lignin remained in solution [56]. The recovered cellulose could be readily hydrolyzed using commercial enzymes at moderate loading levels and generated >90% glucan yields in less than 24 hours, much greater than those yields associated with dilute sulfuric acid pretreatment of the same feedstock [15]. In tandem with other reports, this study has confirmed that, under appropriate conditions, IL pretreatment of switchgrass can be very effective in terms of enhancing the rate of saccharification relative to other pretreatment approaches [15,17,56,58].

Similar results were obtained with miscanthus in a recent paper that evaluated several different hydrophilic imidazolium- and pyridinium-based ILs to determine solubility parameters [59]. Of the 14 different ILs studied, acetate-, chloride- and phosphate-based ILs showed promise. The top performers were [C₁mim][OAc], [C₂mim][OAc], and [C₄mim][OAc] that could dissolve 4–5 wt% of miscanthus. The particle size, presence of water (either from biomass or adsorbed by IL from the atmosphere), process temperature and time, and type of IL were found to be the major determining factors in the rate of biomass dissolution. Based on the empirical data generated, the team used Abraham solvation parameters to develop a correlation that could be used to predict solubility using the quantum chemical model Conductor-like Screening Model–Real Solvents (COSMO-RS). ILs having high hydrogen bond acceptor strength and high polarity more efficiently solubilized miscanthus, in particular those with acetate, chloride and phosphate anions, while other ILs showed little miscanthus solubilization [59]. This approach is an example of the importance of developing predictive computational tools based on validated sets of data, which are capable of screening and selecting novel anions and cations for optimized pretreatment performance.

11.3.2   Agricultural Residues

Agricultural residues, such as corn stover [60], wheat straw [61], sugarcane bagasse [62], and rice straw [63], represent a significant amount of biomass that is potentially available annually for conversion to biofuels on a renewable and sustainable basis. The ILs [C_2mim][OAc] [13] and [C_4mim]Cl [64] have been used to efficiently pretreat corn stover. The general results of reduced lignin and decreased cellulose crystallinty seen in herbaceous materials is also observed for corn stover. It has been shown that [C_2mim][OAc] can effectively pretreat corn stover and provide a fractionated output that can be tailored based on the selection of the anti-solvent as well as avoiding formation of a gel phase during product recovery [13]. Using a series of washes, the residual IL remaining in the biomass was reduced to 0.2 wt%, which is below the inhibitory concentration for yeast fermentation [16]. Approximately 77% of the total carbohydrates could be recovered using this approach, and the recycled ILs were suitable for use in another batch pretreatment [13]. Another report compared the impact of ammonia fiber expansion (AFEX) and [C_2mim][OAc] pretreatment on corn stover [14]. For both pretreatments, more than 70% of theoretical sugar yields were obtained after 48 h using commercially available enzyme cocktails, but the IL pretreatment required less enzyme loading and a shorter hydrolysis time to reach 90% yields. The addition of hemicellulases to the commercial enzyme mixture significantly improved sugar yields obtained for the AFEX process [14].

There was a recent report that combined an ammonia steeping pretreatment with IL pretreatment for the conversion of rice straw into fermentable sugars. Four different ILs were screened, and [C_2mim][OAc] was determined to be the top performer [65]. Results from IL pretreatment alone were then compared with another route in which the rice straw was first treated with ammonia followed by IL pretreatment. The combined IL and ammonia pretreatment recovered 82% of the cellulose with 97% glucose yield, as compared to 84% and 52% for the ammonia or 79% and 76% for the IL pretreatments alone. It was also demonstrated that the IL could be recycled at least 20 times for reuse, a critical aspect in developing a cost-competitive IL pretreatment technology [65]. Another issue is the impact of any residual IL on downstream fermentation and microbial toxicity.

There have been several studies on the pretreatment of wheat straw using a variety of ILs [66–70]. One detailed parametric study used [C_2mim][OAc] for pretreatment of wheat straw using a central composite design, where the variables were temperature, (130–170 °C), time (0.5–5.5 h), and [C_2mim][OAc] concentration (0–100%) [66]. The optimum pretreatment conditions were identified as 158 °C, 49.5% (wt./wt.) [C_2mim][OAc] in water, and 3.6 h that produced 60% sugar yields out of a calculated maximum of 71% [66]. This study highlighted the possibility of using co-solvents to decrease the amount of IL required for effective pretreatment, but in this particular case with lower sugar yields than would be desired for a biorefinery setting. Another study reported the use of 1-ethyl-3-methylimidazolium diethylphosphate ([C_2mim] [DEP]) for pretreatment of wheat straw [69]. IL pretreatment at 130 °C for 30 min produced glucose yields of 54.8% after 12 h of saccharification. The hydrolysates generated were fermented into ethanol using a commercial strain of *Saccharomyces cerevisiae* and reached production levels of 0.43 g/g glucose after 26 h with minimal inhibition observed [69].

11.3.3   Woody Biomass

A good deal of research and development has been conducted into the ability of ILs to process woody biomass, including both softwoods and hardwoods; numerous reports highlight the ability of several ILs to solubilize woody biomass under certain process conditions [45,52,71–83]. For example, the ILs [C_4mim]Cl and 1-allyl-3-methylimidazolium chloride ([C_3mim]Cl) can efficiently solubilize Norway spruce sawdust and pulp derived from southern pine fibers with extensive acetylation [72]. In contrast, the IL1-benzyl-3-methylimidazolium chloride ([Armim]Cl) could form transparent amber solutions from these same woody feedstocks, and the enhanced solubilization of lignin observed for this IL was hypothesized to be due to the presence of the aromatic benzyl ring in the cation and the resultant π-π ring interactions. This highlights that the "tunability" of IL pretreatment performance and chemistry are directly related to the selection of the anion and cation. In both cases, the addition of an anti-solvent generated a product that was efficiently digested by a commercial cellulase cocktail [72].

An extensive study on the impact of the anion in the solubilization of *Pinus radiata* was conducted where the cation was set to [C_4mim]$^+$, and the anions were varied to include trifluoromethanesulfonate, methylsulfate, dimethylphosphate, dicyanamide, chloride, and acetate [74]. It was shown that the anion has a profound impact on the swelling and solvation of pine wood blocks, and that process parameters including water content, temperature, and viscosity all played critical roles in solubilization efficiency. Kamlet–Taft parameters were measured for all of the ILs studied in order to determine the impact of IL polarity, and it was determined that hydrogen bond basicity (β) of the anion correlated with dissolution of pine [74]. It has been hypothesized that the aromatic components in lignin form π-π stacking interactions with the imidazolium cation and that this, combined with the hydrogen bond basicity of the IL, forms a very effective biomass pretreatment solvent.

Another report investigated the impact of [C_2mim][OAc] on *Eucalyptus globulus* [75]. Two-dimensional NMR results indicated that extensive acetylation of xylan, acetylation of the lignin units, selective removal of guaiacyl units (increasing the syringyl:guaiacyl ratio), and decreased β-ether content were among the most dominant changes after IL pretreatment, although complete solubilization of the material did not occur. X-ray diffraction (XRD) measurements indicated that the cellulose recovered was the cellulose II polymorph, and there was a significant enhancement of saccharification efficiency and sugar yield after IL pretreatment [75].

Western red cedar (softwood) and Japanese beech (hardwood) were pretreated using [C_2mim]Cl, and there were significant differences in the dissolution behavior and the composition of the recovered biomass between these two feedstocks. For this pairing of IL and feedstocks, it was reported that the polysaccharides were the first species to be solubilized, followed by the lignin. Both the polysaccharides and the lignin were observed to have a reduction in the molecular weights of the solubilized species as a function of process time and temperature. Gel permeation chromatography (GPC) and gas chromatography-mass spectrometry (GC-MS) indicated that monomeric sugars including mannose, galactose, glucose, arabinose, and xylose were also present. Japanese beech was solubilized to a greater extent than Western red cedar and was attributed to the relatively higher levels of β-O-4 linkages present [79].

11.4   Ionic Liquid Pretreatment: Catalytic Route

In addition to the biochemical approach of liberating fermentable sugars from biomass, another conversion approach (Figure 11.2b) that can be taken is based on the unique chemical environment these ILs provide, coupled with the dissolution of the biopolymers, in order to catalytically hydrolyze the polysaccharides present into fermentable sugars without enzymes. These approaches are particularly intriguing as they eliminate one of the most costly operational expenses within a biorefinery: enzymes [84]. Another compelling feature of these systems is that they can employ biphasic aqueous systems for IL [85,86] and sugar recovery; these systems are tunable and can enable efficient recovery and recycle of the IL with no energy-intensive processes such as distillation. These approaches do however require the development of an efficient and cost-effective means of recovering the sugars and/or fuels from the IL-aqueous mixtures, as well as a means to recover the lignin and IL in order to be cost-competitive and prevent any downstream contamination.

Most of the approaches used in the scientific literature to date are based on variants of anion exchange/capture and/or "desalting" the aqueous-IL mixtures [87], neither of which are commercially viable at a scale relevant for significant biofuel production. Techniques that use liquid-liquid extraction of sugars into a phase that is immiscible with the hydrophilic IL, such as boronic acids in an organic phase [88], may address these challenges. Another limitation of this conversion approach is that it has focused primarily on model substrates (e.g., MCC) and not feedstocks that would be available within a biorefinery setting.

11.4.1   Acid-catalyzed Hydrolysis

There have been several reports on the use of mineral and/or solid acids in selected ILs such as [C$_2$mim]Cl to hydrolyze polysaccharides and produce fermentable sugars and biofuels [87,89–91]. The seminal work by Binder and Raines in 2010 demonstrated that acidolysis (10% w/w HCl) of corn stover in the presence of [C$_2$mim]Cl could generate high yields (71% xylose and 42% glucose) of monosaccharides, and that after recovery these sugars could be efficiently fermented into biofuels [87]. One critical discovery was that the rate and amount of water addition must be carefully controlled in order to achieve high yields of fermentable sugars. Another paper used [C$_1$mim]Cl and HCl to pretreat and hydrolyze three wood species: *Eucalyptus grandis*, southern pine, and Norway spruce pulp. In addition to generating monosaccharides, the recovered IL contained the degradation products 5-hydroxymethylfurfural (HMF), furan-2-carboxylic acid, catechol, methylcatechol, methylguaiacol, acetoguaiacone, and acetol [89]. In a related approach, albeit focused on a different component of plant cell walls, another group demonstrated the cleavage of lignin using the phosphonium-based IL, trihexyltetradecyl phosphonium chloride, and very small amounts of mineral acid on sugarcane bagasse. By varying the temperature from 120 to 150 °C and monitoring the production of ketones from the cleavage of β-aryl ethers, maximum delignification (52%) was found to occur at 150 °C. The presence of the mineral acid was essential for delignification to occur, and it was hypothesized that this IL solubilized lignin fragments but not intact lignin [91].

11.4.2   Metal-catalyzed Hydrolysis

Catalysts based on transition metals have been demonstrated as a means of converting biomass into sugars and/or converting sugars into fuels and chemical intermediates, such as HMF, in the presence of ILs. One system that has achieved considerable attention in the literature is metal chlorides [92,93]. Some initial reports compared the performance of CuCl$_2$ and CrCl$_2$ dissolved in [C$_2$mim]Cl in converting cellulose to HMF at operating temperatures of 80–120 °C [94]. This single-step catalytic process could convert cellulose to HMF at $55.4 \pm 4.0\%$ yields, and the HMF could be selectively extracted and the [C$_2$mim]Cl recovered [94]. The use of metal chloride catalysts has recently been extended to pairing between a primary catalyst (CuCl$_2$) with a second metal chloride (CrCl$_2$, PdCl$_2$, CrCl$_3$ or FeCl$_3$) dissolved in [C$_2$mim]Cl . This combination can substantially accelerate the rate of cellulose depolymerization and conversion under mild conditions (80–120 °C) as compared to single metal chloride systems [95]. In addition to the conversion of cellulose via metal chlorides in ILs, it has also been demonstrated that metal chlorides can also break down lignin in the presence of ILs. Using [C$_4$mim]Cl, metal chlorides (FeCl$_3$, CuCl$_2$, and AlCl$_3$), and water, the hydrolytic cleavage of β-O-4 ether bonds in model lignin was demonstrated [96]. AlCl$_3$ was more efficient in the cleavage of the β-O-4 compared to FeCl$_3$ and CuCl$_2$. After 120 min at 150 °C, *c.* 70% of the β-O-4 bonds present were hydrolyzed in the presence of FeCl$_3$ and CuCl$_2$ while *c.* 80% of β-O-4 were hydrolyzed in the presence of AlCl$_3$ [96]. In addition to metal chlorides, other catalytic systems have been reported. Mesoporous carbon materials functionalized with Ru/CMK-3 were found to be water-tolerant for the hydrolysis of cellulose to glucose with high yield [97]. The layered transition-metal oxide HNbMoO$_6$ was demonstrated to exhibit remarkable catalytic performance for the hydrolysis of polysaccharides (cellobiose, starch, and cellulose) due to accessibility into the strong acidic interlayer of the solid catalyst [98]. Another important aspect of using these catalysts that has not been substantially addressed is their potential role in downstream poisoning of the fermentation organisms.

11.5 Factors Impacting Scalability and Cost of Ionic Liquid Pretreatment

From a performance-based perspective, it is clear that certain ILs have considerable positive attributes as they relate to biomass pretreatment when compared to other pretreatment approaches. The ability to handle a wide range of feedstocks and achieve high sugar yields at low biomass loading and mild process conditions are some of the most important advantages. The primary challenges for ILs are those associated with cost and scalable technology deployment. A thorough systems-level understanding of critical factors that underpin the techno-economics of IL pretreatment must be obtained before scalable commercialization can occur [19]. In order for ILs to be truly competitive and scalable, the following issues must be addressed.

- **Cost of Ionic Liquids.** Of all of the parameters to be discussed, the most challenging to address is the cost of the ILs themselves. The most effective ILs known to date for biomass pretreatment are those based on the imidazolium cations, which are some of the more expensive ILs at the time of writing. Although significant strides have been made, and some of these ILs are now available for under $50/kg, more cost reductions (to a level of *c.* $1–2/kg) are necessary before they can be considered commercially viable for the biofuels industry. This also underscores the need for more exploratory efforts for new anions/cations, alternative IL pretreatment processes, and efficient IL recovery and recycle.
- **Biomass Loading.** Increased biomass loading is a key factor in the overall economics of a biorefinery. The efficiency of certain ILs to partially or fully solubilize biomass may indicate that these solvents could retain their pretreatment efficacy at high biomass loadings. A recent report demonstrated that [C_2mim] [OAc] could efficiently pretreat corn stover at loadings as high as 50% (w/w) [99]. Even at this very high biomass loading, after pretreatment for only 1 h at 125 °C, cellulose crsytallinity was decreased by up to 52% and lignin was reduced by up to 44%. Yields of fermentable sugar reached *c.* 80% for glucose and *c.* 50% for xylose at corn stover loadings of 33% (w/w), with yields of *c.* 55% for glucose and *c.* 34% for xylose at 50% (w/w) biomass loading. Most importantly, similar results were observed for switchgrass, poplar, and maple [99]. These promising results indicate that it may be possible to substantially decrease the amount of IL needed on a w/w basis to effectively pretreat biomass and generate high yields of fermentable sugars.
- **Ionic Liquid Recovery and Recycle.** Due to their higher cost, ILs must be recovered and recycled in order to develop a cost-competitive technology. Although the importance of this aspect of IL biomass pretreatment is understood by almost everyone working in the field, it remains one of the least studied aspects of the technology. One approach was demonstrated that used several different distillation steps to recover the IL, but this approach would likely result in an energy demand that surpasses the amount of energy that could be generated from the biomass [13]. There have also been reports on the development of novel ILs that are easier to recover through distillation, such as 1,1,3,3-tetramethylguanidine (TMG) coupled with carboxylic acids such as formic, acetic, and propionic acids [100]. Another approach is to use aqueous kosmotropic salt solutions that form a three-phase system of the biomass solids and IL-rich and salt-rich phases [101]. This process reduces the amount of water that must be removed through distillation and has the added benefit of higher yields in the conversion of cellulose to glucose when compared to cellulose obtained from biomass pretreated with IL and precipitated with water [101]. IL recovery is also critical to minimize the impacts of residual ILs on the downstream processes such as fermentation.

11.6   Concluding Remarks

ILs are a fascinating and exciting area of new scientific discovery with multiple applications in research and industry, including biomass pretreatment. IL pretreatment is a relatively new technique that has seen significant advances in less than a decade of research and development. This technology will require substantial development before a commercially viable and scalable process can be fully realized. IL pretreatment processes may offer several advantages over other conventional biomass pretreatments in terms of converting crystalline cellulose to an amorphous high-surface-area substrate that, when combined with delignification, provides pretreatment technology that can efficiently process a wide range of lignocellulosic feedstocks. The availability of inexpensive ILs, the development of additional process technologies such as *in situ* chemical or biochemical hydrolysis, and efficient recovery and recycling of ILs are all critical factors that will ultimately decide the fate of this innovative approach to biomass pretreatment.

Acknowledgements

The work conducted by the Joint BioEnergy Institute is supported by the Office of Science, Office of Biological and Environmental Research of the US Department of Energy under contract DE-AC02-05CH11231.

References

1. Beckham, G.T., Matthews, J.F., Peters, B. *et al.* (2011) Molecular-level origins of biomass recalcitrance: decrystallization free energies for four common cellulose polymorphs. *Journal of Physical Chemistry B*, **115** (14), 4118–4127.
2. Gross, A.S. and Chu, J.-W. (2010) On the molecular origins of biomass recalcitrance: the interaction network and solvation structures of cellulose microfibrils. *Journal of Physical Chemistry B*, **114** (42), 13333–13341.
3. Pu, Y., Jiang, N., and Ragauskas, A.J. (2007) Ionic liquid as a green solvent for lignin. *Journal of Wood Chemistry and Technology*, **27** (1), 23–33.
4. Brodeur, G., Yau, E., Badal, K. *et al.* (2011) Chemical and physicochemical pretreatment of lignocellulosic biomass: a review. *Enzyme Research*, **2011**, article ID 787532, 17.
5. Simmons, B.A., Singh, S., Holmes, B.M., and Blanch, H.W. (2010) Ionic liquid pretreatment. *Chemical Engineering Progress*, **106** (3), 50–55.
6. Tadesse, H. and Luque, R. (2011) Advances on biomass pretreatment using ionic liquids: an overview. *Energy & Environmental Science*, **4** (10), 3913–3929.
7. Graenacher, C. (1934) Cellulose Solution. US Patent 1,943,176.
8. Swatloski, R.P., Spear, S.K., Holbrey, J.D., and Rogers, R.D. (2002) Dissolution of cellulose with ionic liquids. *Journal of the American Chemical Society*, **124** (18), 4974–4975.
9. Dadi Anantharam, P., Schall Constance, A., and Varanasi, S. (2007) Mitigation of cellulose recalcitrance to enzymatic hydrolysis by ionic liquid pretreatment. *Applied Biochemistry and Biotechnology*, **137–140** (1–12), 407–421.
10. Dadi Anantharam, P., Varanasi, S., and Schall Constance, A. (2006) Enhancement of cellulose saccharification kinetics using an ionic liquid pretreatment step. *Biotechnology and Bioengineering*, **95** (5), 904–910.
11. Zhao, H., Jones, C.L., Baker, G.A. *et al.* (2009) Regenerating cellulose from ionic liquids for an accelerated enzymatic hydrolysis. *Journal of Biotechnology*, **139** (1), 47–54.
12. Cheng, G., Varanasi, P., Li, C. *et al.* (2011) Transition of cellulose crystalline structure and surface morphology of biomass as a function of ionic liquid pretreatment and its relation to enzymatic hydrolysis. *Biomacromolecules*, **12** (4), 933–941.
13. Dibble, D.C., Li, C., Sun, L. *et al.* (2011) A facile method for the recovery of ionic liquid and lignin from biomass pretreatment. *Green Chemistry*, **13** (11), 3255–3264.

14. Li, C., Cheng, G., Balan, V. *et al.* (2011) Influence of physico-chemical changes on enzymatic digestibility of ionic liquid and AFEX pretreated corn stover. *Bioresource Technology*, **102** (13), 6928–6936.

15. Li, C., Knierim, B., Manisseri, C. *et al.* (2010) Comparison of dilute acid and ionic liquid pretreatment of switchgrass: Biomass recalcitrance, delignification and enzymatic saccharification. *Bioresource Technology*, **101** (13), 4900–4906.

16. Ouellet, M., Datta, S., Dibble, D.C. *et al.* (2011) Impact of ionic liquid pretreated plant biomass on Saccharomyces cerevisiae growth and biofuel production. *Green Chemistry*, **13** (10), 2743–2749.

17. Bokinsky, G., Peralta-Yahya, P.P., George, A. *et al.* (2011) Synthesis of three advanced biofuels from ionic liquid-pretreated switchgrass using engineered Escherichia coli. Proceedings of the National Academy of Sciences of the United States of America, Nov. 28 2011, pp. 1–6.

18. Blanch, H.W., Simmons, B.A., and Klein-Marcuschamer, D. (2011) Biomass deconstruction to sugars. *Biotechnology Journal*, **6** (9), 1086–1102.

19. Klein-Marcuschamer, D., Simmons, B.A., and Blanch, H.W. (2011) Techno-economic analysis of a lignocellulosic ethanol biorefinery with ionic liquid pre-treatment. *Biofuels, Bioproducts & Biorefining*, **5** (5), 562–569.

20. Izgorodina, E.I., Forsyth, M., and MacFarlane, D.R. (2009) On the components of the dielectric constants of ionic liquids: ionic polarization? *Physical Chemistry Chemical Physics*, **11** (14), 2452–2458.

21. Krossing, I., Slattery John, M., Daguenet, C. *et al.* (2006) Why are ionic liquids liquid? A simple explanation based on lattice and solvation energies. *Journal of the American Chemical Society*, **128** (41), 13427–13434.

22. Bentley, T.W., Ebdon, D.N., Kim, E.-J., and Koo, I.S. (2005) Solvent polarity and organic reactivity in mixed solvents: evidence using a reactive molecular probe to assess the role of preferential solvation in aqueous alcohols. *Journal of Organic Chemistry*, **70** (5), 1647–1653.

23. Mashraqui, S.H., Subramanian, S., and Bhasikuttan, A.C. (2007) New ICT probes: synthesis and photophysical studies of N-phenylaza-15-crown-5 aryl/heteroaryl oxadiazoles under acidic condition and in the presence of selected metal ions. *Tetrahedron*, **63** (7), 1680–1688.

24. Poole, C.F. (2004) Chromatographic and spectroscopic methods for the determination of solvent properties of room temperature ionic liquids. *Journal of Chromatography, A*, **1037** (1–2), 49–82.

25. Yuan, M.-S., Liu, Z.-Q., and Fang, Q. (2007) Donor-and-acceptor substituted truxenes as multifunctional fluorescent probes. *Journal of Organic Chemistry*, **72** (21), 7915–7922.

26. Dear, G.J., Plumb, R.S., Sweatman, B.C. *et al.* (2000) Use of directly coupled ion-exchange liquid chromatography-mass spectrometry and liquid chromatography-nuclear magnetic resonance spectroscopy as a strategy for polar metabolite identification. *Journal of Chromatography, B: Biomedical Sciences and Applications*, **748** (1), 295–309.

27. Noij, T.H.M. and van der Kooi, M.M.E. (1995) Automated analysis of polar pesticides in water by online solid phase extraction and gas chromatography using the co-solvent effect. *Journal of High Resolution Chromatography*, **18** (9), 535–539.

28. Street, K.W. Jr. and Acree, W.E. Jr. (1986) The Py solvent polarity scale: binary solvent mixtures used in reversed-phase liquid chromatography. *Journal of Liquid Chromatography*, **9** (13), 2799–2808.

29. Kamlet, M.J. and Taft, R.W. (1976) The solvatochromic comparison method. I. The beta -scale of solvent hydrogen-bond acceptor (HBA) basicities. *Journal of the American Chemical Society*, **98** (2), 377–383.

30. Taft, R.W. and Kamlet, M.J. (1976) The solvatochromic comparison method. 2. The alpha -scale of solvent hydrogen-bond donor (HBD) acidities. *Journal of the American Chemical Society*, **98** (10), 2886–2894.

31. Yokoyama, T., Taft, R.W., and Kamlet, M.J. (1976) The solvatochromic comparison method. 3. Hydrogen bonding by some 2-nitroaniline derivatives. *Journal of the American Chemical Society*, **98** (11), 3233–3237.

32. Lungwitz, R., Strehmel, V., and Spange, S. (2010) The dipolarity/polarizability of 1-alkyl-3-methylimidazolium ionic liquids as function of anion structure and the alkyl chain length. *New Journal of Chemistry*, **34** (6), 1135–1140.

33. Ab Rani, M.A., Brant, A., Crowhurst, L. *et al.* (2011) Understanding the polarity of ionic liquids. *Physical Chemistry Chemical Physics*, **13** (37), 16831–16840.

34. Doherty, T.V., Mora-Pale, M., Foley, S.E. *et al.* (2010) Ionic liquid solvent properties as predictors of lignocellulose pretreatment efficacy. *Green Chemistry*, **12** (11), 1967–1975.

35. Lee, J.-M., Ruckes, S., and Prausnitz John, M. (2008) Solvent polarities and kamlet-taft parameters for ionic liquids containing a pyridinium cation. *The Journal of Physical Chemistry B*, **112** (5), 1473–1476.
36. Lindman, B., Karlstroem, G., and Stigsson, L. (2010) On the mechanism of dissolution of cellulose. *Journal of Molecular Liquids*, **156** (1), 76–81.
37. Bochek, A.M., Petropavlovsky, G.A., and Kallistov, O.V. (1993) Dissolution of cellulose and its derivatives in the same solvent, methylmorpholine-N-oxide and the properties of the resulting solutions. *Cellulose Chemistry and Technology*, **27** (2), 137–144.
38. Eckelt, J., Eich, T., Roeder, T. *et al.* (2009) Phase diagram of the ternary system NMMO/water/cellulose. *Cellulose (Dordrecht, Netherlands)*, **16** (3), 373–379.
39. Hudson, S.M. and Cuculo, J.A. (1980) The solubility of cellulose in liquid ammonia/salt solutions. *Journal of Polymer Science, Polymer Chemistry Edition*, **18** (12), 3469–3481.
40. Hudson, S.M., Cuculo, J.A., and Wadsworth, L.C. (1983) The solubility of cellulose in liquid ammonia/ammonium thiocyanate solution: the effect of composition and temperature on dissolution and solution properties. *Journal of Polymer Science, Polymer Chemistry Edition*, **21** (3), 651–670.
41. Scherer, P.C. Jr. and Hussey, R.E. (1931) Action of sodium on cellulose in liquid ammonia. *Journal of the American Chemical Society*, **53**, 2344–2347.
42. Conte, P., Maccotta, A., De Pasquale, C. *et al.* (2009) Dissolution mechanism of crystalline cellulose in H3PO4 as assessed by high-field NMR spectroscopy and fast field cycling NMR relaxometry. *Journal of Agricultural and Food Chemistry*, **57** (19), 8748–8752.
43. Zhang, Y.H.P. and Lynd Lee, R. (2005) Determination of the number-average degree of polymerization of cellodextrins and cellulose with application to enzymatic hydrolysis. *Biomacromolecules*, **6** (3), 1510–1515.
44. Bentivoglio, G., Roeder, T., Fasching, M. *et al.* (2006) Cellulose processing with chloride-based ionic liquids. *Lenzinger Berichte*, **86**, 154–161.
45. Cuissinat, C., Navard, P., and Heinze, T. (2008) Swelling and dissolution of cellulose. Part IV: Free floating cotton and wood fibres in ionic liquids. *Carbohydrate Polymers*, **72** (4), 590–596.
46. Kosan, B., Dorn, S., Meister, F., and Heinze, T. (2010) Preparation and subsequent shaping of cellulose acetates using ionic liquids. *Macromolecular Materials and Engineering*, **295** (7), 676–681.
47. Remsing, R.C., Swatloski, R.P., Rogers, R.D., and Moyna, G. (2006) Mechanism of cellulose dissolution in the ionic liquid 1-n-butyl-3-methylimidazolium chloride: a 13C and 35/37Cl NMR relaxation study on model systems. *Chemical Communications (Cambridge, United Kingdom)*, (12), 1271–1273.
48. Liu, Z., Remsing, R.C., Moore, P.B., and Moyna, G. (2007) Molecular dynamics study of the mechanism of cellulose dissolution in the ionic liquid 1-n-butyl-3-methylimidazolium chloride. *ACS Symposium Series*, **975** (Ionic Liquids IV), 335–350.
49. Zhang, J., Zhang, H., Wu, J. *et al.* (2010) NMR spectroscopic studies of cellobiose solvation in EmimAc aimed to understand the dissolution mechanism of cellulose in ionic liquids. *Physical Chemistry Chemical Physics*, **12** (8), 1941–1947.
50. Gross, A.S., Bell, A.T., and Chu, J.-W. (2011) Thermodynamics of cellulose solvation in water and the ionic liquid 1-butyl-3-methylimidazolim chloride. *Journal of Physical Chemistry B*, **115** (46), 13433–13440.
51. Liu, H., Sale Kenneth, L., Simmons Blake, A., and Singh, S. (2011) Molecular dynamics study of polysaccharides in binary solvent mixtures of an ionic liquid and water. *The Journal of Physical Chemistry B*, **115** (34), 10251–10258.
52. Lee, S.H., Doherty, T.V., Linhardt, R.J., and Dordick, J.S. (2009) Ionic liquid-mediated selective extraction of lignin from wood leading to enhanced enzymatic cellulose hydrolysis. *Biotechnology and Bioengineering*, **102** (5), 1368–1376.
53. George, A., Tran, K., Morgan, T.J. *et al.* (2011) The effect of ionic liquid cation and anion combinations on the macromolecular structure of lignins. *Green Chemistry*, **13** (12), 3375–3385.
54. Mitchell, R., Vogel, K.P., and Uden, D.R. (2012) The feasibility of switchgrass for biofuel production. *Biofuels*, **3** (1), 47–59.
55. Brown, J.C., Renvoize, S., Chiang, Y.-C. *et al.* (2011) Developing Miscanthus for bioenergy. *RSC Energy and Environment Series*, **3** (Energy Crops), 301–321.

56. Singh, S., Simmons Blake, A., and Vogel Kenneth, P. (2009) Visualization of biomass solubilization and cellulose regeneration during ionic liquid pretreatment of switchgrass. *Biotechnology and Bioengineering*, **104** (1), 68–75.

57. Samayam Indira, P. and Schall Constance, A. (2010) Saccharification of ionic liquid pretreated biomass with commercial enzyme mixtures. *Bioresource Technology*, **101** (10), 3561–3566.

58. Zhao, H., Baker Gary, A., and Cowins Janet, V. (2010) Fast enzymatic saccharification of switchgrass after pretreatment with ionic liquids. *Biotechnology Progress*, **26** (1), 127–133.

59. Padmanabhan, S., Kim, M., Blanch, H.W., and Prausnitz, J.M. (2011) Solubility and rate of dissolution for Miscanthus in hydrophilic ionic liquids. *Fluid Phase Equilibria*, **309** (1), 89–96.

60. Sheehan, J., Aden, A., Paustian, K. *et al.* (2004) Energy and environmental aspects of using corn stover for fuel ethanol. *Journal of Industrial Ecology*, **7** (3–4), 117–146.

61. Saha, B.C., Nichols, N.N., and Cotta, M.A. (2011) Ethanol production from wheat straw by recombinant Escherichia coli strain FBR5 at high solid loading. *Bioresource Technology*, **102** (23), 10892–10897.

62. Kimon, K.S., Alan, E.L., and Sinclair, D.W.O. (2011) Enhanced saccharification kinetics of sugarcane bagasse pretreated in 1-butyl-3-methylimidazolium chloride at high temperature and without complete dissolution. *Bioresource Technology*, **102** (19), 9325–9329.

63. Binod, P., Sindhu, R., Singhania, R.R. *et al.* (2010) Bioethanol production from rice straw: An overview. *Bioresource Technology*, **101** (13), 4767–4774.

64. Geng, X. and Henderson Wesley, A. (2012) Pretreatment of corn stover by combining ionic liquid dissolution with alkali extraction. *Biotechnology and Bioengineering*, **109** (1), 84–91.

65. Nguyen, T.-A.D., Kim, K.-R., Han, S.J. *et al.* (2010) Pretreatment of rice straw with ammonia and ionic liquid for lignocellulose conversion to fermentable sugars. *Bioresource Technology*, **101** (19), 7432–7438.

66. Fu, D. and Mazza, G. (2011) Optimization of processing conditions for the pretreatment of wheat straw using aqueous ionic liquid. *Bioresource Technology*, **102** (17), 8003–8010.

67. Fu, D., Mazza, G., and Tamaki, Y. (2010) Lignin extraction from straw by ionic liquids and enzymatic hydrolysis of the cellulosic residues. *Journal of Agricultural and Food Chemistry*, **58** (5), 2915–2922.

68. Guragain, Y.N., De Coninck, J., Husson, F. *et al.* (2011) Comparison of some new pretreatment methods for second generation bioethanol production from wheat straw and water hyacinth. *Bioresource Technology*, **102** (6), 4416–4424.

69. Li, Q., He, Y.-C., Xian, M. *et al.* (2009) Improving enzymatic hydrolysis of wheat straw using ionic liquid 1-ethyl-3-methyl imidazolium diethyl phosphate pretreatment. *Bioresource Technology*, **100** (14), 3570–3575.

70. Pezoa, R., Cortinez, V., Hyvarinen, S. *et al.* (2010) Use of ionic liquids in the pretreatment of forest and agricultural residues for the production of bioethanol. *Cellulose Chemistry and Technology*, **44** (4–6), 165–172.

71. Xin, Q., Pfeiffer, K., Prausnitz, J.M. *et al.* (2012) Extraction of lignins from aqueous-ionic liquid mixtures by organic solvents. *Biotechnology and Bioengineering*, **109** (2), 346–352.

72. Kilpelaeinen, I., Xie, H., King, A. *et al.* (2007) Dissolution of wood in ionic liquids. *Journal of Agricultural and Food Chemistry*, **55** (22), 9142–9148.

73. Bodirlau, R., Teaca, C.-A., and Spiridon, I. (2010) Enzymatic hydrolysis of Asclepias syriaca fibers in the presence of ionic liquids. *Monatshefte fuer Chemie*, **141** (9), 1043–1048.

74. Brandt, A., Hallett, J.P., Leak, D.J. *et al.* (2010) The effect of the ionic liquid anion in the pretreatment of pine wood chips. *Green Chemistry*, **12** (4), 672–679.

75. Cetinkol, O.P., Dibble, D.C., Cheng, G. *et al.* (2010) Understanding the impact of ionic liquid pretreatment on eucalyptus. *Biofuels*, **1** (1), 33–46.

76. Chen, C. and Li, J. (2010) Synthesis of ionic liquid and its application in Duabanga grandiflora wood powder as green solvent. *Advanced Materials Research (Zuerich, Switzerland)*, **113–116** (Pt. 1, Environment Materials and Environment Management), 407–411.

77. Li, B., Asikkala, J., Filpponen, I., and Argyropoulos, D.S. (2010) Factors affecting wood dissolution and regeneration of ionic liquids. *Industrial & Engineering Chemistry Research*, **49** (5), 2477–2484.

78. Lucas, M., MacDonald, B.A., Wagner, G.L. *et al.* (2010) Ionic liquid pretreatment of poplar wood at room temperature: swelling and incorporation of nanoparticles. *ACS Applied Materials & Interfaces*, **2** (8), 2198–2205.

79. Nakamura, A., Miyafuji, H., and Saka, S. (2010) Liquefaction behavior of Western red cedar and Japanese beech in the ionic liquid 1-ethyl-3-methylimidazolium chloride. *Holzforschung*, **64** (3), 289–294.

80. Shamsuri, A.A. and Abdullah, D.K. (2010) Isolation and characterization of lignin from rubber wood in ionic liquid medium. *Modern Applied Science*, **4** (11), 19–27.

81. Xie, H. and Shi, T. (2010) Liquefaction of wood (Metasequoia glyptostroboides) in allyl alkyl imidazolium ionic liquids. *Wood Science and Technology*, **44** (1), 119–128.

82. Yamashita, Y., Sasaki, C., and Nakamura, Y. (2010) Effective enzyme saccharification and ethanol production from Japanese cedar using various pretreatment methods. *Journal of Bioscience and Bioengineering*, **110** (1), 79–86.

83. Viell, J. and Marquardt, W. (2011) Disintegration and dissolution kinetics of wood chips in ionic liquids: 11th EWLP, Hamburg, Germany, August 16–19, 2010. *Holzforschung*, **65** (4), 519–525.

84. Klein-Marcuschamer, D., Oleskowicz-Popiel, P., Simmons, B.A., and Blanch, H.W. (2012) The challenge of enzyme cost in the production of lignocellulosic biofuels. *Biotechnology and Bioengineering*, **109** (4), 1083–1087.

85. Louros, C.L.S., Claudio, A.F.M., Neves, C.M.S.S. *et al.* (2010) Extraction of biomolecules using phosphonium-based ionic liquids+K3PO4 aqueous biphasic systems. *International Journal of Molecular Sciences*, **11**, 1777–1791.

86. Melgarejo-Torres, R., Torres-Martinez, D., Castillo-Araiza, C.O. *et al.* (2012) Mass transfer coefficient determination in three biphasic systems (water-ionic liquid) using a modified Lewis cell. *Chemical Engineering Journal (Amsterdam, Netherlands)*, **181–182**, 702–707.

87. Binder, J.B. and Raines, R.T. (2010) Fermentable sugars by chemical hydrolysis of biomass. Proceedings of the National Academy of Sciences of the United States of America, Mar. 1 2010, pp. 1–6.

88. Brennan, T.C.R., Datta, S., Blanch, H.W. *et al.* (2010) Recovery of sugars from ionic liquid biomass liquor by solvent extraction. *Bioenergy Research*, **3** (2), 123–133.

89. Li, B., Filpponen, I., and Argyropoulos, D.S. (2010) Acidolysis of wood in ionic liquids. *Industrial & Engineering Chemistry Research*, **49** (7), 3126–3136.

90. Dee, S.J. and Bell, A.T. (2011) A study of the acid-catalyzed hydrolysis of cellulose dissolved in ionic liquids and the factors influencing the dehydration of glucose and the formation of humins. *ChemSusChem*, **4** (8), 1166–1173.

91. Keskar, S., Edye, L., Doherty, W.S., and Bartley, J. (2012) The chemistry of acid catalyzed delignification of sugarcane bagasse in the ionic liquid trihexyl tetradecyl phosphonium chloride. *Journal of Wood Chemistry and Technology*, **32** (1), 71–81.

92. Zhao, H., Holladay, J.E., Brown, H., and Zhang, Z.C. (2007) Metal chlorides in ionic liquid solvents convert sugars to 5-Hydroxymethylfurfural. *Science (Washington, DC, United States)*, **316** (5831), 1597–1600.

93. Wang, P., Yu, H., Zhan, S., and Wang, S. (2011) Catalytic hydrolysis of lignocellulosic biomass into 5-hydroxymethylfurfural in ionic liquid. *Bioresource Technology*, **102** (5), 4179–4183.

94. Su, Y., Brown, H.M., Huang, X. *et al.* (2009) Single-step conversion of cellulose to 5-hydroxymethylfurfural (HMF), a versatile platform chemical. *Applied Catalysis, A: General*, **361** (1–2), 117–122.

95. Su, Y., Brown, H.M., Li, G. *et al.* (2011) Accelerated cellulose depolymerization catalyzed by paired metal chlorides in ionic liquid solvent. *Applied Catalysis, A: General*, **391** (1–2), 436–442.

96. Jia, S.-Y., Cox, B.J., Guo, X.-W. *et al.* (2011) Hydrolytic cleavage of beta -O-4 ether bonds of lignin model compounds in an ionic liquid with metal chlorides. *Industrial & Engineering Chemistry Research*, **50** (2), 849–855.

97. Kobayashi, H., Komanoya, T., Hara, K., and Fukuoka, A. (2010) Water-tolerant mesoporous-carbon-supported ruthenium catalysts for the hydrolysis of cellulose to glucose. *ChemSusChem*, **3** (4), 440–443.

98. Takagaki, A., Tagusagawa, C., and Domen, K. (2008) Glucose production from saccharides using layered transition metal oxide and exfoliated nanosheets as a water-tolerant solid acid catalyst. *Chemical Communications*, (42), 5363–5365.

99. Wu, H., Mora-Pale, M., Miao, J. *et al.* (2011) Facile pretreatment of lignocellulosic biomass at high loadings in room temperature ionic liquids. *Biotechnology and Bioengineering*, **108** (12), 2865–2875.

100. King, A.W.T., Asikkala, J., Mutikainen, I. *et al.* (2011) Distillable acid-base conjugate ionic liquids for cellulose dissolution and processing. *Angewandte Chemie, International Edition*, **50** (28), S/1–S/5.

101. Shill, K., Padmanabhan, S., Xin, Q. *et al.* (2011) Ionic liquid pretreatment of cellulosic biomass: enzymatic hydrolysis and ionic liquid recycle. *Biotechnology and Bioengineering*, **108** (3), 511–520.

12

Comparative Performance of Leading Pretreatment Technologies for Biological Conversion of Corn Stover, Poplar Wood, and Switchgrass to Sugars

Charles E. Wyman[1,2], Bruce E. Dale[3], Venkatesh Balan[3], Richard T. Elander[4], Mark T. Holtzapple[5], Rocío Sierra Ramirez[5,*], Michael R. Ladisch[6,**], Nathan S. Mosier[6], Y.Y. Lee[7], Rajesh Gupta[8], Steven R. Thomas[9,§], Bonnie R. Hames[9,§§], Ryan Warner[10] and Rajeev Kumar[2]

[1] Department of Chemical and Environmental Engineering, University of California, Riverside, USA

[2] Center for Environmental Research and Technology, University of California, Riverside, USA

[3] Department of Chemical Engineering and Materials Science Michigan State University, East Lansing, USA

[4] National Renewable Energy Laboratory, Golden, USA

[5] Department of Chemical Engineering, Texas A&M University, College Station, USA

[6] Laboratory of Renewable Resources Engineering, Purdue University, West Lafayette, USA

[7] Department of Chemical Engineering, Auburn University, USA

[8] Chevron ETC, Houston, USA

[9] Ceres, Inc., Thousand Oaks, USA

[10] DuPont Industrial Biosciences, Palo Alto, USA

* Present address: Department of Chemical Engineering, University of the Andes, Bogota, Columbia
** Dr. Ladisch is at both Mascoma and Purdue
§ Present address: US Department of Energy, Golden, USA
§§ Present address: B. Hames Consulting, Newbury Park, USA

12.1   Introduction

Cellulosic biomass is the only resource from which liquid fuels so vital to transportation can be made at a large scale sustainably and also minimize the conflict with food production [1]. Furthermore, cellulosic biomass costing about $60/dry ton is competitive with petroleum at about $20/barrel on an equivalent energy content basis [2]. Therefore, the primary challenge to competitiveness is low-cost processing of cellulosic biomass to fuels, and biological routes can take advantage of the rapidly evolving tools of biotechnology to radically reduce costs [1,3]. The economic challenge for biological options is to overcome the natural resistance of plants to release of sugars that are the building blocks of cellulose and hemicellulose and typically comprise two-thirds to three-quarters of cellulosic materials. Although modification of plants to facilitate sugar release and better enzyme/organism combinations that can attack cellulosic materials more effectively are often favored to improve performance [1], to date, biomass pretreatment has been essential to achieving the high yields of sugars vital to economic success and is likely to remain an essential step in the overall conversion system [4]. At this point, we are faced with the conundrum that pretreatment is among the most expensive single steps in biological processing [5], but unit production costs are even higher if pretreatment is eliminated, due to the resulting low product yields [6]. It has therefore been stated that the "only step more expensive than pretreatment is no pretreatment" [4]. Full attention must be focused on developing lower-cost pretreatment technologies that can integrate with advanced biological conversion systems and perhaps take advantage of plants that have been modified to facilitate sugar release.

The Biomass Refining Consortium for Applied Fundamentals and Innovation (CAFI) was originally conceived in late 1999 in a meeting in Dallas, Texas among pretreatment leaders interested in working collaboratively. It was formally organized in early 2000 in a Chicago meeting among this team [7]. During the entire CAFI lifetime, the following goals were set:

- Develop data on leading pretreatments using common feedstocks, enzymes, analytical methods, material and energy balance methods and costing methods.
- Seek to understand mechanisms that influence performance and differentiate pretreatments by providing a technology base to facilitate commercial use and identifying promising paths to advance pretreatment technologies that achieve lower costs.
- Train and educate students in biomass conversion technologies.

A key objective was to provide information to help industry select technologies for commercial applications and not to "downselect" pretreatments. Rather, it was vital to provide extensive data on promising options so that others could decide which technologies to employ and avoid a lack of data clouding decisions.

The CAFI team was fortunate to be selected for funding by a new United States Department of Agriculture (USDA) Program called the Initiative for Future Agricultural and Food Systems (IFAFS) through a competitive solicitation released in the spring of 2000. Although the program unfortunately had a short life, the IFAFS approach was unique in that it funded large collaborative projects focused on advancing and applying biomass conversion technologies, consistent with the CAFI spirit. In this inaugural CAFI project that ran from 2000 to 2004, now designated CAFI 1, the emphasis was on comparative data from application of leading pretreatments to a shared source of corn stover, with most of the work focused on performance from just the pretreatment and enzymatic hydrolysis steps [7]. In 2004, the Office of the Biomass Program (OBP) of the US Department of Energy selected the CAFI team for a second project as a result of a competitive solicitation. This project, now known to our team as CAFI 2, applied most of the pretreatment technologies employed in CAFI 1 to poplar wood but with more data developed on enzymatic hydrolysis and fermentation [8]. Following completion of the CAFI 2 work, OBP funded the team to apply the same stable of pretreatment technologies to switchgrass. This latter CAFI 3 project was somewhat broader in scope than prior projects in that, in addition to determining yields from pretreatment and enzymatic hydrolysis, it included such aspects as yield changes over a wider range of enzyme loadings, the impact of enzyme formulation on performance, and the influence of feedstock variables including harvest season and geographic location on results [9].

Numerous features of the three CAFI projects were distinctive. First and essential to its success was the extremely cooperative/collaborative spirit of the CAFI team in working together to provide objective comparative data for leading pretreatment options that were viewed as competitive. Second was use of the same feedstocks and enzymes for all pretreatments, coupled with shared methods and material balance formats to facilitate performance comparisons. Another key attribute was the tracking and reporting of material balance data for each pretreatment to aid in making meaningful performance comparisons. Two vital attributes of the CAFI team were the experience of the investigators in biological conversion and particularly biomass pretreatment. An Agricultural and Industrial Advisory Board who met with the team at six-month intervals to review results was also very valuable in providing feedback, suggestions, and insights that helped guide team activities. Over the CAFI history, the board grew from initially four members to include representatives from the following organizations: Abengoa Bioenergy, Arkion Life Sciences, Aventine Renewable Energy, Cargill, Catchlight Energy, CEA, Ceres, ChevronTexaco, Codexis, DuPont, DDCE, Hercules, John Deere, Lallemand Ethanol Technology, Lignol, LS9, Mascoma, MBI, Mendel Biotechnology, the National Corn Growers Association, NorFalco Sales, Poet, Shell, Synthetic Genomics, the US Department of Agriculture, Verenium, and Weyerhaeuser, as well as Donald Johnson, a retired executive from Grain Processing Corporation.

The core pretreatment leaders of the CAFI projects had extensive expertise and experience in pretreatment and biological conversion, and each had developed technologies that promised high sugar yields at low cost. The pretreatment technologies and lead investigator for each were: ammonia fiber expansion (AFEX), Bruce Dale, Michigan State University; ammonia recycle percolation (ARP) and soaking in aqueous ammonia (SAA), YY Lee, Auburn University; dilute acid (DA), Charles Wyman, University of California, Riverside and Dartmouth College; sulfur dioxide, Jack Saddler, University of British Columbia and Charles Wyman, University of California, Riverside; liquid hot water (LHW), Michael Ladisch and Nathan Mosier, Purdue University; and lime, Mark Holtzapple, Texas A&M University. In addition, Richard Elander and Robert Torget from the National Renewable Energy Laboratory (NREL) along with assistance from Tim Eggeman of Neoterics International provided expertise in dilute acid pretreatment along with key logistical support and economic analyses. For CAFI 2 and 3, Ryan Warner and Colin Mitchinson of Genencor International, later a Danisco Division and now part of DuPont, provided enzymes, and Steve Thomas and then Bonnie Hames of Ceres Corporation ensured supply of three varieties of switchgrass for the CAFI 3 project.

This chapter primarily focuses on comparing maximum sugar yields from application of the CAFI pretreatments followed by enzymatic hydrolysis by baseline cellulase loadings supplemented with β-glucosidase for all three CAFI feedstock types. The chapter includes summaries of compositions of the CAFI feedstocks and information on the enzymes and methods employed for the three projects. The maximum yields of xylose and glucose and their oligomers achieved along with the corresponding compositions of pretreated solids and the combinations of temperatures, times, and catalyst concentrations and types applied to obtain those results are also indicated for each pretreatment with each feedstock. Some information is also given on differences in sugar yields for a second poplar source and two other types of switchgrass.

This chapter does not provide details about the different pretreatment reaction systems employed by the team as these varied somewhat among feedstocks and teams, making succinct summarization too difficult. However, references are provided to papers that report these details for those interested. In addition, Chapter 13 of this book gives more details on how changes in enzyme formulations and loadings impact yields of xylose and glucose and their oligomers. Chapter 15 provides more information on comparative economics of the CAFI pretreatments. It is important to note that any differences in data between this chapter and Chapter 15 for some technologies resulted from the need to perform the economic analysis in parallel with the CAFI research before identification of the best conditions was totally complete. However, any differences should have a small effect on the comparisons. Other information beyond that summarized here can be found in references cited throughout, with a collection of CAFI papers and the associated summaries in three special journal volumes providing more in-depth overviews of each of the three CAFI projects [8–12].

12.2 Materials and Methods

12.2.1 Feedstocks

The NREL procured corn stover for CAFI 1 from a single Harlan, IA source through BioMass AgriProducts. The stover was washed and dried in their commercial operation and then sent to NREL, where it was knife milled to pass through a 1/4-inch round screen and distributed to the team as needed [11].

For CAFI 2, the USDA Northern Research Station in Rhinelander, Wisconsin provided poplar wood. The initial wood was from an Arlington, WI site where it was planted most likely in the spring of 1995 and harvested and shipped in February 2004. However, when it was realized that the amount was insufficient to support the CAFI demand over the length of the project, a second larger tree was obtained for use as the standard feedstock from a farm in Alexandria, Minnesota. This poplar was planted in the spring of 1994 and harvested and shipped to NREL in August 2004. NREL debarked, chipped, and milled both materials at their site Golden, CO using a Mitts and Merrill Model 10×12 knife mill (Saginaw, MI) to pass through a 1/4-inch screen. The resulting materials were kept separate and stored frozen at $-20\,^\circ$C and distributed to CAFI members by NREL as needed [8].

Ceres Corporation provided Alamo, Dacotah, and Shawnee varieties of switchgrass for CAFI 3 [9,13]. Alamo switchgrass was a thick-stemmed southern lowland ecotype of 29° N latitude origin planted in Ardmore, OK (34° latitude and 870 ft elevation). Two batches of Alamo grass were used over the CAFI 3 lifetime: one planted in June 2005 and harvested in December 2006 and a second planted in June 2007 and harvested in November 2007. Shawnee switchgrass is a thin-stemmed northern upland ecotype with a 38° N latitude origin. It was planted in Stillwater, OK (36° N latitude and 960 ft elevation) in June 2005 and harvested in December 2006. Dacotah switchgrass is also a thin-stemmed northern upland ecotype with a 46° N latitude origin but planted in Pierre, SD (44° N latitude and 1420 ft elevation) in December 1999 and harvested in May 2008. All switchgrass was stored as small square bales, dried at $50\,^\circ$C to less than 10% moisture, and knife or hammer milled to a 2–6 mm size range.

The compositions of the corn stover, poplar, and switchgrass employed over the CAFI lifetime are summarized in Table 12.1. Both sources of poplar were distinctive in their higher glucan contents relative to corn stover and switchgrass, while being lower in the hemicellulose structural sugars arabinose, galactose, and xylose. However, the poplar material adopted as a standard was considerably higher in lignin and somewhat higher in ash and extractives than the initial tree, with a consequent reduction in carbohydrates. The baseline Dacotah switchgrass and corn stover were quite similar in polymeric sugar content (cellulose and hemicellulose), but corn stover was about 20% lower in lignin. Three of the switchgrasses contained significantly higher amounts of water-extractable sugars than the other biomass materials employed in the CAFI studies: 9.6% for Alamo 1, 6.9% for Alamo 2, and 8.2% for Shawnee. These values contrast with the sugar contents of 1.2% for corn stover and 0.8% for Dacotah switchgrass. For example, adjusting glucan, xylan, arabinan, lignin, acetyl, ash, and protein contents of Shawnee switchgrass to exclude these soluble sugars resulted in the respective average values also shown in parentheses in Table 12.1: 33.9, 21.8, 3.4, 21.5, 3.2, 4.6, and 4.7%. On this basis, the average structural carbohydrate content for switchgrass is less than 8% different from that in corn stover, while the lignin content is nearly 25% more than for corn stover. The ash content of switchgrass is about 40% less than for corn stover.

Table 12.1 *Compositions of corn stover, poplar, and switchgrass feedstocks used in three CAFI projects. Values in parentheses are adjusted to neglect soluble sugars; UD: undetermined; ND: not detected [8,9,11,13,14].*

Component	Amount, wt% dry basis						
	Corn stover	Standard poplar	Initial poplar	Alamo 1 switchgrass	Alamo 2 switchgrass	Dacotah switchgrass	Shawnee switchgrass
Glucan	36.1 (36.5)	43.8	45.1	29.9 (33.1)	30.9 (33.2)	35.3 (35.6)	30.9 (33.9)
Xylan	21.4 (21.7)	14.9	17.8	20.5 (22.7)	21.6 (23.2)	22.5 (22.7)	20.0 (21.8)
Arabinan	3.5 (3.5)	0.6	0.5	3.4 (3.8)	2.7 (2.9)	3.1 (3.1)	3.1 (3.4)
Mannan	1.8 (1.8)	3.9	1.7	ND	ND	ND	ND
Galactan	2.5 (2.5)	1.0	1.5	ND	ND	ND	ND
Total lignin*	17.2 (17.4)	29.1	21.4	18.8 (20.8)	19.5 (21.0)	22.6 (22.8)	19.7 (21.5)
Protein	4.0 (4.0)	UD	UD	5.4 (6.0)	4.5 (4.8)	1.2 (1.2)	4.3 (4.7)
Acetyl	3.2 (3.2)	3.6	5.7	2.2 (2.4)	2.8 (3.0)	3.6 (3.6)	2.9 (3.2)
Ash	7.1 (7.2)	1.1	0.8	3.9 (4.3)	4.3 (4.6)	3.3 (3.3)	4.2 (4.6)
Uronic acid	3.6 (3.6)	UD	UD	UD	UD	UD	UD
Extractives	UD	3.6	3.4	UD	UD	UD	UD
Soluble sugars	1.2			9.6	6.9	0.8	8.2

*Acid soluble plus insoluble lignin.

12.2.2 Enzymes

Spezyme CP cellulase was provided by Genencor, and supplementary β-glucosidase Novozyme 188 was purchased from Sigma Chemical Company for use in all three CAFI projects. In addition, GC-220 cellulase, Multifect Xylanase, and non-commercial β-glucosidase and beta-xylosidase formulations were provided by Genencor for CAFI 2 and 3. Genencor also provided Accellerase 1000, a whole broth cellulase, for CAFI 3. The baseline enzyme formulation for CAFI 2 and 3 was a combination of β-glucosidase and cellulase at an activity (CBU/FPU) ratio of 2.0. Again, more details are available in papers reporting results from each of the three CAFI projects and in Chapter 13 of this book.

12.2.3 CAFI Pretreatments

Each of the CAFI participants applied their respective pretreatment methods to process these three classes of feedstocks; because details of each method as applied to each feedstock are available in several papers published by the team, information on the pretreatment approaches will not be reported here. However, to facilitate an understanding of the data summarized in this chapter, it is important to note that some of the pretreatment technologies evolved during the three projects. Although improvements in the conditions, configurations, and systems were incorporated in AFEX, dilute sulfuric acid, LHW, and lime pretreatments, the basic concepts remained the same. It should be noted that some papers refer to liquid hot water pretreatment as neutral-pH and controlled-pH pretreatment. In addition, the University of British Columbia (UBC) applied sulfur dioxide pretreatment to corn stover and poplar, but the University of California, Riverside (UCR) applied sulfur dioxide to switchgrass when funding by Natural Resources Canada was unavailable to continue participation by UBC. When the economic analyses that supported the CAFI project determined that the high water use for ARP was detrimental to its competitiveness, Auburn University reduced water use by changing the process configuration to a pretreatment approach they termed SAA. As a result, early CAFI data is for application of ARP and not SAA, while later data is based on results with SAA and not ARP. Another notable difference was the evolution in reaction times for lime pretreatment from the long values (4 weeks; 50–60 °C) applied for CAFI 1 to shorter times (120–240 mins at 120–140 °C) for CAFI 2 and CAFI 3.

12.2.4   Material Balances

A key feature of all three CAFI projects was reporting of material balances for each of the key components to account for recovery of each. The CAFI team adapted the approach of factoring the actual amounts pre-treated to a basis of 100 mass units of dry biomass entering pretreatment, as illustrated in Figure 12.1, and reported the fraction of glucan and xylan that could be accounted for. Only glucan and xylan were tracked because the other sugars could not be measured accurately enough to allow meaningful material balances. It is recommended that all future pretreatment research apply a similar material balance format to facilitate ready comparisons of results.

12.2.5   Free Sugars and Extraction

Unlike corn stover or poplar, Alamo and Shawnee switchgrass contained high amounts of free sugars, and their release and degradation during pretreatment would confuse interpretation of sugar yields from hemi-cellulose and cellulose. Therefore, soluble sugars were removed from these two switchgrass varieties by washing three times with 10 volumes per weight of 80–90 °C deionized (DI) water each time [13]. The liquid from each wash was collected, and the water-extractable soluble free sugars were measured. The material was then dried in an oven at 45 °C. Prior to pretreatment and enzymatic hydrolysis, dried switch-grass was Wiley milled through a 40-mesh (0.422 mm) screen to provide uniform particle sizes across all switchgrass ecotypes. Switchgrass for AFEX pretreatment was milled through a 2-mm screen. All material balances were adjusted to a basis of 100 mass units of dry washed biomass feed to pretreatment to facilitate following the results and comparing pretreatments. The composition of the washed switchgrass is summa-rized in parentheses in Table 12.1. Corn stover also contained some free sugars, and the washed solids composition is also included in this table in parentheses.

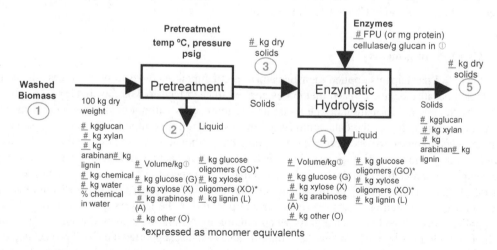

Figure 12.1   *Material balance approach used to track fate of key components measurable in biomass inputs and outputs from pretreatment and enzymatic hydrolysis. For switchgrass, the basis is 100 kg of solids following washing to remove soluble sugars that would otherwise confuse closing material balances to follow fate of structural carbohydrates in cellulose and hemicellulo-ses (See figure in color plate section).*

12.3   Yields of Xylose and Glucose from Pretreatment and Enzymatic Hydrolysis

A distinctive feature of the CAFI team was the objective to identify pretreatment conditions that maximize total glucose plus xylose yields from the combined operations of pretreatment and enzymatic hydrolysis and not just to focus on release of a single sugar from either stage alone. On this basis, overall yields were calculated on the basis of total potential glucose and xylose and not on the percent of each by itself. This approach was adopted because most cellulosic biomass is richer in glucose than xylose and, as a result, glucose yields have greater leverage on overall product yields than xylose. (Note that both sugars are assumed to be fermented equally well to final fuel products.)

In line with this objective, sugar yields were defined by dividing the amount of xylose, glucose, their oligomers, or combinations of these recovered in pretreatment and enzymatic hydrolysis by the maximum potential amount of both sugars. As an example, 100 mass units of dry corn stover has a maximum xylose potential of 24.3 mass units and a maximum glucose potential of 40.1 mass units, resulting in a total xylose plus glucose potential of 64.4 mass units. On this basis, the maximum xylose yield is 24.3/64.4 or 37.7%, the maximum glucose yield is 40.1/64.4 or 62.3%, and the maximum amount of total xylose and glucose is 100%. Data from all three CAFI projects followed this convention.

In addition, both monomers and oligomers in the liquid phases from pretreatment and enzymatic hydrolysis were followed, with each identified in the reported results. Although high recovery of arabinose, galactose, and mannose are also very important to final yields and economic processing, they were not included in the yield calculations because their measurements were not sufficiently accurate to provide meaningful yield values.

12.3.1   Yields from Corn Stover

Table 12.2 reports glucose and xylose yields as a percent of the maximum total potential glucose plus xylose that could be released for each pretreatment (Stage 1) followed by enzymatic hydrolysis (Stage 2) for corn stover, standard poplar, and Dacotah switchgrass. The Stage 1 and 2 designations were adopted to shorten nomenclature and avoid confusion about the sugar sources. Stage 2 enzymatic hydrolysis was carried out with an enzyme loading of about 15 FPU/g of glucan in the original raw feedstocks. However, this was changed to a protein mass basis for switchgrass on advice from Advisory Board members that protein mass could be more accurately measured than activity and that cost would be proportional to total enzyme protein mass. The yields reported in the first entry in each cell of the table are for total sugars released into solution from pretreatment or enzymatic hydrolysis as determined by high-performance liquid chromatography (HPLC) of liquids following post hydrolysis. The values were corrected for sugar losses in post hydrolysis using the established sugar recovery standard approach. The second entry for monomeric xylose or glucose yields was determined by direct analysis of the liquids released from pretreatment or enzymatic hydrolysis. Yields of oligomers alone can therefore be determined by subtracting the yields of monomers from the total xylose or glucose values as determined by post hydrolysis. A single entry in a cell in this table indicates release of just monomers.

Several trends stand out in the yield data. First, we see that overall total xylose plus glucose yields were very high for application of all pretreatments to corn stover, from a high of almost 94.4% with AFEX to a low of just under 87% with lime. Within this range, the order of glucose plus xylose yields from the combined operations of pretreatment and enzymatic hydrolysis with cellulase supplemented with β-glucosidase followed the high to low sequence: AFEX (94.4%), sulfur dioxide (93.9%), dilute sulfuric acid (92.4%), ARP (89.4%), LHW (87.2%), and lime (86.8%). It is important to note that further research and/or changes in process configuration could potentially improve the yields further to close the less than 8% spread between the highest and lowest values.

Table 12.2 Yields of glucose and xylose for each pretreatment (Stage 1) followed by enzymatic hydrolysis (Stage 2) for corn stover, standard poplar, and Dacotah switchgrass [8,9,11]. Stage 1 refers to pretreatment and Stage 2 refers to the enzymatic digestion of washed solids produced in pretreatment. The first value reported in each column is for total sugars released into solution, and the second is just for the monomeric sugar released. A single value indicates release of only monomers. Yields are defined based on the maximum potential sugars that could be captured from corn stover, poplar, and pre-washed Dacotah switchgrass. ND: not determined.

Pretreatment	Xylose yields			Glucose yields			Total sugars		
	Stage 1	Stage 2	Total xylose	Stage 1	Stage 2	Total glucose	Stage 1	Stage 2	Combined total
Corn stover with enzymatic hydrolysis at a loading of 15 FPU/g glucan in raw corn stover									
Max possible			37.7			62.3			100.0
Untreated									
Dilute acid	32.1/31.2	3.2	35.3/34.4	3.9	53.2	57.1	36.0/35.1	56.4	92.4/91.5
Sulfur dioxide	14.7/13.7	20.0	34.7/33.7	2.5/1.7	56.7	59.2/58.4	17.2/15.4	76.7	93.9/92.1
LHW	21.8/0.9	9.0	30.8/9.9	3.5/0.2	52.9	56.4/53.1	25.3/1.1	61.9	87.2/63.0
Lime	9.2/0.3	19.6	28.8/19.9	1.0/0.3	57.0	58.0/57.3	10.2/0.6	76.6	86.8/77.2
ARP	17.8/0	15.5	33.3/15.5		56.1	56.1	17.8/0	71.6	89.4/71.6
AFEX	0.0	34.6/29.3	34.6/29.3	0.0	59.8	59.8		94.4/89.1	94.4/89.1
Standard poplar with enzymatic hydrolysis at a loading of 15 FPU/g glucan in raw poplar									
Max possible			25.7			74.3			100
Untreated									
Dilute acid	16.1	2.4	18.5	17.7	46.6	64.3	33.8	49.0	82.8
Sulfur dioxide	19.2/14.0	2.4	21.6/16.4	2.3	72.0	74.3	21.6/16.3	74.4	95.9/90.7
LHW	15.0/1.0	9.7	24.7/10.7	1.5/0.1	40.0	41.5/40.1	16.5/1.1	49.7	66.2/50.8
Lime	1.2/0.0	18.8/16.8	20.0/16.8	0.2/0.0	71.0/67.2	71.2/67.2	1.5/0.0	89.8/84	91.3/84
ARP	9.6/0.0	8.2/8.0	17.7/8.0	0.4/0.0	36.3	36.7/36.3	10.0/0.0	44.5/44.3	54.5/44.3
AFEX	0.0	13.4	13.4	0.0	39.4	39.4	0.0	52.8	52.8
Dacotah switchgrass with enzymatic hydrolysis at a loading of 30 mg protein/g glucan in raw switchgrass									
Max possible			39.4			60.6			100.0
Untreated	N/A	1.9	1.9	N/A	8.4	8.4	N/A	10.3	10.3
Dilute acid	29.3/27.6	3.4	32.6/31.0	4.3/3.8	42.2	46.5/46.0	33.6/31.4	45.6	79.2/77.0
Sulfur dioxide	28.7/27.2	3.2	31.9/30.4	3.0/1.5	48.3	51.4/49.9	31.7/28.7	51.5	83.2/80.2
LHW	25.9/8.7	5.3/4.2	31.3/13.0	4.1/0.3	47.3	51.4/47.6	30.0/9.0	52.6/51.5	82.6/60.5
Lime	13.6/0.0	22.4/21.6	36.0/21.7	0.9/0.1	54.0/51.0	54.9/51.1	14.5/0.1	76.4/72.6	90.9/72.7
SAA	9.5/0.8	17.8/10.9	27.3/11.8	0.2/0.0	39.8/38.6	40.0/38.6	9.7/0.8	57.6/49.5	67.3/50.3
AFEX	11.1/0.0	25.6/22.6	36.7/22.6	0.8/0.0	47.1	47.9/47.1	11.9/0.0	72.7/69.7	84.6/69.7

Although overall sugar yields were very similar for application of all CAFI pretreatments to corn stover, it is noteworthy that the different pretreatments released sugars in very different patterns. In particular, pretreatment at lower pH with dilute sulfuric acid captured most of the xylose in the pretreatment step and most of the glucose in enzymatic hydrolysis. In addition, we see that most of the solubilized xylose was as monomers with a very low fraction of oligomers. Sulfur dioxide also produced mostly monomers from xylose in pretreatment but left a large fraction behind to be released in enzymatic hydrolysis with high overall yields. As for dilute acid, most of the glucose was left in the solids from pretreatment and released in Stage 2 enzymatic hydrolysis. LHW was similar to dilute acid in that it recovered most of the xylose in pretreatment, but it differed from dilute acid in that the xylose was mostly as oligomers.

Even though high-pH pretreatments also gave high sugar yields, the sugar release patterns are much different from those at low pH. First, the two ammonia-based approaches of AFEX and ARP released essentially no glucose in pretreatment whereas lime recovered a very small fraction of glucose. However, all three achieved very high yields of glucose from the pretreated solids in enzymatic hydrolysis. Xylose release showed an even greater difference from that observed at low pH. First, ARP released similar amounts of xylose in both pretreatment and enzymatic hydrolysis while lime recovered about twice as much xylose in enzymatic hydrolysis as in pretreatment. In both of these pretreatments, almost all of the xylose solubilized in pretreatment was as oligomers. In total contrast to both lime and ARP as well as to the lower-pH pretreatments, AFEX released virtually no xylose in pretreatment of corn stover. Furthermore, about 15% of the xylose that AFEX released in Stage 2 was as oligomers, unlike other pretreatments that produced mostly monomers in enzymatic hydrolysis for the cellulase type and loadings employed.

Table 12.2 shows that total glucose yields from the combined operations of pretreatment and enzymatic hydrolysis of corn stover were very similar for all six CAFI pretreatments, even though there were differences in glucose yield patterns among stages. However, the total xylose yields were 10% or more lower for LHW and lime than for the other four approaches. Higher sugar yields from these two options could be likely achieved by improving xylose recovery through tailoring enzyme formulations and/or reducing hemicellulose degradation in pretreatment.

12.3.2   Yields from Standard Poplar

Table 12.2 also includes glucose and xylose yields for CAFI pretreatments of poplar. Once again, the data are reported as a percent of the maximum total potential glucose plus xylose that could be released for each pretreatment (Stage 1) followed by enzymatic hydrolysis (Stage 2), and include total oligomers plus monomers as well as just monomers. Total sugar yields varied more widely for poplar than for corn stover, with the highest yield being 96% for sulfur dioxide and the lowest being almost 53% for AFEX. Within this range, the order of pretreatment yields from high to low followed the sequence: sulfur dioxide (95.9%), lime (91.3%), dilute sulfuric acid (82.8%), LHW (66.2%), ARP (54.5%), and AFEX (52.8%).

As before, these values must be considered with caution as changes in process configuration or further optimization could shift the balance. For example, dilute acid data for corn stover was produced in batch laboratory tube reactors with tight reaction time control. In contrast, the dilute acid data for poplar was developed at the pilot plant scale using a vertical continuous pretreatment reactor that could not tightly control residence time. Pretreatment of poplar in batch tubes or a horizontal Pandia-type or another reactor design that has better residence time control could potentially improve yields substantially.

In addition to the considerable variation in overall sugar yields from poplar among the CAFI pretreatments and compared to stover, changes in the sugar release patterns were also evident among the technologies with poplar. An exception was pretreatment at lower pH with dilute sulfuric acid that still captured most of the xylose in the pretreatment step, with little left to be released in enzymatic hydrolysis. In addition, most of the solubilized xylose was still as monomers with a very low fraction of oligomers. However, sulfur dioxide pretreatment released most of the xylose in pretreatment in a similar profile to dilute sulfuric acid, although the proportion of xylose released in Stage 1 was higher. Compared to results with corn stover, LHW still produced more xylose in Stage 1 than in Stage 2, but the fraction of total xylose recovered was somewhat lower in pretreatment compared to corn stover results. Once again, most of the xylose recovered in LHW pretreatment of poplar was as oligomers.

Among the higher-pH pretreatments, ARP gave very similar xylose release patterns from poplar as it did with corn stover, with just over half of the total recovered being released in Stage 1. However, the total yield as a percent of potential xylose was somewhat lower for poplar. On the other hand, lime displayed a different split in xylose release for poplar than for corn stover, with almost all of the xylose recovered in enzymatic hydrolysis of pretreated poplar compared to about two-thirds from corn stover. AFEX released no xylose in pretreatment of poplar (as for corn stover), with the result that all of the xylose recovered was still from Stage 2.

Glucose release from dilute acid pretreatment of poplar was noticeably different than from corn stover, with about 30% of the total being released in pretreatment of poplar versus only about 7% from corn stover. This anomaly is likely due to the higher pretreatment temperature employed for poplar and the residence time spread from the vertical pilot plant device that resulted in prolonged reaction times; these could be severe enough to release glucose from cellulose for a portion of the feedstock. On the other hand, glucose yield patterns from sulfur dioxide pretreatment of poplar displayed more similarities to those from corn stover, although with a little higher proportion from Stage 2 for this glucan-rich material. The glucose release pattern for LHW pretreatment of poplar was also similar to that from corn stover, although with a higher relative amount being left to enzymatic hydrolysis. Despite significant differences in overall yields, the distributions of glucose yields were similar for LHW and sulfur dioxide pretreatments. Similarly, lime pretreatment released most of the glucose in enzymatic hydrolysis and also shifted a significantly larger fraction to the second stage compared to corn stover. ARP pretreatment of poplar released almost all of the glucose recovered in Stage 2, a similar glucose yield pattern to ARP of corn stover. AFEX pretreatment of poplar released essentially no glucose in Stage 1 but released all that was recovered in Stage 2, a pattern entirely consistent with that of corn stover. However, the overall glucose yields were much lower from AFEX- pretreated poplar than that experienced with corn stover.

12.3.3 Yields from Dacotah Switchgrass

As shown in Table 12.1, Dacotah switchgrass contained similar amounts of xylan and glucan to corn stover. Dacotah switchgrass also displayed similar yield tendencies for most of the pretreatments to those from corn stover but with some important differences. First, the overall glucose plus xylose yields from combined pretreatment and enzymatic hydrolysis of switchgrass ranged from a high of 90.9% with lime to a low of 67.3% for SAA pretreatment (that replaced ARP to reduce water and energy consumption), a range of 23.6% compared to a range of only about 7.6% for corn stover. Thus, the overall total sugar yields for switchgrass were lower and had a wider spread. In addition, the order of decreasing yields shifted to: lime (90.9%), AFEX (84.6%), sulfur dioxide (83.2%), LHW (82.6%), dilute sulfuric acid (79.2%), and SAA (67.3%) for Dacotah. Lime moved from the lowest yields with corn stover to the highest yields with switchgrass, possibly due to severe conditions applied for the latter, while AFEX, sulfur dioxide, dilute sulfuric acid, and LHW remained in similar relative positions. These results suggest that the superficial similarities in composition between corn stover and Dacotah switchgrass do not translate into similar yields and that other factors must have a considerable bearing on performance. Although SAA yields from Dacotah switchgrass were lower than ARP yields for corn stover, this difference may result from the difference in process configuration and cannot be clearly attributed to differences in biomass recalcitrance.

Xylose release from switchgrass displayed a similar pattern to that from corn stover for dilute sulfuric acid pretreatment in that most of the xylose was produced in pretreatment. Sulfur dioxide also released almost the same amount of xylose in pretreatment as dilute sulfuric acid, with very little left for enzymatic hydrolysis. LHW displayed a similar tendency, albeit with a little less xylose from pretreatment and a little more from enzymatic hydrolysis. As before, LHW produced a large portion of the xylose as oligomers in pretreatment, contrary to the high monomeric xylose yields from dilute acid and sulfur dioxide.

As with corn stover and poplar, the higher pH pretreatments released about twice as much xylose in enzymatic hydrolysis as in pretreatment. For all three of these pretreatments (AFEX, SAA, and lime), most of the xylose recovered in pretreatment was as oligomers with minor to no monomers produced in pretreatment. The overall Stage 1 plus Stage 2 yields of monomeric plus oligomeric xylose from both lime and AFEX were over 10% greater than from lower-pH pretreatments; however, overall xylose yields were about 10% lower from SAA than from the lower-pH approaches. It is interesting to note that AFEX pretreatment released almost a third of the total xylose recovered in pretreatment of switchgrass, in sharp contrast to the essentially zero recovery of xylose from corn stover or poplar in the pretreatment step alone. This difference may result from washing AFEX-pretreated switchgrass but not washing the other two.

With corn stover, all pretreatments produced glucose yields greater than 90% of the maximum possible. In contrast, for switchgrass all yields were less than 90% of the maximum possible (except for lime, which was about 90% of the maximum possible). As summarized in Table 12.2, most of the glucose was released by enzymes in Stage 2 for all pretreatments. However, low-pH pretreatments released about 10% of the total glucose recovered in pretreatment while the high-pH pretreatments produced very low amounts of glucose in that stage. These trends were similar to those observed for both corn stover and poplar. As before, the exception was for dilute sulfuric acid pretreatment of poplar, which released about 30% of the total glucose recovered in Stage 1. This was apparently as a result of the residence time control challenges with the vertical pilot scale pretreatment reactor applied for pretreatment of poplar compared to the better reaction time control for the laboratory systems employed for both corn stover and switchgrass.

12.4   Impact of Changes in Biomass Sources

As mentioned in Section 12.1, to allow ready comparison of the performance of different pretreatments with different feedstocks, all baseline data were gathered for application of the full range of CAFI pretreatments to single sources of corn stover, poplar, and switchgrass. Another source of poplar was however used in CAFI 2 initially until a large enough quantity could be obtained to satisfy all of the different pretreatment needs, and two other sources of switchgrass were employed in CAFI 3 for comparison to the baseline data with Dacotah switchgrass. The compositions of these materials are summarized in Table 12.1. Comparing the composition of the two sources of poplar shows that the lignin was substantially lower at 21.4% for the initial poplar compared to 29.1% for the standard source. Other meaningful composition differences included 17.8% xylan vs. 14.9% and 5.7% acetyl vs. 3.6% for the initial vs. the standard poplar, respectively. Comparing the switchgrass sources shows that the primary difference was in the amount of free sugars in the Alamo and Shawnee feedstocks compared to the Dacotah baseline material. This difference was no doubt due to a winter harvest for Dacotah switchgrass allowing soluble sugars and other constituents to be transported to the roots during senescence, while the Alamo and Shawnee materials were harvested before senescence could occur. Comparison of the compositions in Table 12.1 that were adjusted to neglect soluble compounds reveals that over 10% more lignin but 25% as much protein was found in the Dacotah switchgrass compared to the other two varieties of Alamo and one of Shawnee, while the glucan, xylan, and arabinan levels were quite similar.

Time was too limited to obtain more than just a few performance results for the two poplar trees employed; striking yield differences were evident, however. For example, although overall glucose yields from AFEX pretreatment were only about 40% after 72 hours of enzymatic hydrolysis of standard poplar, they increased to about 60% in the same timeframe for the initial poplar. Furthermore, glucose yields increased to over 80% for 168 hours of enzymatic hydrolysis for AFEX-treated initial poplar, but were stuck at close to 40% even after extended enzymatic hydrolysis of AFEX-treated standard poplar. Interestingly, xylose yields were very similar at close to 40% of theoretical for AFEX pretreatment of either poplar and changed little when enzymatic hydrolysis was extended from 72 to 168 hours. In comparison, sulfur dioxide pretreatment of either poplar variety realized similar high xylose and glucose yields.

There were also differences among glucose yields from the different switchgrass varieties, with some variations more pronounced for some of the pretreatments than for others, as summarized in Table 12.3 [13]. In all cases, glucose yields were lower for Dacotah than for the other two feedstocks. However, glucose yields were always higher from Alamo than from either of the other two after 168 hours of enzymatic hydrolysis. One-hour yields were also highest for dilute sulfuric acid, LHW, and SAA pretreatments of Alamo but slightly higher for Shawnee than Alamo for AFEX and lime. In addition, the 168-hour differences were largest for AFEX and SAA, followed by dilute sulfuric acid and LHW pretreatments. Lime showed the smallest variation in long-term glucose yields among the different feedstocks but still had a difference of about 10% between the highest yield with Alamo switchgrass and the lowest value with Dacotah.

Differences in xylose yields were also very evident among the three switchgrass varieties, but the patterns were different from those displayed for glucose yields. For example, although long-term xylose yields were lower for Dacotah than for the other two switchgrass plants for AFEX, SAA, dilute sulfuric acid, and LHW, lime pretreatment actually gave slightly higher long-term xylose yields from Dacotah and the lowest

Table 12.3 *Glucose and xylose yields from three sources of switchgrass following enzymatic hydrolysis for 1 and 168 hours at cellulase loading of 30 mg protein (15 FPU)/g glucan in raw switchgrass.*

	Alamo		Shawnee		Dacotah		Note
	Glucose yield (%)	Xylose yield (%)	Glucose yield (%)	Xylose yield (%)	Glucose yield (%)	Xylose* yield (%)	
Yields after 1 h of enzymatic hydrolysis							
AFEX	17.5	5.8	18.3	6.6	9.4	5.1	No wash
DA	39.0	26.2	37.4	35.6	23.6	27.7	After hot water wash
LHW	41.1	34.5	34.3	49.6	20.3	52.8	After hot water wash
Lime	19.2	15.2	22.0	17.2	16.0	12.6	After hot water wash
SAA	23.2	15.7	22.8	14.4	15.9	5.5	After hot water wash
Untreated	4.4	0.7	4.8	0.6	5.6	2.5	
Yields after 168 h of enzymatic hydrolysis							
AFEX	83.2	68.0	78.8	68.1	63.1	62.0	No wash
DA	88.8	76.9	88.1	88.7	79.4	71.7	After hot water wash
LHW	92.5	91.1	87.9	85.3	83.9	84.7	After hot water wash
Lime	95.2	80.5	91.3	73.2	88.7	81.4	After hot water wash
SAA	82.9	65.0	81.6	68.5	65.8	47.3	After hot water wash
Untreated	16.60	3.40	16.10	3.20	13.90	3.70	

*Monomeric xylose

yields from Shawnee. Long-term xylose yields were also about 20% lower for application of SAA to Dacotah than that obtained for either Alamo or Shawnee. Long-term xylose yields from pretreatment and enzymatic hydrolysis of Alamo were similar to those from Shawnee for AFEX and SAA, but higher for LHW and lime and lower for dilute sulfuric acid. One-hour xylose yields followed similar trends to those for the long term.

Overall, the results with the two poplar sources and three varieties of switchgrass could suggest that lignin content is a key barrier to high sugar yields. However, there were also differences in xylan and acetyl contents in poplar and protein contents in switchgrass that could play a role. In addition, although glucose yields varied directly with lignin content, xylose yield variations were less tractable. Most likely, a series of resistances influence enzymatic hydrolysis yields, with their importance being affected by plant structure and the removal or alteration of other features that offered greater resistance. More research is needed to more fully understand how yields are impacted with variations in structure of similar plant materials and to determine how the resistances to hydrolysis interact.

12.5 Compositions of Solids Following CAFI Pretreatments

Table 12.4 summarizes the compositions of the solids produced by each CAFI pretreatment at conditions that resulted in the highest glucose plus xylose yields from the combined operations of pretreatment and enzymatic hydrolysis for applications to corn stover, standard poplar, and Dacotah switchgrass. Also included are the compositions of each of the raw feedstocks as a reference point for changes in solids

Table 12.4 *Compositions of untreated and solids resulting from pretreatment of corn stover, standard poplar, and Dacotah switchgrass by CAFI technologies. ND: not determined.*

Component	Glucan	Xylan	Lignin
Corn stover			
Untreated	36.1	21.4	17.2
Dilute acid	59.3	9.3	22.5
Sulfur dioxide	56.9	11.6	23.8
LHW	52.7	16.2	25.2
AFEX	34.4	22.8	18.0
ARP	61.9	17.9	8.8
Lime	56.7	26.4	14.6
Standard poplar			
Untreated	43.8	14.9	29.1
Dilute acid	57.3	2.1	46.1
Sulfur dioxide	55.1	2.5	ND
LHW	58.8	7.0	32.2
AFEX	46.6	15.0	ND
ARP	57.5	13.5	24.8
Lime	53.1	16.8	18.0
Dacotah switchgrass			
Untreated	35.3	22.5	22.6
Dilute acid	50.3	4.5	29.4
Sulfur dioxide	53.9	2.7	27.6
LHW	50.1	2.5	30.6
AFEX	35.9	21.9	24.4
Soaking in ammonia	55.6	21.9	13.9
Lime	53.0	21.5	14.6

make-up. The patterns observed are consistent with what would be expected based on the sugar yield data reported in Table 12.2.

12.5.1   Composition of Pretreated Corn Stover Solids

Examination of the xylan content in each pretreated material shows considerable differences in the amount left in the solids following pretreatment. For corn stover, dilute sulfuric acid reduced xylan levels to the lowest fraction among the pretreatments. Sulfur dioxide and LHW pretreatments also reduced xylan content in the residual solids considerably compared to that in the starting material, but not to such low levels as for dilute acid. ARP displayed a similar xylan composition to that from LHW, while lime pretreatment actually increased the xylan levels to higher values than without pretreatment due to higher lignin removal. As expected from the sugar yield data, AFEX pretreatment left the solids with virtually the same composition as that measured before pretreatment.

Removal of xylan while releasing little glucan in pretreatment resulted in the glucan content increasing for all pretreatments at low pH. The glucan content of the solids from dilute sulfuric acid, sulfur dioxide, and LHW pretreatments therefore ranged from a low of nearly 53% for LHW to close to 60% for dilute acid and sulfur dioxide. Interestingly, even though ARP and lime pretreatments removed less xylan than low -pH treatments, the glucan content rose for both ARP and lime due to removal of a large portion of the lignin. As a result of removing large amounts of both lignin and xylan, ARP produced the highest glucan content in the pretreated solids. Again, no change in solids composition was noticeable for AFEX due to the limited removal of any components from corn stover.

12.5.2   Composition of Pretreated Switchgrass Solids

In light of the closer similarities in sugar yields from switchgrass and corn stover compared to poplar, the solids compositions following pretreatment reported in Table 12.4 are compared for the former two first. The trends are very similar, in that low-pH pretreatments by dilute sulfuric acid, sulfur dioxide, and LHW all reduced xylan content substantially from that of the raw switchgrass, but the resulting xylan fractions were much lower in pretreated switchgrass solids than in corn stover. In addition, LHW pretreatment now changed places with dilute sulfuric acid in achieving the lowest xylan content, while sulfur dioxide had the second-lowest xylan levels for both feedstocks. For switchgrass, all three high-pH pretreatments (SAA, AFEX, and lime) gave nearly identical xylan contents; this is in contrast to corn stover, for which ARP and lime pretreatments reduced the xylan fraction more.

Comparison of glucan contents of the solids from the six pretreatments shows similar consequences to those from pretreatment of corn stover. SAA, similar to ARP, resulted in the highest fraction of glucan while lime gave the third-highest level, the same relative effect as for pretreatment of corn stover. As before, AFEX had no significant effect on glucan content with the result that glucan content in the pretreated solids was the lowest among the pretreatments. Sulfur dioxide pretreatment now gave a higher glucan content than that of dilute sulfuric acid, the opposite of the result for pretreatment of corn stover. LHW pretreatment had the same effect on glucan content in the solids relative to the other pretreatments as for corn stover.

Compared to corn stover, the lignin content in pretreated switchgrass solids reveals more significant changes. The lignin content was still highest for LHW pretreatment but pretreatment by SAA, similar to ARP for corn stover [14], showed the lowest lignin content. Dilute sulfuric acid pretreatment of switchgrass increased the lignin content more than for corn stover, while sulfur dioxide had a similar relative effect as before. Although AFEX had a limited effect on both glucan and xylan content of switchgrass, it did increase the fraction measured as Klason lignin by about 10% compared to raw Dacotah switchgrass. This observation is consistent with current mechanistic understanding of AFEX in which ester linkages between lignin and hemicellulose are attacked via hydrolysis and ammonolyis to generate lower-molecular-weight lignin and hemicellulose fragments. Some of these fragments are translocated to the surface of the biomass during the pressure release, creating additional porosity and increasing enzyme access but also condensing such fragments into Klason lignin-type products [15,16].

12.5.3   Composition of Pretreated Poplar Solids

The patterns in xylan left in the solids following poplar pretreatment were mostly similar to those for corn stover and switchgrass, although there were some exceptions (Table 12.4). Dilute acid resulted in the lowest xylan content, consistent with results for corn stover and not too different from the result of dilute acid pretreatment of switchgrass. Sulfur dioxide pretreatment gave the second-lowest xylan levels, as for both of the other two feedstocks. LHW pretreatment did not reduce xylan content quite as much as dilute sulfuric acid or sulfur dioxide did, consistent with its effects on corn stover but less aggressive than those for switchgrass. The xylan left in the solids following ARP was about twice that from LHW and followed the same trend as for corn stover and switchgrass. Once again, AFEX had little effect on the fraction of xylan left in the solids compared to that in the original standard poplar. Finally, lime pretreatment resulted in the highest levels of xylan of any of the pretreatments when applied to poplar, entirely consistent with the results for corn stover and switchgrass.

Although the trends in xylan content of pretreated poplar solids were quite consistent with those measured following pretreatments of corn stover and switchgrass, the same cannot be said about patterns of glucan content. First, Table 12.4 shows that the glucan levels were the highest following LHW pretreatment, a totally different outcome than the comparatively low values from LHW pretreatment of either corn stover or switchgrass. The high glucan content of solids resulting from dilute sulfuric acid pretreatment of poplar were similar to those from corn stover, while dilute acid did not increase glucan content as much as for switchgrass. By removing significant amounts of both xylan and lignin, ARP once again produced one of the highest glucan levels, being only slightly less than for LHW. For poplar, sulfur dioxide did not have as large an effect on glucan content as it did on switchgrass, but gave a similar relative effect as corn stover. Similarly, lime pretreatment did not increase glucan content from poplar as much as it did for each of the other two feedstocks.

Finally, AFEX increased the glucan content more for pretreatment of poplar then it did for either corn stover or switchgrass, but still resulted in the lowest glucan content of any of the CAFI pretreatments. This result is not unexpected as AFEX is unique among the pretreatments studied by the CAFI; it is a "dry to dry" process and does not require or generate a separate liquid stream following pretreatment. The composition of the biomass following AFEX was, and should be, essentially the same as before pretreatment (unless the AFEX-treated biomass was washed, putting it on the same analytical footing as the other pretreatments). A consequence of this fact is that all of the sugars from AFEX pretreatment exist in a single aqueous stream, unlike the other pretreatments studied which produce sugars in two or more process streams.

12.5.4   Overall Trends in Composition of Pretreated Biomass Solids and Impact on Enzymatic Hydrolysis

The results in Table 12.4 and the discussion in Sections 12.5.1–12.5.3 reveal some important trends in the impacts of the CAFI pretreatments that were generally consistent for corn stover, poplar, and switchgrass. First, low-pH pretreatment by dilute sulfuric acid and sulfur dioxide reduced xylan content to the lowest levels of any pretreatment, while removing relatively little Klason lignin (at least on a net basis). Thus, glucan levels increased significantly from those in the raw feedstocks for both. LHW also increased glucan levels by reducing xylan content considerably but less so than the lower pH approaches while leaving much of the Klason lignin in the solids. Although ARP did not reduce xylan content in the pretreated solids nearly as much as lower pH pretreatments, its glucan content was still high through reducing lignin levels. Lime had the highest xylan content in pretreated solids but still increased glucan content by removing more lignin than other approaches. Finally, AFEX had a small impact on xylan or glucan contents, as expected.

An intriguing outcome is that even though the solids resulting from the CAFI pretreatments differed considerably in composition, all could produce high overall yields from the combination of pretreatment and enzymatic hydrolysis. This outcome was particularly true for corn stover, but the yields were still reasonably consistent for pretreatments of switchgrass. However, not all of the pretreatments were nearly as effective on poplar as on corn stover. These results suggest that there are important features of biomass that may be effectively altered by a variety of pretreatments for some biomass materials, but that there are important differences in the structure of other biomasses that only particular pretreatment approaches can alter sufficiently to improve sugar yields from enzymatic hydrolysis. Furthermore, the fact that both low-pH pretreatments by sulfur dioxide and, to some extent, dilute sulfuric acid and high-pH pretreatment with lime were all effective on all feedstocks suggests that biomass recalcitrance is controlled by a complex set of features that can be disrupted in many ways [14,17,18].

Overall, sugar yields are a complex function of the biomass material used, the pretreatment conditions applied, and both the amount of enzyme applied and the composition of the enzyme mixture used. However, interpretation of these results is complicated because the CAFI project was not directed toward optimizing enzyme combinations for the biomass samples studied. In fact, the amount of enzyme used was held constant for most CAFI studies, as was the composition of the enzyme mixture. Given that acidic pretreatments largely hydrolyze the hemicellulose fraction to monomers while alkaline pretreatments do not [14,19,20], different balances in hemicellulase and cellulase activities are expected to be optimal [21,22]. Time and funding were however insufficient to allow much attention to this aspect.

12.6   Pretreatment Conditions to Maximize Total Glucose Plus Xylose Yields

Table 12.5 summarizes key conditions that resulted in the highest total glucose plus xylose yields when each CAFI pretreatment was coupled with enzymatic hydrolysis for corn stover, standard poplar, and Dacotah switchgrass. It is interesting to note that application of chemicals that spanned nearly the entire pH range were reasonably effective for each feedstock. However, while temperatures as low as 55 °C for lime and 90 °C for AFEX proved capable of achieving high sugar yields from corn stover, the best results were realized at temperatures as high as 160 °C with dilute acid and 190 °C with LHW. On the other hand, higher temperatures were needed for switchgrass, with none of the pretreatments applying a temperature below the 120 °C value for lime. Even higher temperatures were required for the more recalcitrant poplar, with the lowest being 140 °C. Furthermore, application of higher temperatures could not totally overcome the greater recalcitrance of poplar and switchgrass compared to corn stover.

The times to maximum total xylose plus glucose yields spanned a greater range for corn stover than for the other feedstocks, with lower temperature pretreatments requiring more time. At the extremes, 190 °C was applied for only 15 min in the case of LHW while lime pretreatment at 55 °C took about 4 weeks to achieve the highest glucose plus xylose recovery. The significant increase in pretreatment time to maximum yields with dropping temperature is in line with the expectation that the rate constants for pretreatment reactions would likely follow an Arrhenius -type temperature dependence that prolongs times considerably at lower temperatures [23–25]. This tradeoff between time and the exponential dependence on temperature is also reflected in the severity parameter concept. This is often applied to estimate combinations of temperatures and times that can give equivalent yields for LHW or steam explosion pretreatments, and was extended to include the impact of pH in the combined severity parameter when dilute acid is employed. In any event, as temperatures were increased to achieve the highest yields from more recalcitrant switchgrass

Table 12.5 *Pretreatment conditions resulting in highest total glucose plus xylose yields from CAFI pretreatments followed by enzymatic hydrolysis of corn stover, standard poplar, and Dacotah switchgrass [8,9,11]. NS: not studied.*

Component	Reagent used	Percent reagent	Temperature (°C)	Time
Corn stover				
Dilute acid	Sulfuric acid	0.49	160	20 min
Sulfur dioxide	Sulfur dioxide			
LHW	None	0.0	190	15 min
AFEX*	Anhydrous ammonia	1 g/g of biomass	90	5 min
ARP	Aqueous ammonia	15	170	10 min
SAA	Aqueous ammonia	NS		
Lime**	Lime	0.08 g/g biomass	55	4 weeks
Standard poplar				
Dilute acid	Sulfuric acid	2.0	190	1.1 min
Sulfur dioxide	Sulfur dioxide	3.0	190	5 min
LHW	None	0.0	200	10 min
AFEX*	Anhydrous ammonia	2 g/g of biomass	180	10 min
ARP	Aqueous ammonia	15	185	27.5 min
SAA	Aqueous ammonia	NS		
Lime**	Lime	20	140–160	120 min
Dacotah switchgrass				
Dilute acid	Sulfuric acid	1.0	140	40 min
Sulfur dioxide	Sulfur dioxide	5.0	180	10 min
LHW	None	0.0	200	10 min
AFEX*	Anhydrous ammonia	1.5 g/g biomass	150	30 min
SAA	Aqueous ammonia	15	160	60 min
Lime**	Lime	0.3 g/g biomass	120	240 min

*About 97% of ammonia can be recovered after pretreatment and reused.
**Oxygen (or air) are essential for lime pretreatment.

and poplar, the times to maximum yield dropped, consistent with expected effects on rate constants. For example, the time to the best yields for lime pretreatment dropped to 240 min at 120 °C for switchgrass and to 120 min at 140–160 °C for poplar. Although application of more severe conditions resulted in similar yields from all feedstocks for lime pretreatment, little change in temperature was required to achieve high yields for sulfur dioxide pretreatment. The application of more severe conditions was however inadequate to boost yields to the highest levels for other pretreatments of the baseline feedstocks.

12.7   Implications of the CAFI Results

The three CAFI projects convey a number of important messages. First, we see that thermochemical pretreatments spanning the application of a wide range of chemical additives, temperatures, and times can be effective in achieving high yields of xylose and glucose from cellulosic biomass in the combined operations of pretreatment and enzymatic hydrolysis, as can application of high temperatures over short times without additives. In fact, no particular pH or temperature proved essential to realizing high yields from all three feedstocks. However, not all pretreatments were effective with all three feedstocks, and sulfur dioxide and lime proved to be the only pretreatments that realized similar high yields from corn stover, switchgrass, and poplar. In addition to differences in results for these three feedstocks, important differences were also found in sugar yields from different sources of poplar and switchgrass, with only lime and sulfur dioxide pretreatments able to achieve high yields regardless of source. However, it is important to remember that the CAFI team did not have the resources to optimize enzyme compositions and loadings for the pretreatments and biomass samples studied.

Superficial evidence suggests that higher lignin content impeded performance. However, the fact that high yields were achievable with low-pH pretreatments that removed high amounts of xylan, high-pH pretreatments that favored removable of lignin, and AFEX that removed very little, suggests that biomass recalcitrance is more complex than can be explained by a single barrier such as lignin. It must also be kept in mind that the enzyme loadings applied by the CAFI team to achieve these high yields would likely be too costly for commercial success; more work is needed to identify pretreatment approaches that can achieve high yields at lower, more practical enzyme loadings. More in-depth study would therefore be invaluable in defining the key features of cellulosic biomass that are responsible for recalcitrance, and how to alter these features so that low enzyme loadings can effectively deconstruct biomass into sugars.

A vital point that the CAFI project highlights is the importance of developing, closing, and presenting material balances to clearly show the fates of key components and support performance data. Information of this nature should be reported in all applied biomass conversion research papers and presentations, so that others can follow and apply the results with confidence. It is also important to track sugar yields from pretreated biomass against enzyme loadings for favored pretreatment conditions, and to show how these results change with hydrolysis time up to the point where yields level off.

Economic data is not reported here but is available elsewhere, including Chapter 15. However, the overall message from techno-economic evaluations is clear: high yields must be achieved for processes that produce fuels and other commodity products [6,26]. No pretreatment option will be cost effective if yields are low. Furthermore, even though pretreatment is a dominant cost element in the overall process, the CAFI pretreatments were all projected to have similar costs. As a result, the most cost-effective pretreatments to a first approximation will be those that allow the highest sugar yields in combined pretreatment and enzymatic hydrolysis. After high yields are realized, the implications of other process variables such as concentration and residence time on costs can be evaluated.

Although it may be tempting to select one of the pretreatments as the winner and "downselect" others based on CAFI yield and cost data, it is vital to recognize that differences in biomass deconstruction patterns among the pretreatments can have important implications for upstream and downstream operations that must be fully considered. For example, enzyme formulations for specific biomass materials and pretreatments must be optimized, as the set of enzyme activities required to achieve high sugar yields for alkaline pretreatments will almost certainly be different from those required for acidic pretreatments. Another consideration is that fermentative organisms with diauxic effects that favor use of glucose before xylose with resulting low final product yields from xylose would likely realize higher final product yields from pretreatments that release most of the xylose in pretreatment and most of the glucose in enzymatic hydrolysis. In addition, elimination of the need for conditioning of pretreated biomass streams for some pretreatments such as AFEX would have substantial yield and cost benefits. Interactions with pretreatment are also likely to be important for such process steps as particle size reduction, nutrient demands, neutralization costs, enzyme formulations, waste treatment, and utilization of solid residues. Different pretreatments are expected to be better tailored to some feedstocks than others, and vice versa. One size does not fit all, and pretreatment selection must be considered in the context of other technologies employed and overall techno-economic analyses [27].

12.8 Closing Thoughts

The three CAFI projects met the goals originally set by the team in 2000 to provide comparative data that could be readily used by others to judge the merits of different pretreatment approaches. First, data was developed on leading pretreatments using common feedstocks, shared enzymes, identical analytical methods, a consistent and clear material and energy balance approach, and the same cost models. CAFI also provided a complete set of data to help industry select technologies for commercial applications and avoided "downselecting" pretreatments, so that information would be provided on all promising options. In this way, others could decide which pretreatment to employ for their application; neglecting promising pretreatments would only leave open questions about how well those not included would perform in comparison. The team also sought to understand mechanisms that influence performance and differentiate pretreatments to serve as a technology base to facilitate commercial use and identify promising paths to advance pretreatment technologies. Although new insights were gained through this work, the scope of the project was insufficient to dig into the true root causes of the observed differences, and much is left to be done to understand the basis for performance differences.

The keys to success of the projects included using single sources of enzymes and feedstocks and reporting data in a meaningful material balance format that would allow others to readily interpret and apply the results. An even more important factor was the cooperative spirit of the team, comprising institutions from across the United States as well as from Canada in CAFI 2. Major credit of course belongs to the IFAFS Program of the US Department of Agriculture and the Office of the Biomass Program of the US Department of Energy, as this work would have not been possible without their support.

Although the CAFI projects successfully met their objectives, it is clear that a wide diversity of pretreatments could effectively increase enzymatic digestibility. It will therefore be challenging to find a simple relationship between sugar yields and the chemical composition of pretreated solids, choice of chemical, biomass type, reaction temperature, reaction pH, pretreatment chemicals applied, lignin or xylan removal, or other physical and chemical parameters. For example, the classical arguments that enzymatic digestion of pretreated solids can be correlated with just lignin removal [28,29] or just xylan removal [30,31] were not supported by the high yields that resulted from dilute acid, sulfur dioxide, and LHW pretreatments that remove mostly xylan at low pH; from pretreatments with lime and aqueous ammonia that removed lignin at high pH; and from AFEX pretreatment that did not physically remove much, if any, of either. Similarly, contrary to literature hypotheses, the CAFI studies showed no consistent relationship between enzymatic hydrolysis yields and removal of acetyl groups in switchgrass [32,33].

The high yields from lime and AFEX pretreatments of corn stover at low temperatures bring into question the belief that biomass must be pretreated at a temperature above some critical value, such as the glass transition temperature of lignin, to be effective [34]. Although a promising correlation was found between initial hydrolysis rates and maximum cellulase adsorption capacity for most of the CAFI pretreatments, lime-pretreated solids displayed higher yields at their relatively low adsorption capacity than would be predicted by the trends in the yields for the other CAFI pretreatments [35]. This outcome could be due to removal of lignin that is known to non-productively adsorb enzyme, coupled with modifying the lignin left to adsorb less enzyme.

Unfortunately, the CAFI team had insufficient resources to systematically examine these and other features that could explain the basis for high yields over the range of pretreatments applied; further research would be very beneficial to better understand how such a wide range of pretreatment approaches and conditions could produce similar results. Although the CAFI team successfully met its objectives, the project results clearly indicate that much more must be done to understand biomass recalcitrance and the role of pretreatment in deconstructing cellulosic biomass into sugars with high yields.

Acknowledgements

The CAFI projects were made possible through funding by both the United States Department of Agriculture and the United States Department of Energy, and the team is grateful for their support. The first CAFI project directed at corn stover was funded by the USDA Initiative for Future Agricultural and Food Systems (IFAFS) Program as a result of a competitive solicitation through contract 00-52104-9663 to Dartmouth College, where the lead author and project PI was based at the time. The Office of the Biomass Program of the US DOE supported participation by the National Renewable Energy Laboratory (NREL) in this project, with NREL providing enzymes, feedstock, and technical and economic analysis. The second CAFI project that focused on pretreating poplar wood was funded by the US DOE Office of the Biomass Program through a competitive solicitation that resulted in award of contract DE-FG36-04GO14017 to Dartmouth College. The third CAFI project that addressed pretreatment of switchgrass was also funded by the US DOE Office of the Biomass Program though contract DE-FG36-07GO17102 to the University of California, Riverside, where the PI relocated in 2005. Natural Resources Canada provided funding for participation by Dr Jack Saddler and his team at the University of British Columbia for research on sulfur dioxide pretreatment of corn stover and poplar wood in the second CAFI project. We also appreciate the valuable feedback, encouragement, and other interactions with the CAFI Agricultural and Industrial Advisory Board throughout the three projects and appreciate the time and other contributions of board members in helping to guide the projects. It is vital to recognize the participation of all CAFI team leaders, students, and others who were extremely cooperative in fulfilling the CAFI objectives; the success of the projects would not have been possible without the willingness of the team to work objectively together. Finally, but not least, the PI would like to recognize the Ford Motor Company for sponsoring the Chair in Environmental Engineering in the Center for Environmental Research and Technology of the Bourns College of Engineering (CE-CERT) at the University of California, Riverside, that facilitates participation in projects such as these.

References

1. Lynd, L.R., Laser, M.S., Bransby, D. *et al.* (2008) How biotech can transform biofuels. *Nature Biotechnology*, **26** (2), 169–172.
2. Lynd, L.R., Wyman, C.E., and Gerngross, T.U. (1999) Biocommodity engineering. *Biotechnology Progress*, **15** (5), 777–793.
3. Wyman, C.E. (1994) Ethanol from lignocellulosic biomass – technology, economics, and opportunities. *Bioresource Technology*, **50** (1), 3–16.
4. Wyman, C.E. (2007) What is (and is not) vital to advancing cellulosic ethanol. *Trends in Biotechnology*, **25** (4), 153–157.
5. Aden, A., Ruth, M., Ibsen, K. *et al.* (2002) Lignocellulosic biomass to ethanol process design and economics utilizing co-current dilute acid prehydrolysis and enzymatic hydrolysis for corn stover. National Renewable Energy Laboratory, Golden, CO, NREL/TP-510-32438.
6. Eggeman, T. and Elander, R. (2005) Process and economic analysis of pretreatment technologies. *Bioresource Technology*, **96** (18), 2019–2025.
7. Wyman, C.E., Dale, B.E., Elander, R.T. *et al.* (2005) Coordinated development of leading biomass pretreatment technologies. *Bioresource Technology*, **96** (18), 1959–1966.
8. Wyman, C.E., Dale, B.E., Elander, R.T. *et al.* (2009) Comparative sugar recovery and fermentation data following pretreatment of poplar wood by leading technologies. *Biotechnology Progress*, **25** (2), 333–339.
9. Wyman, C., Balan, V., Dale, B. *et al.* (2011) Comparative data on effects of leading pretreatments and enzyme loadings and formulations on sugar yields from different switchgrass sources. *Bioresource Technology*, **102** (24), 11052–11062.
10. Wyman, C.E., Dale, B.E., Elander, R.T. *et al.* (2005) Coordinated development of leading biomass pretreatment technologies. *Bioresource Technology*, **96** (18), 1959–1966.
11. Wyman, C.E., Dale, B.E., Elander, R.T. *et al.* (2005) Comparative sugar recovery data from laboratory scale application of leading pretreatment technologies to corn stover. *Bioresource Technology*, **96** (18), 2026–2032.
12. Elander, R.T., Dale, B.E., Holtzapple, M. *et al.* (2009) Summary of findings from the Biomass Refining Consortium for Applied Fundamentals and Innovation (CAFI): Corn stover pretreatment. *Cellulose*, **16** (4), 649–659.
13. Kim, Y., Mosier, N.S., Ladisch, M.R. *et al.* (2011) Comparative study on enzymatic digestibility of switchgrass varieties and harvests processed by leading pretreatment technologies. *Bioresource Technology*, **102** (24), 11089–11096.

14. Kumar, R., Mago, G., Balan, V., and Wyman, C.E. (2009) Physical and chemical characterizations of corn stover and poplar solids resulting from leading pretreatment technologies. *Bioresource Technology*, **100** (17), 3948–3962.

15. Li, C., Cheng, G., Balan, V. *et al.* (2011) Influence of physico-chemical changes on enzymatic digestibility of ionic liquid and AFEX pretreated corn stover. *Bioresource Technology*, **102** (13), 6928–6936.

16. Chundawat, S.P.S., Donohoe, B.S., da Costa Sousa, L. *et al.* (2011) Multi-scale visualization and characterization of lignocellulosic plant cell wall deconstruction during thermochemical pretreatment. *Energy & Environmental Science*, **4** (3), 973–984.

17. Donohoe, B.S., Vinzant, T.B., Elander, R.T. *et al.* (2011) Surface and ultrastructural characterization of raw and pretreated switchgrass. *Bioresource Technology*, **102** (24), 11097–11104.

18. Kumar, R. and Wyman, C.E. (2010) Key features of pretreated lignocelluloses biomass solids and their impact on hydrolysis, in *Bioalcohol Production: Biochemical Conversion of Lignocellulosic Biomass* (ed. K. Waldon), Woodhead Publishing Limited, Oxford, p. 73–121.

19. da Costa Sousa, L., Chundawat, S.P.S., Balan, V., and Dale, B.E. (2009) 'Cradle-to-grave' assessment of existing lignocellulose pretreatment technologies. *Current Opinion in Biotechnology*, **20** (3), 339–347.

20. Yang, B. and Wyman, C.E. (2008) Pretreatment: the key to unlocking low-cost cellulosic ethanol. *Biofuels, Bioproducts and Biorefining*, **2** (1), 26–40.

21. Gao, D., Chundawat, S.P.S., Krishnan, C. *et al.* (2010) Mixture optimization of six core glycosyl hydrolases for maximizing saccharification of ammonia fiber expansion (AFEX) pretreated corn stover. *Bioresource Technology*, **101** (8), 2770–2781.

22. Kumar, R. and Wyman, C.E. (2009) Effect of xylanase supplementation of cellulase on digestion of corn stover solids prepared by leading pretreatment technologies. *Bioresource Technology*, **100** (18), 4203–4213.

23. Overend, R.P. and Chornet, E. (1987) Fractionation of lignocellulosics by steam-aqueous pretreatment. *Philosophical Transactions of the Royal Society of London Series B: Biological Sciences*, **A321**, 523–536.

24. Chum, H., Johnson, D., Black, S., and Overend, R. (1990) Pretreatment-catalyst effects and the combined severity parameter. *Applied Biochemistry and Biotechnology*, **24–25** (1), 1–14.

25. Chum, H.L., Johnson, D.K., Black, S. *et al.* (1988) Organosolv pretreatment for enzymatic hydrolysis of poplars. 1. Enzyme hydrolysis of cellulosic residues. *Biotechnology and Bioengineering*, **31** (7), 643–649.

26. Hinman, N.D., Schell, D.J., Riley, C.J. *et al.* (1992) Preliminary estimate of the cost of ethanol production for SSF technology. *Applied Biochemistry Biotechnology*, **34/35**, 639–649.

27. Yang, B. and Wyman, C.E. (2008) Pretreatment: the key to unlocking low cost cellulosic ethanol. *Biofuels, Bioproducts, and Biorefining*, **2** (1), 26–40.

28. Chang, V.S. and Holtzapple, M.T. (2000) Fundamental factors affecting biomass enzymatic reactivity. *Applied Biochemistry and Biotechnology*, **84–86**, 5–38.

29. Converse, A.O. (1993) Substrate factors limiting enzymatic hydrolysis, in *Bioconversion of Forest and Agricultural Plant Residues* (ed. J.N. Saddler), CAB International, Wallingford, UK, p. 93–106.

30. Grohmann, K., Torget, R., and Himmel, M. (1985) Optimization of dilute acid pretreatment of biomass. *Biotechnology Bioengineering Symposium*, **15**, 59–80.

31. Grohmann, K., Torget, R., and Himmel, M. (1986) Dilute acid pretreatment of biomass at high solids concentrations. *Biotechnology and Bioengineering Symposium*, **17**, 137–151.

32. Grohmann, K., Mitchell, D.J., Himmel, M.E. *et al.* (1989) The role of ester groups in resistance of plant cell wall polysaccharides to enzymatic hydrolysis. *Applied Biochemistry and Biotechnology*, **20/21**, 45–61.

33. Kong, F., Engler, C.R., and Soltes, E.J. (1992) Effects of cell wall acetate, xylan backbone, and lignin on enzymatic hydrolysis of aspen wood. *Applied Biochemistry Biotechnology*, **34–35**, 23–35.

34. Balan, V., Sousa Lda, C., Chundawat, S.P., Marshall, D., Sharma, L.N., Chambliss, C.K., and Dale, B.E. (2009) Enzymatic digestibility and pretreatment degradation products of AFEX-treated hardwoods (Populus nigra). *Biotechnol Prog.*, **25** (2), 365–375.

35. Shi, J., Ebrik, M.A., Yang, B. *et al.* (2011) Application of cellulase and hemicellulase to pure xylan, pure cellulose, and switchgrass solids from leading pretreatments. *Bioresource Technology*, **102** (24), 11080–11088.

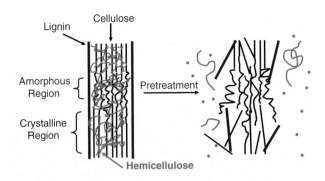

Plate 4.1 Schematic diagram of the effect of pretreatment on cellulose structure and the lignin seal, resulting in exposure of the cellulose and access by cellulolytic enzymes that cause hydrolysis. (Reprinted with permission from Mosier et al. [1] © 2005, Elsevier).

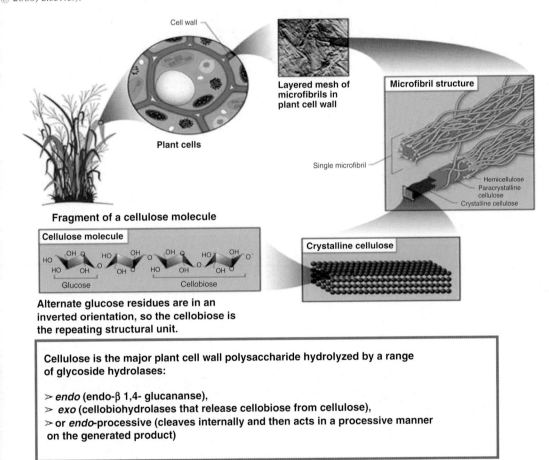

Plate 4.2 Plant cell-wall structure and cellulolytic enzymes. (Adapted from US DOE [5,6], Office of Biological and Environmental Research of the US Department of Energy Office of Science).

Aqueous Pretreatment of Plant Biomass for Biological and Chemical Conversion to Fuels and Chemicals, First Edition. Edited by Charles E. Wyman.
© 2013 John Wiley & Sons, Ltd. Published 2013 by John Wiley & Sons, Ltd.

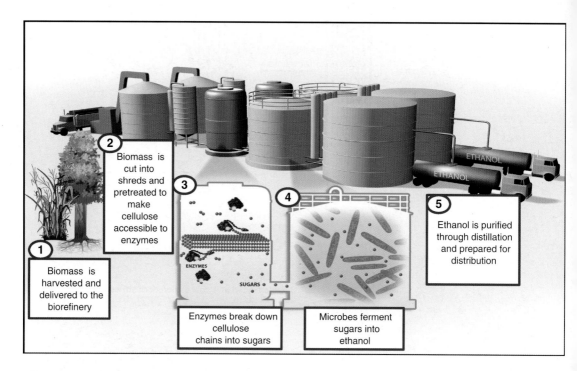

Plate 4.3 *Schematic view of key steps in transforming lignocellulosic materials to ethanol. (Adapted from US DOE [11], Office of Biological and Environmental Research of the US Department of Energy Office of Science).*

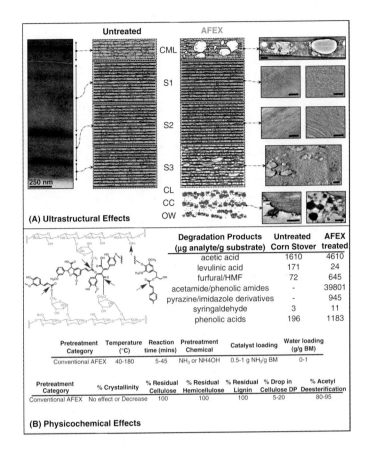

Untreated **AFEX**

CML
S1
S2
S3
CL
CC
OW

250 nm

(A) Ultrastructural Effects

Degradation Products (μg analyte/g substrate)	Untreated Corn Stover	AFEX treated
acetic acid	1610	4610
levulinic acid	171	24
furfural/HMF	72	645
acetamide/phenolic amides	-	39801
pyrazine/imidazole derivatives	-	945
syringaldehyde	3	11
phenolic acids	196	1183

Pretreatment Category	Temperature (°C)	Reaction time (mins)	Pretreatment Chemical	Catalyst loading	Water loading (g/g BM)
Conventional AFEX	40–180	5–45	NH₃ or NH4OH	0.5-1 g NH₃/g BM	0-1

Pretreatment Category	% Crystallinity	% Residual Cellulose	% Residual Hemicellulose	% Residual Lignin	% Drop in Cellulose DP	% Acetyl Deesterification
Conventional AFEX	No effect or Decrease	100	100	100	5-20	80-95

(B) Physicochemical Effects

Plate 9.2 *Ultrastructural and physicochemical impacts of non-extractive AFEX on grass cell walls. (a) Schematic model depicting cell wall structural changes due to AFEX within compound middle lamella (CML), secondary walls (S1-3), cell lumen (CL), cell corners (CC) and outer walls (OW). (b) Figure depicts common chemical linkages cleaved and major cell-wall decomposition products formed during AFEX. Other physicochemical impacts of conventional AFEX are also shown: cell-wall composition, cellulose crystallinity and DP, acetylation, enzyme and microbial digestibility. (Adapted from [2] © 2011, Annual Reviews, Inc., [17] © 2011, Royal Society of Chemistry, [23] © 2011, Elsevier, and [24] © 2009, Elsevier).*

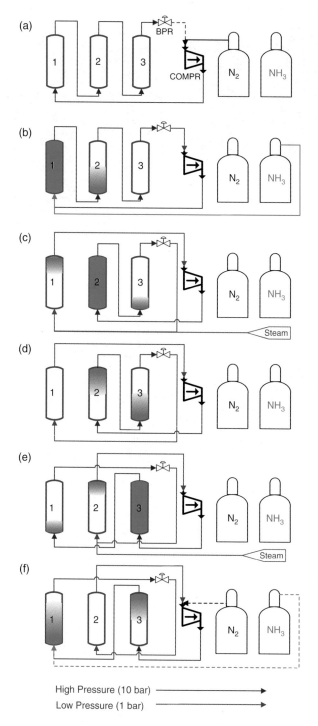

Plate 9.5 *Concept for hybrid AFEX pretreatment and ammonia recovery in a system of three packed beds (1–3) in a step-wise manner (a–f).*

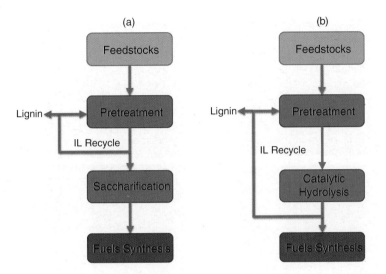

Plate 11.2 *The two general routes of IL-based sugar production and fuel production from lignocellulosic feedstocks: (a) enzymatic and (b) catalytic hydrolysis routes. Lignin is typically a separate process output from the generation of fermentable sugars and/or fuels that can be recovered as a function of IL recycle, dependent on IL pretreatment process conditions used.*

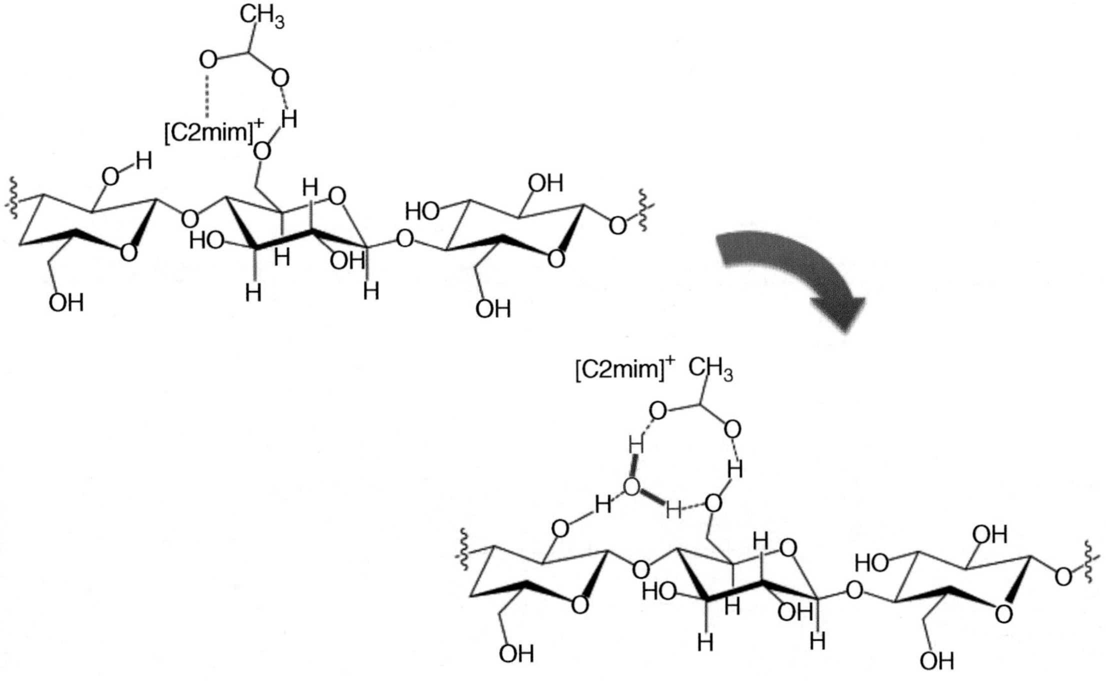

Plate 11.3 *Proposed mechanism of cellulose precipitation from [C$_2$mim][OAc] upon addition of water. (Adapted from [69] © 2009, Elsevier).*

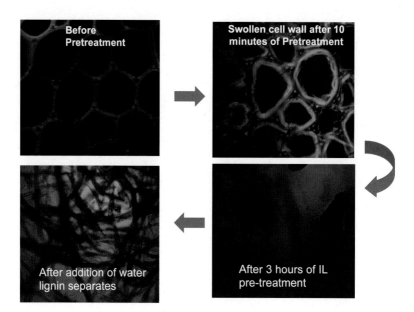

Plate 11.4 *Auto-fluorescence images of plant cell walls of switchgrass before, during, and after IL pretreatment with [C₂mim]* *[OAc]. (Adapted from [74] © 2010, Royal Society of Chemistry).*

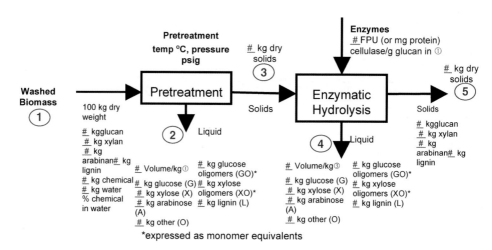

Plate 12.1 *Material balance approach used to track fate of key components measurable in biomass inputs and outputs from pretreatment and enzymatic hydrolysis. For switchgrass, the basis is 100 kg of solids following washing to remove soluble sugars that would otherwise confuse closing material balances to follow fate of structural carbohydrates in cellulose and hemicelluloses.*

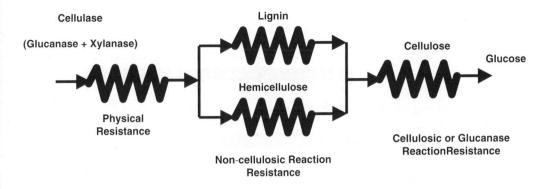

Plate 13.1 *Schematic of resistances encountered by cellulase enzyme in hydrolysis of cellulose within biomass [31].*

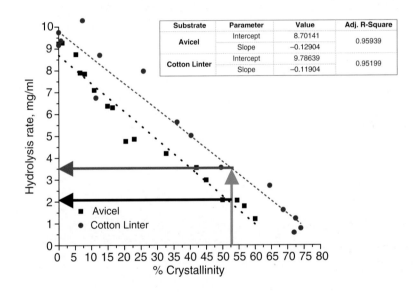

Substrate	Parameter	Value	Adj. R-Square
Avicel	Intercept	8.70141	0.95939
	Slope	−0.12904	
Cotton Linter	Intercept	9.78639	0.95199
	Slope	−0.11904	

Plate 14.4 *Hydrolysis rate versus crystallinity for two cellulose substrates with different initial DPs. Cellulose samples of different crystallinities were prepared at varying phosphoric acid concentrations (Data re-plotted from Bansal et al. [236]).*

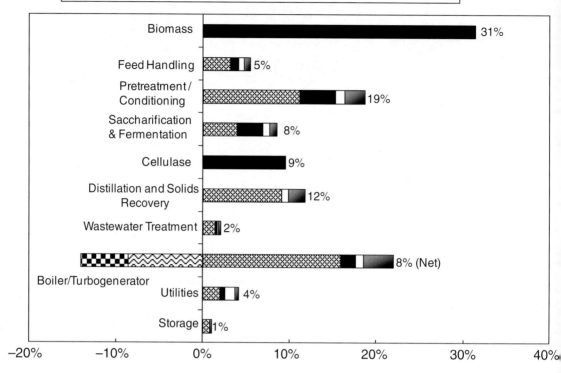

Plate 15.1 *Cost breakdown by process area of cellulosic ethanol using dilute acid pretreatment.*

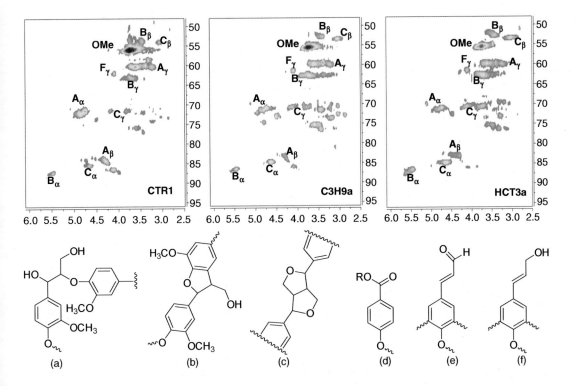

Plate 18.6 *Aliphatic regions of ¹³C/¹H HSQC NMR spectra of alfalfa ball-milled lignins and the identified structures [38]. A: β-O-4 ether linkage; B: phenylcoumaran (β-5/α-O-4); C: resinol (β-β); OMe: methoxyl group; D: p-hydroxybenzoate; E: cinnamaldehyde; F: cinnamyl alcohol; R = H or C; CTR1: wild type; C3H9a: p-coumarate 3-hydroxylase (C3H) transgenic line; HCT3a: hydroxycinnamoyl CoA:shikimate/quinate hydroxycinnamoyl transferase (HCT) transgenic line.*

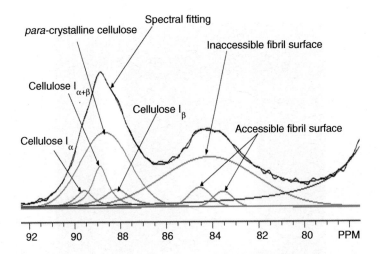

Plate 18.11 *Spectral fitting for the C-4 region of CP/MAS ¹³C-NMR spectrum of native Buddleja davidii cellulose [28]. $I_{\alpha+\beta}$: I(α + β).*

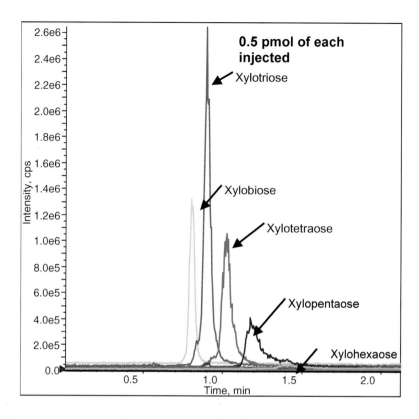

Plate 19.5 *UPLC chromatogram of xylooligosaccharides of DP 2–6. (From Tomkins et al. [93] with permission of the authors).*

Plate 20.3 *Wire-mesh basket and recirculation vessel and pump for small-batch biomass impregnation (National Renewable Energy Laboratory, Golden CO, USA).*

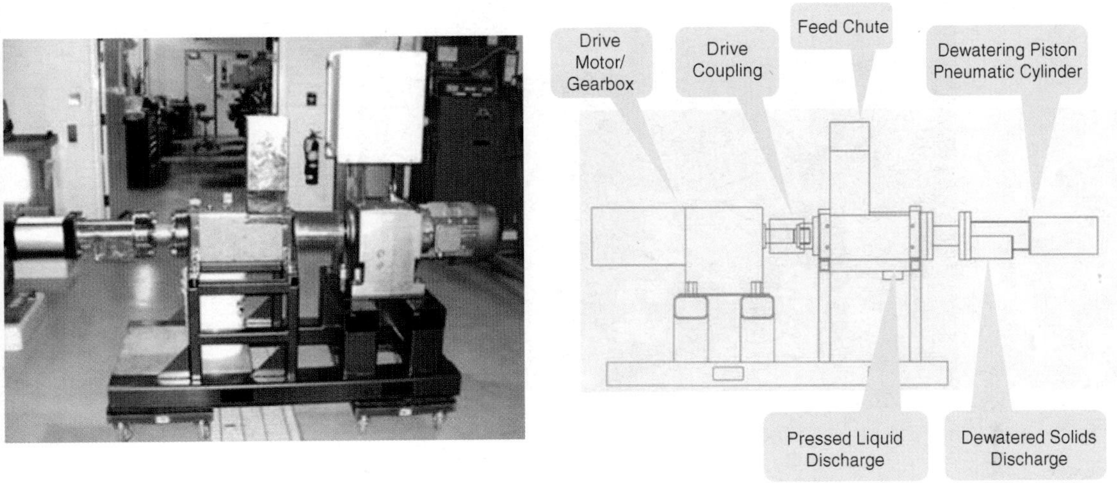
Drive
Motor/
Gearbox
Drive
Coupling
Feed Chute
Dewatering Piston
Pneumatic Cylinder
Pressed Liquid
Discharge
Dewatered Solids
Discharge

Plate 20.4 *Customized dewatering screw press for dewatering pre-impregnated biomass (National Renewable Energy Laboratory, Golden CO, USA).*

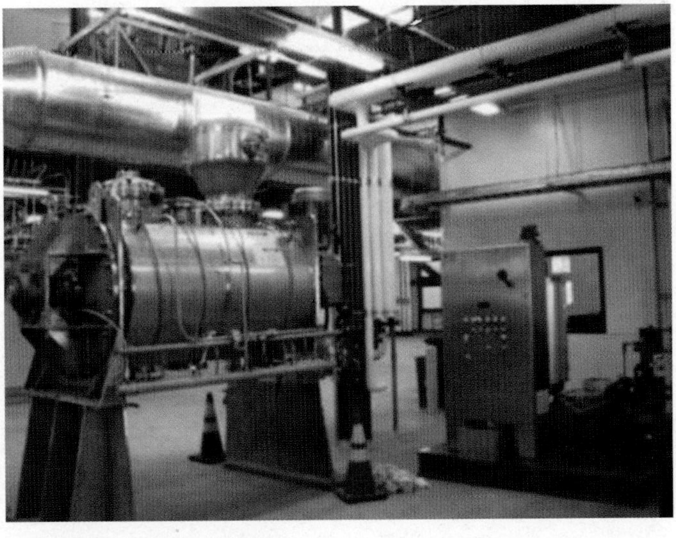

Plate 20.5 *2000 L jacketed high-solids biomass impregnation vessel (National Renewable Energy Laboratory, Golden CO, USA).*

(a) Rotating horizontal reactor system with roller bottles.

(b) Shaking vertical reactor system with modified centrifuge bottles. Photo courtesy of Mascoma Corporation.

Plate 21.1 *Laboratory-scale systems for high-solids enzymatic hydrolysis. (a) Rotating horizontal reactor system with roller bottles. (b) Shaking vertical reactor system with modified centrifuge bottles. (Photo courtesy of Mascoma Corporation).*

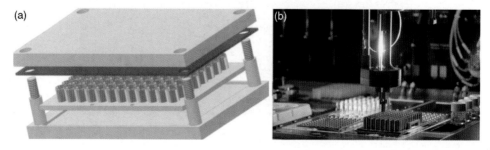

Plate 22.3 *UCR's HTPH reactor system [25] including the original reactor design in which Hastelloy wells with a 250 mg reaction mass were (a) press fit into an aluminum plate clamped between two stainless steel plates during pretreatment, and (b) the updated reactor with larger free-standing Hastelloy wells (450 mg reaction mass) being loaded by a Symyx Core Module.*

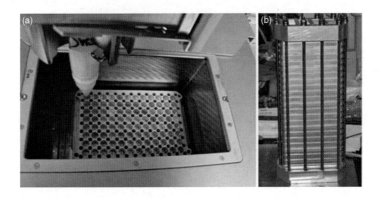

Plate 22.4 *The NREL HTPH reactor system uses the Symyx Powdernium to dispense biomass into (a) the wells of the 96-well reactor plate with a reaction mass of 255 mg per well with (b) 20 reactor plates stacked together in a modified 2-gal Parr reactor for pretreatment. (Reproduced from Decker et al. [12]).*

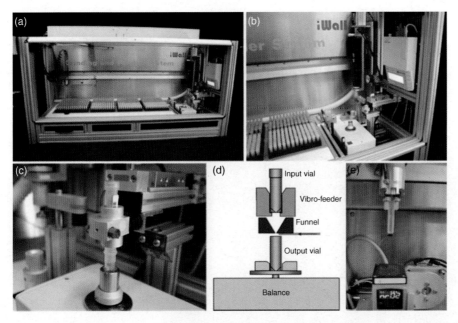

Plate 22.5 *GLBRC's HTPH system, including (a) iWall robotics platform for milling and dispensing. More detailed views of the (b) weighing substation, (c) balance and vibro-feeder dispensing from input (upper) to output (lower) tube, (d) diagram of weighing substation, and (e) bar code scanner substation are also shown. (Reproduced from Santoro et al. [23] with permission from Springer).*

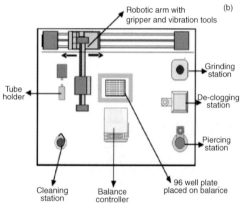

Robotic arm with
gripper and vibration tools

Grinding
station

Tube
holder

De-clogging
station

Piercing
station

Cleaning
station

Balance
controller

96 well plate
placed on balance

Plate 22.6 *HTPH system described by Gomez* et al. *including (a) general view of robotics platform for milling and dispensing and (b) schematic of robot's different substations. (Figure obtained from Gomez* et al. *2010 [22]).*

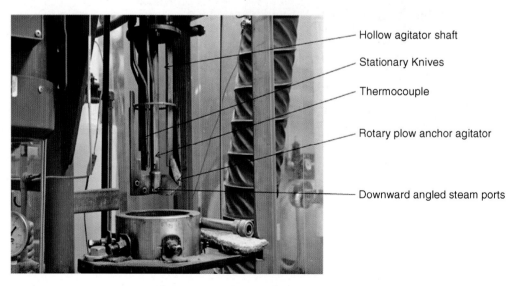

Hollow agitator shaft

Stationary Knives

Thermocouple

Rotary plow anchor agitator

Downward angled steam ports

Plate 23.3 *NREL Zipperclave reactor system.*

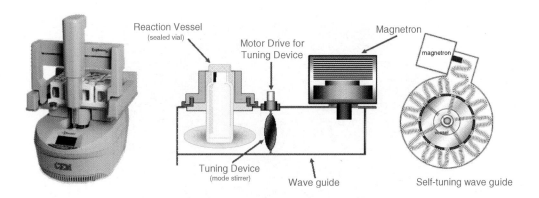

Reaction Vessel
(sealed vial)

Motor Drive for
Tuning Device

Magnetron

magnetron

Tuning Device
(mode stirrer)

Wave guide

Self-tuning wave guide

Plate 23.4   *Single-mode microwave system. (Adapted from http://www.cem.com/).*

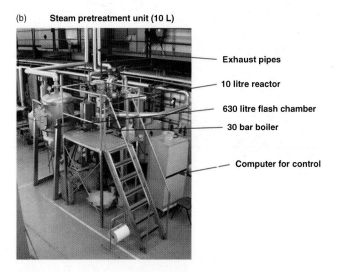

(b)   **Steam pretreatment unit (10 L)**

Exhaust pipes

10 litre reactor

630 litre flash chamber

30 bar boiler

Computer for control

Plate 23.5   *University of Lund 10 L steam explosion reactor. Photo of 10 L steam explosion reactor, flash tank, boiler and computerized controls. (Used with permission of Professor Guido Zacchi, Lund University, Sweden).*

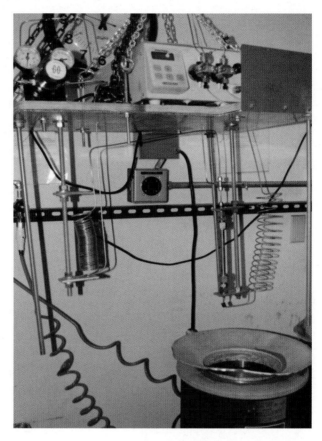

Plate 23.6 *Flowthrough pretreatment reactor system. (Adapted from Yang et al. [111] © 2004, American Chemical Society).*

13

Effects of Enzyme Formulation and Loadings on Conversion of Biomass Pretreated by Leading Technologies

Rajesh Gupta[1] and Y.Y. Lee[2]
[1] *Chevron ETC, Houston, USA*
[2] *Department of Chemical Engineering, Auburn University, USA*

13.1 Introduction

According to a recent economic analysis by the National Renewable Energy Laboratory (NREL, Golden, CO), enzymes are the second-largest cost item (first being the feedstock) in the operating cost of ethanol production from lignocellulosic biomass via the conventional bioconversion scheme that involves pretreatment and enzymatic hydrolysis. The report also points out that reduction of enzymes cost is the biggest challenge facing commercialization of cellulosic ethanol technology. Since it is proportional to the enzyme loading, enzymes cost is highly dependent upon the type of biomass feedstock and pretreatment technology used in the conversion process [1].

The plant cell wall is a composite of multiple carbohydrates, protein, and lignin and this architecture necessitates a variety of enzymes for its degradation [2]. Efficient degradation of biomass polysaccharides requires synergistic action of a broad spectrum of proteins meant to cleave specific substrates present in lignocellulosic biomass. Cellulase, produced by various biomass-degrading microorganisms, is a complex mixture of diverse enzymes which work synergistically to degrade cellulose and hemicellulose. Five endoglucanases (EG-1 to EG-5), two exoglucanases or cellobiohydrolases (CBH-1 and CBH-2), β-glucosidases (β-Gs), and several hemicellulases (including xylanase, β-xylosidases or β-Xs, and others) have been identified in the *T. reesei* cellulase complex [3]. Among these proteins, CBH-1, CBH-2, and EG-1 are the three main proteins in the *T. reesei* cellulase with respective portions of about 60%, 20%, and 12% [4]. As measured by Dien *et al.*, different enzymatic activities are found in commercial cellulase and auxiliary enzyme

Aqueous Pretreatment of Plant Biomass for Biological and Chemical Conversion to Fuels and Chemicals, First Edition.
Edited by Charles E. Wyman.
© 2013 John Wiley & Sons, Ltd. Published 2013 by John Wiley & Sons, Ltd.

preparations used for biomass hydrolysis research [5]. Evidently, none of these enzyme preparations are pure enzyme but contain different activities to target specific bonds. This is also evidence that multiple auxiliary enzymes are important for realizing high yields of sugar from biomass.

The activities of the functional components in the cellulase system are not necessarily balanced at optimal proportions for effective biomass hydrolysis, making this an area of research that can significantly impact production costs. The optimal mix of the enzyme components in cellulase depends on substrate characteristics [6], and optimization of the enzyme mixture can have a direct impact on the process economics (reduction of enzyme cost). A well-balanced enzyme mixture can reduce pretreatment temperature, chemical usage, and reaction time [7,8], thus directly reduces the severity and makes the process safer and more reliable. Low severity in pretreatment will also reduce degradation of biomass and hazardous chemical disposal from the process [9]. Enzyme tailoring can therefore bring about benefits in environmental, financial, and safety aspects of the biofuels industry. This chapter analyzes requirements for various functional components in cellulase enzyme systems and discusses how this is influenced by biomass structure and properties. We also discuss specific functions of these enzymatic components for biomass feedstocks pretreated by various methods and how they interact and affect the effectiveness of the enzymatic digestion of lignocellulosic biomass.

13.2 Synergism among Cellulolytic Enzymes

Cellulolytic enzymes in cellulase comprise three functional protein groups, namely: endoglucanases (EG), exoglucanases (Exo-G), and β-G, which act synergistically to hydrolyze the crystalline molecule of cellulose. The degree of polymerization (DP) of crystalline cellulose varies from approximately 500 to 14 000 depending upon the origin and its respective location in the complex cell wall of biomass [10]. After the exposure of cellulose molecules to enzymatic attack, EG preferentially acts on the amorphous part of the cellulose chain and produces shorter DP cellulose chains as well as cello-oligosaccharides [10,11]. Exo-G acts on the cellulose chain from the reducing (CBH-1) or non-reducing end (CBH-2) processively and releases cellobiose as the main product. Thereafter, β-G hydrolyzes the soluble cellobiose and higher-DP cello-oligosaccharides (COS) from the non-reducing ends to form glucose as the end-product of cellulose hydrolysis. Cellulase is inhibited by cellobiose and adding sufficient β-G also reduces competitive inhibition of cellobiose [12].

Degree of synergy among the enzyme components is defined as the ratio of activity of the enzyme mixture on a particular substrate divided by the sum of the individual activities of these enzyme components on the same substrate [13]. Degree of synergy between EG/Exo-G in cellulose hydrolysis depends upon substrate characteristics. Pretreatment not only alleviates the barrier presented by lignin and hemicelluloses, but also alters the structure of cellulose molecule. Literature information collectively indicates that these structural changes are not limited to only DP and crystallinity of cellulose but include modifications to specific surface area (SSA), fraction of β-glycosidic bonds accessible to enzyme (F_a), fraction of reducing ends (F_{RE}), swelling, and change in molecular orientation [13,14].

Zhang *et al.* [13] delineated a correlation among the aforementioned properties in various pure cellulosic substrates and their hydrolysis behavior with cellulase. Bacterial cellulose and cotton, which have high DP and high crystallinity, exhibited the highest degree of synergy. Filter paper, with high DP and moderate crystallinity, had less degree of synergy. The least synergies were shown by microcrystalline cellulose with its low DP and moderate crystallinity and amorphous cellulose with very low DP and very low crystallinity. Proper understanding of how these characteristics of cellulose affect synergy can be helpful in adjustment of the relative ratio of enzyme components in cellulase. For example, Avicel requires very low EG and low β-G activity while amorphous cellulose requires very low EG but very high β-G activity [10,11]. Cotton-type structures with high DP typically require high EG and high β-G activity [15]. Furthermore, high Exo-G activity is essential for all types of insoluble cellulosic substrates. On the other hand, soluble COS requires high β-G activity [11]. From a different viewpoint, cellulosic substrates with a high F_{RE}/F_a ratio such as Avicel are good choices for measurement of Exo-G activity; substrates with a very low F_{RE}/F_a ratio, such as carboxymethyl cellulose (CMC), are good choices for measurement of EG activity [13].

Efficient mixing and proper enzyme adsorption on its substrate are greatly affected by solid loading during enzymatic hydrolysis [16]. High EG activity is especially important at high substrate loadings because it greatly assists in viscosity reduction, hence promotes uniform mixing and distribution of enzyme. Among EG, Cel7B (EG-1) and Cel5A (EG-2), EG-1 was most effective in reducing viscosity [17]. However, Gao *et al.* [18] observed that both CBH-1 and CBH-2 are indispensable for efficient biomass hydrolysis. Their study also indicated that the synergistic role of EG-1 with other enzyme components was more important at lower protein loadings than at higher protein loadings. One explanation for this observation stated in the study was substrate inhibition at low enzyme loadings. For low enzyme to substrate ratios, the availability of reducing ends for CBH to work on would be very low because there would be a much higher number of glycosidic bonds near CBH molecules compared to the reducing end. Therefore, increasing EG-1 becomes highly effective in enhancing hydrolysis as it generates additional reducing ends [15]. Another explanation for this phenomenon was given by Gupta and Lee [11], who demonstrated that only a minimum activity of EG was sufficient to hydrolyze semi-crystalline cellulose to a low DP crystalline structure, after which EG activity was not required and further hydrolysis was carried out solely by Exo-G.

13.3   Hemicellulose Structure and Hemicellulolytic Enzymes

Feedstocks appropriate for biofuels production come from various sources. The US Department of Energy (DOE) has divided them into a number of different categories including energy crops, forest residue, agriculture residue, and waste materials [19]. These substrates have very different chemical compositions and structures, with hemicellulose molecular structure being one. Due to different hemicellulose structure and composition, the need for hemicellulolytic enzymes in the cellulase enzyme mixture varies with the biomass feedstock to be hydrolyzed. It is therefore vital to understand hemicellulolytic enzyme roles to optimally design "enzyme cocktails."

Hemicellulose is a hetero-linked mixture of various subcomponents including xylans, mannans, arabinans, heteroxylans, galactomannan, glucomannan, galactoglucomannan, arabinoxylan, and acidic residues (glucuronic, acetyl). These are named according to composition and intra-structural bonding. The composition of these hemicellulosic fractions varies from one wood species to another. In general terms, hardwood hemicellulose contains mostly xylans while softwood hemicellulose contains predominately xylan and glucomannan [2,20]. Grasses typically contain 20–40% arabinoxylan (arabino-4-O-methylglucurono-xylan) with varying ratios of arabinan and xylan [21].

Hardwood hemicellulose contains mostly xylans and is laced with groups of 4-O-methylglucuronic acid (Me-GcluU) with (1–2)-glycosidic linkages to xylose units. Many of the —OH groups at the C2 and C3 position of xylose are substituted with an O-acetyl group. Hardwood xylan has two or three branching points with very short chains linked at the C3 position of the backbone. Minor amounts of rhamnose and galacturonic acid are associated with the main chain of hardwood xylan [2,21]. Softwood xylan differs from hardwood xylan by the lack of an acetyl group and by the presence of arabinofuranose units linked by α-(1–3)-glycosidic bonds to the xylan backbone. Softwood xylan has a higher proportion of 4-O-methylglucuronic acid. Higher-molecular-weight xylans in softwood contain increased numbers of arabinose units and more branching points. Arabinofuranose units are esterified with p-coumaric acid and ferulic acid [22].

Wood mannans are more specifically referred to as glucomannans, with the backbone consisting of mannose and glucose units. The ratio of mannose to glucose in glucomannan is about 1.5–2:1 in hardwood and 3:1 in softwood. Softwood contains about 20–25% glucomannan, while hardwood has only 3–5%. Acetyl groups and galactose groups are also attached to glucomannan. On average, one acetyl group per 3–4 hexose units is present. There is evidence of linkages between galacto-gluco-mannans and lignin [20].

Xylanase hydrolyze the β-(1–4) bonds between xylose residues in xylan chains including heteroxylan and xylooligosaccharides (XOS). The affinity and reactivity on XOS for xylanase decrease as DP decreases. Few xylanases are very specific towards xylan molecules, although some hydrolyze cellulose molecules and other may just bind the cellulose. However, reactivity of cellulose molecules with xylanases is very low in comparison to xylan [23,24]. Acetylation of the xylan molecule reduces xylanase reactivity, hence the end-product of heteroxylan reacted with xylanase is a lower DP but highly substituted XOS [25]. Some xylanase, however, can attack heteroxylan chains near a substituted site [26]. The overall reactivity of xylanase can be affected by the presence of auxiliary debranching enzymes α-L-arabinofuranosidase (AFE), acetyl-xylan esterase (AXE), or feruloyl esterase (FE). Similar to β-G, β-xylosidase (β-X) catalyzes hydrolysis of XOS and xylobiose. The reactivity of β-G and β-X increases with a decrease in oligosaccharide DP [27,28]. Further discussion on the role of these enzymes and perspectives on their use in biomass hydrolysis research are presented in Sections 13.5–13.9.

13.4   Substrate Characteristics and Enzymatic Hydrolysis

The overall efficiency of cellulase enzymes is strongly influenced by various factors related to biomass structure (biomass surface area, crystallinity index, and DP of cellulose), composition (lignin/hemicellulose ratio, hemicellulose composition), transport properties (external and internal diffusion resistance, adsorption characteristics), and enzyme reaction kinetic parameters (product and substrate inhibition, enzyme deactivation) [10,29,30]. Figure 13.1 provides a simplified representation of different resistances for cellulose hydrolysis by cellulase. Physical resistance is mainly controlled by transfer of the proteins and successive adsorption onto the biomass surface. Non-cellulosic reaction resistance is related to lignin and hemicellulose structure. These resistances are shown in parallel because they are not independent, and efficient pretreatment benefits can be realized by removing just one of the two resistances. Most pretreatments are designed to reduce and/or alter the structures of lignin or hemicelluloses, consequently enhancing enzyme accessibility to cellulose [8,32]. Cellulose structure such as crystallinity and DP affects cellulosic reaction resistances. Overall, the highest resistance is due to cellulose crystallinity that severely limits the extent and the rates of glucan hydrolysis [33,34]. All of these resistances are further affected by inhibition and other reaction kinetic features of enzymes, as described in this chapter [31].

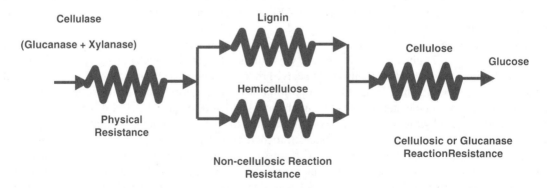

Figure 13.1   *Schematic of resistances encountered by cellulase enzyme in hydrolysis of cellulose within biomass [31]. (See figure in color plate section).*

High enzyme efficiency (defined as sugar yield from biomass/total protein loading) is a prerequisite for an economically viable industrial process. Enzyme efficiency was found to be higher at lower enzyme loadings [35]. This is evident in hydrolysis yields versus mass protein plot with different pretreated feedstocks [36]. Zhu *et al.* [34] indicated that lignin and acetyl contents (the latter being typically proportional to hemicellulose content) affected the amount of enzyme adsorbed onto biomass, but crystallinity controlled enzyme effectiveness. In another study, Zhu *et al.* [36] observed that, for moderate enzyme loadings, biomass hydrolysis rate (Y) depended linearly on the natural logarithm of enzyme loading (E). The slope and intercept of the resulting straight line described by the equation $Y = A \ln (E) + B$ primarily depended on lignin content, acetyl content, carbohydrate content, and cellulose crystallinity. At very low and very high enzyme loadings, enzyme efficiency was much lower than for the moderate loading region. At very low enzyme loadings, physical resistance and unproductive binding of enzyme to lignin could primarily control hydrolysis [37]. At very high enzyme loadings, cellulose hydrolysis did not seem to increase beyond a certain limit [8]. Structural features of hydrolyzed residual cellulose [11] or steric hindrance of enzyme already attached to cellulose might be responsible for this phenomenon [38].

13.5 Xylanase Supplementation for Different Pretreated Biomass and Effect of β-Xylosidase

Several methods have recently been evaluated for pretreatment of corn stover, hybrid poplar, and switchgrass. Pretreatments included dilute acid (DA), SO_2 catalyzed steam explosion, controlled-pH hot water (CHW), lime, ammonia fiber expansion (AFEX), and ammonia recycle percolation (ARP). The effort, undertaken by CAFI (Consortium for Applied Fundamentals and Innovation), was unique as all the substrates were evaluated using a standard protocol and the same commercial enzymes. One of the issues addressed by this group was the effect of xylanase addition on various pretreated substrates. Tables 13.1–13.4 present data compiled from CAFI [41] projects. These pretreatment methods used different reagents that spanned from low pH to high pH and different process conditions that caused diverse structural and physical changes in

Table 13.1 *Change in biomass composition after different pretreatment methods [8,33,39].*

		Corn Stover			Poplar			Switchgrass		
		G^1	X^1	L^1	G	X	L	G	X	L
Untreated		38.3	21.7	20.5	43.8	14.8	29.1	35.6	22.6	21.1
DA	As is	59.3	9.3	22.5	57.3	2.1	46.1	50.3	4.5	29.4
SO_2		56.9	11.6	23.8	55.1	2.5	NR^3	53.9	2.7	27.6
CHW		52.7	16.2	25.2	58.8	7.0	32.2	50.1	2.5	30.6
Lime		56.7	26.4	14.6	53.1	16.8	18.0	53.0	21.5	14.6
ARP		61.9	17.9	8.8	57.5	13.5	24.8	55.6	21.9	13.9
AFEX		34.4	22.8	18.0	46.6	15.0	NR	35.9	22.5	24.4
Untreated		38.3	21.7	20.5	43.8	14.8	29.1	35.6	22.6	21.1
DA	Based upon the	35.8	5.9	13.9	36.0	1.3	28.9	30.4	2.7	17.8
SO_2	untreated biomass	37.1	7.6	15.5	36.4	1.7	NR	33.6	1.7	17.2
CHW		36.0	11.1	17.2	47.6	5.7	26.1	30.1	1.5	18.4
Lime		37.2	17.3	9.6	44.0	13.9	14.9	34.6	14.0	9.5
ARP		37.8	10.4	5.3	41.9	9.8	18.1	34.5	13.6	8.6
AFEX		38.3	21.7	18.6	44.9	14.4	NR	35.9	22.5	24.4

1. G, X and L refer to glucan, xylan, and lignin content in biomass.
2. All the compositions are %wt/wt, oven dry basis.
3. NR: Not reported

Table 13.2 *Effect of xylanase addition on glucan and xylan digestibility for pretreated corn stover [33].*

| | Enzyme loading (mg/g glucan) | | | |
| | C + β-G | | C + β-G + X | |
	14.5 (14 + 0.5)	29(28 + 1)	20 (14 + 0.5 + 5.5)	29 (14 + 0.5 + 14.5)
	Glucan digestibility			
DA	49	56	50	50
SO_2	74	81	80	85
CHW	70	83	75	78
Lime	72	82	73	76
ARP	64	73	74	78
AFEX	64	72	77	82
	Xylan digestibility			
DA	68	73	78	76
SO_2	52	58	61	70
CHW	25	50	45	48
Lime	36	62	47	55
ARP	48	58	66	70
AFEX	61	69	67	69

1. C, β-G and X refer to cellulase, β-glucosidase, and xylanase respectively.
2. Glucan/xylan digestibility is defined as amount (grams) of glucan/xylan solublized from 100 g of glucan/xylan present in substrate during the enzymatic hydrolysis of biomass.

biomass. This dataset provides useful information regarding how the effectiveness of different enzyme formulations varies with properties resulting from various pretreatments.

Table 13.2 shows that supplementation of commercial cellulase with xylanase (both enzymes from Genencor) provides significant benefits for hydrolysis of ARP- and AFEX-pretreated corn stover, but had negligible effect on the performance of DA-pretreated corn stover. These data therefore proved that adjusting enzyme formulation, such as substituting xylanase for a portion of cellulase while keeping the total protein loading constant, improved enzyme efficiency for ARP- and AFEX-pretreated corn stover. However, adding xylanase had limited impact on hydrolysis performance for DA-, controlled-pH hot water (CHW), and lime-pretreated corn stover. The reason is quite clear: ARP and AFEX pretreatment leave large amounts of xylan intact in the biomass that xylanase addition clears away to effectively pave cellulose accessibility. Gupta *et al.* [42] also observed that xylanase supplementation was more beneficial for pretreated biomass with higher residual xylan, as is typically true for alkaline pretreatments that remove more lignin than hemicellulose.

The commercial "cellulase" from *T. reesei* contains hemicellulolytic activities including xylanase, but at much lower levels than present in commercial xylanases [5]. Supplementation with xylanase is therefore an effective tool for boosting overall activity. With ARP- and AFEX-treated corn stover, addition of xylanase enzyme to very low cellulase loading (7.5 filter paper units or FPU) works more effectively than the equivalent protein loading of cellulase. In contrast, with lime-, DA-, and CHW-treated corn stover, even adding very high amounts of xylanase to 7.5 FPU cellulase did not result in greater hydrolysis rates than achieved with 15 FPU of just cellulase with no xylanase addition [33]. This observation is consistent with the noted low levels of xylan in dilute acid and hot-water-treated corn stover (Table 13.1), but not with the very high level of xylan and low level of lignin in corn stover treated with lime. Two factors may partially explain this phenomenon: (1) higher cellulose reactivity in ARP- and AFEX-treated corn stover than for non-ammonium-based pretreatments; and (2) very high resistance to cellulase from residual lignin left intact in lime-, DA-, and CHW-treated corn stover. With SO_2-treated corn stover, digestibility seems independent of enzyme combinations but depends upon the total protein loading [33].

Table 13.3   *Effect of xylanase addition on glucan and xylan digestibility for pretreated hybrid poplar [8].*

| | Enzyme loading (mg/g glucan) | | | |
| | C + β-G | | C + β-G + X | |
	14.5 (14 + 0.5)	29 (28 + 1)	20 (14 + 0.5 + 5.5)	29 (14 + 0.5 + 14.5)
	Glucan digestibility			
DA	53	68	60	66
SO_2	57	61	65	73
CHW	NR	40		
Lime	39	60	48	57
ARP	35	51	40	41
AFEX	15	38	21	23
	Xylan digestibility			
DA	64	100	80	91
SO_2	100	100	100	100
CHW		38		
Lime	39	56	49	59
ARP	40	55	49	54
AFEX	24	45	32	35

1. C, β-G and X refer to cellulase, β-glucosidase, and xylanase respectively.
2. Glucan/xylan digestibility is defined as amount (grams) of glucan/xylan solublized from 100 g of glucan/xylan present in substrate during the enzymatic hydrolysis of biomass.

These trends do not however apply to hybrid poplar (Table 13.3). With hybrid poplar, the same reformulation of the enzyme (substituting part of cellulase with xylanase keeping the total protein loading constant) did not increase glucan digestibility for any of the pretreatments tested except for SO_2 [8]. Although the mechanism is not entirely clear, this observation may be related to the fact that the residual lignin in treated hybrid poplar was much higher than that in pretreated corn stover and switchgrass. For switchgrass however, xylan digestibility of alkali-treated biomass increased after substituting the xylanase in lieu of some of the protein loading in the enzyme mixture (Table 13.4). The primary reason for this is that XOS left unconverted due to low β-X activity in cellulase mixture [43] were eventually hydrolyzed to xylose by supplemental xylanase. Xylanase is also known to promote lignin removal by breaking bonds between lignin and hemicellulose [44], thereby enhancing access by cellulase and reducing unproductive binding of enzymes [45].

Kumar and Wyman [33] defined xylanase leverage as the ratio of percentage increase in glucose release to the percentage increase in xylose release due to addition of xylanase. They found that the xylanase leverage not only varied with type of pretreatment but also with enzyme protein loading. For low enzyme loadings, alkali-treated biomass (AFEX and ARP) typically exhibited higher leverage than DA- or CHW-treated biomass. At high enzyme loadings however, CHW-treated biomass (both corn stover and poplar) showed exceptionally high leverage. In this study, it was concluded that the two main factors affecting xylanase leverage were residual acetyl content in pretreated solids and the concentration of soluble XOS, which were found to be inhibitory to enzyme activity [46].

Reduction in acetyl content was also found to enhance xylanase leverage, indicating that higher acetyl content posed an additional hurdle to cellulose accessibility by enzymes [33]. It has been reported that enzymes (including xylanase) are less active on xylan chains that are highly substituted and contain side groups [25]. The same is true for xylan that originate in the pericarp and germ fractions of corn kernels and are contained in distiller's dried grains with soluble (DDGS), a co-product from corn ethanol production [5]. This xylan is known to be highly substituted and is not easily degraded by fungal hydrolytic enzymes

Table 13.4   *Effect of xylanase addition on glucan and xylan digestibility for pretreated switchgrass [39].*

	Enzyme loading (mg/g glucan)				
	C + β-G			C + β-G + X	
	13.4 (10 + 3.4)	29(28 + 1)	33.4 (30 + 3.4)	13.4 (8.3 + 3.4 + 1.7)	33.4 (15 + 3.4 + 15)
	Glucan digestibility				
DA	53	71	77	58	77
SO$_2$	90	96	91	85	87
CHW	56	72	75	55	75
Lime	62	74	79	63	79
ARP	49	54	69	40	73
AFEX	30	43	43	33	45
	Xylan digestibility				
DA	39	69	55	39	55
SO$_2$	80	NM	80	79	71
CHW	84	79	71	86	72
Lime	47	74	62	52	77
ARP	30	42	50	34	60
AFEX	41	58	51	48	59

1. C, β-G and X refer to cellulase, β-glucosidase, and xylanase respectively.
2. Glucan/xylan digestibility is defined as amount (grams) of glucan/xylan solublized from 100 g of glucan/xylan present in substrate during the enzymatic hydrolysis of biomass.

[2,32]. Removal of acetyl content would be helpful in de-branching of the xylan chain and enhancing xylanase activity on hemicellulose.

Hydrolysis of solids resulting from pretreatment of corn stover by different technologies with a moderate cellulase loading of 16 mg/g glucan resulted in glucose yields of 70–80% for all pretreated solids; however, xylose yields were only 40–70%. It was then found that XOS produced from these solids during enzymatic hydrolysis strongly inhibited cellulase and xylanase activity [40]. Furthermore, cellulase inhibition increased with XOS chain length [46]. Alkaline-pretreated solids (i.e., AFEX, ARP, and lime) generated more XOS than acid-treated solids (i.e., DA and SO$_2$) [43]. These oligomers are believed to be one of the main reasons for higher xylanase leverage with alkali-treated biomass than with acid-treated biomass. As xylanase converts these XOS to xylose monomers in enzymatic hydrolysis of alkali-treated biomass, inhibition to cellulase activity is reduced thus increasing the cellulose and xylose hydrolysis.

Excess xylanase addition may also negatively affect glucanase activity, as found in an optimization study on enzyme reformulation using dilute-acid-treated corn stover [47]. Competition for the same active site between cellulase and xylanase was cited as one possibility. Similar observations were made with lime-pretreated hybrid poplar, for which the maximum hydrolysis yield was obtained with 67% cellulase, 12% β-G, and 24% xylanase at a total enzyme loading of 61 mg/g glucan [35]. In a recent study, xylanase was found to be negatively impacting the simultaneous saccharification and fermentation (SSF) of steam-pretreated biomass [48]. These studies indicated that a careful choice of xylanase loading should be made depending upon the substrate characteristics.

β-X activity is required to convert the XOS and xylobiose generated in enzymatic hydrolysis of biomass. In hydrolysis tests with high-DP XOS using various enzymes (commercial cellulase, β-G, xylanase, and β-X), the highest hydrolysis efficiency was observed with β-X, but even β-X could not hydrolyze 10–20% of the high-DP XOS [46]. Incomplete hydrolysis can be attributed to arabinose side chains in XOS, and complete hydrolysis would require enzymes with debranching activity such as AFE [49].

13.6   Effect of β-Glucosidase Supplementation

Supplementation with β-G is very effective in increasing overall hydrolysis of cellulosic substrates. Berlin *et al.* [50] hydrolyzed organosolv-pretreated hardwood with various cellulase preparations and obtained very different specific conversions. These preparations had a wide range of cellulolytic and hemicellulolytic activities. When these cellulase mixtures were supplemented with exogenous β-G, the range of differences in cellulose-specific conversion became very narrow, reaffirming the negative effect of sugar oligomers on hydrolysis. Because typical cellulase preparations are deficient in β-G activity as well as β-X activity, COS and XOS accumulate during cellulose and xylan hydrolysis, respectively [51]. Gao *et al.* [52] showed that addition of β-G did not improve hydrolysis yields beyond a certain extent. According to Breuil *et al.* [53], although exogenous β-G addition increased the long-term hydrolysis of lignocellulosic biomass, it did not show any significant change in the filter paper activity of the enzyme mixture. Filter paper activity is an indication of initial hydrolysis rate by the cellulase and does not reflect the β-G activity [54]. If substrate does not accumulate much cellobiose and oligosaccharides in the initial phase of hydrolysis reaction, addition of β-G might not affect filter paper activity [55].

With aqueous-ammonia-pretreated switchgrass, addition of β-G to cellulase increased glucan digestibility from 50% to 70% and xylan digestibility from 42% to 53% [56]. The β-G preparation not only hydrolyzed COS but also XOS due to the simultaneous presence of β-X activity [5]. Addition of β-G was especially helpful for hydrolysis at high solids loadings due to reduction of cellulase activity inhibition by cellobiose and oligosaccharides [46,57]. These observations collectively indicate that controlled supplementation with this enzyme can improve enzyme efficiency.

13.7   Effect of Pectinase Addition

Pectinase hydrolyzes the pectin polymer (galactouronan) in biomass hemicellulose, and supplementation with pectinase can enhance hydrolysis of substrates that have high amounts of pectin [58,59]. Commercial pectinase preparations (such as Multifect pectinase from Genencor-Danisco) exhibit much higher β-X and β-G activity than Multifect xylanase [5]. Perhaps because of this, supplementing pectinase also enhances hydrolysis of certain types of substrates that produced high amounts of XOS in enzymatic digestibility with cellulase. It was also shown that addition of pectinase to cellulase-produced equivalent or higher glucose and xylose yields as addition of β-G [59]. In a separate study with CHW- and AFEX-treated DDGS, addition of pectinase to cellulase increased xylan digestibility by 145% and 350%, respectively, while addition of xylanase only increased xylan digestibility by 33% and 55% [5]. Addition of feruloyl esterase (FE) with pectinase and cellulase increased digestibility by an additional 15% (CHW) and 20% (AFEX), plateauing at approximately 80% for both cases.

Similarly, adding pectinase to CHW-treated switchgrass was more effective in improving xylose yields than xylanase addition, but glucose yields were improved more by adding xylanase [60]. CHW-pretreated biomass that contained large amounts of XOS [43] was hydrolyzed better by pectinase than xylanase. Thus, adding pectinase would be very helpful when large amounts of soluble oligomers are produced during enzymatic hydrolysis. The amounts of XOS were measured during hydrolysis of aqueous-ammonia-treated switchgrass with β-G, xylanase, and pectinase additions with constant loading of cellulase. Although xylanase addition resulted in higher hydrolysis rates, pectinase addition reduced XOS levels more than xylanase [59]. In this case, however, adding xylanase produced better results than pectinase in enhancing xylan as well as glucan digestibility due to better action of xylanase on hemicellulose. With ammonia-treated hybrid poplar, pectinase could not match xylanase in enhancing glucan yields [31]. These observations indicate that proper adjustment of pectinase and xylanase loading is important in hydrolysis of XOS as well as reducing recalcitrance due to the presence of hemicellulose.

13.8   Effect of Feruloyl Esterase and Acetyl Xylan Esterase Addition

Unlike *A. niger* and *H. insolens, T. reesei* does not produce FE. Thus, supplementing *T. reesei* cellulase with *A. niger* or *H. insolens* enzymes substantially enhanced digestibility of associated arabinoxylan in grains [5]. It was also reported that FE is produced along with xylanase by different organisms but, in those cases, FE activity is much lower than xylanase activity [61]. As mentioned in the previous section on pectinase, addition of FE to cellulase and pectinase further increased digestibility of DDGS pretreated with CHW (15%) and AFEX (20%) [5]. It was further observed with AFEX-treated hybrid poplar that addition of 1 mg FE to cellulase enhanced glucan digestibility more than the addition of 30 mg of xylanase or 50 mg of pectinase [62]. These results should not however be a benchmark for process applications, as the maximum glucan digestibility was only 35%.

According to Cosgrove [63], arabinoxylan chains bind to cellulose microfibrils and are tied together by ferulic acid esters bonds. It seems that FE increases cellulase access to cellulose by breaking these linkages. Dinis *et al.* [64] also pointed out that xylanase and FE work synergistically to deconstruct the hemicellulose-lignin (H-L) matrix. FE activity was found to promote bio-delignification of wheat and oilseed flax straws by breaking ester-linked ferulic and p-hydroxycinnamic acid structures in the H-L matrix [65,66].

A very high degree of synergy was detected between xylanase and AXE in xylan hydrolysis. This synergy was much more prominent than synergy between xylanase and FE with hot-water-treated corn stover [67]. Recently, Agger *et al.* explained the comparative effect of FE and AXE addition to hydrolysis of arabinoxylan from pretreated corn bran [68]. AXE addition was more effective in releasing xylose from insoluble xylan, while FE was more effective with soluble XOS. They also reported that the acetyl residue is attached directly to the xylan chain whereas the feruloyl residue is not. Feruloyl residues are attached with branched arabinofuran chains and also cross-link with heteroxylan chains [2,68]. In this case, removal of acetyl groups by AXE enhanced endo-xylanase (EX) activity on insoluble arabinoxylan [25]. With soluble XOS, steric hindrance by the feruloyl residue appeared to be interfering with EX, β-X, and AFE activities. Removal of feruloyl residues by FE therefore promoted hydrolysis of XOS [68]. Preference of FE for shorter-chain fragments generated by xylanase was also noticed in a study with oat hulls, in which the synergy between FE and xylanase was particularly important [69].

13.9   Effect of α-L-arabinofuranosidase and Mannanase Addition

Along with xylanase and β-X, AFE plays a key role in efficient hydrolysis of hemicelluloses and especially XOS. Hydrolysis of oat spelt xylan by purified xylanase, β-X, and AFE showed that AFE activity was essential for release of arabinose from hemicellulose. β-X released xylose from XOS, but arabinose side chains were not cleaved without AFE activity [70].

AFE has been found to be very efficient in hydrolyzing XOS that are highly substituted with arabinose [49]. However, it has been shown to hydrolyze only low-molecular-weight L-arabinofuranose containing XOS and not high–molecular-weight insoluble xylans [49,71]. AFE worked synergistically with xylanase in the hydrolysis of the arabinoxylan fraction extracted from wheat bran. AFE alone could not hydrolyze high-molecular-weight arabinoxylan, and xylose/arabinose yields could not be increased by adding xylanase after AFE with xylan. When AFE and xylanase were added simultaneously however, xylose release was 6 times higher and arabinose release was 10 times higher than with xylanase alone [49].

Beukes and Pletschke [72] studied the synergy among cellulosomal mannanase and xylanase with AFE from *T. reesei* on lime-pretreated baggase. They found that a molar ratio of 37.5% AFE, 25% mannanase, and 37.5% xylanase resulted in the highest hydrolysis conversion and highest degree of synergy. This optimum mixture resulted in a 1.5 times higher hydrolysis rate than for an enzyme mixture containing very high amounts of xylanase (75%) and small amounts of AFE (12.5%) and mannanase (12.5%). This result reaffirms the importance of auxiliary enzymes in cleaving side chains that may contain mannose and arabinose residues, thus enhancing the overall hydrolysis of complex hemicellulose polymer. Another important finding in this study was that using these enzymes individually resulted in no significant change in sugar release after 24 h of reaction, whereas the mixture of enzymes sharply increased reducing sugar levels after a certain point in time. The action of one enzyme presumably enhanced access by other enzymes, paving the way for greater hemicellulose hydrolysis. Mannanase activity did not have any appreciable effect on cellulose hydrolysis in pretreated hardwood or softwood [6,50]. Apparently residual mannan in pretreated wood did not significantly restrict enzyme access to cellulose, although synergistic action of mannanase with other hemicellulolytic enzymes can improve digestion of softwood [73].

13.10 Use of Lignin-degrading Enzymes (LDE)

For biological treatment of biomass by white-rot fungi, Dias *et al.* [74] noted that lignin-degrading enzymes such as laccase, manganese-dependent peroxidase, and lignin peroxidase cleaved lignin molecules and raised cellulolytic (Avicelase and CMCase) and xylanolytic (xylanase and FE) activities. These enzymes therefore displayed synergy for biological degradation of biomass. It was also reported that even though LDE delignified biomass, enzymatic hydrolysis was not significantly improved [75]. Although adding LDE to the enzyme cocktail could reduce the lignin barrier, organisms that produce these enzymes grow so slowly that they would not be very efficient for biomass delignification.

13.11 Effect of Inactive Components on Biomass Hydrolysis

Addition of some non-hydrolyzing components (such as certain surfactants, proteins, and polymers) to cellulase can have positive effects on biomass hydrolysis [37,76,77]. However, the extent of benefit depended upon the type of pretreatment and the specific chemical applied [8]. Bura *et al.* [78] showed that addition of BSA (Bovine Serum Albumin: A Protein) to cellulase in hydrolysis of SO_2-treated corn stover was not helpful, but found it effective with ARP-treated poplar (high lignin substrate). Tu and Saddler [79] demonstrated that addition of Tween-80 increased the effective availability of cellulase in supernatant which led to the increased efficiency of enzyme recycle in biomass hydrolysis. The effect of surfactant addition on hydrolysis of biomass was more prominent with higher lignin substrates [80]. Eriksson *et al.* [81] reported that the hydrophobic portion of surfactant bound to lignin, reducing non-productive binding of cellulase to lignin as well as enzyme denaturation, thereby enhancing hydrolysis. Certain polymers, such as polyethylene glycol (PEG), have also been shown to adsorb to lignin by hydrophobic interaction, thus increasing enzymatic digestion of biomass [82]. These observations confirm that non-hydrolytic components, capable of blocking active lignin sites that would otherwise bind cellulase, can be effectively utilized with enzyme cocktails for efficient biomass hydrolysis.

13.12 Adsorption and Accessibility of Enzyme with Different Cellulosic Substrates

As illustrated in Figure 13.1, the barrier for cellulase accessibility to cellulose is one of the main factors controlling hydrolysis. A mechanism conceived by many researchers was that this barrier contributes to slowdown in the hydrolysis rate as cellulose conversion proceeds [83]. With increasing hydrolysis time, the fraction of lignin also increases in the substrate, presenting one of the largest obstacles to cellulose accessibility. It is also shown that enzyme adsorption and activity are inhibited by sugars (monomeric as well as oligomeric) produced in hydrolysis of biomass [84,85]. This would undoubtedly create a negative effect on conversion that would become more significant as hydrolysis progresses [46,84,86].

Pure crystalline cellulose such as Avicel has much less adsorption capacity for enzyme than for many pretreated lignocellulosic substrates [87]. Furthermore, cellulase has a much higher maximum adsorption capacity than β-G on Avicel [87,88]. On the other hand, amorphous cellulose has a higher capacity than crystalline cellulose to adsorb cellulase and β-G [88,89]. SO_2 and lime-pretreated corn stover showed higher cellulase adsorption capacity than DA- or AFEX-pretreated corn stover. Irrespective of the substrate, lime-pretreated biomass always demonstrated high adsorption capacity while AFEX-treated biomass always had low adsorption capacity for cellulase as well as xylanase [88]. High lignin substrates such as hybrid poplar and switchgrass pretreated at very low pH with dilute acid and SO_2 had higher adsorption capacity than when they were pretreated at moderate pH range with hot water [87].

Cellulase adsorption on lignin is due to hydrophobic interactions and, as such, is endothermic. It takes much longer (3 h) for lignin-cellulase adsorption to equilibrate than for pure cellulose-cellulase adsorption (1 h) [90]. Recent work by Várnai *et al.* [91] monitored adsorption of different enzymes during hydrolysis of Avicel and steam-pretreated spruce (SPS) and oxidatively delignified spruce (ODS). With SPS (high lignin and low xylan substrate) and Avicel (no lignin and no xylan substrate), Exo-G was adsorbed spontaneously but did not desorb. With ODS (low lignin and high xylan substrate), Exo-G desorbed back into solution after some time. Xylanase adsorption was found to be more pronounced with SPS than with ODS or Avicel. These observations indicated that Exo-G and xylanase had a higher affinity for lignin than for xylan. β-G adsorption was much lower compared to other enzymes on all substrates [91]. In another study, Berlin *et al.* [92] showed that the hydrolytic activity of β-G was least affected by lignin compared to other enzymes such as cellulase and xylanase. In the same study, it was observed that enzyme inhibition was higher by organosolv lignin than by residual lignin from enzymatic hydrolysis.

Cellulase adsorption on lignin from ethanol-pretreated lodge pole pine has been shown to be much less than on lignin from steam-exploded lodge pole pine [90]. In another study, the lower enzyme adsorption on ethanol-pretreated biomass allowed more effective enzyme recycle than possible for steam-pretreated biomass [79]. Changes in lignin molecular structure therefore appear to vary for different pretreatments. Addition of salt to the hydrolysis solution decreased cellulase adsorption onto lignin and cellulase adsorption onto pure cellulose. This result was attributed to salt ions altering the protein configuration, thus decreasing hydrophobic interaction with lignin [90].

Despite all such observations in the literature, a direct correlation between enzyme adsorption and hydrolytic activity cannot yet be ascertained. Although cellulase adsorption on biomass is a prerequisite for cellulase to be effective, enzyme adsorption on lignin may reduce effective cellulase availability to the structural carbohydrates in biomass.

13.13 Tuning Enzyme Formulations to the Feedstock

Assuming that all biomass-hydrolyzing enzymes have equal cost on a protein mass basis, economics would favor tuning of formulations to minimize the total protein mass required to achieve a given yield for given feedstock characteristics. For example, commercial cellulase contains 60% CBH-1 protein [4], but recently Gao *et al.* [52] showed that the optimal enzyme solution to hydrolyze AFEX-pretreated corn stover required only 28% CBH-1 with the result that only about half of the CBH-1 present in cellulase is utilized for this particular feedstock. The optimal enzyme formulation (also referred to as an "enzyme cocktail" in the industry) was determined to be 28.5% CBH-1, 18% CBH-2, 31% EG-1, 14.1% EX, 4.7% β-G, and 3.8% β-X [52]. In a separate study with AFEX-treated corn stover, the enzyme cocktail was optimized to maximize glucose and xylose release [93]. For maximum glucose release, the optimum mixture was 29% CBH-1, 5% CBH-2, 25% EG-1, 22% EX, 14% β-G, and 5% β-X. For maximum xylose release, however, the optimum cocktail composition identified was different in that the fraction of EX and β-X increased to 34% and 17%, respectively, and the fraction of CBH decreased to 19% [93]. As evident from these studies, the requirement for hemicellulase activity (EX and β-X) is much lower than that for cellulase activity (CBH-1, CBH-2, EG-1, and β-G), possibly due to the amorphous nature and higher accessibility of hemicellulose than cellulose.

In a different study with DA-pretreated corn stover, the proportions of commercial cellulase, β-G, xylanase, and pectinase were optimized using a high-throughput assay and response surface methodology. At the optimized amounts, the total protein loading was reduced by 50% compared to the cellulase only [47]. In this study, synergism between cellulase and xylanase components was shown to be very important. Zhou *et al.* applied purified CBH, EG, and β-G from commercial cellulase to steam-exploded corn stover and determined that the optimized combination of these enzymes was 72% CBH, 35% EG, and 3% β-G, which resulted in a 2.1 times higher hydrolysis yield than possible with the crude cellulase preparation alone [94]. The optimum composition of the enzyme mixture stayed relatively constant as the total enzyme loading was increased. It was also found that hydrolysis was not influenced as much by enzyme loading for the optimum enzyme cocktail as it was for the commercial enzyme [6].

Recent progress in the development of enzyme-related laboratory analytical methods has been remarkable. Examples include a high-throughput microplate system [95–98], Great Lakes Bioenergy Research Center Enzyme Platform (GENPLAT) [93], and automated saccharification assays [99]. Most of these assays use the 96-well plate format for high-throughput and spectrophotometer-based sugar assays for rapid screening. These methods undoubtedly contributed to recent progress in lignocellulosic hydrolysis technology, including mixed enzyme assays and substrate characterization. The repeatability of colorimetric assays based on dinitrosalicylic acid (DNS) or 2,2′-bicinchoninate (BCA) has always been questionable because of different reactivities of sugar oligomers with analytical reagents [100,101]. Bharadwaj *et al.* developed a microfluidic chip-based capillary electrophoresis device for rapid and high-throughput characterization of COS and XOS [102]. This novel assay claimed to overcome the limitation of non-specific colorimetric assays or throughput-limiting HPLC-based assays. Development of screening methods would be particularly useful in optimization of enzyme mixtures.

The optimum enzyme cocktail varies widely with the feedstock species and pretreatment applied, as does hydrolysis outcome for a given enzyme preparation. It was therefore suggested that the target substrate should be used as the basis for reporting gross enzyme activity instead of filter paper or any other model substrate [50]. In their study on hardwood (yellow poplar and red maple) as well as on softwood (Douglas fir and lodge pole pine), Berlin *et al.* demonstrated that specific cellulose conversion had little correlation with filter paper, carboxymethyl cellulose (CMC), and Avicel referenced activities, but correlated well with endogenous β-G and xylanase activities [6,50].

Although optimization of enzyme compositions is attractive for lowering enzyme loadings, preparing this mixture is challenging since it requires separation of enzyme components from the broth, a very costly process carried out by liquid chromatography [52]. Thus, crude enzyme mixtures produced by cellulase-producing organisms need to be further fine-tuned for a particular substrate to attain the benefits of an optimized enzyme cocktail. One approach would be to first identify the optimized enzyme formulation by separating and remixing components to maximize hydrolysis results from a particular feedstock. Once the optimized cocktail composition is known, the commercial enzyme broth composition would need to be adjusted by means such as applying different operating conditions, substrate compositions, and organisms for mass production of hydrolyzing enzymes [98,103–105]. Another approach, perhaps more practical, could be to produce different enzyme activities separately and combine them into an optimized cocktail [5].

13.14 Summary

Formulation of individual hydrolytic enzyme activities in an overall system can reduce total enzyme loadings and potentially costs in a cellulosic biofuels process. Adding enzyme components to the conventional

T. reesei "cellulase" system may further improve protein efficiency. The proportion of activity levels of different enzymes to maximize hydrolysis efficiency depends primarily on hemicellulose structure, characteristics of the cellulose molecule such as crystallinity and DP, and production of oligosaccharides during hydrolysis. The overall protein requirement is also influenced by effective availability of cellulose to cellulase, which is influenced in turn by adsorption of enzyme on lignin (non-productive binding) and carbohydrate molecules. Reactive adsorption of enzyme on the carbohydrate portion of biomass or non-reactive adsorption on lignin are both influenced strongly by biomass composition and especially lignin and acetyl contents [34].

Acetyl content acts as a barrier to cellulase access to cellulose as well as affecting xylanase leverage [33]. The positive effect of adding xylanase (i.e., xylanase leverage) is very high for alkali-pretreated biomass as it contains significant amounts of xylan and a considerable quantity of XOS is produced during enzymatic hydrolysis [43]. EG activity is essential for hydrolysis of cellulose with very high DP [15]. EG activity is also very important at high solids loadings as it effectively reduces cellulose DP and, consequently, the viscosity of the biomass slurry. Oligosaccharides seem to inhibit adsorption of cellulase and its hydrolytic efficiency [46,84]. Enhancement of β-X and β-G activity in cellulase therefore becomes important for biomass hydrolysis, especially when large amounts of XOS and COS are produced. AFE activity is effective in boosting hydrolysis of XOS that contain arabinose side chains [72]. Commercial pectinase also contains large amount of β-X, which obviously enhance hydrolytic activity against XOS [5]. Fortification of the enzyme cocktail with FE and AXE activities facilitates de-branching of hemicellulose and increases xylanase activity as well as the access of cellulase to cellulose [66,67]. The positive role of Exo-G activity, a rate-limiting enzyme in the overall cellulase system, applies to all types of biomass substrates in which glucan is the primary carbohydrate to be hydrolyzed [52,93].

Xylanase supplementation has been proven to enhance hydrolysis efficiency for pretreated biomass that retains significant amounts of hemicellulose, with the exception of lime-pretreated biomass. The exact reason for this difference is still unclear, but it is hypothesized that inhibition by XOS, lower cellulose reactivity, and unproductive binding to the residual lignin may contribute [43]. It was also observed that lime-pretreated biomass has very high enzyme adsorption whereas AFEX-treated biomass has very low enzyme adsorption, although both are based on alkaline treatment and AFEX-treated biomass contains larger amounts of lignin than lime-treated biomass.

Recent pretreatment and hydrolysis studies collectively indicate that unproductive enzyme binding to lignin significantly impeded hydrolysis of acid-treated biomass, whereas inhibition by XOS has a significant effect on hydrolysis of alkali-treated biomass [43,106]. Use of neutral components such as surfactants, proteins, and polymers can enhance biomass hydrolysis by reducing unproductive binding to lignin and enzyme denaturation [81,82].

References

1. Kazi, F.K. (2010) Iowa State U, ConocoPhillips, National Renewable Energy L. Techno-economic analysis of biochemical scenarios for production of cellulosic ethanol. Golden, CO: National Renewable Energy Laboratory; Available from: http://purl.access.gpo.gov/GPO/LPS124651.
2. Saha, B. (2003) Hemicellulose bioconversion. *Journal of Industrial Microbiology & Biotechnology*, **30** (5), 279–291.
3. Clarke, A.J. (ed.) (1997) Enzymology of biodegradation of cellulose and hemicellulose, in *Biodegradation of cellulose: Enzymology and Biotechnology*, Technomic publishing co., Inc., Lancaster, Basel.
4. Goyal, A., Ghosh, B., and Eveleigh, D. (1991) Characteristics of fungal cellulases. *Bioresource Technology*, **36** (1), 37–50. doi: 10.1016/0960-8524(91)90098-5

5. Dien, B.S., Ximenes, E.A., O'Bryan, P.J. *et al.* (2008) Enzyme characterization for hydrolysis of AFEX and liquid hot-water pretreated distillers' grains and their conversion to ethanol. *Bioresource Technology*, **99** (12), 5216–5225. doi: 10.1016/j.biortech.2007.09.030

6. Berlin, A., Gilkes, N., Kilburn, D. *et al.* (2005) Evaluation of novel fungal cellulase preparations for ability to hydrolyze softwood substrates – evidence for the role of accessory enzymes. *Enzyme and Microbial Technology*, **37** (2), 175–184. doi: 10.1016/j.enzmictec.2005.01.039

7. Lloyd, T.A. and Wyman, C.E. (2005) Combined sugar yields for dilute sulfuric acid pretreatment of corn stover followed by enzymatic hydrolysis of the remaining solids. *Bioresource Technology*, **96** (18), 1967–1977.

8. Kumar, R. and Wyman, C.E. (2009) Effects of cellulase and xylanase enzymes on the deconstruction of solids from pretreatment of poplar by leading technologies. *Biotechnology Progress*, **25** (2), 302–314.

9. Gupta, R. and Lee, Y.Y. (2010) Pretreatment of corn stover and hybrid poplar by sodium hydroxide and hydrogen peroxide. *Biotechnology Progress*, **26** (4), 1180–1186.

10. Zhang, Y.-H.P. and Lynd, L.R. (2004) Toward an aggregated understanding of enzymatic hydrolysis of cellulose: Noncomplexed cellulase systems. *Biotechnology and Bioengineering*, **88** (7), 797–824.

11. Gupta, R. and Lee, Y.Y. (2009) Mechanism of cellulase reaction on pure cellulosic substrates. *Biotechnology and Bioengineering*, **102** (6), 1570–1581.

12. Howell, J.A. and Stuck, J.D. (1975) Kinetics of solka floc cellulose hydrolysis by trichoderma viride cellulase. *Biotechnology and Bioengineering*, **17** (6), 873–893

13. Zhang, P.Y.H., Himmel, M.E., and Mielenz, J.R. (2006) Outlook for cellulase improvement: Screening and selection strategies. *Biotechnology Advances*, **24** (5), 452–481. doi: 10.1016/j.biotechadv.2006.03.003

14. Fengel, D. and Wegener, G. (1984) Reactions in alkaline medium, in *Wood:Chemistry, Ultrastructure, Reaction*, Walter de Gruyter, Berlin. New York, p. 296–318.

15. Valjamae, P., Pettersson, G., and Johansson, G. (2001) Mechanism of substrate inhibition in cellulose synergistic degradation. *European Journal of Biochemistry*, **268** (16), 4520–4526.

16. Kristensen, J.B., Felby, C., and Jorgensen, H. (2009) Yield-determining factors in high-solids enzymatic hydrolysis of lignocellulose. *Biotechnology for Biofuels*, **2** (11), 10.

17. Szijártó, N., Siika-aho, M., Sontag-Strohm, T., and Viikari, L. (2011) Liquefaction of hydrothermally pretreated wheat straw at high-solids content by purified Trichoderma enzymes. *Bioresource Technology*, **102** (2), 1968–1974. doi: 10.1016/j.biortech.2010.09.012

18. Gao, D., Chundawat, S., Liu, T. *et al.* (2010) Strategy for identification of novel fungal and bacterial glycosyl hydrolase hybrid mixtures that can efficiently saccharify pretreated lignocellulosic biomass. *BioEnergy Research*, **3** (1), 67–81.

19. DOE (2013) Feedstock Related Links. http://www1.eere.energy.gov/biomass/feedstocks_links.html.

20. Alen, R. (2000) *Structure and Chemical Composition of Wood*. Papermaking Science and Technology series. Published in cooperation with the Finnish Paper Engineers Association and TAPPI; p. 12–57.

21. Clarke, A.J. (ed.) (1997) Chemistry and structure of cellulose and heteroxylan, in *Biodegradation of Cellulose: Enzymology and Biotechnology*, Technomic Publishing co., Inc., Lancaster, Basel.

22. Fengel, D. and Wegener, G. (eds) (1984) Hemicellulose, in *Wood:Chemistry, Ultrastructure, Reaction*, Walter de Gruyter, Berlin. New York, p. 106–131.

23. Fournier, R.A., Frederick, M.M., Frederick, J.R., and Reilly, P.J. (1985) Purification and characterization of endo-xylanases from Aspergillus Niger. III. An enzyme of PL 365. *Biotechnology and Bioengineering*, **27** (4), 539–546.

24. Hall, J., Hazlewood, G.P., Huskisson, N.S. *et al.* (1989) Conserved serine-rich sequences in xylanase and cellulase from Pseudomonas fluorescens subspecies cellulosa: internal signal sequence and unusual protein processing. *Molecular Microbiology*, **3** (9), 1211–1219.

25. Wood, T.M. and McCrae, S.I. (1986) The effect of acetyl groups on the hydrolysis of ryegrass cell walls by xylanase and cellulase from trichoderma koningii. *Phytochemistry*, **25** (5), 1053–1055. doi: 10.1016/S0031-9422(00)81552-6

26. Frederick, M.M., Kiang, C.-H., Frederick, J.R., and Reilly, P.J. (1985) Purification and characterization of endo-xylanases from Aspergillus niger. I. Two isozymes active on xylan backbones near branch points. *Biotechnology and Bioengineering*, **27** (4), 525–532.

27. Matsuo, M. and Yasui, T. (1984) Purification and some properties of β-xylosidase from Emericella nidulans. *Agricultural and Biological Chemistry*, **48**, 1853–1869.

28. Doorslaer, V., Kersters-Hilderson, H., and DeBruyne, C.K. (1985) Hydrolysis of β-D-xylo-oligosaccharides by β-D-xylosidase from Bacillus pumilus. *Carbohydrate Research*, **140**, 342–346.

29. Hsu, T.-A. (1996) Pretreatment of biomass, in *Handbook on Bioethanol* (ed. C.E. Wyman), Taylor & Francis, Washington, DC, p. 179–212.

30. Kumar, R., Mago, G., Balan, V., and Wyman, C.E. (2009) Physical and chemical characterizations of corn stover and poplar solids resulting from leading pretreatment technologies. *Bioresource Technology*, **100** (17), 3948–3962. doi: 10.1016/j.biortech.2009.01.075

31. Gupta, R. (2008) Alkaline pretreatment of biomass for ethanol production and understanding the factors influencing the cellulose hydrolysis. PhD thesis. Auburn University, Auburn, AL.

32. Grabber, J.H., Hatfield, R.D., and Ralph, J. (1998) Diferulate cross-links impede the enzymatic degradation of non-lignified maize walls. *Journal of the Science of Food and Agriculture*, **77** (2), 193–200.

33. Kumar, R. and Wyman, C.E. (2009) Effect of xylanase supplementation of cellulase on digestion of corn stover solids prepared by leading pretreatment technologies. *Bioresource Technology*, **100** (18), 4203–4213

34. Zhu, L., O'Dwyer, J.P., Chang, V.S. *et al.* (2008) Structural features affecting biomass enzymatic digestibility. *Bioresource Technology*, **99** (9), 3817–3828. doi: 10.1016/j.biortech.2007.07.033.

35. Sierra, R. and Holtzapple, M.T. (April 18, 2008) CAFI-2 Project Progress Report, Texas A&M University. Limited access file at http://www.eng.auburn.edu/cafi/Reports%20a.htm.

36. Zhu, L., O'Dwyer, J.P., Chang, V.S. *et al.* (2010) Multiple linear regression model for predicting biomass digestibility from structural features. *Bioresource Technology*, **101** (13), 4971–4979. doi: 10.1016/j.biortech.2009.11.034

37. Yang, B. and Wyman, C.E. (2009) Lignin Blocking Treatment of Biomass and Uses Thereof, U.S. Patent 7,604,967.

38. Yang, B., Willies, D.M., and Wyman, C.E. (2006) Changes in the enzymatic hydrolysis rate of Avicel cellulose with conversion. *Biotechnology and Bioengineering*, **94** (6), 1122–1128

39. Falls, M., Sierra, R.*et al.* (eds) (2009) Optimization of enzyme formulation on pretreated switchgrass. AIChE Annual Meeting; Nashville, TN.

40. Kumar, R. and Wyman, C.E. (2009) Effect of enzyme supplementation at moderate cellulase loadings on initial glucose and xylose release from corn stover solids pretreated by leading technologies. *Biotechnology and Bioengineering*, **102** (2), 457–467.

41. CAFI Consortium for Applied and Fundamental Innovation. http://www.eng.auburn.edu/cafi/.

42. Gupta, R., Kim, T., and Lee, Y. (2008) Substrate dependency and effect of xylanase supplementation on enzymatic hydrolysis of ammonia-treated biomass. *Applied Biochemistry and Biotechnology*, **148** (1), 59–70.

43. Garlock, R.J.Balan, V. (eds) *et al.* (2009) Comparative material balances around leading pretreatment technologies for the conversion of switchgrass to soluble sugars. AIChE Annual Meeting; Nashville, TN.

44. Valls, C., Gallardo, O., Vidal, T. *et al.* (2010) New xylanases to obtain modified eucalypt fibres with high-cellulose content. *Bioresource Technology*, **101** (19), 7439–7445. doi: 10.1016/j.biortech.2010.04.085.

45. Mansfield Shawn, D. and Esteghlalian Ali, R. (eds) (2003) Applications of biotechnology in the forest products industry, in *Applications of Enzymes to Lignocellulosics*, American Chemical Society, p. 2–29.

46. Qing, Q. and Wyman, C.E. (2011) Hydrolysis of different chain length xylooliogmers by cellulase and hemicellulase. *Bioresource Technology*, **102** (2), 1359–1366.

47. Berlin, A., Maximenko, V., Gilkes, N., and Saddler, J. (2007) Optimization of enzyme complexes for lignocellulose hydrolysis. *Biotechnology and Bioengineering*, **97** (2), 287–296.

48. Shen, F., Kumar, L., Hu, J., and Saddler, J.N. (2011) Evaluation of hemicellulose removal by xylanase and delignification on SHF and SSF for bioethanol production with steam-pretreated substrates. *Bioresource Technology*, **102** (19), 8945–8951. doi: 10.1016/j.biortech.2011.07.028

49. Guerfali, M., Gargouri, A., and Belghith, H. (2011) Catalytic properties of Talaromyces thermophilus [alpha]-l-arabinofuranosidase and its synergistic action with immobilized endo-[beta]-1,4-xylanase. *Journal of Molecular Catalysis B: Enzymatic*, **68** (2), 192–199. doi: 10.1016/j.molcatb.2010.11.003

50. Berlin, A., Gilkes, N., Kilburn, D. *et al.* (2006) Evaluation of cellulase preparations for hydrolysis of hardwood substrates. *Applied Biochemistry and Biotechnology*, **129–132**, 528–545.

51. Haltrich, D., Laussamayer, B., Steiner, W. *et al.* (1994) Cellulolytic and hemicellulolytic enzymes of sclerotium rolfsii: Optimization of the culture medium and enzymatic hydrolysis of lignocellulosic material. *Bioresource Technology*, **50** (1), 43–50. doi: 10.1016/0960-8524(94)90219-4

52. Gao, D., Chundawat, S.P., Krishnan, C. *et al.* (2010) Mixture optimization of six core glycosyl hydrolases for maximizing saccharification of ammonia fiber expansion (AFEX) pretreated corn stover. *Bioresource Technology*, **101** (8), 2770–2781.

53. Breuil, C., Chan, M., Gilbert, M., and Saddler, J.N. (1992) Influence of [beta]-glucosidase on the filter paper activity and hydrolysis of lignocellulosic substrates. *Bioresource Technology*, **39** (2), 139–142. doi: 10.1016/0960-8524(92)90132-H

54. Ghosh, T.K. (1987) Measurement of cellulase activities. *Pure and Applied Chemistry*, **59** (2), 257–268.

55. Coward-Kelly, G., Aiello-Mazzari, C., Kim, S. *et al.* (2003) Suggested improvements to the standard filter paper assay used to measure cellulase activity. *Biotechnology and Bioengineering*, **82** (6), 745–749.

56. Pallapolu, R. and Lee, Y.Y. (July 2 (2009)) CAFI-3 Project Progress Report, Auburn University. http://www.eng .auburn.edu/cafi/Reports%20a.htm.

57. Holtzapple, M., Cognata, M., Shu, Y., and Hendrickson, C. (1990) Inhibition of Trichoderma reesei cellulase by sugars and solvents. *Biotechnology and Bioengineering*, **36** (3), 275–287.

58. Kashyap, D.R., Vohra, P.K., Chopra, S., and Tewari, R. (2001) Applications of pectinases in the commercial sector: a review. *Bioresource Technology*, **77** (3), 215–227. doi: 10.1016/S0960-8524(00)00118-8

59. Pallapolu, R., Kang, L., and Lee, Y.Y. (eds) (2010) Effect of enzyme mix on saccharification of SAA treated switchgrass. AIChE Annual Meeting; 2010 Nov 07–12; Salt lake City, UT.

60. Kim, Y., Mosier, N., and Ladisch, M.R. (March 26, 2009) CAFI-3 Project Progress Report, LORRE, Purdue University. http://www.eng.auburn.edu/cafi/Reports%20a.htm.

61. Mandalari, G., Bisignano, G., Lo Curto, R.B. *et al.* (2008) Production of feruloyl esterases and xylanases by Talaromyces stipitatus and Humicola grisea var. thermoidea on industrial food processing by-products. *Bioresource Technology*, **99** (11), 5130–5133. doi: 10.1016/j.biortech.2007.09.022

62. Balan, V., Chundawat, S., and Dale, B. (August 3, 2006) CAFI-2 Project Progress Report, BCRL, Michigan State University. http://www.eng.auburn.edu/cafi/Reports%20a.htm.

63. Cosgrove, D.J. (2005) Growth of the plant cell wall. *Nature Reviews. Molecular Cell Biology*, **6** (11), 850–861. doi: 10.1038/nrm1746

64. Dinis, M.J., Bezerra, R.M.F., Nunes, F. *et al.* (2009) Modification of wheat straw lignin by solid state fermentation with white-rot fungi. *Bioresource Technology*, **100** (20), 4829–4835. doi: 10.1016/j.biortech.2009.04.036

65. Sun, R., Lawther, J.M., and Banks, W.B. (1997) A tentative chemical structure of wheat straw lignin. *Industrial Crops and Products*, **6** (1), 1–8. doi: 10.1016/S0926-6690(96)00170-7

66. Tapin, S., Sigoillot, J.-C., Asther, M., and Petit-Conil, M. (2006) Feruloyl esterase utilization for simultaneous processing of nonwood plants into phenolic compounds and pulp fibers. *Journal of Agricultural and Food Chemistry*, **54** (10), 3697–3703. doi: 10.1021/jf052725y

67. Selig, M.J., Knoshaug, E.P., Adney, W.S. *et al.* (2008) Synergistic enhancement of cellobiohydrolase performance on pretreated corn stover by addition of xylanase and esterase activities. *Bioresource Technology*, **99** (11), 4997–5005. doi: 10.1016/j.biortech.2007.09.064

68. Agger, J., Viksø-Nielsen, A., and Meyer, A.S. (2010) Enzymatic xylose release from pretreated corn bran arabinoxylan: differential effects of deacetylation and deferuloylation on insoluble and soluble substrate fractions. *Journal of Agricultural and Food Chemistry*, **58** (10), 6141–6148. doi: 10.1021/jf100633f

69. Yu, P., McKinnon, J.J., Maenz, D.D. *et al.* (2002) Enzymic release of reducing sugars from oat hulls by cellulase, as influenced by aspergillus ferulic acid esterase and trichoderma xylanase. *Journal of Agricultural and Food Chemistry*, **51** (1), 218–223. doi: 10.1021/jf020476x.

70. Carapito, R., Carapito, C., Jeltsch, J-M, and Phalip, V. (2009) Efficient hydrolysis of hemicellulose by a Fusarium graminearum xylanase blend produced at high levels in Escherichia coli. *Bioresource Technology*, **100** (2), 845–850. doi: 10.1016/j.biortech.2008.07.006.

71. Henrissat, B. and Bairoch, A. (1996) Updating the sequence-based classification of glycosyl hydrolases. *Biochemical Journal*, **316** (2), 695–696.

72. Beukes, N. and Pletschke, B.I. (2010) Effect of lime pre-treatment on the synergistic hydrolysis of sugarcane bagasse by hemicellulases. *Bioresource Technology*, **101** (12), 4472–4478. doi: 10.1016/j.biortech.2010.01.081

73. Várnai, A., Huikko, L., Pere, J. *et al.* (2011) Synergistic action of xylanase and mannanase improves the total hydrolysis of softwood. *Bioresource Technology*, **102** (19), 9096–9104.

74. Dias, A.A., Freitas, G.S., Marques, G.S.M. *et al.* (2010) Enzymatic saccharification of biologically pre-treated wheat straw with white-rot fungi. *Bioresource Technology*, **101** (15), 6045–6050. doi: 10.1016/j.biortech.2010.02.110

75. Shi, J., Sharma-Shivappa, R.R., Chinn, M., and Howell, N. (2009) Effect of microbial pretreatment on enzymatic hydrolysis and fermentation of cotton stalks for ethanol production. *Biomass and Bioenergy*, **33** (1), 88–96. doi: 10.1016/j.biombioe.2008.04.016

76. Yang, B. and Wyman, C.E. (2011) Lignin Blockers and Uses Thereof, U. S. Patent 7,875,444.

77. Yang, B. and Wyman, C. (2006) BSA treatment to enhance enzymatic hydrolysis of cellulose in lignin containing substrates. *Biotechnology and Bioengineering*, **94** (4), 611–617.

78. Bura, R., Chandra, R., and Saddler, J.N. (April 5, 2007) CAFI-2 Project Progress Report, Forest Products Biotechnology Group, University of British Columbia. http://www.eng.auburn.edu/cafi/Reports%20a.htm.

79. Tu, M. and Saddler, J. (2010) Potential enzyme cost reduction with the addition of surfactant during the hydrolysis of pretreated softwood. *Applied Biochemistry and Biotechnology*, **161** (1), 274–287.

80. Tu, M., Zhang, X., Paice, M. *et al.* (2009) The potential of enzyme recycling during the hydrolysis of a mixed softwood feedstock. *Bioresource Technology*, **100** (24), 6407–6415. doi: 10.1016/j.biortech.2009.06.108

81. Eriksson, T., Börjesson, J., and Tjerneld, F. (2002) Mechanism of surfactant effect in enzymatic hydrolysis of lignocellulose. *Enzyme and Microbial Technology*, **31** (3), 353–364. doi: 10.1016/S0141-0229(02)00134-5

82. Börjesson, J., Engqvist, M., Sipos, B., and Tjerneld, F. (2007) Effect of poly(ethylene glycol) on enzymatic hydrolysis and adsorption of cellulase enzymes to pretreated lignocellulose. *Enzyme and Microbial Technology*, **41** (1–2), 186–195. doi: 10.1016/j.enzmictec.2007.01.003

83. Mansfield, S.D., Mooney, C., and Saddler, J.N. (1999) Substrate and enzyme characteristics that limit cellulose hydrolysis. *Biotechnology Progress*, **15** (5), 804–816.

84. Shi, J., Ebrik, M.A., Yang, B. *et al.* (2011) Application of cellulase and hemicellulase to pure xylan, pure cellulose, and switchgrass solids from leading pretreatments. *Bioresource Technology*, **102** (24), 11080–11088. doi: 10.1016/j.biortech.2011.04.003

85. Qing, Q., Yang, B., and Wyman, C.E. (2010) Xylooligomers are strong inhibitors of cellulose hydrolysis by enzymes. *Bioresource Technology*, **101** (24), 9624–9630. doi: 10.1016/j.biortech.2010.06.137

86. Kumar, R. and Wyman, C.E. (2008) An improved method to directly estimate cellulase adsorption on biomass solids. *Enzyme and Microbial Technology*, **42** (5), 426–433. doi: 10.1016/j.enzmictec.2007.12.005

87. Shi, J.Ebrik, M.A. *et al.* (eds) (2009) Properties of cellulase and non-cellulase enzymes and their interactions with switchgrass processed by leading pretreatment technologies. AIChE Annual Meeting; Nashville, TN.

88. Kumar, R. and Wyman, C.E. (July 25, 2007) CAFI-2 Project Progress Report, Dartmouth College. http://www.eng.auburn.edu/cafi/Reports%20a.htm.

89. Hong, J., Ye, X., and Zhang, Y.H.P. (2007) Quantitative determination of cellulose accessibility to cellulase based on adsorption of a nonhydrolytic fusion protein containing CBM and GFP with its applications. *Langmuir*, **23** (25), 12535–12540. doi: 10.1021/la7025686

90. Tu, M., Pan, X., and Saddler, J.N. (2009) Adsorption of cellulase on cellulolytic enzyme lignin from lodgepole pine. *Journal of Agricultural and Food Chemistry*, **57** (17), 7771–7778.

91. Várnai, A., Viikari, L., Marjamaa, K., and Siika-aho, M. (2011) Adsorption of monocomponent enzymes in enzyme mixture analyzed quantitatively during hydrolysis of lignocellulose substrates. *Bioresource Technology*, **102** (2), 1220–1227. doi: 10.1016/j.biortech.2010.07.120

92. Berlin, A., Balakshin, M., Gilkes, N. *et al.* (2006) Inhibition of cellulase, xylanase and [beta]-glucosidase activities by softwood lignin preparations. *Journal of Biotechnology*, **125** (2), 198–209. doi: 10.1016/j.jbiotec.2006.02.021.

93. Banerjee, G., Car, S., Scott-Craig, J.S. *et al.* (2010) Synthetic enzyme mixtures for biomass deconstruction: Production and optimization of a core set. *Biotechnology and Bioengineering*, **106** (5), 707–720

94. Zhou, J., Wang, Y.-H., Chu, J. *et al.* (2009) Optimization of cellulase mixture for efficient hydrolysis of steam-exploded corn stover by statistically designed experiments. *Bioresource Technology*, **100** (2), 819–825. doi: 10.1016/j.biortech.2008.06.068

95. Chundawat, S.P.S., Balan, V., and Dale, B.E. (2008) High-throughput microplate technique for enzymatic hydrolysis of lignocellulosic biomass. *Biotechnology and Bioengineering*, **99** (6), 1281–1294.

96. Studer, M.H., DeMartini, J.D., Brethauer, S. *et al.* (2010) Engineering of a high-throughput screening system to identify cellulosic biomass, pretreatments, and enzyme formulations that enhance sugar release. *Biotechnology and Bioengineering*, **105** (2), 231–238.

97. Berlin, A., Maximenko, V., Bura, R. *et al.* (2006) A rapid microassay to evaluate enzymatic hydrolysis of lignocellulosic substrates. *Biotechnology and Bioengineering*, **93** (5), 880–886.

98. King, B.C., Donnelly, M.K., Bergstrom, G.C. *et al.* (2009) An optimized microplate assay system for quantitative evaluation of plant cell wall–degrading enzyme activity of fungal culture extracts. *Biotechnology and Bioengineering*, **102** (4), 1033–1044.

99. Gomez, L., Whitehead, C., Barakate, A. *et al.* (2010) Automated saccharification assay for determination of digestibility in plant materials. *Biotechnology for Biofuels*, **3** (1), 1–12.

100. Zhang, Y.H.P. and Lynd, L.R. (2005) Determination of the number-average degree of polymerization of cellodextrins and cellulose with application to enzymatic hydrolysis. *Biomacromolecules*, **6** (3), 1510–1515. doi: 10.1021/bm049235j

101. Sengupta, S., Jana, M.L., Sengupta, D., and Naskar, A.K. (2000) A note on the estimation of microbial glycosidase activities by dinitrosalicylic acid reagent. *Applied Microbiology and Biotechnology*, **53** (6), 732–735.

102. Bharadwaj, R., Chen, Z., Datta, S. *et al.* (2010) Microfluidic glycosyl hydrolase screening for biomass-to-biofuel conversion. *Analytical Chemistry*, **82** (22), 9513–9520. doi: 10.1021/ac102243f

103. Mamma, D., Kourtoglou, E., and Christakopoulos, P. (2008) Fungal multienzyme production on industrial by-products of the citrus-processing industry. *Bioresource Technology*, **99** (7), 2373–2383. doi: 10.1016/j.biortech.2007.05.018

104. Kang, S.W., Park, Y.S., Lee, J.S. *et al.* (2004) Production of cellulases and hemicellulases by Aspergillus niger KK2 from lignocellulosic biomass. *Bioresource Technology*, **91** (2), 153–156. doi: 10.1016/S0960-8524(03)00172-X

105. Oliveira, L.A., Porto, A.L.F., and Tambourgi, E.B. (2006) Production of xylanase and protease by Penicillium janthinellum CRC 87M-115 from different agricultural wastes. *Bioresource Technology*, **97** (6), 862–867. doi: 10.1016/j.biortech.2005.04.017

106. Yang, B.H. and Wyman C.E.N. (2004) Lignin-blocking treatment of biomass and uses thereof United States patent 20040185542.

14

Physical and Chemical Features of Pretreated Biomass that Influence Macro-/Micro-accessibility and Biological Processing

Rajeev Kumar[1,3] and Charles E. Wyman[1,2,3]

[1] *Center for Environmental Research and Technology, University of California, Riverside, USA*
[2] *Department of Chemical and Environmental Engineering, University of California, Riverside, USA*
[3] *BioEnergy Science Center, Oak Ridge, USA*

14.1 Introduction

Lignocellulosic biomass containing about 55–65 wt% sugars in the form of polymeric structural carbohydrates can be a sustainable feedstock for production of a variety of sugars that can in turn be converted into fuels and chemicals [1–3]. For biological deconstruction of structural carbohydrates to sugars, tiny protein molecules called enzymes need to reach appropriate substrates to depolymerize them. In native plants however, carbohydrates are trapped in a complex matrix comprising non-sugar constituents and polymers such as lignin, which forms a strong and complex sheath around carbohydrates and presents challenges to effective and economical release of sugars from biomass [1,2,4,5]. Prior to biological conversion, some kind of treatment of biomass is therefore needed to make the carbohydrates accessible [6,7]. This pretreatment can be purely mechanical, thermal, chemical, biological, or combinations of these, as reviewed elsewhere [8–10]. Thermochemical pretreatments, however, are considered leading options due to their comparatively short reaction time, effectiveness, and lower energy requirements compared to mechanical options [11–13]. Nonetheless, thermochemical pretreatments typically need

Aqueous Pretreatment of Plant Biomass for Biological and Chemical Conversion to Fuels and Chemicals, First Edition.
Edited by Charles E. Wyman.
© 2013 John Wiley & Sons, Ltd. Published 2013 by John Wiley & Sons, Ltd.

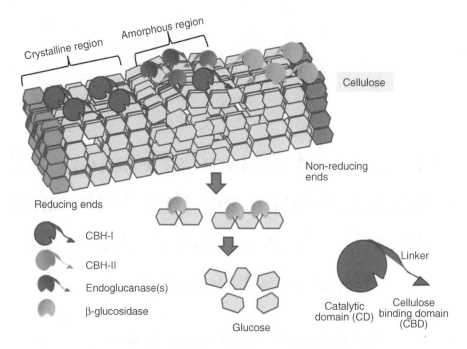

Figure 14.1   *A simplified schematic of cellulose hydrolysis by a typical non-complexed cellulase system mediated by cellobiohydrolases (CBH I and CBH II) and endoglucanases (EG I to EG V) adsorption, cellobiose and cellodextrin production, and their catalysis to glucose by β-glucosidase. Also shown is one of the cellulase components with different domains.*

some mechanical treatment such as size reduction prior to or after pretreatment to realize acceptable performance [14,15].

Enzymatic conversion (saccharification/hydrolysis) of structural carbohydrates is a heterogeneous phenomenon which, unlike homogeneous reactions, involves two major steps: enzyme adsorption on the substrate surface and hydrolysis of the polymers to form shorter chained molecules [16–18], as depicted in Figure 14.1. Compared to the hydrolysis step, enzyme adsorption is fast and usually completed within a few hours (*c.* 2 hours compared to 72–120 hours for hydrolysis, as shown in Figure 14.2) [18,19]. Most carbohydrases, at least in the fungi family that includes cellulases and hemicellulases, are composed of two distinct domains connected by a peptide linker: a cellulose or carbohydrate binding domain (CBM) and a catalytic domain (CD) [20,21], schematically shown in Figure 14.1. Although adsorption is largely mediated by CBMs, catalytic domains also participate in adsorption [22–25].

The literature suggest that several physical and chemical features of pretreated biomass impact its biological conversion including surface area/porosity; hemicellulose and lignin contents; particle size; and cellulose crystallinity, type, and chain length [18,26–28]. However, we believe that all these features are aspects, directly or indirectly, of two main factors that most control conversion: (1) enzyme accessibility to their appropriate substrates; and (2) enzyme effectiveness once adsorbed on the surface. Because a few cellulolytic and xylanolytic enzymes such as β-glucosidase and β-xylosidase catalyze homogenous reactions [29,30], some of the above-mentioned features do not seem to play any direct role in controlling accessibility of these enzymes to their substrates to the best of our knowledge. Nevertheless, they can affect their effectiveness [31–33]. Enzyme accessibility can be further categorized into macro- and micro-accessibility. In this chapter, the impact of physical and chemical features on cellulase macro-

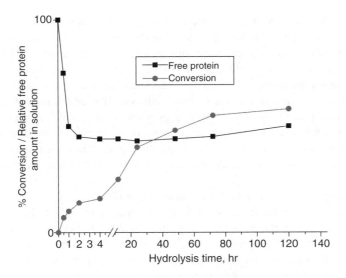

Figure 14.2 *A typical profile of cellulose hydrolysis progression and relative amounts of free protein in solution over hydrolysis time.*

and micro-accessibility and enzyme effectiveness once adsorbed (or in the solution for enzymes which do not participate in adsorption) are reviewed.

14.2 Definitions of Macro-/Micro-accessibility and Effectiveness

Except for a few soluble lower-molecular-weight glucose oligomers such as cellobiose and higher cellodextrins [34], enzymatic hydrolysis of cellulose is a heterogeneous reaction that requires enzymes to adsorb onto a cellulose surface with specific/common binding sites [17,35–38]. The availability of cellulose binding sites to cellulase is called accessibility, and accessibility has been defined both in comparative and quantitative terms. The quantitative definition is often expressed in terms of the maximum adsorption capacity (σ_{max} in μg or mg/g biomass) of cellulase (or its individual components) per gram of biomass or glucan and can be estimated by fitting multiple points of cellulase adsorption data to a non-linear Langmuir model [16–19]. Adsorption data is commonly collected by equilibrating various concentrations of enzymes with a given mass of biomass at non-hydrolytic temperatures, for example 4 °C, to minimize reaction during the measurement [36,39]. The comparative definition is often used to represent the change in accessibility and/or relative accessibility by performing adsorption with a single or multiple concentrations of enzymes, with adsorption expressed as μg or mg protein/g biomass (or glucan) [40–42].

Here we postulate a new concept based on dividing accessibility, at least for real biomass, into two stages: macro and micro. Although techniques to quantify the relative role of each in governing overall accessibility are not yet available due to the complexity of cellulosic biomass, pure cellulose (e.g., Avicel) can be pictured as being completely macro-accessible; it has no obvious obstacles (such as lignin and hemicelluloses) to prevent cellulase enzymes from reaching the cellulose surface, provided physiochemical conditions such as solids concentration and pH are optimized to not impede accessibility [16,43,44]. On the other hand, the hemicellulose/lignin sheath in cellulosic biomass interferes with cellulase reaching cellulose, limiting the ability for enzymes to attach to cellulose [4,45,46]. Carbohydrates in untreated and pretreated real biomass can therefore be pictured as having varying degrees of macro-accessibility compared to pure cellulose or hemicelluloses, depending on the amount and nature of residual lignin and/or hemicellulose. Once enzymes reach the cellulose surface, a second stage of accessibility (micro-accessibility) can be pictured in terms of how readily the enzymes can reach cellulose binding/catalytic sites buried within the compact, semi-crystalline cellulose structure. Amorphous cellulose can be thought of as representing the highest degree of macro- and micro-accessibility in that cellulase enzymes should be able to reach all cellulose chains. From the cell-wall physiology point of view, cellulase access to cellulose microfibrils and elementary fibrils within cellulose microfibers, respectively, can be referred to as macro- and micro-accessible. Emergence of new tools to measure cellulase–cellulose interactions, such as that developed by Hong *et al.* [39], would be very valuable in clarifying how macro- and micro-accessibility influence cellulose hydrolysis.

Enzyme effectiveness can be defined by the increase in hydrolysis rate/yields with little or no change in accessibility quantified by either the maximum adsorption capacity (MAC), estimated by non-linear fit of adsorption data to the Langmuir model, or the change in protein adsorption per unit biomass. For example, although somewhat contentious, temperature is believed to have limited impact on enzyme access to substrate as shown in the literature by a negligible change in enzyme adsorption with temperature [47]. However, temperature has a great impact on enzyme effectiveness. As another example, most inhibitors such as cellobiose and xylan oligomers have been shown to greatly impact enzyme effectiveness [48–52], but limited studies have indicated limited impact on accessibility. However, Kumar and Wyman showed that sugars, and especially cellobiose, can cause enzyme desorption even at low concentrations [36], a result substantiated in a study by Kristensen *et al.* [53].

14.3   Features Influencing Macro-accessibility and their Impacts on Enzyme Effectiveness

14.3.1   Lignin

Role in Accessibility

Based on limited data, lignin removal does not appear to make morphological changes in the cellulose structure such as cellulose chain length [28,54,55] or cellulose crystallinity; the effects on cellulose macro-accessibility therefore appear more important than micro-accessibility. In plants, the amounts of p-hydroxyphenyl (*H*), guaiacyl (*G*), and syringyl (*S*) lignin varies with species and cell wall, and the secondary cell wall is generally found to be more lignified than the primary cell wall [56–58]. Woody biomass, dicots, and especially softwoods generally have higher lignin content than agricultural residues and herbaceous energy crops, that is, monocots [59–61]. Unfortunately, conventional analysis as acid-soluble and acid-insoluble (Klason) lignin by gravimetric or other methods does not distinguish the impact of lignin from individual cell walls on cellulose digestibility, limiting knowledge of its role.

Lignin appears to be a major impediment to cellulose accessibility and ultimately to polysaccharide saccharification, as most studies in the literature reported that enzymatic conversion of polysaccharides is enhanced by down-regulation of lignin in genetically modified plants and/or delignification of hardwood/softwood and other lignocellulosics [61–70]. Others have found no correlation or a negative correlation between lignin content/removal and cellulose digestibility [4,41,71–74] but, in most of these studies, other effects accompanying delignification or the thermal treatment used for lignin removal makes it difficult to interpret the outcomes.

Although removing lignin appears critical to making biomass amenable to enzymatic hydrolysis in some situations, this conclusion may not be universally applicable as some feedstocks with equal or somewhat less lignin than hardwoods or softwoods, such as corn stover, are susceptible to biological catalysis without significant lignin removal. For example, in a collaborative project jointly funded by DOE and USDA, we found that corn stover solids prepared by ammonia fiber expansion (AFEX) pretreatment that removed little if any lignin were equally or more susceptible to enzymes as solids prepared by pretreatments that removed a significant amount of lignin [75,76]. Although conditions vary with the type of biomass and pretreatment catalyst type and loading, most leading pretreatments were effective in the range of 140–200 °C; soaking in aqueous ammonia (SAA) and lime were exceptions that were quite effective at lower temperatures [77–79]. These temperatures are well above the glass transition temperature (T_g) of lignin of 80–130 °C, recognizing that the latter depends on the lignin state and measurement methods [80,81]. However, some plasticizers such as glycerol have been reported to lower the lignin T_g[82]. Therefore, lignin melting and relocation appears to occur during most leading thermochemical pretreatments and seems particularly certain for steam explosion and dilute-acid pretreatments in several studies [46,83–86] and in a recent study for AFEX pretreatment of corn stover [57,87].

The impact of lignin removal from biomass on cellulase accessibility to cellulose is not entirely clear. Based on hypothetical models of the intricate lignin-hemicellulose-cellulose structure presented in the literature, hemicellulose appears to form a sheath around cellulose chains, and lignin in turn surrounds hemicellulose. It can therefore be hypothesized that lignin does not directly restrict cellulase accessibility to cellulose to the extent that hemicellulose does, although access to hemicellulose is strongly limited by lignin. Unfortunately, experimental data to support such hypotheses are scarce.

Several studies have reported that lignin removal affects digestibility of hemicellulose more than that of cellulose [71,88–93]. For example, adding crude cellulase at a loading of 25 IFPU/g (international filter paper units) glucan to hybrid poplar and sugar cane bagasse that had lignin selectively removed by peracetic acid pretreatment resulted in low glucan digestibility, unless supplemented with an industrial-grade xylanase SP431 or SP431 supplemented with Novozyme®188 β-glucosidase [89]. Nevertheless, as recently discovered, this result could be due to strong cellulase inhibition by xylooligomers released during enzymatic hydrolysis, which can be relieved by these supplementary enzymes [50–52,94]. In a recent study with highly purified enzyme components and alkaline-peroxide-delignified corn stover, lignin appeared to have a more direct impact on xylan than glucan accessibility [95]. In another paper at that time, Kumar and Wyman showed that selective removal of lignin from corn stover using peracetic acid did not significantly increase cellulase accessibility to cellulose, as measured by purified Cel7A adsorption. Instead, lignin removal appeared to more directly enhance xylan accessibility, which in turn affected cellulose accessibility, as evidenced by a much higher increase in digestion of xylan than of glucan and a linear relation between the percentage increase in xylan and glucan removal [17,96]. Thus, although lignin may not appear to affect cellulose accessibility directly, lignin removal accelerates xylan removal via both biological and thermal means [97], thereby reducing occlusion of glucan chains by hemicellulose [40,95] and apparently greatly enhancing cellulose macro-accessibility. Lignin removal can also alter other biomass features such as increasing Brunauer–Emmett–Teller (BET) surface area and biomass crystallinity [98,99]. However, more investigation is needed to clarify the role of lignin in cellulose macro-/micro-accessibility.

Role in Enzymes Effectiveness

Leading thermochemical pretreatments result in partial removal and/or dislocation of lignin, and some of the lignin solubilized during pretreatment has been shown to precipitate back on cellulose fibers in droplet form [46,83]. The literature suggests that lignin droplets deposited on cellulose may interact with water, as one study showed that hydrophobic surfaces at a macroscopic level do not repel but attract water [100] to form a boundary layer impeding cellulase movement and limiting cellulose accessibility [46,84,101]. However, the source of these droplets was not possible to determine with the conventional NREL two-step acid hydrolysis method often employed for K-lignin determination, as some hemicellulose during pretreatment degrades to lignin-like compounds called humins (pseudo-lignin, char) and may also deposit onto cellulose

[97,102–104]. Enzymes can also unproductively bind to lignin, and possibly to hemicellulose-derived humins, to limit enzyme effectiveness [33,84,105–112]. Unproductive adsorption of cellulase or other enzymes on lignin is hypothesized to be due to hydrophobic interactions between the two [113–116]. Proteins, and especially cellulase that is highly hydrophobic due to clusters of closely located non-polar residues on its surface [117–121], tend to adsorb strongly on hydrophobic surfaces [100,114]; that attachment results in conformational changes and consequently irreversible adsorption and deactivation [122–125]. In line with this, hydrolysis yields and free enzyme concentrations have been reported to increase with increased cellulase hydrophilicity [122,123].

Lignin can impact performance in other ways. First, lignin linkages with cellulose [126–128] should presumably impede the processive action of enzyme components. In addition, lignin breakdown products generally inhibit fermentation and cellulase effectiveness [84,107,129–134]. Moreover, lignin and its derivatives were also reported to precipitate with proteins [135–137]. Overall, lignin may reduce the amount of active enzyme available for cellulose hydrolysis, but its influence on the effectiveness of adsorbed cellulase needs more study.

14.3.2 Hemicellulose

Role in Accessibility

Based on lignocellulosic biomass data and models in the literature, hemicellulose appears to play a direct and vital role in cellulase accessibility to cellulose. Hemicellulose, the other major carbohydrate in biomass, is composed of combinations of hexose and pentose sugars, namely arabinoxylan, arabinogalactan, glucomannan, galactoglucomannan, and xyloglucan [138,139]. Furthermore, the chains of these hemicellulose compounds are often substituted with uronic and glucuronic acids (one per ten xylopyranosyl residues in hardwoods) and acetate (grasses and agricultural residues) [60,140]. Although it would greatly help in designing better pretreatment and enzyme cocktails, unfortunately the exact amounts of these complex hemicellulose polymers cannot be determined due to the complex structure of biomass and limitations in current invasive/non-invasive compositional analysis tools. Xylan, a polymer of xylose, is the major constituent in hemicellulose (>85%) for most lignocellulosic feedstocks except softwoods, for which mannan dominates. Because hemicellulose removal by thermochemical pretreatments is generally not highly selective, its removal is often accompanied by other changes such as a reduction in the cellulose degree of polymerization (DP), creation/enhancement of nanopores within cellulose fibers, and changes in cross-sectional radius of the crystalline cellulose fibril [28,85,141–143], with these effects heavily impacting cellulose micro-accessibility. However, the scale of these contributions to cellulose micro-accessibility appears to be smaller than the increase in cellulose macro-accessibility; hemicellulose removal is therefore attributed to be mostly responsible for enhancing biomass macro-accessibility.

Hemicellulose has been pictured as obstructing cellulase access to cellulose by forming a sheath around glucan chains [40,94,96,99,144–147], with several studies showing a direct relationship between cellulose digestibility and hemicellulose removal [4,68,71,148–154]. However, as reported by Torget *et al.* for short rotation woody crops (silver maple, sycamore, and black locust) and corn residues (stover and cobs), hemicellulose removal was complete at 140 °C and 160 °C for all these substrates. However, woods and stover solids prepared by dilute acid pretreatment at 160 °C were more digestible than when prepared at 140 °C, suggesting that hemicellulose removal is not the sole factor impacting cellulose conversion [155]. In addition, some reports do not postulate any role for hemicellulose removal in changing cellulose digestibility [156–158]. Unfortunately, as discussed above, hemicellulose alteration during pretreatment also disrupts other biomass components [28,158–162]. Reports on impact of selective hemicellulose removal on cellulose accessibility are scarce, making it challenging to draw firm conclusions about the degree to which it controls access of enzymes to cellulose.

In addition, some argue that hemicellulose may actually be a marker signaling disruption of the far less soluble lignin and that lignin disruption (and/or lignin-carbohydrate complex) could be the key to greater digestion [68,164–167]. In contrast to traditional beliefs, AFEX pretreatment (which was earlier believed to enhance digestibility without removing any components) has recently been shown to solubilize hemicellulose as oligomers that deposited back onto the biomass due to ammonia evaporation, thus enhancing cellulose accessibility [38,57].

Jeoh *et al.* [4,40] reported that cellulose accessibility increased with xylan removal by dilute acid pretreatment of corn stover, as measured by the adsorption of fluorescent labeled Cel7A (CBHI). Furthermore, several reports showed significant effects of xylanase supplementation to cellulase on digestion of glucan as well as xylan, and linear relationships between xylan and glucan digestion suggested that xylan removal affects cellulose accessibility [94,96,145,147,168–171]. Although further investigation is needed, a recent study showed that solubilized hemicellulose oligomers, which can make up a large percentage of the solubles produced by enzymatic hydrolysis and pretreatment for lower enzymes loadings and some pretreatments [50], can negatively impact cellulose accessibility through reduced cellulase adsorption [172].

The xylan backbone of hemicellulose is often substituted with uronic (methyl) and glucuronic acids (up to 6 wt% of dry biomass, and collectively termed glucuronoxylan), acetate (up to 5 wt% of dry biomass), and/or arabinose (up to 3.5 wt% of dry biomass, collectively termed arabinoxylan). Although xylan removal has been reported to affect cellulose accessibility in several studies, the effect of side-chain removal on enzymatic hydrolysis has received little attention. In particular, almost no studies have been directed at thermochemical removal/modification of arabinose; this is hypothesized to be directly ester-bonded with lignin and to significantly affect lignin-carbohydrate complex (LCC) linkages, thus enhancing accessibility and enzymes effectiveness. In recent studies, however, it has been shown that arabinose side-chain modification is one of the reasons why AFEX pretreatment is more effective for grasses than other kinds of biomass [60]. Other side-chain components found in almost all types of biomass are as the anhydrous form of acetic acid called acetyl groups or acetate. It is often presumed that most acetyl groups are on the xylan backbone, but some have also been reported to be acetylated for some plant species of lignin [173,174]. Acetyl removal from the xylan backbone was shown to enhance xylan hydrolysis due to increased xylan access, but not to directly impact glucan conversion [175]. Kumar and Wyman showed that selective deacetylation by the method developed by Kong *et al.* [176] enhanced cellbiohydrolase-I (CBHI) adsorption, increased the initial rate, and produced greater digestibility of cellulose and xylan, indicating increased cellulose accessibility and/or enzymes effectiveness [17,94]. This result led to the hypothesis that acetyl groups may restrict cellulase accessibility to cellulose by inhibiting productive binding through increasing the diameter of cellulose and/or changing its hydrophobicity [177].

Most of the studies reported above looked at the impact of hemicellulose removal on accessibility/digestibility for either agricultural residues or hardwoods, both of which have highly substituted xylan as a major hemicellulose component. Fortunately, the high dosages of cellulase and other commercial enzyme preparations often used, such as Novozyme 188, have a broad enough range of activities to hydrolyze carbohydrate polymers such as these completely [178]. However, hemicellulose in softwood, dried distiller's grains with solubles (DDGS), and corn fiber contains a wider range of components including (galacto)glucomannan, arabinogalactan, and arabinoxylan [179]. A multiplicity of enzymes is therefore required to realize high sugar yields [180], particularly at low enzyme loadings. In line with this reasoning, complexities in hemicellulose composition could be additional factors beyond the high lignin content often cited that result in higher enzyme loadings and much lower glucose yields from cellulose for softwoods compared to hardwoods or agricultural residues [181].

Role in Effectiveness

Removing hemicellulose not only enhances cellulose accessibility but greatly improves enzyme effectiveness. We recently showed for the first time that removing hemicellulose during pretreatment can reduce cellulase/xylanase inhibition by soluble xylooligomers generated during enzymatic hydrolysis [50,94]. Extension of these studies by Qing *et al.* and Qing and Wyman showed that xylooligomers inhibit cellulase even more than the well-established powerful inhibitor cellobiose at equal concentrations [51,52]. Previously, xylooligomers were demonstrated to inhibit endoxylanase action [182]. Although the effect of xylan removal on cellulase efficiency is still not completely established, it can be hypothesized to interfere with the processive action of Cel7A and reduce enzyme availability by binding cellulase unproductively [183,184].

During pretreatments, and especially batch pretreatments, hemicellulose degrades to furfural that further (and possibly hemicellulose directly) degrades to carbon-rich compounds called humins or pseudo-lignin [103,104,185]. For example, as shown in Figure 14.3, data adapted from Liu and Wyman [165] for batch hydrothermal pretreatment of corn stover at different temperatures showed that the amount of lignin removed with xylan solubilization increased and then decreased, possibly due to hemicellulose (xylan in this case) degradation to humins. The decrease in amount of lignin removed with xylan solubilized was sharp for pretreatment at high temperatures, possibly due to humins formation unless solubilized lignin precipitates more at high temperatures. Hemicellulose-derived humins, as recently discovered by Kumar *et al.* [103], can deposit onto (and/or co-exist with) cellulose and can affect both cellulose accessibility and enzymes effectiveness.

For enzymatic hydrolysis of lignocellulosics, deacetylation and removal of other side chains may indirectly affect cellulase effectiveness through removing bonds/linkages to xylose that xylanase could not otherwise hydrolyze, thereby making xylanase more effective [138,182,186–195] and in turn increasing cellulose digestibility [170,196,197]. Although the role of acetyl groups and other side chains is unclear, removal of such side chains during pretreatment would surely reduce enzyme requirements and enhance both xylan and glucan digestions as well [94,145,189]. Although not yet established experimentally, we

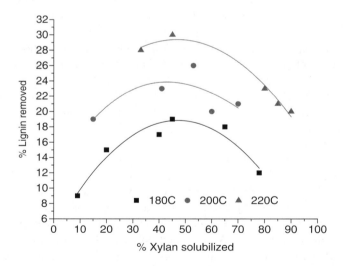

Figure 14.3   *Relationship between the amount of lignin removed (% of original) and xylan solubilized for batch hydrothermal pretreatment of corn stover performed at various temperatures and times (Data adapted from figure 14.6 in Liu and Wyman [165]).*

believe that removing arabinose substitution from the xylan backbone should impact accessibility and/or enzymes effectiveness due to its direct linkages with lignin.

14.4   Features Influencing Micro-accessibility and their Impact on Enzymes Effectiveness

14.4.1   Cellulose Crystallinity (Structure)

Role in Accessibility

Cellulose, which accounts for 25 to nearly 50% (w/w; dry basis) of lignocellulosic biomass, is composed of glucose molecules covalently bonded to one another in linear chains that are linked to other parallel cellulose chains in a highly ordered structure by hydrogen bonds [29,198–200]. However, glucose molecules in real cellulosic biomass are often collectively referred to as "glucan" because current compositional assays cannot accurately distinguish how much of the glucose entity is contained in cellulose versus hemicellulose (e.g., in xyloglucan, glucomannan). There are different allomorphs of cellulose (i.e., cellulose I, II, III [III$_I$ and III$_{II}$], and IV [IV$_I$ and IV$_{II}$]) that contain highly ordered crystalline and less-ordered amorphous regions [201–203]. Cellulose I, that has two polymorphs Iα and Iβ [204], is found in most of the plant cell wall with the amounts and type varying with anatomy (e.g., stalks, leaves, nodes, and internodes) and position in the plant cell wall [57,201,205]. For amorphous cellulose, the total number of hydrogen bonds per repeat units is 5.3 as compared to 9 in crystalline cellulose (Iα). Also, the cohesive energy density in the crystalline form is much higher than for amorphous cellulose, suggesting that non-crystalline forms of cellulose should be more reactive [206–208]. Although various methods (e.g., x-ray diffraction or XRD, nuclear magnetic resonance or NMR, Fourier transform infrared or FTIR spectroscopy [205,209]) have been applied to estimate cellulose crystallinity that indirectly indicates the extent of hydrogen bonding, most methods estimate cellulose bulk average crystallinity and do not distinguish differences in crystallinity (or the extent of hydrogen bonding) between/among different cell walls which may adversely affect cellulase accessibility/digestibility [40]. As an example, in their study Kataoka and Kondo showed that cellulose in the secondary cell wall is more crystalline than in primary walls [210,211].

Cellulase adsorption onto cellulose and cellulose moisture uptake are often reported to increase with decreasing crystallinity, indicating enhanced cellulose accessibility [16,212–214]. For example, Joeh *et al.* [40] showed that crystallinity greatly reduced adsorption of cellobiohydrolase I (Cel7A; CBHI), leading to a decreased extent of hydrolysis. In another study, although increasing crystallinity was reported to reduce the adsorption capacity of cellulose for complete cellulase [215], the maximum adsorption of purified exo- and endocellulases of *Irpex lecteus* protein could be inversely related to cellulose crystallinity [216,217], and more cellulose binding domains from cellulase adsorbed on amorphous than crystalline cellulose [218]. Working with a single source of cellulose, Hall *et al.* found that the cellulase adsorption increased as crystallinity decreased to an extent and then remained constant. Furthermore, the crystallinity value at adsorption saturation was a function of cellulase loading [219]. Nevertheless, contrary to the many other studies, when working with pure cellulose and lignocellulosic substrates Goel and Ramachandran did not find any correlation between crystallinity and cellulase adsorption as measured by enzyme activities in solution [220]. Interestingly, Banka *et al.* showed that adsorption of a non-hydrolytic protein increased with crystallinity [221].

Overall, the greater cellulase adsorption capacity of amorphous cellulose would lead one to expect amorphous regions to have greater hydrolysis rates and yields (by a factor of 2–25) than for crystalline areas [29,39,40,218,222–229]. The ordered structure of crystalline cellulose can therefore impact the ability of cellulase to access cellulose based on the mechanism that a layer of cellulose must be removed before enzymes can attack layers [208,230–232] and active sites lying underneath [223,233–235]. The observed slowdown in rates with increasing cellulose crystallinity, such as for the

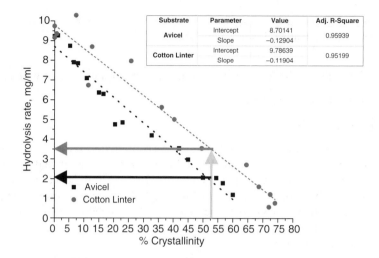

Substrate	Parameter	Value	Adj. R-Square
Avicel	Intercept	8.70141	0.95939
	Slope	−0.12904	
Cotton Linter	Intercept	9.78639	0.95199
	Slope	−0.11904	

Figure 14.4   *Hydrolysis rate versus crystallinity for two cellulose substrates with different initial DPs. Cellulose samples of different crystallinities were prepared at varying phosphoric acid concentrations (Data re-plotted from Bansal et al. [236]) (See figure in color plate section).*

example in Figure 14.4 for untreated and phosphoric-acid-treated Avicel and cotton linter [236], is consistent with this hypothesis [219,226,237,238].

However, others have observed the opposite effect for real biomass: hydrolysis performance improves with crystallinity [162,239,240]. Unfortunately, results with real biomass may be misinterpreted because removal of amorphous lignin and/or hemicellulose can increase biomass crystallinity and enhance digestibility. In some cases, cellulose crystallinity was considered to have no effect on hydrolysis rates [27,239,241–246]. One explanation could be that greater moisture uptake does not necessarily mean enhanced hydrolysis rates, as the cellulase molecular weight is about 3000 times that of water.

If hydrolysis rates are much slower for crystalline regions and amorphous regions are preferentially hydrolyzed, the classical question arises as to why crystallinity does not increase over the course of cellulose hydrolysis as a result of more rapid removal of amorphous cellulose [225,247]. However, no significant change in crystallinity has been measured over the course of cellulose hydrolysis [219,248–252]. Working with a bleached softwood Kraft pulp (albeit at a high enzyme loading of about 166 IU/g glucan), Pu *et al.* showed preferential action by cellulase on non-crystalline cellulose regions, cellulose Iα polymorph, and para-crystalline cellulose compared to cellulose Iβ polymorph and highly crystalline cellulose regions only during the initial phase of hydrolysis, during which cellulose crystallinity increased to some extent [253]. However, after the initial phase, all regions were similarly susceptible to enzymatic hydrolysis with little change in crystallinity. Consistent with this observation, Hall *et al.*, Cateto *et al.* and Penttila *et al.* recently showed similar findings in their respective studies [219,252,254]. The following points may help understand this conundrum and possible mechanisms [16].

1. High enzyme loadings, as often employed to determine the impact of biomass features and other factors on hydrolysis, may lead to misinterpretation by saturating the substrate and impacting other biomass features with enzymes activities other than cellulase in commercial preparations. In particular, fungal cellulases have >50% of their total protein designed to attack and deconstruct crystalline cellulose, with the result that high enzyme loadings would make it hard to differentiate the effect of crystallinity (or modest changes in crystallinity, as often misinterpreted) on hydrolysis.

2. As demonstrated in recent studies, cellulolytic components not only function as hydrolytic agents but can simultaneously disrupt cellulose structure to a significant extent [255–261]. Thus, during hydrolysis, the action(s) of individual monocomponent enzymes is likely obfuscated by concurrent modification by complementing enzymes [255]. Park *et al.* addressed this phenomenon by stating that: "*if the enzymes work ablatively on cellulose microfibril surfaces, consuming the less ordered surface layers of cellulose, then internal ordered cellulose chains will become surface chains with decreased order, so that conversion of amorphous cellulose results in production of more 'amorphous cellulose' and a further decrease in cellulose CI*" [205].
3. For real biomass, crystallinity should not be confused with absolute cellulose crystallinity as real biomass has amorphous components other than cellulose [28,209,242].
4. Almost all the characterization methods require treatment before analysis such as drying or coating that may disturb the structure of biomass [29,209]. Also different analytical methods give different crystallinity values; hence the outcome must be interpreted with caution [29,199,205,209,262].

Nonetheless, a better understanding of cellulase behavior at the micro level and advanced analytical tools would help to understand the role of crystallinity/structure in controlling cellulose micro-accessibility.

Role in Enzymes Effectiveness

From the previous section, it appears that cellulose crystallinity plays a very important role in cellulose accessibility; however, cellulose crystallinity would likely impact the effectiveness of adsorbed cellulase components as well. The latter became more evident from a first-of-its-kind of study by Hall *et al.* in which they showed that cellulase adsorption increased with a reduction in crystallinity only up to a point. However, the hydrolysis rate kept increasing with further reductions in crystallinity while adsorption remained constant [219]. This study directly supports the hypothesis we stated earlier that crystallinity impacts both accessibility and effectiveness [16,263]. Furthermore, the literature showed that cellulose crystallinity affects synergism between cellulase components [29,219,232,264–271]. Because crystalline cellulose is highly hydrophobic, it can irreversibly bind cellulase; this lowers enzyme effectiveness as surface-bound CBHI may lose up to 70% of its activity in only 10 minutes [272]. Crystallinity can therefore impact enzyme effectiveness.

Other than cellulose bulk crystallinity, the cellulose types and crystalline cellulose polymorphs may also affect enzyme effectiveness. Igarashi *et al.* showed that the maximum cellulase adsorption capacity on cellulose Iβ was approximately 1.5 times that for cellulose Iα, although the rate of cellobiose generation from cellulose Iβ was lower than that from cellulose Iα [273,274]. In another study, Igrashi *et al.* showed that the activation of cellulose I to cellulose III$_I$ by ammonia treatment resulted in 5 times more sugar generation, with twice the amount of cellulase adsorbed [275]. Consistent with this, by employing molecular dynamics simulations Chundawat *et al.* recently showed that ammonia treatment converts cellulose I to cellulose III by redistributing the hydrogen bonds to enhance cellulose accessibility and enzyme effectiveness [276].

14.4.2 Cellulose Chain Length/Reducing Ends

Role in Accessibility

Information on the effect of cellulose degree of polymerization (DP) on cellulose accessibility is very limited. However, based on catalytic preferences of cellulase exoglucanases [277–281], it appears that the greater the number of reducing (non-reducing) ends, the greater should be cellulose accessibility. The ambiguity surrounding the effect of cellulose chain length on accessibility to enzymes is partly due to its close association with cellulose crystallinity and particle size. For example, mechanical pretreatments such as ball milling generally reduce particle size, crystallinity, and DP [215,226,228,282,283] for pure cellulose and could also affect lignin structure in real biomass. On the other hand, thermochemical pretreatments (especially at low pH) by such options as steam explosion, dilute acid, and other chemical/physical treatments such as commonly used acid-chlorite delignification [54,55], significantly affect cellulose DP in addition to several other features, again making it difficult to differentiate control [7,28,284–286]. Due to the complexity of real biomass and close association with other physiochemical characteristics, it is therefore difficult to differentiate the impact of cellulose DP on accessibility.

Although conclusive studies directly showing the impact of cellulose chain length on accessibility are scarce, Kaplan *et al.* [287] found a significant drop in cellulase adsorption and hydrolysis of altered cellulose following photochemical degradation, probably due to a decrease in cellulose DP and some ring opening for weathered cotton cellulose. Although cellulase adsorption was not measured to determine its impact, a few other studies indirectly showed the impact of cellulose chain length on cellulose accessibility. For instance, working with wheat straw and bagasse Puri and Pearce [239,245] showed that a reduction in cellulose DP improved hydrolysis, but the impact of features other than crystallinity was not studied. Using highly crystalline bacterial cellulose (BC), Väljamäe *et al.* showed the impact of cellulose DP on both CBHI (Cel7A) and endoglucanase I (EGI) (Cel7B) activity [232]. Treating highly crystalline BC with 1 M hydrochloric acid resulted in little change in cellulose crystallinity, but a major reduction in DP (from 2620 to 150 in 40 min). The reduction in DP increased CBHI relative activity from 37% to 100% but reduced EGI relative activity from 100% to 50%. However, longer incubation of BC with HCl dropped the DP from 150 to 114, with a somewhat negative impact on CBHI and a positive effect on EGI relative activities [232]. For subcritical water pretreatment of Avicel cellulose, Kumar *et al.* reported that microcrystalline cellulose DP decreased dramatically from *c.* 320 to *c.* 100 at temperatures above 300 °C, along with a slight transformation of cellulose I to cellulose II but no change in crystallinity, apparently resulting in increased hydrolysis rates and yields [288]. Hu *et al.* and Hu and Ragauskas and showed that the enzymatic digestibility of cellulose in leaves was much higher than in internodes for hydrothermal pretreatment despite similar cellulose and lignin structures and profiles except cellulose DP, which was much lower for cellulose in leaves [285,289]. Cellulose decrystallization via ball milling is often accompanied by particle size and cellulose chain length reduction, making it difficult to isolate the effect of either feature. However, Bansal *et al.* applied different phosphoric acid concentrations to alter the crystallinity of two different cellulose types: Avicel (CrI *c.* 60%) and cotton linter (CrI *c.* 72%). Their data, re-plotted in Figure 14.4, implied that cotton linter with lower DP of about 180 [217] had a higher cellulose hydrolysis rate than Avicel (DP *c.* 350) at any given crystallinity, assuming that phosphoric acid treatment had no or similar relative effects on chain length for both cellulose types. From correlations in Figure 14.4 it appears that at 0% crystallinity, the enzymatic hydrolysis rate for cotton linter and Avicel would be 9.78 mg/mL and 8.70 mg/mL, respectively.

Kasprzyk *et al.* showed that gamma radiation below 120 kGy did not have a significant impact on cellulose morphology and crystallinity [290]. Consistent with this, Katsumata *et al.* applied gamma radiation of 100 kGy to sapwood and showed enhanced termite feeding activity, mainly due to decreased cellulose DP resulting from gamma radiation [291]. In another study, Knappert and coworkers developed a qualitative relationship between cellulose DP and digestibility [292]. However, Sinistyn *et al.* contradicted these conclusions by showing that a reduction in DP of cotton linters by application of gamma irradiation while keeping crystallinity index (CI) constant had a negligible impact on hydrolysis rates [226]. Furthermore, a kinetic study by Zhang and Lynd indicated that reducing cellulose DP had less effect on accelerating hydrolysis rates than increasing accessibility of β-glycosidic bonds as measured by the maximum amount of cellulase adsorbed on cellulose [293]. However, the possibility of cellulose chain length (DP) impacting accessibility was not discussed. On a different note, cellulose surface properties (free energy [gS $= 37.56 + 0.02$ DP] and polarity [P $= 11.88 - 0.02$ DP $+ 9.10$ DP^2] were related to cellulose chain length [294] and both were correlated to increase with cellulose DP [294].

Role in Effectiveness

Other than impacting accessibility, cellulose DP can also affect enzyme effectiveness; relevant literature is once more limited, however. Hypothetically, the lower the DP, the more reducing and non-reducing ends are available, and more CBHI/II would be expected to be able to work at one time while exposing more sites for endoglucanases attack. However, as cellulose deconstruction is assumed to be a surface phenomenon and the mechanism is often characterized in terms of a peeling action [252,295], the surface availability of reducing/non-reducing ends can be limited [234,235] and affected by other structural features, as discussed above. Nidetzky *et al.* found that, for soluble cellulose, the initial degradation velocity of cello-oligosaccharides by CBHI increased with DP below cellohexose but was constant for higher DP [281]. Similar DP effects of soluble cellodextrins on CBHII and EGI activity have been reviewed elsewhere [29]. Furthermore, a decrease in β-glucosidase activity with increasing DP has been reported [296,297]. To the authors' knowledge, not much information is available on the effect of insoluble cellulose DP on the catalytic efficiency of cellulose, except that higher DP could result in higher synergy between CBHI and EGI [268,293,298,299]. However, Gupta and Lee showed that cellulase could hydrolyze non-crystalline cellulose (NCC) with high yields when cellulose DP was large, because the CBHI component of the cellulase system cannot recognize substrates with a DP below 10; this led to the result that cello-oligosaccharides with DP < 10 and > 3 are not hydrolyzed [35]. Furthermore, cellulose DP may affect the processivity index, with full processivity of CBHI possibly not realized for short chains [35,232]. Overall, studies of the effect of DP and crystallinity on enzymatic digestibility demonstrated that the susceptibility of pretreated substrates to enzymatic hydrolysis could not be easily predicted from differences in their cellulose DP and crystallinity [239,300], likely due to the complexity of real cellulosic substrates.

14.5 Concluding Remarks

In this review chapter based on the information in literature and findings in our lab on strong correlations between maximum cellulase (complete or a component) adsorption capacity and hydrolysis rates/yields [17,106,301,302], such as for the example in Figure 14.5, it was hypothesized that enzymatic hydrolysis of cellulose in pretreated biomass is controlled by two main factors: enzyme accessibility to cellulose and enzyme effectiveness on cellulose. However, accessibility can further be visualized in two categories: macro-accessibility and micro-accessibility. The factors influencing macro- (mainly lignin and hemicellulose) and micro-accessibility (cellulose crystallinity and type and chain length) most and their impacts on enzymes effectiveness discussed in this chapter are summarized in Table 14.1. Due to complexity of actual cellulosic biomass materials, it is difficult to assign some features to a single category as they can likely affect both macro- and micro-accessibility.

Some other biomass features that are also believed to affect enzymatic hydrolysis such as biomass porosity and pore volume, biomass and cellulose surface area, and particle size [72,161,303] were not included due to their lower impact on hydrolysis for real biomass, their non-exclusive behavior, or inadequate evidence to support our hypothesis. For example, several studies showed that biomass porosity/pore volume, which should not be confused with micro/nanopores or cracks within cellulose microfibrils [142,304,305], play a major role in accessibility and enzymes effectiveness due to the molecular weights/sizes of cellulolytic components. Although pores/porosity appear to play a vital role, unpretreated cell walls have a porous structure to support plant growth that no doubt changes during pretreatment, and an increase in porosity is

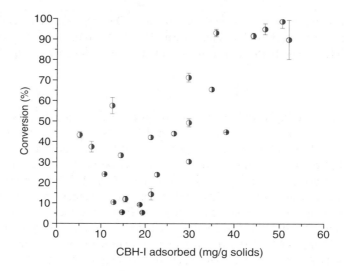

Figure 14.5   *The 24-h cellulose conversion (%) versus CBH-I adsorption (mg/g solids) for Avicel cellulose and poplar solids prepared by various leading pretreatments and those solids hydrolyzed to different extents (Data adapted from Kumar and Wyman [42]).*

often concomitant with changes in other biomass structural features such as hemicellulose and pectin removal during pretreatment [150,306,307]. It is therefore difficult to ascertain the influence of porosity on accessibility and enzyme effectiveness. Other than that, the methods employed to measure porosity and pore volumes do not appear to be reliable enough. However, although difficult to measure for real biomass, cracks and micro/nanopores may also exist within cellulose microfibrils but, according to the current picture of cellulase taking cellulose apart layer by layer, cracks and nanopores within microfibrils may have a limited effect on accessibility but could impact effectiveness. For example, Tanka *et al.* showed that cellulase components trapped within pores may have low synergism and thus lower hydrolysis yields/rates [305]. On the contrary, working with several types of pure cellulose Gama *et al.* showed that cellulase does not enter these micropores [304]. Consistent with this, Penttila *et al.* recently reached the same conclusion, namely: *"enzymes act on the surface of cellulose bundles and are unable to penetrate into the nanopores of wet cellulose"* [254].

Table 14.1   *Summary of pretreated biomass key chemical and physical features and their role in cellulose macro-/micro-accessibility and enzyme effectiveness.*

Biomass chemical/physical features	Role in cellulose accessibility		Role in enzyme effectiveness
	Macro	Micro	
Lignin	Major (apparently indirect[a])	Minor, if any	Significant
Hemicellulose	Major (direct)	Minor, if any	Major
Acetyl/arabinan/other substitution on xylan backbone	Significant[a]	Minor, if any	Significant
Crystallinity	Minor, if any	Major	Major
Cellulose chain length/ DP	Minor, if any	Possibly significant[a]	Possibly significant

[a] Hypothesis needs more research.

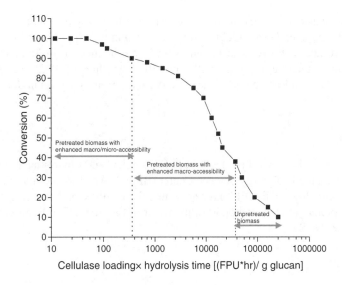

Figure 14.6 *Hypothetical depiction of cellulose conversion versus cellulase loading × hydrolysis time (FPU hr/g glucan) with changes in macro-/micro-accessibility in pretreated biomass.*

Biomass particle size is often suggested to affect enzymatic hydrolysis by increasing the surface area for greater enzyme adsorption, based on data with pure celluloses [220,308–310]. However, particle size does not appear to play a plausible role for real biomass unless one of the other major features (either lignin or hemicellulose) impacting cellulose macro-accessibility is altered [311]. In addition, data on the effect of particle size should be interpreted cautiously as mechanical size reduction may also affect other structural features.

Understanding the effect of surface area, often deemed responsible for effective enzymatic hydrolysis, is again marred by the measurement techniques applied and interpretation of the data. For example, Gupta and Lee showed that BET surface area of NCC almost doubled compared to the original untreated cellulose; however, hydrolysis rates for NCC were almost 10 times higher [35]. Together, we believe that the features described here contribute directly or indirectly to cellulose accessibility and enzyme effectiveness, but the extent of each is still debatable.

The key question, however, is which accessibility categories or biomass features and their alteration affect cellulose hydrolysis the most. The choice will depend strongly on biomass type and the desired outcome. For instance, impacts on either accessibility category can achieve reasonable sugar yields with a somewhat high enzyme loading; however, they must both be changed to realize acceptably high sugar yields with very low enzyme loadings and short processing times, as illustrated in Figure 14.6.

Most leading pretreatments, except for those at low pH, seem to affect features in the macro-accessibility category and apparently enhance biomass digestibility by changing mostly cellulose macro-accessibility through altering/removing either hemicellulose or lignin or both. Although solids pretreated by these techniques are highly digestible, high and yet economically unfeasible enzyme loadings are required to realize these high yields. Conversely, other pretreatments such as concentrated phosphoric acid (COSLIF) [312,313] and ionic liquids [314–316] appear to enhance both macro- and micro-accessibility and thus require less enzyme and processing time compared to pretreatments that impact only macro-accessibility However, changing only micro-accessibility (such as through ball milling to alter cellulose structure/crystallinity) does not lead to effective conversions at low enzyme loadings (personal communication with Dr Mark Holtzapple, Texas A&M, 2011). In addition, COSLIF/ionic-liquid pretreatments that strongly address micro-accessibility need to remove/relocate hemicellulose/lignin (i.e., address macro-accessibility) to realize higher sugar yields with low FPU hr/g glucan [199,209,317,318]. Furthermore, even within the macro-accessibility category, the effectiveness of removing lignin or hemicellulose depends on biomass types.

Acknowledgements

We gratefully acknowledge the support by the Office of Biological and Environmental Research in the DOE Office of Science through the BioEnergy Science Center (BESC). We are grateful to the Center for Environmental Research & Technology (CE-CERT) for providing the facilities and equipment. We would also like to thank Ford motor company for their support of the Chair in Environmental Engineering at the University of California Riverside (UCR).

References

1. Lynd, L.R., Cushman, J.H., Nichols, R.J., and Wyman, C.E. (1991) Fuel ethanol from cellulosic biomass. *Science*, **251** (4999), 1318–1323.
2. Wyman, C.E. and Goodman, B.J. (1993) Biotechnology for production of fuels, chemicals, and materials from biomass. *Applied Biochemistry and Biotechnology*, **39/40**, 41–59.
3. Lynd, L.R., Wyman, C.E., and Gerngross, T.U. (1999) Biocommodity engineering. *Biotechnology Progress*, **15** (5), 777–793.
4. Jeoh, T., Johnson, D.K., Adney, W.S., and Himmel, M.E. (2005) Measuring cellulase accessibility of dilute-acid pretreated corn stover. *Preprints of Symposia – American Chemical Society Division of Fuel Chemistry*, **50** (2), 673–674.
5. Wyman, C. and Hinman, N. (1990) Ethanol- Fundamentals of production from renewable feedstocks and use as a transportation fuel. *Applied Biochemistry and Biotechnology*, **24–25** (1), 735–753.
6. Sun, Y. and Cheng, J. (2002) Hydrolysis of lignocellulosic materials for ethanol production: a review. *Bioresource Technology*, **83** (1), 1–11.
7. Chandra, R.P., Bura, R., Mabee, W.E. *et al.* (2007) Substrate pretreatment: the key to effective enzymatic hydrolysis of lignocellulosics? *Advances in Biochemical Engineering/Biotechnology*, **108**, 67–93.
8. Mosier, N., Wyman, C., Dale, B. *et al.* (2005) Features of promising technologies for pretreatment of lignocellulosic biomass. *Bioresource Technology*, **96** (6), 673–686.
9. da Costa Sousa, L., Chundawat, S.P.S., Balan, V., and Dale, B.E. (2009) 'Cradle-to-grave' assessment of existing lignocellulose pretreatment technologies. *Current Opinion in Biotechnology*, **20** (3), 339–347.
10. Kumar, P., Barrett, D.M., Delwiche, M.J., and Stroeve, P. (2009) Methods for pretreatment of lignocellulosic biomass for efficient hydrolysis and biofuel production. *Industrial & Engineering Chemistry Research*, **48** (8), 3713–3729.
11. Sun, Y. and Cheng, J.Y. (2002) Hydrolysis of lignocellulosic materials for ethanol production: a review. *Bioresource Technology*, **83** (1), 1–11.
12. Holtzapple, M.T., Humphrey, A.E., and Taylor, J.D. (1989) Energy requirements for the size reduction of poplar and aspen wood. *Biotechnology and Bioengineering*, **33** (2), 207–210.
13. Zhu, J.Y. and Pan, X.J. (2010) Woody biomass pretreatment for cellulosic ethanol production: Technology and energy consumption evaluation. *Bioresource Technology*, **101** (13), 4992–5002.
14. Zhu, J.Y., Pan, X.J., Wang, G.S., and Gleisner, R. (2009) Sulfite pretreatment (SPORL) for robust enzymatic saccharification of spruce and red pine. *Bioresource Technology*, **100** (8), 2411–2418.
15. Falls, M. and Holtzapple, M. (2011) Oxidative lime pretreatment of alamo switchgrass. *Applied Biochemistry and Biotechnology*, **165** (2), 506–522.

16. Kumar, R. and Wyman, C.E. (2010) Key features of pretreated lignocelluloses biomass solids and their impact on hydrolysis, in *Bioalcohol Production: Biochemical Conversion of Lignocellulosic Biomass* (ed. K. Waldon), Woodhead Publishing Limited, Oxford, p. 73–121.

17. Kumar, R. and Wyman, C.E. (2009) Cellulase adsorption and relationship to features of corn stover solids produced by leading pretreatments. *Biotechnology and Bioengineering*, **103** (2), 252–267.

18. Zhang, Y.-H.P. and Lynd, L.R. (2004) Toward an aggregated understanding of enzymatic hydrolysis of cellulose: Noncomplexed cellulase systems. *Biotechnology and Bioengineering*, **88** (7), 797–824.

19. Lynd, L.R., Weimer, P.J., van Zyl, W.H., and Pretorius, I.S. (2002) Microbial cellulose utilization: fundamentals and biotechnology. *Microbiology and Molecular Biology Reviews*, **66** (3), 506–577.

20. Henrissat, B. (1994) Cellulases and their interaction with cellulose. *Cellulose*, **1** (3), 169–196.

21. Bayer, E.A., Chanzy, H., Lamed, R., and Shoham, Y. (1998) Cellulose, cellulases and cellulosomes. *Current Opinion in Structural Biology*, **8** (5), 548–557.

22. Jung, H., Wilson, D.B., and Walker, L.P. (2002) Binding of Thermobifida fusca CDCel5A, CDCel6B and CDCel48A to easily hydrolysable and recalcitrant cellulose fractions on BMCC. *Enzyme and Microbial Technology*, **31** (7), 941–948.

23. Palonen, H., Tenkanen, M., and Linder, M. (1999) Dynamic interaction of Trichoderma reesei cellobiohydrolases Cel6A and Cel7A and cellulose at equilibrium and during hydrolysis. *Applied and Environmental Microbiology*, **65** (12), 5229–5233.

24. Woodward, J., Jp, B., Br, E., and Ka, A. (1994) Papain digestion of crude Trichoderma reesei cellulase: purification and properties of cellobiohydrolase I and II core proteins. *Biotechnology and Applied Biochemistry*, **19** (2), 141–153.

25. Palonen, H., Tjerneld, F., Zacchi, G., and Tenkanen, M. (2004) Adsorption of Trichoderma reesei CBH I and EG II and their catalytic domains on steam pretreated softwood and isolated lignin. *Journal of Biotechnology*, **107** (1), 65–72.

26. Converse, A.O. (1993) Substrate factors limiting enzymatic hydrolysis, in *Bioconversion of Forest and Agricultural Plant Residues* (ed. J.N. Saddler), CAB International, Wallingford, UK, p. 93–106.

27. Mansfield, S.D., Mooney, C., and Saddler, J.N. (1999) Substrate and enzyme characteristics that limit cellulose hydrolysis. *Biotechnology Progress*, **15** (5), 804–816.

28. Kumar, R., Mago, G., Balan, V., and Wyman, C.E. (2009) Physical and chemical characterizations of corn stover and poplar solids resulting from leading pretreatment technologies. *Bioresource Technology*, **100** (17), 3948–3962.

29. Zhang, Y.H. and Lynd, L.R. (2004) Toward an aggregated understanding of enzymatic hydrolysis of cellulose: noncomplexed cellulase systems. *Biotechnology and Bioengineering*, **88** (7), 797–824.

30. Poutanen, K. and Puls, J. (1988) Characteristics of Trichoderma reesei β-xylosidase and its use in the hydrolysis of solubilized xylans. *Applied Microbiology and Biotechnology*, **28** (4), 425–432.

31. Willies, D. (2007) An Investigation into the Adsorption of Enzymes and BSA Protein onto Lignocellulosic Biomass Fractions and the Benefit to Hydrolysis of Non-Catalytic Additives. M.Sc. thesis, Thayer School of Engineering, Dartmouth College, USA.

32. Qing, Q. and Wyman, C. (2011) Supplementation with xylanase and beta-xylosidase to reduce xylo-oligomer and xylan inhibition of enzymatic hydrolysis of cellulose and pretreated corn stover. *Biotechnology for Biofuels*, **4** (1), 18.

33. Yang, B. and Wyman, C.E. (2006) BSA treatment to enhance enzymatic hydrolysis of cellulose in lignin containing substrates. *Biotechnology and Bioengineering*, **94** (4), 611–617.

34. Percival Zhang, Y.H., Himmel, M.E., and Mielenz, J.R. (2006) Outlook for cellulase improvement: Screening and selection strategies. *Biotechnology Advances*, **24** (5), 452–481.

35. Gupta, R. and Lee, Y.Y. (2009) Mechanism of cellulase reaction on pure cellulosic substrates. *Biotechnology and Bioengineering*, **102** (6), 1570–1581.

36. Kumar, R. and Wyman, C.E. (2008) An improved method to directly estimate cellulase adsorption on biomass solids. *Enzyme and Microbial Technology*, **42** (5), 426–433.

37. Jeoh, T., Wilson, D.B., and Walker, L.P. (2002) Cooperative and competitive binding in synergistic mixtures of thermobifida fuscaCellulases Cel5A, Cel6B, and Cel9A. *Biotechnology Progress*, **18** (4), 760–769.

38. Gao, D., Chundawat, S.P.S., Uppugundla, N. *et al.* (2011) Binding characteristics of Trichoderma reesei cellulases on untreated, ammonia fiber expansion (AFEX), and dilute-acid pretreated lignocellulosic biomass. *Biotechnology and Bioengineering*, **108** (8), 1788–1800.

39. Hong, J., Ye, X., and Zhang, Y.H. (2007) Quantitative determination of cellulose accessibility to cellulase based on adsorption of a nonhydrolytic fusion protein containing CBM and GFP with its applications. *Langmuir*, **23** (25), 12535–12540.

40. Jeoh, T., Ishizawa, C.I., Davis, M.F. *et al.* (2007) Cellulase digestibility of pretreated biomass is limited by cellulose accessibility. *Biotechnology and Bioengineering*, **98** (1), 112–122.

41. Ishizawa, C., Jeoh, T., Adney, W. *et al.* (2009) Can delignification decrease cellulose digestibility in acid pretreated corn stover? *Cellulose*, **16** (4), 677–686.

42. Kumar, R. and Wyman, C.E. (2009) Does change in accessibility with conversion depend on both the substrate and pretreatment technology? *Bioresource Technology*, **100** (18), 4193–4202.

43. Wang, W., Kang, L., Wei, H. *et al.* (2011) Study on the decreased sugar yield in enzymatic hydrolysis of cellulosic substrate at high solid loading. *Applied Biochemistry and Biotechnology*, **164** (7), 1139–1149.

44. Kyriacou, A., Neufeld, R.J., and MacKenzie, C.R. (1988) Effect of physical parameters on the adsorption characteristics of fractionated Trichoderma reesei cellulase components. *Enzyme and Microbial Technology*, **10** (11), 675–681.

45. Donohoe, B.S., Selig, M.J., Viamajala, S. *et al.* (2009) Detecting cellulase penetration into corn stover cell walls by immuno-electron microscopy. *Biotechnology and Bioengineering*, **103** (3), 480–489.

46. Donohoe, B.S., Decker, S.R., Tucker, M.P. *et al.* (2008) Visualizing lignin coalescence and migration through maize cell walls following thermochemical pretreatment. *Biotechnology and Bioengineering*, **101** (5), 913–925.

47. Tatsumoto, K., Baker, J.O., Tucker, M.P. *et al.* (1988) Digestion of pretreated aspen substrates. Hydrolysis rates and adsorptive loss of cellulase enzymes. *Applied Biochemistry and Biotechnology*, **18**, 159–174.

48. Hong, J., Ladisch, M.R., Gong, C.S. *et al.* (1981) Combined product and substrate inhibition equation for cellobiase. *Biotechnology and Bioengineering*, **23** (12), 2779–2788.

49. Holtzapple, M., Cognata, M., Shu, Y., and Hendrickson, C. (1990) Inhibition of Trichoderma reesei cellulase by sugars and solvents. *Biotechnology and Bioengineering*, **36** (3), 275–287.

50. Kumar, R. and Wyman, C.E. (2009) Effect of enzyme supplementation at moderate cellulase loadings on initial glucose and xylose release from corn stover solids pretreated by leading technologies. *Biotechnology and Bioengineering*, **102** (2), 457–467.

51. Qing, Q., Yang, B., and Wyman, C.E. (2010) Xylooligomers are strong inhibitors of cellulose hydrolysis by enzymes. *Bioresource Technology*, **101** (24), 9624–9630.

52. Qing, Q. and Wyman, C.E. (2011) Hydrolysis of different chain length xylooliogmers by cellulase and hemicellulase. *Bioresource Technology*, **102** (2), 1359–1366.

53. Kristensen, J., Felby, C., and Jorgensen, H. (2009) Yield-determining factors in high-solids enzymatic hydrolysis of lignocellulose. *Biotechnology for Biofuels*, **2** (1), 11.

54. Hubbell, C.A. and Ragauskas, A.J. (2010) Effect of acid-chlorite delignification on cellulose degree of polymerization. *Bioresource Technology*, **101** (19), 7410–7415.

55. Hallac, B.B. and Ragauskas, A.J. (2011) Analyzing cellulose degree of polymerization and its relevancy to cellulosic ethanol. *Biofuels, Bioproducts and Biorefining*, **5** (2), 215–225.

56. Dhugga, K.S. (2007) Maize biomass yield and composition for biofuels. *Crop Science*, **47** (6), 2211–2227.

57. Chundawat, S.P.S., Donohoe, B.S., da Costa Sousa, L. *et al.* (2011) Multi-scale visualization and characterization of lignocellulosic plant cell wall deconstruction during thermochemical pretreatment. *Energy & Environmental Science*, **4** (3), 973–984.

58. Grabber, J.H., Ralph, J., and Hatfield, R.D. (1997) Quideau S. p-Hydroxyphenyl, guaiacyl, and syringyl lignins have similar inhibitory effects on wall degradability. *Journal of Agricultural and Food Chemistry*, **45** (7), 2530–2532.

59. Chundawat, S.P.S., Beckham, G.T., Himmel, M.E., and Dale, B.E. (2010) Deconstruction of lignocellulosic biomass to fuels and chemicals. *Annual Review of Chemical and Biomolecular Engineering*, **2** (6), 1–25.

60. Venkatesh, B., Leonardo da Costa, S., Shishir, P.S.C. *et al.* (2009) Enzymatic digestibility and pretreatment degradation products of AFEX-treated hardwoods Populus (Populus nigra). *Biotechnology Progress*, **25** (2), 365–375.

61. Studer, M.H., DeMartini, J.D., Davis, M.F. *et al.* (2011) Lignin content in natural Populus variants affects sugar release. *Proceedings of the National Academy of Sciences*, **108** (15), 6300–6305.
62. Chang, V.S. and Holtzapple, M.T. (2000) Fundamental factors affecting biomass enzymatic reactivity. *Applied Biochemistry and Biotechnology*, **84–86**, 5–37.
63. Sudo, K., Matsumura, Y., and Shimizu, K. (1976) Enzymic hydrolysis of woods. I. Effect of delignification on the hydrolysis of woods by Trichoderma viride cellulase. *Mokuzai Gakkaishi*, **22** (12), 670–676.
64. Cunningham, R.L., Detroy, R.W., Bagby, M.O., and Baker, F.L. (1981) Modifications of wheat straw to enhance cellulose saccharification by enzymic hydrolysis. *Transactions of the Illinois State Academy of Science*, **74** (3–4), 67–75.
65. Avgerinos, G.C. and Wang, D.I.C. (1983) Selective solvent delignification for fermentation enhancement. *Biotechnology and Bioengineering*, **25** (1), 67–83.
66. Várnai, A., Siika-aho, M., and Viikari, L. (2010) Restriction of the enzymatic hydrolysis of steam-pretreated spruce by lignin and hemicellulose. *Enzyme and Microbial Technology*, **46** (3–4), 185–193.
67. Koullas, D.P., Christakopoulos, P.F., Kekos, D. *et al.* (1993) Effect of alkali delignification on wheat straw saccharification by fusarium oxysporum cellulases. *Biomass and Bioenergy*, **4** (1), 9–13.
68. Yang, B. and Wyman, C.E. (2004) Effect of xylan and lignin removal by batch and flowthrough pretreatment on the enzymatic digestibility of corn stover cellulose. *Biotechnology and Bioenginnering*, **86** (1), 88–95.
69. Fu, C., Mielenz, J.R., Xiao, X. *et al.* (2011) Genetic manipulation of lignin reduces recalcitrance and improves ethanol production from switchgrass. *Proceedings of the National Academy of Sciences*, **108** (9), 3803–3808.
70. Kumar, L., Chandra, R., and Saddler, J. (2011) Influence of steam pretreatment severity on post-treatments used to enhance the enzymatic hydrolysis of pretreated softwoods at low enzyme loadings. *Biotechnology and Bioengineering*, **108** (10), 2300–2311.
71. Kim, S.B., Um, B.H., and Park, S.C. (2001) Effect of pretreatment reagent and hydrogen peroxide on enzymatic hydrolysis of oak in percolation process. *Applied Biochemistry and Biotechnology*, **91–93**, 81–94.
72. Wong, K.K.Y., Deverell, K.F., Mackie, K.L. *et al.* (1988) The relationship between fiber-porosity and cellulose digestibility in steam-exploded Pinus radiata. *Biotechnology and Bioengineering*, **31** (5), 447–456.
73. Draude, K.M., Kurniawan, C.B., and Duff, S.J.B. (2001) Effect of oxygen delignification on the rate and extent of enzymatic hydrolysis of lignocellulosic material. *Bioresource Technology*, **79** (2), 113–120.
74. Ohgren, K., Bura, R., Saddler, J., and Zacchi, G. (2007) Effect of hemicellulose and lignin removal on enzymatic hydrolysis of steam pretreated corn stover. *Bioresource Technology*, **98** (13), 2503–2510.
75. Wyman, C.E., Dale, B.E., Elander, R.T. *et al.* (2005) Comparative sugar recovery data from laboratory scale application of leading pretreatment technologies to corn stover. *Bioresource Technology*, **96** (18), 2026–2032.
76. Elander, R., Dale, B., Holtzapple, M. *et al.* (2009) Summary of findings from the biomass refining consortium for applied fundamentals and innovation (CAFI): corn stover pretreatment. *Cellulose*, **16** (4), 649–659.
77. Kim, T. and Lee, Y. (2005) Pretreatment of corn stover by soaking in aqueous ammonia. *Applied Biochemistry and Biotechnology*, **124** (1), 1119–1131.
78. Kim, S. and Holtzapple, M.T. (2005) Lime pretreatment and enzymatic hydrolysis of corn stover. *Bioresource Technology*, **96** (18), 1994–2006.
79. Kim, T.H. and Lee, Y.Y. (2007) Pretreatment of corn stover by soaking in aqueous ammonia at moderate temperatures. *Applied Biochemistry and Biotechnology*, **137–140**, 81–92.
80. Abe, A., Dusek, K., Kobayashi, S. *et al.* (2010) Lignin structure, properties, and applications, in *Biopolymers*, Springer, Berlin/Heidelberg, p. 1–63.
81. Glasser, W.G., Barnett, C.A., Muller, P.C., and Sarkanen, K.V. (1983) The chemistry of several novel bioconversion lignins. *Journal of Agricultural and Food Chemistry*, **31** (5), 921–930.
82. Bouajila, J., Dole, P., Joly, C., and Limare, A. (2006) Some laws of a lignin plasticization. *Journal of Applied Polymer Science*, **102** (2), 1445–1451.
83. Michalowicz, G., Toussaint, B., and Vignon, M.R. (1991) Ultrastructural changes in poplar cell wall during steam explosion treatment. *Holzforschung*, **45** (3), 175–179.
84. Selig, M.J., Viamajala, S., Decker, S.R. *et al.* (2007) Deposition of lignin droplets produced during dilute acid pretreatment of maize stems retards enzymatic hydrolysis of cellulose. *Biotechnology Progress*, **23** (6), 1333–1339.

85. Foston, M. and Ragauskas, A.J. (2010) Changes in lignocellulosic supramolecular and ultrastructure during dilute acid pretreatment of Populus and switchgrass. *Biomass and Bioenergy*, **34** (12), 1885–1895.

86. Donaldson, L.A., Wong, K.K.Y., and Mackie, K.L. (1988) Ultrastructure of steam-exploded wood. *Wood Science and Technology*, **22** (2), 103–114.

87. Li, C., Cheng, G., Balan, V. *et al.* (2011) Influence of physico-chemical changes on enzymatic digestibility of ionic liquid and AFEX pretreated corn stover. *Bioresource Technology*, **102** (13), 6928–6936.

88. Beveridge, R.J. and Richards, G.N. (1975) Digestion of polysaccharide constituents of tropical pasture herbage in the bovine rumen. VI. Investigation of the digestion of cell-wall polysaccharides of spear grass and of cotton cellulose by viscometry and by x-ray diffraction. *Carbohydrate Research*, **43** (1), 163–172.

89. Teixeira, L., Linden, J., and Schroeder, H. (1999) Alkaline peracetic acid pretreatments of biomass for ethanol production. *Applied Biochemistry and Biotechnology*, **77** (1), 19–34.

90. Morrison, I.M. (1983) The effect of physical and chemical treatments on the degradation of wheat and barley straws by rumen liquor-pepsin and pepsin-cellulase systems. *Journal of the Science of Food and Agriculture*, **34** (12), 1323–1329.

91. Ford, C.W. (1983) Effect of particle size and delignification on the rate of digestion of hemicellulose and cellulose by cellulase in mature pangola grass stems. *Australian Journal of Agricultural Research*, **34** (3), 241–248.

92. Mes-Hartree, M., Yu, E.K.C., Reid, I.D., and Saddler, J.N. (1987) Suitability of aspenwood biologically deligni-fied with Pheblia tremellosus for fermentation to ethanol or butanediol. *Applied Microbiology and Biotechnology*, **26** (2), 120–125.

93. Prabhu, K.A. and Maheshwari, R. (1999) Biochemical properties of xylanases from a thermophilic fungus, Mela-nocarpus albomyces, and their action on plant cell walls. *Journal of Biosciences*, **24** (4), 461–470.

94. Kumar, R. and Wyman, C.E. (2009) Effect of xylanase supplementation of cellulase on digestion of corn stover solids prepared by leading pretreatment technologies. *Bioresource Technology*, **100** (18), 4203–4213.

95. Selig, M., Vinzant, T., Himmel, M., and Decker, S. (2009) The effect of lignin removal by alkaline peroxide pretreatment on the susceptibility of corn stover to purified cellulolytic and xylanolytic enzymes. *Applied Biochemistry and Biotechnology*, **155** (1), 94–103.

96. Kumar, R. and Wyman, C.E. (2009) Effects of cellulase and xylanase enzymes on the deconstruction of solids from pretreatment of poplar by leading technologies. *Biotechnology Progress*, **25** (2), 302–314.

97. Schwald, W., Brownell, H.H., and Saddler, J.N. (1988) Enzymatic hydrolysis of steam treated aspen wood: influ-ence of partial hemicellulose and lignin removal prior to pretreatment. *Journal of Wood Chemistry and Technol-ogy*, **8** (4), 543–560.

98. Zhao, X.-B., Wang, L., and Liu, D.-H. (2008) Peracetic acid pretreatment of sugarcane bagasse for enzymatic hydrolysis: a continued work. *Journal of Chemical Technology & Biotechnology*, **83** (6), 950–956.

99. Yoshida, M., Liu, Y., Uchida, S. *et al.* (2008) Effects of cellulose crystallinity, hemicellulose, and lignin on the enzymatic hydrolysis of miscanthus sinensis to monosaccharides. *Bioscience, Biotechnology, and Biochemistry*, **72** (3), 805–810.

100. van Oss, C.J. (1995) Hydrophobicity of biosurfaces – origin, quantitative determination and interaction energies. *Colloids and Surfaces, B: Biointerfaces*, **5** (3/4), 91–110.

101. Matthews, J.F., Skopec, C.E., Mason, P.E. *et al.* (2006) Computer simulation studies of microcrystalline cellulose I[beta]. *Carbohydrate Research*, **341** (1), 138–152.

102. Sannigrahi, P., Kim, D.H., Jung, S., and Ragauskas, A. (2011) Pseudo-lignin and pretreatment chemistry. *Energy & Environmental Science*, **4** (4), 1306–1310.

103. Kumar, R., Hu, F., Sannigrahi, P., Jung, S., Ragauskas, A.J., Wyman, C.E. (2013) Carbohydrate derived-pseudo-lignin can retard cellulose biological conversion. *Biotechnology and Bioengineering*, **110** (3), 737–53.

104. Hu, F., Jung, S., Ragauskas, A. (2012) Pseudo-lignin formation and its impact on enzymatic hydrolysis. *Biore-source Technology*, **117** (0), 7–12.

105. Berlin, A., Balakshin, M., Gilkes, N. *et al.* (2006) Inhibition of cellulase, xylanase and beta-glucosidase activities by softwood lignin preparations. *Journal of Biotechnology*, **125** (2), 198–209.

106. Kumar, R. and Wyman, C.E. (2009) Access of cellulase to cellulose and lignin for poplar solids produced by leading pretreatment technologies. *Biotechnology Progress*, **25** (3), 807–819.

107. Excoffier, G., Toussaint, B., and Vignon, M.R. (1991) Saccharification of steam-exploded poplar wood. *Biotechnology and Bioengineering*, **38** (11), 1308–1317.
108. Berlin, A., Gilkes, N., Kurabi, A. *et al.* (2005) Weak lignin-binding enzymes: a novel approach to improve activity of cellulases for hydrolysis of lignocellulosics. *Applied Biochemistry and Biotechnology*, **121–124**, 163–170.
109. Sewalt, V.J.H., Glasser, W.G., and Beauchemin, K.A. (1997) Lignin impact on fiber degradation. 3. Reversal of inhibition of enzymic hydrolysis by chemical modification of lignin and by additives. *Journal of Agricultural and Food Chemistry*, **45** (5), 1823–1828.
110. Mandels, M. and Reese, E.T. (1965) Inhibition of cellulases. *Annual Review of Phytopathology*, **3**, 85–102.
111. Wu, Z. and Lee, Y.Y. (1997) Ammonia recycled percolation as a complementary pretreatment to the dilute-acid process. *Applied Biochemistry and Biotechnology*, **63–65**, 21–34.
112. Jørgensen, H. and Olsson, L. (2006) Production of cellulases by Penicillium brasilianum IBT 20888–Effect of substrate on hydrolytic performance. *Enzyme and Microbial Technology*, **38** (3–4), 381–390.
113. Tilton, R.D., Robertson, C.R., and Gast, A.P. (1991) Manipulation of hydrophobic interactions in protein adsorption. *Langmuir*, **7** (11), 2710–2718.
114. Kongruang, S., Bothwell, M.K., McGuire, J. *et al.* (2003) Assaying the activities of Thermomonospora fusca E5 and Trichoderma reesei CBHI cellulase bound to polystyrene. *Enzyme and Microbial Technology*, **32** (5), 539–545.
115. Bai, G., Goncalves, C., Gama, F.M., and Bastos, M. (2008) Self-aggregation of hydrophobically modified dextrin and their interaction with surfactant. *Thermochimica Acta*, **467** (1–2), 54–62.
116. Chundawat, S.P., Venkatesh, B., and Dale, B.E. (2007) Effect of particle size based separation of milled corn stover on AFEX pretreatment and enzymatic digestibility. *Biotechnology and Bioengineering*, **96** (2), 219–231.
117. Suvajittanont, W., Bothwell, M.K., and McGuire, J. (2000) Adsorption of Trichoderma reesei CBHI cellulase on silanized silica. *Biotechnology and Bioengineering*, **69** (6), 688–692.
118. Andreaus, J., Azevedo, H., and Cavaco-Paulo, A. (1999) Effects of temperature on the cellulose binding ability of cellulase enzymes. *Journal of Molecular Catalysis B: Enzymatic*, **7**, 233–239.
119. Reinikainen, T., Teleman, O., and Teeri, T.T. (1995) Effects of pH and high ionic strength on the adsorption and activity of native and mutated cellobiohydrolase I from Trichoderma reesei. *Proteins*, **22** (4), 392–403.
120. Halder, E., Chattoraj, D.K., and Das, K.P. (2005) Adsorption of biopolymers at hydrophilic cellulose-water interface. *Biopolymers*, **77** (5), 286–295.
121. Karlsson, M., Ekeroth, J., Elwing, H., and Carlsson, U. (2005) Reduction of irreversible protein adsorption on solid surfaces by protein engineering for increased stability. *The Journal of Biological Chemistry*, **280** (27), 25558–25564.
122. Park, J.-W., Park, K., Song, H., and Shin, H. (2002) Saccharification and adsorption characteristics of modified cellulases with hydrophilic/hydrophobic copolymers. *Journal of Biotechnology*, **93** (3), 203–208.
123. Kajiuchi, T., Park, J.W., and Moon, H.Y. (1993) Adsorption control of cellulase onto cellulose by modification with amphiphilic copolymer. *Journal of Chemical Engineering of Japan*, **26** (1), 28–33.
124. Borjesson, J., Engqvist, M., Sipos, B., and Tjerneld, F. (2007) Effect of poly(ethylene glycol) on enzymatic hydrolysis and adsorption of cellulase enzymes to pretreated lignocellulose. *Enzyme and Microbial Technology*, **41** (1–2), 186–195.
125. Palonen, H. (2004) Role of Lignin in Enzymatic Hydrolysis of Lignocellulose. Ph.D. thesis, University of Helsinki, Finland.
126. Jin, Z., Katsumata, K.S., Bach, T. *et al.* (2006) Covalent linkages between cellulose and lignin in cell walls of coniferous and nonconiferous woods. *Biopolymers*, **83** (2), 103–110.
127. Karlsson, O. and Westermark, U. (1996) Evidence for chemical bonds between lignin and cellulose in kraft pulps. *Journal of Pulp and Paper Science*, **22** (10), J397–J401.
128. Kotelnikova, N.E., Shashilov, A.A., and Yongfa, H. (1993) Effect of the presence of lignin on the structure and reactivity to hydrolysis of lignincarbohydrate complexes of poplar wood obtained by sulfate pulping. *Wood Science and Technology*, **27** (4), 263–269.
129. Garcia-Aparicio, M.P., Ballesteros, I., Gonzalez, A. *et al.* (2006) Effect of inhibitors released during steam-explosion pretreatment of barley straw on enzymatic hydrolysis. *Applied Biochemistry and Biotechnology*, **129–132**, 278–288.

130. Kaya, F., Heitmann, J.A. Jr., and Joyce, T.W. (1999) Effect of dissolved lignin and related compounds on the enzymic hydrolysis of cellulose model compound. *Cellulose Chemistry and Technology*, **33** (3–4), 203–213.

131. Paul, S.S., Kamra, D.N., Sastry, V.R.B. *et al.* (2003) Effect of phenolic monomers on biomass and hydrolytic enzyme activities of an anaerobic fungus isolated from wild nil gai (Boselaphus tragocamelus). *Letters in Applied Microbiology*, **36** (6), 377–381.

132. Weil, J.R., Dien, B., Bothast, R. *et al.* (2002) Removal of Fermentation Inhibitors Formed during Pretreatment of Biomass by Polymeric Adsorbents. *Industrial & Engineering Chemistry Research*, **41** (24), 6132–6138.

133. Ximenes, E., Kim, Y., Mosier, N. *et al.* (2011) Deactivation of cellulases by phenols. *Enzyme and Microbial Technology*, **48** (1), 54–60.

134. Kim, Y., Ximenes, E., Mosier, N.S., and Ladisch, M.R. (2011) Soluble inhibitors/deactivators of cellulase enzymes from lignocellulosic biomass. *Enzyme and Microbial Technology*, **48** (4–5), 408–415.

135. Kawamoto, H., Nakatsubo, F., and Murakami, K. (1992) Protein-adsorbing capacities of lignin samples. *Mokuzai Gakkaishi*, **38** (1), 81–84.

136. Makkar, H.P.S., Dawra, R.K., and Singh, B. (1987) Protein precipitation assay for quantitation of tannins: determination of protein in tannin-protein complex. *Analytical Biochemistry*, **166** (2), 435–439.

137. Kawamoto, H., Nakatsubo, F., and Murakami, K. (1990) Synthesis of condensed tannin derivatives and their protein-precipitating capacity. *Journal of Wood Chemistry and Technology*, **10** (1), 59–74.

138. Shallom, D. and Shoham, Y. (2003) Microbial hemicellulases. *Current Opinion in Microbiology*, **6** (3), 219–228.

139. Wyman, C.E., Decker, S.R., Himmel, M.E. *et al.* (2005) Hydrolysis of cellulose and hemicelluloses, in *Polysaccharides: Structural Diversity and Functional Versatility* (Dumitriu, S., ed.), CRC Press, Boca Raton, 995–1033.

140. Maxim, S., Rajeev, K., Haitao, Z., and Steven, H. (2011) Novelties of the cellulolytic system of a marine bacterium applicable to cellulosic sugar production. *Biofuels*, **2**, 59–70.

141. Sannigrahi, P., Ragauskas, A., and Miller, S. (2008) Effects of two-stage dilute acid pretreatment on the structure and composition of lignin and cellulose in loblolly pine. *BioEnergy Research*, **1** (3), 205–214.

142. Pingali, S.V., Urban, V.S., Heller, W.T. *et al.* (2010) Breakdown of cell wall nanostructure in dilute acid pretreated biomass. *Biomacromolecules*, **11** (9), 2329–2335.

143. Wan, J., Wang, Y., and Xiao, Q. (2010) Effects of hemicellulose removal on cellulose fiber structure and recycling characteristics of eucalyptus pulp. *Bioresource Technology*, **101** (12), 4577–4583.

144. Ding, S.-Y. and Himmel, M.E. (2006) The maize primary cell wall microfibril: a new model derived from direct visualization. *Journal of Agricultural and Food Chemistry*, **54** (3), 597–606.

145. Selig, M.J., Knoshaug, E.P., Adney, W.S. *et al.* (2008) Synergistic enhancement of cellobiohydrolase performance on pretreated corn stover by addition of xylanase and esterase activities. *Bioresource Technology*, **99** (11), 4997–5005.

146. Himmel, M.E., Ding, S.Y., Johnson, D.K. *et al.* (2007) Biomass recalcitrance: engineering plants and enzymes for biofuels production. *Science*, **315** (5813), 804–807.

147. Berlin, A., Maximenko, V., Gilkes, N., and Saddler, J. (2007) Optimization of enzyme complexes for lignocellulose hydrolysis. *Biotechnology and Bioengineering*, **97** (2), 287–296.

148. Zhu, Y., Lee, Y.Y., and Elander, R.T. (2005) Optimization of dilute-acid pretreatment of corn stover using a high-solids percolation reactor. *Applied Biochemistry and Biotechnology*, **121–124**, 1045–1054.

149. Allen, S.G., Schulman, D., Lichwa, J. *et al.* (2001) A comparison of aqueous and dilute-acid single-temperature pretreatment of yellow poplar sawdust. *Industrial & Engineering Chemistry Research*, **40** (10), 2352–2361.

150. Ishizawa, C.I., Davis, M.F., Schell, D.F., and Johnson, D.K. (2007) Porosity and its effect on the digestibility of dilute sulfuric acid pretreated corn stover. *Journal of Agricultural and Food Chemistry*, **55** (7), 2575–2581.

151. Palonen, H., Thomsen, A.B., Tenkanen, M. *et al.* (2004) Evaluation of wet oxidation pretreatment for enzymatic hydrolysis of softwood. *Applied Biochemistry and Biotechnology*, **117** (1), 1–17.

152. Kabel, M.A., Bos, G., Zeevalking, J. *et al.* (2007) Effect of pretreatment severity on xylan solubility and enzymatic breakdown of the remaining cellulose from wheat straw. *Bioresource Technology*, **98** (10), 2034–2042.

153. Grohmann, K., Torget, R., and Himmel, M. (1986) Optimization of dilute acid pretreatment of biomass. In *Proceedings of 7th Biotechnology and Bioengineering Symposium*, **15**, 59–80.

154. Um, B.-H., Karim, M.N., and Henk, L.L. (2003) Effect of sulfuric and phosphoric acid pretreatments on enzymatic hydrolysis of corn stover. *Applied Biochemistry and Biotechnology*, **105–108**, 115–125.

155. Torget, R., Walter, P., Himmel, M., and Grohmann, K. (1991) Dilute-acid pretreatment of corn residues and short-rotation woody crops. *Applied Biochemistry and Biotechnology*, **28–29** (1), 75–86.

156. Tsao, G.T., Ladisch, M., Ladisch, C. *et al.* (1978) Fermentation substrates from cellulosic materials: production of fermentable sugars from cellulosic materials. *Annual Reports on Fermentation Processes*, **2**, 1–21.

157. Millett, M.A., Baker, A.J., and Satter, L.D. (1975) Pretreatments to enhance chemical, enzymatic, and microbiological attack of cellulosic materials. *Biotechnology and Bioengineering Symposium*, 193–219.

158. Fan, L., Lee, Y.-H., and Gharpuray, M. (1982) The nature of lignocellulosics and their pretreatments for enzymatic hydrolysis. *Advances in Biochemical Engineering*, **23**, 157–187.

159. Chum, H.L., Johnson, D.K., Black, S.K. *et al.* (1988) Organosolv pretreatment for enzymatic hydrolysis of poplar. I. Enzyme hydrolysis of cellulosic residues. *Biotechnology and Bioengineering*, **31**, 643–649.

160. Iyer, P.V. and Lee, Y.Y. (1999) Product inhibition in simultaneous saccharification and fermentation of cellulose into lactic acid. *Biotechnology Letters*, **21** (5), 371–373.

161. Grethlein, H.E. (1984) Pretreatment for enhanced hydrolysis of cellulosic biomass. *Biotechnology Advances*, **2** (1), 43–62.

162. Grethlein, H.E. (1985) Effect of pore size distribution on the rate of enzymatic hydrolysis of cellulosic substrates. *Bio/Technology*, **3** (2), 155–160.

163. Maloney, M.T., Chapman, T.W., and Baker, A.J. (1985) Dilute acid hydrolysis of paper birch: Kinetics studies of xylan and acetyl-group hydrolysis. *Biotechnology and Bioengineering*, **27** (3), 355–361.

164. Liu, C. and Wyman, C.E. (2004) Effect of the flow rate of a very dilute sulfuric acid on xylan, lignin, and total mass removal from corn stover. *Industrial & Engineering Chemistry Research*, **43** (11), 2781–2788.

165. Liu, C. and Wyman, C.E. (2003) The effect of flow rate of compressed hot water on xylan, lignin, and total mass removal from corn stover. *Industrial & Engineering Chemistry Research*, **42** (21), 5409–5416.

166. Liu, C. and Wyman, C.E. (2004) Impact of fluid velocity on hot 4water only pretreatment of corn stover in a flowthrough reactor. *Applied Biochemistry and Biotechnology*, **113–116**, 977–987.

167. Liu, C. and Wyman, C.E. (2005) Partial flow of compressed-hot water through corn stover to enhance hemicellulose sugar recovery and enzymatic digestibility of cellulose. *Bioresource Technology*, **96** (18), 1978–1985.

168. Gupta, R., Kim, T., and Lee, Y. (2008) Substrate dependency and effect of xylanase supplementation on enzymatic hydrolysis of ammonia-treated biomass. *Applied Biochemistry and Biotechnology*, **148** (1), 59–70.

169. Beukes, N., Chan, H., Doi, R.H., and Pletschke, B.I. (2008) Synergistic associations between Clostridium cellulovorans enzymes XynA., ManA and EngE against sugarcane bagasse. *Enzyme and Microbial Technology*, **42** (6), 492–498.

170. García-Aparicio, M., Ballesteros, M., Manzanares, P. *et al.* (2007) Xylanase contribution to the efficiency of cellulose enzymatic hydrolysis of barley straw. *Applied Biochemistry and Biotechnology*, **137–140** (1), 353–365.

171. Murashima, K., Kosugi, A., and Doi, R.H. (2003) Synergistic effects of cellulosomal xylanase and cellulases from Clostridium cellulovorans on plant cell wall degradation. *Journal of Bacteriology*, **185** (5), 1518–1524.

172. Shi, J., Ebrik, M.A., Yang, B. *et al.* (2011) Application of cellulase and hemicellulase to pure xylan, pure cellulose, and switchgrass solids from leading pretreatments. *Bioresource Technology*, **102** (24), 11080–11088.

173. Del Rio, J.C., Marques, G., Rencoret, J. *et al.* (2007) Occurrence of naturally acetylated lignin units. *Journal of Agricultural and Food Chemistry*, **55** (14), 5461–5468.

174. del Río, J.C., Gutiérrez, A., and Martínez, Á.T. (2004) Identifying acetylated lignin units in non-wood fibers using pyrolysis-gas chromatography/mass spectrometry. *Rapid Communications in Mass Spectrometry*, **18** (11), 1181–1185.

175. Selig, M., Adney, W., Himmel, M., and Decker, S. (2009) The impact of cell wall acetylation on corn stover hydrolysis by cellulolytic and xylanolytic enzymes. *Cellulose*, **16** (4), 711–722.

176. Kong, R., Engler, C.R., and Soltes, E.J. (1992) Effects of cell-wall acetate xylan backbone and lignin on enzymatic hydrolysis of aspen wood. *Applied Biochemistry and Biotechnology*, **34**, 23–35.

177. Pan, X., Gilkes, N., and Saddler, J.N. (2006) Effect of acetyl groups on enzymatic hydrolysis of cellulosic substrates. *Holzforschung*, **60**, 398–401.

178. Dien, B.S., Ximenes, E.A., O'Bryan, P.J. *et al.* (2008) Enzyme characterization for hydrolysis of AFEX and liquid hot-water pretreated distillers grains and their conversion to ethanol. *Bioresource Technology*, **99** (12), 5216–5225.

179. Gubitz, G.M., Haltrich, D., Latal, B., and Steiner, W. (1997) Mode of depolymerisation of hemicellulose by various mannanases and xylanases in relation to their ability to bleach softwood pulp. *Applied Microbiology and Biotechnology*, **47** (6), 658–662.

180. Dien, B.S., Li, X.L., Iten, L.B. *et al.* (2006) Enzymatic saccharification of hot-water pretreated corn fiber for production of monosaccharides. *Enzyme and Microbial Technology*, **39** (5), 1137–1144.

181. Berlin, A., Gilkes, N., Kilburn, D. *et al.* (2006) Evaluation of cellulase preparations for hydrolysis of hardwood substrates. *Applied Biochemistry and Biotechnology*, **129–132**, 528–545.

182. Suh, J.-H. and Choi, Y.-J. (1996) Synergism among endo-xylanase, β-xylosidase, and acetyl xylan esterase from Bacillus stearothermophilus. *Journal of Microbiology and Biotechnology*, **6** (3), 173–178.

183. Chernoglazov, V.M., Ermolova, O.V., and Klesov, A.A. (1988) Adsorption of high-purity endo-1,4-b-glucanases from Trichoderma reesei on components of lignocellulosic materials: cellulose, lignin, and xylan. *Enzyme and Microbial Technology*, **10** (8), 503–507.

184. Tenkanen, M., Buchert, J., and Viikari, L. (1995) Binding of hemicellulases on isolated polysaccharide substrates. *Enzyme and Microbial Technology*, **17** (6), 499–505.

185. Samuel, R., Foston, M., Jaing, N. *et al.* (2011) HSQC (heteronuclear single quantum coherence) 13C-1H correlation spectra of whole biomass in perdeuterated pyridinium chloride-DMSO system: An effective tool for evaluating pretreatment. *Fuel*, **90** (9), 2836–2842.

186. Grohmann, K., Mitchell, D.J., Himmel, M.E. *et al.* (1989) The role of ester groups in resistance of plant cell wall polysaccharides to enzymic hydrolysis. *Applied Biochemistry and Biotechnology*, **20–21**, 45–61.

187. Fernandes, A.C., Fontes, C.M.G.A., Gilbert, H.J. *et al.* (1999) Homologous xylanases from Clostridium thermocellum: evidence for bi-functional activity, synergism between xylanase catalytic modules and the presence of xylan-binding domains in enzyme complexes. *Biochemical Journal*, **342** (1), 105–110.

188. Tenkanen, M., Siika-aho, M., Hausalo, T. *et al.* (1996) Synergism of xylanolytic enzymes of Trichoderma reesei in the degradation of acetyl-4-O-methylglucuronoxylan. Biotechnology in the Pulp and Paper Industry: Recent Advances in Applied and Fundamental Research. Proceedings of the International Conference on Biotechnology in the Pulp and Paper Industry, 6th, Vienna, June 11–15, 1995, pp. 503–508.

189. Kormelink, F.J.M. and Voragen, A.G.J. (1992) Combined action of xylan-degrading and accessory enzymes on different glucurono-arabino xylans. *Progress in Biotechnology*, **7**, 415–418.

190. Wood, T.M. and McCrae, S.I. (1986) The effect of acetyl groups on the hydrolysis of ryegrass cell walls by xylanase and cellulase from Trichoderma koningii. *Phytochemistry*, **25** (5), 1053–1055.

191. Mitchell, D.J., Grohmann, K., Himmel, M.E. *et al.* (1990) Effect of the degree of acetylation on the enzymic digestion of acetylated xylans. *Journal of Wood Chemistry and Technology*, **10** (1), 111–121.

192. Anand, L. and Vithayathil, P.J. (1996) Xylan-degrading enzymes from the thermophilic fungus Humicola lanuginosa (Griffon and Maublanc) Bunce: Action pattern of xylanase and [beta]-glucosidase on xylans, xylooligomers and arabinoxylooligomers. *Journal of Fermentation and Bioengineering*, **81** (6), 511–517.

193. Rivard, C.J., Adney, W.S., Himmel, M.E. *et al.* (1992) Effects of natural polymer acetylation on the anaerobic bioconversion to methane and carbon dioxide. *Applied Biochemistry and Biotechnology*, **34–35**, 725–736.

194. Grabber, J.H., Hatfield, R.D., and Ralph, J. (1998) Diferulate cross-links impede the enzymatic degradation of non-lignified maize walls. *Journal of the Science of Food and Agriculture*, **77** (2), 193–200.

195. Glasser, W.G., Ravindran, G., Jain, R.K. *et al.* (1995) Comparative enzyme biodegradability of xylan, cellulose, and starch derivatives. *Biotechnology Progress*, **11**, 552–557.

196. Yu, P., McKinnon, J.J., Maenz, D.D. *et al.* (2003) Enzymic release of reducing sugars from oat hulls by cellulase, as influenced by aspergillus ferulic acid esterase and trichoderma xylanase. *Journal of Agricultural and Food Chemistry*, **51** (1), 218–223.

197. Tabka, M.G., Herpoel-Gimbert, I., Monod, F. *et al.* (2006) Enzymatic saccharification of wheat straw for bioethanol production by a combined cellulase xylanase and feruloyl esterase treatment. *Enzyme and Microbial Technology*, **39** (4), 897–902.

198. Jarvis, M. (2003) Chemistry: cellulose stacks up. *Nature*, **426** (6967), 611–612.

199. Rollin, J.A., Zhu, Z., Sathitsuksanoh, N., and Zhang, YH.P. (2011) Increasing cellulose accessibility is more important than removing lignin: A comparison of cellulose solvent-based lignocellulose fractionation and soaking in aqueous ammonia. *Biotechnology and Bioengineering*, **108** (1), 22–30.

200. Mittal, A., Katahira, R., Himmel, M., and Johnson, D. (2011) Effects of alkaline or liquid-ammonia treatment on crystalline cellulose: Changes in crystalline structure and effects on enzymatic digestibility. *Biotechnology for Biofuels*, **4** (1), 41.

201. Hayashi, J., Sufoka, A., Ohkita, J., and Watanabe, S. (1975) The confirmation of existences of cellulose IIII, IIIII, IVI, and IVII by the X-ray method. *Journal of Polymer Science: Polymer Letters Edition*, **13** (1), 23–27.

202. Chanzy, H., Henrissat, B., Vincendon, M. *et al.* (1987) Solid-state 13C-N.M.R. and electron microscopy study on the reversible cellulose I--> cellulose IIII transformation in Valonia. *Carbohydrate Research*, **160** (0), 1–11.

203. O'Sullivan, A. (1997) Cellulose: the structure slowly unravels. *Cellulose*, **4** (3), 173–207.

204. Atalla, R.H. and VanderHart, D.L. (1984) Native cellulose: A composite of two distinct crystalline forms. *Science*, **223**, 283–285.

205. Park, S., Baker, J., Himmel, M. *et al.* (2010) Cellulose crystallinity index: measurement techniques and their impact on interpreting cellulase performance. *Biotechnology for Biofuels*, **3** (1), 10.

206. Mazeau, K. and Heux, L. (2003) Molecular dynamics simulations of bulk native crystalline and amorphous structures of cellulose. *The Journal of Physical Chemistry B*, **107** (10), 2394–2403.

207. Zhao, H., Kwak, J.H., Wang, Y. *et al.* (2006) Effects of crystallinity on dilute acid hydrolysis of cellulose by cellulose ball-milling study. *Energy Fuels*, **20** (2), 807–811.

208. Zhao, H., Kwak, J.H., Zhang, Z.C. *et al.* (2007) Studying cellulose fiber structure by SEM, XRD, NMR and acid hydrolysis. *Carbohydrate Polymers*, **68** (2), 235–241.

209. Sathitsuksanoh, N., Zhu, Z., Wi, S., and Percival Zhang, Y.H. (2011) Cellulose solvent-based biomass pretreatment breaks highly ordered hydrogen bonds in cellulose fibers of switchgrass. *Biotechnology and Bioengineering*, **108** (3), 521–529.

210. Kataoka, Y. and Kondo, T. (1996) Changing cellulose crystalline structure in forming wood cell walls. *Macromolecules*, **29** (19), 6356–6358.

211. Kataoka, Y. and Kondo, T. (1998) FT-IR microscopic analysis of changing cellulose crystalline structure during wood cell wall formation. *Macromolecules*, **31** (3), 760–764.

212. Mihranyan, A., Llagostera, A.P., Karmhag, R. *et al.* (2004) Moisture sorption by cellulose powders of varying crystallinity. *International Journal of Pharmaceutics*, **269** (2), 433–442.

213. Bertran, M.S. and Dale, B.E. (1986) Determination of cellulose accessibility by differential scanning calorimetry. *Journal of Applied Polymer Science*, **32**, 4241–4253.

214. Luo, X., Zhu, J., Gleisner, R., and Zhan, H. (2011) Effects of wet-pressing-induced fiber hornification on enzymatic saccharification of lignocelluloses. *Cellulose*, **18** (4), 1055–1062.

215. Lee, S.B., Shin, H.S., and Ryu, D.D.Y. (1982) Adsoprtion of Cellulase on cellulose: effect of physiochemical properties of cellulose on adsoprtion and rate of hydrolysis. *Biotechnology and Bioengineering*, **XXIV**, 2137–2153.

216. Hoshino, E., Kanda, T., Sasaki, Y., and Nisizawa, K. (1992) Adsorption mode of exo- and endo-cellulases from irpex lacteus (Polyporus tulipiferae) on cellulose with different crystallinities. *Journal of Biochemistry (Tokyo)*, **111** (5), 600–605.

217. Hoshino, E. and Kanda, T. (1997) Scope and mechanism of cellulase action on different cellulosic substrates. *Oyo Toshitsu Kagaku*, **44** (1), 87–104.

218. Pinto, R., Carvalho, J., Mota, M., and Gama, M. (2006) Large-scale production of cellulose-binding domains. Adsorption studies using CBD-FITC conjugates. *Cellulose*, **13** (5), 557–569.

219. Hall, M., Bansal, P., Lee, J.H. *et al.* (2010) Cellulose crystallinity – a key predictor of the enzymatic hydrolysis rate. *FEBS Journal*, **277** (6), 1571–1582.

220. Goel, S.C. and Ramachandran, K.B. (1983) Studies on the adsorption of cellulase on lignocellulosics. *Journal of Fermentation Technology*, **61** (3), 281–286.

221. Banka, R.R. and Mishra, S. (2002) Adsorption properties of the fibril forming protein from Trichoderma reesei. *Enzyme and Microbial Technology*, **31** (6), 784–793.

222. Lynd, L.R. (1996) Overview and evaluation of fuel ethanol from cellulosic biomass:Technology, economics, the environment, and policy. *Annual Review of Energy and the Environment*, **21**, 403–465.

223. Zhang, Y.H. and Lynd, L.R. (2005) Determination of the number-average degree of polymerization of cellodextrins and cellulose with application to enzymatic hydrolysis. *Biomacromolecules*, **6** (3), 1510–1515.

224. Ryu, D.D.Y. and Lee, S.B. (1986) Enzymic hydrolysis of cellulose: determination of kinetic parameters. *Chemical Engineering Communications*, **45** (1–6), 119–134.

225. Ooshima, H., Sakata, M., and Harano, Y. (1983) Adsorption of cellulase from Trichoderma viride on cellulose. *Biotechnology and Bioengineering*, **25** (12), 3103–3114.

226. Sinitsyn, A., Gusakov, A., and Vlasenko, E. (1991) Effect of structural and physico-chemical features of cellulosic substrates on the efficiency of enzymatic hydrolysis. *Applied Biochemistry and Biotechnology*, **30** (1), 43–59.

227. Meunier-Goddik, L. and Penner, M.H. (1999) Enzyme-catalyzed saccharification of model celluloses in the presence of lignacious residues. *Journal of Agricultural and Food Chemistry*, **47** (1), 346–351.

228. Caulfield, D.F. and Moore, W.E. (1974) Effect of varying crystallinity of cellulose on enzymic hydrolysis. *Wood Science*, **6** (4), 375–379.

229. Christakopoulos, P., Koullas, D.P., Kekos, D. *et al.* (1991) Direct conversion of straw to ethanol by Fusarium oxysporum: Effect of cellulose crystallinity. *Enzyme and Microbial Technology*, **13** (3), 272–274.

230. Fan, L.T., Lee, Y.-H., and Beardmore, D.H. (1980) Major chemical and physical features of cellulosic materials as substrates for enzymic hydrolysis. *Advances in Biochemical Engineering*, **14**, 101–117.

231. Lee, Y.-H. and Fan, L.T. (1983) Kinetic studies of enzymatic hydrolysis of insoluble cellulose: (II). Analysis of extended hydrolysis times. *Biotechnology and Bioengineering*, **25**, 939–966.

232. Väljamäe, P., Sild, V., Nutt, A. *et al.* (1999) Acid hydrolysis of bacterial cellulose reveals different modes of synergistic action between cellobiohydrolase I and endoglucanase I. *European Journal of Biochemistry*, **266** (2), 327–334.

233. Teeri, T.T. (1997) Crystalline cellulose degradation: new insight into the function of cellobiohydrolases. *Trends in Biotechnology*, **15** (5), 160–167.

234. Kongruang, S. and Penner, M.H. (2004) Borohydride reactivity of cellulose reducing ends. *Carbohydrate Polymers*, **58** (2), 131–138.

235. Kongruang, S., Han, M., Breton, C., and Penner, M. (2004) Quantitative analysis of cellulose-reducing ends. *Applied Biochemistry and Biotechnology*, **113** (1), 213–231.

236. Bansal, P., Hall, M., Realff, M.J. *et al.* (2010) Multivariate statistical analysis of X-ray data from cellulose: A new method to determine degree of crystallinity and predict hydrolysis rates. *Bioresource Technology*, **101** (12), 4461–4471.

237. Fan, L.T., Gharpuray, M.M., and Lee, Y. (1981) Evaluation of pretreatments for enzymatic conversion of agricultural residues. *Biotechnology and Bioengineering Symposium*, **11**, 29–45.

238. Sasaki, T., Tanaka, T., Nanbu, N. *et al.* (1979) Correlation between x-ray diffraction measurements of cellulose crystalline structure and the susceptibility to microbial cellulase. *Biotechnology and Bioengineering*, **21** (6), 1031–1042.

239. Puri, V.P. (1984) Effect of crystallinity and degree of 7polymerization of cellulose on enzymic saccharification. *Biotechnology and Bioengineering*, **26** (10), 1219–1222.

240. Maeda, R.N., Serpa, V.I., Rocha, V.A.L. *et al.* (2011) Enzymatic hydrolysis of pretreated sugar cane bagasse using Penicillium funiculosum and Trichoderma harzianum cellulases. *Process Biochemistry*, **46** (5), 1196–1201.

241. Converse, A.O. (1993) Substrate factors limiting enzymatic hydrolysis, in *Bioconversion of Forest and Agricultural Plant Residue* (ed. J.N. Saddler), CAB International, Wallingford, Oxon, UK, p. 93–106.

242. Kim, S. and Holtzapple, M.T. (2006) Effect of structural features on enzyme digestibility of corn stover. *Bioresource Technology*, **97** (4), 583–591.

243. Rivers, D.B. and Emert, G.H. (1988) Factors affecting the enzymatic hydrolysis of municipal-solid-waste components. *Biotechnology and Bioengineering*, **31** (3), 278–281.

244. Rivers, D.B. and Emert, G.H. (1988) Factors affecting the enzymatic hydrolysis of bagasse and rice straw. *Biological Wastes*, **26**, 85–95.

245. Puri, V.P. and Pearce, G.R. (1986) Alkali-explosion pretreatment of straw and bagasse for enzymic hydrolysis. *Biotechnology and Bioengineering*, **28** (4), 480–485.

246. Gharpuray, M.M., Lee, Y.H., and Fan, L.T. (1981) Pretreatment of wheat straw for cellulose hydrolysis. Proceedings of the Annual Biochemical Engineering Symposium 11th, pp. 1–10.

247. Paralikar, K.M. and Betrabet, S.M. (1977) Electron-diffraction technique for determination of cellulose crystallinity. *Journal of Applied Polymer Science*, **21** (4), 899–903.

248. Puls, J. and Wood, T.M. (1991) The degradation pattern of cellulose by extracellular cellulases of aerobic and anaerobic microorganisms. *Bioresource Technology*, **36** (1), 15–19.

249. Lenz, J., Esterbauer, H., Sattler, W. *et al.* (1990) Changes of structure and morphology of regenerated cellulose caused by acid and enzymatic hydrolysis. *Journal of Applied Polymer Science*, **41** (5–6), 1315–1326.

250. Boisset, C., Chanzy, H., Henrissat, B. *et al.* (1999) Digestion of crystalline cellulose substrates by the clostridium thermocellum cellulosome: structural and morphological aspects. *The Biochemical Journal*, **340** (Pt 3), 829–835.

251. Chen, Y., Stipanovic, A., Winter, W. *et al.* (2007) Effect of digestion by pure cellulases on crystallinity and average chain length for bacterial and microcrystalline celluloses. *Cellulose*, **14** (4), 283–293.

252. Cateto, C., Hu, G., and Ragauskas, A. (2011) Enzymatic hydrolysis of organosolv Kanlow switchgrass and its impact on cellulose crystallinity and degree of polymerization. *Energy & Environmental Science*, **4** (4), 1516–1521.

253. Pu, Y., Ziemer, C., and Ragauskas, A.J. (2006) CP/MAS 13C NMR analysis of cellulase treated bleached softwood kraft pulp. *Carbohydrate Research*, **341** (5), 591–597.

254. Penttila, P.A., Valrnai, A., Leppanen, K. *et al.* (2010) Changes in submicrometer structure of enzymatically hydrolyzed microcrystalline cellulose. *Biomacromolecules*, **11** (4), 1111–1117.

255. Mansfield, S.D. and Meder, R. (2003) Cellulose hydrolysis – the role of monocomponent cellulases in crystalline cellulose degradation. *Cellulose*, **10** (2), 159–169.

256. Xiao, Z., Gao, P., Qu, Y., and Wang, T. (2001) Cellulose-binding domain of endoglucanase III from Trichoderma reesei disrupting the structure of cellulose. *Biotechnology Letters*, **23** (9), 711–715.

257. Wang, L., Zhang, Y., and Gao, P. (2008) A novel function for the cellulose binding module of cellobiohydrolase I. *Science in China Series C: Life Sciences*, **51** (7), 620–629.

258. Sinnott, M.L. (1998) The cellobiohydrolases of Trichoderma reesei: a review of indirect and direct evidence that their function is not just glycosidic bond hydrolysis. *Biochemical Society Transactions*, **26** (2), 160–164.

259. Himmel, M.E., Ruth, M.F., and Wyman, C.E. (1999) Cellulase for commodity products from cellulosic biomass. *Current Opinion in Biotechnology*, **10** (4), 358–364.

260. Hall, M., Bansal, P., Lee, J.H. *et al.* (2011) Biological pretreatment of cellulose: Enhancing enzymatic hydrolysis rate using cellulose-binding domains from cellulases. *Bioresource Technology*, **102** (3), 2910–2915.

261. Arantes, V. and Saddler, J. (2010) Access to cellulose limits the efficiency of enzymatic hydrolysis: the role of amorphogenesis. *Biotechnology for Biofuels*, **3** (1), 4.

262. Park, S., Johnson, D., Ishizawa, C. *et al.* (2009) Measuring the crystallinity index of cellulose by solid state 13C nuclear magnetic resonance. *Cellulose*, **16** (4), 641–647.

263. Kumar, R. (2008) Enzymatic Hydrolysis of Cellulosic Biomass Solids Prepared by Leading Pretreatments and Identification of Key Features Governing Performance. Ph.D. thesis, Thayer School of Engineering, Dartmouth College, Hanover, NH, USA.

264. Nidetzky, B., Hayn, M., Macarron, R., and Steiner, W. (1993) Synergism of Trichoderma reesei cellulases while degrading different celluloses. *Biotechnology Letters*, **15** (1), 71–76.

265. Hoshino, E., Shiroishi, M., Amano, Y. *et al.* (1997) Synergistic actions of exo-type cellulases in the hydrolysis of cellulose with different crystallinities. *Journal of Fermentation and Bioengineering*, **84** (4), 300–306.

266. Valjamae, P. (2002) The Kinetics of Cellulose Enzymatic Hydrolysis: Implications of the Synergism Between Enzymes. Ph.D. thesis, Uppsala University, Sweden.

267. Tarantili, P.A., Koullas, D.P., Christakopoulos, P. *et al.* (1996) Cross-synergism in enzymic hydrolysis of lignocellulosics: Mathematical correlations according to a hyperbolic model. *Biomass and Bioenergy*, **10** (4), 213–219.

268. Henrissat, B. (1994) Cellulases and their interaction with cellulose. *Cellulose*, **1**, 169–196.

269. Kanda, T., Wakabayashi, K., and Nisizawa, K. (1980) Modes of action of exo- and endo-cellulases in the degradation of celluloses I and II. *Journal of Biochemistry (Tokyo)*, **87** (6), 1635–1639.

270. Murashima, K., Kosugi, A., and Doi, R.H. (2002) Synergistic effects on crystalline cellulose degradation between cellulosomal cellulases from clostridium cellulovorans. *Journal of Bacteriology*, **184** (18), 5088–5095.

271. Henrissat, B., Driguez, H., Viet, C., and Schulein, M. (1985) Synergism of cellulases from trichoderma reesei in the degradation of cellulose. *Nature Biotechnology*, **3** (8), 722–726.

272. Ma, A., Hu, Q., Qu, Y. *et al.* (2008) The enzymatic hydrolysis rate of cellulose decreases with irreversible adsorption of cellobiohydrolase I. *Enzyme and Microbial Technology*, **42** (7), 543–547.

273. Igarashi, K., Wada, M., Hori, R., and Samejima, M. (2006) Surface density of cellobiohydrolase on crystalline celluloses. A critical parameter to evaluate enzymatic kinetics at a solid-liquid interface. *FEBS Journal*, **273** (13), 2869–2878.

274. Igarashi, K., Wada, M., and Samejima, M. (2006) Enzymatic kinetics at a solid-liquid interface: hydrolysis of crystalline celluloses by cellobiohydrolase. *Cellulose Communications.*, **13** (4), 173–177.

275. Igarashi, K., Wada, M., and Samejima, M. (2007) Activation of crystalline cellulose to cellulose III(I) results in efficient hydrolysis by cellobiohydrolase. *FEBS Journal*, **274** (7), 1785–1792.

276. Chundawat, S.P.S., Bellesia, G., Uppugundla, N. *et al.* (2011) Restructuring the crystalline cellulose hydrogen bond network enhances its depolymerization rate. *Journal of the American Chemical Society*, **133** (29), 11163–11174.

277. Teeri, T.T., Koivula, A., Linder, M. *et al.* (1995) Modes of action of two Trichoderma reesei cellobiohydrolases. *Progress in Biotechnology*, **10**, 211–224.

278. Christina Divne, J.S.h., and Teeri, T.T., and Jones, T. Alwyn (1998) High-resolution crystal structures reveal how a cellulose chain is bound in the 50 a long tunnel of cellobiohydrolase i from Trichoderma reesei. *Journal of Molecular Biology*, **275**, 309–325.

279. Teeri, T.T. (1997) Crystalline cellulose degradation:new insight into the function of cellobiohydrolases. *Tibtech*, **15**, 160–167.

280. Beldman, G., Searle-Van Leeuwen, M.F., Rombouts, F.M., and Voragen, F.G. (1985) The cellulase of Trichoderma viride. Purification, characterization and comparison of all detectable endoglucanases, exoglucanases and beta-glucosidases. *European Journal of Biochemistry*, **146** (2), 301–308.

281. Nidetzky, B., Zachariae, W., Gercken, G. *et al.* (1994) Hydrolysis of cellooligosaccharides by Trichoderma reesei cellobiohydrolases: Experimental data and kinetic modeling. *Enzyme and Microbial Technology*, **16** (1), 43–52.

282. Schwanninger, M., Rodrigues, J.C., Pereira, H., and Hinterstoisser, B. (2004) Effects of short-time vibratory ball milling on the shape of FT-IR spectra of wood and cellulose. *Vibrational Spectroscopy*, **36** (1), 23–40.

283. Oh, K.D. and Kim, C. (1987) A study on enzymic hydrolysis of cellulose in an attrition bioreactor. *Korean Journal of Chemical Engineering*, **4** (2), 105–112.

284. Martinez, J.M., Reguant, J., Montero, M.A. *et al.* (1997) Hydrolytic pretreatment of softwood and almond shells. Degree of polymerization and enzymatic digestibility of the cellulose fraction. *Industrial & Engineering Chemistry Research*, **36** (3), 688–696.

285. Hu, Z., Foston, M., and Ragauskas, A.J. (2011) Comparative studies on hydrothermal pretreatment and enzymatic saccharification of leaves and internodes of alamo switchgrass. *Bioresource Technology*, **102** (14), 7224–7228.

286. Hallac, B.B., Sannigrahi, P., Pu, Y. *et al.* (2010) Effect of ethanol organosolv pretreatment on enzymatic hydrolysis of buddleja davidii stem biomass. *Industrial & Engineering Chemistry Research*, **49** (4), 1467–1472.

287. Kaplan, A.M., Mandels, M., Pillion, E., and Greenberger, M. (1970) Resistance of weathered cotton cellulose to cellulase action. *Applied Microbiology*, **20** (1), 85–93.

288. Kumar, S., Gupta, R., Lee, Y.Y., and Gupta, R.B. (2010) Cellulose pretreatment in subcritical water: Effect of temperature on molecular structure and enzymatic reactivity. *Bioresource Technology*, **101** (4), 1337–1347.

289. Hu, Z. and Ragauskas, A.J. (2011) Hydrothermal pretreatment of switchgrass. *Industrial & Engineering Chemistry Research*, **50** (8), 4225–4230.

290. Kasprzyk, H., Wichlacz, K., and Borysiak, S. (2004) The effect of gamma radiation on the supramolecular structure of pine wood cellulose in situ revealed by X-ray diffraction. *Electronic Journal of Polish Agricultural Universities*, **7** (1).

291. Katsumata, N., Yoshimura, T., Tsunoda, K., and Imamura, Y. (2007) Resistance of gamma-irradiated sapwood of *Cryptomeria japonica* to biological attacks. *Journal of Wood Science*, **53** (4), 320–323.

292. Knappert, D., Grethlein, H., and Converse, A. (1980) Partial acid hydrolysis of cellulosic materials as a pretreatment for enzymatic hydrolysis. *Biotechnology and Bioengineering*, **XXI**, 1449–1463.

293. Zhang, Y.H.P. and Lynd, L.R. (2006) A functionally based model for hydrolysis of cellulose by fungal cellulase. *Biotechnology and Bioengineering*, **94** (5), 888–898.

294. Xu, Y., Ding, H.-g., and Shen, Q. (2007) Influence of the degree of polymerization on the surface properties of cellulose. *Xianweisu Kexue Yu Jishu*, **15** (2), 53–56.

295. Ramos, L.P., Nazhad, M.M., and Saddler, J.N. (1993) Effect of enzymatic-hydrolysis on the morphology and fine-structure of pretreated cellulosic residues. *Enzyme and Microbial Technology*, **15** (10), 821–831.

296. Lee, Y.H. and Fan, L.T. (1980) Properties and mode of action of cellulase. *Advances in Biochemical Engineering*, **17**, 101–129.

297. Wilson, C.A., McCrae, S.I., and Wood, T.M. (1994) Characterisation of a β-glucosidase from the anaerobic rumen fungus Neocallimastix frontalis with particular reference to attack on cello-oligosaccharides. *Journal of Biotechnology*, **37** (3), 217–227.

298. Okazaki, M. and Moo-Young, M. (1978) Kinetics of enzymic hydrolysis of cellulose: analytical description of a mechanistic model. *Biotechnology and Bioengineering*, **20** (5), 637–663.

299. Okazaki, M., Miura, Y., and Moo-Young, M. (1981) Synergistic effect of enzymic hydrolysis of cellulose. *Biotechnology Advances*, **2**, 3–8.

300. Ramos, L.P., Breuil, C., and Saddler, J.N. (1993) The use of enzyme recycling and the influence of sugar accumulation on cellulose hydrolysis by Trichoderma cellulases. *Enzyme and Microbial Technology*, **15** (1), 19–25.

301. Lee, Y.H. and Fan, L.T. (1981) Kinetics of enzymic hydrolysis of insoluble cellulose: experimental observation on initial rates. *Biotechnology Advances*, **2**, 9–14.

302. Ryu, D.D.Y. and Lee, S.B. (1982) Enzymic hydrolysis of cellulose: effects of structural properties of cellulose on hydrolysis kinetics. *Enzyme Engineering*, **6**, 325–333.

303. Chandra, R.P., Esteghlalian, A.R., and Saddler, J.N. (2008) Assessing substrate accessibility to enzymatic hydrolysis by cellulases, in *Characterization of Lignocellulosic Materials* (ed. T.Q. Hu), Blackwell Publishing Ltd., Oxford, UK, 60–80.

304. Gama, F.M., Teixeira, J.A., and Mota, M. (1994) Cellulose morphology and enzymic reactivity: a modified solute exclusion technique. *Biotechnology and Bioengineering*, **43** (5), 381–387.

305. Tanaka, M., Ikesaka, M., Matsuno, R., and Converse, A.O. (1988) Effect of pore size in substrate and diffusion of enzyme on hydrolysis of cellulosic materials with cellulases. *Biotechnology and Bioengineering*, **32** (5), 698–706.

306. Kim, T.H., Kim, J.S., Sunwoo, C., and Lee, Y.Y. (2003) Pretreatment of corn stover by aqueous ammonia. *Bioresource Technology*, **90** (1), 39–47.

307. Lin, K.W., Ladisch, M.R., Voloch, M. *et al.* (1985) Effect of pretreatments and fermentation on pore size in cellulosic materials. *Biotechnology and Bioengineering*, **27** (10), 1427–1433.

308. Dasari, R. and Eric Berson, R. (2007) The effect of particle size on hydrolysis reaction rates and rheological properties in cellulosic slurries. *Applied Biochemistry and Biotechnology*, **137–140** (1), 289–299.

309. Kumakura, M. (1986) Adsorption of cellulase by various substances. *Journal of Materials Science Letters*, **5** (1), 78–80.

310. Mooney, C.A., Mansfield, S.D., Tuohy, M.G., and Saddler, J.N. (1997) The effect of lignin content on cellulose accessibility and enzymic hydrolysis of softwood pulps. Biological Sciences Symposium; 1997 Oct. 19–23; San Francisco, pp. 259–265.

311. Goel, S.C. and Ramachandran, K.B. (1983) Comparison of the rates of enzymatic hydrolysis of pretreated rice straw and bagasse with celluloses. *Enzyme and Microbial Technology*, **5** (4), 281–284.

312. Zhu, Z., Sathitsuksanoh, N., Vinzant, T. *et al.* (2009) Comparative study of corn stover pretreated by dilute acid and cellulose solvent-based lignocellulose fractionation: Enzymatic hydrolysis, supramolecular structure, and substrate accessibility. *Biotechnology and Bioengineering*, **103** (4), 715–724.

313. Zhang, Y.-H.P., Ding, S.-Y., Mielenz, J.R. *et al.* (2007) Fractionating recalcitrant lignocellulose at modest reaction conditions. *Biotechnology and Bioengineering*, **97** (2), 214–223.

314. Hua, Z., Gary, A.B., and Janet, V.C. (2010) Fast enzymatic saccharification of switchgrass after pretreatment with ionic liquids. *Biotechnology Progress*, **26** (1), 127–133.

315. Zhao, H., Jones, C.L., Baker, G.A. *et al.* (2009) Regenerating cellulose from ionic liquids for an accelerated enzymatic hydrolysis. *Journal of Biotechnology*, **139** (1), 47–54.

316. Dadi, A.P., Varanasi, S., and Schall, C.A. (2006) Enhancement of cellulose saccharification kinetics using an ionic liquid pretreatment step. *Biotechnology and Bioengineering*, **95** (5), 904–910.
317. Seema, S., Blake, A.S., and Kenneth, P.V. (2009) Visualization of biomass solubilization and cellulose regeneration during ionic liquid pretreatment of switchgrass. *Biotechnology and Bioengineering*, **104** (1), 68–75.
318. Dadi, A., Schall, C., and Varanasi, S. (2007) Mitigation of cellulose recalcitrance to enzymatic hydrolysis by ionic liquid pretreatment. *Applied Biochemistry and Biotechnology*, **137–140** (1), 407–421.

15

Economics of Pretreatment for Biological Processing

Ling Tao, Andy Aden* and Richard T. Elander

National Renewable Energy Laboratory, Golden, USA

15.1 Introduction

While many of the preceding chapters focused on pretreatment fundamentals, this chapter will put pretreatment technologies into a more applied setting. Specifically, the importance of pretreatment of lignocellulosic feedstocks in the overall economics of the biorefinery will be discussed along with key parameters that influence pretreatment economics. Understanding the economics of pretreatment for biological conversion processes is crucially important for several reasons; it often represents a significant fraction of the overall cost of conversion and pretreatment impacts many of the downstream unit operations within the process, thus influencing their performance and economics as well.

This chapter will demonstrate several process and economic aspects of pretreatment:

- the fraction of overall processing costs represented by pretreatment;
- the impact of pretreatment on downstream unit operations and their respective economics;
- the role key process parameters play on pretreatment economics and economic tradeoffs within the biorefinery;
- comparative analyses between multiple chemical pretreatments; and
- the need for assessing future economic biomass pretreatment technologies.

15.2 Importance of Pretreatment

While specific technology combinations are plentiful, most biological conversion processes of lignocellulosic feedstocks include some variation of the steps: feedstock handling, pretreatment, enzymatic

*Present address: URS Corporation, Denver, USA

Aqueous Pretreatment of Plant Biomass for Biological and Chemical Conversion to Fuels and Chemicals, First Edition.
Edited by Charles E. Wyman.
© 2013 John Wiley & Sons, Ltd. Published 2013 by John Wiley & Sons, Ltd.

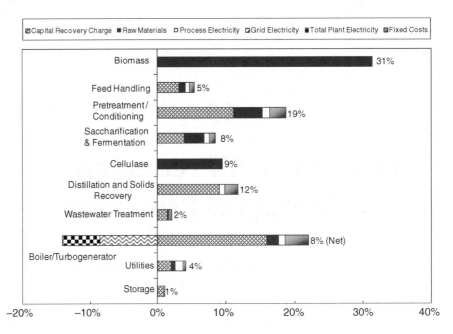

Figure 15.1   *Cost breakdown by process area of cellulosic ethanol using dilute acid pretreatment (See figure in color plate section).*

hydrolysis (or saccharification), fermentation, product recovery and purification, by-product utilization, and waste treatment. In addition, these processes are often designed with significant recycle of water and other process streams within the biorefinery.

Pretreatment of biomass for biofuel production is a crucial step. Its primary role is to disrupt the matrix of polymeric compounds that are physically and chemically bonded within lignocellulosic biomass cell-wall structures. These compounds include cellulose microfibrils, lignin, and hemicellulose. Pretreatment has a significant impact not only on enzymatic hydrolysis, fermentation, and downstream processing [1], but also on overall process economics and sustainability. Typically, hydrolysis yields in the absence of pretreatment are <20% of theoretical yields, whereas sugar yields after pretreatment can exceed 90% of theoretical [2]. It is noted that hydrolysis yields are sugar yields, while the fermentation yields are mainly focused on ethanol yields from fermentable sugars. Figure 15.1 is a chart taken from a conceptual design for a cellulosic corn stover ethanol process utilizing dilute acid pretreatment in this study, similarly shown in NREL design reports [3]. It depicts the overall processing costs by individual section and clearly shows that pretreatment can often represent a significant fraction of the overall processing cost.

Many pretreatment technologies also provide secondary functionalities that aid in the overall conversion process. For example, acid-based pretreatments also serve to hydrolyze and depolymerize hemicellulosic polymers (xylan, mannan, arabinan, and galactan) as well as some of the cellulose. Some of the lignin is also solubilized by acid treatment. Ammonia-based pretreatments, on the other hand, can decrease the crystallinity of structures, swell fibers, and begin to delignify biomass; they do not hydrolyze components, however. These additional functionalities of pretreatment are important factors within the overall process economics context.

In many of these biological processes, cellulase enzymes selectively convert cellulose polymers to shorter chain monomers (glucose) and oligomers (soluble short-length sugar polymers). These enzyme cocktails also have multiple functionalities, and often possess activity for hydrolyzing xylan and debranching hemicellulose linkages. While the primary role of pretreatment is disruption, deconstruction therefore happens through a combination of size reduction, pretreatment, and enzymatic hydrolysis.

15.3 History of Pretreatment Economic Analysis

The economics of pretreatment for biological processes has been studied for at least 25 years. For example, analysis at the Solar Energy Research Institute (SERI, now NREL) investigated economics for organosolv, wet oxidation, and steam explosion pretreatments for woody feedstocks [4]. During this time, fundamental experiments for cellulase enzyme systems were ongoing and little to no pilot-scale data were available. As time progressed, researchers explored a number of pretreatment paradigms that have helped to shape the technologies seen today.

In the most general sense, pretreatment has been classified into at least three broad categories: biological, physical (or mechanical), and chemical (along with combinations such as physiochemical). Biological pretreatment typically refers to the use of fungi for biological degradation of biomass, particularly wood. However, this approach has been shown to have limited industrial applicability due to relatively slow processing rates with longer pretreatment time and yield losses to by-products [5]. Physical approaches typically refer to milling of biomass down to very small particle size in order to facilitate enzymatic hydrolysis by increasing the surface area of cellulose. The most common milling technologies include ball mills, knife mills, and hammer mills, the choice of which depends on the specific type of biomass being processed.

Several studies [6–9] have shown that significant amounts of energy are required in these mills to reduce the particle size enough to significantly impact enzymatic hydrolysis, perhaps even greater than the theoretical energy content of the biomass. In one study, it was found that biomass size reduction below 1 mm did not improve the cellulase reaction rate [10]. Recent studies have however shown that, if the particle size was further reduced, the enzymatic hydrolysis digestibility greatly improved if the particle size ranged from 10 nm to 1 μm [8]. Significant energy consumption was a financial concern for mechanical pretreatment for commercial-scale development if applied alone. However, less intense mechanical pretreatment is applicable prior to chemical pretreatment and can also be used after it. Recent research has revealed improved yields at very-low-severity dilute-acid pretreatment conditions [6,7,9].

Radiation is another type of pretreatment that can be classified as physical in nature and there have been a number of studies [11–13] on this particular subject. In general, moderate to significant breakdown of biomass has been observed with these technologies, depending upon type of radiation and specific dosage. Concerns with energy costs and safety risks of these technologies have however prevented them from advancing to larger scale.

Due to limitations from biological and physical pretreatments, chemical and physiochemical pretreatments became the most commonly studied pretreatment technologies. During much of the 1990s, two pretreatment paradigms emerged. The first was total chemical pretreatment and hydrolysis without use of enzymes. The second was chemical pretreatment coupled with enzymatic hydrolysis. The very high cost of enzymes was the primary driver for investigating total chemical pretreatment. For example, both concentrated acid and two-stage dilute-acid approaches were developed. The two-stage process attempted to maximize hydrolysis yields of hemicellulose and cellulose separately. However, these processes suffered from lower yields compared to enzymatic approaches, never achieving much greater yields than 60% of theoretical [14].

Total chemical hydrolysis (no enzymes) as well as chemical pretreatment coupled with enzymatic hydrolysis are both being pursued commercially. Blue Fire Renewables, for example, is a company that uses concentrated sulfuric acid hydrolysis. Virdia (formerly HCl Cleantech) is another such company, who instead uses hydrochloric acid. Companies pursuing chemical pretreatment with enzymatic hydrolysis include Abengoa Bioenergy, POET, etc.

15.4 Methodologies for Economic Assessment

A number of methodologies have been used to estimate pretreatment and overall process economics for biomass conversion to liquid fuels, often mirroring established engineering economic approaches used in the chemicals and petroleum refining industries. It is important that both capital costs and operating costs are appropriately represented in these analyses. In order to ascertain the initial feasibility of pretreatment economics, a conceptual level of cost modeling has been used. This allows the overall process to be modeled without the use of site-specific data but with larger ranges of uncertainty. These estimates are best used for relative comparisons against technological variations or process improvements. Use of absolute values without a detailed understanding of the basis behind them can be misleading.

For the case of an existing industrial process that is adding or modifying technology, a "breakeven" analysis is typical. This methodology is used to calculate how long it will take for additional revenues to equal the added cost of the new technology. For example, if someone is interested in a technology that will increase the yields (and therefore revenue) of their primary product by 25% and the installed capital cost of this technology is known, then the time to pay back or "break even" on this investment can be calculated. While this methodology is sufficient for the paradigm of existing industrial processes, it is insufficient for new biorefineries.

Another common approach for existing processes is an annualized cost of production. For processes where much of the capital in the refinery has already been depreciated, the annual cost of production only needs to consider operating costs over a single year. This approach is typically used to represent what it costs a company to produce a product and often does not factor in either capital cost or any revenue or profit.

More common for so-called "greenfield" projects are cash-flow economic analyses. Cash-flow economic analyses are useful because they capture both capital and operating costs over a specific time period from initial ground-breaking and construction, through plant start-up, and through continuous operation. In this fashion, the overall profitability of a project can be quantified as a net present value (NPV), a return on investment (ROI), or an internal rate of return (IRR).

The economic analyses presented in this chapter are built upon a modified discounted cash-flow rate of return (DCFROR) methodology [3,15]. Using this methodology, a minimum ethanol selling price (MESP) is calculated as the minimum cost the fuel must be sold for in order for the NPV of the project to be zero or greater at a certain discount rate. For these analyses, the discount rate is roughly equivalent to the internal rate of return of the project. For conceptual analyses of this type, factored estimates are typically used to project the total project investment (TPI). This is further based upon the calculation of total capital investment. This is appropriate for the conceptual level of design where process flow diagrams (PFDs) are generated. Piping and instrument diagrams (P&ID) are typically only required when additional detail and accuracy become of value, such as for a site-specific project that is being designed for eventual construction.

An added level of engineering rigor was brought to pretreatment economic analysis when detailed process simulation was coupled and integrated with the economic analysis. NREL began using this approach to guide research and development (R&D) in the mid–late 1990s. Advanced process simulation tools and software (e.g., Aspen Plus [16], Pro Sim, ChemCAD) had been developed to simulate steady-state chemical production and refining processes and were soon leveraged to model conceptual biorefineries. Using detailed thermodynamic physical properties and equations, material and energy balance data for biorefining processes could be generated relatively easily. These data were then used to size capital equipment and estimate material flow rates for cost estimation. Key to these tools was the development of customized physical property data (e.g., heat of combustion and heat capacity) for non-conventional biomass compounds such as lignin, cellulose, xylan, and protein [17].

15.5   Overview of Pretreatment Technologies

As described previously, mechanical and biological pretreatment technologies have not emerged as leading industrial candidates for various reasons. Chemical pretreatment technologies, however, have received considerable R&D attention. These can be further classified into the following categories: acidic (both strong and weak acid), alkaline, and solvent-based.

15.5.1   Acidic Pretreatments

Acid-based pretreatments target hydrolysis of significant fractions of hemicellulose and small fractions of cellulose. Depending upon the reaction conditions, varying amounts of oligomeric (soluble) and monomeric sugars are produced, as are varying amounts of by-products such as aldehydes. Organic acid (acetic) is liberated from hemicellulose during pretreatment. This acid and the aldehydes are known fermentation inhibitors and can impact downstream conversion even in relatively dilute concentrations. Because of the relatively high temperatures used, lignin will undergo a phase transition. Strong acids have been preferred over weak acids, and concentrated acid and dilute acid pretreatment technologies have been developed. Sulfuric acid has been a common choice simply because of its lower cost compared to other strong acids, such as hydrochloric or nitric. Liquid hot water (LHW) pretreatment could be regarded as a weak acid technology in the context of this discussion of acid pretreatment.

Resistant metallurgy is needed for wetted parts of strong acid reactor systems to provide corrosion protection. In fact, corrosion studies [3] have shown dilute acid to be more corrosive than concentrated acid. Metallurgical considerations can lead to high costs for these systems, and nickel-based alloys (e.g., Incoloy, Hastelloy) provide a good balance between cost and functionality. Concentrated acid systems typically necessitate acid recovery, using evaporation and/or chromatographic separations. If dilute acid concentrations are used, it may be more economical to neutralize the acid than to attempt recovery. However, this can impact many of the downstream operations, including wastewater treatment and lignin utilization sections. R&D is now focusing on using milder conditions in the reactor, with lower acid concentrations and temperatures, but longer residence time.

Neutral-pH pretreatments such as LHW can control pH in order to minimize hydrolysis. While this approach shifts hydrolysis burden onto enzymes, it also minimizes side reactions to known inhibition products.

15.5.2   Alkaline Pretreatments

Alkaline pretreatments break bonds between lignin and carbohydrates and disrupt lignin structure, making carbohydrates more accessible to enzymatic hydrolysis [1]. Alkaline pretreatment normally results in better delignification than acid pretreatment by cleaving lignin-hemicellulose linkages and swelling cellulose [10], with virtually no hydrolysis of the cellulose or hemicellulose. Ammonia and lime are the most common alkaline pretreatments and are discussed here. Sodium hydroxide (caustic) is another important pretreatment option but is not discussed in detail. In lime pretreatment, the presence of an oxidative agent enhances delignification significantly.

Solvent recovery is often important for these systems, particularly if chemicals are used in large amounts or high concentrations. While technologies for solvent recovery are not typically complex for these systems, they do add cost to the overall pretreatment system. However, reactor metallurgy requirements are often less costly than acid-based pretreatments and downstream processes (including water recycling) become less expensive due to lesser salt formation since most of the catalysts are removed upstream.

15.5.3   Solvent-based Pretreatments

A wide range of solvent-based pretreatments have been researched that can be further classified into two subcategories: ionic liquids and organic solvent extraction (organosolv). In ionic liquid pretreatments, the inter- and intra-molecular hydrogen bonds of cellulose are disrupted and replaced by hydrogen bonding between ionic liquid anion and carbohydrate hydroxyl groups [18]. This results in significant solubilization of lignin and cellulose. Cellulose surface areas are increased, leading to increased enzyme accessibility [19]. Depending on temperature (20–150 °C), time (1–48 hours), the nature of ionic liquids ([Amin], [Bmin] or [Emin] chloride or acetate),[1] and biomass particle size distribution, up to 25% of cellulose can be dissolved in the ionic liquids to form a homogenous solution [20,21]. After dissolution, the cellulose is precipitated from the ionic liquid, creating multiple phases. Decantation and filtration can be used to separate the phases. Washing of the solids (water and/or buffer solution) may be needed in order to remove ionic liquids prior to enzymatic hydrolysis or fermentation, because ionic liquids are known to be toxic to fermentation strains. A relatively purified lignin could be sold in higher-value markets to improve the overall process economics.

Several potential drawbacks currently exist for ionic liquid pretreatments. Because of the extremely high cost of these solvents, solvent recovery is crucial for minimizing cost. Evaporation for the solvent recovery system is known to be energy intensive. The recovery of IL also uses multiple technologies and is therefore relatively complex; drying is another potentially energy-intensive step. Further R&D is required to improve hemicellulose yields for ionic liquid pretreatment to be economical. Extensive studies are also needed to improve ionic liquid recycling, energy integration, solid liquid separation, and drying.

Organosolv pretreatments were originally derived from pulping processes with goals typically being delignification and fractionation of cellulose, hemicellulose, and lignin fractions. Solvent systems have included low boiling point alcohols (e.g., methanol, ethanol), high boiling point alcohols (e.g., glycols), and other classes of organic solvents (e.g., ketones, ethers, phenols). The cost of using organic solvents to pretreat biomass is still very high, mainly due to the high price of chemicals and, in some cases, high energy consumption [22].

In organosolv pretreatments, hemicellulose and lignin are dissolved in the organophilic phase, but much of the cellulose remains insoluble [22]. Reaction times for organosolv pretreatments range from minutes to a few hours, depending on the solvent-chemical combination and reaction temperature. The pretreated solids need to be washed with organic solvents, then with water extensively to avoid lignin loss, inhibition to enzymatic hydrolysis, and downstream processing. A high degree of solvent recovery is also necessary for economic purposes, to avoid the high cost of adding make-up solvents. Inorganic pretreatment chemicals are either recovered or discarded, likely adding significant variable costs to the overall process. Cellulose is sent to enzymatic hydrolysis and then combined with hemicellulose sugars prior to fermentation. Organosolv pretreatment results in up to 90% cellulose to glucose yields if it is combined with effective enzymatic hydrolysis [23,24].

15.6   Comparative Pretreatment Economics

As discussed Chapter 12, the Consortium for Applied Fundamentals and Innovation (CAFI) was formed in 2000 [25] to collaboratively develop and publish data on leading biomass pretreatment options. Building upon methodologies and tools described above, technoeconomic analysis (TEA) was an important tool for R&D facilitation within the collaborative projects conducted by the CAFI team. The first CAFI project

---

[1] [Amin]chloride is 1-alkyl-3-methyl-imidazolium chloride, [Bmin]chloride is 1-butyl-3-methyl-imidazolium chloride and [Emin]chloride is 1-ethyl-3-methyl-imidazolium chloride, [Amin]acetate is 1-alkyl-3-methyl-imidazolium acetate, [Bmin]acetate is 1-butyl-3-methyl-imidazolium acetate and [Emin]acetate is 1-ethyl-3-methyl-imidazolium acetate.

(CAFI 1) applied various pretreatment processes to a single corn stover feedstock to determine glucose and xylose yields from pretreatment and subsequent enzymatic hydrolysis. TEA analysis for each pretreatment process was conducted using these reported sugar yields [26]. The second phase of this effort (CAFI 2) examined these pretreatments and their performance on poplar, a woody feedstock. The CAFI project team recently extended this approach to switchgrass in the CAFI 3 project. When the experimental results of these CAFI projects were integrated into the TEA models, comparative process economic data for six pretreatment processes across three feedstocks resulted, with particular focus on estimating the total capital investment and MESP.

15.6.1 Modeling Basis and Assumptions for Comparative CAFI Analysis

A common modeling platform was applied for these comparative TEA studies [27] to provide a consistent process and economic framework for comparison. The base model was sized to handle 2000 dry metric tons per day of biomass, operating for 8400 hours per year, using an approach similar to the NREL 2002 design report [3]. Six separate models were developed, one for each of the different pretreatment technologies: dilute acid (DA); SO_2; LHW; ammonia fiber expansion (AFEX); soaking in aqueous ammonia (SAA); and lime. For each of these separate models, the pretreatment sections were modeled based upon conceptual engineering designs developed jointly between NREL modelers and CAFI researchers. One limitation to the CAFI project, however, was the lack of process optimization for each of the pretreatments. It is conceivable that process economics for many of these processes could improve through further optimization.

A short description of the process models employed in the CAFI studies follows. In the feed handling section, the feedstock was delivered to the plant gate in its respective transported form. Minimal storage and size reduction were applied to prepare the biomass for pretreatment. Following pretreatment, the material was washed and/or neutralized prior to enzymatic hydrolysis, with the pH adjusted using either acid or base depending upon the pretreatment pH. For enzymatic hydrolysis, a purchased enzyme cocktail was assumed, consisting of either cellulase only or combined cellulase/hemicellulase activity. The pretreated biomass was loaded at 20% total solids if not defined otherwise, the physical upper limit for processing in a conventional stirred tank reactor. The hydrolyzed material was transferred to the fermentation section, where recombinant *Zymomonas mobilis* bacteria converted sugars into ethanol. The enzymatic hydrolysis and fermentation process assumed here is separate hydrolysis and fermentation (known as SHF), since enzymatic hydrolysis is operated at elevated temperature. The combined residence time for hydrolysis and fermentation was 7 days. Low-cost nutrients (corn steep liquor or CSL and diammonium phosphate or DAP) were assumed to be effective. Unless otherwise stated oligomers were not considered fermentable sugars in these models, with the impact of their conversion shown in subsequent sensitivity analyses. The fermentation yields were assumed to be the same for all pretreatments with glucose fermented to ethanol at 92% of theoretical efficiency and xylose to ethanol at 85% of the theoretical efficiency. The other sugars (arabinose, mannose, and galactose) were not fermented, an assumption that should be verified as subsequent R&D continues.

The resulting beer was separated into ethanol, water, and residual solids (stillage) through a combination of distillation, molecular sieve dehydration (to break the azeotrope), and liquid/solid separation. Solids (lignin residue, unconverted solids) were sent to the onsite combustor while the liquid (water) was treated and recycled within the process. Anaerobic and aerobic wastewater treatment was applied to reduce the chemical oxygen demand (COD) and total dissolved solids (TDS) of the water. Methane formed by anaerobic digestion was burned to provide heat and power for the biorefinery, with a fluidized bed combustor coupled with turbogenerator producing steam along with electricity. The ash from the combustor was land filled. In this combined heat and power (CHP) configuration, excess electricity was available for sale as a co-product back to the power grid after meeting refinery demands. Other ancillary process areas for costing included utilities (cooling and chilled water systems) and chemical and product storage.

Table 15.1 Raw material costs in US dollars (2007 prices).

Raw materials cost	Price ($2007)
Corn stover	69.60/dry ton
Hybrid poplar	67.55/dry ton
Switchgrass	69.60/dry ton
CSL	205.00/ton
Cellulase enzyme	5.00/kg protein
Diammonium phosphate	182.30/ton
Propane	5.90/ton
Sulfuric acid (93%)	32.10/dry ton
NH_3	300.00/ton
Lime	89.80/ton
SO_2	400.00/ton
Water	0.40/ton
Power (by-product credit)	0.04/kWh

The variable operating costs were estimated based on material and energy balance calculations using Aspen Plus simulations, similar process modeling for NREL biochemical conversion design report [3]. Raw materials included biomass feedstocks, pretreatment and neutralization chemicals, nutrients (corn steep liquor, diammonium phosphate, potassium salts), wastewater treatment chemicals and polymers, with their unit costs listed in Table 15.1. Onsite utilities included steam, power, water, and nitrogen gas. All costs were on a constant year-2007 dollars basis. For this analysis, the enzyme cost was assumed to be uniformly $5.00/kg protein, and the enzyme loading was calculated for each CAFI project approaches based on mg protein per gram cellulose. This enzyme price did not reflect any commercial enzyme costs but was solely an estimated contribution of enzyme cost to the operating cost of ethanol production. Enzyme cost is an important element in raw material costs, and is discussed below.

Salaries were inflated to 2007 dollars using NREL's 2002 design basis [3]. In addition to salary, general overheads were calculated as 60% of the total salary and covered items such as safety, general engineering, general plant maintenance, and payroll overheads. Annual maintenance materials were estimated at 2% of the total project investment, and property insurance and local property tax were estimated at 1.5% of the total project investment based on standard literature assumptions [15].

Major capital equipment was sized using material and energy balance data, as well as equipment retention time. Equipment costs were developed using a number of sources, including past vendor quotations (for more specialized equipment) from the corn stover ethanol biochemical design report [3], the CAFI 1 process designs [26], the United State Department of Agriculture (USDA) corn ethanol model [28], and the NREL equipment database, as well as costing software estimates (for simpler equipment such as distillation columns, pumps, and tanks), chemical engineering textbooks [15], and other database information. Equipment costs for areas outside of pretreatment were derived from the NREL 2002 design case using the power law to adjust for changes in capacity. The scaling exponent for the power law was obtained from the NREL 2002 design case [3] for most of the equipment. For equipment not listed in the NREL 2002 design case and for which vendor's guidance was not available, the exponent term was assumed to be 0.60. Standard NREL factors [3] were used to obtain the total project investment from the purchased equipment costs. However, pretreatment reactor installation factors were re-evaluated to reflect both proper scaling-up and materials of construction considerations, based on common engineering heuristics.

The discounted cash-flow calculation in this study assumed 100% equity financing with 2.5 years for construction plus 0.5 year for start-up. The plant life was 20 years, the income tax 39%, and working capital 5% of fixed cost investment (FCI). The MESP was the minimum price that ethanol must realize to generate a NPV of zero for a 10% IRR, making the MESP higher than the cash cost of production. The cost year was 2007 (US dollars), with equipment quotations obtained in earlier or later years inflated or deflated to this year using chemical indices. Several sensitivity cases were analyzed for comparison with the base case scenario for each pretreatment.

For CAFI 1, the average cellulase enzyme activity was assumed to be 31.2 FPU/mL (FPU: filter paper unit) of solution. β-glucosidase loadings were based on results with Novozyme 188, a supplementary enzyme used by the CAFI team and purchased from Sigma Chemical Company (Cat. No. C6150, Lot No. 11K1088) [29]. The cellulase loadings were taken as 15 FPU/g glucan and supplemented with 30 carboxypeptidase B units (CBU) of β-glucosidase/g glucan [29]. 15 FPU/g glucan in untreated corn stover was taken to be equivalent to a 58 mg/g glucan protein loading for the CAFI 1 technoeconomic analysis [26].

For CAFI 2, cellulase enzyme Spezyme CP provided by Genencor International had a filter paper activity of 59 FPU/mL and a protein concentration of 123 mg protein/mL. Another cellulase, GC-220, had corresponding values of 89 FPU/mL and 184 mg protein/mL, respectively. The other enzymes and their respective protein contents were beta-glucosidase at 32 mg protein/mL, multifect xylanase at 41 mg protein/mL, and b-xylosidase at 85 mg protein/mL. The baseline enzyme formulation was a combination of β-glucosidase and cellulase at a CBU/FPU ratio of 2.0 [27]. A cellulase loading of 15 FPU/g glucan and β-glucosidase loading of 30 CBU/g glucan were used, equivalent to 29.0 mg protein/g glucan [30].

For CAFI 3, enzymes were added at 15 FPU cellulase in Spezyme CP plus 30 CBU β-glucosidase in Novozyme 188 per g glucan of untreated raw switchgrass (equivalent to 27 mg protein/g glucan). Since the enzyme activity improved over the years, 27 mg protein per gram glucan was used for consistency in the economics analysis. Enzyme costs are compared in Figure 15.2 assuming a cost of $5.00/kg protein. It is noted that the enzyme cost varies from $0.37 to $0.86 per gallon ethanol, depending on pretreatment technologies, but averages around $0.45/gal ethanol produced (Figure 15.2) with the highest cost being for ammonia recycle percolation (ARP) and SAA pretreatment of hybrid poplar and switchgrass due to the low yields.

Pretreatment performance varied significantly for the three feedstocks used in the CAFI project: corn stover, hybrid poplar, and switchgrass. As shown in Table 15.2, hybrid poplar had the highest cellulose and lignin contents on a dry weight basis but also had the highest moisture level. Because acetate content in the corn stover studied in the CAFI 1 project was relatively high, acetate removal by ion exchange chromatography before pretreatment might significantly enhance performance and yields but was not considered in this study for process design for any of the three feedstocks.

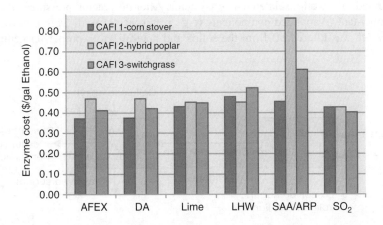

Figure 15.2 *Enzyme cost per gallon ethanol for CAFI projects with various pretreatment methods.*

Table 15.2 Feedstock compositions (in percent) for CAFI projects.

Feedstock composition	CAFI 1 Corn stover	CAFI 2 Hybrid poplar	CAFI 3 Dacotah switchgrass
Cellulose	33	43	35
Xylan	22	15	23
Arabinan	4	1	3
Mannan	1	4	0
Galactan	1	1	1
Lignin	11	29	23
Extractives	8	4	10
Ash	6	1	3
Acetate	5	4	2
Protein	2	0	1
Soluble solids[a]	6	0	0
Moisture	15	50	15

[a] Unknown soluble solids are calculated by difference to close the mass balance.

15.6.2 CAFI Project Comparative Data

Glucose and xylose can be released first in pretreatment (also designated as Stage 1 in Figure 15.3) and then in enzymatic hydrolysis (Stage 2 in Figure 15.3). Sugar is fermented following enzymatic hydrolysis of washed solids from each pretreatment. Washing the pretreated solids may not be desirable for all pretreatments but was included in this study for consistency. The process conditions for all the pretreatments studied in the CAFI project are summarized in Table 15.3. It is important to recognize that none of the CAFI pretreatments were optimized; as a result, the ethanol yields and process economics presented here may not represent the best that could be done for the pretreatments studied in the CAFI projects.

15.6.3 Reactor Design and Costing Data

A horizontal reactor was chosen for dilute acid pretreatment. Biomass entered through a system of screw-feeders, and steam was then directly injected into the reactor to rapidly achieve reaction temperature, typically in the range of 150–200 °C. For the base case, biomass was loaded at 30% total solids, and minimal heat loss was assumed for a well-insulated reactor systems. After the reactor, pressure was dropped to ambient via a blow-down tank. Vaporized components (e.g., water, acetic acid) were sent to onsite treatment while the hydrolyzate was filtered off from the solids and neutralized to proper fermentation conditions.

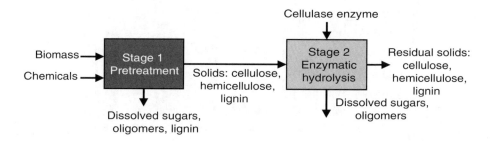

Figure 15.3 Schematic process of pretreatment followed by enzymatic hydrolysis used in the CAFI projects.

Table 15.3 *Summary of process conditions used in the CAFI projects.*

Method	Project	Temp (°C)	Residence time (min)	Total solids (wt%)	Catalyst	Catalyst loading (g/100 g DBP)	Catalyst recycle	Dilution water and steam (g/100 DBP)
AFEX	CAFI 1	90	5	47%	Anhydrous NH_3	100	Yes	—
	CAFI 2	180	30	42%	Concentrated NH_3	100	Yes	—
	CAFI 3	150	30	30%	Concentrated NH_3	152	Yes	63
DA	CAFI 1	160	20	25%	H_2SO_4	2	No	285
	CAFI 2	190	1.1	25%	H_2SO_4	2	No	198
	CAFI 3	140	40	30%	H_2SO_4	9	No	206
Lime	CAFI 1	55	4 wk	48%	CaO	8	Yes	—
	CAFI 2	160	120	30%	$Ca(OH)_2$	39	Yes	224
	CAFI 3	120	240	20%	$Ca(OH)_2$	100	Yes	282
					O_2	100 psi	Yes	
LHW	CAFI 1	190	15	16%	Water	N/A	Yes	425
	CAFI 2	200	10	15%	Water	N/A	Yes	467
	CAFI 3	200	10	20%	Water	N/A	Yes	382
ARP	CAFI 1	170	10	25%	Aqueous NH_3	15	Yes	270
	CAFI 2	185	27.5	25%	Aqueous NH_3	15	Yes	185
SAA	CAFI 3	160	60	30%	Aqueous NH_3	135	Yes	80
SO_2	CAFI 1	—	—	—	—	—	—	—
	CAFI 2	190	5	25%	SO_2	3	No	197
	CAFI 3	180	10	30%	SO_2	5	No	210

Costs were obtained from vendor quotations. Reactor size is calculated based on pretreatment conditions listed in Table 15.3.

Gaseous sulfur dioxide (SO_2) enhances glucose and xylose yields in a manner similar to that of liquid acid pretreatment. Because purchased costs (and safety concerns) for SO_2 would be relatively high if shipped, onsite production of SO_2 is believed to be more cost-effective for use at an industrial scale. For the economic models applied here, an impregnation step was used prior to the primary reactor, and the reactor pressure was held just at the bubble point of the mixture. Most of the other design elements were identical to those for dilute acid pretreatment.

For LHW, biomass was fed (pumped) at 15–20% total solids through an indirect heat exchanger to bring the reaction temperature up to 200 °C. A long tubular reactor was employed as a simple approach to achieve a residence time of 10–20 minutes. The reactor effluent was then cooled by cross-exchange and then entered a blow-down tank to vent non-condensable gases.

For this study, anhydrous ammonia was employed for AFEX pretreatment at moderate temperatures (*c.* 90–120 °C) and high pressures (up to 400 psig). The pretreatment process flow diagram is depicted in Figure 15.4. After a short reaction time, the pressure was rapidly released to physically disrupt biomass. Although virtually no hydrolysis or lignin removal took place, disruption was sufficient to enhance enzymatic hydrolysis. The ratios of ammonia to dry biomass were in the range of 1.0–1.6. No vendor quotations were used to estimate the reactor capital costs, but a cost of $1M was assumed for the reactor alone. The vapors released upon depressurization contained concentrated ammonia and were compressed and condensed. For the CAFI 1 project [26], ammonia recovery was enhanced by a more complex drying and distillation system. However, an absorber system was applied for the economic study for CAFI 3 [31] to simplify recovery. Costs for the recovery system were obtained from a variety of costing software and databases such as Aspen IPE. Most of the pH adjustment prior to enzymatic hydrolysis was provided by washing the solids with water.

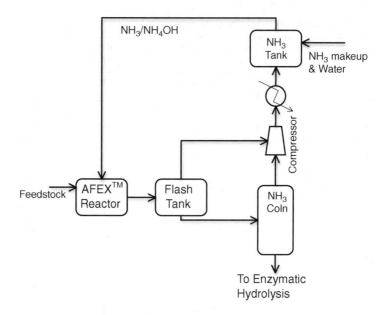

Figure 15.4 *Process flow diagram of AFEX pretreatment process.*

For CAFI 1, ARP at 150–180 °C applied ammonia concentrations of 10–15%. However, the approach evolved to SAA to improve the economics, with a relatively simple pretreatment reactor configuration applied to solubilize about 70% of the lignin. The SAA process (shown in Figure 15.5) applied a wider temperature range (60–180 °C) and lower ammonia concentrations (1.35 g NH3/g dry biomass). For the base case, the reaction temperature was set at 160 °C, pressure at 350 psig, and

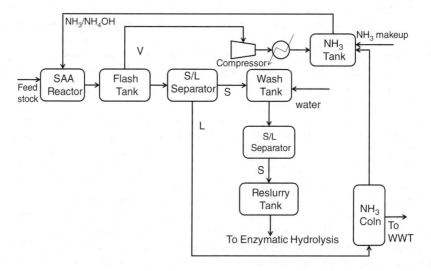

Figure 15.5 *Process flow diagram of SAA pretreatment process.*

residence time at 60 minutes, with the solids loading being 20% total and the ammonia concentration 30%. Screw-type reactors (similar to those assigned to dilute acid pretreatment) were chosen but made of stainless steel to reduce cost. Steam was directly injected to obtain the target reaction temperature. After the reactor, most of the ammonia was condensed and recycled to the reactor. The solids were then washed using a belt filter and sent to enzymatic hydrolysis. The wash liquor was sent to a recovery column where ammonia was further recovered and recycled. The bottoms of the recovery column were sent to wastewater treatment (WWT). Total ammonia recovery was taken to be 98%, with 2% lost to either reaction with acetic acid to form ammonium acetate or carried downstream with the solids. Since much of the lignin was assumed to be solubilized, it was washed from the solids and eventually ended up downstream at the residue combustion section.

For lime pretreatment, high-pressure oxygen (100 psi) was obtained from an onsite air separation unit using pressure swing adsorption (PSA). It was added to the pretreatment reactor which was maintained at constant pressure to blanket the head space of the vessel. Lime pretreatment can be performed over a wide range of temperatures, ranging from ambient to 160 °C. The design for lime has changed significantly since the first CAFI project. In CAFI 1, pretreatment was assumed to be performed in a "pile" with long residences times (up to 8 weeks) similar to biological pretreatment. Subsequently, in CAFIs 2 and 3, higher temperature (above 120 °C) and much reduced residence times (4 hours) were applied to allow reaction in a closed reactor. A process flow diagram of the lime pretreatment process applicable to the CAFI 2 and 3 projects is shown in Figure 15.6. 25% of the xylan was released during pretreatment, primarily in the form of oligomeric xylose. After pretreatment, the pretreated biomass was washed. Calcium carbonate from this system follows the solids downstream through hydrolysis and fermentation to end up in the residue combustor. The ash from the combustor contains CaO which is then separated and slaked to produce lime which can be recycled to the pretreatment reactor. Approximately 20% CaO loss was assumed from the ash-lime decanter.

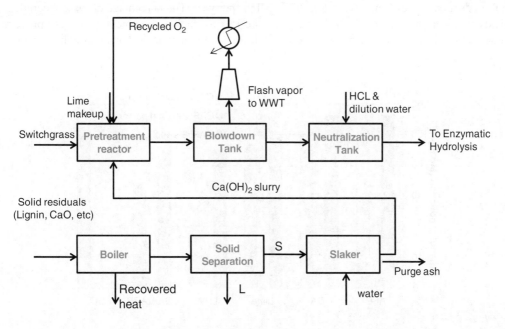

Figure 15.6 *Process flow diagram of lime pretreatment process.*

Table 15.4   *The monomer sugar and total sugar yields and resulting percentage of theoretical ethanol yields of CAFI projects.*

Method	Monosugar yield%			Total sugar yield%			% of theoretical ethanol yield		
	CAFI 1	CAFI 2	CAFI 3	CAFI 1	CAFI 2	CAFI 3	CAFI 1	CAFI 2	CAFI 3
AFEX	94	53	76	94	53	79	66	42	58
DA	92	83	76	92	83	78	58	64	57
Lime	77	84	70	87	91	87	57	67	53
LHW	63	52	61	87	69	83	57	41	46
SAA/ARP	72	44	52	89	54	69	65	35	39
SO$_2$	—	91	79	—	96	82	—	71	60

15.6.4   Comparison of Sugar and Ethanol Yields

The total sugar yields and monomer sugar yields from glucan and xylan for each stage are summarized in Table 15.4 for the CAFI projects. In this TEA study, only monomeric glucose and xylose were assumed to be fermentable to ethanol. As a result, overall ethanol yields and associated annual production levels were strongly dependent on monomeric sugar yields. Ethanol production was expressed in terms of the percent of maximum theoretical ethanol yields (Table 15.4) and million gallons of ethanol annual production (Figure 15.7). Because higher ethanol production generally results in lower MESP, the MESP values are highly dependent on monomeric sugar yields. For instance, the ARP/SAA pretreatment has the highest MESP (representing ethanol costs on a per gallon basis) as a result of lower total monomeric sugar yields for both hybrid poplar and switchgrass feedstocks.

Based on the feedstock compositions shown in Table 15.2, the theoretical ethanol production shown in Table 15.5 ranges from 84.4 to 87.8 million gallons per year if all the sugars in cellulose and hemicellulose were converted to ethanol. However, if only glucose and xylose were converted to ethanol, the theoretical ethanol production ranges from 76 to 78 million gallons per year. The percentage of these theoretical ethanol yields ranged from 35% to 71%, as shown in Figure 15.7 for each pretreatment method. Improvement of theoretical ethanol yields should be possible through improved enzyme systems that convert more of the

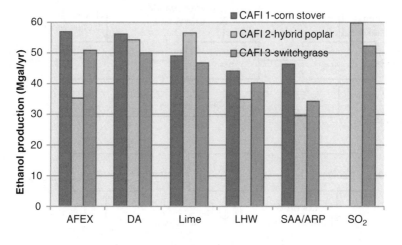

Figure 15.7   *Annual ethanol production of pretreatment studied in the CAFI projects.*

Table 15.5   *Theoretical ethanol production for 2000 dry MT/day feed of corn stover, hybrid poplar and switchgrass used in the CAFI.*

	Theoretical ethanol production (MGal/yr)	Theoretical ethanol production from glucose and xylose (MGal/yr)
CAFI 1 Corn stover	86	76
CAFI 2 Hybrid poplar	84	77
CAFI 3 Switchgrass	88	78

oligomers to monomers and through strain development to enable utilization of the minor sugars (arabinose, mannose and galactose).

Sugar and ethanol yields from CAFI 1 using corn stover have the least variability across all five pretreatments. On the other hand, for CAFI 2 with hybrid poplar, sugar yields vary significantly with DA, lime, and SO_2 having significantly higher yields than from ARP, AFEX, and LHW. For CAFI 3 with switchgrass, pretreatment by AFEX, DA, lime, and SO_2 have similar sugar and ethanol yields, but SAA and LHW pretreatments have significant lower yields, as shown in Table 15.4 and Figure 15.7.

Some pretreatments such as DA and lime showed little difference in sugar yields among the three feedstocks, and both have slightly better ethanol yields with hybrid poplar than with corn stover or switchgrass because of the higher cellulose content. However, AFEX pretreatment achieved better sugar yields from corn stover and switchgrass than from hybrid poplar. Since alkaline pretreatments solublize lignin, they are more effective on agricultural residues and herbaceous crops than on woody biomass that contains more lignin. It is noted that the primary mechanism for alkaline pretreatment is thought to be lignin solubilization with little or no hemicellulose dissolution, whereas dilute acid pretreatment dissolves both hemicellulose and lignin. Given lignin solubility limits, pretreatment effectiveness decreases with increasing lignin content for alkaline pretreatments. The high lignin content (*c.* 29%) of hybrid poplar therefore likely contributes to low sugar or ethanol yields from AFEX or ARP/SAA pretreatments. Ethanol yields are relatively high for corn stover among all the pretreatment technologies, while ethanol yields from switchgrass fall between stover and wood as shown in both Table 15.4 and Figure 15.7. Further, it was found that SAA pretreatment may be more effective with fall harvest switchgrass feedstocks and less effective in disrupting the lignin structure for switchgrass harvested in the spring.

15.6.5   Comparison of Pretreatment Capital Costs

The direct capital cost for pretreatment is strongly dependent on processing conditions such as reaction temperature, residence time, solids levels, and pretreatment chemical recovery strategies. As mentioned above, the process conditions are summarized in Table 15.3 for the pretreatment processes included in the CAFI projects, although these were not optimized for ethanol production or process economics. The reaction temperatures and residence times were defined by the CAFI research teams developing each technology based on extensive bench-scale studies. The total solids levels were defined based on both bench-scale process conditions and through reasonable engineering judgment for a conceptual commercial-scale continuous process configuration. The pretreatment chemical loading for each case was based on bench-scale data provided by respective CAFI project teams. The solids and pretreatment chemical loadings of each pretreatment impact the size of the pretreatment reactors, the final concentration of ethanol in the fermentation broth, plant steam usage, and plant power balance, as well as capital costs of downstream processes of fermentation and ethanol recovery areas. Further optimization of catalyst loading, as well as other process conditions for commercial scale operation, is therefore needed for a more detailed engineering design.

Table 15.6   *Summary of pretreatment capital and total project investment of CAFI projects.*

2007$	Pretreatment reactor cost ($M)			Pretreatment Capital ($M)			Total Capital ($M)			Total project($M)		
Method	CAFI 1	CAFI 2	CAFI 3	CAFI 1	CAFI 2	CAFI 3	CAFI 1	CAFI 2	CAFI 3	CAFI 1	CAFI 2	CAFI 3
AFEX	17	26	22	35	41	40	164	197	200	298	359	363
DA	22	34	34	34	46	45	162	185	192	294	337	349
Lime	6	32	25	30	64	57	124	214	224	226	390	407
LHW	0.3	4	4	6	20	20	157	169	179	285	308	325
SAA/ARP	6	19	33	34	53	87	160	174	203	290	317	368
SO$_2$	—	25	23	—	39	35	—	181	187	—	329	340

Table 15.6 summarizes installed pretreatment reactor costs, pretreatment area capital costs in million dollars ($M), total capital for the entire cellulosic ethanol processes and TPI costs for pretreatment technologies studied in the CAFI efforts. Total pretreatment area capital costs include pretreatment reactors, conditioning/neutralization equipment, and any required recovery and recycle systems. Equipment was sized using stream data from the Aspen Plus simulations [16]. Most pretreatment reactor costs were referenced from the CAFI 1 project with corn stover [26] and CAFI 3 project with switchgrass [31], with appropriate scaling. Total capital costs represented the process equipment, while the TPI included indirect costs such as field expenses, home office and construction fees, project contingencies, start-up, and permits. The indirect cost represented 45% of TPI in the study.

Pretreatment capital costs were found to range widely for the study (Table 15.6) from $6M (LHW in CAFI 1) to $87M (SAA in CAFI 3). SAA pretreatment was more expensive due to long residence time. Lime pretreatments were also more expensive than other pretreatments due to higher costs for lime recovery using an expensive slaking operation and oxygen PSA unit, as well as higher cost of lime pretreatment reactors. Although a lime pretreatment reactor by itself was not expensive compared to the dilute acid pretreatment reactor, the number of reactors required was significant due to the longer residence time (4 hours). DA and SO$_2$ pretreatment reactor costs contributed to over 75% of their pretreatment capital cost due to the high cost of materials of construction and no recycling of pretreatment chemicals. For AFEX, SAA, and lime pretreatments, significant portions of cost were associated with the recovery of pretreatment chemicals, as one-pass use of pretreatment chemicals was unrealistic at the high pretreatment chemical usage for these processes. However, the design of these recovery and recycle systems was preliminary, and further development of efficient recovery configurations may lower costs for these pretreatments.

In comparing capital investments across CAFIs 1, 2, and 3, pretreatment, enzymatic hydrolysis, fermentation, and recovery sections are responsible for slightly less than half of the total direct fixed capital of the plant [31], regardless of pretreatment technologies. The boiler system to recover heating value from residual biomass and lignin is about one-third of total direct fixed capital cost.

15.6.6   Comparison of MESP

Figure 15.8 summarizes MESP expressed in cost per gallon of ethanol produced for all the pretreatments analyzed in the CAFI projects. MESP is further broken down into contributions to the overall MESP by feedstock, other variable inputs, fixed costs, depreciation, income tax, and return on capital. The feedstock costs were derived from DOE's Office of the Biomass Program (OBP) multiyear program plan (MYPP) [32] for a dry ton cost delivered to the gate of an ethanol plant. The other variable costs account for costs for

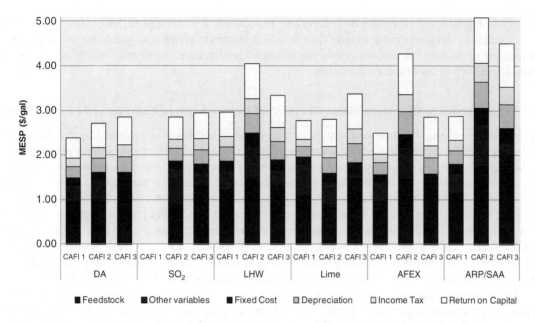

Figure 15.8 *The MESP of all the cases studied in CAFI, in comparison with ethanol annual productions.*

pretreatment chemicals, enzymes, nutrients, and other chemicals as well as net electricity generation credits, and are positive even after taking electricity by-product credits. Fixed costs include labor, maintenance, insurance, and other costs not tied to production rate. The depreciation taken as the TPI over 20 years contributed $0.24–0.58 per gallon ethanol to the MESPs. The income tax rate was assumed to be 39%, and a 10% discount rate (or internal rate of return, IRR) was assumed for the discounted cash-flow analysis to calculate rate of return on a per gallon basis. Higher IRR resulted in higher required return of capital cost, as well as higher MESPs [3]. The MESP contribution from return on capital was high, even for a 10% IRR, and ranged from $0.43/gal for lime pretreatment to $1.04/gal for ARP pretreatment.

The estimated MESP ranged from $2.39/gal (DA in CAFI 1) to $5.10/gal (SAA in CAFI 3), with the MESPs below $3.00/gal for most of the pretreatments in the CAFI projects. For the cases with MESP higher than $3.00/gal, the sugar and ethanol yields were relatively low. Although all the cases were capital intensive, it is obvious that the MESP results were more dependent on ethanol yields than on capital investments. Fermenting oligomeric sugars to ethanol or converting them to fermentable monomers after pretreatment would therefore significantly improve the MESP for several pretreatments.

15.7 Impact of Key Variables on Pretreatment Economics

Pretreatment represents a significant fraction of overall costs of conversion and impacts many of the downstream portions of the process. It is therefore important to understand the many key variables that impact pretreatment economics, as the following sections, while not exhaustive, illustrate.

15.7.1 Yield

As demonstrated, yield represents one of the single-most important factors that influence conversion economics, with yields combining those from pretreatment, enzymatic hydrolysis, and fermentation.

It therefore includes the efficiency of hydrolyzing hemicellulose as well as cellulose and the metabolic yield of ethanol from fermentable sugars. Therefore, it is important to optimize yields across the process as a whole and to examine interactions among pretreatment, conditioning, enzymatic hydrolysis, and fermentation, as opposed to any of these in isolation.

15.7.2   Conversion to Oligomers/Monomers (Shift of Burden between Enzymes and Pretreatment)

The importance of sugar distribution between monomers and oligomers is governed by the downstream fermenting organism. Other than a few exceptions, most of the leading co-fermenters (both yeast and bacteria) are capable of only fermenting monomers and cannot ferment sugar oligomers to ethanol. As a result, maximizing monomeric sugars becomes vital to maximizing yields. For acid-catalyzed reactions, this becomes a bit of a "tight-rope" balancing act in that the kinetic reactions must be controlled such that material is not under- or over-reacted. In other words, high-severity processing leads to higher conversion to sugar degradation products such as aldehydes, while lower-severity processing can lead to higher percentages of material left as insolubles or oligomers (or both). A research strategy currently employed is to convert oligomers to monomers thermally through holding at temperature for a couple of hours. An alternate strategy is to increase the concentration of activities that hydrolyze oligomers in enzyme cocktails.

A key benefit of using enzymes in these processes is their high selectivity. Cellulase enzymes, for example, can be pictured as a mixture of proteins with at least three distinct functionalities that each promote a particular reaction. First, endoglucanase (EG) enzymes break cellulose chains randomly to expose chain ends. Exoglucanase (cellobiohydrolase) enzymes then work processively from those ends to cleave the long chains into dimer units called cellobiose. Finally, β-glucosidase enzymes break the dimers into monomers. Not only do these systems possess the ability to convert oligomers into monomers, but they do not produce by-products. Xylanase and other hemicellulase enzymes can function with similar selectivity for attacking hemicellulose.

The strategy behind LHW pretreatment is to maximize the enzymatic advantage by minimizing the severity of pretreatment, to leave the majority of the hydrolysis burden on the enzyme system. Current advanced commercial cellulase products possess some xylanase activity; however, further development of advanced enzyme cocktails is required for systems that could benefit from more hemicellulase activity. Enhancing enzyme performance for the many combinations of feedstocks, pretreatment technologies, and processing conditions is a complex challenge.

15.7.3   Biomass Loading/Concentration

Biomass loading to pretreatment can have a large impact on pretreatment costs. First, the feeding systems for getting loose, low-density biomass into a pressurized reactor can be complex and costly, with screw-feeders typical for such pretreatment reactors. However, the biomass concentration in a given system also plays a significant role in that more dilute systems must have larger equipment for the significant volumes of liquid present, adding to the capital cost of the entire conversion process. Higher-solids-concentration systems can process the same mass of dry biomass with less capital cost.

Most of the pretreatment systems in the CAFI comparative study loaded biomass into pretreatment to meet specifications for 20–30% total solids. When solids are loaded in more dilute proportions than this, the material will flow more easily and in some instances may be pumpable. However, previous techno-economic studies have shown that cost savings associated with processing higher solids slurries are significant, particularly if high conversion yields can be maintained. Biomass materials at solids concentrations beyond 30% no longer resemble slurries but instead look like moist or slightly wetted biomass. Significant mass transfer barriers exist in these regimes, and significant processing energy can be needed to disrupt biomass fibers.

15.7.4   Chemical Loading/Recovery/Metallurgy

The amount of chemicals used for biomass pretreatment also impacts the economics in several ways. First of all, it will define the overall pH of the system which impacts physical responses and chemical reactions. For example, even with dilute strong acid systems, relatively little chemical loading can result in very acidic solutions with a pH of 1–2. The pH itself also has an impact on the materials of construction required for the reactor. Alloys far more expensive than stainless steel are often required in acid systems (for corrosion protection), whereas stainless steel provides sufficient corrosion protection for alkaline systems.

Secondly, chemicals loaded have a direct economic effect. Clearly larger volumes of chemicals results in higher operating costs, and the cost and loading percentage of these chemicals will ultimately determine whether chemical recovery and recycle is needed. Unless dilute proportions (<5% by weight) are needed, there is a high probability of chemical recovery being the economically preferred option, but even dilute portions of very expensive reagents may still require recovery. Where solvents and pretreatment chemicals are used in greater proportions, the efficiency of recovering these chemicals will be a primary economic driver.

Dilute sulfuric acid pretreatments will typically not employ chemical recovery since it is considered a lost-cost acid. However concentrated acid systems require recovery. Liquid hot water does not utilize further pretreatment chemicals, thus negating recovery needs. For higher-cost acids such as nitric or hydrochloric recovery systems are often advantageous, but the economics will depend upon concentrations used. However, all of the alkaline CAFI systems modeled included a solvent recovery system because of their higher costs and/or greater amounts. For example, ammonia is considerably more expensive than sulfuric acid. On the other hand, while lime is relatively inexpensive, the ability to regenerate, recover, and recycle it cannot be ignored. Because organosolv and ionic liquid systems employ even more expensive solvents, efficient recovery is paramount.

Chemical loadings also have important implications on operations downstream from pretreatment. First of all, the need for subsequent neutralization chemicals will bring additional cost and also bring additional compounds into the system which must be disposed of, treated, or sold. In line with this, downstream operations such as wastewater treatment and residue combustion can require significant design changes to handle salts and other compounds. Some of the pretreatment and neutralization chemicals also have known inhibitory or even toxic effects on the biological systems downstream, and sufficient quantities can negatively impact performance by hurting yields and possibly killing organisms. Subsequently, minimizing added chemicals is important. Adapting organisms to environments with these chemicals present is one potential strategy to reduce inhibition.

For the example of a dilute sulfuric acid system, neutralization is used to adjust the pH to 5.0 to be compatible with the enzymes, and more than stoichiometric amounts of neutralization chemicals can have detoxification benefits for the organisms. In cases where lime has been used, this is referred to as "overliming." This reaction between sulfuric acid and lime produces gypsum or calcium sulfate that is relatively insoluble and precipitates out of solution for easy recovery. Sugar losses from side reactions also result from neutralization with lime. However, these salts also present a large volume of impure by-products that must be disposed of with associated costs of disposal and, more importantly, life cycle and sustainability implications. While gypsum can be used for products such as wallboard, significant clean-up of this bio-derived gypsum would be needed to meet performance specifications. In addition, any calcium salts not recovered will travel downstream. Because calcium becomes less soluble with increasing temperature, handling precautions are essential to prevent these salts from plating out on reboilers and other equipment. A different neutralization chemical, such as ammonium hydroxide, can change the conversion system drastically in that the ammonium sulfate salts that result are soluble and do not precipitate out, and recent R&D

has reduced sugar losses for this system. However, the impact of these salts on organism performance is not yet well characterized. Also of concern is the concentration of these salts entering wastewater treatment and residue combustion, and significant costs may be incurred in processing these soluble salts downstream. Ultimately, minimizing the use of added chemicals for pretreatment and neutralization in biomass conversion systems will have important cost and sustainability benefits.

15.7.5   Reaction Conditions: Pressure, Temperature, Residence Time

The reaction conditions required for biomass conversion have cost implications for pretreatment, particularly on the reactor design elements. Specifically, parameters such as temperature and pressure, and kinetic parameters such as residence time, are of utmost importance and are some of the key variables explored through bench and pilot scale R&D. In fact, several of these parameters can be combined into an overall "severity" factor to describe pretreatment conditions [33]. While many techniques are available to achieve temperature in a reactor, direct steam injection and cross-exchange with hot process streams are two of the most common. As higher reaction temperatures are required, more energy is clearly needed to achieve these temperatures. Pressure has an even more pronounced impact on the inner workings of the reactor. Because biomass slurries at industrially relevant moisture levels are not pumpable by typical means, they require energy (mostly electric power) to feed pressurized reactor systems. Pressure can come when mid- or high-pressure steam is directly injected, but must be obtained by other means when steam is not fed live. Novel systems involving supercritical conditions (e.g., water, CO_2) can have significant costs associated with achieving the extremely high pressures needed.

Residence time continues to play an important role in overall conversion economics, in that longer residence times require larger and more costly equipment. This was clearly shown for lime pretreatment with a 4-hour residence time. In the past, much of the R&D for dilute acid focused on high-temperature (190 °C) systems with very short residence times (<5 minutes). However, research is now focusing on lower temperatures with longer residence times in order to lower pretreatment severity. In addition, residence time control has been shown to significantly impact conversion yields of xylan to xylose during pretreatment, with tremendous impact on overall economics.

15.7.6   Reactor Orientation: Horizontal/Vertical

Reactor design directly impacts pretreatment economics. While pressure, temperature, and residence time are extremely important, other elements including metallurgy, moving parts, seals, and controls also impact costs. Underlying these factors is the reactor orientation, whether vertical or horizontal, and combinations of vertical and horizontal reactors have been explored for two-stage reactions. There are tradeoffs with each orientation. Vertical reactors will typically be lower in cost than horizontal reactors because they are gravity assisted in moving solids from top to bottom. Thus, they typically have fewer or no moving internals for conveying biomass, which helps to lower costs. Horizontal systems on the other hand often have larger footprints and higher costs associated with conveying biomass through the reactor. However, horizontal systems have the added advantage of biomass residence time control and variable speed drives can improve residence time control, minimizing over-cooking and undercooking. Consequently, horizontal reactors have better ability to maximize yields compared to vertical systems.

15.7.7   Batch versus Continuous Processing

Also important is consideration of batch versus continuous systems. While batch pretreatment is commonplace at the laboratory and bench scale, both batch and continuous systems have been employed at pilot and larger scales. Continuous systems allow higher throughputs and are more easily scalable than batch systems. Batch systems are limited in scale by their ability to achieve high steam and chemical penetration, uniform mixing and cooking, as well as other aspects. Turnaround time is also an important consideration when choosing between batch and continuous systems.

15.8   Future Needs for Evaluation of Pretreatment Economics

While the economics of biomass pretreatment are fairly well understood for well-developed systems, novel systems would benefit from further investigation. Companies who are commercializing biological conversion technologies employ a number of pretreatments, including those based on dilute acid, concentrated acid, lime, liquid hot water (including steam pretreatment), and ammonia. In many cases, however, performance data are not available to the research community, and more detailed design and costing of these systems would be beneficial. Industry will benefit as R&D continues to lower the cost and severity of these pretreatment systems.

The CAFI series of projects provided an important look at how different pretreatment technologies perform for three different feedstocks: corn stover, switchgrass, and poplar wood. However, many other biomass sources exist, including straws (wheat, rice), bagasse, miscanthus, various hardwoods and softwoods, and woody residues. Generating more pretreatment data on multiple feedstocks is clearly needed. More detailed costing is also required for many of the non-acid systems as many of these costs are not yet based upon vendor designs and quotations for these specific applications. This additional design and cost detail will improve the accuracy of economic estimates for biomass conversion, including additional details on chemical and solvent recovery systems.

Although the feasibility of many pretreatment technologies has been demonstrated, the robustness of these systems should also be a focus of R&D. More pilot pretreatment facilities now exist than ever before, both in the US and internationally. Using these facilities to generate long-term data at a larger scale will demonstrate the robustness of technologies for commercial operation.

Advanced enzymes would favorably impact pretreatment economics. As these enzymes become more efficient in hydrolyzing both cellulose and hemicellulose, they will allow less severe pretreatment conditions to be used which can lower the costs. The enzyme costs have been drastically reduced over the past 10 years, with commercial preparations available. However, this remains an active area for R&D that can lead to additional large cost savings.

The impact of pretreatment and neutralization chemicals on downstream elements will continue to be an area for active investigation. Chemicals and resulting salts can have significant impact on enzymatic hydrolysis and fermentation through potentially toxic or inhibitory impacts. Beyond biological processing, these chemicals and salts can also impact the design and operation of downstream operations such as distillation, residue utilization, combustion, and even wastewater treatment. Cost savings in a particular process area can be offset by cost increases in other areas for various chemical combinations. This demonstrates the importance of designing and simulating integrated processes that include recycle to understand these trade-offs in cost and performance.

More novel pretreatment concepts, such as those using ionic liquids or supercritical conditions, have not been modeled or developed to the same degree that others have. As such, there is significant need for R&D to investigate the potential of these pretreatment technologies, particularly in the context of the entire conversion process and not in isolation. Solvent recovery and cost will be key R&D targets for solvent-based pretreatments, but it remains to be seen how well these will compete with the more developed systems. Further research and development to improve hemicellulose yields is required for ionic liquid pretreatments to be economical. Extensive studies are also needed to improve ionic liquid recycling, energy integration, solid liquid separation, and feedstock drying.

15.9   Conclusions

A broad perspective on pretreatment economics within an overall conversion context has been presented in this chapter. A large array of pretreatment technology options exist and are currently being researched and developed. There are many performance and cost tradeoffs for these varying technologies, and only through continued R&D will these tradeoffs be more fully understood. Process economics, especially for promising pretreatment technologies studied in the CAFI projects, were summarized in this work. The six pretreatment technologies varied greatly in terms of their process designs and projected total capital investments. Pretreatment capital costs are based on past research, existing databases, and engineering judgments; they include not only pretreatment reactor costs, but also pretreatment chemical recovery and recycle equipment costs for the processes that use high pretreatment chemical loadings. More accurate cost estimates for each pretreatment will require specific vendor quotations based on pretreatment process conditions and pretreatment chemical recovery requirements. Overall ethanol yields, which are largely based on overall sugar yields achieved in pretreatment and subsequent enzymatic hydrolysis, are the single-most important factor in determining projected MESP for each pretreatment. Many key variables and parameters can significantly impact pretreatment economics and, while many have been discussed in this chapter, R&D will generate data to guide development of the biomass conversion industry. Sensitivity analysis is critical to help identify the most important design variables.

Acknowledgements

This research was funded under the Office of the Biomass Program of the United States Department of Energy through contract number DE-FG36-04GO14017. We would like to acknowledge the generous help from each CAFI project PI to establish the design basis for each pretreatment method.

References

1. Galbe, M. and Zacchi, G. (2007) Pretreatment of lignocellulosic materials for efficient bioethanol production. *Biofuels*, **108**, 41–65.
2. Lynd, L.R. (1996) Overview and evaluation of fuel ethanol from cellulosic biomass: Technology, economics, the environment, and policy. *Annual Review of Energy and the Environment*, **21**, 403–465.
3. Aden, A., Ruth, M. *et al.* (2002) Lignocellulosic biomass to ethanol process design and economics utilizing co-current dilute acid prehydrolysis and enzymatic hydrolysis for corn stover. Report NREL/TP-510-32438. http://www.nrel.gov/docs/fy02osti/32438.pdf.
4. Chum, H.L., Douglas, L.J., Black, S.K., and Overend, R.P. (1985) Pretreatment catalyst effects and the combined severity parameter. *Applied Biochemistry and Biotechnology*, **24** (5), 1–14.
5. Hsu, T.A. (ed.) (1996) *Handbook on Bioethanol Production and Utilization*, Taylor & Francis, Washington.
6. Burke, M.J., Saville, B. *et al.* (2008) *Treatment of Lignocellulosic Materials Ulitizing Disc Refining and Enzymatic Hydrolysis*, S. B. Inc., p. 31.
7. South, C.R., Garant, H. *et al.* (2008) *Combined Thermochemical Pretreatment and Refining of Lignocellulosic Biomass*, M. Corporation.
8. Teramoto, Y., Tanaka, N. *et al.* (2008) Pretreatment of eucalyptus wood chips for enzymatic saccharification using combined sulfuric acid-free ethanol cooking and ball milling. *Biotechnology and Bioengineering*, **99** (1), 75–85.
9. Chen, X., Tao, L., *et al.* (2012) Improved ethanol yield and reduced Minimum Ethanol Selling Price (MESP) by modifying low severity dilute acid pretreatment with deacetylation and mechanical refining: 1) Experimental. *Biotechnology for Biofuels*, **5** (1), 60.
10. Shill, K., Padmanabhan, S. *et al.* (2011) Ionic liquid pretreatment of cellulosic biomass: enzymatic hydrolysis and ionic liquid recycle. *Biotechnology and Bioengineering*, **108** (3), 511–520.

11. Kumakura, M. and Kaetsu, I. (1983) Effect of radiation pretreatment of bagasse on enzymatic and acid hydrolysis. *Biomass*, **3**, 199–208.

12. Yang, C.P., Shen, Z.Q., Yu, G.C., Wang, J.L. (2008) Effect and aftereffect of gamma radiation pretreatment on enzymatic hydrolysis of wheatstraw. *Bioresource Technology*, **99** (14), 6240–6245.

13. Keshwani, D.R., Burns, J.C. *et al.* (2007) Microwave Pretreatment of Switchgrass to Enhance Enzymatic Hydrolysis. Conference Presentations and White Papers: Biological Systems Engineering. Paper 35. University of Nebraska, Lincoln. ASABE paper #077127.

14. Nguyen, Q., Tucker, M.P., Keller, F.A., Eddy, F.P. (2000) Two-stage dilute acid pretreatment of softwoods. *Applied Biochemistry and Biotechnology*, **84** (1–9), 561–576.

15. Peters, M. and Timmerhaus, K. (1991) *Plant Design and Economics for Chemical Engineers*, McGraw-Hill, New York City.

16. AspenPlus™ (2007) *Release 7.2*, Aspen Technology Inc., Cambridge MA.

17. Wooley, R.J., Ruth, M., Sheehan, J., Ibsen, K., Majdeski, H., Galvez, A. (1999) Lignocellulosic biomass to ethanol process design and economics utilizing co-current dilute acid prehydrolysis and enzymatic hydrolysis current and futuristic scenarios. NREL TP-580-26157. http://www.nrel.gov/docs/fy99osti/26157.pdf.

18. Remsing, R.C., Swatloski, R.P. *et al.* (2006) Mechanism of cellulose dissolution in the ionic liquid 1-n-butyl-3-methylimidazolium chloride: a C-13 and Cl-35/37 NMR relaxation study on model systems. *Chemical Communications* (12), 1271–1273.

19. Li, C.L., Knierim, B. *et al.* (2010) Comparison of dilute acid and ionic liquid pretreatment of switchgrass: Biomass recalcitrance, delignification and enzymatic saccharification. *Bioresource Technology*, **101** (13), 4900–4906.

20. Fort, D.A., Remsing, R.C. *et al.* (2007) Can ionic liquids dissolve wood? Processing and analysis of lignocellulosic materials with 1-n-butyl-3-methylimidazolium chloride. *Green Chemistry*, **9** (1), 63–69.

21. Li, C.Z. and Zhao, Z.K.B. (2007) Efficient acid-catalyzed hydrolysis of cellulose in ionic liquid. *Advanced Synthesis & Catalysis*, **349**, 1847–1850.

22. Zhao, X.B., Cheng, K.K. *et al.* (2009) Organosolv pretreatment of lignocellulosic biomass for enzymatic hydrolysis. *Applied Microbiology and Biotechnology*, **82** (5), 815–827.

23. Holtzapple, M.T. and Humphrey, A.E. (1984) The effect of organosolv pretreatment on the enzymatic-hydrolysis of poplar. *Biotechnology and Bioengineering*, **26** (7), 670–676.

24. Pan, X.J., Xie, D. *et al.* (2005) Strategies to enhance the enzymatic hydrolysis of pretreated softwood with high residual lignin content. *Applied Biochemistry and Biotechnology*, **121**, 1069–1079.

25. Elander, R.T., Dale, B.E. *et al.* (2009) Summary of findings from the biomass refining consortium for applied fundamentals and innovation (CAFI): corn stover pretreatment. *Cellulose*, **16** (4), 649–659.

26. Eggeman, T. and Elander, R.T. (2005) Process and economic analysis of pretreatment technologies. *Bioresource Technology*, **96** (18), 2019–2025.

27. Wyman, C.E., Dale, B.E. *et al.* (2009) Comparative sugar recovery and fermentation data following pretreatment of poplar wood by leading technologies. *Biotechnology Progress*, **25** (2), 333–339.

28. Tao, L. and Aden, A. (2009) The economics of current and future biofuels. *In Vitro Cellular & Developmental Biology-Plant*, **45** (3), 199–217.

29. Wyman, C.E., Dale, B.E. *et al.* (2005) Coordinated development of leading biomass pretreatment technologies. *Bioresource Technology*, **96** (18), 1959–1966.

30. Kumar, R. and Wyman, C.E. (2009) Effects of cellulase and xylanase enzymes on the deconstruction of solids from pretreatment of poplar by leading technologies. *Biotechnology Progress*, **25** (2), 302–314.

31. Tao, L., Aden, A. *et al.* (2011) Process and technoeconomic analysis of leading pretreatment technologies for lignocellulosic ethanol production using switchgrass. *Bioresource Technology*, **102** (24), 11105–11114.

32. MYPP (2011) Biomass Multi-Year Program Plan Office of the Biomass Program, Energy Efficiency and Renewable Energy. US Department of Energy. http://www1.eere.energy.gov/biomass/pdfs/mypp_april_2011.pdf.

33. Chum, H., Johnson, D. *et al.* (1990) Pretreatment-catalyst effects and the combined severity parameter. *Applied Biochemistry and Biotechnology*, **24–25** (1), 1–14.

16

Progress in the Summative Analysis of Biomass Feedstocks for Biofuels Production

Foster A. Agblevor[1] and Junia Pereira[2]

[1] *Department of Biological Engineering, Utah State University, Logan, USA*
[2] *Department of Biological Systems Engineering, Virginia Polytechnic Institute and State University, Blacksburg, USA*

16.1   Introduction

The analysis of biomass is critical to the development of the emerging bio-based economy because design information, material and energy balances, yields of products, and the evaluation of economic and technical feasibility of technologies depends on accurate analytical data. The analysis of lignocellulosic biomass has been investigated extensively in the past, but methods vary depending on the feedstocks and the end application. The wide variation in chemical compositions of biomass feedstocks explains the existence of a large variety of analytical methods.

All biomass feedstocks are composed of three structural polymers; lignin, cellulose, and hemicellulose. Cellulose is composed of long chains of glucose arranged in microfibrils formed from about 24–40 cellulose chains [1–5], depending on the sample source. The microfibrils are embedded in the matrix of lignin and hemicelluloses [1]. The degree of crystallinity of cellulose also differs for various biomass feedstocks [1].

Unlike the homogeneous chemical composition of cellulose for all biomass feedstocks, the chemical composition of lignin and hemicelluloses vary significantly for different biomass feedstocks [6–8]. Lignins are formed by the enzyme-initiated dehydrogenative polymerization of three primary precursors: trans-coniferyl, trans-sinapyl and trans-p-coumaryl alcohols [1,7]. Although there are other types of structures

Aqueous Pretreatment of Plant Biomass for Biological and Chemical Conversion to Fuels and Chemicals, First Edition.
Edited by Charles E. Wyman.
© 2013 John Wiley & Sons, Ltd. Published 2013 by John Wiley & Sons, Ltd.

associated with lignin such as aromatic carboxylic acids in ester-like combination, they are minor. Lignins can be classified into three major lignin types: guaiacyl (G), syringyl (S), and p-hydroxyphenyl (H) [7]. Differences in the composition of lignins from various biomass feedstocks stem from differences in the proportions of these three types of lignins. Softwoods (also known as gymnosperms or coniferous woods) have a preponderance of G lignin, while hardwoods (also known as angiosperm or deciduous woods) have large fractions of both S and G lignins. Grasses on the other hand contain a higher fraction of H lignin in addition to the S and G lignins [7].

Hemicellulose is a heteropolymer and its components also vary according to biomass species. For instance, softwoods have predominantly galactoglucomannan, arabinoglucuronoxylan and arabinogalactan [8], whereas grasses and hardwoods have mostly glucuronoxylan- and glucomannan-type hemicelluloses. Seed coats and other agricultural residues have more complex hemicellulose components such as arabinoxylan and other combinations [6,8,9].

These three major biomass polymers are packed in a highly ordered pattern during the biosynthesis of cell walls. In addition to these structural polymers, biomass feedstocks contain other compounds such as proteins, extractives, starch, soluble sugars, waxes, uronic acids, ash, and other extraneous materials [6,8]. The content of any of these compounds in the biomass feedstock depends on the biomass species, geographic region, time and method of harvest, age of biomass, and post-harvest storage method [10].

The chemical composition of biomass also differs in various parts of the biomass such as stem, bark, branches, roots, and leafs [6–9]. Analysis of biomass is therefore very challenging, especially when a good mass closure is needed. Determination of the summative composition of the biomass without considering the morphological distribution of the main groups or individual components represents a common type of analysis.

Because of the complexity and variability of biomass feedstocks, there are many biomass analysis methods with each geared towards specific needs. Each set of analytical methods was developed to meet the needs of specific industries and to accommodate unique features of each biomass type. The emerging biomass industry uses American Society for Testing and Materials (ASTM) International standards or combinations thereof, as well as the National Renewable Energy Laboratory (NREL) Laboratory Analytical Procedures (LAPs), available from the NREL website (http://www.nrel.gov). The pulp and paper industry uses the Technical Association of Pulp and Paper Institute (TAPPI) standards for hardwood and softwood analysis, and the forage industry uses the Association of Official and Analytical Chemists International (AOAC) standards based on acid and neutral detergent methods. The TAPPI methods were developed foremost for analyzing mature woody biomass feedstocks for the pulp and paper industry and are optimized for such purposes. The forage and feed industries developed the AOAC methods that use detergent and other methods for measuring the digestibility of food and feed. These methods are therefore optimized for herbaceous feedstocks and foods. The pulp and paper and the food and feed industries methods cover a broad range of biomass feedstocks. However, in the case of the emerging biofuels industry, the feedstocks include herbaceous and woody biomass: short-rotation energy crops and agricultural residues whose analyses do not fit within the above-described classes. Methods are therefore needed that are applicable to these broad classes of feedstocks.

The standard analytical methods tend to target measuring biomass features that are most relevant to that industry. For the pulp and paper industry, cellulose is the most valuable component, whereas digestible fiber is the most valuable feature for the feed and food industry. In the case of the biofuels industry, the component targeted depends on the processing technology applied: polysaccharides are the most valuable for biochemical conversion, whereas all components are equally important for thermochemical conversion. In this chapter, we summarize methods for analysis of various types of biomass feedstocks with references provided for more comprehensive information. The discussion that follows is based on the analysis of various fractions of the biomass.

16.2   Preparation of Biomass Feedstocks for Analysis

The summative analysis of biomass feedstocks is strongly influenced by preparation before the analysis including material sampling, milling, and drying. Woody feedstocks are heterogeneous with statistically significant variations in chemical composition within a tree and between trees [6,8]. In the case of herbaceous biomass, there is variation between young and matured plants with respect to the proportion of leaves, stems, and branches [11,12]. In case of agricultural and forestry residues, in addition to the above factors, other extraneous materials such as soil and other plant materials must be accounted for in addition to the targeted plant. All these factors must be considered to ensure that the compositional data is representative of the entire feedstock under consideration. Both ASTM (ASTM E 1757) [13] and NREL LAP [14] have standardized procedures for biomass sample preparation; although TAPPI also has a standard method [15], it is limited to woody biomass feedstocks.

For the ASTM and NREL LAP methods [13,14], there are three main preparation procedures according to the nature of the feedstock referred to as methods A, B, and C. Method A is suitable for field-collected materials such as woody, herbaceous, and agricultural wastes and for pretreated wet biomass materials that are capable of becoming moldy. Woody samples must first be available as chips of a nominal $5 \times 5 \times 0.6$ cm ($2 \times 2 \times 1/4$ in) or less and twigs not exceeding 0.6 cm (1/4 in) diameter. Herbaceous materials may be processed as whole straw. It is recommended that wastepaper should be shredded into pieces less than 1 cm (1/2 in) wide [13,14]. Furthermore, it is recommended that twigs, straw, and wastepaper should not exceed 61 cm (24 in) in length to facilitate handling. Methods B and C are recommended for very moist feedstocks, samples that would not be stable during prolonged exposure to ambient conditions, or for drying materials when room conditions deviate from the ambient conditions [13,14]. These test methods are also suitable for handling small samples of biomass (<20 g).

In Method A the samples need to be milled, preferably with a knife mill, to facilitate easy filtration after hydrolysis; hammer-milled samples tend to bridge and block filters during filtration of the hydrolysates [13,16]. After milling, the samples need to be sieved to provide a relatively uniform material for analysis. The sieves need to be shaken for 15 ± 1 min, after which the sieves are removed and the fraction retained on the 20 mesh sieve (+20 mesh fraction) should be re-milled. The fraction retained on the 80 mesh sieve (−20/+80 mesh fraction) should be used for compositional analysis. To ensure uniformity of the samples, the fractions collected from various sieves are mixed thoroughly in a riffler. The samples are then stored for analysis.

For feedstocks that are very wet such as fermentation residues, pretreated biomass or sludges, Method B is recommended. The material is first washed three times with distilled water and then dried in a convection oven at 45 ± 3 °C for 36–48 hours. For small quantities (<20 g) containing material that would not pass through a 20 mesh screen, the particle size is reduced by knife-milling the entire sample with an intermediate sized knife-mill with a 1 mm mesh screen. For larger quantities (>20 g) containing material that would not pass through a 20 mesh screen, the particle size is reduced by knife-milling the entire sample and then sieved as described in Method A.

For materials such as olive oil wastewater residue that are very wet and heat sensitive and cannot be prepared by Method B, the feedstocks are lyophilized using a suitable freeze-drier under Method C [13,14]. Then, for small quantities (<20 g) material that would not pass through a 20 mesh screen, the particle size is reduced by knife-milling the entire sample using an intermediate-sized knife-mill equipped with a 1 mm mesh screen. For larger quantities (>20 g) of material that would not pass through a 20 mesh screen, the particle size is reduced by knife-milling the entire sample followed by sieving.

16.3   Determination of Non-structural Components of Biomass Feedstocks

16.3.1   Moisture Content of Biomass Feedstocks

Moisture content is an ubiquitous, variable component of any biomass sample, and reporting data on a moisture-free basis is very important to avoid confusion biased by the moisture content of the material. Moisture is not considered a structural component of biomass and can change with storage and handling of biomass samples. The determination of moisture content corrects biomass samples to an oven-dried solids mass basis that is constant for a particular sample. This method is also called total solids content to account for materials that have very large amounts of moisture, such as acid-pretreated biomass or fermentation residues.

Several moisture-determination methods are described in the literature, including oven-dry [17–19], Karl Fischer [20], and infrared [17] methods. The most frequently used methods, which have been standardized by ASTM (E-1756) [17], TAPPI (T412 om-06) [18], and NREL LAP [19], are the convective oven drying and the loss-on-drying infrared methods. The infrared method is particularly unsuitable for materials that contain large amounts of free sugars or proteins that caramelize or turn brown under direct infrared heating elements. In such cases, convective oven drying is recommended; the moisture content of a biomass sample is the amount of mass lost during drying of the sample at 105 °C to constant mass. An inherent error of these methods and any oven-drying procedure is the loss of volatile substances other than water during drying as a result of extractives migrating to the surface of the biomass with unknown amounts of volatile organic compounds evaporating [21–24]. Larnøye [24] reported that the loss of resin acids during oven drying of pine wood ranged from 16 to 67 wt%, depending on the type of acid.

The Karl Fischer method [20,25,26] dissolves the water in ground wood in methanol during a 20-min extraction process. The moisture content is then determined by Karl–Fischer titration of the methanol solution. This method does not suffer from the loss of volatile constituents of the biomass feedstock.

Nuclear magnetic resonance (NMR) spectroscopy [27] has also been used for absolute moisture determination without loss of volatile compounds, but this method is tedious and costly and therefore not recommended for standard routine laboratory analysis.

16.3.2   Determination of Ash in Biomass

All biomass feedstocks contain intrinsic inorganic materials, especially potassium, calcium, and phosphorous, that the plant uses for growth. Additionally, extraneous materials such as soil and silica collected with the biomass during harvesting also contribute to the ash. These inorganic materials, both intrinsic and extraneous, are called ash and are not structural biopolymers. The ash content is an approximate measure of the mineral content and other inorganic matter in biomass and is used in conjunction with other assays to determine the total composition of biomass samples. Some of the ash components such as potassium compounds can be easily extracted with suitable solvents. The ash content is determined by dividing the mass of residue remaining after dry oxidation (oxidation at 575 ± 25 °C) by the original mass of biomass. All results are reported relative to the 105 °C oven-dried mass of the sample. Ash determination is described in all major standardized procedures such as the ASTM International (ASTM E 1755) [28], NREL LAP [29], and TAPPI (T211 om-07) [30].

16.3.3   Protein Content of Biomass

Many types of biomass used as feedstocks for conversion to fuels and chemicals contain protein and other nitrogen-containing materials which must be measured as part of a comprehensive biomass analysis [31–33]. Protein in biomass is not always measured directly. In many cases the nitrogen content of a biomass sample is measured and then applied to estimate the protein content using an appropriate nitrogen factor (NF) [31–33]. Many standard methods recommend applying an NF of 6.25 for all types of biomass except wheat grains, for which an NF of 5.70 is recommended [31–33]. For woody biomass feedstocks, the amount of nitrogenous compounds and proteins are so low that, for all practical purposes, they can be considered negligible [6,7]. Currently, only the NREL LAP method has a specific protocol for protein determination in biomass feedstocks [34]. The AOAC method (AOAC 984) [31] and NREL LAP methods for measuring proteins use similar conversion factors as those reported by Mossé [32] and Mossé *et al.* [33].

16.3.4   Extractives Content of Biomass

The extractives are a broad class of non-structural compounds found in all types of biomass and include proteins, waxes, fats, inorganic salts, phytosterols, resins, and non-volatile hydrocarbons, which play various physiological roles in the biomass feedstocks [6,8]. The composition of extractives vary with biomass species (woody, herbaceous, agricultural residues), among different morphological structures of the biomass, and with age of the biomass before harvest [6,8,11,12]. Extractive compounds generally tend to interfere with determination of biomass structural components and also influence biomass processing to higher-value products. It is therefore essential to remove these compounds using polar and non-polar solvents before analyzing the polymer constituents of biomass.

In the USA pulp and paper industry, the TAPPI test method [35] has been developed for hardwoods, softwoods, and pulp. The method involves sequential extraction of the non-polar compounds with benzene or toluene, followed by benzene/ethanol extraction, and finally the extraction of the polar compounds with water [35]. The solvents are vacuum-evaporated and the yield calculated. The extractions are usually carried out in a Soxhlet extractor with up to 8 hours or more needed to complete extraction. For health reasons, acetone has been recommended to replace benzene for these extractions and other methods use dichloromethane for extraction. However, the benzene methods produce the highest yield of extractives [35].

In the animal feed and food industries, the neutral detergent fiber (NDF) method [36] is an indirect measurement of the extractives content of herbaceous biomass. The goal of this method is to determine the fiber content of the feed (cellulose, hemicelluloses, and lignin) by extracting the non-structural components (sugars, starch, and other components) that are soluble in neutral detergent (sodium lauryl sulfate, EDTA, pH 7). The residue after this extraction is termed the neutral detergent fiber [36]. The NDF soluble fraction can therefore be considered as the extractives fraction of the biomass.

In the emerging biofuels industry, ASTM [37] and NREL LAP [38] methods are commonly employed. Both of them use smaller amounts of biomass for analysis and the extraction solvent (95% ethanol) is different from those used in the TAPPI procedure. The methods are considered safer and can remove most of the extractives from biomass. These procedures also use the Soxhlet extractor for determination of the extractives.

For the pulp and paper and feed industries, the extractives determination methods do not have any negative influence on the mass balance with respect to the product (fiber in the feed or wood pulp) yields [39,40], and data are often expressed in terms of an extractives-free biomass basis. However, in the case of the biofuels industry, the extractives determination method may have a negative impact on the mass balance with respective to product yield (e.g., ethanol) in some feedstocks. For example, feedstocks such as corn stover, switchgrass, and fescues have significant quantities of fermentable sugars such as fructose, glucose, sucrose, and their oligomers in the extractives in addition to other compounds [39–41]. Yields of products based on extractives-free biomass will therefore underestimate the true yields of products from such feedstocks. Studies by Chen *et al.* [39] showed that the water-soluble extract from corn stover contains as much as 4–12 wt% sugar monomers and 4–12 wt% oligomeric sugars. Thammasouk *et al.* [40] also reported that yields of Klason lignin, glucose, and other sugars for switchgrass, fescue, and corn stover on an extractives-free basis were much lower than expected, probably because some of the sugars were extracted by the ethanol and water extractions and about 40% of the total solids extracted measured as Klason lignin. Prudence is therefore required when reporting yields of various products on extractives-free basis. The recommended approach for herbaceous biomass is to report on both extractives-free and whole-biomass bases. This problem is not important with woody biomass feedstocks however, because their extractives contain negligible amounts of sugars [6,8].

16.4   Quantitative Determination of Lignin Content of Biomass

Lignin, one of the major structural components of all lignocellulosic biomass, is an amorphous polymer of phenylpropane units and includes industrial as well as native lignins. Industrial lignins such as those isolated from pulp and paper or biofuels processes have different compositions from native lignins because the separation process alters the structure of native lignin [6–8]. However, the discussion here is confined to native lignins found in biomass feedstocks. There are three types of native lignins: syringyl (S), guaiacyl (G), and p-hydroxyphenyl (H). All three types of lignins occur in all lignocellulosic biomass [6,7], but the proportions of each unit vary according to the biomass source. Although lignin appears very attractive as a potential source of aromatic compounds, few applications have been found for this polymer because of its complexity [7]. Consequently, lignin is mostly used as boiler fuel in the pulp and paper industry and could find similar application in the bioethanol industry. In the animal feed industry, lignin is not digestible by animals and is therefore considered to be more of a liability than an asset. However, in thermochemical processing, lignin will generally be valued as an energy-dense component of the feedstock [42].

Because lignin acts like glue cementing cellulose and hemicelluloses together to form the strong lignocellulosic structures, it also limits access to the polysaccharides [1,3,6]. Further, because there is no direct method of determining cellulose and hemicelluloses content in biomass feedstocks, lignin must be removed to gain access to the cellulose and hemicelluloses in biomass. It is also important to know the lignin content of the biomass in order to achieve a good mass closure.

Currently, all the known primary analytical methods for the determination of lignins rely on the degradation of the lignin component or hydrolysis of the polysaccharides. Rapid analytical methods that do not require degradation of lignin or hydrolysis of polysaccharides also need primary standards for calibration [10,43,44]; degradation methods are therefore critical for biomass analysis. Similar to the extractives determination methods discussed in the previous section, there are several methods for determining the lignin content of lignocellulosic biomass. Standard methods are industry driven; the pulp and paper industry has therefore devised TAPPI procedures, the feed industry employs AOAC analytical methods, and the biofuels industry relies on ASTM and NREL LAP methods.

The oldest and the most common method for quantitative determination of lignin in woods and pulps is the Klason analysis [45]. This method has been extensively modified for various applications because of differences in the properties of biomass materials. In the Klason lignin method [45], the extractives-free wood is first reacted with 72 wt% sulfuric acid at room temperature followed by heating with dilute acid to hydrolyze the polysaccharides to soluble fragments (mostly monosaccharides). The solid residue is then washed several times with distilled water and dried to a constant weight. This method is excellent for softwoods, but has some serious shortcomings when applied to hardwoods and herbaceous biomass [46–49]. For hardwoods, a fraction of the lignin is soluble in acid and the method therefore underestimates the native lignin content of the feedstock. The soluble lignin fraction is therefore determined using ultraviolet (UV) spectrophotometry in the wavelength range 200–205 nm. However, furfural and hydroxymethylfurfural also absorb UV in the same range and interfere with measurements [48,50]. One major challenge for acid-soluble lignin determination is the variation of absorptivity of soluble lignin from different hardwood feedstocks [7]. Softwoods, however, do not show major variation in the UV absorption of the soluble compounds and lignin solubility in the acid is low [7].

The Klason lignin method is also unsuitable for herbaceous biomass and agricultural residues because of interference from proteins and ash in biomass. For herbaceous biomass, some of the protein tends to condense on the lignin, resulting in overestimation of the lignin content [46]. Some studies have however shown that the condensation of protein on lignin is non-linear, in that the protein condensation levels off at some concentration [46]; it therefore may be possible to estimate how much protein is incorporated into lignin. Protein content determinations are however questionable because it is generally determined by multiplying the nitrogen content by a factor of 6.25. This result tends to overestimate the protein content and underestimate biomass lignin content as other sources of nitrogen such as nucleic acid are also associated with lignin [46].

Lignin also condenses on the silica in agricultural residues, thus increasing the apparent lignin content of such feedstock [49,50]. Corrections have to be made through other methods such as ashing the residue after the Klason lignin determination and using the difference as the lignin content.

The shortcomings associated with the Klason lignin method discussed in this section have necessitated development of other methods to meet the needs of industrial groups such as the animal feed industry, which uses herbaceous biomass. The acid detergent method [36] has become the most important approach in the feed industry for determining fiber digestibility in herbaceous biomass. This method is based on the concept that plant cells can be divided into less digestible cell walls (cellulose, lignin, and hemicelluloses) and more digestible cell materials (starch and sugars). An acid detergent (cetyl trimethyl ammonium bromide in $1\,M\,H_2SO_4$) is used as the first step in the determination of the lignin content of biomass because the acid detergent fiber (ADF) is a combination of cellulose and lignin.

To avoid interference from extractives, a sequential method such as that described by Van Soest and Roberston [47] is most appropriate. In this method, about $100\,mL$ of neutral detergent solution is added to about $0.5\,g$ of dried ground herbaceous biomass at room temperature and refluxed for $60\,min$ from the onset of boiling. After refluxing, the mixture is filtered through a coarse porosity crucible, washed three times with boiling water, washed twice with cold acetone and then dried overnight at $100\,°C$. The dry material is the NDF which can be considered as extractives-free biomass. After weighing the NDF, about $200\,mL$ acid detergent solution is added to it and the resulting mixture is refluxed for $60\,min$. Following filtration, the residue is washed with boiling water followed by acetone and then dried at $105\,°C$ for several hours to a constant weight. The resulting solids are designated as ADF. To determine the lignin content of biomass, the ADF is treated with $12\,M\,H_2SO_4$ at room temperature and allowed to drain through the crucible. Additional concentrated acid is added to the crucible over a period of 3 hours at room temperature to complete the hydrolysis. The acid-insoluble residue is then washed with hot water and acetone and then dried overnight at $100\,°C$ followed by cooling in a dessicator. The residue is weighed and then ashed at $550\,°C$ for 2 hours. The difference between the ash and the residue is considered the lignin content of the biomass.

The major difference between this method and the Klason lignin determination is the sequence of acid addition. In the ADF method, biomass is first treated with dilute sulfuric acid and then with concentrated acid. In contrast, for the Klason lignin determination, biomass is first treated with concentrated sulfuric acid and in the second step the oligosaccharides are hydrolyzed with dilute acid.

Comparative studies of these two methods [46] have shown significant differences between the lignin values. The authors concluded that protein condensation on Klason lignin was limited; however, the ADF solution dissolved some of the lignin fraction and therefore the lignin determined by this method underestimated the true lignin content of herbaceous biomass. The Klason lignin determination with correction for the limited protein condensation was considered to be more representative of the true lignin content in herbaceous biomass. Additionally, the ADF method does not address the acid-soluble lignin question.

In the emerging biofuels industry, both herbaceous and woody biomass feedstocks are processed to biofuels. As discussed in Section 16.4 above, a unified method is needed to determine the lignin content of biomass feedstocks. In addition to the raw biomass feedstocks (herbaceous, agricultural residues, and woody), the lignin content must also be measured for pretreated biomass produced in biological processing. To address the shortcomings noted above and the large sample sizes used in the analysis, ASTM method E 1721 [48] was developed in the early 1990s and later modified as the NREL LAP method [50]. The ASTM method [48] was developed through a "round robin" consensus [49] using modifications of the method of Theander [51] and Theander and Westlund [52]. This method is similar to the Klason lignin approach described above with a couple of exceptions: the mass of biomass used is 0.3 g, and the ash content is usually determined at 575 °C for subtraction from the acid-insoluble residue. These methods also measure the acid-soluble lignin content of biomass through UV absorption at a wavelength of either 240 nm or 320 nm. For samples high in protein, correction is made by determining the protein content independently and then subtracting it from the total insoluble lignin. The lignin content determined in this way therefore addresses all the shortcomings of the other methods and is also applicable to the large diversity of biomass feedstocks that could potentially be used for the production of biofuels.

16.5 Quantitative Analysis of Sugars in Lignocellulosic Biomass

The sugars in biomass feedstocks originate from two sources: polysaccharides from plant cell walls and sugars from the non-structural components (extractives) in cell walls. The contribution of sugars from non-structural components in most biomass is minimal except for sweet sorghum, sugar cane, corn stover, and other herbaceous biomass [41]. Most sugar analysis methods developed are therefore focused on plant cell-wall polysaccharides (cellulose and hemicelluloses). The analysis of biomass sugars is important to several industries including pulp and paper, animal feed, biofuels, and forestry; for the latter, cellulose degradation products generated by fires are monitored using sugar analysis methods [53,54]. Because of this strong interest, several wet chemical analyses as well as rapid instrumental methods have been developed [11,43,44,49]. This chapter is focused on the wet chemical analysis methods.

Holocellulose (hemicellulose + cellulose) and cellulose can be determined by polysaccharide hydrolysis with most modern chemical analytical methods based on chemical degradation of the polysaccharides followed by either derivatization and gas chromatography (GC) analysis or high-performance liquid chromatography (HPLC) analysis. There are also colorimetric methods such as the Miller method [55] and Nelson–Somogyi method [56] that quantitatively measure the total reducing sugars released. Browning [57] described in detail the separation of sugars using a paper chromatographic method. In all the methods, the goal is to determine the total cellulose and hemicellulose content of the biomass as well as other minor constituents such as uronic acids and acetyl compounds.

16.5.1 Holocellulose Content of Plant Cell Walls

The holocellulose method [16,57–59] is mostly used by the pulp and paper industry; there has not been much interest from the biomass conversion community. Since holocellulose is the sum of cellulose and hemicellulose, it should correlate well with either ethanol conversion or total sugar content of the biomass feedstock in cases where there are minimal amounts of sugars in the non-structural component of the biomass. A shortcoming of this method is the loss of some carbohydrates and retention of some lignin in the product. It may be worthwhile for the biomass conversion community to evaluate this method of biomass analysis.

16.5.2   Monoethanolamine Method for Cellulose Determination

Nelson and Leming [60] reported a monoethanolamine method for determining the total cellulose in lignocellulosic biomass. In this approach, lignocellulosic biomass (3 g) is refluxed with 100 mL of mono-ethanolamine at 170 °C for 3 hours. Water is then added to the mixture after it is cooled down to room temperature. The mixture is then decanted and filtered through a sintered glass crucible and the solid residue is washed with hot water (60 °C). The cellulose is then bleached with 10 mL H_2SO_4 (10% v/v solution) and 10 mL NaOCl (24 g/L solution) for 5 min at room temperature. The mixture is filtered and several sequential bleaching steps followed until all the lignin is removed (the appearance of rose color indicates the presence of lignin). The cellulose is filtered and then washed with boiling water and acetic acid. The final product is dried at 105 °C to a constant mass and the cellulose content is calculated based on the initial mass of lignocelluloses. Although this method has not been utilized by the emerging biofuels community, it merits consideration as a way to estimate the amount of cellulose in biomass that could be converted to biofuels.

16.6   Chemical Hydrolysis of Biomass Polysaccharides

The polysaccharide content of biomass can be determined as holocellulose or through the breakdown of polysaccharides into monosaccharides, followed by analysis of these sugars that can in turn be used to calculate the polymeric sugar content. The calculations normally produce reasonably accurate indications of cellulose and hemicellulose in the biomass feedstocks. The most common monosaccharides obtained from degradation of the cell wall are glucose, xylose, mannose, galactose, and arabinose. In the case of corn stover and other herbaceous biomass feedstocks, fructose and sucrose are also present but most of these originate from the extractives fraction of the biomass. Plants belonging mainly to the *Compositae* and *Graminae* families store polymers of fructose such as inulin in significant quantities as a reserve storage material instead of starch [61,62].

The glucosidic bonds in the biomass polymers are cleaved by either acid hydrolysis [49,63] or methanolysis [64,65]. Extractives-free lignocellulosic materials are hydrolyzed with strong mineral acids or trifluoroacetic acid (TFA) which degrades the crystalline structure of cellulose. In contract, methanolysis is faster than hydrolysis, but has limited quantitative applicability because it is only effective on non-crystalline cellulose.

16.6.1   Mineral Acid Hydrolysis

The most common mineral acid used in the hydrolysis of polysaccharides is sulfuric acid because of the ease of acid removal as either calcium or barium sulfate after the reaction. The complete hydrolysis of the cell-wall polysaccharides is accomplished in two steps. Primary hydrolysis involves treatment of the extractives-free sample with 72 wt% sulfuric acid for 1 h at 30 °C. A secondary hydrolysis dilutes the sulfuric acid to 4 wt% followed by refluxing for 4 h or heating at 121 °C (15 psi) for 1 h. The TAPPI method [45] is similar to the above, but the ASTM E1721 [48] and NREL LAP methods [50] use slightly modified versions of the above approach. In both the ASTM and LAP methods, 300 mg of extractives-free biomass is used for the analysis instead of the 1 g used in the TAPPI method. Other researchers also have developed microtechniques for the hydrolysis procedure [66–68].

For herbaceous biomass in the feed industry, the hydrolysis approach is quite different. In the primary hydrolysis step, biomass is treated with boiling acid detergent solution (cetyl trimethylammonium bromide in 1 M H_2SO_4) for 1 h instead of the concentrated acid hydrolysis at the low temperatures used in the TAPPI, ASTM, and LAP procedures. In the secondary hydrolysis step, concentrated sulfuric acid (12 M H_2SO_4) is used at room temperature. The acid detergent fiber is treated with $12\,M\,H_2SO_4$ through periodic additions and draining of acid over a 3 h period [52,69].

In case of herbaceous feedstocks that contain significant amounts of starch, Theander and Westlund [52] recommended first treating the sample with alpha amylase to remove the starch. The starch-free residue can then be hydrolyzed with the concentrated sulfuric acid as described in the ASTM and NREL LAP methods. The sulfuric acid concentration for hydrolysis varies from 72% to 64%, and the treatment time and temperatures also vary. Most of these methods trace their origins to Seaman *et al.* [63] and Moore and Johnson [70], and various researchers have since improved these techniques [49–52,66,67,71].

In comparative studies of microcrystalline cellulose using 64 wt% sulfuric acid for the hydrolysis [69], 25 wt% acid insoluble materials (apparent lignin) was recovered after hydrolysis, suggesting incomplete hydrolysis of the microcrystalline cellulose. When the hydrolysis time was increased from 2 h to 19 h, glucose released increased from 72.73% to 86.23%. When similar studies were conducted on newspaper, glucose did not change suggesting complete hydrolysis of the substrate. These results imply that microcrystalline cellulose cannot be used as a reference substrate for the biomass hydrolysis studies.

16.6.2   Trifluoroacetic Acid (TFA)

TFA has been proposed to improve hydrolysis of plant cell-wall polysaccharides [72,73] with major advantages including minimal loss of monosaccharides, short hydrolysis time (1 h), and removal of acid by evaporation. TFA can be used in methods with variations that depend on the cellulose content of the feedstock. TFA procedures can be classified as low-lignin and high-lignin methods, based on the hydrolysis severity [72]. In a variation for low-lignin feedstocks [72], the extractives-free sample is first steeped in 80% TFA at room temperature for 15 min to swell the material. The swollen material is then successively refluxed at 15 min intervals until followed by a final hydrolysis step in which TFA is diluted with water to a 30% concentration and refluxed for 30 min to complete the hydrolysis. The hydrolysate is then evaporated under vacuum and the residue is suspended in 5 mL water and analyzed for sugar content. For water and alkali-soluble polysaccharides, the samples are refluxed with 2 M TFA for 1 h.

In comparative studies using the low-lignin TFA method, only 32.9% of the microcrystalline cellulose was hydrolyzed to glucose [69]. Similarly, when the high-lignin TFA method (100% TFA) was applied to microcrystalline cellulose by the same authors, the glucose yield was still very low. The authors also detected cellobiose and polysaccharides in the hydrolysates, suggesting that incomplete hydrolysis of microcrystalline cellulose and correction factors were applied to account for the incomplete hydrolysis [72]. Furthermore, hemicellulose monomers produced in this method were also very low. Studies with paper and pulps [74] also showed limited hydrolysis. Other studies showed that TFA did not hydrolyze cellulose from oat straw even after 8 days in solution at $37\,^\circ C$ [75]. The TFA method is therefore most suitable for non-crystalline materials; for crystalline materials however, repeated hydrolysis steps are needed to ensure complete hydrolysis.

16.6.3   Methanolysis

The methanolysis method offers an alternative to acid hydrolysis, presenting the advantage of greater stability of released methyl glycosides [64] and simultaneous analysis of acidic and neutral sugars by capillary GC [76] or HPLC [65,77,78].

The cleavage of 4-O-methyl D-glucuronic acid units in hemicellulose hydrolysates using methanolysis is more difficult because glucosic bonds between the uronic acid and the monossachiride units are more resistant than those between neutral monosaccharides. Severe acid conditions are therefore needed for cleavage of these bonds, with resulting degradation of some of the uronic acid [77]. In addition, polysaccharide bonds in highly crystalline materials such as wood pulp are not completely cleaved unless samples are heated in 2 M hydrochloric acid in anhydrous methanol for 3 h at 100 °C. The difficulty in converting crystalline polymers to methyl glycosides results in underestimating the total sugar content in biomass; coupling of enzyme hydrolysis with partial acid hydrolysis has been suggested to facilitate the complete determination of the total sugars [79].

It is clear from the techniques described in this chapter (sulfuric acid, TFA, ADF, methanolysis, monoethanolamine) that no single method can be universally applied to hydrolyze biomass. Some of this difficulty stems from the heterogeneity of the substrate. Methods are therefore needed to optimize the release of monomeric sugars that can be readily measured by HPLC. The need is particularly critical because other modern instrumental methods rely heavily on standards derived from cleavage of the glucosidic bonds for calibration.

16.7   Analysis of Monosaccharides

Several methods are used in the quantification of monosaccharides released from biomass including colorimetry, GC, HPLC, gas chromatography/mass spectrometry (GC/MS), capillary electrophoresis (CE), anion exchange chromatography (AEC), high-pH anion-exchange chromatography with pulsed amperometric detection (HPAE/PAD), and proton NMR spectroscopy (^{1}H-NMR). The most important and frequently used methods are outlined in the following sections, along with references to facilitate access to more details on each.

16.7.1   Colorimetric Analysis of Biomass Monosaccharides

Although colorimetric methods such as dinitrosalicylic acid (DNS) described by Miller [55] and the Nelson–Somogyi reducing sugar assay [56] are rapid, they cannot distinguish between the individual monosaccharides; this makes it difficult to ascertain the contents of cellulose and hemicelluloses in biomass. Furthermore, these methods do not detect non-reducing sugars such as fructose; sugar concentrations will therefore be underestimated for herbaceous feedstocks that contain significant amounts of these sugars. As a result, colorimetric methods are primarily useful for monitoring the overall rate of reaction of sugars in a mixture, such as the rate of sugar release in enzymatic hydrolysis of pure substrates such as microcrystalline cellulose [39].

16.7.2   Gas Chromatographic Sugar Analysis

Modern methods of sugar analysis use GC extensively [50,51,77,80–82]; however, because neutral sugars are not volatile, they are derivatized before GC analysis. Derivatization reduces interactions with the stationary phase [83] and improves the separation and quantification of the sugars. The three common derivatization methods are: trimethylsilylation (TMS) [84,85]; methanolysis [64,76]; and reduction of monosaccharides to alditols followed by acetylation to alditol acetates [49–52]. All three methods usually produce accurate results [84,85] but each has its own shortcomings [86]. Several detectors coupled with gas chromatographic methods have been used to detect and quantify the sugars, including flame ionization detection (FID) and mass selective detector (GC/MS) [84]. The GC/MS method is particularly useful for detection and quantification of trace sugars and has been used successfully to analyze sugars in environmental samples [84].

Silylation of Monosaccharides

The TMS method takes only 5 min, making it the fastest derivatization procedure for monosaccharides. Several TMS agents are available either singly or in combinations (see Table 16.1) and internal standards

Table 16.1 *Silylation reagents for monosaccharide derivatization.*

Reagent	Abbreviation	Structure
Trimethylchlorosilane	TMCS	$(CH_3)_3SiCl$
Hexamethyldisilazane	HMDS	$(CH_3)_3SiNHSi(CH_3)_3$
N,O-Bis(trimethylsilyl)-trifluoroacetamide	BSFTA	$CF_3[OSi(CH_3)_3]=N[Si(CH_3)_3]$
N,O-Bis(trimethylsilyl)-acetamide	BSA	$CH_3C[OSi(CH_3)_3]=N[Si(CH_3)_3]$
Trimethylsilyimidazole	TMSI	$Si(CH_3)_3(CH_3N_2)$
N-methyl-N-(trimethylsilyl)trifluoroacetamide	MSTFA	$CF_3(CO)N(CH_3)[Si(CH_3)_3]$
N-methyl-N-(tert-butyl-dimethylsilyl)trifluoroaceetamide	MTBSFTA	$CF_3(CO)N(CH_3)[Si(C_4H_9)(CH_3)_2]$
N-Trimethylsilyl-N-diethylamine	TMSDEA	$Si(CH_3)_3N(C_2H_5)_2$

such as xylitol or mannitol are added to the sugar mixtures during the derivatization [84,85]. Because a separate derivative is formed for each anomeric form of the sugar, complete resolution of all peaks in the mixture is sometimes difficult however, especially if a packed column is used for GC analysis [85]; capillary columns however improve the separation of the sugar anomers. FIDs are commonly used for detection and quantification of the sugars. GC/MS has also been successfully applied to the identification and quantification of sugar derivatives [84].

Methanolysis of Monosaccharides

The methanolysis procedure was described in detail in Section 16.6.3 and will not be repeated here. That procedure can prepare methyl glucosides which can then be analyzed by either GC or HPLC.

Alditol Acetates of Monosaccharides

In this method, monosaccharides are reduced to alditols followed by acetylation to alditol acetates and then chromatographed. This approach has been extensively used to analyze biomass hydrolysates [50–52,84,86,87]. In one method, 100 μL of 12 M NH$_4$OH was added to 1.0 mL of hydrolysate resulting from sulfuric acid hydrolysis of biomass in a 10 mL test tube and thoroughly mixed. The monosaccharides were reduced by the addition of 100 μL of freshly prepared 3 M aqueous ammonium hydroxide containing 15 mg of potassium borohydride and left for 1 h in a water bath at 40 °C. After the reaction, 100 μL glacial acetic acid is added and thoroughly mixed. Solutions of about 0.5 mL are then transferred into screw-cap test tubes and 0.5 mL 1-methylimidazole is added to each followed by 5 mL of acetic anhydride. Next, 1.0 mL of absolute alcohol was added and after 10 min the tubes were placed in a water bath at room temperature. After that 5 mL of water was added, followed by 5 mL of 7.5 M KOH after 3 min. The samples are thoroughly mixed, and the tubes are capped and hand-shaken for 10 min. The upper layer of ethyl acetate solution of the alditol acetates is transferred into sampling vials for GC analysis. Reaction with potassium borohydride eliminates the anomeric center by converting them into alditol and thus simplifying the chromatogram (because each alditol produces one chromatographic peak). GC analysis is carried out using capillary columns and flame ionization detector identification and quantification of the sugars. Loss factors are also calculated by analyzing losses of standard sugars run in parallel with the hydrolysates. Although this procedure can be very accurate and give excellent resolution of the sugar peaks as alditol acetate, it is very time consuming and prone to errors; highly skilled technicians are therefore required to perform this procedure.

16.8   Gas Chromatography-Mass Spectrometry (GC/MS)

The GC/MS method for biomass analysis requires derivatization of sugars as described in Sections 16.7.2. The method is particularly useful for environmental samples that contain very complex mixtures of polar and non-polar compounds including sugars; since sample concentrations are usually very low, sensitive methods are required for the analysis [84]. The GC/MS methodology is more conclusive than the GC method which relies on relative retention times of monosaccharides. Medeiros and Simoneit [84] applied this method to analyze 50 sugar standards and sugars in environmental samples by derivatization with N,O-bis-(trimethylsilyl) trifluoroacetamide (BASTFA) containing 1% trimethylcholorosilane (TMCS) and pyridine. Sugar solutions as low as $120-200 \, \mu g \, mL^{-1}$, monosaccharides, anhydrosugars, sugar alcohols, and disaccharides were successfully analyzed. The only drawback to this method is that the monosaccharides of pentoses and hexoses had two GC peaks due to the 1ά- and 1β-configurations of the OH group on the pyrano-ring or furano-ring. However, sugar alcohols and anhydrosugars did not have isomers in the chromatograms. Fragmentation patterns and retention times are important to identify and quantify the sugars because of the similarities in the fragmentation patterns. Because most significant biomass sugars usually have higher concentrations, the problems associated with environmental samples may not be encountered in biofuels sugars analysis.

16.9   High-performance Liquid Chromatographic Sugar Analysis

Another important method for biomass sugars analysis is high performance liquid chromatography (HPLC). Unlike the gas chromatographic method, the HPLC method does not require derivatization of the hydrolyzed sugars. However, because chemical and physical properties vary only slightly within the different classes of sugars, HPLC methods depend on differences in conformation, configuration, and column type for separation of sugars; as a result, no single HPLC column or method is capable of separating all sugars. The major challenges of separating sugars by HPLC is finding suitable columns that provide baseline resolution and cost-effective detectors to detect low sugar concentrations similar to or equivalent to the GC method. The two commonly employed methods are partition chromatography on cation exchange resins with refractive index detection [88,89] and high-performance anion-exchange chromatography with pulsed amperometric detection. The latter is also called high-pH anion-exchange chromatography with pulsed amperometric detection (HPAE/PAD) [90–95].

   Bonn and Bobleter [88] evaluated various HPLC column materials for the analysis of monomeric and oligomeric sugars in biomass hydrolysates and fermentation solutions. Four stationary phases were investigated to determine their ability to separate sugars and analysis times. An amino-bonded material (Nucloeosil® 5-NH$_2$) was shown to be effective for monomeric and oligomeric sugars and a Ca-loaded ion-exchanged material with relatively large particle diameter (20 μm Aminex® HPX-42A, Bio-Rad) had good resolution for gluco-oligomers and separated hydroxymethylfurfural in 40.5 min. Ca-loaded ion exchange with smaller particles (8 μm u-spherogel) was also suitable for analysis of sugars. A strong acid-exchanged material (Aminex® HPX-87H, Bio-Rad) was not suitable for the oligomeric carbohydrates, but was applicable to monomeric sugars, ethanol, and furfural. Subsequent studies have applied variations of this method including use of Pb-loaded ion-exchange columns (BP-100 Benson polymeric carbohydrate column, Supelcogel™ Pb) which provide the highest resolution and the best selectivity for monosaccharides, including excellent separation of xylose, galactose, and mannose [96]. For woody biomass, which is composed of mostly glucose, xylose, arabinose, galactose, and mannose, these columns are capable of separating the sugars. However, for herbaceous biomass, such as corn stover and sugar cane bagasse, fructose and some of the minor sugars tend to co-elute and cause disagreement between conversion and predicted results from sugar analysis [97]. These ion-exchange columns also require neutralization and sulfate removal using barium hydroxide, and filtration and concentration of the biomass hydrolysates to improve detection and quantification of sugars [90].

The NREL LAP [50] carbohydrate and lignin method and the ASTM E1758 [97] standard methods use similar columns for HPLC sugar determination. In addition to the challenge of finding suitable columns for sugar separation, sugar detection presents another major problem. The most common and least expensive detector for sugar analysis is refractive index (RI). However, although RI detectors can identify most sugars and are relatively inexpensive, their detection limits are several-fold higher than that of gas chromatographic detectors [49]. Thus, in "round robin" tests on biomass analysis [49], it was concluded that for low sugar concentrations of mannose, galactose, and arabinose, the gas chromatographic method was superior.

An alternative liquid chromatographic method is HPAE/PAD. This method also has lengthy chromatographic run times as well as lengthy times for column conditioning and re-equilibration [91,92,98]. Other limitations include difficulty in resolving minor sugars such as arabinose, galactose, mannose, and xylose [90,99,100]. The tailing-off of some of the sugars such as mannose reduces the accuracy of quantification of these sugars. Modified versions of this method, including solid-phase extraction (SPE) and a reverse-gradient method with acetate loading during column conditioning, have been reported to reduce the overall process to only a few minutes and provide an improved baseline resolution of the sugars [90]. Further, the improved method does not require extensive preparation to remove the sulfates before injection into the column [90].

Other researchers have developed gradient methods that provide excellent resolution of neutral and acidic sugars and reduced analysis time of citrus hydrolysates using HPAE/PAD [93,94]. However, the PAD detector is more expensive than the RI detector. The PAD detector is also very sensitive; injection

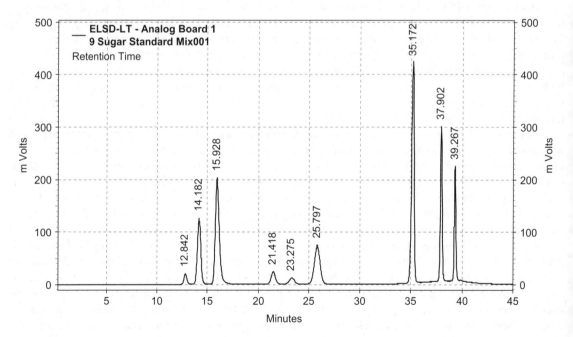

Figure 16.1   *Chromatogram of standard sugar mixture on Prevail^{TM} carbohydrate column showing their retention times (RTs) in minutes. RT12.84, arabinose; RT14.12, xylose; RT15.92, fructose; RT21.41, mannose; RT23.27, galactose; RT25.79, glucose; RT35.17, inositol(internal standard); RT 37.9, sucrose; RT 39.26, cellobiose [100].*

amounts are limited by the linear range of the detector, and 50–100-fold dilutions of the hydrolysates are required for analysis.

To address the major problems associated with sugar resolution, detection, and long chromatography times as well as poor quantification, Agblevor *et al.* [95] developed a modified method based on partition chromatography using a Prevail™ carbohydrate column and evaporative light scattering detector (ELSD). This method not only reduced the analysis times, but also showed improved baseline resolution of glucose, fructose, xylose, arabinose, mannose, and galactose as well as dissaccharides such as cellobiose and sucrose. The method was also very effective for analysis of biomass hydrolysates [101]. A typical chromatogram from the application of this method is shown in Figure 16.1. The Prevail™ carbohydrate method required minimal preparation of biomass hydrolysates and short analysis times, but required a gradient method to maximize baseline peak resolutions. Additionally, the method cannot use the common RI detector because of gradient elution.

16.10   NMR Analysis of Biomass Sugars

Both proton (^{1}H-NMR) and carbon-13 (^{13}C-NMR) NMR have been routinely applied to qualitative analysis of biomass hydrolysates and other biomass components [102]. ^{1}H-NMR is a relatively rapid method compared to ^{13}C-NMR, and has been used extensively to analyze various compounds [103–106]. This method was recently applied to quantitative analysis of biomass hydrolysates [107,108]. Although the results of the test showed significant variations in the compositional estimates by various laboratories [107], the method holds some promise as another analytical tool for biomass hydrolysates in that it requires minimal preparation time before analysis. The major drawback is that NMR spectroscopic equipment is very specialized and expensive, and not readily available to average biomass analytical laboratories.

16.11   Conclusions

Although extensive research had been applied to analysis of biomass sugars, no universally simple method can be readily applied to unambiguously analyze biomass feedstocks. Analytical methods are simpler for woody biomass because of the limited number of sugars present. However, although significant progress has been made in the field for herbaceous feedstocks, comprehensive analysis that can support excellent mass balances in conversion reactions still remain a challenge. To date, the gas chromatographic method is still superior, but requires extensive derivatization and other sample preparations. HPLC methods promise rapid analysis, but no universal column can be used for all feedstocks. Current developments with the Prevail™ carbohydrate column and the evaporative light scattering detector have the potential to finally resolve and significantly speed biomass sugar analysis to complement rapid instrumental methods of analysis that rely on wet chemical methods for calibration.

References

1. Perez, S. and Mackie, B. (2001) Structure and morphology of cellulose, http://www.cermav.cnrs.fr/glyco3d/lessons/cellulose/.
2. Bergenstrahle, M., Mathews, J., Crowley, M., and Brady, J. (2010) Cellulose crystal structure and force fields. International Conference on Nanotechnology for Forest Products Industry, September 27–29, 2010, Otaniemi, Espoo, Finland.
3. O'Sullivan, A.C. (1997) Cellulose: structure slowly unravels. *Cellulose*, **4**, 173–207.
4. Frey-Wyssling, A. (1954) The fine structure of cellulose microfibrils. *Science*, **119**, 80–82.

5. Fernandes, A.N., Thomas, L.H., Altaner, C.M. *et al.* (2011) Nanostructure of cellulose microfibrils in spruce wood. *Proceedings of the National Academy of Sciences of the United States of America*, **108** (47), E1195–E1203.

6. Fengel, D. and Wegener, G. (1989) *Wood Chemistry, Ultrastructure and Reactions*, Walter de Gruyter, New York.

7. Sarkanen, K.V. and Ludwig, C.H. (eds) (1971) *Lignins, Occurrence, Formation, Structure, and Reactions*, John Wiley, New York, NY.

8. Sjostrom, E. (1992) *Wood Chemistry, Fundamentals and Applications*, 2nd edn, Academic Press, New York, NY.

9. Schadel, C. (2009) Cell-wall hemicellulose as mobile carbon stores in plants. PhD Thesis, Basel University, Basel, Switzerland

10. Wiselogle, A.E., Agblevor, F.A., Johnso, D.K. *et al.* (2020) Compositional changes during storage of large round switchgrass bales. *Bioresource Technology*, **56** (1), 103–109.

11. Agblevor, F.A., Evans, R.J., and Johnson, K.D. (1994) Molecular-beam mass spectrometric analysis of ligno-cellulosic materials 1: Herbaceous biomass. *Journal of Analytical and Applied Pyrolysis*, **30** (2), 125–144.

12. Chenney, J.H., Johnson, K.D., Volenec, J.J., and Anliker, K.S. (1988) Chemical composition of herbaceous grass, legume species grown for maximum biomass production. *Biomass*, **17** (4), 215–238.

13. ASTM E1757 (2008) Standard practice for preparation of biomass for compositional analysis, in *Annual Book of ASTM Standards*, vol. **11.05**, ASTM International, West Conshohocken, PA.

14. Hames, B., Ruiz, R., Scarlata, C. *et al.* (2008) Preparation of samples for compositional analysis, Laboratory Analytical Procedures (LAP). Technical Report, NREL/TP-510-42620.

15. T264-cm-07 (2007) Preparation of wood for chemical analysis, in *TAPPI Test Methods*, TAPPI Press, Norcross, GA.

16. Milne, T.A., Brennan, A.H., and Glen, B. (1990) *Sourcebook of Methods of Analysis for Biomass and Biomass Conversion Processes*, Elsevier Science, New York, NY.

17. ASTM E1756 (2008) Standard test method for determination of total solids in biomass, in *Annual Book of ASTM Standards*, vol. **11.05**, ASTM International, West Conshohocken, PA.

18. T412 om-06 (2006) Moisture in pulp, paper and paperboard, in *TAPPI Test Methods*, TAPPI Press, Norcross, GA.

19. Sluiter, A., Hames, B., Hyman, D. *et al.* (2008) Determination of total solids in biomass and total dissolved solids in liquid process samples, Laboratory Analytical Procedure (LAPs), Technical Report, NREL/TP-510-42621.

20. Lohse, G. and Dietrich, H.H. (1972) Investigations on the suitability of Karl Fischer's titration method for determining wood moisture content. *Holz als Roh- und Werkstoff*, **57** (4), 125–129.

21. Shmulsky, R. (2000) Influence of lumber dimension on VOC emission from oven-drying loblolly pine lumber. *Forest Products Journal*, **50** (3), 63–66.

22. McDonald, A.G., Gifford, J.S., Dare, P.H., and Steward, D. (1999) Characterization of the condensate generated from vacuum-drying of radiate pine wood. *Holz als Roh- und Werkstoff*, **57** (4), 251–258.

23. Risholm-Sundman, M., Lundgren, M., Vestin, E., and Herder, P. (1998) Emission of acetic acid and other volatiles organic compounds from different species of solid wood. *Holz als Roh- und Werkstoff*, **56** (2), 125–129.

24. Larnøye, E. (2008) Mass loss evaluation of wood; are the results correct? Proceedings, 4th Meeting of the Nordic-Baltic Network in Wood Materials Science and Engineering, November 13–14, 2008, Riga, Latvia.

25. Resch, H. and Ecklund, A. (1963) Moisture content determination of wood with highly votalile contents. *Forest Products Journal*, **13**, 481–482.

26. Kollman, F. and Hockele, G. (1962) Comparison of procedure for the determination of moisture content in wood. *Holz als Roh- und Werkstoff*, **20** (12), 461–473.

27. Hartley, I.D., Kamke, F.A., and Peemoeller, H. (1994) Absolute moisture content determination of aspen wood below the fiber saturation point using pulsed NMR. *Holzforschung*, **48** (6), 474–479.

28. ASTM E1755 (2008) Standard method for ash in biomass, in *Annual Book of ASTM Standards*, vol. **11.05**, ASTM International, West Conshohocken, PA.

29. Sluiter, A., Hames, B., Hyman, D. *et al.* (2008) Determination of ash in biomass, Laboratory Analytical Procedure (LAP), Technical Report, NREL/TP-510-42622.

30. T211 om-07 (2007) Ash in wood, pulp, paper, paperboard: combustion at 525 °C, in *TAPPI Test Methods*, TAPPI Press, Norcross, GA.

31. AOAC 984 (1990) Protein (crude) determination in animal feed: copper catalyst Kjeldahl method, in *Official methods of Analysis*, 15th edn, AOAC.

32. Mossé, J. (1990) Nitrogen to protein conversion factor for 10 cereals and 6 legumes or oilseeds – a reappraisal of its definition and determination – variation according to species and to seed protein content. *Journal of Agricultural and Food Chemistry*, **38** (1), 18–24.

33. Mossé, J., Huet, J.C., and Baudet, J. (1985) The amino-acid composition of wheat-grain as a function of nitrogen-content. *Journal of Cereal Science*, **3** (2), 115–130.

34. Hames, B., Scarlata, C., and Sluiter, A. (2008) Determination of protein content in biomass, Laboratory Analytical Procedure (LAP), Technical Report, NREL/TP-510-42625.

35. T204 cm-07 (2008) Test method for solvent extractives in wood and pulp, in *TAPPI Test Methods*, TAPPI Press, Norcross, GA.

36. Van Soest, P.J. and Wine, R.H. (1967) Use of detergent in the analysis of fibrous feeds. IV. Determination of plant cell wall constituents. *Journal of the Association of Official Analytical Chemists*, **48**, 785–790.

37. ASTM E1690 (2008) Standard test method for determination of ethanol extractives in biomass, in *Annual Book of ASTM Standards*, vol. **11.05**, ASTM International, West Conshohocken, PA.

38. Sluiter, A., Ruiz, R., Scarlata, C. *et al.* (2008) Determination of extractives in biomass, Laboratory Analytical Procedure (LAP), Technical Report, NREL/TP-510-42619.

39. Chen, S.F., Mowery, R.A., Scarlata, J., and Chambliss, C.K. (2007) Compositional analysis of water-soluble materials in corn stover. *Journal of Agricultural and Food Chemistry*, **55**, 5912–5918.

40. Thammasouk, K., Tandjo, D., and Penner, M.H. (1997) Influence of extractives on the analysis of herbaceous biomass. *Journal of Agricultural and Food Chemistry*, **45**, 437–443.

41. Agblevor, F.A., Hames, B.R., Schell, D., and Chum, H.L. (2007) Analysis of biomass sugars using a novel HPLC method. *Applied Biochemistry and Biotechnology*, **136**, 310–236.

42. Agblevor, F.A., Beis, S., Mante, O., and Abdoulmoumine, A. (2010) Fractional catalytic pyrolysis of hybrid poplar wood. *Industrial & Engineering Chemistry Research*, **49**, 3533–3538.

43. Sanderson, M.A., Agblevor, F., Collins, M., and Johnson, D.K. (1996) Compositional analysis of biomass feedstocks by near infrared reflectance spectroscospy. *Biomass and Bioenergy*, **11** (5), 365–370.

44. Hames, B.R., Thomas, S.R., Sluiter, A.D. *et al.* (2003) Rapid biomass analysis. *Applied Biochemistry and Biotechnology*, **105–108**, 5–16.

45. T222 om-11 (2011) Test method for acid insoluble lignin in wood and pulp, in *TAPPI Test Methods*, TAPPI Press, Norcross, GA.

46. Hatfield, R.D., Jung, H.J.G., Ralph, J. *et al.* (1994) A comparison of the insoluble residues produced by the Klason lignin and acid detergent lignin procedures. *Journal of the Science of Food and Agriculture*, **65**, 51–58.

47. Van Soest, P.J. and Robertson, J.B. (1980) Systems of analysis for evaluating feeds, in *Standardization of Analytical Methodology for Feeds* (eds W.J. Pigden, C.C. Balch, and M. Graham), IDRC, Ottawa, ON, Canada.

48. ASTM E1721 (2008) Standard test method for determination of acid-insoluble residue in biomass, in *Annual Book of ASTM Standards*, vol. **11.05**, ASTM International, West Conshohocken, PA.

49. Milne, T.A., Chum, H.L., Agblevor, F.A., and Johnson, D.K. (1992) Standardized analysis of biomass sugars. *Biomass & Bioenergy*, **21** (1–6), 341–366.

50. Sluiter, A., Hames, B., Ruiz, R. *et al.* (2011) Determination of structural carbohydrates and lignin in biomass, Laboratory Analytical Procedure (LAP), Technical Report, NREL/TP-510-42618.

51. Theander, O. (1991) Chemical analysis of lignocelluloses materials. *Animal Feed Science and Technology*, **32**, 35–44.

52. Theander, O. and Westlund, E.A. (1986) Studies on dietary fibre. 3. Improved procedures for analysis of dietary fibre. *Journal of Agricultural and Food Chemistry*, **34**, 330–336.

53. Wan, E.C.H. and Yu, J.Z. (2007) Analysis of sugars and sugar polyols in atmospheric aerosols by chloride attachment in liquid chromatography/negative ion electrospray mass spectrometry. *Environmental Science & Technology*, **41**, 2459–2466.

54. Engling, G., Carrico, C.M., Kreidenweis, S.M. *et al.* (2006) Determination of levoglucosan in biomass combustion aerosol by high-performance anion-exchange chromatography with pulsed amperiometric detection. *Atmospheric Environment*, **40** (2), 299–311.

55. Miller, G.L. (1959) Use of dinitrosalicylic acid reagent for determination of reducing sugar. *Analytical Chemistry*, **31** (3), 426–428.

56. Green, F. III, Clausen, C.A., and Highley, T.L. (1989) Adaptation of the Nelson-Somogyi reducing sugar assay to a microassay using microtiter plates. *Analytical Biochemistry*, **182**, 197–199.

57. Browning, B.L. (1967) *Methods in Wood Chemistry*, vol. **2**, Wiley Interscience, New York.

58. Rabemanolontsoa, H. and Saka, S. (2012) Holocellulose determination in biomass, in *Zero-Carbon Energy, Kyoto, 2011, Green Energy and Technology, Part II* (ed. T. Yao), Springer, New York, pp. 135–140.

59. Easty, D.B. and Thompson, N.S. (1991) Wood analysis, in *Wood Structure and Composition, International Fiber Science and Technology 11* (ed. M. Lewin and I.S. Goldstein), Marcel Dekker, New York, NY.

60. Nelson, L. and Leming, J.A. (1957) Evaluation of monomethanolamine method for cellulose determination for agricultural residues. *TAPPI Journal*, **40** (10), 846–850.

61. Carpita, N.C., Kanabus, J., and Housely, T.L. (1989) Linkage structure of fructans and fructan oligomers from *Triticum aestivum* and *Festuca arundinacea* leaves. *Journal of Plant Physiology*, **134**, 162–168.

62. Van Loo, J., Coussement, P., Leenheer, L.D. *et al.* (1995) On the presence of inulin and oligofructose as natural ingredients in western diet. *Critical Reviews in Food Science and Nutrition*, **35**, 525–552.

63. Seaman, J.F., Moore, W.E., Mitchell, R.L., and Millet, M.A. (1954) Techniques for the determination of pulp constituents by quantitative paper chromatography. *TAPPI Journal*, **37**, 336–343.

64. Chambers, R.E. and Clamp, J.R. (1971) An assessment of methanolysis and other factors used in the analysis of carbohydrate containing materials. *The Biochemical Journal*, **125**, 1009–1018.

65. Cheetham, N.H.W. and Sirimanne, P. (1983) Methanolysis studies of carbohydrates, using HPLC. *Carbohydrate Research*, **112**, 1–10.

66. DeMartini, J.D., Studer, M.H., and Wyman, C.E. (2011) Small-scale and automatable high-throughput compositional analysis of biomass. *Biotechnology and Bioengineering*, **108** (2), 306–312.

67. Sluiter, J.B., Ruiz, R.O., Scarlata, C.J. *et al.* (2010) Compositional analysis of lignocellulosic feedstocks. 1. Review and description of methods. *Journal of Agricultural and Food Chemistry*, **58** (16), 9043–9053.

68. Templeton, D.W., Scarlata, C.J., Sluiter, J.B., and Wolfrum, E.J. (2010) Compositional analysis of lignocellulosic feedstocks. 2. Method uncertainties. *Journal of Agricultural and Food Chemistry*, **58** (16), 9054–9062.

69. Foyle, T., Jennings, L., and Mulcahy, P. (2007) Compositional analysis of lignocellulosic materials: Evaluation of methods used for sugar analysis of waste paper and straw. *Bioresource Technology*, **98**, 3026–3036.

70. Moore, W.E. and Johnson, D.B. (1967) Determination of wood sugars, in *Procedures for the Chemical Analysis of Wood and Wood Products*, Forest Products Laboratory, Forest Service, US Department of Agriculture, Madison, WI.

71. Grohmann, K. and Bothast, R. (1997) Saccharification of corn fibre by combined treatment with dilute sulfuric acid and enzymes. *Process Biochemistry*, **32** (5), 405–415.

72. Fengel, D. and Wegner, G. (1979) Hydrolysis of polysaccharides with trifluoroacetic acid and its application to rapid wood analysis, in *Hydrolysis of Cellulose: Mechanisms of Enzymatic and Acid Catalysis Advances in Chemistry Series No 181* (eds R.D. Brown and L. Jurasek), American Chemical Society, Washington DC.

73. Albersheim, P., Nevins, D.J., English, P.D., and Kaar, A. (1967) A method for the analysis of sugars in plant cell-wall polysaccharides by gas-liquid chromatography. *Carbohydrate Research*, **5** (3), 340–345.

74. Gutleben, W., Unterholzner, V., Volger, D., and Zemann, A. (2004) Characterization of carbohydrates in paper and pulps using anion exchange chromatography and principal component analysis. *Michrochim Acta*, **146**, 111–117.

75. Morrison, I.M. and Stewart, D. (1998) Plant cell wall fragments released on solubilization in trifluoroacetic acid. *Phytochemistry*, **49** (6), 1555–1563.

76. Chaplin, M.F. (1982) A rapid and sensitive method for the analysis of carbohydrate components in glycoproteins using gas-liquid chromatography. *Analytical Biochemistry*, **123**, 336–341.

77. Quemener, B. and Thibault, J.P. (1990) Assessment of methanolysis for determination of sugars in pectins. *Carbohydrate Research*, **206**, 277–287.

78. Quemener, B., Lahaye, M., and Thibault, J.F. (1993) Studies on the simultaneous determination of acidic and neutral sugars of plant cell wall materials by HPLC of their methyl glucosides after combine methanolysis and enzymic prehydrolysis. *Carbohydrate Polymers*, **20**, 87–94.

79. Quemener, B., Lahaye, M., and Bobib-Dubigeon, C. (1997) Sugar determination in ulvans by chemical-enzymatic method coupled to high performance anion exchange chromatography. *Journal of Applied Phycology*, **9**, 179–188.

80. ASTM E 1821 (2009) Determination of carbohydrates in biomass by gas chromatography, in *Annual Book of ASTM Standards*, vol. **11.05**, ASTM International, West Conshohocken, PA.
81. T249 cm-09 (2009) Carbohydrate composition of extractives-free wood and wood pulp by gas-liquid chromatography, in *TAPPI Test Methods*, TAPPI Press, Norcross, GA.
82. Agblevor, F.A., Chum, H.L., and Johnson, D.K. (1993) Compositional analysis of NIST biomass standards from the IEA whole feedstock round robin, in *Energy from Biomass and Wastes*, vol. **16** (ed. D.L. Klass), Institute of Gas Technology, Chicago, IL.
83. Biermann, C.J. and McGinnis, G.D. (1989) *Analysis of Carbohydrates by GLC and MS*, CRC Press, Boca Raton, FL.
84. Madeiros, P. and Simoneit, R.T. (2007) Analysis of sugars in environmental samples by gas chromatography-mass spectrometry. *Journal of Chromatography. A*, **1141** (2), 271–278.
85. Robards, K. and Whitelaw, M. (1986) Chromatography of monosaccharides and disaccharides. *Journal of Chromatography. A*, **373**, 81–110.
86. Agblevor, F.A., Batz, S., and Trumbo, J. (2003) Composition and ethanol potential of cotton gin residues. *Applied Biochemistry and Biotechnology*, **105–108**, 219–230.
87. Blakeney, A.B., Harris, P.J., Henry, R.J., and Stone, B.A. (1983) A simple and rapid preparation of alditol acetates for monosaccharide analysis. *Carbohydrate Research*, **113**, 291–299.
88. Bonn, G. and Bobleter, O. (1984) HPLC-analysis of plant biomass hydrolysis and fermentation solutions. *Chromatographia*, **18** (8), 445–448.
89. Kaar, W.E. and Brink, D.L. (1991) The complete analysis of wood polysaccharides using HPLC. *Journal of Wood Chemistry and Technology*, **11**, 447–463.
90. Davis, M.W. (1998) A rapid modified method for compositional carbohydrate analysis of lignocellulosics by high pH anion-exchanged chromatography with pulsed amperometric detection (HPAEC/PAD). *Journal of Wood Chemistry and Technology*, **18** (2), 235–252.
91. Pettersen, R.C. and Schwandt, V.H. (1991) Wood sugar analysis by anion chromatography. *Journal of Wood Chemistry and Technology*, **11**, 495.
92. Worral, J.J. and Anderson, K.M. (1993) Ample preparation for analysis of wood sugars by anion chromatography. *Journal of Wood Chemistry and Technology*, **13**, 429.
93. Clarke, A.J., Sarabia, V., Keenleyside, W. *et al.* (1991) The compositional analysis of bacterial extracellular polysaccharides by high performance anion-exchange chromatography. *Analytical Biochemistry*, **199**, 68–74.
94. Widmer, W. (2010) An improved method for analysis of biomass sugars and galacturonic acid by anion exchanged chromatography. *Biotechnology Letters*, **32**, 435–438.
95. Agblevor, F.A., Murden, A., and Hames, B.R. (2004) Improved method of biomass sugars analysis using high-performance liquid chromatography. *Biotechnology Letters*, **26**, 1207–1210.
96. Baker, J.O. and Himmel, M.E. (1986) Separation of sugar anomers by aqueous chromatography on calcium- and lead- form ion-exchange columns: application to anomeric analysis of enzymes reaction products. *Journal of Chromatography. A*, **357**, 161–181.
97. ASTM E 1758 (2008) Determination of carbohydrates in biomass by high performance liquid chromatography, in *Annual Book of ASTM Standards*, vol. **11.05**, ASTM International, West Conshohocken, PA.
98. Adsul, M.G., Ghule, J.E., Shaikh, H. *et al.* (2006) Enzymatic hydrolysis of delignified bagasse polysaccharides. *Carbohydrate Polymers*, **62** (1), 6–10.
99. Laver, M.L. and Wilson, K.P. (1993) Determination of carbohydrates in wood pulp products. *TAPPI Journal*, **76** (6), 155.
100. Suzuki, M., Sakamoto, R., and Aoyagi, T. (1995) Rapid carbohydrate analysis of wood pulps by ion chromatography. *TAPPI Journal*, **78** (7), 174.
101. Agblevor, F.A., Hames, B.R., Schell, D., and Chum, H.L. (2007) Analysis of biomass sugars using novel HPLC method. *Applied Biochemistry and Biotechnology*, **136**, 309–326.
102. Agblevor, F.A., Fu, J., Hames, B., and McMillan, J.D. (2004) Identification of microbial inhibitory functional groups in corn stover hydrolysate by carbon-13 nuclear magnetic resonance spectroscopy. *Applied Biochemistry and Biotechnology*, **119**, 97–120.

103. Srokol, Z., Bouche, A.-G., Estrik, A. *et al.* (2004) Hydrothermal upgrading of biomass to biofuel; studies on some monosaccharide model compounds. *Carbohydrate Research*, **339** (10), 1717–1726.
104. Pan, X., Arato, C., Gilkes, N. *et al.* (2005) Biorefining of softwoods using ethanol organosolv pulping: Preliminary evaluation of process streams for manufacture of fuel-grade ethanol and co-products. *Biotechnology and Bioengineering*, **90** (4), 473–481.
105. Liu, T., Lin, L., Sun, Z. *et al.* (2010) Bioethanol fermentation by recombinant *E. coli* FBR5 and its robust mutant FBHW using hot water wood extract hydrolyzates as substrate. *Biotechnology Advances*, **28** (5), 602–608.
106. Zhao, H., Halladay, J.E., Brown, H., and Zhang, Z.C. (2007) Metal chlorides in ionic liquid solvents convert sugars to 5-hydroxymethylfurfural. *Science*, **316** (5831), 1597–1600.
107. Mittal, A., Scott, G.M., Amidon, T.E. *et al.* (2009) Quantitative analysis of sugars in wood hydrolysates with H NMR during the authohydrolysis of hardwoods. *Bioresource Technology*, **100**, 6398–6406.
108. Kiemle, D.J., Stipanovic, A.J., and Mayo, K.E. (2004) *Proton NMR methods in the compositional characterization of polysaccharides in Hemicelluloses: Science and Technology, ACS Symposium Series 864* (eds P. Gatenholm and M. Taenkanen), American Chemical Society, Washington DC.

17

High-throughput NIR Analysis of Biomass Pretreatment Streams

Bonnie R. Hames

B Hames Consulting, Newbury Park, USA

17.1 Introduction

Pretreatment methods for conversion of plant biomass to fuels and valuable chemicals begin with a chemically complex feedstock which is fractionated into liquid and solid phases, presenting a significant challenge from the perspective of analytical chemistry. Traditional analytical methods have been developed at the National Renewable Energy Laboratory (NREL) to support pretreatment research and development activities, and are available for download from http://www.nrel.gov/biomass/analytical_procedures.html. The NREL methods are widely used in academic and industrial settings and are currently being vetted as international standards through joint efforts of the American Society for Testing and Materials (ASTM, http://www.astm.org/) and the International Standards Organization (ISO, http://www.iso.org/). The main limitations of these methods are their per-sample cost and complexity, which delay results for days or even weeks.

This chapter focuses on the development, validation and application of high-throughput (HTP) methods for monitoring pretreatment processes. The HTP methods described here are rapid analysis methods based on a combination of near-infrared (NIR) spectroscopy and advanced multivariate analysis techniques. HTP rapid analysis methods are capable of characterizing complex substrates with excellent mass closure, delivering the precision and accuracy needed to understand pretreatment processes [1]. Once calibrated and validated, the simplicity and low per-sample cost of these methods provide a realistic path to levels of information that would have been too expensive to pursue using traditional methods. For example, using traditional wet chemical methods, a full mass closure characterization of feedstock, pretreated solids and pretreatment liquor for a batch of 10 samples would cost around $30 000 and require days of sample preparation and analysis. In contrast, rapid analysis methods could provide the same quality of data on a thousand samples for the same budget of $30 000 with a throughput of minutes instead of days. Calibration and

Aqueous Pretreatment of Plant Biomass for Biological and Chemical Conversion to Fuels and Chemicals, First Edition.
Edited by Charles E. Wyman.
© 2013 John Wiley & Sons, Ltd. Published 2013 by John Wiley & Sons, Ltd.

validation are expensive, often costing $300 000 or more, but this cost is recovered quickly with a saving of $2970 per sample.

The ability to gather high-quality data on thousands of samples on a limited analytical budget is an essential factor in exploring the technical and economic effects of feedstock variability on pretreatment and all downstream steps. Comprehensive data on thousands of samples provide an opportunity to evaluate feedstocks as a statistical population. This big-picture approach should provide a much more complete understanding of the complex behavior of plant polymers during pretreatment and lead to new methods of producing high-quality pretreated substrates from a diverse slate of potential biomass feedstocks [2]. As pretreatment moves from fundamental research to a commercial scale, the methods that supported fundamental research will need to be simplified and their speed increased to provide quality data for on-line process control as demonstrated widely in commercial systems in other applications [3].

The biggest challenge for analytical methods designed to support pretreatment is the complexity of the starting feedstocks. All types of plant-based feedstocks are chemically and biologically diverse representing the cumulative effects of many factors including genetics, soil, agronomy, weather, harvest time, harvest method, storage conditions and duration, drying, and particle size reduction [4,5]. The incoming biomass feedstocks will contain a combination of structural matter derived from the plant cell walls, and non-structural matter including salts, sucrose and a complex mixture of plant metabolites.

Ideally, optimization of pretreatment would begin with a thorough characterization of the starting feedstock and include tracking the chemical and physical changes that occur as each component of the feedstock is partitioned into soluble and insoluble pretreatment fractions that are compatible with downstream processes. The sheer number of constituents to measure and track through pretreatment illustrates the complexity of this challenge. The cell-wall material represents a complex three-dimensional matrix of natural polymers including cellulose, hemicellulose, lignin, and, for some feedstocks, protein [4]. Many feedstocks will also have a significant amount of structural silica and other inorganic materials, commonly reported as ash. Tracking only feedstock carbohydrates through the process of depolymerization during pretreatment would require individual calibrations for quantifying glucose, xylose, mannose, arabinose, galactose, sucrose, maltose, cellobiose, and such common carbohydrate degradation products as 5- hydroxymethylfurfural (HMF) and furfural [6].

Tracking lignin, protein, and inorganic salts requires another complex set of methods and assumptions about their initial chemical structure and chemical reactions that occur during pretreatment. It is in this complexity that the true value of modern HTP methods becomes apparent. These methods are designed to pull information about the concentration of all of these constituents from a single near-infrared spectrum. The time required for a full comprehensive compositional analysis is simply the time required for sample preparation and spectral collection, with a few seconds for spectral processing through the multivariate analysis software.

17.2 Rapid Analysis Essentials

Like all analytical methods, rapid analysis methods require calibration, validation and quality monitoring. The calibration process differs from traditional methods in the application of multivariate analysis methods. Traditional analytical methods, such as high-performance liquid chromatography (HPLC), physically separate and resolve individual analytes. It is this physical separation to isolate individual constituents that make wet chemical methods slow and labor-intensive. Calibrations for these methods are based on direct relationships between the concentration of isolated analytes and simple detector response. In contrast, the quantitative analysis methods described here use NIR spectra collected from intact samples. Individual components of interest are resolved from whole sample spectra using complex mathematical algorithms and advanced statistics. The analysis is non-destructive, meaning that all samples are available for additional testing or confirmation using traditional methods, if desired. These important differences require different modes of calibration, validation, method development, and quality assurance.

Five essential components in the development of a rapid analysis method are discussed in the following sections: rapid, spectroscopic technique; calibration and validation samples; quality calibration data for each calibration sample; multivariate analysis to resolve complex sample spectra; validation of new methods; and standard reference materials and protocols for ongoing quality assurance/quality control (QA/QC).

17.2.1 Rapid Spectroscopic Techniques

The first step in rapid analysis method development is choosing an HTP technique that will be the basis of the final methods. Many different spectroscopic techniques can be used as the basis for rapid analysis methods, including mid-range IR [7], NIR, mass spectra [8], Raman [9], nuclear magnetic resonance (NMR) [10], optical acoustics [11], and so on. The technique selected must contain information about the constituents found in biomass feedstocks. The examples presented here are all based on NIR spectroscopy where there is a richness of information on the chemical bonds commonly found in plant biomass, such as C—H, C—O, C—C, C—N, and N—H. NIR spectra also include signals from first, second, and third overtones of fundamental vibrations in these bonds, a redundancy which can be useful when developing chemometric methods [3]. The NIR spectrum contains little or no direct information about inorganic molecules, so high-throughput methods for measuring these constituents may require the use of one of the other techniques listed above. Most of the procedures necessary for the development of high-throughput models for monitoring biomass pretreatment are independent of the spectroscopic methods selected. The examples used in the chapter will all use NIR spectroscopy, but the principles described here are relevant to any of the HTP techniques listed above.

Other important criteria include the robust nature of the instruments, the complexity of sample preparation, sample presentation options, and sample throughput. This selection determines the operator training requirements, per-sample cost, and throughput of the final methods.

The availability of accessories for sample presentation can also be very important when developing methods for pretreatment monitoring, since this task requires analysis of feedstocks, process solids, and process liquids. With some feedstocks and processes, it may even be possible to build methods capable of analyzing pretreatment slurries [12]. In NIR spectroscopy, temperature control is necessary for the analysis of liquid samples and slurries. The best instrument and accessory choices will depend on the feedstock choice and methods used in pretreatment. Gathering a few representative samples for use in instrument demonstrations prior to instrument selection is strongly advised. For maximum flexibility in the application of chemometric methods, a NIR spectrometer with a sampling interval of at least 2 nm and detectors that cover the full range of the NIR spectrum of approximately 800–2500 nm (12 500–4000 cm^{-1}) may be needed. (NIR spectrometers meeting these requirements, with accessories for liquid, solid, and slurry analysis, are available from a variety of manufacturers including Foss, Bruker Optics, Analytical Spectral Devices, Perten Intstuments, and others.) Scanning accessories and sample presentation modes will vary by manufacturer.

When an HTP method has been selected, the next step is the development of robust spectroscopic sampling and protocols. This is easily accomplished with a few large, stable samples that can be analyzed numerous times while instrument options are varied. At this point decisions can be made about sample presentation. Will samples be solids, liquids, or slurries? Will they be analyzed by transmission, reflectance, transflectance, or other options depending on instrument choices? What sample preparation requirements are necessary? Do feedstock samples need to be milled or dried? Do pretreated liquids need to be filtered or diluted? Do pretreated solids need to be washed, dried, or milled? The goals are maximum reproducibility, maximum throughput, minimum sample preparation, and minimum analysis time. When these parameters are defined and uniform and reproducible analysis can be demonstrated, the selection of calibration and validation samples can begin.

17.2.2   Calibration and Validation Samples

Traditional quantitative analytical methods are calibrated using purified samples of individual analytes. This calibration approach will not work for rapid analysis methods. To be useful in calibrating rapid analysis methods, the spectrum of the pure, isolated components would need to be identical to the spectrum of that component as it exists in the intact plant matrix. Standards of this type are not available for the components of interest in lignocellulosic feedstocks. The complex, three-dimensional matrix of natural polymers found in lignocellulosic materials makes it difficult, if not impossible, to separate individual components in their native form without modification or degradation. Additionally, the NIR spectrum of whole biomass contains information about the interactions between individual components, both covalent and ionic, which is lost upon isolation. When developing rapid analysis methods, a completely different approach to calibration is required.

Calibration and validation sets for analytical methods based on multivariate analysis are a collection of diverse and intact samples. The selection criteria are the same for calibration and validation samples, so for this discussion they will be considered as one calibration collection that will be divided into calibration and validation sets as method development progresses. The requirements for the calibration collection are that the samples are similar enough to each other to represent a single population, but diverse enough to allow independent resolution of each constituent of interest.

The calibration samples must be similar in composition and structure to the samples to be evaluated. For example, calibration sets for methods that will evaluate corn stover must contain corn stover samples, but they might also contain other herbaceous feedstocks of similar chemical composition and structure such as switchgrass or sorghum. Feedstocks such as hardwoods, softwoods and grain have major differences in composition and structure from each other and from grasses. These differences can be clearly distinguished by NIR spectroscopy, and multivariate analysis will view them as separate populations. Each of these feedstock populations will require separate calibration collections when building rapid analysis methods. Because pretreatment chemically alters the cell-wall structures in biomass feedstocks, pretreated materials are easily distinguished from feedstock as a separate spectroscopic population; for this reason, separate calibration samples are usually required for feedstocks and pretreated materials [13].

Within each sample population, the calibrations samples must also represent wide ranges of diversity for each constituent. Ideally the concentration ranges for each constituent in the calibration collection would represent the diversity found in nature. At a minimum, the calibration set must include samples with much higher and lower concentrations than the values anticipated in the samples to be evaluated by the finished method. The calibration set must also exhibit independent variability of individual constituents. Mutivariate analysis incorporates concepts of experimental design and, in that sense, requires a calibration set that allows constituents to be mathematically resolved based on their NIR spectra. An example of how this is accomplished would be using a calibration set that includes some samples with high lignin and low cellulose, some with high lignin and high cellulose, and some with low lignin and high cellulose.

Independent variability is needed to identify structural patterns relating to individual constituents. Spiking calibration samples with purified components to modify their composition will not work for rapid analysis methods. As mentioned earlier, the spectra of purified components are different from that of components bound in the cell-wall matrix. However, it may be possible to generate new levels of compositional diversity by anatomical fractionation or blending of natural samples, for example, removing only leaf material from a corn stover sample lowers ash and protein and raises carbohydrate levels. The polymers are still in their native form, but are now present in new proportions. As in experimental design, the quality of the

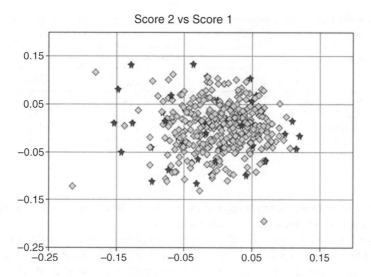

Figure 17.1 *Plot of PCA scores for 250 potential calibration samples (gray diamonds) and selection of 40 samples for method calibration (black stars).*

rapid analysis method is determined by the quality of this statistical resolution. These variability requirements are often the biggest challenge in locating appropriate calibration samples. With a highly diverse calibration set whose composition ranges span those of the population in nature, an excellent rapid analysis method can be built with around 100 samples. More commonly, it takes around 300 samples to represent an entire population and adequately resolve all constituents of interest [14].

To aid in the identification of samples that would expand or improve calibrations, most multivariate analysis software packages include principal component analysis (PCA) options that allow selection of calibration samples based only on spectroscopic variance. This process begins by scanning a large number of potential calibration samples using the HTP techniques developed earlier. Surprisingly, evaluating potential calibration samples for compositional diversity does not require prior information about the chemical composition of the individual samples. Selection can be based entirely on the spectroscopic signatures of the underlying independent constituent variance. Figure 17.1 shows 250 samples of herbaceous feedstocks relative to two principal components that represent spectroscopic differences in the sample population. PCA tools have been used to identify and remove redundant samples, leaving 40 samples that are statistically unique relative to significant sources of spectroscopic variance in the larger population. In this example, they are plotted relative to the larger sample population and identified with black stars. The reasonable assumption is made that spectroscopic differences reflect important chemical differences. The compositional ranges and compositional diversity of the entire population is captured in 40 individual samples. The calibration of the rapid analysis method can now proceed with 40 samples instead of 250, allowing considerable savings in the cost of the next essential element: generation of calibration data.

17.2.3 Quality Calibration Data for Each Calibration Sample

Rapid analysis methods require compositional data for each calibration sample, obtained using a traditional wet chemical method. These data will be used in a data matrix to resolve complex spectra into constituent information. Rapid analysis methods often match the precision and accuracy of the wet chemical methods,

Table 17.1 *Three calibration samples with the same hemicellulose content but significant differences in hemicellulose structure, reflected in their NIR spectra.*

Constituent	Concentration (% dry weight)		
	Sample A	Sample B	Sample C
Hemicellulose	25	25	25
Xylan	18	15	12
Arabinan	4	6	7
Acetyl	3	4	6

but they cannot be better than the methods used to generate this calibration matrix. For this reason it is wise to employ the best available methods and analysts for this task.

When selecting calibration methods it is also important to choose methods and data that reflect chemical structures in the calibration samples and calibration data should reflect constituents that vary independently. An example of this is seen in Table 17.1. Separating hemicellulose into individual components such as xylan, arabinan, acetyl, and so on and building separate methods for each of these constituents will allow the multivariate analysis software to find the spectroscopic patterns relating to each component as they vary independently in the calibration set. It is much less likely that the multivariate analysis software can locate any spectroscopic patterns that vary with a total hemicellulose number, because the same value for hemicellulose content will be assigned to three very different spectroscopic patterns.

Another example is seen in liquid sample measurements of soluble oligomeric sugars. Lab-based HPLC methods measure monomeric sugars and total sugars, after oligomer hydrolysis. Since the NIR spectra contain information on monomers and oligomers, these are the values to use in the calibration data matrix. The oligomeric form of each sugar can be calculated from the HPLC data as the difference between the monomeric and total sugar, after adjustment to an anhydro basis.

The portfolio of analytical methods developed at NREL meet the criteria listed here and have been used to calibrate rapid analysis methods for a variety of bioenergy feedstocks and pretreated materials [1,3,5,10,11,13,15,16]. Generating primary calibration data is usually the most expensive and slowest step in rapid analysis method development. It is important to employ very strict quality control limits and to document precision and accuracy for each constituent that is measured. Precision and accuracy information will be used during multivariate analysis to avoid over-fitting of models. In most cases, the precision and accuracy of rapid analysis methods closely match those of the calibration methods. For this reason, investing in the highest-quality data available will bring an immediate and long-lasting return.

One valuable quality assessment tool is full mass closure, where the concentrations of individual constituents are converted to an anhydro basis and the sum of all measured constituent concentrations is calculated. The anhydro adjustments account for water added during polymer hydrolysis. For the NREL portfolio methods, mass closures of 98% for feedstocks and 95% for pretreated solids are expected. This summative check ensures that no major constituents are missed and that errors associated with counting mass in more than one category are minimized [17,18].

Mass closures for rapid analysis methods will also closely reflect those of the calibration data. In HTP mode, each constituent is measured independently from the same NIR spectra for a comprehensive compositional profile. As sample mass closure improves, so will mass balance from feedstocks to pretreated solids and liquids. Tables 17.2–17.4 contain a list of constituents commonly measured for full mass closure analysis of feedstocks and pretreated materials [6]. Separate lists are shown for herbaceous materials, hardwood, and softwood feedstocks to reflect differences in the structure of the native polymers and the complexity of

Table 17.2 *Constituents found in herbaceous feedstocks and pretreated materials.*

Constituent	Herbaceous feedstock	Pretreated solids	Pretreated liquids
Cellulose/glucan	X	X	
Glucose			X
Cellobiose			X
Hemicellulose			
Xylose/xylan	X	X	X
Arabinose/arabinan	X	X	X
Galactose/galactan	X	X	X
Mannose/mannan	X	X	X
Glucose/glucan	X	X	X
Acetic acid/acetyl groups	X	X	X
Ferulic acids/ferulate	X	X	X
Uronic acids/uronyl groups	X	X	X
Lignin	X	X	X
Nitrogen	X	X	X
Protein	X	X	
Nitrates			X
Starch	X		
Maltose			X
Ash	X	X	
Extractives	X		
Sucrose	X		X

the plant cell walls. The liquid and solid phases following pretreatment are commonly separated prior to compositional analysis and are listed separately.

Tables 17.2–17.4 illustrate the analytical complexity of tracking plant polymers as they hydrolyze and partition between liquid and solid phases during pretreatment. The number and nature of constituents to measure varies according to the cell-wall structure of the biomass feedstock, with softwood processes requiring 25 calibrations, hardwood processes requiring 31 calibrations, and herbaceous processes requiring 44 calibrations for comprehensive mass closure. Tracking all of these constituents using conventional wet

Table 17.3 *Constituents found in hardwood feedstocks and pretreated materials.*

Constituent	Hardwood feedstock	Pretreated solids	Pretreated liquids
Cellulose/glucan	X	X	
Glucose			X
Cellobiose			X
Hemicellulose			
Xylose/xylan	X	X	X
Arabinose/arabinan	X	X	X
Galactose/galactan	X	X	X
Mannose/mannan	X	X	X
Glucose/glucan	X	X	X
Acetic acid/acetyl groups	X	X	X
Uronic acids/uronyl groups	X	X	X
Lignin	X	X	X
Ash	X	X	
Extractives	X		

Table 17.4   *Constituents found in softwood feedstocks and pretreated materials.*

Constituent	Softwood feedstock	Pretreated solids	Pretreated liquids
Cellulose/glucan	X	X	
Glucose			X
Cellobiose			X
Hemicellulose			
Mannose/mannan	X	X	X
Glucose/glucan	X	X	X
Galactose/galactan	X	X	X
Xylose/xylan	X	X	X
Arabinose/arabinan	X	X	X
Lignin	X	X	X
Ash	X	X	
Extractives	X		

chemical methods is well beyond most analytical budgets, so shortcuts are often taken to improve through-put and reduce costs. Fewer constituents are measured, assumptions are made about partitioning and yields, decisions are made using incomplete information, and risks associated with constituents that are not tracked cannot be assessed. With rapid analysis methods, all of these measurements can be made in minutes from NIR spectra of just three materials: feedstocks, pretreated liquids, and pretreated solids. More information is available for process optimization, economic evaluations, and risk assessments. This information would have been too costly to obtain using traditional methods, so the value of this information should also be considered when justifying the expense of calibrating and validating HTP rapid analysis methods.

For most materials derived from plant biomass, challenges in representative sampling can introduce ana-lytical uncertainty. The impact of sampling differences can be minimized when building a multivariate analysis model by performing chemical analysis on the exact sample that was scanned by NIR. The non-destructive nature of NIR spectroscopy allows this type of sample hand-off and ensures the closest match between spectroscopic and compositional data.

Once a NIR spectrum has been taken of each of the calibration samples, the stored calibration spectra can be used as many times as needed to build new models. There is no need to wait for all of the constituent measurements to begin building rapid analysis methods. New constituent models can be added any time that data are available. For this reason it is a good idea to store samples of the feedstocks and pretreated materi-als that have been used for method calibration for future updates and method expansions. Sample stability during storage can easily be assessed by rescanning stored samples and comparing data and mass closure with previous measurements.

17.2.4   Multivariate Analysis to Resolve Complex Sample Spectra

Once calibration and validation samples are collected, scanned, and characterized for the constituents of interest, the actual building of rapid analysis methods proceeds quickly. The spectroscopic and constituent data are imported into multivariate analysis (MVA) software for generation of calibration equations. Many MVA software options are available. Four of the most popular are WinISI (http://www.winisi.com/), OPUS (http://www.brukeroptics.com/opus.html), Unscrambler (http://www.camo.com/rt/Products/Unscrambler/unscrambler.html) and Grams IQ (http://gramssuite.com/Downloads/). Some software choices are deter-mined by the selection of a spectrometer when MVA analysis package is part of the spectral collection software (Bruker OPUS and Foss WinISI). Other spectrometer manufacturers rely on MVA equations developed in separate, specialized chemometric software such as Unscrambler and Grams IQ. Each of these software options has strengths and limitations. User-friendly data visualization tools are nice, but are often paired with mathematical and statistics packages with limited access and "black box" choices.

Building robust rapid analysis methods incorporates elements of advance mathematical treatments for reducing spectroscopic noise and scatter effects, identification and removal of outlier samples, and advanced linear algebra to combine the spectra and data matrices into scores and loadings that form the basis of the complex linear equation that is the essence of the rapid analysis model [19,20]. When building multivariate analysis methods, it is wise to seek the advice of a chemometrician or an expert in multivariate analysis for chemical applications for guidance through the complex mathematical options.

The Unscrambler software by CAMO is the tool most widely used for MVA by serious chemometricans and statisticians. This software package interfaces with most spectral and analytical file formats for easy import and export of data. The Unscrambler software offers many options for data manipulation and mathematical treatments. This is a serious software tool, without pretty graphics, but with unlimited, transparent access to equations and data. The tools available in this software allow for the most complete statistical analysis of MVA equations during method development. Methods passing this rigorous analysis are much more likely to be accurate, precise and robust.

All multivariate analysis software will translate the data and spectroscopic matrices into a series of scores and loadings. Loadings are spectroscopic patterns that reflect the chemical structure of the constituent being measured. Scores are coefficients that reflect the concentration of that constituent in each calibration sample. The calibration population is described by a series of principle components, with separate sets of scores and loadings [14].

The first principal component (PC) contains the most information about the constituent to be measured. The loading for the first PC will reflect chemical bonds that are statistically most significant for the identification of the constituent to be measured. Individual calibration samples will be assigned scores that reflect the fraction of their spectrum that is described by this loading pattern. This value is compared to the measured constituent concentration and the variance is reported as the calibration error for that PC. The information used in generating the first PC is then subtracted from each spectrum and a new loading pattern is generated that describes the most significant patterns in the remaining spectroscopic data. The second PC will contain a bit less of constituent chemistry but will often fine-tune the distinction between samples. New scores are generated, constituent concentrations are estimated, and method errors are again compared to measured calibration data.

The generation of new scores and loadings continues in this manner until the loading represents spectroscopic patterns that do not contain information about the constituent of interest and the cumulative errors in concentration estimates increase instead of decrease. In this manner, the software will define the calibration population and recommend a number of principal components to be incorporated into the final method for each constituent. This recommendation is based entirely on a minimization of the calculated errors of calibration and validation.

Modern software and data visualization tools can make this complex process seem simple (in many cases, too simple) and great care should be taken to ensure that the models do not over-fit the data. There are two simple rules that should guide the selection of an appropriate number of principal components:

- the only valid metric for the "best" method is robust performance over time that can be supported by independent validation; and
- rapid analysis methods cannot be more accurate than the methods used to validate them.

Software prompts and chemometric guidelines often suggest the selection of principal components based strictly on minimization of method errors [14,19,20]. Using error minimization as the sole criteria can lead to over-fit models that violate these two rules. Over-fit methods will rarely perform with the accuracy and precision claimed [19]. To avoid this situation, it may be necessary to ignore the software recommendation and manually select the number of principal components based on knowledge of the calibration samples and the analytical methods used to generate the calibration data. A useful guide for selection of the number of principal components is to aim for a model where the method error and average residuals reflect the accuracy and precision of the primary method used to generate the calibration data [14,19].

Quantitative analysis methods for biomass feedstocks and biomass-derived materials are commonly built using partial least-squares (PLS) chemometric techniques, also known as projection to latent structures [19,20]. PLS-1 methods are used to build separate equations for each constituent. In these methods all spectroscopic information is available to be used in calculating each constituent equation. Most chemometric software will employ this technique. Summative analysis is accomplished by applying multiple constituent equations to the same spectroscopic data. Some software packages offer a second option, PLS-2, which builds equations for the analysis of multiple constituents and restricts the use of the same spectroscopic features to one constituent. Both methods have been successfully employed in building rapid analysis methods for biomass conversion applications.

Rapid analysis methods are whole spectral techniques, meaning that they use all or most of the available spectroscopic data. They are designed to pull the important information from spectra, so peak picking is not necessary or advised [21]. Mathematical treatments are often applied to spectra prior to multivariate analysis to minimize the significance of variance in the calibration spectra not related to composition such as baseline offset (due to path length differences) and scatter (due to particle size differences). The combination of first derivative, standard normal variate, and detrend mathematical treatments has been shown to work well in rapid analysis methods for feedstocks and process intermediates [1,3,5,6,8,13,16,17,22].

17.2.5   Validation of New Methods

Building a rapid analysis equation is an iterative process of generation, validation, and improvement. Mathematical options are selected. Samples are added and/or deleted. An appropriate number of principal components are selected, and the precision and accuracy of the method are calculated [19]. Figures 17.2 and 17.3 illustrate the characteristics of a robust model. In Figure 17.2, data for the calibration set are plotted, comparing lab measurements of a fermentation product to full cross-validation estimates for each sample in the calibration set. Cross-validation is designed to be a conservative measurement of the performance of the final method. In this process, each sample is evaluated using a model built while that sample has been excluded from the calibration set. Figure 17.3 shows a similar comparison between calibration data and the evaluation of a set of samples that are not part of the calibration. The error calculated for the NIR/PLS model is 305 mg/mL which closely matches that of the primary lab method at 300 mg/mL. The R^2 values around 0.98 reflect the excellent agreement between the calibration lab method and the rapid analysis method, demonstrating that rapid analysis methods can offer significant savings in time and per-sample cost without sacrificing data quality.

Once they are calibrated and validated, analysis using these methods is as simple as sample preparation, scanning, and applying the method equations to the spectroscopic data. The equations use the stored loading patterns for each PC to generate scores for the unknown samples, which are then used to estimate the concentration of individual constituents.

17.2.6   Standard Reference Materials and Protocols for Ongoing QA/QC

Like all analytical methods, rapid analysis methods require ongoing performance monitoring for calibrations and instruments. The simplest methods for monitoring both the PLS equation performance and the NIR spectrometer is through use of a well-characterized reference standard. This standard is scanned at the

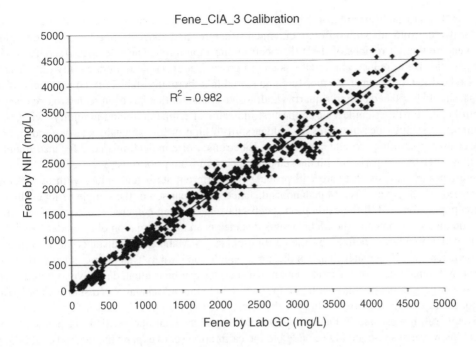

Figure 17.2   *Comparison of lab values and full cross-validation values for the calibration set.*

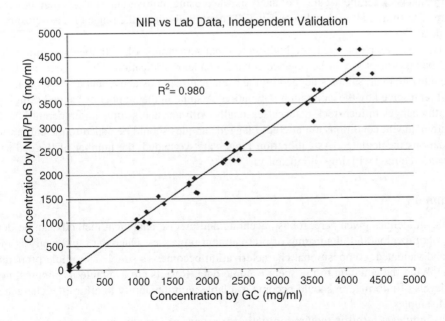

Figure 17.3   *Comparison of lab values and NIR/PLS results for an independent test set.*

beginning and end of each spectroscopic run and tracked by rigorous control charting. It is usually sufficient to monitor two constituents with different chemical structures and concentration levels, such as glucan and lignin, to ensure that all regions of the NIR spectrum are monitored. The reference standard should be of uniform particle size to reduce scatter effects and improve spectroscopic reproducibility [23]. The samples should also be dried to less than 5% moisture and stored in a thermally stable, dry environment. If possible, the sample should be placed behind a quartz window in a container that has been sealed to exclude moisture. The control chart for this sample will be the first indicator of temperature and humidity swings in the lab that could affect sample measurements. A sudden shift to high or low values is a strong indicator of light intensity changes that often precede lamp failure. Electronic noise indicating a need for instrument maintenance will be seen by comparison with the standard spectrum from prior runs [24].

All multivariate analysis software will provide two different statistical values with every constituent value reported. The first value is a Mahalanobis distance, a comparison of the sample being evaluated to the calibration population [25]. If the calibration spectra have been mean-centered and z-scored as part of multivariate equation development, the Mahalanobis distance is an approximation of a standard deviation from the mean of the population defined by the calibration set. A Mahalanobis distance of 0.0 represents the population mean. A value greater than 3 indicates a population outlier. This metric is particularly valuable in monitoring biomass-derived materials. Unknown samples can be evaluated relative to both internal controls and the larger sample population. It is important to note that a Mahalanobis distance value greater than 3 does not mean that the reported value for the constituent is wrong; it is only a flag indicating that the sample is not well represented in the calibration and the value should be verified independently. Samples flagged in this way can subsequently be added to the calibration set to expand the method calibration range.

The second statistical value is a measure of the differences between the sample being analyzed and the most similar calibration sample. This value allows quick identification of samples that represent levels of independent variability between constituents not represented in the current calibration. Samples flagged in this manner are valuable for their potential to improve the accuracy of multivariate analysis equations.

An additional check for the accuracy of analysis when using multivariate analysis methods is summative mass closure. The sum of all measured constituents should fall within the range of mass closure seen in the calibration data.

Rapid analysis is often deployed as a high-throughput screening tool, with the expectation that 1–2% of the samples being evaluated will be selected for independent verification. These independent verification batches should contain all samples flagged by the statistical methods above and at least 1% of the samples that are well represented by the calibration population. A spike in the number of flagged samples should be expected with changes in feedstock batches seasonally with unusual swings in temperature and humidity, and when pretreatment conditions are altered. Flagged samples should be added to the calibration set, and method equations recalculated. As calibration ranges are expanded, the number of flagged samples and method updates required will drop significantly.

17.3 Summary

Following the guidelines given here, robust, accurate and precise analytical methods can be developed to support the optimization and monitoring of pretreatment processes. Once the rapid analysis methods are developed and validated, compositional characterization becomes as simple as sample preparation, scanning, and applying the equation to the spectroscopic data. Results are available in minutes, not hours or days. New levels of information should become available with the ability to afford the characterization of thousands of samples.

Although significant effort is required initially for proper calibration, once in place, routine compositional analysis using rapid analysis methods usually requires much less technical knowledge and training than traditional wet chemical methods, making these methods very user-friendly and accessible. When calibrated using appropriate samples and high-quality data, rapid analysis methods can perform well in many environments with most operators and for decades with minimal updates [22].

References

1. Hames, B.R., Thomas, S.R., Sluiter, A.D. *et al.* (2003) Rapid biomass analysis. New tools for compositional analysis of corn stover feedstocks and process intermediates from ethanol production. *Applied Biochemistry and Biotechnology*, **105–108**, 5–16.
2. Perlack, R.D., Stokes, B.J. *et al.* (2011) US Department of Energy. US Billion-Ton Update: Biomass Supply for a Bioenergy and Bioproducts Industry. ORNL/TM-2011/224. Oak Ridge National Laboratory, Oak Ridge, TN. 227 p.
3. Burns, D.A. and Ciurczak, E.W. (eds) (2001) *Handbook of Near-Infrared Analysis*, Part IV. Applications, 2nd edn, Marcel Dekker, Inc., New York, NY, 419–802.
4. Hopkins, W.G. and Hüner, Norman P.A. (2008) *Introduction to Plant Physiology*, 4th edn, Wiley, NJ.
5. Templeton, D.W., Sluiter, AmieD., Hayward, T.K. *et al.* (2009) Assessing corn stover composition and sources of variability via NIRS. *Cellulose*, **16** (4), 621–639.
6. Hames, B.R. (2009) Chapter 11, Biomass compositional analysis for energy applications, in *Biofuels: Methods and Protocols, Series: Methods in Molecular Biology*, **581** (ed. Mielenz, J.), Humana.
7. Martin, M.E., Newman, S.D., Aber, J.D., and Congalton, R.G. (1998) Determining forest species composition using high spectral resolution remote sensing data. *Remote Sensing of Environment*, **85** (3), 249–254.
8. Kelley, S.S., Rowell, R.M., Davis, M. *et al.* (2004) Rapid analysis of the chemical composition of agricultural fibers using near-infrared spectroscopy and pyolysis molecular beam mass spectrometry. *Biomass and Bioenergy*, **27** (1), 77–88.
9. Adapa, P., Karunakaran, C., Tabil, L., and Schoenau, G. (2009) Potential applications of infrared and raman spectromicroscopy for agricultural biomass. *Agricultural Engineering International: the CIGRE Journal*, **X**, 1081.
10. Hedenström, M., Wiklund-Lundström, S., Oman, T. *et al.* (2009) Identification of lignin and polysaccharide modifications in poplus wood by chemometric analysis and 2D NMR spectra from dissolved cell walls. *Molecular Plant*, **2** (5), 933–942.
11. Jones, R.W., Meglen, R.R., Hames, B.R., and McClelland, J.F. (2002) Chemical analysis of wood chips in motion using thermal-emission mid-infrared spectroscopy with projection to latent structures regression. *Analytical Chemistry*, **74** (2), 453–457.
12. Friesen, W.E. (1996) Qualitative analysis of oil sand slurries using on-line NIR spectroscopy. *Applied Spectroscopy*, **50**, 1535–1540.
13. Wolfrum, E.J. and Sluiter, A.D. (2009) Improved multivariate calibration models for corn stover feedstock and dilute-acid pretreated corn stover. *Cellulose*, **16**, 567–576.
14. Massart, D.L., Vandeginste, B.G.M., Deming, S.M. *et al.* (1998) *Chemometrics: A Textbook (Data Handling in Science and Technology)*, Elsevier.
15. Kruse, T.K., Alexiades, A., East, G. *et al.* (2009) Small-scale Enzymatic Conversion Screens to Assist in the Development of Improved Energy Crop Varieties. Oral Presentation at 31st Symposium on Biotechnology for Fuels and Chemicals, San Francisco, CA, May 3–6, 2009, To request copies of presentation slides go to http://www.ceres.net.
16. Kruse, T., Alexiades, A., East, G. *et al.* (2009) Rapid conversion analysis of switchgrass feedstocks using NIR spectroscopy. Poster presentation at the 31st Symposium on Biotechnology for Fuels and Chemicals, San Francisco, CA, May 3–6, 2009, To request copies of poster presentation go to http://www.ceres.net.
17. Sluiter, J.B., Ruiz, R.O., Scarlata, C.J. *et al.* (2010) Compositional analysis of lignocellulosic feedstocks. 1. Review and description of methods. *Journal of Agricultural and Food Chemistry*, **58** (16), 9043–9053.
18. Templeton, D.W., Scarlata, C.J., Sluiter, J.B., and Wolfrum, E.J. (2010) Compositional analysis of lignocellulosic feedstocks. 2. Method uncertainties. *Journal of Agricultural and Food Chemistry*, **58** (16), 9054–9062.
19. Geladi., P. and Kowalski, B.R. (1986) Partial least squares regression: a tutorial. *Analytical Chemica Acta*, **185**, 1–17.

20. Burns, D.A. and Ciurczak, E.W. (eds) (2001) *Handbook of Near-Infrared Analysis*, 2nd edn, Marcel Dekker, New York.
21. DiFoggio, R. (1995) Examination of some misconceptions about near-infrared analysis. *Applied Spectroscopy*, **49** (1), 67–75.
22. Marten, G.C., Shenk, J.S., and Barton, F.E. (eds) (1989) Near infrared reflectance spectroscopy (NIRS): analysis of forage quality, in *Agriculture Handbook 643*, United States Department of Agriculture, Washington, D.C.
23. Theissing, H.H. (1950) Macrodistribution of light scattered by dispersions of spherical dielectric particles. *Journal of the Optical Society of America*, **40** (4), 232–242.
24. Wang, Y. and Kowalski, B.R. (1992) Calibration transfer and measurement stability of near- infrared spectrometers. *Applied Spectroscopy*, **46**, 764–771.
25. Gnanadesikan, R. and Kettenring, J.R. (1972) The mahalanobis distance. *Chemometrics and Intelligent Laboratory Systems*, **50**, 1–18.

18

Plant Biomass Characterization: Application of Solution- and Solid-state NMR Spectroscopy

Yunqiao Pu[1,3]**, Bassem Hallac**[2,3] **and Arthur J. Ragauskas**[1,2,3]

[1] *Georgia Institute of Technology, Atlanta, USA*
[2] *School of Chemistry and Biochemistry, Georgia Institute of Technology, Atlanta, USA*
[3] *BioEnergy Science Center, Oak Ridge, USA*

18.1 Introduction

Releasing fermentable sugars from lignocellulosic materials remains challenging due to resistance of plants to breakdown. A pretreatment stage is required to reduce this recalcitrance, which is considered to be the most intensive operating cost component of cellulosic ethanol production. There are different features that make plant biomass resistant to chemical and biological degradation, such as lignin content/structure, lignin-carbohydrate complexes (LCCs), hemicellulose content, as well as cellulose ultrastructure and degree of polymerization (DP). Research on this subject is therefore focused on understanding the effects of pretreatment technologies on the reduction of biomass recalcitrance as well as on fundamental structural characteristics of biomass that impact pretreatment and subsequent enzymatic hydrolysis. Improving our fundamental knowledge of pretreatment technologies will lead to significant advances in the field of sustainable low-cost cellulosic biofuels production [1].

Nuclear magnetic resonance (NMR) spectroscopy is a powerful tool for detailed structural elucidation of the major constitutes of plant biomass, lignin, hemicellulose, and cellulose [2–8]. Many of the structural details/characteristics of biopolymers (especially lignin) in native and transgenic plant biomass we know of today were revealed with NMR measurements, in addition to key information about lignin/hemicellulose synthesis in plants as well as plant genetic engineering [2,3]. This chapter focuses on the application of solution- and solid-state NMR spectroscopy techniques to characterize the structural features of cellulose

Aqueous Pretreatment of Plant Biomass for Biological and Chemical Conversion to Fuels and Chemicals, First Edition.
Edited by Charles E. Wyman.
© 2013 John Wiley & Sons, Ltd. Published 2013 by John Wiley & Sons, Ltd.

and lignin during aqueous pretreatment of plant biomass for biological and chemical conversion to fuels and chemicals. Specifically, the most commonly employed solution-state NMR techniques including one-dimensional (1D) ^1H, ^{13}C, and ^{31}P NMR and two-dimensional (2D) heteronuclear single quantum coherence (HSQC) will be discussed for lignin structure characterization. Solid-state cross-polarization/magic angle spinning (CP/MAS) ^{13}C NMR will also be reviewed as the technique used for analysis of the crystallinity and ultrastructure of plant cellulose.

18.2 Plant Biomass Constituents

Plant biomass, including woody and herbaceous lignocellulosics, is a natural biocomposite primarily composed of three major biopolymers (i.e., cellulose, hemicellulose, and lignin), usually with minor amounts of inorganics and extractives. In a plant cell wall, these polymers typically interact with each other physically and chemically to form an intricate three-dimensional network structure. Plants have a complicated and dynamic cell wall which is generally composed of three anatomical regions, that is, middle lamella, the primary cell wall, and the secondary cell wall. The cell types and chemical compositions vary among different species as well as among various regions of the same plant.

The predominant polysaccharide in plant biomass is cellulose, which is a linear homopolymer of (1 → 4)-linked β-D-glucopyranosyl units with the degree of polymerization varying from 300 to *c.* 15 000 [9,10]. The hydroxyl groups on these glucopyranosyl units have a strong tendency to form intra- and intermolecular hydrogen bonds among the linear glucan chains, which stiffen the chains and can facilitate cellulose aggregations to form highly ordered or crystalline cellulose fibril structures. Most plant celluloses also contain varying degrees of amorphous domains that are more amenable to chemical and enzymatic attack.

After cellulose, the next major polysaccharide in plant biomass is hemicellulose. Hemicelluloses generally refer to a group of mixed heteroglycans of pentoses and hexanoses which link together in a plant with a DP of *c.* 70–200 and frequently have branching and substitution groups [11,12]. The major hemicelluloses in softwoods include glucomannans and arabinoglucuronoxylan, while in many hardwood and herbaceous plants the predominant hemicellulose is glucuronoxylan.

Compared to cellulose and hemicellulose, lignin does not have a distinct chemical structure. Lignin is an amorphous and irregular polyphenolic biopolymer that is synthesized by enzymatic dehydrogenative polymerization of phenylpropanoid monolignols. Three types of phenylpropane units are generally considered as basic building blocks for biosynthesis of protolignin: coniferyl, sinapyl, and *p*-coumaryl alcohol (Figure 18.1), which correspond to the guaiacyl (G), syringyl (S) and *p*-hydroxyphenyl (H) structures of lignin, respectively [13–15]. Softwood lignin contains predominantly guaiacyl and minor amounts of

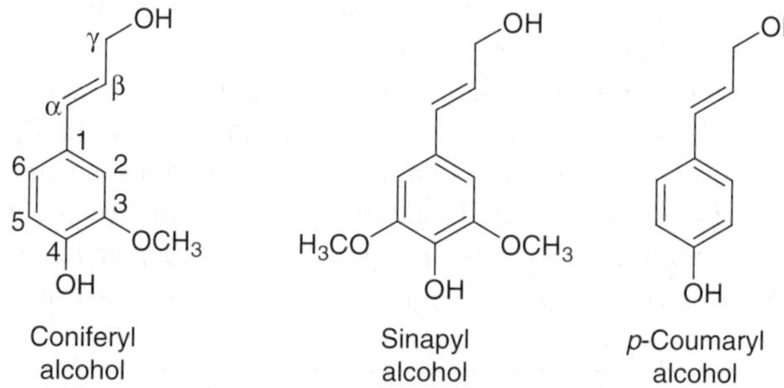

Figure 18.1   *Phenylpropanoid units involved in lignin biosynthesis [13].*

p-hydroxyphenyl units, while hardwood lignin is primarily composed of guaiacyl and syringyl units. Lignin in herbaceous plants generally contains all three types of monolignol units, with a core structure composed mainly of guaiacyl and syringyl units and incorporated peripheral groups such as *p*-hydroxycinnamic acid and ferulic acid units [14,15].

18.3   Solution-state NMR Characterization of Lignin

Lignin is considered as one of the most recalcitrant components in plant cell walls and protects plants against microbial and enzymatic deconstruction. Lignin is intimately associated with carbohydrate components in the cell walls of vascular plants, forming an amorphous network embedding microfibril cellulosic materials. The lignin macromolecule is primarily connected through carbon-carbon and carbon-oxygen ether bonds among the building blocks of phenylpropane monomers [13]. The structure of lignin is complex, irregular, and highly heterogeneous, with no regular extended repeating unit structures observed. Compared to other biopolymers, the structural determination of lignin is more challenging due to its complexity and the difficulty in isolation of a highly representative and structurally unchanged lignin sample from plant species. Although the exact structure of protolignin in a plant is still not fully understood, advances in spectroscopic methods (especially NMR techniques) and computational modeling have enabled scientists to elucidate the predominant structural features of lignin, such as inter-unit linkages and their relative abundances in plant biomass. Some common inter-unit linkages identified in lignin, such as β-O-4, α-O-4/β-5 (phenylcoumaran), β-β (pinoresinol), dibenzodioxocin, and 4-O-5, are presented in Figure 18.2. The relative proportions of such units are usually dependent on the biomass species as well as the processing methods employed [11,16]. The discovery of dibenzodioxocin in softwood lignins at a level of *c.* 10–15% in the early 1990s has initiated a flurry of research efforts that led to detection/identification of several new subunit structures such as spirodienone [17–21].

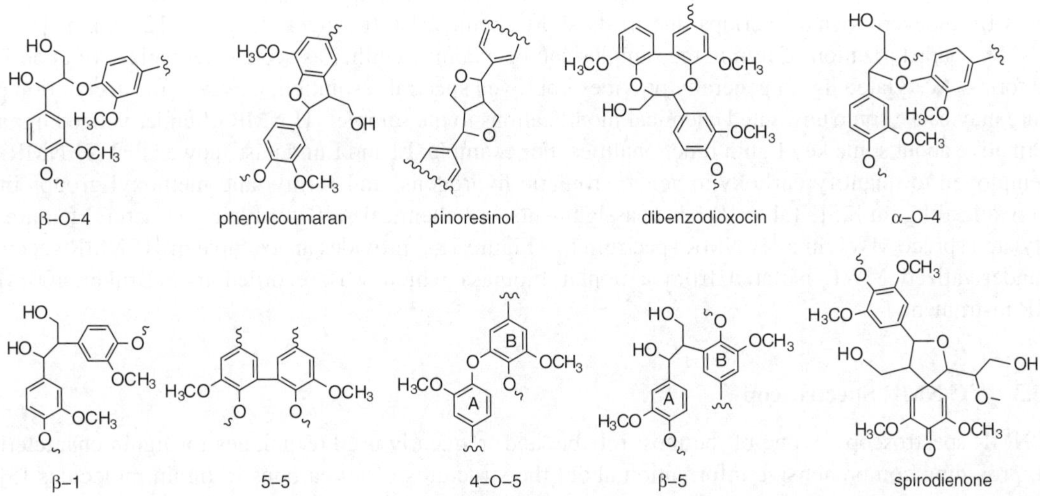

Figure 18.2   *Typical inter-unit linkages in plant lignins.*

18.3.1   Lignin Sample Preparation

No applicable method has yet been developed that can be regarded as ideal for the isolation of a highly representative native lignin with unaltered structures from plant biomass materials. To date, the most commonly used methods for lignin isolation are ball-milled lignin isolation developed by Bjorkman and cellulolytic enzyme lignin (CEL) preparation [22–24]. The Bjorkman method involves extensive milling of the plant materials followed by extraction with neutral solvents of dioxane-water at room temperature [22]; the resulting isolated material is usually referred to as milled wood lignin (MWL) when dealing with woody biomass. The milling is carried out either in a non-swelling medium such as toluene or in the dry state, usually under inert atmospheric gases such as nitrogen or argon in the milling jar. This method generally offers a low or moderate lignin yield (up to *c.* 25%), depending upon various plant species. Despite its low/ moderate yield and possible structural alteration during milling, this methodology is widely accepted as a typical lignin isolation method, affording a lignin sample with minimal structural alteration and the closest native representation.

CEL preparation treats finely ground (i.e., ball-milled) plant powders with cellulolytic enzymes prior to solvent extraction to partially remove polysaccharides [23]. The enzymatically treated plant meal is then successively extracted with 96% and 50% aqueous dioxane, yielding two lignin fractions. Compared to the traditional Bjorkman procedure, the CEL procedure offers a significantly improved lignin yield (up to 55%) for various species, which is probably more representative of total lignin in a plant. However, this preparation usually suffers from higher carbohydrate contamination (*c.* 10–12% for a spruce wood compared to less than 5% in the Bjorkman procedure) of the lignin and is a more tedious and time-consuming process [24].

18.3.2 ^{1}H NMR Spectroscopy

^{1}H NMR spectroscopy is a valuable technique for characterization and classification of lignin structural features. Earlier work on the characterization of lignin by NMR mainly relied on ^{1}H NMR spectroscopy, and extensive databases of ^{1}H NMR chemical shifts were established for lignin model compounds and functional groups [6,8]. One advantage of ^{1}H NMR is that the ^{1}H nucleus is the most abundant among the nuclei that can be detected by NMR, thus giving a high signal to noise (S/N) ratio in a short experimental time (typically within several minutes). The drawback of ^{1}H NMR for lignin analysis is that it usually suffers from severe signal overlaps due to its short chemical shift ranges (i.e., δ *c.* 12–0 ppm) [8]. For ^{1}H NMR characterization of lignin, the samples can be examined either as acetate derivatives or underivatized forms. Acetylated lignin generally provides improved spectral resolution; however, the acetylation procedure may cause some unwanted chemical modifications to the sample. ^{1}H NMR of underivatized lignin is informative about some key lignin functionalities. For example, Li and Lundquist showed that ^{1}H NMR can be employed to quantify carboxylic acids, aromatic hydrogens, and formyl and methoxyl groups in an underivatized lignin [25]. Table 18.1 lists assignments and chemical shifts of typical structural features of acetylated spruce MWL in a ^{1}H NMR spectrum [8]. Figure 18.3 provides an example of ^{1}H NMR spectrum of underivatized MWL isolated from a poplar biomass which was recorded in a Bruker 400-MHz NMR instrument.

18.3.3 ^{13}C NMR Spectroscopy

^{13}C NMR spectroscopy is one of the most reliable and frequently used techniques for lignin characterization, providing comprehensive information about the structures of all carbons in lignin molecules [5]. It benefits from a broader spectral window (i.e., δ *c.* 240–0 ppm) in comparison to ^{1}H NMR, with better resolution and less overlap of signals. A routine qualitative ^{13}C NMR spectrum of lignin is generally recorded

Table 18.1 *Typical signals assignment and chemical shifts in the* ^1H *NMR spectrum of acetylated spruce lignin using deuterated chloroform as solvent [8].*

δ (ppm)	Assignment
1.26	Hydrocarbon contaminant
2.01	Aliphatic acetate
2.28	Aromatic acetate
2.62	Benzylic protons in β-β structures
3.81	Protons in methoxyl groups
4.27	H_γ in several structures
4.39	H_γ in, primarily, β-O-4 structures and β-5 structures
4.65	H_β in β-O-4 structures
4.80	Inflection possibly due to H_α in pinoresinol units and H_β in noncyclic benzyl aryl ethers
5.49	H_α in β-5 structures
6.06	H_α in β-O-4 structures (H_α in β-1 structures)
6.93	Aromatic protons (certain vinyl protons)
7.41	Aromatic protons in benzaldehyde units and vinyl protons on the carbon atoms adjacent to aromatic rings in cinnamaldehyde units
7.53	Aromatic protons in benzaldehyde units
9.64	Formyl protons in cinnamaldehyde units
9.84	Formyl protons in benzaldehyde units

with a pulse angle in the range 30–60°, a pulse delay of *c.* 0.5–2 s, and the transient number of about 10 000–20 000. Quantitative information about specific functional groups and structures present in lignin can be also estimated when the ^{13}C NMR spectrum is recorded under quantitative requirement conditions, although the recording of spectra is time consuming. In general, a quantitative ^{13}C NMR spectrum with a satisfactory signal-to-noise ratio is obtained at *c.* 50 °C using a 90° pulse, a pulse delay of about 12 s, an inversed gated decoupling pulse sequence, and thousands of transient acquisition numbers. The total experiment time is much longer than typical ^1H NMR, usually being up to 24–36 h.

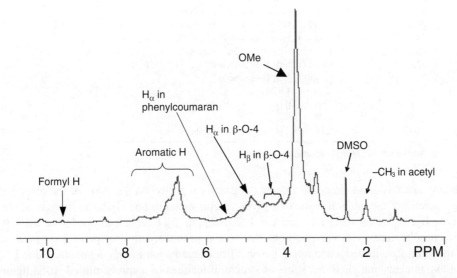

Figure 18.3 *An example of* ^1H *NMR spectrum of a poplar mill-wood lignin using DMSO as solvent.*

Table 18.2 *Typical chemical shifts and signal assignments for a spruce milled wood lignin (MWL) in a* [13]*C NMR spectrum [5,26,27]..*

δ (ppm)	Assignment
193.4	C=O in Ar—CH=CH—CHO; C=O in Ar—CO—CH(—OAr)—C—
191.6	C=O in Ar—CHO
169.4	Ester C=O in R'—O—CO—CH$_3$
166.2	C=O in Ar—COOH; Ester C=O in Ar—CO—OR
156.4	C-4 in H-units
152.9	C-3/C-3' in etherified 5-5 units; C-α in Ar—CH=CH—CHO units
152.1	C-3/C-5 in etherified S units and B ring of 4-O-5 units[a]
151.3	C-4 in etherified G units with α-C=O
149.4	C-3 in etherified G units
149.1	C-3 in etherified G type β-O-4 units
146.8	C-4 in etherified G units
146.6	C-3 in non-etherified G units (β-O-4 type)
145.8	C-4 in non-etherified G units
145.0	C-4/C-4' of etherified 5-5 units
143.3	C-4 in ring B of β-5 units[a]; C-4/C-4' of non-etherified 5-5 units
134.6	C-1 in etherified G units
132.4	C-5/C-5' in etherified 5-5 units
131.1	C-1 in non-etherified 5-5 units
129.3	C-β in Ar—CH=CH—CHO
128.0	C-α and C-β in Ar-CH=CH—CH$_2$OH
125.9	C-5/C-5' in non-etherified 5-5 units
122.6	C-1 and C-6 in Ar—CO—C—C units
119.9	C-6 in G units
118.4	C-6 in G units
115.1	C-5 in G units
114.7	C-5 in G units
111.1	C-2 in G units
110.4	C-2 in G units
86.6	C-α in G type β-5 units
84.6	C-β in G type β-O-4 units (threo)
83.8	C-β in G type β-O-4 units (erythro)
71.8	C-α in G type β-O-4 units (erythro)
71.2	C-α in G type β-O-4 units (threo); C-γ in G type β-β
63.2	C-γ in G type β-O-4 units with α-C=O
62.8	C-γ in G type β-5, β-1 units
60.2	C-γ in G type β-O-4 units
55.6	C in Ar—OCH$_3$
53.9	C-β in β-β units
53.4	C-β in β-5 units
40-15	CH$_3$ and CH$_2$ in saturated aliphatic chain

[a] see Figure 18.2 for structures; Ar: aromatic.

Quantitative data on lignin structural features is usually reported on the basis of the phenylpropane by calculating the ratio of the integral value of a given carbon signal to one-sixth the integral of the aromatic carbons whose signals are located in the range *c.* 102–162 ppm [5]. Corrections need to be made if a sample contains structural features with vinyl carbons that appear in this region. Figure 18.4 shows a quantitative ^{13}C NMR spectrum for a milled wood lignin isolated from a hardwood *Buddleja davidii*. Table 18.2 summarizes signal assignments and chemical shifts of structural features of a spruce milled wood lignin in a ^{13}C NMR measured using deuterated dimethyl sulfoxide (DMSO) as solvent [5,26,27].

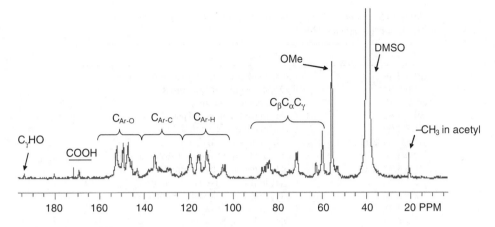

Figure 18.4 *Quantitative* ^{13}C *NMR spectrum of a milled wood lignin isolated from a hardwood* Buddleja davidii *[28]. Ar: aromatic; OMe: methoxyl; DMSO: dimethyl sulfoxide.*

Hallac *et al.* [28] employed quantitative ^{13}C NMR to characterize structural features of milled wood lignin isolated from a shrub *Buddleja davidii* as a potential feedstock for biofuels production, and reported that the lignin had an approximate guaiacyl/syringyl (G/S) ratio of 81:19 with no H-units observed. They further investigated the chemical transformations of *Buddleja davidii* lignin during ethanol organosolv pretreatment using ^{13}C NMR together with other NMR techniques, and demonstrated quantitative changes of lignin structural features such as substantive decrease of β-O-4 linkages after the pretreatments [29]. Using ^{13}C NMR, Samuel *et al.* [30] observed a 36% decrease in β-O-4 linkages and an S/G ratio decrease from 0.80 to 0.53 for ball-milled lignin isolated from Alamo switchgrass after dilute acid pretreatment. Sannigrahi *et al.* investigated lignin isolated from loblolly pine before and after ethanol organosolv pretreatment and reported a *c.* 50% decrease in β-O-4 linkages utilizing quantitative ^{13}C NMR analysis, suggesting that acid-catalyzed cleavage of β-O-4 linkages was a major mechanism for lignin cleavage [31,32].

18.3.4 HSQC Correlation Spectroscopy

Although 1D ^{1}H and ^{13}C NMR are very efficient tools for lignin structural analysis, these classical 1D NMR techniques usually suffer from signal overlaps in spectra. With advancements in NMR instrumentation, these issues are now addressed with a host of 2D and 3D NMR techniques, among which HSQC is the most commonly applied. Heteronuclear multidimensional correlation NMR experiments can not only increase the sensitivity of ^{13}C nuclei by polarization transfer but also separate overlapped signals which usually occur in 1D spectra, making ^{1}H-^{13}C correlation methods very efficient for lignin structural analysis. Indeed, the use of 2D NMR has been instrumental in advancing the analysis of lignin structure, especially in discovering new lignin subunits and the presence of lignin-carbohydrate complexes such as dibenzodioxocin and spirodienone structures [4,17–21]. However, it has limitations in that it is not quantitative and a spectral overlap of lignin functionality still occurs. The application of these techniques to lignin isolated from native as well as genetically altered plants was recently summarized by Ralph *et al.* [33].

HSQC is the most frequently collected 2D NMR spectrum that is used not only for structural identification but also for estimation of the relative abundance of inter-unit linkages as well as S/G ratios in lignin. Table 18.3 summarizes chemical shifts and assignments of cross-peaks of typical inter-unit linkages and/or subunits in HSQC spectra of lignin when using DMSO as solvent. Although this approach

Table 18.3   *Chemical shifts and assignment of ^{13}C-^1H correlation signals in HSQC spectra of lignin [34–39].*

δ_C/δ_H (ppm)	Assignment[a]
53.1/3.44	C_β/H_β in phenylcoumaran substructure (B)
53.6/3.03	C_β/H_β in resinol substructure (C)
55.7/3.70	C/H in methoxyl group
59.8/3.62	C_γ/H_γ in β-O-4 ether linkage (A)
61.7/4.09	C_γ/H_γ in cinnamyl alcohol (F)
62.3/4.08,3.95	C_γ/H_γ in dibenzodioxocin
62.8/3.76	C_γ/H_γ in phenylcoumaran substructure (B)
71.1/3.77, 4.13	C_γ/H_γ in resinol substructure (C)
71.4/4.76	C_α/H_α in β-O-4 linked to a G unit (A)
72.1/4.86	C_α/H_α in β-O-4 linked to a S unit (A)
76.0/4.81	C_α/H_α in benzodioxane
78.2/4.00	C_β/H_β in benzodioxane
81.4/5.1	C_β/H_β in spirodienone substructure
83.7/4.31	C_β/H_β in β-O-4 linked to a G unit (A)
84.2/4.69	C_α/H_α in dibenzodioxocin
84.7/4.7	C_α/H_α in spirodienone substructure
85.2/4.63	C_α/H_α in resinol substructure (C)
86.3/4.13	C_β/H_β in β-O-4 linked to a S unit (A)
86.6/4.08	C_β/H_β in dibenzodioxocin
87.0/5.52	C_α/H_α in phenylcoumaran substructure (B)
103.8/6.70	$C_{2,6}/H_{2,6}$ in syringyl units (S)
105.5/7.3	$C_{2,6}/H_{2,6}$ in oxidized syringyl (S') units with $C_\alpha = O$
111.0/6.98	C_2/H_2 in guaiacyl units (G)
114.8/6.73	$C_{3,5}/H_{3,5}$ in p-hydroxyphenyl units (H)
115.1/6.72, 6.98	C_5/H_5 in guaiacyl units
119.1/6.80	C_6/H_6 in guaiacyl units
128.0/7.17	$C_{2,6}/H_{2,6}$ in p-hydroxyphenyl units
128.2/6.75	C_β/H_β in cinnamaldehyde unit (E)
128.3/6.45	C_α/H_α in cinnamyl alcohol (F)
128.3/6.25	C_β/H_β in cinnamyl alcohol (F)
130.6/7.65, 7.87	$C_{2,6}/H_{2,6}$ in p-hydroxybenzoate units (D)
153.6/7.62	C_α/H_α in cinnamaldehyde unit (E)

[a] G: guaiacyl; S: syringyl; S″ = oxidized syringyl with C_α=O; H: p-hydroxyphenyl; A: β-O-4 ether linkage; B: β-5/α-O-4 phenylcoumaran; C: resinol (β-β); D: p-hydroxybenzoate; E: cinnamaldehyde; F: cinnamyl alcohol.

is not considered quantitative, it has been widely employed in a semi-quantitative way to offer relative comparisons of inter-unit linkage levels in biomass lignins, such as alfalfa, eucalyptus, and poplar [34–37]. The well-resolved α-carbon contours in various linkages are generally used for volume integration, and the relative abundance of each respective inter-unit linkage is then calculated as the percentage of integrals of total linkages [34,37].

Pu *et al.* [38] recently investigated structural characteristics of lignin isolated from wild-type alfalfa, p-coumarate 3-hydroxylase (C3H) down-regulated, and hydroxycinnamoyl CoA:shikimate/quinate hydroxycinnamoyl transferase (HCT) down-regulated alfalfa transgenic lines. The HSQC spectra of lignin in wild-type and transgenic alfalfa shown in Figures 18.5 and 18.6 illustrate ^{1}H-^{13}C correlation signals in aromatic regions and aliphatic side-chain ranges, respectively. The HSQC spectra demonstrated that C3H and HCT down-regulated alfalfa lignins showed significantly more p-hydroxyphenyl units along with a substantial

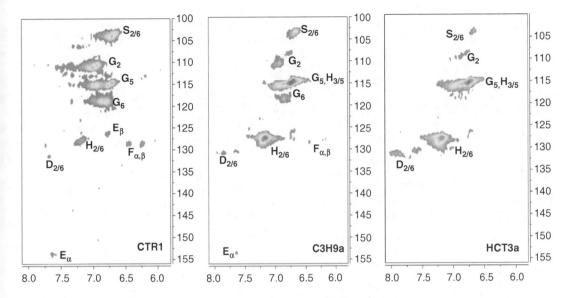

Figure 18.5 *Aromatic regions of $^{13}C/^{1}H$ HSQC NMR spectra of alfalfa ball-milled lignins [38]. G: guaiacyl; S: syringyl; H: p-hydroxyphenyl; D: p-hydroxybenzoate; E: cinnamaldehyde; F: cinnamyl alcohol. CTR1: wild type; C3H9a: p-coumarate 3-hydroxylase (C3H) transgenic line; HCT3a: hydroxycinnamoyl CoA:shikimate/quinate hydroxycinnamoyl transferase (HCT) transgenic line.*

decrease in the intensities of syringyl and guaiacyl correlations. Compared to the wild-type plant, the C3H and HCT transgenic lines had an increased phenylcoumaran and resinol linkages in lignins. Using HSQC NMR, Ralph *et al.* [34] also characterized the structures of acetylated alfalfa lignins isolated from severely down-regulated C3H transgenic lines. They reported structural differences in inter-unit linkage distribution with a decrease in β-aryl ether units, which were accompanied by relatively higher levels of phenylcoumarans and resinols. In addition, Ralph *et al.* [34] provided data for coupling and cross-coupling propensities of *p*-coumaryl alcohol and *p*-hydroxyphenyl units in C3H-deficient alfalfa lignin and revealed that (1) β-ether units were from G, S, and *p*-hydroxyphenyl but with relatively low levels of the G- (especially) and S-units, and (2) the phenylcoumarans were almost entirely from *p*-hydroxyphenyl units. Moinuddin *et al.* [39] investigated lignin structures of *Arabidopsis thaliana* COMT mutant *Atomt1* using HSQC NMR and suggested that β-O-4 linkage frequency in the lignin isolates of both the *Atomt1* mutant and wild-type line was conserved.

18.3.5 ^{31}P NMR Spectroscopy

An approach to deal with the limitations of the general 1D ^{1}H and ^{13}C NMR and 2D correlation NMR techniques is to "selectively tag" specific functional groups in lignin with an NMR active nucleus through derivatization and then analyze the derivatized lignin with NMR. Phosphorous reagents have been employed to tag hydroxyl groups or quinone structures in lignin for determining their concentration by ^{31}P NMR [28–30,40–43]. Hydroxyl groups, especially free phenoxy groups, are among the most important functionalities affecting physical and chemical properties of lignin. These functional groups exhibit a prominent role in defining reactivity of lignin to promote cleavage of inter-unit linkages and/or oxidative degradation during pretreatment processes. The traditional wet chemistry methods employed to determine

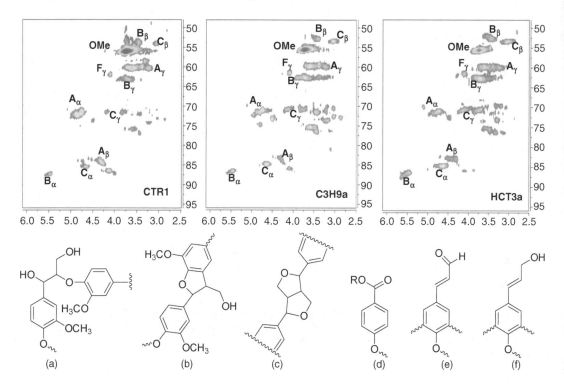

Figure 18.6 *Aliphatic regions of* $^{13}C/^{1}H$ *HSQC NMR spectra of alfalfa ball-milled lignins and the identified structures [38]. A: β-O-4 ether linkage; B: phenylcoumaran (β-5/α-O-4); C: resinol (β-β); OMe: methoxyl group; D: p-hydroxybenzoate; E: cinnamaldehyde; F: cinnamyl alcohol; R = H or C; CTR1: wild type; C3H9a: p-coumarate 3-hydroxylase (C3H) transgenic line; HCT3a: hydroxycinnamoyl CoA:shikimate/quinate hydroxycinnamoyl transferase (HCT) transgenic line (See figure in color plate section).*

hydroxyl contents in lignin typically involve time-consuming and/or laborious multistep derivatizations [43]. The ^{31}P NMR method has been shown to be very effective for determining the presence of hydroxyl groups in lignin. It can provide quantitative information in a single spectrum for various types of major hydroxyl groups including aliphatic, carboxylic, guaiacyl, syringyl, C_5-substituted phenolic hydroxyls, and p–hydroxyphenyls in a relatively short experimental time and with small sample size requirements. The quantitative information gained from this technique has been verified against other techniques such as GC, ^{1}H NMR, ^{13}C NMR, Fourier transform infrared (FTIR), and wet chemistry methods during an international "round robin" lignin study [44,45]. Compared to ^{1}H NMR, the large range of chemical shifts for ^{31}P nucleus generates a better separation and resolution of signals. In addition, the 100% natural abundance of the ^{31}P and its high sensitivity renders ^{31}P NMR a rapid analytical tool in comparison to ^{13}C NMR.

The ^{31}P NMR technique usually involves treating lignin samples with the phosphorylation reagent 2-chloro-4,4,5,5 tetramethyl-1,3,2-dioxaphospholane (TMDP) to phosphorylate the labile hydroxyl protons in lignin according to the reaction outlined in Figure 18.7. The ^{31}P NMR spectrum of this derivatized sample is then recorded with an internal standard such as cyclohexanol or N-hydroxy-5-norbornene-2,3-dicarboximide. The ^{31}P NMR spectrum contains well-separated peaks corresponding to the various types of hydroxyl groups present in lignin. The fact that these peaks are well separated is very important in making it possible to distinguish between regions containing various types of hydroxyl groups and allow for their

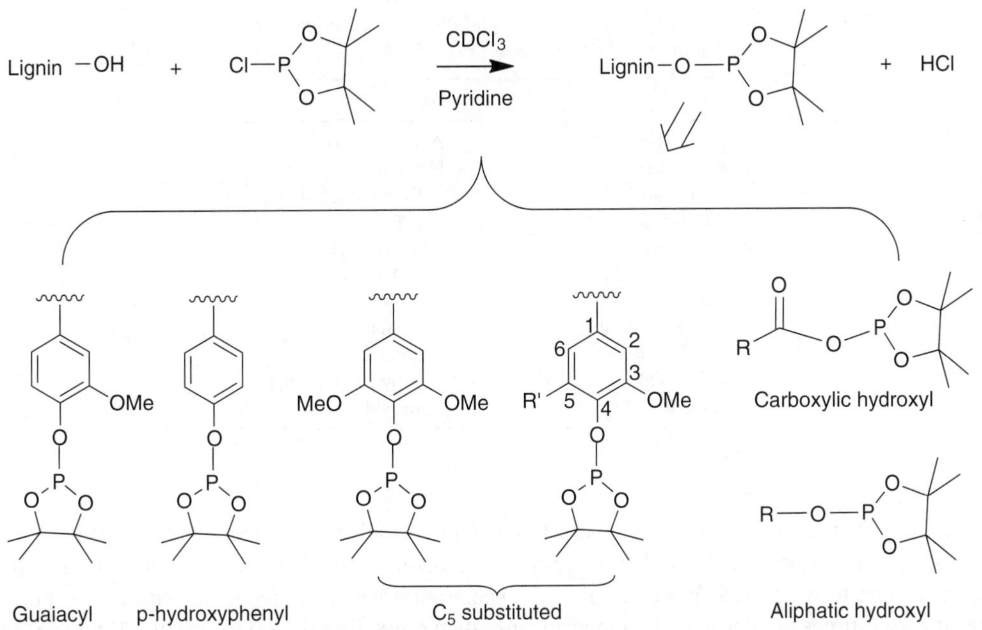

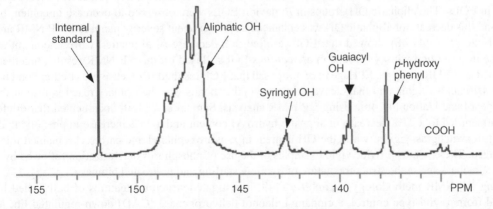

Figure 18.7 *Reaction of labile hydroxyls present in lignins with 2-chloro-4,4,5,5 tetramethyl-1,3,2-dioxaphospholane. R: Lignin side-chain; R′: lignin unit.*

accurate integration. Figure 18.8, which is a ^{31}P NMR spectrum of a hardwood lignin derivatized with 2-chloro-4,4,5,5 tetramethyl-1,3,2-dioxaphospholane, illustrates the well-separated signals arising from the various hydroxyl groups in lignin. A compilation of typical assignments and chemical shifts/integration ranges of hydroxyl groups in lignin using TMDP/^{31}P NMR analysis is shown in Table 18.4 [46].

For ^{31}P NMR analysis of lignin, the solvent employed is usually a mixture of anhydrous pyridine and deuterated chloroform (*c.* 1.6 : 1.0 v/v) containing a relaxation agent (i.e., chromium (III) acetylacetonate) and an internal standard. An accurately weighed dried lignin sample (10–25 mg) is dissolved in a NMR

Figure 18.8 *Quantitative ^{31}P NMR spectrum of a hardwood lignin derivatized with TMDP using N-hydroxy-5-norbornene-2,3-dicarboximide as internal standard.*

Table 18.4 *Typical chemical shifts and integration regions for lignins in a ^{31}P NMR spectrum [46].*

δ (ppm)	Assignment
145.4–150.0	Aliphatic OH
137.6–144.0	Phenols
a) 140.2–144.5	C_5 substituted phenols
c. 143.5	β–5
c. 142.7	Syringyl
c. 142.3	4–O–5
c. 141.2	5–5
b) 139.0–140.2	Guaiacyl
c) *c.* 138.9	Catechol
d) *c.* 137.8	*p*–hydroxyphenyl
133.6–136.0	Carboxylic OH

solvent mixture (0.50 mL). TMDP reagent (*c.* 0.05–0.10 mL) is added and stirred for a short period of time at room temperature. Since the derivatization reagent is moisture sensitive, all efforts need to be directed at reducing exposure to water. Quantitative ^{31}P NMR spectra are generally recorded with a long pulse delay which is at least 5 times greater than the longest spin-lattice relaxation time (i.e., T_1) of ^{31}P nucleus to allow phosphorus nuclei to reach thermal equilibrium prior to a subsequent pulse. Chromium (III) acetylacetonate in a solvent system is generally used as a relaxation agent to shorten the spin-lattice relaxation time of phosphorus nuclei. Typically, a 25 s pulse delay is considered appropriate for quantitative ^{31}P NMR analysis of lignin. In addition, an inverse gated decoupling pulse is employed to eliminate nuclear Overhauser effects for quantification. Using a 90° pulse and the conditions above, 128–256 acquisitions (*c.* 1–2 h) at room temperature are sufficient to acquire a spectrum with a satisfactory S/N ratio.

Sannigrahi *et al.* [31,32] employed ^{31}P NMR to characterize ball-milled lignin isolated from loblolly pine and reported that ethanol organosolv pretreatment (EOP) of loblolly pine resulted in an ethanol organosolv lignin (EOL) with a higher content of guaiacyl phenolic, *p*-hydroxyl phenolic, and carboxylic hydroxyl groups. Hallac *et al.* [28,29] applied ^{31}P NMR to determine the hydroxyl content of *Buddleja davidii* lignin during ethanol organosolv pretreatments with various pretreatment severities. Compared to milled wood lignin from native *B. davidii*, the amount of phenolic OH, both condensed and guaiacyl, increased significantly in EOLs. The aliphatic OH groups in *B.* davidii EOLs were observed to decrease in content by 41–59%, and this decrease of aliphatic OH was enhanced as pretreatment severity increased. ^{31}P NMR analysis by El Hage *et al.* [47,48] showed that EOP resulted in a decrease of aliphatic hydroxyl content and an increase in phenolic hydroxyl groups in *Miscanthus* EOLs. Based on the ^{31}P NMR results, together with ^{13}C NMR and FTIR analysis, El Hage *et al.* proposed that EOP resulted in extensive aryl-ether bond hydrolysis of *Miscanthus* lignin and that cleavage of α-aryl ether bonds was the primary reaction responsible for lignin depolymerization [47,48]. Using ^{31}P NMR analysis, Samuel *et al.* [30] documented that dilute-acid pretreatment led to a 27% decrease in aliphatic hydroxyl content and a 25% increase in phenolic hydroxyl content in switchgrass lignin, while the OH content in *p*-hydroxyphenyl and carboxyl remained relatively unchanged. These results provide vivid examples of the use of phosphorus derivatization followed by NMR analysis to characterize the structural nature of lignin in starting and pretreated biomass.

Using ^{31}P NMR methodology, Akim *et al.* [49] investigated structural features of ball-milled lignins isolated from a wild-type control, a cinnamyl alcohol dehydrogenase (CAD) down-regulated line, and a caffeic acid/5-hydroxyferulic acid O-methyl transferase (COMT) down-regulated transgenic poplar. According to the ^{31}P NMR results, Akim *et al.* [49] documented that moderate CAD down-regulation (70%

deficient) resulted in no drastic changes in structures of poplar lignin. More severe CAD depletion for 6-month old poplar led to a slight increase in the amount of condensed phenolic hydroxyls, which the authors suggested was indicative of a higher degree of cross-linked lignin. Compared to the wild-type control, COMT down-regulation (90% deficient) yielded a poplar lignin with a lower content of syringyl and aliphatic OH group as well as an increased guaiacyl phenolic OH amount, while *p*-hydroxyphenyl and carboxylic OH content was observed to remain unchanged after COMT down-regulation [49].

18.4 Solid-state NMR Characterization of Plant Cellulose

18.4.1 CP/MAS ^{13}C NMR Analysis of Cellulose

Since the early 1980s, the solid-state CP/MAS ^{13}C NMR technique has been widely applied for investigation of structural features of cellulose, providing not only information of crystallinity index but also enabling a thorough investigation of the ultrastructure of cellulose [50,51]. Cellulose is a linear polymer made up of β-D glucopyranose units covalently linked by $1 \rightarrow 4$ glycosidic bonds, with the degree of polymerization varying with the biomass source [10,52]. Figure 18.9 illustrates the molecular structure of cellulose [52].

The large number of hydroxyl groups on cellulose chains form intra- and inter-molecular hydrogen bonds, resulting in the crystalline structure of cellulose. Cellulose also has a less-ordered structure called amorphous cellulose. Native crystalline cellulose (cellulose I) has been shown to co-exist in the form of two allomorphs (i.e., I_α and I_β) [50]. The degree of cellulose crystallinity has been studied over the years for many species, and a term called crystallinity index (CrI) is widely used to represent the relative proportion of crystalline cellulose to the total cellulose present in a material. The two most common techniques used to measure this value are X-ray diffraction (XRD) and solid-state CP/MAS ^{13}C NMR. XRD is based on the concept that the X-ray scattering can be divided into two components due to crystalline and amorphous structures [53]. Table 18.5 presents the degree of crystallinity of several cellulose samples measured by XRD [10,11,53].

The cellulose sample for solid-state CP/MAS ^{13}C NMR is usually ground and packed into a MAS rotor which is then inserted into a MAS probe and spun at a frequency of 5–10 kHz. CP/MAS ^{13}C NMR measurements are generally carried out with a 90° proton pulse, 0.8–2.0 ms contact pulse, 4 s recycle delay, and 2024–8192 scans for a good S/N ratio. The spectra are usually recorded on moist samples (*c.* 30–60% water content) to increase signal resolution. The solid-state CP/MAS ^{13}C NMR technique to measure the crystallinity of cellulose is based on the intensity of the two peaks in the C-4 region (δ *c.* 80–92 ppm): the first peak corresponds to the crystalline structure ($\delta = 86$–92 ppm), whereas the amorphous region is located in the range of $\delta = 80$–86 ppm. This method is referred to as the C-4 peak separation NMR technique. In order for this technique to be accurate, hemicelluloses and lignin must be removed from the cellulose samples because they interfere with the area of the amorphous region [54]. Figure 18.10 shows a typical CP/MAS ^{13}C NMR spectrum of a cellulose sample isolated from switchgrass. Cellulose crystallinity values for various lignocellulosic materials measured by this method are summarized in Table 18.6 [7,28,31,55,56].

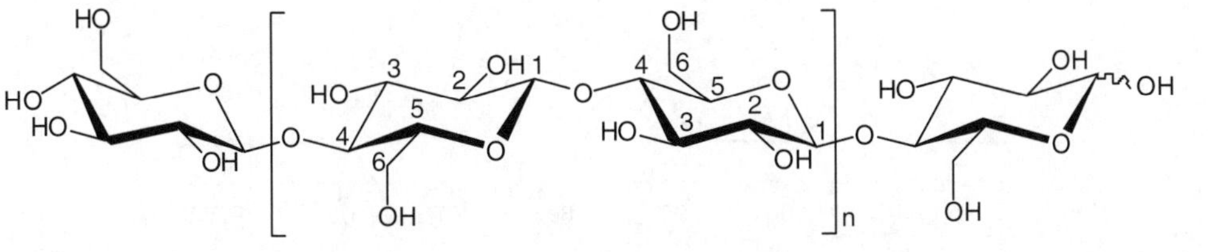

Figure 18.9 *Molecular structure of cellulose [52].*

Table 18.5   *Crystallinity index of some cellulosic materials determined by XRD [10,11,53].*

Sample	Crystallinity index (%)
Cotton linters	50–63
Wood pulp	50–70
Viscose rayon	27–40
Regenerated cellulose film	40–45
Avicel	50–60
Cotton	81–95
Algal cellulose	>80
Bacterial cellulose	65–79
Ramie	44–47

Table 18.6   *Crystallinity index of some cellulosic materials measured by C-4 peak separation NMR technique [7,28,31,55,56].*

Sample	Crystallinity index (%)
Hybrid poplar	63
Loblolly pine	63
Switchgrass alamo	44
Buddleja davidii	55
Spruce	48
Birch	36

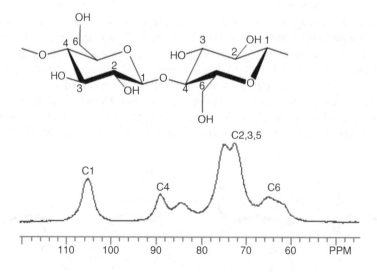

Figure 18.10   *CP/MAS ^{13}C -NMR spectra of cellulose isolated from switchgrass.*

In a recent study by Park *et al.*, a new NMR technique to measure cellulose crystallinity was introduced [57]. This novel method is based on digitally subtracting the spectrum of standard amorphous cellulose from the original spectrum. The authors believe that this method is straightforward, easier than XRD and C-4 peak separation methods, and could be applied to cellulose with any level of crystallinity [57].

The current interest in determining the degree of crystallinity of cellulose is to correlate the relationship between changes in CrI during pretreatment to the efficiency of enzymatic hydrolysis in the biological process of converting cellulosic biomass to biofuels. It is believed that cellulose crystallinity is a key property contributing to biomass recalcitrance [58]. Lowering cellulose crystallinity should therefore make it more readily digestible by cellulase. However, it has been suggested in various studies that changes in crystallinity after pretreatment and its effect on the enzymatic hydrolysis are related to the nature of the original material as well as the pretreatment technologies employed. Therefore, crystallinity index itself may not provide enough information to explain enzymatic hydrolysis behavior. Greater importance needs to be assigned to the ultrastructure of cellulose and not just to crystallinity in general.

18.4.2 Cellulose Crystallinity

In the realm of the utilization of cellulose as a material and as a feedstock for biofuels production, many studies have examined the effect of pretreatment on the crystallinity of cellulose. Table 18.7 [56–63] summarizes the CrI of cellulose of various untreated and pretreated biomass after some pretreatment processes such as ammonia fiber expansion (AFEX), ammonia recycled percolation (ARP), controlled pH, dilute sulfuric acid, lime, SO_2, ozone (O_3), carbon dioxide explosion (CE), alkaline explosion (AE), and organosolv.

As to be expected, each pretreatment had a different effect on cellulose crystallinity. Some pretreatment technologies caused a reduction in cellulose crystallinity, some showed no effect on crystallinity, and others exhibited an increase in crystallinity. Low-pH pretreatments generally enriched biomass crystallinity, while all high-pH pretreatments had less effect and even reduced biomass crystallinity in some instances. It also appears that the effect of the same pretreatment is biomass-dependent. For instance, controlled-pH pretreatment (i.e., controlling the pH at near neutral conditions) reduced the crystallinity index of corn stover, while it caused an increase in crystallinity of poplar. In general, dilute acid, lime, CO_2 explosion, alkaline explosion, and SO_2 pretreatments increased cellulose crystallinity due to the fact that amorphous cellulose degrades more easily than the more stable crystalline cellulose during pretreatment. Pretreatments such as AFEX, ARP, and ethanol organosolv are capable of reducing the crystallinity of cellulose, suggesting possible decrystallization of cellulose. Furthermore, the resulting crystallinity of pretreated biomass depends on the pretreatment conditions employed. Table 18.8 shows that lodgepole pine cellulose crystallinity increased when the pretreatment severity increased from condition set 1 to 3, causing an enrichment in the crystalline form of cellulose due to selective hydrolysis of amorphous cellulose during pretreatment [64].

Since cellulose crystallinity is thought to be a key property contributing to plant recalcitrance, many studies have focused on establishing a correlation between crystallinity and enzymatic hydrolysis of cellulose. A study by Jeoh *et al.* demonstrated the effect of cellulose crystallinity on cellulase accessibility [58]. Amorphous cellulose samples were prepared from Avicel and filter paper as follows. Cellulose was dissolved in a dimethylsulfoxide-paraformaldehyde solution, and then regenerated by slow addition of the cellulose solution to a solution of 0.2 M sodium alkoxide in methanol/i-propanol (1 : 1). This procedure has been demonstrated to produce amorphous cellulose without altering the degree of polymerization (DP) and reducing end-group concentration of the starting cellulose [58,65]. The resulting amorphous forms of cellulose were found to be significantly more digestible by cellulase than the original crystalline forms. For both Avicel and filter paper, the extent of cellulose hydrolysis increased from 10% to 80% for crystalline and amorphous forms, respectively [58]. Specifically, the bound cellulase concentrations on the amorphous

Table 18.7 *Crystallinity index of several untreated and pretreated biomass cellulose using various pretreatment technologies.*

Biomass	Pretreatment	Crystallinity index (%)
Corn stover[a,] [59]	—	50.3
	Ammonia fiber expansion	36.3
	Ammonia recycled percolation	25.9
	Controlled pH	44.5
	Dilute Acid	52.5
	Lime	56.2
Poplar[a,] [59]	—	49.9
	Ammonia fiber expansion	47.9
	Ammonia recycled percolation	49.5
	Controlled pH	54.0
	Dilute acid	50.6
	Lime	54.5
	SO_2	56.5
Bagasse[a,] [60]	—	37
	O_3	38
	CO_2 explosion	57
	Alkaline explosion	62
Wheat straw[a,] [60]	—	35
	O_3	34
	CO_2 explosion	56
	Alkaline explosion	53
Eucalyptus regnans [a,] [60]	—	37
	O_3	40
	CO_2 explosion	53
Pinus radiata [a,] [60]	—	34
	O_3	36
Switchgrass[a,] [61]	—	46.1
	Lime	51.9
Loblolly pine[b,] [31,62]	—	62.5
	Dilute acid	69.9
	Ethanol organosolv	53
Buddleja davidii [b,] [63]	—	55
	Ethanol organosolv	49

[a] CrI measured by X-ray diffraction.
[b] CrI measured by CP/MAS ^{13}C NMR.

Table 18.8 *Crystallinity index of ethanol-organosolv-pretreated substrates prepared from lodgepole pine under various conditions.*

Condition	Temperature (°C)	Time (min)	Sulfuric acid dosage (%)	Concentration of ethanol (%)	Crystallinity index (%)
1	170	60	0.76	65	75
2	170	60	1.10	65	78
3	180	60	1.10	65	85

forms of both filter paper and Avicel were significantly higher than on the crystalline forms (i.e., 0.1 μmoles/g of remaining cellulose for the crystalline form and 1.5 μmoles/g of remaining cellulose for the amorphous form) [58]. The maximum extents of binding on the amorphous forms increased by a factor of 15 over that of the original forms. The change in crystallinity of the cellulose samples may therefore have allowed increased access to cellulase [58]. It could therefore be inferred that the increased access may have contributed to the increased cellulose hydrolysis rates observed. Another study by Zhu *et al.* illustrated the relationship between biomass digestibility and crystallinity. Hybrid poplar was treated with varying amounts of peracetic acid and KOH to generate samples with different crystallinity [66]. The results clearly indicated that the enzymatic digestibility of the biomass increased with decreasing biomass crystallinity, suggesting that amorphous cellulose is more accessible to enzymatic digestibility.

18.4.3   Cellulose Ultrastructure

To further understand the effects of pretreatments on biomass enzymatic digestibility, the changes in the ultrastructure of cellulose and how it is affected by pretreatment need to be investigated. The ultrastructure of cellulose has long been studied by CP/MAS ^{13}C NMR. Atalla and VanderHart were among the first to apply this NMR technique and concluded that native cellulose in plants has two crystalline allomorphs: cellulose I$_\alpha$ and cellulose I$_\beta$ [50]. Cellulose I$_\alpha$, a one-chain triclinic unit cell, is the dominant form in bacterial and algal cellulose; cellulose I$_\beta$, a monoclinic two-chain unit cell, is dominant in higher plants such as cotton, ramie, and wood [67]. By annealing, the meta-stable cellulose I$_\alpha$ can be converted to the thermodynamically more stable cellulose I$_\beta$ [68]. Nishiyama *et al.* proposed that slippage of the glucan chains is the most likely mechanism for conversion of cellulose I$_\alpha$ to cellulose I$_\beta$ [69]. Solid-state CP/MAS ^{13}C NMR of cellulose has been shown to be a convenient analytical technique to characterize several other forms of cellulose such as *para*-crystalline cellulose, and two non-crystalline forms: amorphous cellulose at accessible and inaccessible fibril surfaces [54,67,70,71]. *para*-Crystalline cellulose is the form that is less ordered than cellulose I$_\alpha$ and cellulose I$_\beta$, but more ordered than amorphous cellulose [67]. Accessible fibril surfaces are those in contact with water/solvent, while the inaccessible fibril surfaces are fibril-fibril contact surfaces and surfaces resulting from distortions in the fibril interior [31]. In order to analyze and quantify these various crystalline allomorphs and amorphous domains, Larsson *et al.* developed a model and methodology based on non-linear spectral fitting with a combination of Lorentzian and Gaussian functions [70,72]. Figure 18.11 shows the spectral fitting of C-4 region of the solid-state CP/MAS ^{13}C NMR spectrum of cellulose isolated from *Buddleja davidii*, and the assignments of signals and fitting parameters are presented in Table 18.9 [28].

Table 18.9   *Assignments of signals in the C-4 region of CP/MAS ^{13}C NMR spectrum obtained from Buddleja davidii cellulose.*

Assignments	Chemical shift (ppm)	FWHH^a (Hz)	Intensity (%)	Line type
Cellulose I$_\alpha$	89.6	96	4.2	Lorentz
Cellulose I$_{\alpha+\beta}$	88.9	85	8.7	Lorentz
Para-crystalline cellulose	88.7	258	32.9	Gauss
Cellulose I$_\beta$	88.2	142	6.5	Lorentz
Accessible fibril surface	84.6	116	3.9	Gauss
Inaccessible fibril surface	84.1	482	41.1	Gauss
Accessible fibril surface	83.6	101	2.7	Gauss

^{a} FWHH: Full width at half-height.

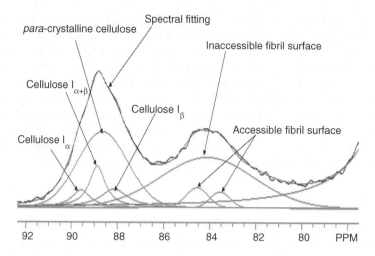

Figure 18.11   *Spectral fitting for the C-4 region of CP/MAS ^{13}C-NMR spectrum of native Buddleja davidii cellulose [28]. $I_{\alpha+\beta}$: I($\alpha + \beta$) (See figure in color plate section).*

A study by Pu *et al.* [67] showed that cellulose I_α, *para*-crystalline, and amorphous celluloses of fully bleached softwood (pine) kraft pulp are more susceptible to enzymatic hydrolysis than cellulose I_β, which is the more ordered and stable crystal structure. Since cellulose I_β is the predominant form in wood, it is of great importance for pretreatment technologies to convert the I_β allomorph to the more easily digestible *para*-crystalline and amorphous celluloses. In another study, Sannigrahi *et al.* [62] showed that the CrI of ethanol-organosolv-pretreated loblolly pine cellulose increased by 53% after treatment with cellulase and that the relative intensities of amorphous and *para*-crystalline cellulose decreased, suggesting that enzymes selectively degrade these forms of cellulose. In contrast, Cateto *et al.* [73] reported recently that the crystallinity of ethanol-organosolv-pretreated Kanlow switchgrass remained approximately constant upon enzymatic hydrolysis by cellulase. The explanation for such results was hypothesized to be by the effect of the synergistic action of endo- and exoglucanases on the removal of the outer layers of the cellulose crystallite in order to gain access to the inner layers, characteristic of a "peeling-off" type mechanism [73]. It therefore appears that the impact of enzymatic hydrolysis on the crystallinity of cellulose is dependent, to some extent, on the nature/structure of the cellulosic substrate.

The effects of dilute acid and organosolv pretreatment on the ultrastructure of various cellulosic materials are summarized in Table 18.10 [7,31,62,63]. The data indicate that the main difference between

Table 18.10   *Variations in the ultrastructure of cellulose after pretreatment [7,31,62,63].*

Species	Pretreatment	Changes in ultrastructure of cellulose[a]		
		I_β	*Para*-crystalline	Amorphous
Loblolly pine	Dilute acid	+	−	−
Loblolly pine	Ethanol organosolv	+	−	+
Switchgrass	Dilute acid	+	+	−
Buddleja davidii	Ethanol organosolv	−	+	+

[a]+ indicates increase and − indicates decrease in the specific cellulosic structure.

pretreatments is the ability of ethanol organosolv pretreatment to increase the amorphous regions of pine and *Buddleja davidii* cellulose, which could possibly improve enzymatic hydrolysis. Dilute-acid pretreatment increased cellulose I_β, an observation that can be explained by the thermal conversion of cellulose I_α to the more stable cellulose I_β, or simply an enrichment in the crystalline structure after hydrolysis of the amorphous structures. The results clearly show that the variations in cellulose ultrastructure are pretreatment dependent and related to the nature of the origin of the cellulosic material.

18.5 Future Perspectives

The solution- and solid-state NMR techniques presented in this chapter offer powerful and effective tools for analysis of lignin and cellulose for biomass characterization. Given the importance and need for thorough analysis of the fundamental structures of plant biomass as well as its conversion chemistry in aqueous pretreatments for biological and chemical conversion to fuels and chemicals, these methods can and will continue to have broad applicability for researchers involved in biomass conversion to second- and third-generation biofuels in the future.

Acknowledgements

This work was supported and performed as part of the BioEnergy Science Center (BESC). The BioEnergy Science Center is a US Department of Energy Bioenergy Research Center supported by the Office of Biological and Environmental Research in the DOE Office of Science.

References

1. Yang, B. and Wyman, C.E. (2008) Pretreatment: the key to unlocking low-cost cellulosic ethanol. *Biofuels, Bioproducts, and Biorefining*, **2**, 26–40.
2. Ralph, J. and Landucci, L.L. (2010) NMR of lignins, in *Lignin and Lignans* (eds C. Heitner, D.R. Dimmel, and J.A. Schmidt), CRC Press, Boca Raton, Fla, pp. 137–244.
3. Anterola, A.M. and Lewis, N.G. (2002) Trends in lignin modification: a comprehensive analysis of the effects of genetic manipulations/mutations on lignification and vascular integrity. *Phytochemistry*, **61** (3), 221–294.
4. Balakshin, M.Y., Capanema, E.A., and Chang, H.M. (2007) MWL fraction with a high concentration of lignin-carbohydrate linkages: isolation and 2D NMR spectroscopic analysis. *Holzforschung*, **61**, 1–7.
5. Robert, D. (1992) Carbon-13 nuclear magnetic resonance spectroscopy, in *Methods in Lignin Chemistry* (eds S.Y. Lin and C.W. Dence), Springer-Verlag, New York, NY, pp. 250–273.
6. Ludwig, C., Nist, B., and McCarthy, J.L. (1964) The high resolution nuclear magnetic resonance spectroscopy of protons in acetylated lignins. *Journal of the American Chemical Society*, **86**, 1196–1202.
7. Samuel, R., Pu, Y., Foston, M., and Ragauskas, A.J. (2010) Solid-state NMR characterization of switchgrass cellulose after dilute acid pretreatment. *Biofuels*, **1**, 85–90.
8. Lundquist, K. (1992) Proton (^{1}H) NMR spectroscopy, in *Methods in Lignin Chemistry* (eds S.Y. Lin and C.W. Dence) Springer-Verlag, New York, NY, pp. 242–249.
9. Ragauskas, A.J., Williams, C.K., Davison, B.H. *et al.* (2006) The path forward for biofuels and biomaterials. *Science*, **311**, 484–489.
10. Klemm, D., Heublein, B., Fink, H.-B., and Bohn, A. (2005) Cellulose: Fascinating biopolymer and sustainable raw material. *Angewandte Chemie-International Edition*, **44**, 3358–3393.
11. Pu, Y., Zhang, D., Singh, P.M., and Ragauskas, A.J. (2007) The new forestry biofuels sector. *Biofuels, Bioproducts, and Biorefining*, **2**, 58–73.
12. Harris, P.J. and Stone, B.A. (2008) Chemistry and molecular organization of plant cell walls, in *Biomass Recalcitrance: Deconstructing the Plant Cell Wall for Bioenergy* (ed. M.E. Himmel), Blackwell, pp. 61–93.

13. Sjöström, E. (1993) *Wood Chemistry: Fundamentals and Applications*, 2nd edn, Academic Press, New York, NY.
14. Boerjan, W., Ralph, J., and Baucher, M. (2003) Lignin biosynthesis. *Annual Review of Plant Biology*, **54**, 519–546.
15. Davin, L.B. and Lewis, N.G. (2005) Lignin primary structures and dirigent sites. *Current Opinion in Biotechnology*, **16**, 407–415.
16. Chakar, F.S. and Ragauskas, A.J. (2004) Review of current and future softwood kraft lignin process chemistry. *Industrial Crops and Products*, **20**, 131–141.
17. Brunow, G., Kilpelainen, I., Sipila, J. *et al.* (1998) Oxidative coupling of phenols and the biosynthesis of lignin, in *ACS Symposium Series, 697 (Lignin and Lignan Biosynthesis)*, American Chemical Society.
18. Karhunen, P., Rummakko, P., Sipila, J., and Brunow, G. (1995) Dibenzodioxocins: a novel type of linkage in softwood lignins. *Tetrahedron Letters*, **36**, 169–170.
19. Kukkola, E.M., Koutaniemi, S., Pollanen, E. *et al.* (2004) The dibenzodioxocin lignin substructure is abundant in the inner part of the secondary wall in Norway spruce and silver birch xylem. *Planta*, **218**, 497–500.
20. Zhang, L. and Gellerstedt, G. (2001) NMR observation of a new lignin structure, a spiro-dienone. *Chemical Communications*, **24**, 2744–2745.
21. Zhang, L., Gellerstedt, G., Ralph, J., and Lu, F. (2006) NMR studies on the occurrence of spirodienone structures in lignins. *Journal of Wood Chemistry and Technology*, **26**, 65–79.
22. Bjorkman, A. (1954) Isolation of lignin from finely divided wood with neutral solvents. *Nature*, **174**, 1057–1058.
23. Chang, H.M., Cowling, E.B., Brown, W. *et al.* (1975) Comparative studies on cellulolytic enzyme lignin and milled wood lignin of sweetgum and spruce. *Holzforschung*, **29**, 153–159.
24. Wu, S. and Argyropoulos, D.S. (2003) An improved method for isolating lignin in high yield and purity. *Journal of Pulp and Paper Science*, **29**, 235–240.
25. Li, S. and Lundquist, K. (1994) A new method for the analysis of phenolic groups in lignins by ^{1}H-NMR spectrometry. *Nordic Pulp and Paper Research Journal*, **3**, 191–195.
26. Drumond, M., Aoyama, M., Chen, C.-L., and Robert, D. (1989) Substituent effects on C-13 chemical shifts of aromatic carbons in biphenyl type lignin model compounds. *Journal of Wood Chemistry and Technology*, **9**, 421–411.
27. Pan, X., Lachenal, D., Neirinck, V., and Robert, D. (1994) Structure and reactivity of spruce mechanical pulp lignins IV: ^{13}C-NMR spectral studies of isolated lignins. *Journal of Wood Chemistry and Technology*, **14**, 483–506.
28. Hallac, B.B., Sannigrahi, P., Pu, Y. *et al.* (2009) Biomass characterization of Buddleja davidii: A potential feedstock for biofuel production. *Journal of Agricultural and Food Chemistry*, **57**, 1275–1281.
29. Hallac, B.B., Pu, Y., and Ragauskas, A.J. (2010) Chemical transformations of Buddleja davidii lignin during ethanol organosolv pretreatment. *Energy & Fuels*, **24**, 2723–2732.
30. Samuel, R., Pu, Y., Raman, B., and Ragauskas, A.J. (2010) Structural characterization and comparison of switchgrass lignin before and after dilute acid pretreatment. *Applied Biochemistry and Biotechnology*, **162**, 62–74.
31. Sannigrahi, P., Ragauskas, A.J., and Miller, S.J. (2008) Effects of two-stage dilute acid pretreatment on the structure and composition of lignin and cellulose in loblolly pine. *Bioenergy Research*, **1**, 205–214.
32. Sannigrahi, P., Ragauskas, A.J., and Miller, S.J. (2010) Lignin structural modifications resulting from ethanol organosolv treatment of loblolly pine. *Energy & Fuels*, **24**, 683–689.
33. Ralph, J., Marita, J.M., Ralph, S.A. *et al.* (1999) Solution state NMR of lignins, in *Advances in Lignocellulosics Characterization*, Tappi Press, Atlanta, GA.
34. Ralph, J., Akiyama, T., Kim, H. *et al.* (2006) Effects of coumarate 3-hydroxylase down-regulation on lignin structure. *The Journal of Biological Chemistry*, **281**, 8843–8853.
35. Rencoret, J., Marques, G., Gutierrez, A. *et al.* (2008) Structural characterization of milled wood lignins from different eucalypt species. *Holzforschung*, **62**, 514–526.
36. del Río, J.C., Rencoret, J., Marques, G. *et al.* (2008) Highly acylated (acetylated and/or p-coumaroylated) native lignins from diverse herbaceous plants. *Journal of Agricultural and Food Chemistry*, **56**, 9525–9534.
37. Stewart, J.J., Akiyama, T., Chapple, C. *et al.* (2009) The effects on lignin structure of overexpression of ferulate 5-hydroxylase in hybrid poplar. *Plant Physiology*, **150**, 621–635.
38. Pu, Y., Chen, F., Ziebell, A. *et al.* (2009) NMR characterization of C3H and HCT down-regulated alfalfa lignin. *Bioenergy Research*, **2**, 198–208.

39. Moinuddin, S.G.A., Jourdes, M., Laskar, D.D. *et al.* (2010) Insights into lignin primary structure and deconstruction from Arabidopsis thaliana COMT (caffeic acid O-methyl transferase) mutant Atomt1. *Organic and Biomolecular Chemistry*, **8**, 3928–3946.

40. Zawadzki, M., Runge, T.M., and Ragauskas, A.J. (2000) Facile detection of ortho- and para-quinone structures in residual kraft lignin by phosphorus-31 NMR spectroscopy. *Journal of Pulp and Paper Science*, **26**, 102–106.

41. Granata, A. and Argyropoulos, D.S. (1995) 2-Chloro-4,4,5,5-tetramethyl-1,3,2-dioxaphospholane, a reagent for the accurate determination of the uncondensed and condensed phenolic moieties in lignins. *Journal of Agricultural and Food Chemistry*, **43**, 1538–1544.

42. Argyropoulos, D.S. (2010) Heteronuclear NMR Spectroscopy of Lignins, in *Lignin & Lignans; Advances in Chemistry* (eds C. Heitner, D. Dimmel, and J. Schmidt), CRC Press, pp. 245–265.

43. Lai, Y.-Z. (1992) Determination of phenolic hydroxyl groups, in *Methods in Lignin Chemistry* (eds S.Y. Lin and C. W. Dence), Springer-Verlag, New York, NY, pp. 423–434.

44. Faix, O., Andersons, B., Argyropoulos, D.S., and Robert, D. (1995) Quantitative determination of hydroxyl and carbonyl groups of lignins – an overview. Proceedings of 8th International Symposium on Wood and Pulping Chemistry, Gummerus Kirjapaino Oy, Jyvaskyla, Finland, Vol. 1, pp. 559–566.

45. Faix, O., Argyropoulos, D.S., Robert, D., and Neirinck, V. (1994) Determination of hydroxyl groups in lignins evaluation of ^{1}H-, ^{13}C-, ^{31}P-NMR, FTIR and wet chemical methods. *Holzforschung*, **48**, 387–394.

46. Zawadzki, M. (1999) Quantitative determination of quinone chromophore changes during ECF bleaching of kraft pulp. Ph.D. thesis, Institute of Paper Science and Technology, Atlanta.

47. El Hage, R., Brosse, N., sannigrahi, P., and Ragauskas, A. (2010) Effects of process severity on the chemical structure of Miscanthus ethanol organosolv lignin. *Polymer Degradation and Stability*, **95**, 997–1003.

48. El Hage, R., Brosse, N., Chrusciel, L. *et al.* (2009) Characterization of milled wood lignin and ethanol organosolv lignin from miscanthus. *Polymer Degradation and Stability*, **94**, 1632–1638.

49. Akim, L.G., Argyropoulos, D.S., Jouanin, L. *et al.* (2001) Quantitative phosphorus-31 NMR spectroscopy of lignins from transgenic poplars. *Holzforschung*, **55**, 386–390.

50. Atalla, R.H. and VanderHart, D.L. (1984) Native cellulose: A composite of two distinct crystalline forms. *Science*, **223**, 283–285.

51. Atalla, R.H. (1999) Celluloses, in *Comprehensive Natural Products Chemistry: Carbohydrates and their Derivatives Including Tannins, Cellulose, and Related Lignins*, vol **3** (ed. B.M. Pinto), Elsevier, Amsterdam, pp. 529–598.

52. Hallac, B.B. and Ragauskas, A.J. (2011) Analyzing cellulose degree of polymerization and its relevancy to cellulosic ethanol. *Biofuels, Bioproducts, and Biorefining*, **5**, 215–225.

53. Zhang, Y.P. and Lynd, L.R. (2004) Toward an aggregated understanding of enzymatic hydrolysis of cellulose: Non-complexed cellulase systems. *Biotechnology and Bioengineering*, **88**, 797–824.

54. Wickholm, K., Larsson, P.T., and Iversen, T. (1998) Assignment of non-crystalline forms in cellulose I by CP/MAS carbon-13 NMR spectroscopy. *Carbohydrate Research*, **312**, 123–129.

55. Sannigrahi, P., Ragauskas, A.J., and Tuskan, G.A. (2010) Poplar as a feedstock for biofuels: A review of compositional characteristics. *Biofuels, Bioproducts, and Biorefining*, **4**, 209–226.

56. Foston, M., Hubbell, C.A., Davis, M., and Ragauskas, A.J. (2009) Variations in cellulosic ultrastructure of poplar. *Bioenergy Research*, **2**, 193–197.

57. Park, S., Johnson, D.K., Ishizawa, C.I. *et al.* (2009) Measuring the crystallinity index of cellulose by solid state ^{13}C nuclear magnetic resonance. *Cellulose*, **16**, 641–647.

58. Jeoh, T., Ishizawa, C.I., Davis, M.F., Himmel, M.E., Adney, W.S. and Johnson, D.K. (2007) Cellulase digestibility of pretreated biomass is limited by cellulose accessibility. *Biotechnology and Bioengineering*, **98**, 112–122.

59. Kumar, R., Mago, G., Balan, V., and Wymand, C.E. (2009) Physical and chemical characterizations of corn stover and poplar solids resulting from leading pretreatment technologies. *Bioresource Technology*, **100**, 3948–3962.

60. Puri, V.P. (1984) Effect of crystallinity and degree of polymerization of cellulose on enzymatic saccharification. *Biotechnology and Bioengineering*, **26**, 1219–1222.

61. Chang, V.S. and Holtzapple, M.T. (2000) Fundamental factors affecting biomass enzymatic reactivity. *Applied Biochemistry and Biotechnology*, **84–86**, 5–37.

62. Sannigrahi, P., Miller, S.J., and Ragauskas, A.J. (2010) Effects of organosolv pretreatment and enzymatic hydrolysis on cellulose structure and crystallinity in Loblolly pine. *Carbohydrate Research*, **345**, 965–970.

63. Hallac, B.B., Sannigrahi, P., Pu, Y. *et al.* (2010) Effect of ethanol organosolv pretreatment on enzymatic hydrolysis of Buddleja davidii stem biomass. *Industrial & Engineering Chemistry Research*, **49**, 1467–1472.

64. Pan, X., Xie, D., Yu, R.W., and Saddler, J.N. (2008) The bioconversion of mountain pine beetle-killed lodgepole pine to fuel ethanol using the organosolv process. *Biotechnology and Bioengineering*, **101**, 39–48.

65. Schroeder, L.R., Gentile, V.M., and Atalla, R.H. (1986) Nondegradative preparation of amorphous cellulose. *Journal of Wood Chemistry and Technology*, **6**, 1–14.

66. Zhu, L., O'Dwyer, J.P., Chang, V.S. *et al.* (2008) Structural features affecting biomass enzymatic digestibility. *Bioresource Technology*, **99**, 3817–3828.

67. Pu, Y., Ziemer, C., and Ragauskas, A.J. (2006) CP/MAS ^{13}C NMR analysis of cellulase treated bleached softwood kraft pulp. *Carbohydrate Research*, **341**, 591–97.

68. Yamamoto, H. and Horii, F. (1993) CPMAS carbon-13 NMR analysis of the crystal transformation induced for Valonia cellulose by annealing at high temperatures. *Macromolecules*, **26**, 1313–1317.

69. Nishiyama, Y., Sugiyama, J., Chanzy, H., and Langan, P. (2003) Crystal structure and hydrogen bonding system in cellulose I_α from synchrotron x-ray and neutron fiber diffraction. *Journal of the American Chemical Society*, **125**, 14300–14306.

70. Larsson, P.T., Wickholm, K., and Iversen, T. (1997) A CP/MAS ^{13}C NMR investigation of molecular ordering in celluloses. *Carbohydrate Research*, **302**, 19–25.

71. Larsson, P.T., Hult, E.L., Wickholm, K. *et al.* (1999) CP/MAS ^{13}C-NMR spectroscopy applied to structure and interaction studies on cellulose I. *Solid State Nuclear Magnetic Resonance*, **15**, 31–40.

72. Larsson, P.T., Westermark, U., and Iversen, T. (1995) Determination of the cellulose Iα allomorph content in a tunicate cellulose by CP/MAS ^{13}C-NMR spectroscopy. *Carbohydrate Research*, **278**, 339–343.

73. Cateto, C., Hu, G., and Ragauskas, A.J. (2011) Enzymatic hydrolysis of organosolv Kanlow switchgrass and its impact on cellulose crystallinity and degree of polymerization. *Energy & Environmental Science*, **4**, 1516–1521.

19

Xylooligosaccharides Production, Quantification, and Characterization in Context of Lignocellulosic Biomass Pretreatment

Qing Qing[1], Hongjia Li[2,3,4,*], Rajeev Kumar[2,4] and Charles E. Wyman[2,3,4]

[1] *Pharmaceutical Engineering & Life Science, Changzhou University, Changzhou, China*
[2] *Center for Environmental Research and Technology, University of California, Riverside, USA*
[3] *Department of Chemical and Environmental Engineering, University of California, Riverside, USA*
[4] *BioEnergy Science Center, Oak Ridge, USA*

19.1 Introduction

19.1.1 Definition of Oligosaccharides

Oligosaccharides, also termed sugar oligomers, refer to short-chain polymers of monosaccharide units connected by α and/or β glycosidic bonds. In structure, oligosaccharides represent a class of carbohydrates between polysaccharides and monosaccharides, but the range of degree of polymerization (DP, chain length) spanned by oligosaccharides has not been consistently defined. For example, the Medical Subject Headings (MeSH) database of the US National Library of Medicine defines oligosaccharides as carbohydrates consisting of 2–10 monosaccharide units; in other literature, sugar polymers with DPs of up to 30–40 have been included as oligosaccharides [1–3].

*Present address: DuPont Industrial Biosciences, Palo Alto, USA

Aqueous Pretreatment of Plant Biomass for Biological and Chemical Conversion to Fuels and Chemicals, First Edition.
Edited by Charles E. Wyman.
© 2013 John Wiley & Sons, Ltd. Published 2013 by John Wiley & Sons, Ltd.

Table 19.1   *Lignocellulosic feedstocks that have heteroxylans as dominant hemicellulose types.*

Plant group	Examples	Wall type	
		Primary cell wall	Secondary cell wall
Hardwood	Poplar	Xyloglucan	4-O-methyl-glucuronoxylan
Energy grasses	Switchgrass	Glucuronoarabinoxylan	Glucuronoarabinoxylan
Agricultural residues	Corn stover	Glucuronoarabinoxylan	Glucuronoarabinoxylan

19.1.2   Types of Oligosaccharides Released during Lignocellulosic Biomass Pretreatment

Oligosaccharides exist naturally in plant tissues, but their amounts are small compared to cell-wall structural polysaccharides, such as cellulose and hemicellulose [4]. During pretreatment of lignocellulosic biomass, most of the insoluble hemicellulose is removed from the surface of cellulose microfibrils and broken into various soluble oligosaccharides. However, the amounts and structures vary with pretreatment types and severity. The majority of oligosaccharides released during lignocellulosic biomass pretreatment are hydrolysis products of hemicellulose, and the types of oligosaccharides (composition, DP, and substitution) depend on the structure and composition of the corresponding hemicellulose.

Hemicellulose refers to several amorphous polysaccharides found in the plant cell-wall matrix that have β-(1–4)-linked backbones with an equatorial configuration [5], which are commonly categorized into several groups such as xyloglucans, heteroxylans, (galacto) glucomannans, and arabinogalactans [6–8]. For example, glucuronoarabinoxylan refers to one type of heteroxylans which have a backbone of β-(1–4)-xylosyl residues with a few short side chains that mainly contain arabinosyl residues and glucuronic acid residues, but could also contain other sugars or sugar acid residues [4]. The number of side chains and the side-chain residues composition vary with biomass and cell-wall types and life stage of the same plant. Dominant forms of hemicellulose polysaccharides in major lignocellulosic biomass feedstocks, except softwoods, are xyloglucans and "heteroxylans" as listed in Table 19.1. In lignocellulosic biomass feedstocks, the mass fraction of secondary cell walls based on total plant dry weight is much greater than that of primary cell walls [4]. Thus, xylooligosaccharides (XOs) from heteroxylans hydrolysis are the predominant type of oligosaccharides released during pretreatment.

19.1.3   The Importance of Measuring Xylooligosaccharides

Understand Plant Cell-wall Structure and its Role in Biomass Recalcitrance

In general, plant cell walls represent an enormous source of complex polysaccharides that can be broken down to monosaccharides for potential conversion into biofuels and chemicals. The framework of plant cell walls is cellulose, a highly ordered, water-excluding natural crystalline polymer of glucose molecules joined by β-(1–4)-glycosidic bonds, with its chains connected by many intra/inter-chain hydrogen bonds. Outside the framework, cellulose microfibrils and hemicellulose are intimately interlocked with one another and often with lignin, both covalently and non-covalently [5]. The hydrophobic association of cell-wall polysaccharides and lignin, termed the lignin-carbohydrate complex (LCC), is an important part of plant cell-wall defense and has been recognized as the main barrier for economic deconstruction of cell-wall polysaccharides [5,9–11]. Such collective resistance, which plants and plant materials pose to deconstruction by microbes and enzymes, is defined as "biomass recalcitrance" [5,12]. Although the aspect(s) most responsible for biomass recalcitrance to conversion are not clear, a better understanding of cell-wall polysaccharides compositions and structures would greatly facilitate advanced process designs that achieve more effective breaking of such defenses with lower cost, as well as aid in production of less recalcitrant plants using genetic tools. For example, through comparison of glucuronoxylan (GX) structures in poplar wood, Lee *et al.* found transgenic reduction of GX in secondary cell wall reduced recalcitrance of wood to cellulase digestion [13].

Unfortunately, direct characterization of cell-wall polysaccharides is difficult because of the heterogeneous and complex nature of cell walls. Thus, using either enzymes or chemicals to break down cell-wall polysaccharides followed by characterizing the corresponding oligosaccharides and monomers has been an effective way to study cell-wall polysaccharides structures and their possible roles in biomass recalcitrance. Effective structural studies normally contain two parts. First, optimized enzymatic or chemical treatment methods are applied to extract certain types of polysaccharides from the insoluble cell wall in which they are held. For example, heteroxylans are typically extracted by 4% KOH whereas heteroglucans may require 24% KOH [5,14]. The isolated polysaccharides or fragments are then purified and broken down into oligosaccharides for detailed characterization. Important structural information about hemicellulose polysaccharides can be determined, such as the glycosyl residue composition, the glycosyl linkage composition, the sequence of glycosyl residues in both the backbone and side chains, and non-carbohydrate substituents through characterizing hemicellulose oligosaccharides [4].

Engineer Reaction Pathways for Economic Deconstruction of Structural Cell-wall Polysaccharides

As a feedstock for fuels and chemicals production, lignocellulosic biomass has many benefits such as not competing for food and feed supply, low production costs, and wide availability over a range of locations and climates [15,16]. Utilization of cell-wall carbohydrates makes lignocellulosic biomass a promising renewable feedstock for large-scale conversion into liquid fuels and organic chemicals. Different reaction pathways have been devised to break down cell-wall polysaccharides in lignocellulosic biomass into monomeric sugars: thermal, chemical, biological, and/or a combination of these. In lignocellulosic ethanol production for example, cellulase (which is a synergistic combination of several proteins) in combination with hemicellulases and other accessory enzymes degrades cellulose and residual hemicellulose into glucose and xylose. However, pretreatments have proven to be essential to open up the rigid biomass structure through removing or altering hemicellulose and lignin and loosening the structure of cellulose, enhancing access of enzymes to their respective substrates. Hemicellulose polysaccharide chains can be broken into oligosaccharides and then further hydrolyzed to monosaccharides, especially during low to neutral pH pretreatments, which in turn can react to degradation products as described in the following:

$$\text{Polysaccharides}_{(solid)} \rightarrow \text{Hemicellulose Oligosaccharides}_{(aq)}$$
$$\rightarrow \text{Monosaccharides}_{(aq)} \rightarrow \text{Degradation products}_{(aq)}$$

Employing harsh pretreatment conditions can reduce macro-barriers to enzymes reaching cellulose and improve micro-accessibility of cellulose to enzymes through changes in its crystal structure and degree of polymerization and result in better conversion to sugars [17–19]. However, such conditions also degrade xylooligosaccharides and xylose into by-products such as furfural [20,21], resulting in sugar losses and formation of inhibitors to enzymes and microbes for sugars fermentation [22,23]. Pathway optimization is therefore needed to achieve the highest sugar recovery for economical processing. For that reason, qualitative and quantitative measurements of xylooligosaccharides are important because they are essential for detailed studies of hemicellulose hydrolysis kinetics and degradation mechanisms. Such studies can also play a key role in engineering effective pretreatment technologies to achieve high sugar recovery with good economics.

It is also important to note that xylooligosaccharides have recently been shown to have a strong negative effect on cellulase activity in decomposing cell-wall polysaccharides into fermentable sugars [24,25]. Quantitative analysis and characterization of xylooligosaccharides, including improved purification and characterization techniques, facilitate the understanding of xylooligosaccharide inhibition mechanisms and development of strategies for reducing inhibition.

Oligosaccharides for High-value-added Products

Xylooligosaccharides have been shown to have important prebiotic properties and thus great potential for use in medicinal, food, and health products [26]. Xylooligosaccharides for such uses are mainly derived from lignocellulosic biomass by enzymatic and/or chemical hydrolysis to remove hemicellulose polysaccharides (mainly heteroxylans in the case of cellulosic biomass) from the surface of cellulose and break them into water-soluble xylooligosaccharides. Separation technologies then isolate and purify these xylooligosaccharides into desired DP ranges for prebiotic applications. This fast-growing market for xylooligosaccharides creates great opportunities to process xylan-rich pretreatment hydrolyzates in cellulosic biorefineries into high-value products which could improve conversion economics.

19.2   Xylooligosaccharides Production

Xylooligosaccharides are usually produced from xylan-rich lignocellulosic materials (LCM) by autohydrolysis from heating in water or steam, chemical treatments in dilute aqueous solutions of mineral acids [27,28], direct enzymatic hydrolysis of susceptible lignocellulosic materials [29–31], or chemical fractionation of a suitable LCM to isolate (or solubilize) xylan with further enzymatic hydrolysis to XOs [32]. Typical raw materials for XOs production are hardwoods (e.g., birchwood, beechwood), corn cobs, straws, bagasse, rice hulls, malt cakes, and bran [26]. In recent years, the fast-growing functional food market and the increasing number of other industrial applications are encouraging identification of renewable and cheap xylan sources instead of hardwood xylan for XOs production. As a result, agricultural residues such as cotton stalks, tobacco stalks, and wheat straw have also been intensively studied [27].

19.2.1   Thermochemical Production of XOs

Thermochemical production of XOs is usually accomplished by steam, dilute mineral acids, or dilute alkaline solutions. The single-step production of XOs by reaction with steam or water through hydronium-catalyzed degradation of xylan is known as autohydrolysis, hydrothermolysis, or water prehydrolysis [26]. Autohydrolysis takes place at slightly acidic (pH ≤ 4) conditions created by acetic acid released by partial cleavage of acetyl groups in the plant cell wall. A considerable fraction of acetyl and uronic acid groups remain attached to the oligosaccharides, giving them distinctive characteristics like very high solubility in water [33]. In autohydrolysis treatment, XOs behave as typical reaction intermediates whose concentration depends mainly on the tradeoff between breakdown of polymeric hemicellulose in biomass to XOs and their further decomposition to monomeric xylose. Therefore, reaction severity (R_o) influences the concentrations of total XOs as well as of monomeric xylose that could be achieved in hydrolysate and is often represented by a single parameter that combines temperature, time, and reaction pH [34]:

$$R_o = t \exp\left(\frac{T - 100}{14.75}\right) - pH$$

Medium-severity conditions are usually preferred to balance formation of oligosaccharides against their degradation and maximize XO concentration [35]. However, the degree of polymerization DP (or molecular weight) distribution in XOs mixtures generally depends on not only the treatment severity but also on the substrate and its concentration during treatment [33]. In a study by Nabarlatz *et al.*, comparative assessment of six agricultural residues of different botanic origin showed that characteristics of the raw material played a major role in the yield and composition of XOs. However, their yield also depended on the initial content of acetyl groups since their cleavage liberated acetic acid, which in turn catalyzed xylan depolymerization into XOs [33]. In the initial stage of autohydrolysis, hydronium ions were generated through autoionization (dissociation) of water under high temperature or pressure. However, as the reaction proceeded, cleavage of acetyl groups from the xylan backbone formed acetic acid and was believed to contribute more hydronium ions. Although adding acids beyond that released naturally from biomass can facilitate xylan or hemicellulose degradation, XO yields will generally be reduced by generating more monomeric xylose than without added acid. Controlling the temperature and reaction time can also influence XO characteristics such as the acetyl content and the molar mass distribution [36], although the nature of the raw material plays a significant role [33].

Autohydrolysis has the advantage of eliminating corrosive chemicals for extraction and hydrolysis of xylan, but requires equipment that can be operated at temperatures and pressures as high as or higher than acid or alkali treatments. Besides xylan degradation, several concurrent processes occur including extractives removal, solubilization of acid-soluble lignin, and solubilization of ash, all of which contribute to undesired non-saccharide compounds in liquors from autohydrolysis processing. The molecular weight distribution of XOs produced by autohydrolysis after solvent extraction contains a large proportion of high-DP compounds (MW 1000–3000 g/mol) and a much smaller fraction of low-DP compounds (MW < 300 g/mol) [33]. In addition, autohydrolysis at mild temperatures does not modify cellulose and lignin substantially, allowing their recovery for further processing and utilization.

XOs can also be produced by hydrolytic processes either in basic or dilute acidic media. Dilute sulfuric acid (0.1–0.5 M) is most commonly used for acid production of XOs. The DP distribution of the XOs depends on acid concentration, temperature, and reaction time, but the yield of monosaccharides also depends on the structure and composition of xylan [27]. A major disadvantage of acid hydrolysis is low yields of oligomers compared to monomers in addition to production of furfural and other degradation products. However, this disadvantage could be controlled by shortening the reaction time, reducing acid concentration, or removing these by-products by adsorption chromatography and membrane separation. A major advantage of acid hydrolysis is simple, rapid kinetics; for example, dilute acid hydrolysis requires much less reaction time compared to enzymatic hydrolysis to achieve the same xylan to XOs conversion (a few minutes compared to several hours) [27].

Figure 19.1 summarizes the xylan reaction pathway. The depolymerization of xylan has been described as combined reactions of fast-reacting and slow-reacting fractions, which are first decomposed into high-molecular-weight XOs [28]. As the hydrolytic degradation reaction proceeds, high-molecular-weight XOs

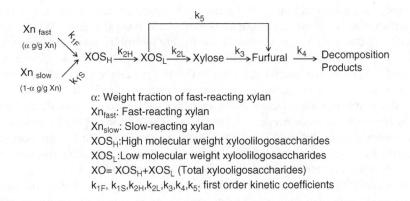

α: Weight fraction of fast-reacting xylan
Xn_{fast}: Fast-reacting xylan
Xn_{slow}: Slow-reacting xylan
XOS_H: High molecular weight xyloolilogosaccharides
XOS_L: Low molecular weight xyloolilogosaccharides
$XO = XOS_H + XOS_L$ (Total xylooligosaccharides)
k_{1F}, k_{1S}, k_{2H}, k_{2L}, k_3, k_4, k_5: first order kinetic coefficients

Figure 19.1   *Xylan reaction pathway in autohydrolysis to oligomers, xylose, furfural, and degradation products. (Adapted from Parajo et al. [28] © 2004, Elsevier).*

are converted into lower-molecular-weight XOs which are further depolymerized to xylose; xylose is then degraded to furfural and many unidentified degradation products. In some cases, low-molecular-weight XOs can be directly degraded to furfural or other degradation products [21,28]. First-order kinetics with Arrhenius-type dependence on temperature are usually adequate to describe reaction rates profiles, with the weight fraction of fast-reacting xylan, the pre-exponential factors of the kinetic coefficients involved in the reaction, and the corresponding activation energies determined by fitting the data to the kinetic model. In the study by Kumar and Wyman with purified xylooligosaccharides degradation at different pH values, they showed that all the XOs disappeared at higher rates compared to monomeric xylose, and the ratio of XOs disappearance rate constants to xylose degradation rate constant increased with decreasing pH. In addition, the direct degradation of low-DP XOs (mainly DP 2 and 3) to undesired products was significant for hydrothermal reactions but could be minimized by adding acid [21].

Alternatively, depolymerized hemicellulose may be extracted from lignocellulosic materials by strong alkali solutions (for example, a solution of KOH, NaOH, $Ca(OH)_2$, ammonia, or a mixture of these compounds). However, the extractability of depolymerized hemicellulose varies with the alkali type and isolation conditions used for different plants. In general, alkaline treatment disrupts the cell wall of lignocellulosic materials by dissolving hemicelluloses and lignin, hydrolyzing uronic and acetic esters, swelling the cellulose, decreasing cellulose crystallinity, and cleaving the α-ether linkages between lignin and hemicelluloses as well as the ester bonds between lignin and/or hemicelluloses and hydroxycinnamic acids, such as *p*-coumaric and ferulic acids. The depolymerized xylan therefore loses acetyl groups and uronic acids by saponification during extraction and has very limited solubility in neutral aqueous solutions [37]. Alkali processing of xylan-containing materials is favored by the pH stability of this polymer, and solubilized xylan and xylan degradation products can be recovered by precipitation with organic compounds (including acids, alcohols or ketones) [26]. However, xylan or soluble XOs obtained from alkali extraction require dilute acid or enzymatic treatment to break them down further to lower-DP XOs [26].

19.2.2   Production of XOs by Enzymatic Hydrolysis

XOs can also be produced by enzymatic hydrolysis of xylan-containing materials. However, because the xylan-lignin complex is naturally resistant to enzyme attack, current commercial processes are usually carried out in a two-stage sequence: (1) alkaline extraction followed by (2) enzymatic hydrolysis. In most plant materials, xylan is a heteropolymer with homopolymeric backbone composed of β-1, 4-linked xylose units and various branching units including L-arabinose, D-glucuronic acid, 4-O-methyl glucuronic acid, D-galacturonic acid, ferulic acid, coumaric, and acetic acid residues and, to a lesser extent, L-rhamnose, L-fucose, and various O-methylated neutral sugars [38]. Consequently, synergistic action of different enzymes is needed to completely hydrolyze these complex xylan structures. Generally, endo-β-1, 4-xylanases degrade xylan by attacking the β-1, 4-bonds between xylose units to produce XOs, and β-xylosidase converts lower-DP XOs into monomeric xylose. In order to maximize production of XOs and minimize xylose production, enzyme mixtures with low endoxylanase and/or β-xylosidase activity are desirable. Debranching enzymes such as α-L-arabinofuranosidase, α-glucuronidase, and several esterases are needed to cleave xylan side groups [39,40] and can be dissolved in the reaction media or immobilized. They can also be produced *in situ* by microorganisms such as fungi and bacteria that make multiple endoxylanase isoenzymes, reflecting the need for xylanases with specificities that are capable of acting on different substrates [41].

In contrast to autohydrolysis and chemical treatment methods, enzymatic hydrolysis avoids production of undesirable by-products or high amounts of monosaccharides, or require high-pressure or high-temperature equipment. However, enzymatic methods usually require much longer reaction times than acid hydrolysis or autohydrolysis. In addition, xylanase with different substrate specificities produces different hydrolysis end-products, and control of production of XOs with a desired DP range can be more difficult. On the other hand, acid hydrolysis of xylan randomly hydrolyzes glycosidic bonds between adjacent xylose units. Acid hydrolysis is therefore more practical for production of XOs in the DP range of 2–15 [42]. A study of the hydrolysis patterns of purified endoxylanase on birchwood, beechwood, and oat spelt xylans indicated that xylotriose (X3) is the shortest XOs released by xylanase [41]. Xylotriose and xylotetraose (X4) fragments are believed to be inaccessible to xylanase enzymes, probably due to substitution with arabinosyl residues. Commercial xylanase preparations are often low in β-xylosidase activity, resulting in xylobiose accumulation (X_2) [43]. Similarly, commercial cellulase preparations are usually low in β-xylosidase activity; that deficiency, coupled with the high inhibition of cellulase by xylooligosaccharides, has recently been shown to be an important contributor to reduced hydrolysis of xylooligosaccharides to xylose [43] as well as cellulose to glucose [25].

19.3 Xylooligosaccharides Separation and Purification

XOs from thermochemical or enzymatic treatment usually contain a wide DP range of oligomers and possibly other compounds as stated in the previous section. To produce more pure XO fractions used in food or pharmaceutical industries, the hydrolysis liquor must be refined by removing monosaccharides or non-saccharide compounds to obtain the highest possible XO content or a given DP range. Purification and separation of XOs from autohydrolysis liquor is complicated and may require multistage processing for reaction and/or fractionation. Depending on the degree of purity desired, a sequence of several physicochemical treatments may be needed [44].

19.3.1 Solvent Extraction

Solvent extraction is frequently applied to recover XOs and also applied to pre-extract interfering components before chemical or enzymatic treatment to simplify purification. Vacuum evaporation may be applied first to concentrate the crude XOs solution produced by hydrothermal processing and remove volatile components. Then, as shown in Figure 19.2, solvent extraction can remove non-saccharide compounds to yield both a refined aqueous phase and a solvent-soluble fraction that mainly contains most of the phenolics and extractive-derived compounds. The recovery yields and the degree of purification depend on the solvent employed for extraction, with ethanol, acetone, and 2-propanol the most common choices to refine crude XOs solution [44,46,47]. However, lignocellulosic materials used for XOs production may contain stabilizing non-saccharide components, especially comparatively high proportions of uronic groups and/or compounds in autohydrolysis liquors that are influenced by the XOs substitution pattern [44]. A study of solvent extraction of freeze-dried solids by 2-propanol, acetone, and ethanol showed that the highest purities were achieved with ethanol, although at the expense of lower recovery yields [44].

19.3.2 Adsorption by Surface Active Materials

Adsorption by surface active materials has been used in combination with other treatment steps to separate oligosaccharides from monosaccharides or remove other undesired compounds. The most widely used adsorbents for purification of XOs liquors include activated charcoal, acid clay, bentonite, diatomaceous earth, aluminum hydroxide or oxide, titanium, silica, and porous synthetic materials [26]. For example, Pellerin *et al.* used activated charcoal followed by elution with ethanol to fractionate XOs based on their molecular weight [48]. In the first stage, XOs were retained by activated charcoal and then released according to DP by changing the ethanol concentration during elution [48]. Zhu *et al.* [49] employed the same approach to purify oligosaccharides from aqueous-ammonia-pretreated corn stover and cobs. In this case,

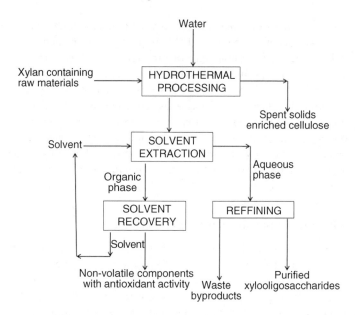

Figure 19.2   *Hydrothermal treatment coupled with solvent extraction for production of purified XOs from xylan-containing lignocellulosic materials. (Adapted from Moure et al. [45] © 2006, Elsevier).*

1–10% w/w activated carbon was added to the supernatant containing oligosaccharides, and the mixture was subsequently eluted with a solution containing 0–50% ethanol in water. The highest XOs yield was achieved for elution with 15–30% ethanol, but only the total oligosaccharides concentration was measured by traditional post-hydrolysis with 4% sulfuric acid at 121 °C for 1 hour and not the concentrations of each oligosaccharide DP fraction [50].

Montane *et al.* proposed that activated carbon treatment of raw XOs solutions obtained by autohydrolysis of lignocellulosic materials is feasible for removal of extractives, lignin-derived compounds, and carbohydrate degradation products [51]. Selective adsorption of lignin products compared to carbohydrates was favored by three commercial activated carbons at slightly acidic pH. The results also showed that selectivity towards lignin adsorption was higher when the carbon used was highly microporous and had smaller mesopore diameters, a low volume of mesopores, and a low concentration of basic surface groups to favor adsorption of lignin derivatives [51].

19.3.3   Chromatographic Separation Techniques

Although all the methods outlined above could be used to refine and concentrate XOs solutions, the resulting XOs solution may not be sufficiently pure. On the other hand, high-purity XO fractions have been produced at the analytical scale by chromatographic separations. For example, samples from hydrothermally treated lignocellulosic materials were fractionated by anion-exchange or size-exclusion chromatography [52–54]. However, techniques such as ^{13}C NMR [55], matrix-assisted laser desorption/ionization-time of flight (MALDI-TOF), and nanospray mass spectrometry have usually been employed for refining samples before structural characterization of XOs [53]. Katapodis *et al.* employed size-exclusion chromatography (SEC) in combination with other techniques for purification of feruloylated oligosaccharides [56]. Jacobs *et al.* purified hemicellulose-derived products from hydrothermal microwave treatments of flax shive by employing ion-exchange chromatography and/or SEC in combination with enzymatic processing [57].

Industrially, charcoal chromatography is preferred for sugar purifications due to its higher loading capacity than for other separation methods. However, it is difficult to separate XOs with high DPs, and acidic oligosaccharides (XOs with uronic acid substituents) would overlap with simpler XOs on the chromatograph. As a result, Dowex 1-X4 anion-exchange resin in the acetate form was used before charcoal chromatography to avoid overlapping. Due to its low efficiency and time-consuming operation, this method is less satisfactory in separation of high-purity XOs. In addition, continuous operation was not applied because of the gradient elution mode used and requirement for column regeneration.

Gel permeation chromatography (GPC) with cross-linked polyacrylamide and cross-linked dextran beads has been successfully applied for fractionating oligosaccharides since the 1960s. Sugars from mannose through mannoheptose were separated from each other using Sephadex G-25, but longer oligomers were not well resolved [58].

Unlike Sephadex, Bio-Gel is composed of polyacrylamide which is not susceptible to microbial degradation and does not leak carbohydrates during elution. Pontis applied Bio-Gel P-2 to separate sucrose through the heptaoligosaccharide of fructosan but, once again, larger fractosans were not separated well [59]. Havlicek and Samuelson applied a Bio-Gel P-2 column to separate XOs with DP ranging from 2 to 18 from acid-pretreated birchwood xylan hydrolysate after removing acidic saccharides with ion-exchange resin [60]. Korner *et al.* [61] fractionated XOs up to DP 7 using a Bio-Gel P-4 column operated at 40 °C with 0.05 M Tris/HCl buffer (pH 7.8) at a flow rate of 30 mL/h. Under such conditions, the series of XOs were eluted according to size exclusion principles, whereas acidic saccharides composed of xylose and uronic acids were separated according to partition principles, resulting in xyloheptose being superimposed on glucuronosylxylose. Because Bio-Gel and Toyopearl gels are known to be resistant to the permeation of acidic saccharides into pores of the gel particles, acidic saccharides are eluted near the void fraction with distilled water as eluent. Distilled water eliminates the need to remove buffer salts after separation, and the XOs fractions collected after separation could be easily concentrated by evaporation. Furthermore, columns filled with Bio-Gel with different pore sizes could be combined in series to maximize separation purity. Sun *et al.* [42] used three combinations of two columns connected in series to isolate xylose and XOs with DPs ranging from 2 to 15. Bio-Gel P-4 and Toyopearl HW-40F columns provided good resolution, and chromatography with Bio-Gel P-4 and P-2 columns also achieved separation of XOs up to DP 15. However, the resolution of the latter was slightly lower than that of Bio-Gel P-4 and Toyopearl HW-40 columns. In contrast, Toyopearl HW-50 and HW-40 columns connected in series could only separate XOs up to DP 8 (Figure 19.3).

Gel-permeation chromatography is a widely used separation technique that can be easily adapted for an auto-preparative system using an auto-sampler (or injection pump), a refractive index detector, and an automatic fraction collector that responds to the detector signal. Depending on the Bio-Gel pore size, relatively high-purity XOs fractions with different DP ranges can be collected, and more columns can be connected in series to further improve separation performance. The main disadvantage of GPC for separation of oligosaccharides is its relatively high cost. Thus, although GPC purification is frequently used to obtain fractions of XOs for structural characterization and the degree of purification of the different DP fractions is relatively good, GPC does not tend to be cost effective for large-scale production of XOs.

19.3.4 Membrane Separation

Membrane technology, mainly ultrafiltration and nanofiltration, is currently seen as the most promising downstream strategy for industrial manufacture of high-purity and concentrated oligosaccharides. Ultrafiltration separates oligosaccharides from higher-molecular-weight compounds or fractionates oligosaccharides of different DP. On the other hand, nanofiltration can concentrate liquors and/or remove undesired

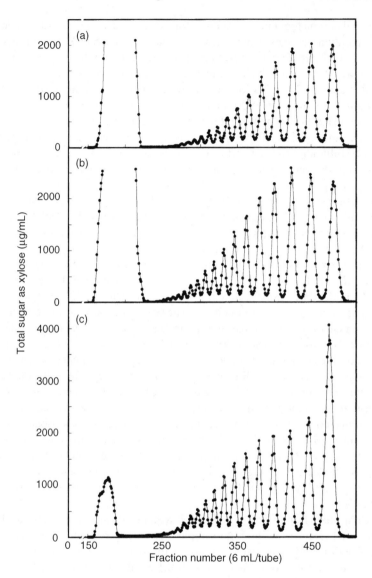

Figure 19.3   *Elution profiles of hydrolysis products of three kinds of xylans on BioGel P-4 and Toyopearl HW-40F columns connected in series: (a) cottonseed xylan; (b) birchwood xylan; and (c) oat spelt xylan. (Adapted from Sun* et al. *[42] © 2002, Elsevier).*

low-molecular-weight compounds such as monosaccharides or phenolics, enabling purification of oligosaccharide mixtures [62]. Compared to the other purification methods discussed above, membrane separation has a number of advantages including low energy requirements, easily manipulated critical operational variables, and relatively easy scale-up [63,64]. Certain operational variables including pressure, temperature, feed flow rate, and agitation can impact properties of the solute and membrane and physical aspects directly related to diffusion and convection of the solute, which in turn affect the overall process efficiency. In addition, oligosaccharide inhibition of enzymatic reactions can be reduced by continuous or semi-continuous product removal by membranes [62]. However, membrane separation performance can be affected by structural characteristics of oligosaccharides, including the types of monosaccharides, substitutions and linkages in oligomer structure, and the final molecular weight and extent of branching. Furthermore, oligosaccharide solubility has a major impact on membrane separation performance [62].

Membrane separations have been used for preparing various concentrations of several oligosaccharides, including fructose oligosaccharides [65], maltooligosaccharides [66], soybean oligosaccharides [67], pecticoligosaccharides [68], and chitooligosaccharides [69]. However, applications of membranes to refining XOs-containing solutions are limited in the literature; some publications have dealt with the processing of solutions/slurry resulting from hydrolytic treatment followed by enzymatic reaction. Recently, some studies successfully applied membranes to XOs produced by enzymatic hydrolysis or autohydrolysis of xylan-containing materials [47,70]. Yuan *et al.* employed nanofiltration membranes for concentrating XOs obtained by enzymatic hydrolysis of xylan from steamed corn cobs, whereas concentration and fractionation of XOs by sequential membrane-based steps has been employed in multistage purification processes [71].

Although microfiltration and ultrafiltration are well-established separation processes for purifying oligosaccharides from high-molecular-weight enzymes and polysaccharides, commercial streams often contain low-molecular-weight sugars that are undesirable or do not contribute to beneficial properties of the higher-molecular-weight oligosaccharides. In a study by Akpinar *et al.* [31], ultrafiltration was used to separate and purify XOs from hydrolysate generated by enzymatic hydrolysis of cotton stalk xylan. The hydrolysate was first filtered through a 10 kDa molecular weight cut-off membrane to remove high-molecular-weight polysaccharides and enzymes, followed by filtration through a 1–3 kDa molecular weight cut-off membrane to further fractionate the XOs. Permeate from the 1 and 3 kDa membranes contained mixtures of different DPs, with 43.3% and 81.6% reported to have DPs higher than 5, respectively. Although chromatography is still the principal choice, Leiva and Guzman reported that nanofiltration membranes can concentrate or purify oligosaccharide mixtures [72] as an alternative to more expensive chromatographic techniques. The molecular weight cut-off of nanofiltration membranes is in the range of 200–1000 Da, combining ultrafiltration and reverse osmosis separation properties. However, despite its promise for industrial-scale purification and concentration of oligosaccharide mixtures, its performance for fractionation of oligosaccharide mixtures has not yet been convincingly proven.

19.3.5 Centrifugal Partition Chromatography

Centrifugal purification chromatography (CPC), a method based on countercurrent chromatography, has recently been proposed for XOs purification [73]. Separation is based on the differences in partitioning behavior of components between two immiscible liquids. CPC uses a so-called "hydrostatic mode" resulting from constant centrifugal force intensity and direction for separation. Therefore, the mobile phase penetrates the stationary phase either by forming droplets or by jets stuck to the channel walls, broken jets, or atomization. The intensity of agitation of both phases depends on centrifugal force intensity, mobile phase flow rate, and solvent physical properties. Compared to countercurrent chromatography, stationary phase retention is less sensitive to physical properties of the solvent systems such as viscosity, density, and interfacial tension [74]. Similar to other chromatographic techniques, the CPC method is able to separate compounds with a broad range of molecular weights. In addition, samples can be recovered by flushing the system since the stationary phase is also a liquid. In contrast to other chromatographic techniques such as GPC, CPC could be used for preparative separation or purification because of its large stationary phase volume [75].

CPC has been widely used as an efficient purification and separation tool for many compounds including flavonoids, flavonolignans, and macrolide antibiotics [75]. Shibusawa *et al.* [76] employed CPC to purify apple-derived catechin oligosaccharides by operating in an ascending mode with a solvent system containing equal volumes of hexane, methyl acetate, acetonitrile, and water. Apple catechin oligosaccharides up to DP 10 were successfully fractionated by this method. However, the total mass and corresponding purity of each DP fraction were not reported.

Lau *et al.* [75] used CPC to separate and purify xylan-derived oligosaccharides from birchwood xylan. A CPC solvent system containing dimethyl sulfoxide (DMSO), tetrahydrofuran (THF), and water in a 1:6:3 volumetric ratio, respectively, was chosen for its ability to dissolve XOs of different DPs. Monomeric xylose and XOs up to DP 5 (xylopentose) were collected with this separation system with relatively high purity for DP 1 and 2 (higher than 85%) and relatively low purity for other DPs (lower than 55%).

19.4   Characterization and Quantification of Xylooligosaccharides

19.4.1   Measuring Xylooligosaccharides by Quantification of Reducing Ends

With few exceptions, each oligosaccharide chain has a reducing end on its terminal sugar residue. Because the aldehyde or ketone group of this terminal sugar residue is not fixed into a ring structure, it is free to undergo oxidation-reduction reactions with chemical reagents to form products that can be detected by colorimetric methods. By measuring the number of reducing ends in a sample, the total number of oligosaccharide chains can be determined. Colorimetric methods to measure monosaccharides as well as oligosaccharides employ a UV-Vis spectrophotometer, a simple and inexpensive instrument. Although these methods are still used today, different types of sugars cannot be differentiated.

The most widely used method for colorometric quantification of reducing ends is the dinitrosalicylic acid (DNS) assay, which was first developed to determine the concentration of monosaccharides [77–79] and then applied to quantify the total numbers of oligosaccharide chains in aqueous solution [80,81]. In the DNS assay, 3, 5-dinitrosalicylic acid reacts with sugar reducing ends to form red-brown 3-amino-5-nitrosalicylate, quantified by comparison of its absorbance at 560 nm or above [77–79] to that with pure sugar calibration standards. The volumetric concentration of reducing ends can therefore be calculated by determining the intensity of color formation of 3-amino-5-nitrosalicylate. However, the equivalence between amino-nitrosalicylate produced and the number of reducing ends varies for different sugars, suggesting that the DNS assay can only be accurate for evaluation of a single sugar [77,82]. Other methods in this category, such as the arsenomolybdate (ARS; also known as Nelson-Somogyi assay) assay [83], the p-hydroxybenzoic acid hydrazide (PAHBAH) assay [84,85], and the phenol-sulfuric acid assay [86,87] are also used to measure reducing sugars but have similar response variance issues with different sugars as the DNS assay. When these colorimetric methods were applied to measure reducing ends of xylooligosaccharides, they responded differently to xylooligosaccharides of different DPs. For example, the ARS assay showed less reactivity to higher-DP xylooligosaccharides, while the DNS assay showed the opposite trend [81]. The reason, however, is not well understood.

19.4.2   Characterizing Xylooligosaccharides Composition

The determination of structure of oligosaccharides released from biomass hydrolysis first requires knowledge of what monosaccharide components are present and in what amounts. This can be achieved by enzymatic or chemical decomposition of oligosaccharides into their monosaccharide building blocks followed by identification and quantification of each component by gas chromatography (GC) or high-performance liquid chromatography (HPLC) [4]. GC methods require multistep formation of volatile derivatives of monosaccharides prior to analysis, with two derivatization methods routinely used: formation of alditol acetates or TMS methyl glycosides [4].

Although GC methods have the advantage of baseline sugar resolution [50], HPLC is more widely used for analysis because sugar derivatization is not required. In standard biomass analytical procedures developed by the National Renewable Energy Laboratory (NREL), HPLC employing a refractive index (RI) detector is the default tool for determining total component monosaccharides released from post hydrolysis of oligosaccharides with 4 wt% sulfuric acid at 121 °C for 1 hour [88]. Two columns, both from Bio-Rad, are commonly used in this application. The HPX-87P column can separate all common biomass sugars (cellobiose, glucose, xylose, galactose, arabinose, and mannose) with high resolution. However, retention times (RT) for xylose, mannose, and galactose on the HPX-87H column are close (within 0.1 min); this often results in a single peak, depending on column conditions. Considering that heteroxylans are the dominant form of hemicellulose in most lignocellulosic feedstocks (refer to Section 19.1.2), the amounts of galactose and mannose in oligosaccharides released from biomass pretreatment are low. The HPX-87H column is therefore widely used to measure the glycosyl composition of xylooligosaccharides because it provides stable and near-baseline resolution of glucose, xylose, and arabinose.

19.4.3 Direct Characterization of Different DP Xylooligosaccharides

As discussed in Section 19.1.3, there is a significant and increasing demand for reproducible, fast, and simple methods to characterize and quantify XOs released from biomass pretreatment to better understand the decomposition mechanisms of hemicellulose in major lignocellulosic feedstocks. To date, several methodologies for qualitative and quantitative analysis of XOs have been developed, which can be grouped into the following categories: HPLC, high-performance anion-exchange chromatography (HPAEC), and capillary electrophoresis (CE).

HPLC

Li *et al.* [89] quantitatively analyzed XOs derived from hydrothermal pretreatment of oat spelt xylan at 200 °C for 15 min with a 5 wt% solid loading. A Waters model 717 chromatography system, equipped with a RI detector and a Bio-Rad Aminex HPX-42A ion-moderated partition (IMP) column was used. At a flow rate of 0.2 mL/min and a column temperature of 85 °C, xylooligosaccharides up to DP 10 were separated, but the baseline for the IMP chromatogram was difficult to resolve, especially for DP higher than 5 as shown in Figure 19.4. Commercial low-DP XOs standards (xylobiose, xylotriose, xylotetraose, and xylopentaose) were used to calibrate the IMP system for quantification of xylooligosaccharides with DPs in that same range of 2–5. It was also shown that xylooligosaccharides of DP 2–5 could be quantified by taking the ratio of peak heights of each XO to the peak height for xylose and multiplying this ratio by the concentration of the latter. These results confirmed that peak height followed concentrations closely for xylooligosaccharides with DP less than 5 for an RI detector. This approach was extended to quantifying XOs with higher DPs from 6 to 10; however, accuracy could not be confirmed due to the lack of standards.

Ohara *et al.* [90] used a cation-exchange column (Sugar KS-802; Showa Denko, Tokyo) with RI detector to characterize xylooligosaccharides up to DP 6, which were prepared by enzymatic hydrolysis of birchwood xylan with endoxylanase. The column temperature was 60 °C and the mobile phase was water with a flow rate of 0.6 mL/min.

For HPLC systems, an evaporative light scattering detector (ELSD) was reported to be more sensitive and thus provide better baseline stability than RI-based detectors for measuring oligosaccharides [91]. Yu and Wu [92] identified gluco-oligosaccharides from DP 2 to DP 6 using a Prevail carbohydrate ES column with an ELSD 3300 detector from Alltech. Other literature on specific characterization of xylooligosaccharides using ELSD, however, is scarce.

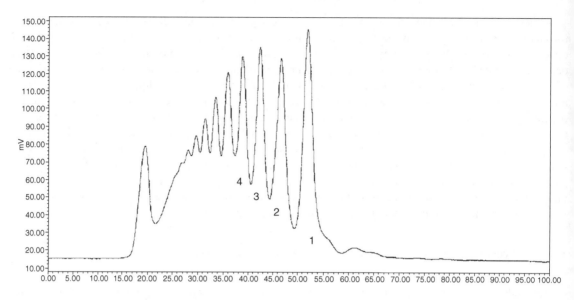

Figure 19.4   *IMP chromatogram of xylooligosaccharides derived from hydrothermal pretreatment of oat spelt xylan. 1: DP1; 2: DP2; 3: DP3; 4: DP4. (From Li et al. [89] with permission from Springer).*

Recently, a Waters Acquity ultrahigh-performance liquid chromatography (UPLC) equipped with a BEH HILIC (unbonded ethylene bridged hybrid efficient hydrophilic interaction chromatography) column and 4000 QTrap MS detector was applied to characterize xylooligosaccharides by Tomkins *et al.* [93]. As the chromatogram in Figure 19.5 shows, the approach was very sensitive, with detection of xylooligosaccharides at concentrations of about 1 pmol and very fast within only 2 min needed to separate xylooligosaccharides of DP up to 6.

HPAEC

The advent of HPAEC featuring pulsed amperometric detection (HPAEC-PAD) in the 1980s provided a highly sensitive and selective tool for separation and detection of complex carbohydrates without derivatization. The recognition by Johnson in 1986 that oligosaccharides could be detected by PAD greatly enhanced the popularity of HPAEC [94]. HPAE-CPAD is often classified as an HPLC method. However, HPAEC-PAD technology is discussed in some detail here because of its novel ability to analyze and characterize XOs over a wide DP range. The unique advantages of HPAEC were first described in the paper by Rocklin and Pohl in 1983 [95]. The oligosaccharides were separated in strong alkaline eluents (pH > 13), where their hydroxyl groups were deprotonated and thus rendered anionic. The number of hydroxyl groups in a single oligosaccharide molecule varies with DP, resulting in various weakly acidic properties which the HPAEC uses to separate oligosaccharides.

The PAD detection mechanism consists primarily of a three-step potential waveform with a frequency of 1–2 Hz. When analyte molecules are absorbed on the oxidation-free surface of the gold working electrode illustrated in Figure 19.6 [96], a detection potential (E_1) appropriate for the analyte properties as well as oxidation mechanism is applied first, and then the analyte molecules are oxidized. Thus, the anodic signal current can be measured in this step. Following detection, the electrode surface is usually oxidatively cleaned by a positive potential (E_2) and then reactivated by a negative

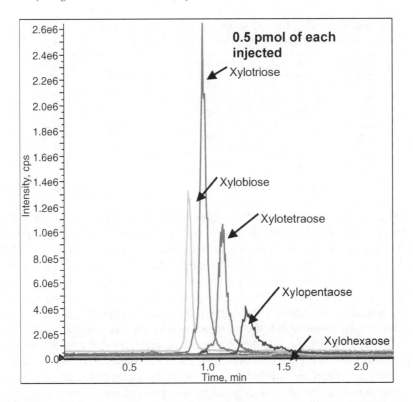

Figure 19.5 *UPLC chromatogram of xylooligosaccharides of DP 2–6. (From Tomkins et al. [93] with permission of the authors). (See figure in color plate section).*

potential (E_3). Alternatively, the cleaning potential (E_2) could be negative as demonstrated for effectively minimizing electrode wear [97,98]. In particular, a waveform suggested by Dionex Technical Note 21 has been widely used for characterizing monosaccharides and oligosaccharides, as shown in Table 19.2 [97].

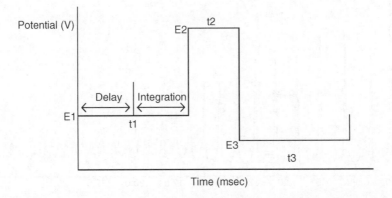

Figure 19.6 *Diagram of the pulse sequence for carbohydrate detection on a PAD detector. (Reproduced with permission of Dionex Corporation).*

Table 19.2   *Waveform of PAD for carbohydrates analysis using the Dionex IC system. (Reproduced with permission of Dionex Corporation).*

Time (ms)	Potential (V)	Integration
0	+0.1	
200	+0.1	Begin
400	+0.1	End
410	−2.0	
420	−2.0	
430	+0.6	
440	−0.1	
500	−0.1	

Dionex Corporation recently advanced its oligosaccharides profiling using HPAEC-PAD with its CarboPac PA-100 and CarboPac PA-200 columns. Several publications have successfully profiled the DP distribution of the oligosaccharides amylopectin, maltodextrin, and inulin up to DP 60 [3,99–103]. However, HPAEC-PAD characterization of xylooligosaccharides derived from hemicellulose in lignocellulosic biomass is difficult due to their low solubility at room temperature and resulting precipitation of higher-DP oligosaccharides [104]. In addition, the heterogeneous glycosyl residue composition of side chains and different linkage substitutions also limit HPAEC-PAD. Yang and Wyman [2] successfully separated XOs released from hydrothermal pretreatment of corn stover using a Dionex DX-600 module with a CarboPac PA100 column. The mobile phase was operated in a gradient mode (50–450 mM of sodium acetate, NaAc) through 150 mM NaOH [103] with the same waveform shown in Table 19.2. As the chromatogram shows in Figure 19.7, xylooligosaccharides with DPs up to 13 were separated well with near baseline resolution. Peaks suspected to represent higher-DP xylooligosaccharides could also be detected, but the separation was relatively poor.

The PAD response of these xylooligosaccharides is believed to depend on the size and spatial structure of analyte molecules [105,106] and vary with DP in this situation. Koch *et al.* showed that relative

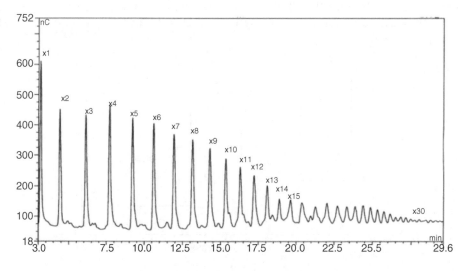

Figure 19.7   *Dionex IC chromatogram of xylooligosaccharides released from hydrothermal pretreatment of corn stover. (From Yang and Wyman [2] with permission of Elsevier).*

electrochemical responses of amylopectin oligosaccharides increased with DP based on molar concentrations but decreased with DP based on mass concentrations [100]. The variable detection behavior is one of the disadvantages of this technique and results in the need for sugar standards for accurate quantification of each DP fraction. However, standards of xylooligosaccharides are only available for DPs below 6 (Megazyme International Ireland Ltd., Ireland) and even then are very expensive.

In the case of HPAEC-PAD for oligosaccharides with the same glycosidic linkages, smaller DP molecules elute first followed by larger ones; however, the order can change when different linkage variants are mixed. For example, Morales *et al.* showed that isomaltohexaose eluted before maltotriose [107]. In fact, factors like charge, molecular size, sugar composition, and glycosidic linkages can impact chromatographic separation [108]. The effects of these factors therefore must be considered when measuring oligosaccharides released by lignocellulosic biomass, particularly for biomass with highly heterogeneous cell-wall polysaccharides. In this situation, additional analytical techniques following HPAEC-PAD are required, such as mass spectroscopy (MS) or nuclear magnetic resonance spectroscopy (NMR).

Capillary Electrophoresis

Capillary electrophoresis (CE) has been successfully applied to separate a wide range of xylooligosaccharide compounds. Since the mid-1990s, application of CE to characterize xylooligosaccharides released from plant cell-wall polysaccharides has been reported in several publications [109–112]. However, due to the lack of charged groups and chromophores, applications to separation of important oligosaccharides by CE are limited [113,114]. Kabel *et al.* applied CE-LIF (laser-induced fluorescence detector) successfully to separate xylooligosaccharides derived from hydrothermally treated *Eucalyptus* woods [115]. In this case, xylooligosaccharides were derivatized with 9-aminopyrene-1,4,6-trisulfonate (APTS), which attached to the reducing end of oligosaccharide molecules to provide a fluorescent APTS tag as well as three negative charges. As shown by the LIF-electropherograms in Figure 19.8, a series of xylooligosaccharides up to DP

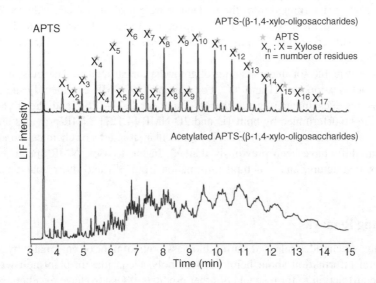

Figure 19.8 *LIF-electropherograms of APTS derivatized β-(1, 4)-xylooligosaccharides (top) and (less diluted) O-acetylated β-(1, 4)-xylooligosaccharides (bottom) obtained from hydrothermally treated* Eucalyptus *wood (* is maltose internal standard). (From Kabel et al. [115] with permission from Elsevier).*

17 that had been derivatized with APTS were separated with very high resolution and much better than similar xylooligosaccharide samples that were not derivatized. Coupled with MS, the minor peaks between the major peaks were identified as linear 1, 4-β-xylooligosaccharides with a different structure.

Although derivatization leads to improved sensitivity and resolution with CE, different reactivity of derivatizing reagents and formation of several by-products result in control problems for consistent preparation of analytes [116]. High-pH buffer and other detection techniques have been used to avoid derivatization, but successful application of oligosaccharides profiling has not been shown.

19.4.4   Determining Detailed Structures of Oligosaccharides by MS and NMR

The analytical techniques reviewed in Sections 19.4.1–19.4.3 are effective for characterizing glycosyl residue compositions for oligosaccharides as well as DP profiling of oligosaccharides with the same type of glycosyl linkages. However, they cannot provide detailed structural information for oligosaccharides such as glycosyl linkage composition, the sequence of glycosyl residues, and the anomeric configuration. MS and NMR are needed to characterize such structural features for oligosaccharides.

MS has proven to be valuable for several aspects of structural characterization of oligosaccharides. With different combinations of ionizations and analyzers, MS is often coupled to chromatography techniques such as GC-MS, HPLC-MS, and HPAEC-MS, but many challenges remain. For example, HPAEC could provide good separation of oligosaccharides without the need for derivatization, but the eluent used in HPAEC contains a high concentration of salt which limits use with MS [115]. Regardless of whether MS is online or offline, good separation of oligosaccharides prior to MS analysis will always facilitate structural characterization. Thus, preparative columns such as size exclusion [52] and ion exchange are also used to isolate oligosaccharides into different fractions for offline MS analysis. Currently, electrospray ionization (ESI) and matrix-assisted laser desorption (MALDI) [52] are the most common ionization sources used for xylooligosaccharides characterization in combination with tandem MS analyzers [117–122]. Usually, one type of MS analysis is better for a particular type of oligosaccharide; for example, MALDI-TOF (time-of-flight) MS allows routine determination of the molecular weight of oligosaccharides containing more than 10 glycosyl residues [4]. As the MALDI-TOF mass spectra in Figure 19.9 show, xylooligosaccharides from *Eucalyptus* wood with different chain lengths were characterized with additional information on the degree of acetylation [123].

NMR has proven valuable for understanding oligosaccharide structures. The most-used isotopes in oligosaccharides characterization are ^{1}H and ^{13}C. For example, ^{1}H-NMR can identify the anomeric configuration of glucosyl residues in an oligosaccharide fragment [124–126], and the glycosyl sequence of oligosaccharides can be determined by both 1D and 2D NMR [4,127]. NMR provides an effective method for quick and accurate characterization of specified molecular structures or chemical bonds when the corresponding chemical shifts have been previously defined. In most cases, NMR spectra are insufficient to analyze an unknown structure, and structural information from MS and other analytical techniques must be also used.

19.5   Concluding Remarks

Characterizing oligosaccharides released during biomass pretreatment or enzymatic hydrolysis can reveal important structural information about hemicellulose polysaccharides in plant cell walls and how they change during deconstruction to form sugars or other products. XOs also have excellent potential for applications in pharmaceutical, agriculture, and food industries. XOs can be produced by chemical or enzymatic methods on an industrial scale from lignocellulosic materials. Chemical methods are preferred to produce XO mixtures with a wide DP range, while enzymatic methods are preferred in the food or pharmaceutical

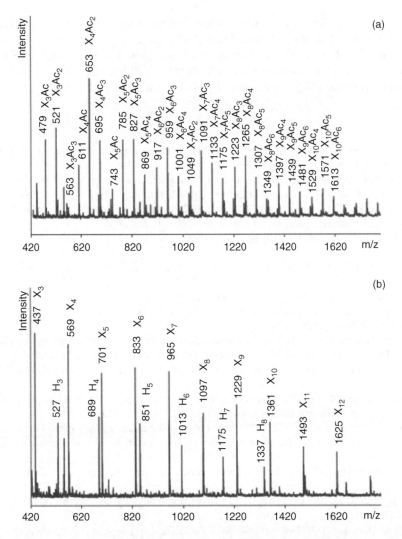

Figure 19.9 *MALDI-TOF mass spectra of the neutral xylooligosaccharides obtained from* Eucalyptus *wood hydrolysate (a) before and (b) after saponification (X = xylose; Ac = acetyl-group; H = hexose). (From Kabel et al. [123] with permission from Elsevier).*

industries to reduce formation of degradation products. With the growing importance of making fuels from cellulosic biomass and the increasing demand for xylooligosaccharides, more opportunities are emerging to process xylan-rich pretreatment hydrolyzate in a cellulosic biorefinery into high-value products that could further lower the cost of cellulosic biofuels. However, separation technologies are needed to produce high-purity XO fractions that span desired DP ranges for industry applications or characterization.

Acknowledgements

We gratefully acknowledge the support by the Office of Biological and Environmental Research in the DOE Office of Science through the BioEnergy Science Center (BESC). The authors would like to thank Professor

Eugene A. Nothnagel at the Botany and Plant Science Department of University of California, Riverside and Dr Bruce A. Tomkins at Oak Ridge National Laboratory for valuable discussions and comments on this chapter. We are grateful to the Center for Environmental Research & Technology (CE-CERT) for providing the facilities and equipment. We would also like to thank the Ford Motor Company for their support of the Chair in Environmental Engineering at the University of California Riverside (UCR).

References

1. Abballe, F., Toppazzini, M., Campa, C. *et al.* (2007) Study of molar response of dextrans in electrochemical detection. *Journal of Chromatography A*, **1149** (1), 38–45.
2. Yang, B. and Wyman, C.E. (2008) Characterization of the degree of polymerization of xylooligomers produced by flowthrough hydrolysis of pure xylan and corn stover with water. *Bioresource Technology*, **99** (13), 5756–5762.
3. Ronkart, S.N., Blecker, C.S., Fourmanoir, H. *et al.* (2007) Isolation and identification of inulooligosaccharides resulting from inulin hydrolysis. *Analytica Chimica Acta*, **604** (1), 81–87.
4. Albersheim, P., Darvill, A., Roberts, K. *et al.* (2010) *Plant Cell Walls: From Chemistry to Biology*, Garland Science, New York, NY.
5. Himmel, M.E. (2008) *Biomass Recalcitrance: Deconstructing the Plant Cell Wall for Bioenergy*, Blackwell Publishing, Oxford.
6. Kabyemela, B.M., Adschiri, T., Malaluan, R.M., and Arai, K. (1999) Glucose and fructose decomposition in subcritical and supercritical water: Detailed reaction pathway, mechanisms, and kinetics. *Industrial & Engineering Chemistry Research*, **38** (8), 2888–2895.
7. Shallom, D. and Shoham, Y. (2003) Microbial hemicellulases. *Current Opinion in Microbiology*, **6** (3), 219–228.
8. Wyman, C.E., Decker, S.R., Himmel, M. *et al.* (2005) Hydrolysis of cellulose and hemicellulose, in *Polysaccharides: Structural Diversity and Functional Versatility*, 2nd edn (ed. S. Dumitriu), Marcel Dekker, New York, p. xvii p. 1204.
9. Kumar, R. and Wyman, C.E. (2009) Access of cellulase to cellulose and lignin for poplar solids produced by leading pretreatment technologies. *Biotechnology Progress*, **25** (3), 807–819.
10. Studer, M.H., DeMartini, J.D., Davis, M.F. *et al.* (2011) Lignin content in natural Populus variants affects sugar release. *Proceedings of the National Academy of Sciences of the United States of America*, **108** (15), 6300–6305.
11. Wyman, C.E. (2007) What is (and is not) vital to advancing cellulosic ethanol. *Trends in Biotechnology*, **25** (4), 153–157.
12. Lynd, L.R., Wyman, C.E., and Gerngross, T.U. (1999) Biocommodity Engineering. *Biotechnology Progress*, **15** (5), 777–793.
13. Lee, C.H., Teng, Q. *et al.* (2009) Down-regulation of PoGT47C expression in poplar results in a reduced glucuronoxylan content and an increased wood digestibility by cellulase. *Plant & Cell Physiology*, **50** (6), 1075–1089.
14. Azadi, P., Naran, R., Black, S., and Decker, S.R. (2009) Extraction and characterization of native heteroxylans from delignified corn stover and aspen. *Cellulose*, **16** (4), 661–675.
15. Wyman, C.E. (2003) Potential synergies and challenges in refining cellulosic biomass to fuels, chemicals, and power. *Biotechnology Progress*, **19** (2), 254–262.
16. Somerville, C., Youngs, H., Taylor, C. *et al.* (2010) Feedstocks for lignocellulosic biofuels. *Science*, **329** (5993), 790–792.
17. Wyman, C.E. (1999) Biomass ethanol: technical progress, opportunities, and commercial challenges. *Annual Review of Energy and the Environment*, **24**, 189–226.
18. Wyman, C.E., Dale, B.E., Elander, R.T. *et al.* (2005) Coordinated development of leading biomass pretreatment technologies. *Bioresource Technology*, **96** (18), 1959–1966.
19. Kumar, R., Mago, G., Balan, V., and Wyman, C.E. (2009) Physical and chemical characterizations of corn stover and poplar solids resulting from leading pretreatment technologies. *Bioresource Technology*, **100** (17), 3948–3962.

20. Selig, M.J., Viamajala, S., Decker, S.R. *et al.* (2007) Deposition of lignin droplets produced during dilute acid pretreatment of maize stems retards enzymatic hydrolysis of cellulose. *Biotechnology Progress*, **23** (6), 1333–1339.

21. Kumar, R. and Wyman, C.E. (2008) The impact of dilute sulfuric acid on the selectivity of xylooligomer depolymerization to monomers. *Carbohydrate Research*, **343** (2), 290–300.

22. Hahn-Hagerdal, B. and Palmqvist, E. (2000) Fermentation of lignocellulosic hydrolysates. II: inhibitors and mechanisms of inhibition. *Bioresource Technology*, **74** (1), 25–33.

23. Ladisch, M., Mosier, N., Wyman, C. *et al.* (2005) Features of promising technologies for pretreatment of lignocellulosic biomass. *Bioresource Technology*, **96** (6), 673–686.

24. Kumar, R. and Wyman, C.E. (2009) Effect of enzyme supplementation at moderate cellulase loadings on initial glucose and xylose release from corn stover solids pretreated by leading technologies. *Biotechnology and Bioengineering*, **102** (2), 457–467.

25. Qing, Q., Yang, B., and Wyman, C.E. (2010) Xylooligomers are strong inhibitors of cellulose hydrolysis by enzymes. *Bioresource Technology*, **101** (24), 9624–9630.

26. Vazquez, M.J., Alonso, J.L., Dominguez, H., and Parajo, J.C. (2000) Xylooligosaccharides: manufacture and applications. *Trends in Food Science & Technology*, **11** (11), 387–393.

27. Akpinar, O., Erdogan, K., and Bostanci, S. (2009) Production of xylooligosaccharides by controlled acid hydrolysis of lignocellulosic materials. *Carbohydrate Research*, **344** (5), 660–666.

28. Parajo, J.C., Garrote, G., Cruz, J.M., and Dominguez, H. (2004) Production of xylooligosaccharides by autohydrolysis of lignocellulosic materials. *Trends in Food Science & Technology*, **15** (3–4), 115–120.

29. de Menezes, C.R., Silva, I.S., Pavarina, E.C. *et al.* (2009) Production of xylooligosaccharides from enzymatic hydrolysis of xylan by the white-rot fungi Pleurotus. *International Biodeterioration & Biodegradation*, **63** (6), 673–678.

30. Brienzo, M., Carvalho, W., and Milagres, A.M.F. (2010) Xylooligosaccharides production from alkali-pretreated sugarcane bagasse using xylanases from thermoascus aurantiacus. *Applied Biochemistry and Biotechnology*, **162** (4), 1195–1205.

31. Akpinar, O., Ak, O., Kavas, A. *et al.* (2007) Enzymatic production of xylooligosaccharides from cotton stalks. *Journal of Agricultural and Food Chemistry*, **55** (14), 5544–5551.

32. Teng, C., Yan, Q.J., Jiang, Z.Q. *et al.* (2010) Production of xylooligosaccharides from the steam explosion liquor of corncobs coupled with enzymatic hydrolysis using a thermostable xylanase. *Bioresource Technology*, **101** (19), 7679–7682.

33. Nabarlatz, D., Ebringerova, A., and Montane, D. (2007) Autohydrolysis of agricultural by-products for the production of xylo-oligosaccharides. *Carbohydrate Polymers*, **69** (1), 20–28.

34. Lloyd, T.A. and Wyman, C.E. (2005) Combined sugar yields for dilute sulfuric acid pretreatment of corn stover followed by enzymatic hydrolysis of the remaining solids. *Bioresource Technology*, **96** (18), 1967–1977.

35. Jacobsen, S.E. and Wyman, C.E. (2002) Xylose monomer and oligomer yields for uncatalyzed hydrolysis of sugarcane bagasse hemicellulose at varying solids concentration. *Industrial & Engineering Chemistry Research*, **41** (6), 1454–1461.

36. Nabarlatz, D., Farriol, X., and Montane, D. (2004) Kinetic modeling of the autohydrolysis of lignocellulosic biomass for the production of hemicellulose-derived ligosaccharides. *Industrial & Engineering Chemistry Research*, **43** (15), 4124–4131.

37. Nabarlatz, D., Farriol, X., and Montane, D. (2005) Autohydrolysis of almond shells for the production of xylooligosaccharides: product characteristics and reaction kinetics. *Industrial & Engineering Chemistry Research*, **44** (20), 7746–7755.

38. Sun, R.C., Tomkinson, J., Ma, P.L., and Liang, S.F. (2000) Comparative study of hemicelluloses from rice straw by alkali and hydrogen peroxide treatments. *Carbohydrate Polymers*, **42** (2), 111–122.

39. Uffen, R.L. (1997) Xylan degradation: a glimpse at microbial diversity. *Journal of Industrial Microbiology & Biotechnology*, **19** (1), 1–6.

40. Den Haan, R. and Van Zyl, W.H. (2003) Enhanced xylan degradation and utilisation by Pichia stipitis overproducing fungal xylanolytic enzymes. *Enzyme and Microbial Technology*, **33** (5), 620–628.

41. Milagres, A.M.F., Magalhaes, P.O., and Ferraz, A. (2005) Purification and properties of a xylanase from Ceriporiopsis subvermispora cultivated on Pinus taeda. *FEMS Microbiology Letters*, **253** (2), 267–272.

42. Sun, H.J., Yoshida, S., Park, N.H., and Kusakabe, I. (2002) Preparation of (1 → 4)-beta-D-xylooligosaccharides from an acid hydrolysate of cotton-seed xylan: suitability of cotton-seed xylan as a starting material for the preparation of (1 → 4)-beta-D-xylooligosaccharides. *Carbohydrate Research*, **337** (7), 657–661.

43. Qing, Q. and Wyman, C.E. (2011) Hydrolysis of different chain length xylooliogmers by cellulase and hemicellulase. *Bioresource Technology*, **102** (2), 1359–1366.

44. Vazquez, M.J., Garrote, G., Alonso, J.L. *et al.* (2005) Refining of autohydrolysis liquors for manufacturing xylooligosaccharides: evaluation of operational strategies. *Bioresource Technology*, **96** (8), 889–896.

45. Moure, A., Gullon, P., Dominguez, H., and Parajo, J.C. (2006) Advances in the manufacture, purification and applications of xylo-oligosaccharides as food additives and nutraceuticals. *Process Biochemistry (Barking, London, England)*, **41** (9), 1913–1923.

46. Vegas, R., Alonso, J.L., Dominguez, H., and Parajo, J.C. (2005) Manufacture and refining of oligosaccharides from industrial solid wastes. *Industrial & Engineering Chemistry Research*, **44** (3), 614–620.

47. Swennen, K., Courtin, C.M., Van der Bruggen, B. *et al.* (2005) Ultrafiltration and ethanol precipitation for isolation of arabinoxylooligosaccharides with different structures. *Carbohydrate Polymers*, **62** (3), 283–292.

48. Pellerin, P., Gosselin, M., Lepoutre, J.P. *et al.* (1991) Enzymatic production of oligosaccharides from corncob xylan. *Enzyme and Microbial Technology*, **13** (8), 617–621.

49. Zhu, Y.M., Kim, T.H., Lee, Y.Y. *et al.* (2006) Enzymatic production of xylooligosaccharides from corn stover and corn cobs treated with aqueous ammonia. *Applied Biochemistry and Biotechnology*, **130** (1–3), 586–598.

50. Sluiter, J.B., Ruiz, R.O., Scarlata, C.J. *et al.* (2010) Compositional analysis of lignocellulosic feedstocks. 1. Review and description of methods. *Journal of Agricultural and Food Chemistry*, **58** (16), 9043–9053.

51. Montane, D., Nabarlatz, D., Martorell, A. *et al.* (2006) Removal of lignin and associated impurities from xylooligosaccharides by activated carbon adsorption. *Industrial & Engineering Chemistry Research*, **45** (7), 2294–2302.

52. Aad, G., Abat, E., Abdallah, J. *et al.* (2008) The ATLAS experiment at the CERN large hadron collider. *Journal of Instrumentation*, **3**, 1–380.

53. Kabel, M.A., Carvalheiro, F., Garrote, G. *et al.* (2002) Hydrothermally treated xylan rich by-products yield different classes of xylo-oligosaccharides. *Carbohydrate Polymers*, **50** (1), 47–56.

54. Kabel, M.A., Kortenoeven, L., Schols, H.A., and Voragen, A.G.J. (2002) *In vitro* fermentability of differently substituted xylo-oligosaccharides. *Journal of Agricultural and Food Chemistry*, **50** (21), 6205–6210.

55. Christakopoulos, P., Katapodis, P., Kalogeris, E. *et al.* (2003) Antimicrobial activity of acidic xylo-oligosaccharides produced by family 10 and 11 endoxylanases. *International Journal of Biological Macromolecules*, **31** (4–5), 171–175.

56. Katapodis, P., Vardakou, M., Kalogeris, E. *et al.* (2003) Enzymic production of a feruloylated oligosaccharide with antioxidant activity from wheat flour arabinoxylan. *European Journal of Nutrition*, **42** (1), 55–60.

57. Jacobs, A., Palm, M., Zacchi, G., and Dahlman, O. (2003) Isolation and characterization of water-soluble hemicelluloses from flax shive. *Carbohydrate Research*, **338** (18), 1869–1876.

58. Stewart, T.S., Mendersh, P.b., and Ballou, C.E. (1968) Preparation of a mannopentaose mannohexaose and mannoheptaose from saccharomyces cerevisiae mannan. *Biochemistry-Us*, **7** (5), 1843.

59. Pontis, H.G. (1968) Separation of fructosans by gel filtration. *Analytical Biochemistry*, **23** (2), 331.

60. Havlicek, J. and Samuelson, O. (1972) Chromatography of oligosaccharides from xylan by various techniques. *Carbohydrate Research*, **22** (2), 307.

61. Körner, H.-U., Gottschalk, D., Wiegel, J., and Puls, J. (1984) The degradation pattern of oligomers and polymers from lignocelluloses. *Analytica Chimica Acta*, **163**, 55–66.

62. Meyer, A.S., Pinelo, M., and Jonsson, G. (2009) Membrane technology for purification of enzymatically produced oligosaccharides: Molecular and operational features affecting performance. *Separation and Purification Technology*, **70** (1), 1–11.

63. Cano, A. and Palet, C. (2007) Xylooligosaccharide recovery from agricultural biomass waste treatment with enzymatic polymeric membranes and characterization of products with MALDI-TOF-MS. *Journal of Membrane Science*, **291** (1–2), 96–105.

64. Czermak, P., Ebrahimi, M., Grau, K. *et al.* (2004) Membrane-assisted enzymatic production of galactosyl-oligo-saccharides from lactose in a continuous process. *Journal of Membrane Science*, **232** (1–2), 85–91.

65. Li, W.Y., Li, J.D., Chen, T.Q. *et al.* (2005) Study on nanofiltration for purifying fructo-oligosaccharides II. Extended pore model. *Journal of Membrane Science*, **258** (1–2), 8–15.

66. Slominska, L. and Grzeskowiak-Przywecka, A. (2004) Study on the membrane filtration of starch hydrolysates. *Desalination*, **162** (1–3), 255–261.

67. Kim, S., Kim, W., and Hwang, I.K. (2003) Optimization of the extraction and purification of oligosaccharides from defatted soybean meal. *International Journal of Food Science and Technology*, **38** (3), 337–342.

68. Iwasaki, K. and Matsubara, Y. (2000) Purification of pectate oligosaccharides showing root-growth-promoting activity in lettuce using ultrafiltration and nanofiltration membranes. *Journal of Bioscience and Bioengineering*, **89** (5), 495–497.

69. Jeon, Y.J. and Kim, S.K. (2000) Production of chitooligosaccharides using an ultrafiltration membrane reactor and their antibacterial activity. *Carbohydrate Polymers*, **41** (2), 133–141.

70. Vegas, R., Luque, S., Alvarez, J.R. *et al.* (2006) Membrane-assisted processing of xylooligosaccharide-containing liquors. *Journal of Agricultural and Food Chemistry*, **54** (15), 5430–5436.

71. Yuan, Q.P., Zhang, H., Qian, Z.M., and Yang, X.J. (2004) Pilot-plant production of xylo-oligosaccharides from corncob by steaming, enzymatic hydrolysis and nanofiltration. *Journal of Chemical Technology and Biotechnology*, **79** (10), 1073–1079.

72. Leiva, M.H.L. and Guzman, M. (1995) Formation of oligosaccharides during enzymatic-hydrolysis of milk whey permeates. *Process Biochemistry*, **30** (8), 757–762.

73. Marchal, L., Legrand, J., and Foucault, A. (2003) Centrifugal partition chromatography: A survey of its history, and our recent advances in the field. *Chemical Record*, **3** (3), 133–143.

74. Armstrong, D.W. (1988) Theory and use of centrifugal partition chromatography. *Journal of Liquid Chromatography & Related Technologies*, **11** (12), 2433–2446.

75. Lau, C.S., Bunnell, K.A., Clausen, E.C. *et al.* (2011) Separation and purification of xylose oligomers using centrifugal partition chromatography. *Journal of Industrial Microbiology & Biotechnology*, **38** (2), 363–370.

76. Shibusawa, Y., Yanagida, A., Shindo, H., and Ito, Y. (2003) Separation of apple catechin oligomers by CCC. *Journal of Liquid Chromatography & Related Technologies*, **26** (9–10), 1609–1621.

77. Miller, G.L. (1959) Use of dinitrosalicylic acid reagent for determination of reducing sugar. *Analytical Chemistry*, **31** (3), 426–428.

78. Sumner, J.B. and Graham, V.A. (1921) Dinitrosalicylic acid: A reagent for the estimation of sugar in normal and diabetic urine. *The Journal of Biological Chemistry.*, **47** (1), 5–9.

79. Sumner, J.B. and Noback, C.V. (1924) The estimation of sugar in diabetic urine, using dinitrosalicylic acid. *The Journal of Biological Chemistry*, **62** (2), 287–290.

80. Bailey, M.J., Biely, P., and Poutanen, K. (1992) Interlaboratory testing of methods for assay of xylanase activity. *Journal of Biotechnology*, **23** (3), 257–270.

81. Jeffries, T.W., Yang, V.W., and Davis, M.W. (1998) Comparative study of xylanase kinetics using dinitrosalicylic, arsenomolybdate, and ion chromatographic assays. *Applied Biochemistry and Biotechnology*, **70–2**, 257–265.

82. Rivers, D.B., Gracheck, S.J., Woodford, L.C., and Emert, G.H. (1984) Limitations of the dns assay for reducing sugars from saccharified-lignocellulosics. *Biotechnology and Bioengineering*, **26** (7), 800–802.

83. Somogyi, M. (1952) Notes on sugar determination. *The Journal of Biological Chemistry*, **195** (1), 19–23.

84. Lever, M. (1972) New reaction for colorimetric determination of carbohydrates. *Analytical Biochemistry*, **47** (1), 273.

85. Lever, M. (1977) Carbohydrate determination with 4-hydroxybenzoic acid hydrazide (Pahbah) – effect of bismuth on reaction. *Analytical Biochemistry*, **81** (1), 21–27.

86. Dubois, M., Gilles, K., Hamilton, J.K. *et al.* (1951) A colorimetric method for the determination of sugars. *Nature*, **168** (4265), 167.

87. Saha, S.K. and Brewer, C.F. (1994) Determination of the concentrations of oligosaccharides, complex type carbohydrates, and glycoproteins using the phenol sulfuric-acid method. *Carbohydrate Research*, **254**, 157–167.

88. Sluiter, A., Hames, B., Ruiz, R. *et al.* (2006) *Determination of Sugars, Byproducts, and Degradation Products in Liquid Fraction Process Samples*, National Renewable Energy Laboratory, Golden, CO, USA.

89. Li, X., Converse, A.O., and Wyman, C.E. (2003) Characterization of molecular weight distribution of oligomers from autocatalyzed batch hydrolysis of xylan. *Applied Biochemistry and Biotechnology*, **105**, 515–522.

90. Ohara, H., Owaki, M., and Sonomoto, K. (2006) Xylooligosaccharide fermentation with Leuconostoc lactis. *Journal of Bioscience and Bioengineering*, **101** (5), 415–420.

91. Alltech. (2005) Carbohydrate Analysis-Prevail Carbohydrate ES HPLC Columns and ELSD. Alltech, Brochure No 467A.

92. Yu, Y. and Wu, H.W. (2009) Characteristics precipitation of glucose oligomers in the fresh liquid products obtained from the hydrolysis of cellulose in hot-compressed water. *Industrial & Engineering Chemistry Research*, **48** (23), 10682–10690.

93. Tomkins, B.A., Van Berkel, G.J., Emory, J.F., and Tschaplinski, T.J. (2010) Development and application of ultra-performance liquid chromatography/mass spectrometric methods for metabolite characterization and quantitation. DOE BioEnergy Science Center Annual Retreat; June 21; Asheville, NC2010.

94. Johnson, D.C. (1986) Carbohydrate detection gains potential. *Nature*, **321** (6068), 451–452.

95. Rocklin, R.D. and Pohl, C.A. (1983) Determination of carbohydrates by anion exchange chromatography with pulsed amperometric detection. *Journal of Liquid Chromatography*, **6** (9), 1577–1590.

96. Dionex. (2004) Analysis of carbohydrates by high-performanceanion-exchange chromatography with pulsedamperometric detection (HPAE-PAD). Dionex Technical Note 20. http://www.dionex.com/en-us/webdocs/5023-TN20_LPN032857-04.pdf.

97. Dionex. (1998) Optimal settings for pulsed amperometric detectionof carbohydrates using the Dionex ED40 electrochemical detector. Dionex Technical Note 21. http://www.dionex.com/en-us/webdocs/5050-TN21_LPN034889-03.pdf.

98. Jensen, M.B. and Johnson, D.C. (1997) Fast wave forms for pulsed electrochemical detection of glucose by incorporation of reductive desorption of oxidation products. *Analytical Chemistry*, **69** (9), 1776–1781.

99. Hanashiro, I., Abe, J., and Hizukuri, S. (1996) A periodic distribution of the chain length of amylopectin as revealed by high-performance anion-exchange chromatography. *Carbohydrate Research*, **283**, 151–159.

100. Koch, K., Andersson, R., and Aman, P. (1998) Quantitative analysis of amylopectin unit chains by means of high-performance anion-exchange chromatography with pulsed amperometric detection. *Journal of Chromatography A*, **800** (2), 199–206.

101. Koizumi, K., Fukuda, M., and Hizukuri, S. (1991) Estimation of the distributions of chain-length of amylopectins by high-performance liquid-chromatography with pulsed amperometric detection. *Journal of Chromatography*, **585** (2), 233–238.

102. Koizumi, K., Kubota, Y., Tanimoto, T., and Okada, Y. (1989) High-performance anion-exchange chromatography of homogeneous D-Gluco-oligosaccharides and D-Gluco-polysaccharides (polymerization degree-greater-than-or-equal-to-50) with pulsed amperometric detection. *Journal of Chromatography*, **464** (2), 365–373.

103. Dionex. (2003) Determination of plant-derived neutral oligo- and polysaccharides. Application Note 67. http://www.dionex.com/en-us/webdocs/5039-AN67_IC_plant_saccharides_Apr03_LPN1474.pdf.

104. Gray, M.C., Converse, A.O., and Wyman, C.E. (2007) Solubilities of oligomer mixtures produced by the hydrolysis of xylans and corn stover in water at 180 degrees C. *Industrial & Engineering Chemistry Research*, **46** (8), 2383–2391.

105. Cataldi, T.R.I., Campa, C., and De Benedetto, G.E. (2000) Carbohydrate analysis by high-performance anion-exchange chromatography with pulsed amperometric detection: The potential is still growing. *Fresenius Journal of Analytical Chemistry*, **368** (8), 739–758.

106. Paskach, T.J., Lieker, H.P., Reilly, P.J., and Thielecke, K. (1991) High-performance anion-exchange chromatography of sugars and sugar alcohols on quaternary ammonium resins under alkaline conditions. *Carbohydrate Research*, **215** (1), 1–14.

107. Morales, V., Sanz, M.L., Olano, A., and Corzo, N. (2006) Rapid separation on activated charcoal of high oligosaccharides in honey. *Chromatographia*, **64** (3–4), 233–238.

108. Gohlke, M. and Blanchard, V. (2008) Separation of N-glycans by HPLC. *Methods in Molecular Biology*, **446**, 239–254.

109. Khandurina, J. and Guttman, A. (2005) High resolution capillary electrophoresis of oligosaccharide structural isomers. *Chromatographia*, **62**, S37–S41.

110. Rydlund, A. and Dahlman, O. (1997) Oligosaccharides obtained by enzymatic hydrolysis of birch kraft pulp xylan: Analysis by capillary zone electrophoresis and mass spectrometry. *Carbohydrate Research*, **300** (2), 95–102.

111. Rydlund, A. and Dahlman, O. (1997) Rapid analysis of unsaturated acidic xylooligosaccharides from kraft pulps using capillary zone electrophoresis. *HRC-Journal of High Resolution Chromatography*, **20** (2), 72–76.

112. Sartori, J., Potthast, A., Ecker, A. *et al.* (2003) Alkaline degradation kinetics and CE-separation of cello- and xylooligomers. Part I. *Carbohydrate Research*, **338** (11), 1209–1216.

113. Arentoft, A.M., Michaelsen, S., and Sorensen, H. (1993) Determination of oligosaccharides by capillary zone electrophoresis. *Journal of Chromatography A*, **652** (2), 517–524.

114. Zemann, A., Nguyen, D.T., and Bonn, G. (1997) Fast separation of underivatized carbohydrates by coelectroosmotic capillary electrophoresis. *Electrophoresis*, **18** (7), 1142–1147.

115. Kabel, M.A., Heijnis, W.H., Bakx, E.J. *et al.* (2006) Capillary electrophoresis fingerprinting, quantification and mass-identification of various 9-aminopyrene-1,4,6-trisulfonate-derivatized oligomers derived from plant polysaccharides. *Journal of Chromatography A*, **1137** (1), 119–126.

116. Lee, Y.H. and Lin, T.I. (1996) Determination of carbohydrates by high-performance capillary electrophoresis with indirect absorbance detection. *Journal of Chromatography B-Biomedical Applications*, **681** (1), 87–97.

117. Harvey, D.J. (1999) Matrix-assisted laser desorption/ionization mass spectrometry of carbohydrates. *Mass Spectrometry Reviews*, **18** (6), 349–450.

118. Reis, A., Coimbra, M.A., Domingues, P. *et al.* (2004) Fragmentation pattern of underivatised xylo-oligosaccharides and their alditol derivatives by electrospray tandem mass spectrometry. *Carbohydrate Polymers*, **55** (4), 401–409.

119. Reis, A., Pinto, P., Coimbra, M.A. *et al.* (2004) Structural differentiation of uronosyl substitution patterns in acidic heteroxylans by electrospray tandem mass spectrometry. *Journal of the American Society for Mass Spectrometry*, **15** (1), 43–47.

120. Reis, A., Domingues, M.R.M., Domingues, P. *et al.* (2003) Positive and negative electrospray ionisation tandem mass spectrometry as a tool for structural characterisation of acid released oligosaccharides from olive pulp glucuronoxylans. *Carbohydrate Research*, **338** (14), 1497–1505.

121. Reis, A., Domingues, M.R.M., Ferrer-Correia, A.J., and Coimbra, M.A. (2003) Structural characterisation by MALDI-MS of olive xylo-oligosaccharides obtained by partial acid hydrolysis. *Carbohydrate Polymers*, **53** (1), 101–107.

122. Reis, A., Coimbra, M.A., Domingues, P. *et al.* (2002) Structural characterisation of underivatised olive pulp xylo-oligosaccharides by mass spectrometry using matrix-assisted laser desorption/ionisation and electrospray ionisation. *Rapid Communications in Mass Spectrometry*, **16** (22), 2124–2132.

123. Kabel, M.A., Schols, H.A., and Voragen, A.G.J. (2002) Complex xylo-oligosaccharides identified from hydrothermally treated Eucalyptus wood and brewery's spent grain. *Carbohydrate Polymers*, **50** (2), 191–200.

124. Hoffmann, R.A., Geijtenbeek, T., Kamerling, J.P., and Vliegenthart, J.F.G. (1992) H-1-Nmr study of enzymatically generated wheat-endosperm arabinoxylan oligosaccharides – structures of hepta-saccharides to tetradeca-saccharides containing 2 or 3 branched xylose residues. *Carbohydrate Research*, **223**, 19–44.

125. Kormelink, F.J.M., Hoffmann, R.A., Gruppen, H. *et al.* (1993) Characterization by H-1-Nmr spectroscopy of oligosaccharides derived from alkali-extractable wheat-flour arabinoxylan by digestion with Endo-(1-]4)-Beta-D-Xylanase-iii from aspergillus-awamori. *Carbohydrate Research*, **249** (2), 369–382.

126. York, W.S., Vanhalbeek, H., Darvill, A.G., and Albersheim, P. (1990) The structure of plant-cell walls .29. Structural-analysis of xyloglucan oligosaccharides by H-1-Nmr spectroscopy and fast-atom-bombardment mass-spectrometry. *Carbohydrate Research*, **200**, 9–31.

127. Hoffmann, R.A., Leeflang, B.R., de Barse, M.M.J. *et al.* (1991) Characterisation by 1H-n.m.r. spectroscopy of oligosaccharides, derived from arabinoxylans of white endosperm of wheat, that contain the elements → 4)[α-l-Araf-(1-ar3)]-β-d-Xylp-(1 → or → 4)[α-l-Araf-(1 → 2)][α-lAraf-(1 → 3)]-β-d-Xylp-(1 →. *Carbohydrate Research*, **221** (1), 63–81.

20

Experimental Pretreatment Systems from Laboratory to Pilot Scale

Richard T. Elander

National Renewable Energy Laboratory, Golden, USA

20.1 Introduction

Pretreatment processes that facilitate the deconstruction of lignocellulosic biomass have been under development for several decades. In general, pretreatment refers to some form of thermal, chemical, and/or mechanical operation to render the resulting biomass in a form that is significantly less recalcitrant than native biomass to saccharification by cellulolytic enzyme systems. Depending on the type of pretreatment process and the specific pretreatment conditions, this decreased recalcitrance is achieved by solubilization of certain insoluble biomass components (such as hemicellulose or lignin), decreased order of resistant structural arrangements in biomass (i.e., by disruption of the crystalline arrangement of cellulose), increased available surface area via mechanical and/or chemical means, or some combination of these phenomena.

Biomass pretreatment approaches are designed to work in concert with enzymatic hydrolysis processes to effectively saccharify structural carbohydrates in native biomass to monomer sugars that can then be converted to biofuels. Often, such processes target the production of cellulosic ethanol via fermentation routes, although there is growing interest and technological advancement toward "infrastructure-compatible" hydrocarbon biofuels that can be produced via either biochemical or catalytic routes. As such processes generally require soluble sugars and/or other soluble carbon compounds as an intermediate, the pretreatment and enzymatic hydrolysis processes developed for cellulosic ethanol can serve as technology platforms for processes to produce these types of biofuels.

Modern pretreatment approaches have evolved from the strict thermochemical hydrolysis processes developed in the early 20th century. These processes used different acid hydrolysis schemes to produce soluble sugars and were employed commercially in some countries during wartime periods [1]. Most of these processes utilized concentrated acids (primarily sulfuric or hydrochloric acid) at relatively low temperatures (under 100 °C) or dilute acids (again, typically sulfuric or hydrochloric acid) at much higher

Aqueous Pretreatment of Plant Biomass for Biological and Chemical Conversion to Fuels and Chemicals, First Edition.
Edited by Charles E. Wyman.
© 2013 John Wiley & Sons, Ltd. Published 2013 by John Wiley & Sons, Ltd.

temperatures (often above 200 °C). Such conditions generally resulted in relatively low recoverable sugar yields (less than 60% of theoretical yield) due to sugar degradation reactions that produce aldehydes and organic acids, along with other undesirable compounds. Also, the large amounts of chemicals used caused such processes to be costly, either from the requirements for pressurized corrosion-resistant reactors capable of processing solid materials or the economic requirement to recover and recycle the acid.

The development of cellulase enzyme systems since the 1970s has dramatically changed the context of the thermochemical hydrolysis step in the conversion of lignocellulosic biomass to sugars. Instead of exclusively producing soluble sugars from biomass, this step can now be viewed as a true pretreatment step whose purpose is to prepare the biomass for subsequent enzymatic hydrolysis to generate monomeric sugars. This process scenario has several inherent advantages over relying on just thermochemical hydrolysis, including lower-temperature and milder-pH conditions leading to less-expensive and less-complex reactor systems and significantly reduced loss of resulting sugars to degradation products. Within this context, several unique pretreatment approaches have been developed and investigated across a variety of biomass feedstock types. Published review articles and comparative studies are available that provide a general overview of various pretreatment approaches [1–11]. Across these various pretreatment approaches, the actual reaction conditions including temperature, pressure, residence time, solids concentration, and corrosion potential vary widely. In Table 20.1, some of these reaction conditions for pretreatment categories that are mature enough for pilot-scale development are presented. It should be noted that specific reaction conditions required are dependent on ancillary considerations, such as feedstock type and amount and type of enzymes used in subsequent enzymatic hydrolysis.

While no one particular pretreatment process can presently be viewed as the "ideal" approach for all feedstocks or for all process circumstances, desired properties of an ideal pretreatment process have been identified [2]. Such an ideal pretreatment process would include the following features:

- produces a highly digestible pretreated solid;
- does not significantly degrade solubilized carbohydrates;
- does not significantly inhibit subsequent fermentation steps;
- requires little or no feedstock size reduction;
- can work in reactors of reasonable size and moderate cost;
- produces no solid-waste residues;
- is simple and practical; and
- is effective at high solids loadings.

Depending on the actual pretreatment approach used and the specific pretreatment reaction conditions, biomass carbohydrate polymers and lignin are solubilized to different extents during pretreatment. Therefore, the resulting composition of the pretreated solid and liquid fractions can differ greatly. For example, dilute sulfuric acid pretreatment processes can solubilize nearly all of the hemicellulose but very little lignin or cellulose. Other pretreatment approaches such as alkaline processes are more effective at solubilizing lignin, but leave extensive amounts of the hemicellulose fraction as an insoluble component of the pretreated solids. These factors greatly impact the relative composition of the pretreated solids and the requirements for effective enzymatic saccharification in subsequent processing steps. As the ultimate goal of an integrated pretreatment-enzymatic hydrolysis "system" is to achieve high recovered yields of biomass carbohydrates in monomeric sugar form, the effectiveness of enzyme systems used will greatly determine the pretreatment reaction conditions and associated design of pretreatment reactor systems.

There are considerable economic drivers to reduce the overall severity of the pretreatment operation [25], including lower-cost reactor materials of construction, lower temperature and/or residence time, lower losses of resulting sugars to degradation products, and lower requirements to adequately "condition" pretreatment hydrolysates for subsequent fermentation. Less-aggressive pretreatment conditions will generally

Table 20.1   *Pretreatment reaction conditions used for various pretreatment approaches that have been operated at a pilot scale.*

Pretreatment category	Pretreatment conditions and process considerations	Pretreatment mechanism	Ref.
Auto-catalyzed (water, steam)	• High-temperature/pressure conditions (often >200 °C) to achieve good enzymatic digestibility of cellulose • Often generates high proportion of solubilized hemicellulose in oligomeric form • Some potential sugar degradation losses, with resulting hydrolyzate toxicity consideration • Low corrosion potential (stainless steel reactor construction) • Explosive decompression option is challenging to operate in continuous large-scale systems	• Moderate to extensive hemicellulose hydrolysis via mild acid autohydrolysis (enables enzymatic digestion of cellulose) • Limited lignin solubilization (except in flowthrough mode) • Reduced particle size, increased surface area (especially upon explosive decompression)	[12–15]
Oxidative (wet oxidation, oxygen/air, ozone)	• High-temperature conditions (often >180 °C) with further overpressurization (compressed air/oxygen) • Oxidative agent process considerations (compressed air/oxygen generation and usage levels of peroxide/ozone) • Often generates high proportion of solubilized hemicelluloses in oligomeric form • Some potential sugar degradation losses (primarily organic acids), with resulting hydrolyzate toxicity consideration • Low corrosion potential (stainless steel reactor construction)	• Partial hemicellulose hydrolysis (primarily generates oligomers) • Partial lignin solubilization/re-arrangement • Enhanced cellulose enzymatic digestibility from partial hemicellulose/lignin solubilization	[16–19]
Organic solvent (including co-catalysts)	• High temperature (usually above 150 °C), with high pressure due to use of volatile organic solvents • Requires economically viable solvent recovery, including consideration of residual solvent toxicity • Recovery of value-added products/energy from soluble lignin • Need for demonstrated continuous, pilot-scale operation at high solids loading (and associated solvent recovery) • Process development tailored to biomass component fractions	• Limited to extensive hemicellulose hydrolysis (with acid co-catalysts) • Limited cellulose hydrolysis/solubilization • Extensive lignin solubilization • Fractionates biomass (organic and aqueous phases)	[20–22]
Acidic (sulfuric, nitric, phosphoric, SO₂)	• High-temperature (usually above 150 °C) and high-pressure conditions • Corrosion potential may require expensive alloys for pretreatment reactor (esp. sulfuric acid) • Potential sugar degradation losses • Toxicity of hydrolyzate due to sugar degradation products and/or release of soluble compounds from biomass • Generation of insoluble neutralization products (some cases)	• Extensive hemicellulose hydrolysis with high yield of hemicellulosic sugars (enables enzymatic digestion of cellulose) • Limited lignin solubilization (except in flowthrough mode) • Reduced particle size, increased surface area	[23–26]

(continued)

Table 20.1 *(Continued)*

Pretreatment category	Pretreatment conditions and process considerations	Pretreatment mechanism	Ref.
Alkaline (ammonia, lime NaOH)	• Extensive process development at pilot-scale at relevant conditions (high solids loading, pre-processing of feedstock, etc.) • Lower temperature (<100 °C) to higher temperature (>180 °C) conditions, depending on chemical used • High-pressure conditions (especially ammonia processes) • Often requires enzymatic conversion of hemicellulosic polymers/oligomers to monomers • Stainless steel reactor construction is acceptable • Chemical containment and recovery/recycle requirements • Generally less effective on woody feedstocks (aggressive alkaline pulping chemistry does not preserve hemicellulosic sugars)	• Swelling and decrystallization (allomorphism/peeling) of cellulose • Partial or no hemicellulose solubilization (high proportion of oligomers) • Lignin re-arrangement and deconstruction/solubilization	[27–33]

result in less sugar release (primarily from hemicellulose) in the pretreatment step, and can result in reduced enzymatic digestibity of cellulose in the pretreated solids. This will shift more of the hydrolytic sugar production requirement from the pretreatment step to the enzymatic hydrolysis step and will have an impact on the amount and type of enzymes required to achieve high sugar yields from both cellulose and hemicellulose in less-severely pretreated biomass.

Given these general process considerations, this chapter will discuss design features for both laboratory-scale and pilot-scale pretreatment reactor systems. As there is a wide range of operating conditions across various pretreatment approaches and even within a given pretreatment approach (as shown in Table 20.1), it is important that pretreatment reactor selection, either for bench-scale or pilot-scale applications, consider a range of expected process conditions. The discussion of laboratory-scale pretreatment equipment will focus on simple batch systems, as continuous systems at the bench scale are difficult to implement and often require components that are not scalable or require unrealistically small feedstock particle sizes. Emphasis will be placed on pretreatment equipment that can operate under process-relevant conditions, such as at reasonably high solids loadings.

Although batch-mode pilot-scale pretreatment systems are of some interest, most pilot-scale pretreatment systems are designed for continuous operation as commercial-scale pretreatment systems are expected to typically employ continuous reactors. Such systems become particularly challenging to design and operate effectively at high solids loadings (above 15% insoluble solids), as heat and mass transfer considerations and control of residence time to tight tolerances are critical design elements that must be adequately addressed. Fortunately, large-scale continuous pulping reactors that have been developed and widely utilized in the pulp and paper industry are being modified for use in biomass pretreatment applications [34], although certain unique challenges for biomass pretreatment (such as control of residence times typically on the order of minutes as opposed to hours in pulping processes) must be considered.

Design features of the various components of continuous pilot-scale biomass pretreatment reactors and associated ancillary equipment will be discussed, including feedstock preparation and handling, pretreatment chemical introduction, pressurized biomass feeding equipment, pretreatment reactor conveyance systems and residence time control, pretreatment reactor discharge devices, blow-down tank design, and flash recovery systems. Examples of integrated continuous pilot-scale pretreatment reactor systems that incorporate these various equipment components are provided.

20.2 Laboratory-scale Pretreatment Equipment

Biomass pretreatment has been performed in various types of batch bench-scale reactors for many decades, beginning with early acid hydrolysis concepts. Much of the published pretreatment research is conducted in stirred-reactor pressure vessels or bomb-type reactors, often with high amounts of added liquid to aid in mass and heat transfer. Such studies may provide valuable insights into the effectiveness and potential of various pretreatment schemes, even if they are not performed in a manner that generates commercially relevant performance data. While a wide variety of laboratory-scale pretreatment reactor systems are available and are discussed more fully in Chapters 22 and 23, there are some general features that are common to all such units given the broad assumption that a pretreatment process is performed at an elevated temperature (above 100 °C), at an elevated pressure (above atmospheric pressure), for a relatively short time (generally less than 60 min), for the generation of commercially relevant process data, and with a wetted biomass slurry that is typically at a moisture content of 40% or higher. Such a reactor system must provide the features described in the following sections.

20.2.1 Heating and Cooling Capability

Virtually all pretreatment processes and conditions that are commonly considered require heating biomass to a specified elevated temperature within a prescribed period of time, maintaining a targeted single temperature or temperature profile followed by rapid cooling of the pretreated biomass to moderate temperatures (generally below 80 °C) for subsequent processing steps. Many heating approaches are used, including indirect electrical resistance heating, indirect heating via jackets or other heat-exchange devices using steam or other heat-transferring fluids, direct injection with steam or other previously heated process-compatible fluids, and heating by microwaves or other electromagnetic sources. Ideally, whatever heating strategy is employed will quickly achieve the targeted desired temperature (within a few seconds to a few minutes) as pretreatment reactions can occur during the heating period and can confound interpretation of process data and reaction kinetics.

At the end of the pretreatment reaction hold period, a number of cooling strategies can also be considered including indirect cooling via a coolant liquid or vapor passed through a jacket or other heat-exchange device, external cooling by immersion of the reactor vessel into an ice bath or similar quenching fluid, depressurization of the vapor head space of the pretreatment vessel via a condenser or similar device, or rapid decompression and evacuation of the entire pretreatment reactor contents via a rapidly opened discharge valve into a suitable flash receiver. Again, rapid cooling to temperatures where pretreatment reactions cease is desirable to avoid a confounding "shoulder" in the pretreatment temperature profile during the cooling stage.

20.2.2 Contacting of Biomass Particles with Water and/or Pretreatment Chemicals

In most processes, biomass pretreatment involves some degree of structural carbohydrate hydrolysis, particularly of the hemicellulosic carbohydrate fraction. As hydrolysis reactions require water, it is important that water is well distributed throughout the fine structure of the biomass, in addition to any additional pretreatment chemicals that catalyze or otherwise participate in pretreatment reactions. Provisions to properly distribute water and any pretreatment chemicals are therefore an integral part of any pretreatment system.

As many biomass feedstocks will be stored in a dry state (less than 15% moisture content) to prevent microbial spoilage during long storage periods, the "impregnation" of dry biomass with water and pretreatment chemicals becomes particularly challenging, especially at high solids loadings. In very dilute systems (<10% insoluble solids content prior to pretreatment), it is relatively easy to fully disperse water and pretreatment chemicals throughout the biomass particle interstitial volume due to the presence of a continuous liquid medium. At high solids loadings, there will not be a continuous liquid medium and biomass particles may not be fully saturated with liquid. Since it is desirable from a process economic standpoint to conduct pretreatment at high solids loadings, suitable methods to contact biomass particles with water and pretreatment chemicals becomes essential.

In low-solids-loadings bench-scale pretreatment systems, high-pressure reactor vessels with integrated stirred impellers of a proper design are generally adequate and are commonly used. For high-solids-loadings bench-scale pretreatment systems, several of which are described in Chapter 23, adequate mass transfer to distribute water and pretreatment chemicals within short pretreatment reaction times may not be provided by such stirred-tank reactor systems, even when using specialized high-solids-loadings mixing impellers.

Under these circumstances, prior contacting of water and pretreatment chemicals before heating to pretreatment temperatures is often practiced. Such processes, often referred to as "pre-impregnation," are carried out in devices designed to adequately contact biomass particles, water, and pretreatment chemicals at high solids concentration for relatively long times, sometimes with the use of moderate temperatures (below 80 °C) and/or vacuum or pressure cycles to aid in delivery of water and pretreatment chemical to the interstitial spaces deep within biomass particles. Alternatively, soaking of biomass particles with the proper concentration of pretreatment chemicals in water for an appropriate period of time, followed by a dewatering operation to generate pre-impregnated biomass at a relatively high solids concentration for delivery to a high-solids bench-scale pretreatment reactor, can be employed. Several studies that investigate the effect and mechanisms of biomass particles impregnation [26,35–38] generally indicate that inadequate attention to effective biomass pre-impregnation when conducting pretreatment experiments in laboratory-scale reactors at high solids loadings will result in less-than-optimal pretreatment process performance.

20.2.3　Mass and Heat Transfer

It is important to avoid reactant and product concentration gradients during pretreatment in order to achieve good pretreatment process performance. In the case of pretreatment processes that target high degrees of hemicellulose hydrolysis (acidic and steam/hot water pretreatment), such gradients that are the result of poor mass transfer can result in variable amounts of less-reacted biomass as well as over-pretreated biomass, generating unwanted sugar degradation products. In low-solids pretreatment reactors, adequate mass transfer can be provided by mixing via a properly designed impeller. In high-solids pretreatment reactors, it is usually important to effectively pre-contact the biomass with water and pretreatment chemicals in a pre-impregnation step, as described above. After pre-impregnation, a high-solids pretreatment process at bench scale can be performed in mixed pressure vessels using properly designed high-solids impeller systems or in unmixed steam-explosion-type reactors where steam injected via multiple ports can provide even heating and temperature profiles. An example of this is the 4 L steam explosion reactor at the National Renewable Energy Laboratory (Golden CO, USA) [26,38], where steam addition flow is split in order to balance the steam addition to the top and the bottom of the pretreatment chamber, as shown in Figure 20.1. The split flow not only allows for good mixing of the biomass particles by providing a fluidization effect throughout the reactor, but also aids in effective heat transfer by avoiding the formation of a condensation layer on the surface of the stagnant biomass bed which can cause poor heat-transfer and resulting temperature gradients during relatively short pretreatment reaction times.

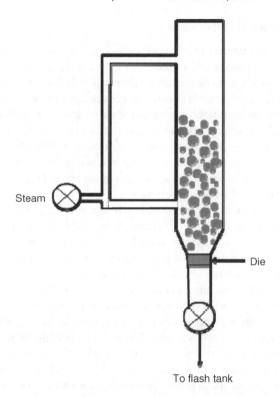

Figure 20.1 *Schematic of 4 L steam explosion pretreatment reactor at National Renewable Energy Laboratory (Golden, CO, USA). The split-flow steam addition allows the steam flows to be balanced to achieve a partial fluidization of biomass particles, aiding both mass and heat transfer.*

20.2.4 Proper Materials of Construction

Laboratory-scale pretreatment reactors should be constructed from materials that ensure adequate corrosion protection for the pretreatment chemicals and concentrations used. In particular, pretreatments that utilize mineral acids such as H_2SO_4 or HCl should be conducted using corrosion-resistant metals, such as various grades of Hastelloy, zirconium, or other corrosion-resistant alloys [25]. Alternatively, glass-lined reactors can be used, but may suffer from wear and cracking especially when used in abrasive high-solids reaction conditions. When high solids loadings are used, reliance upon any corrosion-resistant passivation layer on metal surfaces is inadequate, as such a layer can easily be scoured away by abrasive biomass.

Even though corrosion rates from standard corrosion testing are typically calculated based upon continual long-term use of equipment as typically encountered in commercial-scale operations, laboratory-scale reactors may be used much less frequently and would therefore have a longer effective lifetime. However, there are other considerations that point to use of highly-corrosion-resistant alloys when mineral acids are used in pretreatment. Released corrosion products, such as chromium ions from various grades of stainless steel, have been found to catalyze the degradation of sugars to unwanted aldehydes and organic acids at fairly low concentrations, including ancillary stainless steel tubing and fittings that are exposed to pretreatment conditions even when the pretreatment reactor is constructed of an appropriately corrosion-resistant alloy [39]. This can cause significantly lower pretreatment sugar yields than would otherwise be expected.

20.2.5   Instrumentation and Control Systems

The most significant process parameters to measure and control in a bench-scale pretreatment reactor system are reactor temperature and pressure. Most pretreatment reaction conditions include fairly high temperatures (up to about 200 °C) for short residence times (generally less than 20 min, and sometimes as short as 1–2 min). In order to be able to interpret pretreatment conversion data using reaction kinetics tools, it is desirable to quickly heat the pretreatment reactor contents to desired reaction temperatures and also to quickly cool the resulting pretreated material to temperatures such that the contributions to pretreatment reactions during such heat-up and cool-down periods are minimized. This also allows for better translation to pilot-scale and/or commercial-scale pretreatment systems, which are likely to use a continuous reactor design where the pretreatment reactor itself is always maintained at the targeted temperature and the heating and cooling of entering and exiting biomass is very fast.

Multiple temperature probes should be included in any bench-scale pretreatment system to understand (and minimize, if possible) any temperature gradients across all dimensions of the pretreatment reactor. Temperature probes should be designed and positioned to ensure that measured temperatures are indicative of the biomass pretreatment reaction medium and not the reactor walls, which may heat or cool more slowly due to high heat capacity of the fairly thick-walled pressure vessels.

In many cases, it may be more convenient and useful to control and monitor pretreatment temperature by measuring and controlling the pressure in the pretreatment reactor. In many pretreatment chemistries using steam or pressurized hot water either alone or with non-volatile pretreatment chemicals, it is straightforward to relate saturated steam pressure to temperature. When using volatile pretreatment chemicals such as ammonia the temperature–pressure relationship is complicated by high vapor pressures of such chemicals.

Irrespective of whether temperature or pressure is used as the controlling variable, data collection systems that permit rapid recording of these parameters during the entire sequence of heating, maintaining, and cooling of the biomass pretreatment reaction medium is an important requirement. Depending on heating and cooling profiles and the overall pretreatment residence time, appropriate data collection cycle times may be as short as a few seconds.

20.2.6   Translating to Pilot-scale Pretreatment Systems

Laboratory-scale pretreatment systems that can be operated at high solids loadings (>20% insoluble solids) and can achieve rapid heating to reaction temperatures (in less than 1 min) and then rapid cooling to temperatures at which pretreatment reactions cease (also in less than 1 min) will generate data that will be more easily translatable to pilot-scale pretreatment systems. Examples of such laboratory-scale systems include: small-diameter tubular reactors using biomass pre-impregnated with pretreatment chemicals that utilize rapid-heating sand baths or oil baths and rapid-cooling ice baths; and high-solids steam-injection reactors, either with high-solids mixing impellers (permitting injection of pretreatment chemicals during or after reactor heating and pressurization), or without impellers (requiring proper pre-impregnation of biomass with pretreatment chemicals) with rapid flash-cooling capabilities for reaction quenching.

20.3   Pilot-scale Batch Pretreatment Equipment

There is no universally accepted throughput or volumetric capacity standard that defines the size range of pilot-scale pretreatment reactors. In general, pilot-scale equipment should be (1) of a design that the mechanical reactor components and associated ancillary equipment are similar to what is expected in commercial-scale systems, but (2) at the minimum size necessary to generate relevant data and provide operational validation and experience that can lead to successful equipment and integrated process scale-up. For pretreatment of biomass materials, this consideration is highly related to the biomass particle size and shape range expected in a commercial-scale process. Biomass particle sizing is often determined by feedstock harvesting and collection equipment and long-term storage considerations, and can vary widely across different feedstock types. For example, trees would likely be processed into typical pulping-type wood chips while herbaceous grasses and many other types of agricultural residues may be coarsely chopped and gathered into bales of various shapes and sizes. Further size reduction is undesirable, as there can be significant energy inputs required to achieve small feedstock particle sizes, particularly for woody feedstocks [25,40]. Some amount of size reduction may be necessary due to the heat and mass transfer requirements associated with relatively short pretreatment residence times that are typically used.

As pilot-scale pretreatment systems are intended to provide scale-up data and mimic the process capabilities and performance of envisioned commercial-scale pretreatment systems, the choice of equipment configuration for a pilot plant should take into account commercial-scale considerations. Depending on the envisioned size of a particular commercial-scale facility and its associated pretreatment equipment, either batch or continuous biomass pretreatment processes and equipment are conceivable. A similar situation exists in the pulp and paper industry, where both batch and continuous pulping equipment are used in commercial application. Generally, that industry has trended toward continuous pulping equipment and processes in recent decades, particularly in larger facilities with capacities well in excess of 2000 dry tons per day.

A study conducted for the National Renewable Energy Laboratory compared the cost and other considerations for large-scale lignocellulosic ethanol commercial processing based on either batch or continuous pretreatment equipment and processes [41]. This study concluded that at both the 1000 and 2000 ton of dry feedstock per day plant size, a continuous pretreatment system was favored from an overall cost and operational standpoint. While batch pretreatment reactor designs are available from pulping equipment suppliers, the large plant size and short cycle time per batch (corresponding to short pretreatment times) requires a large number of batch reactors. For a 2000 ton of dry feedstock per day plant size, a scenario that requires 16 batch reactors was developed. This would compare to 2 reactor systems for a continuous pretreatment process. Greater labor requirements and operational complexity are also likely with the batch pretreatment process, along with greater steam demands due to cooling of reactors between cycles. There are also concerns about reactor fatigue in batch reactors due to the very short cycle times that would require up to 120 heating and cooling cycles per day. Similar batch systems have been used in pulping processes, but cycle times for pulping processes are much longer (200–350 min), which limits the typical number of heating and cooling cycles to about 5 per day. For these reasons, most commercial-scale lignocellulosic ethanol process designs are based on continuous pretreatment reactor systems. This is not only true for dilute acid pretreatment, but also for steam/hot water pretreatments and some alkaline pretreatments such as ammonia fiber expansion (AFEX). It is believed that ongoing R&D and pilot-scale development work can improve pretreatment performance in continuous reactor systems that achieve similar performances as seen in batch reactors. However, there may be scenarios where batch pretreatment systems are worthy of consideration for commercial-scale applications. In certain regions where local agricultural practices and feedstock transportation infrastructure favor smaller overall plant sizes and more diverse feedstock types, batch processes may be more suitable.

There are a number of proposed designs of batch pretreatment processes and types of equipment available that are potentially practical. Several reactor designs that have been developed and utilized for pulping applications could be modified for use in shorter-cycle biomass pretreatment applications. One particular design that is viewed as most applicable to biomass pretreatment applications is the Bauer Rapid Cycle Digester [41]. This design has primarily been utilized for wood chips, but could probably be used for other biomass feedstock provided that the feedstock particle size is not too small. A drawing of a 240 ft^3 batch reactor based upon this design concept is provided in Figure 20.2.

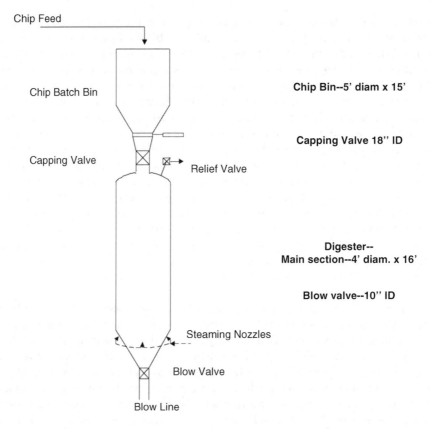

Figure 20.2   *Conceptual design of a Bauer Rapid Cycle Digester for use as a batch pretreatment reactor.*

For commercial-scale purposes, this reactor size would require 16 units to handle a 2000 ton of dry feedstock per day plant size based upon a short dilute acid pretreatment cycle time of about 9 minutes. In this system, feedstock is charged via a batching bin through a capping valve into the main digester body. In this context the feedstock would be pre-impregnated with the pretreatment catalyst and possibly pre-steamed to remove air prior to charging. In the reactor (digester), steam would be introduced via a properly designed series of valves and nozzles to achieve effective heat transfer and held for the prescribed pretreatment time. At the end of the pretreatment time, the pretreated solids would be rapidly discharged via a fast-acting discharge valve and flashed into a flash receiver. Conceptually, this is similar to the much smaller steam-explosion type reactors used in many research laboratories and could be designed as an appropriately sized pilot-scale pretreatment reactor vessel. In commercial applications, carbon or stainless steel with acid-resistant linings could be utilized, but it may be more practical to utilize an appropriate acid-resistant alloy for a pilot-scale unit. The flash receiver would also need to be constructed from acid-resistant materials and should include a condenser for recovery of flash vapors.

Other batch pretreatment system designs suitable for pilot-scale operation are manufactured by various equipment vendors, such as Andritz Inc. (Glens Falls, NY, USA and Lachine, Quebec, Canada). These designs often consist of a high-solids steam pretreatment (with or without an acid pretreatment chemical), followed by a pressurized mechanical disc refiner to achieve further size reduction and additional surface area generation of softened, pretreated biomass and discharge via a blow valve to an atmospheric cyclone for separation of flash vapors and collection of pretreated solids. Such systems are currently manufactured in sizes ranging from 5 to 20 kg/hr, operating on a batch-wise basis. In addition, other pilot-scale batch pretreatment reactor systems using high-solids paddle mixers have been described [42].

20.4 Pilot-scale Continuous Pretreatment Equipment

Pretreatment system equipment components and the overall pretreatment process configurations for pilot-scale and commercial-scale reactors operated in a continuous-flow manner are significantly different to batch systems of similar overall capacity. These differences are found in the pre-processing of raw biomass, feeding of properly prepared feedstock to the pretreatment reactor, chemical/water/steam introduction into the reactor, conveyance of biomass through the reactor, and reactor discharge and pretreated slurry collection systems. The following continuous pretreatment reactor system considerations are discussed thoroughly in the following sections, with particular emphasis on pilot-scale applications:

- feedstock handling and size reduction;
- pretreatment chemical and water addition;
- pressurized continuous pretreatment feeder equipment;
- pretreatment reactor throughput and residence time control;
- reactor discharge devices; and
- blow-down vessel and flash vapor recovery.

20.4.1 Feedstock Handling and Size Reduction

Biomass feedstocks must be properly sized prior to introduction into pretreatment reactors in order to achieve good heat and mass transfer necessary for effective and consistent pretreatment across biomass particles and within individual biomass particles. While this is true for batch pretreatment systems, for which studies have shown the importance of batch reactor sizing and dimensions for effective heat transfer within tubular reactors [43] and pretreatment chemical mass transfer related to feedstock particle size and geometry [36], there are additional considerations in continuous pretreatment reactors. In continuous systems, biomass particles must be actively transferred to the pretreatment equipment via various staging devices such as live-bottom hoppers and conveyance devices, such as screw conveyors or cleated-belt conveyors. In pilot-scale systems, the physical dimensions of such conveyance devices may be of such a size that requires smaller biomass particles than larger commercial-scale systems. For instance, a continuous pretreatment system with a nominal dry biomass throughput rate of 5 kg/hr may use various types of screw conveyors in feedstock hopper live-bottom dischargers and pressurized feeders with a pitch distance between screw flights ranging from about 100 mm to as low as 25 mm. Biomass feedstock particles that are larger (or even slightly smaller) than those distances in their maximum dimension could face difficulty in properly entering into and being conveyed by these devices in the intended manner.

Mechanical processes for reducing biomass feedstock particles size should generally be performed to the minimum extent possible, as large amounts of energy are needed to produce small biomass particles. This is particularly true for woody feedstocks. The required particles size necessary for effective pretreatment is governed by the pretreatment residence time, as shorter residence times can require smaller particle sizes in order to achieve the necessary heat and mass transfer within that short time. If the feedstock is thoroughly pre-impregnated with pretreatment chemicals and water prior to being introduced into a continuous pretreatment reactor, then it may be possible to effectively utilize larger feedstock particles. Larger feedstock particles can potentially be utilized if particle size reduction is performed in a manner such that the vascular structure of the biomass is maintained and is utilized as a transport system to help deliver pretreatment chemicals and steam to the interior of the feedstock particles. This may allow for increased size of feedstock particles along the direction of these vascular structures [36]. The development of the standard wood chip for pulp and paper manufacture and the associated large-scale equipment to produce such wood chips clearly takes advantage of the vascular structure in wood to allow for a significantly longer dimension in the direction of the vascular structure [44]. Despite this, pre-impregnation of wood chips with pulping chemicals is often performed as a separate operation prior to chip feeding to commercial-scale continuous pulping reactors, even though thermochemical pulping process residence times are usually significantly longer than those of most biomass pretreatment processes being developed.

While one consideration for determining appropriate biomass particle size is related to the physical dimensions of conveyance systems in continuous pretreatment systems (screw conveyor pitch, pressurized biomass feeder dimensions, etc.), the influence of biomass particle size on conversion yields obtained in pretreatment and enzymatic hydrolysis operations is generally a more important consideration. As a stand-alone pretreatment process, size reduction by various means such as ball milling, hammer milling, and attrition milling has been found to increase the enzymatic digestibility of otherwise un-pretreated biomass feedstocks such a poplar wood [45], spruce wood [46], corn stover [47], and cotton gin trash [48]. However, this effect has not been universally seen and resulting sugar yields are usually limited to no more than 50% of theoretical, even with average particles sizes usually much less than 1 mm [40]. This is typically much lower than sugar yields achieved upon enzymatic hydrolysis of thermochemically pretreated biomass using larger particle sizes (often 3–15 mm).

The most common approaches used for milling biomass feedstock prior to a thermochemical pretreatment step involve either a cutting action, such as in a knife mill, or an impact action, such as in a hammer mill. Both of these milling approaches utilize the concept of introducing larger feedstock particles into a perforated-wall cylindrical drum, typically with round holes of a prescribed diameter although square or rectangular holes/slots are sometimes used. The milling implements are located within the drum and rotate at the necessary speed to effectively reduce the size of incoming biomass particles. In the case of a knife mill, the rotating blades cut through the feedstock particles. Biomass particles remain within the drum until a given particle becomes small enough to pass through a perforation in the drum. It is possible for particles with a length dimension that is longer that the perforation diameter to pass through a perforation if properly oriented. A hammer hill works under a similar mechanism, but size reduction is achieved by the force of a hammer or other impacting or pulverizing device, as opposed to the cutting action of a knife mill.

Using either technique, highly rigid woody biomass generally requires 5–10 times greater energy input to mill biomass to pass through a screen of diameter 2–4 mm compared to herbaceous energy crops or agricultural residues [40]. Although both types of mills can be used on different biomass types, it is more common for hammer mills to be used on woody biomass while knife mills are often used on herbaceous energy crops and agricultural residues. Woody biomass, particularly when pre-sized as typical wood pulping chips and partially dried to less than 50% moisture content, is somewhat brittle and can split along the wood grain and then across the wood grain as narrower "pin chips" are formed upon the impinging impact within a hammer mill.

Herbaceous energy crops and agricultural residues generally consist of flatter 2D structures as compared to a more 3D structure of wood chips. In a hammer mill, the pulverizing action on flatter, less brittle stems and leaves more commonly found in these types of biomass feedstocks is not particularly effective at reducing particle size in a uniform way. As these types of feedstocks are commonly harvested and stored in a relatively dry state (<15% moisture content) to resist spoilage during long storage periods, stringy biomass strands may tend to form nest-like aggregates or wrap around rotating shafts as a result. Additionally, excessive amounts of fine-particulate "dust" can be generated from such feedstocks being processed in a hammer mill. This dust can cause explosion concerns and require excessively large and expensive dust mitigation and collection equipment, resulting in some loss in the amount of incoming biomass that is ultimately delivered to the conversion process. Finally, the pulverizing nature of impact mills has been speculated to cause a crushing and collapse of the vascular and pore structure of these biomass types, which could reduce the ability of pretreatment chemicals and enzymes to effectively utilize the plant's vascular structure as a transport mechanism to achieve good distribution of these hydrolysis catalysts and to allow for effective diffusion of reaction products back into the bulk fluid medium [49]. These considerations require a hammer mill or other type of impact device to be properly operated, particularly with such feedstock types. A knife mill imparts a cutting action as opposed to an impact action, potentially better preserving these vascular structures. However, knife mills may require greater maintenance, as cutting surfaces must be kept sharp or milling performance will decline.

20.4.2 Pretreatment Chemical and Water Addition

Pretreatment processes that use chemical catalysis in liquid form (e.g., dilute mineral/organic acids or dilute/aqueous alkali agents) or in gaseous form (e.g., anhydrous ammonia, sulfur dioxide, carbon dioxide, or oxygen/air) are generally more effective if such chemicals are well-distributed throughout and within biomass feedstock particles. This is particularly true for pretreatment processes conducted at high solids loadings, where mass transfer of pretreatment chemicals to biomass particles may be impacted by the lack of a free liquid phase, and for very short residence time pretreatment conditions, where time is limited for sufficient distribution of pretreatment chemicals during the short reaction time [36,50]. One approach for addressing this situation involves proper sizing of feedstock particles, as discussed in Section 20.4.1.

A pre-impregnation step can be used to ensure good distribution of pretreatment chemicals prior to introducing biomass feedstocks to a continuous pilot-scale pretreatment system. Pre-steaming and pre-impregnation systems are commonly used in commercial-scale continuous pulping reactors to ensure proper contacting of pulping chemicals with wood pulping chips [44]. In biomass pretreatment processes, proper pre-impregnation with pretreatment chemicals may be even more important as residence times are usually shorter than those used in pulping processes and pretreatment kinetic reactions can result in the production of undesired sugar-degradation compounds (e.g., aldehydes and organic acids) from sugars that are generated during pretreatment. The generation of these degradation products not only reduces the availability of desired sugar intermediates for subsequent conversion to biofuel products of interest, but these compounds can also be inhibitory to fermentative microorganisms that convert biomass-derived sugars to biofuel products or desired intermediates [18,51].

The pretreatment "severity index," first developed as a lumped reaction parameter encompassing the combined effect of reaction time and temperature for steam pretreatments [52,53] and since modified for dilute acid pretreatment to include an acidity contribution in a "combined severity factor" [38,54], assumes that all biomass particles experience the same time and temperature profile and, in the case of the combined severity factor, are at the targeted acid concentration with no pH gradients. Proper pre-impregnation processes can help ensure that biomass entering the pretreatment reactor does not contain significant pretreatment chemical concentration gradients, but steam injection and distribution systems within the pretreatment reactor must be designed in a manner to minimize heat and chemical concentration gradients as a result of condensing steam that dilutes the pre-impregnation chemical concentration unevenly. This can cause an uneven pattern of pretreatment severity within the reactor, leading to inconsistent pretreatment performance. If the pre-impregnation process results in pores within biomass particles being fully saturated with water and pretreatment chemicals prior to steam injection, steam can condense quickly on biomass particle exterior surfaces without effectively penetrating pore structures. This leads to slower heat transfer to the interior of biomass particles and uneven pretreatment chemical concentrations (higher in the interior of particles and lower on exterior surfaces). Higher sugar degradation product formation rates can result in the

Figure 20.3   *Wire-mesh basket and recirculation vessel and pump for small-batch biomass impregnation. (National Renewable Energy Laboratory, Golden CO, USA). (See figure in color plate section).*

interior regions of lower moisture content, which slows the diffusion of soluble pretreatment hydrolysis products to the bulk liquid medium, further exacerbating sugar degradation losses [38].

Published studies have shown that in dilute acid pretreatment of various feedstocks, a properly implemented pre-impregnation step can enhance high-solids pretreatment xylose yields by about 10% as compared to poorly conceived methods of contacting dilute sulfuric acid with biomass particles [26,35,37,38,55]. Dye penetration studies [36] on corn stalk sections have demonstrated the ability of properly sized stalk particles to utilize plant tissue vascular structures to transport dissolved substances into interior regions, along with the benefit of using vacuum to hasten such transport.

An example of a small-batch pre-impregnation system located at the National Renewable Energy Laboratory (Golden, CO, USA) is shown in Figure 20.3. This system [38] consists of a Hastelloy C-276 wire-mesh basket to which up to 15 kg (dry matter basis) of appropriately milled biomass is added. Pretreatment chemicals diluted with water to the appropriate concentration (accounting for further dilution anticipated from condensate upon injection of steam to the pretreatment reactor) are batched in a 200 L vessel. The biomass-containing basket is suspended in this vessel and a pump is used to recirculate the pretreatment chemical solution, ensuring thorough soaking of the biomass particles that are retained within the wire-mesh basket. The recirculation operation is typically conducted from a minimum of 30 min to 3 hours, depending on the wetting characteristics of the biomass. After recirculation, the wire-mesh basket is lifted from the pretreatment chemical vessel and allowed to drain until dripping ceases.

Typically, the pre-impregnated biomass is at a *c.* 20% solids loading at this point and can be further dewatered in a hydraulic press or a continuous screw press, as shown in Figure 20.4. This dewatering step not only removes excess liquid to prepare the pre-impregnated biomass to a reasonably low moisture content for high-solids pretreatment, but the pressing action also serves to further force pretreatment chemicals deep into the vascular and pore structure of the biomass particles. The dewatering screw press incorporates an electric motor with a variable-frequency drive coupled to the compression screw. A feed chute with a wide mesh screen guard allows acid-soaked or -pretreated biomass to be safely fed into the compression screw. The discharge chute allows the dewatered biomass to be collected a low-profile collection tub.

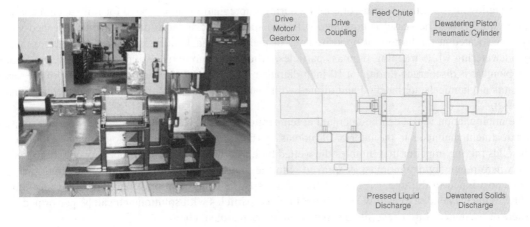

Figure 20.4 *Customized dewatering screw press for dewatering pre-impregnated biomass. (National Renewable Energy Laboratory, Golden CO, USA). (See figure in color plate section).*

The liquor squeezed from the biomass during dewatering is collected separately and can be quantified and analyzed for extractives and other compounds solubilized and removed during pretreatment chemical impregnation. The amount of liquor squeezed from the biomass is controlled by adjusting the amount of back pressure exerted on a cone-shaped choke in the discharge from the compression screw. Back pressure on the choke is easily adjusted using an air pressure regulator to a piston connected to the choke device, allowing for a wide range of dewatering efficiency range of operations. Typically, pre-impregnated biomass is dewatered to achieve a total solids content ranging from 40 to 60% before feeding to a pretreatment reactor.

A larger, specially designed, high-solids impregnation jacketed mixing vessel at NREL is shown in Figure 20.5. This 2000 L vessel (American Process Systems/Eirich Machines Co., Gurnee, IL, USA) is

Figure 20.5 *2000 L jacketed high-solids biomass impregnation vessel. (National Renewable Energy Laboratory, Golden CO, USA). (See figure in color plate section).*

jacketed, allowing for either heating or cooling during impregnation, and is rated for full vacuum and pressures up to 50 psig. The vessel is equipped with two specially designed 6-inch-diameter screened valves, which are located at the bottom at opposite ends of the vessel. Each valve is covered with 2 mm mesh to allow dewatering while trapping biomass particles. After the biomass is partially dewatered, pre-impregnated biomass is discharged through a 10-inch-diameter valve centrally located at the bottom of the vessel. Pretreatment chemicals are batched and diluted with water to the proper concentration in a separate batching vessel.

The impregnation vessel can be operated at temperatures up to 80 °C to enable more thorough transport of pretreatment chemicals to the interior of biomass particles. This transport is also enabled by the use of a high-solids mixing impeller that tumbles the wetted biomass. The system can be operated in a higher-solids mode where relatively dry biomass is sprayed with the pretreatment chemical solution via multiple spray nozzles positioned in the upper region of the vessel spaced along its length while the biomass is being tumbled. This is intended to evenly contact the biomass particles with solution and can be performed under pressure or vacuum to further facilitate penetration of pretreatment chemicals.

In this mode of operation, the amount of liquid added to the biomass is typically not enough to fully saturate the biomass and does not result in a liquid-free slurry, so draining of excess liquid from the pre-impregnated biomass via the screened valve ports is not performed. The partially wetted biomass is then discharged from the pre-impregnation vessel and can be further dewatered using the continuous screw press, if desired.

Another mode of operation involves soaking the biomass particles in an excess volume of pretreatment chemical solution (typically 5–10% solids concentration). In this mode, slurried biomass is mixed at a controlled temperature for a specified time (possibly under elevated pressure or vacuum conditions) and is then partially dewatered by the screened drain ports. Draining can be facilitated by pressurizing the vessel while the screened ports are opened and/or providing vacuum on the drained liquid collection tubing. In addition to being operated in a standard chemical impregnation mode, additional pre-processing steps can be performed in this equipment to remove undesired compounds such as acetate, ash, or other easily solubilized extractives prior to pretreatment chemical pre-impregnation.

Chemical pre-impregnation can also be facilitated by the pressing action of certain continuous feeding devices, as discussed in the following sections.

20.4.3   Pressurized Continuous Pretreatment Feeder Equipment

A number of equipment designs for continuously or semi-continuously feeding as-received or pre-impregnated biomass to pilot-scale continuous reactors have been developed. Many of these designs were initially developed for commercial-scale continuous pulping reactors for the pulp and paper industry, while other designs were developed for high-pressure thermochemical conversion processes on coal or biomass feedstocks, such as gasification or pyrolysis [56,57]. These feeder devices can be divided into five general categories: lock-hopper; rotary valve; piston-compression; screw-compression; and particle pump. The key features of each continuous feeding device are summarized in Table 20.2 and are discussed in detail in the following sections.

Lock-hopper Feeders

A lock-hopper feeder is a system where two reciprocating valves isolate a hopper, allowing intermittent charging, pressurization of hopper contents using steam or a pressurized inert gas, and discharging of biomass either directly into a pressurized continuous pretreatment reactor or into a pressurized metering bin, where an injection screw directly feeds the reactor system. Conceptually, a lock-hopper feeder is similar in

Table 20.2 *Continuous pressure-sealing biomass feeder devices.*

Feeder category	Pressure-sealing mechanism	Other considerations
Lock-hopper	• Ball or gate valves at top and bottom of hopper	• Simple design, no compression of biomass structure • Pressure-cycling fatigue, not true continuous feeding
Rotary valve	• Rotating, sealed valve vanes or pockets	• Compact, no compression of biomass structure • Biomass bridging, limited allowable pressures
Screw-compression	• Compressed biomass plug via conical screw housing	• Good pressure-sealing capability (moderate pressure) • High power consumption and abrasive wear
Piston-compression	• Compressed biomass plug via reciprocating piston	• Good pressure-sealing capability (high pressure) • Moderate power consumption and abrasive wear
Particle pump	• Steam condensate front in lightly compressed biomass plug	• Low power consumption, low wear • Requires precise control of condensate front

design to the batch pretreatment reactor shown in Figure 20.2, with the "capping valve" acting as the upper lock-hopper valve and the "blow valve" acting as the lower lock-hopper valve.

The upper valve opens, allowing for feedstock to be charged via gravity under atmospheric pressure. Once charged, the upper valve is closed and steam or other pressurized gas is used to pressurize the chamber. Once pressurized, the lower valve is opened and material is discharged via gravity (with a possible chase of steam or pressurized gas) into a pressurized continuous pretreatment reactor or into a pressurized metering bin.

Lock-hopper feeders are often used in applications where relatively dry biomass is required or compaction of biomass into densified formats is not desired, such as gasification. They have been used in very large (exceeding 2000 tons per day) coal gasification processes at pressures up to 1500 psig and at scales of up to several tons per day at pressures up to 400 psig [57].

Lock-hopper feeders offer the general advantage of a simple design with few moving parts, of being capable of accepting a wide range of feedstock particle sizes provided that flowability through the valves and hopper is maintained, and of providing a method of feeding solids into a pressurized reactor without actually compressing or compacting the feedstock, which can require high power consumption. However, lock-hopper valves are subject to wear due to high-pressure differentials and the abrasive nature of many biomass feedstocks; system components can experience fatigue due to pressure cycling (and associated temperature cycling in many pretreatment processes); and complex control strategies are associated with valve cycling, hopper charging and pressurization, hopper content discharging, and depressurization. Also, lock-hopper feeders do not result in truly continuous feed rates unless multiple parallel units are used to feed a reactor system and are sequenced in a manner to achieve an overall continuous feed rate.

Rotary Valve Feeders

Rotary valve feeders are known by several different names, including star, pocket, lock, and asthma feeders or valves. These types of feeders are commonly used in commercial-scale (1000–5000 dry tons/day) continuous vertical pulping reactor applications. In most of these applications, pulping reaction conditions require a pressure differential of less than 150 psig, although higher-pressure systems requiring slurry addition at relatively low solids loadings have been developed [56]. Rotary valve feeders operate under the principle of a rotating pocket or set of pockets within a circular housing separated by vanes. Feedstock is introduced at atmospheric pressure (or at low pressure from a pre-steaming bin or pre-impregnation operation), typically into the upper pocket. As the valve rotates, the pockets that were initially open to accept fresh feedstock rotate into the valve housing and are sealed off by close contact of the vanes to the valve housing. As these valves continue to rotate, they reach a steam inlet where high-pressure steam is injected into the pocket to increase the pressure of the pocket contents. This initial steam injection is often recovered steam from the back side of the valve (after discharge of pressurized biomass), captured during depressurizing of the discharged valve pockets. Multiple steam injections are usually made as the pockets rotate to gradually build pressure to reach or slightly exceed the reactor pressure. Once the pocket is fully pressurized and rotates to the discharge position, the valve contents discharge via gravity to the reactor, possibly with a flowing steam chase to fully evacuate any sticky biomass solids from the pocket. The emptied pocket continues its rotation, passing through a pressure discharging section (where evacuated steam is released and is potentially recovered in the pressure-building side of the valve) until it completes a full cycle and returns to the atmospheric-pressure pocket-filling position.

A simpler variation is known as an asthma feeder, which consists of only one pocket that first receives incoming feedstock under zero or very low pressure and then rotates to a pocket-emptying position, where flowing steam helps to discharge contents to the reactor. This specific type of rotary valve has been used in sawdust and other non-wood chip biomass types [44] and may therefore be well suited to a variety of biomass feedstocks in a pretreatment reactor feeding application.

Rotary valve feeders of various designs offer the general advantages of good feed rate control, good pressure-sealing ability (up to about 150 psig), compact size, low power consumption, utility on a variety of feedstock types, and the ability to capture and re-use pressurizing fluids (primarily steam). However, rotary valve feeders are somewhat susceptible to bridging and incomplete solids discharge (particularly with resinous or otherwise sticky biomass particles), pressure-sealing problems due to abrasive wear on pocket vanes and the associated seal with the valve body, and the limitation to pressure differentials of less than 150 psig.

Screw-compression Feeders

Screw-compression feeders have been commonly used in continuous pilot-scale biomass pretreatment systems. Often referred to as plug-screw feeders, these devices operate under the principle of using a screw conveyor to compress incoming biomass particles into a densified plug, which serves as a pressure-sealing barrier [58]. Feedstock, typically either pre-impregnated with pretreatment chemicals or pre-wetted/pre-steamed, enters the screw conveyor chamber via a gravity-discharging atmospheric-pressure hopper. Dry biomass feedstock particles (<20% moisture content) are not well suited for forming consistent pressure-holding plugs and produce excessive friction, which can result in biomass charring, abrasion of screw conveyor and feeder housing components, and excessive power requirements.

Screw conveyor flights advance the feedstock forward into a conical throat section. This conical section is lined with slotted or screened perforations to allow any liquid that is expressed from the densified biomass to seep out. If the expressed liquid is not removed, flooding can result in the plug-forming zone, which can cause the plug to disintegrate. This expressed liquid can also be reintroduced into the pretreatment reactor, ideally by being sprayed onto the expanding plug as it is being discharged. In a manner similar to a sponge that is compressed and then released, the discharged plug can readily adsorb the reintroduced liquid, serving as an excellent pretreatment chemical impregnation method.

The conical throat usually contains anti-rotation bars or slots to help direct the feedstock forward through the feeder rather than rotating with the feeder shaft. The conical throat ends at a cylindrical plug pipe, where the feedstock has been sufficiently compressed to form the pressure-holding plug. A pneumatically controlled piston or conical choke pushes against the forward-moving plug, helping to provide integrity to the plug and also a dampening of the explosive pressure release (known as a blowback) that can occur if the pressure-sealing plug is compromised. A variation of a standard plug-screw feeder, known as a modular screw device, combines size reduction and pressure sealing in a single unit using axial

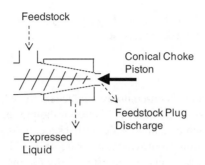

Figure 20.6 *A plug-screw feeder arrangement for use with a continuous pilot-scale pretreatment reactor system.*

compression forces [34]. A schematic depiction of a typical plug-screw feeder arrangement for use with a continuous pilot-scale pretreatment reactor system is shown in Figure 20.6.

Screw-compression feeders of various designs offer the general advantages of truly continuous feeding, good pressure-sealing capability (up to a pressure differential of about 200 psig), compact size with versatile installation options, and a large number of existing designs and manufacturers, and has been specifically used in pilot-scale biomass pretreatment applications. However, these devices can be susceptible to bridging at the feeder inlet (especially for wet, sticky, or fluffy feedstocks), are vulnerable to jamming caused by tramp material lodging in the rotating screw, require high power usage due to frictional and compression forces (with associated erosion of key components that may need periodic replacement or refurbishing), along with difficulty processing sticky or stringy feedstocks, which have a tendency to become packed between screw flights or wound around the screw shaft.

Piston-compression Feeders

Piston-compression feeders are similar in concept to screw-compression feeders, but utilize a reciprocating piston mechanism to compress incoming biomass particles into a densified plug which serves as a pressure-sealing barrier. Several variations of piston-compression feeders have been developed for continuous pilot-scale coal and biomass gasification processes, including modified concrete boom pumping devices [57].

For biomass pretreatment applications, perhaps the most common piston-compression feeder is the StakeTech/Sunopta Bioprocess reciprocating piston feeder (now Mascoma Canada, Mississauga, ON, Canada). This system delivers biomass to the face of a reciprocating piston via a screw conveyor that penetrates through the piston body. The piston compresses the delivered biomass particles into a densified plug that pushes against a pneumatically controlled piston or conical choke. The compression by the piston causes a very dense biomass plug to be formed, allowing for differential pressures as high as 450 psig in pilot-scale pretreatment operations on woody and agricultural residue feedstocks [59–61]. The pressure differentials possible with this feeder device are significantly higher than for typical screw-compression feeders, so this feeder is well suited for continuous steam explosion at high temperatures and pressures without added pretreatment chemicals (i.e., autohydrolysis).

Piston-compression feeders offer the general advantages of short cycling operation that results in near-continuous feeding, good pressure sealing at high-pressure differentials, compact size, potential suitability for sticky or stringy feedstocks, and consistent feedstock moisture due to uniform compression during plug formation. However, the mechanism to deliver feedstock to the plug-forming zone at the face of the piston is complex and potentially costly, power consumption can be high due to the compression required to achieve high operating pressure differentials, and frictional forces may cause abrasive wear that requires periodic replacement or refurbishment of key feeder components.

Particle Pump Feeders

While screw-compression feeders and piston-compression feeders force biomass into a densified plug as the primary means of providing a pressure-holding seal between the pretreatment reactor and upstream feedstock delivery systems, a particle pump or flow-feeder utilizes a combination of low-density biomass compression and steam condensate sealing to provide a pressure seal [62]. As a low-density biomass plug is pushed forward via a cylindrical screw or piston against a mildly forcing disintegration choke device, steam condensate forms a liquid "front" from the pressurized side of the pretreatment reactor. Proper maintenance of the biomass feeding rate and the compaction of the biomass in the feeder cylinder influence how far the steam condensation front penetrates the biomass, with the goal of achieving a stationary location of the condensation front. This liquid barrier provides the pressure holding seal, which can be monitored via a series of temperature or pressure sensors positioned along the length of the cylindrical feed tube [62].

As compared to high-compression biomass feeder devices, this type of feeder system offers the potential advantages of lower power consumption, less abrasive wear on feeder components, ability to process feedstocks of larger and more varied particle size and shape, and less compression and disruption of biomass cellular and vascular structure. However, successful performance of this device is highly dependent upon the balancing of feed rates of feedstock and steam and requires precise and rapid control to maintain a stationary pressure-sealing condensation front. Also, these devices typically operate better at high moisture contents (30–45% solids loading in the feeder zone), so the ability to achieve very high pretreatment solids loadings (>30% total solids) may be limited.

20.4.4   Pretreatment Reactor Throughput and Residence Time Control

Once the biomass is processed though a pressure-sealing feeder system, it must flow though the reaction zone of the pretreatment reactor. In continuous reactor systems, the rate of movement of the pretreating biomass from the discharge of the feeder system to the pressure-releasing reactor discharge system determines the pretreatment residence time. As residence time is a key parameter of pretreatment effectiveness, good control of residence time in a continuous reactor is very important. Ideally, plug-flow behavior will exist within a properly designed continuous pilot-scale pretreatment reactor, with all biomass particles experiencing the same residence time and temperature profile. In practice, there will be some deviation from ideal plug-flow behavior, so minimization of any residence time distribution is desired.

The basic orientation of the pretreatment reactor system and the mechanical means by which biomass is moved through the pretreatment reaction zone will ultimately determine the residence time profile. In general, there are two basic reactor configurations that have been developed and utilized as continuous pilot-scale pretreatment reactors [34]: horizontal and vertical.

A horizontal or inclined reactor tube or series of tubes with a screw conveyor is a common design. Various screw conveyor designs are possible within this configuration, including continuous-flight, interrupted or cut-flight, or extrusion-flight options. A continuous-flight screw is designed to minimize mixing of materials across screw flights and, when volumetric fill levels in the reactor tube are adequate, minimize back-mixing of material within the pitch space between screw flights. Interrupted-flight screws, and especially extrusion-type screws, result in significant mixing of material across screw flights and other implements installed on the screw shaft. While this provides good mixing and uniformity of material, significant back-mixing and deviation from ideal plug-flow occurs. For this reason, a continuous-flight screw is preferred when precise residence time control is desired.

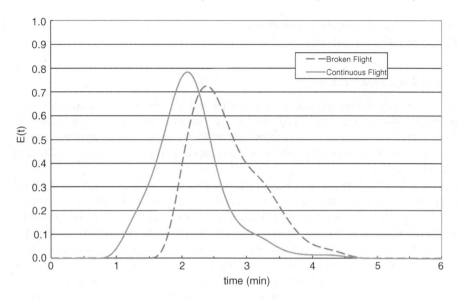

Figure 20.7 *RTD profiles for a horizontal screw pretreatment reactor using broken-flight screw conveyor and continuous-flight screw conveyor at conveyor rotation rate of 49 rpm on untreated corn stover at ambient temperatures using NaCl tracer [63].*

Residence time distribution (RTD) studies have been performed to compare the RTD profiles of a continuous-flight screw conveyor to a interrupted-flight screw conveyor in a single-tube horizontal pretreatment reactor operated at ambient conditions using water-impregnated corn stover, with injection of a 15 s pulse of NaCl-solution-impregnated corn stover. The resulting RTD profile for the continuous-flight screw is less dispersed and achieves a shorter mean residence time than the interrupted-flight screw, as shown in Figure 20.7 [63]. While this study was not conducted under pretreatment reaction conditions, where particle size and rheological changes during progression through the pretreatment reactor may change material motion and associated residence time distribution profiles, a number of methods to measure RTD profiles in horizontal screw reactor using *in situ* methods have been proposed which may be applicable to continuous biomass pretreatment systems [64–66].

A vertical reactor orientation is typically a simpler and therefore less expensive pretreatment reactor design option, especially when applied to larger demonstration-scale and commercial-scale installations [34]. These types of reactor systems are commonly used in large continuous pulping operations in the pulp and paper industry. Feedstock (often pre-steamed or pre-impregnated with chemicals) is introduced via a pressure-sealing feeder into the top of the reactor, which is at the appropriate temperature and pressure. Residence time is controlled by maintaining the accumulated level of biomass in the reactor at a certain height, with smaller heights corresponding to shorter residence heights and larger heights corresponding to longer residence times. Due to the thick walls of the high-pressure-rated reactor, the level is often measured by use of nuclear sources (i.e., gamma ray) and detectors. Once a biomass particle reaches the bottom of the reactor, it exits the reactor via a discharging arrangement. The rate at which a volumetric element of pretreated biomass is removed from the reactor should match the rate at which a same-sized volumetric element enters the top of the reactor to achieve steady-state residence time conditions, accounting for any change in bulk packing density of the biomass as a result of the pretreatment reaction. Ideally, there is no forward or backward mixing of biomass particles within a cross-sectional slice of the reactor column in order to achieve a consistent residence time although, in practice, there will be deviations.

In general, a vertical reactor arrangement is a simpler and less expensive design than the horizontal reactor with a screw conveyor, but does not offer such precise residence time control.

20.4.5   Reactor Discharge Devices

Effective discharge of pretreated biomass is essential for reliable and consistent operation of a continuous pretreatment reactor system. Just as the feeding system to the pretreatment reactor requires a pressure-sealing capability, the discharge system must continuously or near-continuously remove pretreated biomass at a targeted rate into a low-pressure or atmospheric-pressure receiver while maintaining steady temperature and pressure within the reactor.

Perhaps the simplest discharge system is a single, actuated valve (usually a full-port ball valve) that can be partially opened to allow discharge across the partially open valve and/or an inline orifice. This is often referred to as a "blow" valve. Alternatively, a quick-acting open-close valve, such as a poppet valve, can be used. By properly throttling or quickly cycling the valve and keeping the discharge pipe upstream of the valve full (often by use of a short-screw conveyor), pressure loss across the single discharge valve can be minimized, although periodic near-complete or complete closure of the valve may be required to allow the reactor pressure to be maintained within a narrow range. The discharge of pretreated biomass across the partially opened valve and optional orifice can also provide an explosive shearing of the pretreated biomass, which can increase its surface area and result in enhanced enzymatic digestibility. As such, this type of discharge system is often used in pretreatment reactors designed to act as continuous steam-explosion systems [59–61].

Another common continuous pretreatment reactor discharge system involves two actuated valves that are installed in series and are sequenced in a reciprocating manner [34]. This mode of operation more closely maintains the reactor pressure since both valves are never open at the same time. The valve sequencing is controlled with valve opening/closing times and time between valve sequencing adjusted to match the targeted throughput rate and achieve good pretreated slurry transfer through the valves and into the flash vessel. If needed, a steam chase can be introduced between the two discharge valves to assist the evacuation of pretreated slurry to the flash vessel. This may be needed during reactor start-up or when processing mildly pretreated or sticky materials. The steam chase control valve is briefly pulsed when the second (downstream) discharge valve is opened to help sweep the material forward through the discharge piping. A pressurized mechanical refiner can be installed inline immediately before the discharge valve(s) to further reduce the particle size of pretreated biomass as a means of enhancing enzymatic hydrolysis effectiveness.

A number of other discharge mechanisms have been described [34], most of which have been applied in pulping reactor operations and could be utilized in continuous pretreatment reactor systems. For a horizontal reactor system, a screw discharger can be used with a stoker to charge one or more discharge valves, which is an effective manner of delivering high-solids-content materials directly to the pressure-releasing valve. For a vertical reactor system, a properly designed rotating paddle or sweeper that consistently charges a discharge valve port located on the side of the reactor vessel near the bottom or a center-mounted discharge valve port at the bottom of a dished or otherwise sloped reactor bottom not only ensures efficient delivery of pretreated solids to the discharge port, but can help reduce residence time distribution by delivering pretreated solids to the discharge port immediately upon reaching the reactor bottom.

20.4.6   Blow-down Vessel and Flash Vapor Recovery

Once the pretreated biomass passes through the reactor discharger arrangement, it is typically delivered to a blow-down or flash vessel. This vessel serves as a collection tank for pretreated slurry as it exits the continuous pretreatment reactor. It should also allow for removal of vapors that flash from the pretreated slurry as it rapidly depressurizes and cools upon exiting through the discharger arrangement. A common blow-down vessel arrangement includes a tangential entry port and a large headspace to encourage vapor disengagement from the pretreated slurry, in a manner similar to a cyclone separator [44]. The pretreated slurry falls to the bottom of the vessel, where it collects and can be periodically or continuously discharged from the blow-down vessel using a live-bottom screw discharge arrangement.

Mechanical internals such as rotating knock-down paddles are often useful in preventing the accumulation of sticky pretreated biomass slurry on the walls of the blow-down vessel. Since the blow-down vessel is immediately downstream of the high-pressure pretreatment reactor, proper pressure-relief devices (e.g., a rupture disk whose discharge piping is vented to a safe location) should be provided in order to prevent over-pressurization, as could happen if a malfunction in the pressure-sealing discharge arrangement occurs. The flashed vapors exit via a suitably large port and can be condensed via an indirect chilled water condenser.

Besides water vapor, the flash vapor can contain volatile chemical species such as furfural, along with some of the acetic acid released from acetyl content of the biomass during pretreatment. As these compounds may be inhibitory to downstream conversion operations, such as fermentation of biomass-derived sugars to ethanol [18,51] or other biofuels products or intermediates, the removal of at least a portion of those compounds is beneficial. The removal of flash vapor also increases the solids concentration in the pretreated biomass, which may be beneficial if high-solids concentration downstream processing is desired.

For overall and individual component mass balance purposes, it is important to collect the condensed flash vapor and analyze its chemical composition. In smaller continuous pilot-scale systems, it is appropriate to collect the entire volume of condensate over a specified period of time while larger pilot-scale systems would more commonly utilize a flash-vapor-flow meter/totalizer and a compositional analysis of a representative sample of condensate for mass balancing purposes. It is also possible to utilize the heat content in the flash vapor for indirect heating of desired incoming process streams. This may not be feasible or desired in pilot-scale systems, but would be an important heat integration opportunity for envisioned commercial-scale continuous biomass pretreatment systems [25].

20.5 Continuous Pilot-scale Pretreatment Reactor Systems

20.5.1 Historical Development of Pilot-scale Reactor Systems

Early pilot, demonstration, and small commercial plants for hydrolyzing biomass to sugars with subsequent conversion to fuels and chemicals were constructed and operated in the first half of the 20th century. These facilities were based on early acid hydrolysis processes, as enzymatic approaches for hydrolyzing pretreated cellulosic biomass had not yet been developed. There were facilities developed during wartime in Germany and the former Soviet Union, with early efforts in the United States starting in the 1940s [67,68]. By the 1970s, enzymatic hydrolysis technologies had developed to the point where initial pilot-scale pretreatment (primarily based on dilute-acid or steam pretreatment technologies) and downstream enzymatic hydrolysis and fermentation technologies were implemented, often conducted in a simultaneous saccharification and fermentation (SSF) mode [69,70]. There were various designs of continuous acid hydrolysis or pretreatment reactors developed for use in these pioneering facilities, which have been thoroughly described elsewhere [67–70].

Beginning in the 1980s, more modern continuous pilot-scale pretreatment systems have been designed and operated. Several continuous steam-explosion reactors using high-pressure piston-compression feeders have been operated in Canada [59], France, [60,61] and Italy [71]. These systems employ the Staketech/Sunopta (now Mascoma Canada, Mississauga, ON, Canada) piston-compression feeder with a horizontal screw conveyor pretreatment reactor, as discussed in the previous section. Another continuous pilot plant reactor using an inclined screw conveyor tube with a particle pump-type feeder has recently been installed in Denmark. This single-stage unit has a throughput of 100 kg/hr and has been run using wheat straw as the feedstock at a variety of acidic, alkaline, wet oxidation, and steam explosion conditions at temperatures up to 200 °C. [72]. This system served as the basis for a larger two-stage pretreatment reactor installation, with inter-stage washing to remove a portion of solubilized hemicellulosic sugars after the first stage. This larger demonstration-scale biorefinery has since been constructed and operated in Denmark (www.inbicon.com). Additional continuous single-stage and two-stage pilot-scale pretreatment systems using plug-screw compression feeders have been described [34]. There are numerous other pilot-scale, demonstration-scale, and now small commercial-scale continuous pretreatment reactor systems that have been installed or are being designed in conjunction with second-generation biofuels facilities. While the existence of these systems is often indicated by company websites and press releases, there is generally little revealed in the way of pretreatment process or reactor design details. Given the proprietary nature of most of these projects, detailed information on the pretreatment systems is not widely available.

An overall continuous pilot-scale system consists of many or all of the individual components discussed in Section 20.4, designed in an integrated manner to allow for safe and reproducible operations. Smaller continuous pilot-scale pretreatment units may be skid-mounted and portable, not requiring special facility layout provisions. An example of this is a 200 kg dry biomass/day horizontal screw pretreatment reactor at NREL, as shown in Figure 20.8. Pre-impregnated milled biomass (passing through a 0.5 inch or smaller screen) which is dewatered to *c.* 50% moisture content is metered at a prescribed feed rate to a pressure-sealing plug-screw feeder and into a continuous-flight horizontal screw reactor, where steam is injected to achieve the desired temperature. This reactor is equipped with a steam jacket to help reduce the required steam injection in order to achieve higher solids loadings. Residence time in the reactor is controlled by the feed rate, discharge rate, and rotational speed of the screw conveyor. The pretreated biomass is continuously discharged via two reciprocating full-port ball valves and enters a flash receiver, where the pretreated biomass cools, and flash vapors are vented to a condenser and then collected.

In order to improve the logistics of handling and moving low-density biomass feedstocks to, through, and from larger continuous pilot-scale pretreatment systems (500 dry kg/day or larger), it is important to

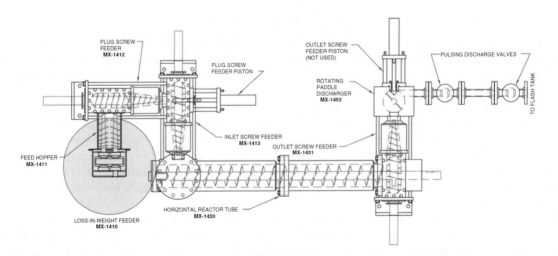

Figure 20.8   *Schematic drawing of 200 kg/day horizontal pretreatment reactor installed at the National Renewable Energy Laboratory (Equipment vendor: Metso Inc., Norcross, GA, USA).*

consider the overall equipment layout and the facility in which it is installed. One of the simplest considerations involves the use of gravity flow, especially in the feedstock handling, pretreatment, and enzymatic saccharification systems operated at high solids loadings. Ideally, such a pilot plant can be constructed on multiple levels, where pre-processed biomass that has already been appropriately cleaned and/or size-reduced will be raised to feed hoppers using pneumatic, screw, cleated-belt, or drag-chain conveyors that supply the pretreatment reactor, which is located at the highest level of the pilot plant. The biomass is fed into the pretreatment reactor and is typically expelled into an appropriate flash tank. A live-bottom feeding mechanism at the bottom of the flash tank can remove the high-solids pretreated slurry, which can be gravity fed downward into conditioning equipment and/or high-solids enzymatic hydrolysis reactors if no conditioning is needed or if conditioning is intended to be performed after enzymatic hydrolysis. Once the high-solids slurry is sufficiently liquefied by action of the hydrolytic enzymes to a point that it is flowable and pumpable, appropriately designed transfer pumps can then be used to transfer the liquefied slurry to subsequent unit operations.

20.5.2 NREL Gravity-flow Reactor Systems

The principal of gravity flow has been applied in a pilot plant expansion at NREL. This facility began operations in 2011 and includes two new continuous pilot-scale pretreatment reactor systems, each with dry feedstock processing rates of between 500 and 1000 kg/day. Each system is integrated with a continuous knife-mill that has interchangeable screen openings ranging from 0.25 inch to 1 inch. The milling systems can be operated in a stand-alone manner, where milled feedstock can be packaged for later use, or in a continuous manner where the output from the mill is continuously fed to its companion pretreatment system at a feed rate that matches the pretreatment throughput rate. The milling facilities are located at the lowest level of the facility, with feedstock to the pretreatment systems delivered to the uppermost level of the facility via a pneumatic blower and tube arrangement. Dust generated in the milling operation can be removed via a cyclone separator located above each pretreatment system, with dust collected in a baghouse filter. As milled feedstock is delivered to the top of each pretreatment reactor, it is delivered via a horizontal screw conveyor to a metering weigh-belt. The speed of the weigh-belt conveyor automatically adjusts to deliver the specified biomass feed rate.

Metered biomass drops via gravity into a pug-mill mixer, which is a high-solids horizontal paddle mixer with liquid-injection nozzles, where pretreatment chemicals and/or water are sprayed onto the biomass at controlled addition rates as it enters the pug-mill mixer. The pug-mill mixer is designed to thoroughly mix the biomass and the added liquids as the wetted biomass continuously moves to the discharge end of this device and drops, via gravity, into a continuous pretreatment reactor system. Alternatively, previously pre-impregnated biomass can be delivered via an optional feedstock metering device directly to the pug-mill mixer without additional water or pretreatment chemical addition. In this case, the pug-mill mixer is simply a transfer device to deliver pre-impregnated biomass to the pretreatment system.

Horizontal-tube Pretreatment System

The first pretreatment system was provided by Metso Inc. (Norcross, GA, USA) and consists of a series of four horizontal-tube screw conveyors, as shown in Figure 20.9. Pre-impregnated feedstock enters the system directly from the pug-mill mixer via gravity to a small live-bottom surge hopper to the pressure-sealing feeder. This feeder is a screw-compression plug-screw feeder device, which compresses the pre-impregnated biomass into a pressure-sealing plug against an opposing choke piston with adjustable pneumatic back-pressure. This also serves to partially dampen the effects of "blowback," which results from a loss of the pressure-sealing plug. Vent piping directs the blowback slurry and steam to a safe location.

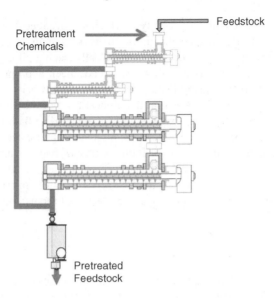

Figure 20.9   *Schematic drawing of 1 ton/day multi-tube horizontal pretreatment reactor installed at the National Renewable Energy Laboratory (Equipment vendor: Metso Inc., Norcross, GA, USA).*

The performance of the plug-screw feeder is of critical importance to reliable operation of the entire pretreatment system and can be monitored by recording the electric current to the drive motor. Fluctuations in the measured current can be an indication of inconsistent processing rates through the plug-screw feeder, which can lead to a blowback if not quickly corrected.

Liquid that is expressed from the pre-impregnated biomass upon formation of the pressure-sealing plug passes through screens in the plug-screw feeder and exits the feeder housing. It can be re-injected into the pressurized zone of the pretreatment reactor immediately past the plug-screw feeder and be re-adsorbed into the biomass. Upon exiting the plug-screw feeder, the biomass is now within the high-pressure zone of the reactor and quickly heats to the desired reaction temperature via steam injection. Steam is delivered from a boiler via pressure control valves and a steam flowmeter to various injection points in the reactor system. The biomass is processed through a shredding device to break up any hard clumps formed in the plug-screw feeder and is conveyed upwards through a short vertical impregnation section, where additional chemicals, water, and/or steam can be added.

From this point, the biomass slurry enters the system of four steam-jacketed horizontal screw conveyor tubes. As shown in Figure 20.9, these tubes are arranged in series and are of different lengths and diameters. There are a number of different possible options to bypass one or more tubes, depending on the pretreatment chemistry used and the residence time desired. Corrosive acidic pretreatments are typically conducted at shorter residence times (less than 20 min), so the first two smaller tubes are constructed of acid-resistant alloy (Hastelloy C-2000, Haynes International, Inc., Kokomo, IN, USA). Using either just the first tube or the first two tubes in series, acidic pretreatment residence times can range from 1 to 20 min depending on throughput rate and screw conveyor rotational speeds. Various geometrically designed spool sections and flange end-caps allow for easy reconfiguration of the reactor tubes to provide versatile operating scenarios.

The final two larger tubes are constructed of stainless steel and provide options for longer residence times that are more suitable for steam/hot water (autohydrolysis) and mild alkali pretreatment conditions.

When all four tubes are connected in series, pretreatment residence times of 60–120 min can be achieved. An additional option of bypassing the first three tubes exists, allowing for pre-impregnated feedstock to be fed directly to the fourth tube which provides a residence time of 20–60 min.

Irrespective of which reactor tube configuration is used, pretreated slurry is delivered to a horizontal screw discharger and two reciprocating full-port ball valves. The pretreated slurry is depressurized as it passes through these valves and enters a blow-down flash vessel. Vent piping allows for flashed vapors to pass through a flowmeter into a chilled-water condenser, from which condensate can be sampled and analyzed for mass-balancing purposes. Pretreated slurry, typically at a solids loading between 20% and 40% total solids, is continuously removed from the bottom of the flash vessel via a live-bottom screw discharger, where it passes via gravity through a diverter valve arrangement into collection drums, high-solids enzymatic hydrolysis reactors, or a solid-liquid separator feed tank.

Vertically Oriented Pretreatment System

The second pretreatment system was provided by AdvanceBio Systems LLC (Milford, OH, USA) and consists of a series of two vertically oriented reactor units, as shown in Figure 20.10. This system utilizes a similar parallel set of pneumatic feedstock delivery, cyclone separator, screw conveyor,

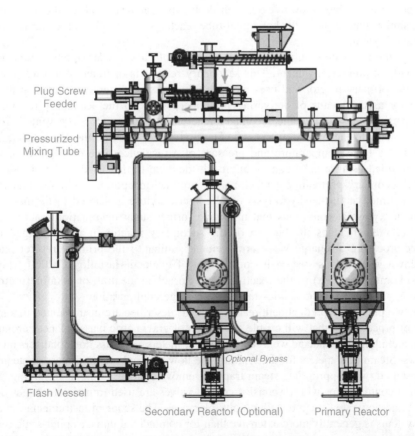

Figure 20.10 *Schematic drawing of 1 ton/day vertical pretreatment reactor installed at the National Renewable Energy Laboratory (Equipment vendor: AdvanceBio Systems LLC., Milford, OH, USA).*

weigh-belt feeder, and pug-mill mixer components as described for the horizontal-tube pretreatment system. It also utilizes a screw-compression plug-screw feeder device of similar (but not identical) design to the Metso reactor plug-screw feeder. Blowback vent piping and provisions for monitoring the motor current of the plug-screw feeder is also provided, as are plug-screw feeder-expressed liquid collection and re-injection provisions.

Upon exiting the plug-screw feeder, the biomass is within the high-pressure zone of the reactor and quickly heats to the desired reaction temperature via steam injection. Steam is delivered from a boiler via pressure-control valves and a steam flow meter to various injection points in the reactor system. The biomass drops from the plug-screw feeder into a horizontal screw impregnation tube where steam and possibly additional chemicals and/or water are added. The biomass slurry only spends a short time (a few minutes or less) in this impregnation tube, where is drops via gravity into the primary vertical reactor.

The residence time in the vertical reactor is controlled by maintaining the accumulated level of biomass in the reactor at a height corresponding to the targeted residence time using gamma sources and detectors. Once a biomass particle reaches the bottom of the reactor, it exits the reactor via a discharge rotating paddle and valve arrangement. The nominal residence time range in this primary reactor ranges over 10–40 min, although it can be reduced to about 5 min by installing an optional volume-occupying insert.

Once a pretreated slurry element reaches the bottom of the reactor, it is pushed by a rotating paddle into discharge piping and is depressurized via two reciprocating full-port discharge valves. Typically, discharged biomass is expelled into a blow-down vessel with flash vapor recovery and live-bottom pretreated slurry discharge in a similar manner as in the horizontal-tube reactor system, as described in Section Horizontal-tube Pretreatment System.

Alternatively, the pretreated slurry can be discharged into a secondary pretreatment reactor via an alternate discharge piping arrangement. This secondary reactor is operated at a lower temperature and pressure than the primary reactor and uses the resulting pressure differential between the primary and secondary reactor as the means of delivering pretreated slurry to the secondary reactor. As the pretreated slurry partially flashes into the secondary reactor, any flash vapors are vented from the top of the secondary reactor to the blow-down vessel. The pretreated slurry in the secondary reactor is additionally treated at a lower temperature and pressure of a residence time of 10–30 min, with the residence time controlled using a level-control approach identical to that used in the primary reactor. The purpose of the secondary pretreatment reactor is to provide an optional mild secondary pretreatment stage, primarily aimed at further hydrolysis of oligomeric xylose generated in the first stage to additional monomeric xylose at conditions that minimize further sugar degradation reactions. Once a pretreated slurry element reaches the bottom of the secondary reactor, it is discharged by a rotating paddle and reciprocating discharge valve arrangement similar to the primary reactor and is expelled to the blow-down vessel. This system is constructed of cladded Hastelloy C-2000 (Haynes International, Inc., Kokomo, IN, USA) in the reaction zone, so acidic, neutral, or alkaline pretreatment conditions can be applied in both the single-stage or two-stage configuration.

Each of the two pilot-scale pretreatment reactor systems described are instrumented in a manner to permit collection of process data that will enable generation of gravimetric mass balances around each major unit operation, including accurate and well-calibrated flow meters and mass flow totalizers on liquid, steam, and slurry flows. Steam flow meters for determining the flow rate and totalized mass of steam are properly located and installed, with appropriate steam traps to remove condensate in steam piping. Properly calibrated and tuned control loops with responsive control valves are used to maintain targeted temperature, pressure, mass, and flow rate setpoints within tight tolerances. The scope of instrumentation in these pilot-scale reactor systems is generally more extensive than for commercial units, as pilot-scale reactors require collection of extensive data to evaluate the effect of changing process conditions and equipment configurations on numerous performance parameters.

20.6 Summary

This chapter has discussed design features for both laboratory-scale and pilot-scale pretreatment reactors. As there is a wide range of operating conditions across various pretreatment approaches and even within a given pretreatment approach, it is important that pretreatment reactor selection, either for bench-scale or pilot-scale applications, consider a range of expected process conditions. The discussion of laboratory-scale pretreatment equipment has focused on simple batch systems, as continuous systems at the bench scale are difficult to implement and often require components that are not scalable or require unrealistically small feedstock particle sizes. Within these limitations, emphasis has been focused on bench-scale pretreatment equipment that can operate under process-relevant conditions, such as at reasonably high solids loadings. Such pretreatment equipment should provide for rapid heating and cooling capabilities, provisions (either prior to or inside the reactor) to adequately contact biomass particles of appropriate size with water, steam, and/or pretreatment chemicals, good heat and mass transfer characteristics during the pretreatment reaction time, adequate materials of construction to ensure corrosion resistance (needed to provide reactor system longevity, safe operation, avoidance of corrosion products that may influence reaction chemistry kinetics), and good instrumentation and control systems for accurate data collection and reproducible operation.

At the pilot scale, loosely defined as pretreatment equipment of a size and design that the mechanical reactor components and associated ancillary equipment are similar to (albeit smaller than) that found in commercial-scale systems, either batch or continuous pretreatment systems can be considered. Since pilot-scale equipment is intended to be scalable to anticipated commercial-scale processes, emphasis has been placed on discussion of continuous pilot-scale systems, as it is widely believed that continuous pretreatment reactors will be utilized at scales above 1000 dry tons of biomass per day in future commercial biorefineries. Examples of batch pilot-scale pretreatment reactor systems have been provided, which generally consist of a biomass and pretreatment chemical addition system, a pressure-rated pretreatment reactor (usually injected with steam to quickly achieve targeted pretreatment reaction temperatures and pressures), and a means of quickly cooling and de-pressurizing the pretreated biomass slurry, often by discharge of the batch pretreatment reactor via a quick-opening valve to a receiving vessel with venting and condensing of volatile flash vapors.

Most pilot-scale pretreatment systems are designed for continuous operation, as commercial-scale pretreatment systems are expected to typically employ continuous reactors. Such systems become particularly challenging to design and operate effectively at high solids loadings (above 15% insoluble solids), as heat and mass transfer considerations and control of residence time to tight tolerances are critical design elements that must be adequately addressed. Large-scale continuous pulping reactors that have been developed and widely utilized in the pulp and paper industry have served as starting design basis for biomass pretreatment systems, although certain unique challenges for biomass pretreatment, such as control of residence times typically on the order of minutes (as opposed to hours in pulping processes) must be considered. The design considerations and broad features of the various elements of continuous pilot-scale biomass pretreatment reactors and associated ancillary equipment have been discussed, including feedstock preparation and handling, pretreatment chemical injection and contacting, pressurized biomass feeding equipment, pretreatment reactor conveyance systems and residence time control, pretreatment reactor discharge devices, blow-down tank design, and flash recovery systems.

The features of several feeder devices have been described, including lock-hoppers, rotary valve feeders, screw-compression feeders, piston-compression feeders, and particle pumps. These feeder devices are critical to the safe and reliable operation of continuous reactor systems, as the particular challenge of continuously adding solid matter to a pressurized reactor is not trivial. The selection of a particular feeder device is somewhat dependent on required operating pressures and temperatures, as some of these feeder systems have practical limitations in this regard. The mechanical means of how particular feeder devices provide the pressure-sealing capability should also be considered as some systems cause extreme compression of the incoming feedstock, which may be desired in some cases and not in other situations.

The control of residence time in a continuous pilot-scale pretreatment reactor is extremely important. Most pretreatment reaction times are relatively short (often less than 20 min) and certain pretreatment chemistries can cause undesirable sugar degradation reactions if residence times exceed certain limits. Ideally, plug-flow behavior will exist within a properly designed continuous pilot-scale pretreatment reactor, with all biomass particles and fluids experiencing the same residence time and temperature profile throughout the reactor. In practice, there will be some deviation from ideal plug-flow behavior, so minimization of any residence time distribution is desired.

There are two basic reactor configurations that have been developed and utilized as continuous pilot-scale pretreatment reactors. A horizontal or inclined reactor tube or series of tubes with a screw conveyor is a common option. Properly designed screw conveyors can achieve residence time control within relatively tight tolerances, but such systems are complex and expensive and require properly designed pressure seals for rotating-screw conveyor shafts. A vertical reactor orientation is typically a simpler and therefore less expensive pretreatment reactor design option, especially in larger demonstration-scale and commercial-scale contexts. Residence time in the pressurized reaction zone is controlled by maintaining the accumulated level of biomass in the reactor at a certain height, with smaller heights corresponding to shorter residence heights and larger heights corresponding to longer residence times. However, these can suffer from deviations in residence time due to uneven movement of biomass particles across a reactor cross-section as individual volumetric biomass elements move toward the bottom of the vertical reactor, where they are discharged to a pretreated slurry receiver.

Finally, examples of integrated continuous pilot-scale pretreatment reactor systems that incorporate these various equipment components were provided. These include both horizontally and vertically oriented systems with different pressure-sealing feeder and discharge options. The integration of desirable design features such as gravity-assisted movement of solid biomass into, through, and from the pretreatment system was discussed, along with the requirements for extensive instrumentation and sophisticated control strategies as these pilot-scale reactors require large amounts of data to be collected to evaluate the effect of changing process conditions and equipment configurations on numerous performance parameters.

In conclusion, biomass pretreatment approaches are generally intended to work in concert with enzymatic hydrolysis processes to effectively saccharify structural carbohydrates in native biomass to monomer sugars that can then be converted to biofuels. Pretreatment reactor systems at both the laboratory and pilot scale have been developed and continue to evolve in sophistication and reliability. The selection of the type of pretreatment system is largely dependent on whether the purpose of the system is a scaled-up demonstration of previously developed pretreatment process conditions within a narrow range of operating conditions, or if a flexible pilot-scale reactor capable of operating under a variety of pretreatment chemistries and operational conditions across a wide range of biomass feedstock types is desired. In either case, it is anticipated that the function of the pretreatment operation will continue to evolve as advances in the efficacy and cost of enzyme preparations to further saccharify pretreated biomass influence pretreatment reaction requirements.

Acknowledgements

The author acknowledges funding from the Office of the Biomass Program of the United States Department of Energy through contract number DE-FG36-04GO14017). Additionally, the author acknowledges numerous discussions with various pretreatment reactor equipment providers over many years, which has helped to frame the author's understanding of the desirable features and best practices for pretreatment equipment reactor design and operation.

References

1. Johnson, D.K. and Elander, R.T. (2007) Pretreatments for enhanced digestibility of feedstocks, in *Biomass Recalcitrance* (ed. M.E. Himmel), Blackwell, W. Sussex, UK, pp. 436–453.
2. Mosier, N., Wyman, C., Dale, B. *et al.* (2005) Features of promising technologies for pretreatment of lignocellulosic biomass. *Bioresource Technology*, **96**, 673–686.
3. Sun, Y. and Cheng, J. (2002) Hydrolysis of lignocellulosic materials for ethanol production: a review. *Bioresource Technology*, **83**, 1–11.
4. Hsu, T.A. (1996) Pretreatment of biomass, in *Handbook on Bioethanol, Production and Utilization* (ed. C.E. Wyman), Taylor & Francis, Washington, D.C., pp. 179–211.
5. Duff, S.J.B. and Murray, W.D. (1996) Bioconversion of forest products industry waste cellulosics to fuel ethanol: a review. *Bioresource Technology*, **55**, 1–33.
6. McMillan, J.D. (1994) Pretreatment of lignocellulosic biomass, in *Enzymatic Conversion of Biomass for Fuels Production* (eds M.E. Himmel, J.O. Baker, and R.P. Overend), American Chemical Society Symposium Series 566, Washington D.C., pp. 292–324.
7. Wyman, C.E., Dale, B.E., Elander, R.T. *et al.* (2005) Coordinated development of leading biomass pretreatment technologies. *Bioresource Technology*, **96**, 1959–1966.
8. Wyman, C.E., Dale, B.E., Elander, R.T. *et al.* (2005) Comparative sugar recovery data from laboratory scale application of leading pretreatment technologies to corn stover. *Bioresource Technology*, **96**, 2026–2032.
9. Eggeman, T. and Elander, R.T. (2005) Process and economic analysis of pretreatment technologies. *Bioresource Technology*, **96**, 2019–2025.
10. Elander, R.T., Dale, B.E., Holtzapple, M. *et al.* (2009) Summary of findings from the Biomass Refining Consortium for Applied Fundamentals and Innovation (CAFI): Corn stover pretreatment. *Cellulose*, **16**, 649–659.
11. Wyman, C.E., Dale, B.E., Elander, R.T. *et al.* (2009) Comparative sugar recovery and fermentation data following pretreatment of poplar wood by leading technologies. *Biotechnology Progress*, **25**, 333–339.
12. Brownell, H.H. and Saddler, J.N. (1987) Steam pretreatment of lignocellulosic material for enhanced enzymatic hydrolysis. *Biotechnology and Bioengineering*, **29**, 228.
13. Mok, W.S.L. and Antal, M.J. Jr. (1992) Uncatalyzed solvolysis of whole biomass hemicellulose by hot compressed liquid water. *Industrial & Engineering Chemistry Research*, **31**, 1157.
14. Mosier, N., Hendrickson, N., Ho, N. *et al.* (2005) Optimization of pH controlled liquid hot water pretreatment of corn stover. *Bioresource Technology*, **96**, 1986–1993.
15. Liu, C. and Wyman, C.E. (2003) The effect of flow rate of compressed hot water on xylan, lignin, and total mass removed from corn stover. *Industrial & Engineering Chemistry Research*, **42**, 5409.
16. Saha, B.C. and Cotta, M.A. (2006) Ethanol production from alkaline peroxide pretreated enzymatically saccharified wheat straw. *Biotechnology Progress*, **22**, 449.
17. Varga, E., Schmidt, S.A., Reczey, K., and Thomsen, A.B. (2003) Pretreatment of corn stover using wet oxidation to enhance enzymatic digestibility. *Applied Biochemistry and Biotechnology*, **104**, 37–50.
18. Klinke, H.B., Ahring, B.K., Schmidt, A.S., and Thomsen, A.B. (2002) Characterization of degradation products from alkaline wet oxidation of wheat straw. *Bioresource Technology*, **82**, 15–26.
19. Sorensen, A., Teller, P.J., Hilstrom, T., and Ahring, B.K. (2008) Hydrolysis of *Miscanthus* for bioethanol production using dilute acid presoaking combined with wet explosion pre-treatment and enzymatic treatment. *Bioresource Technology*, **99**, 6602–6607.
20. Bozell, J.J., Black, S.K., Myers, M. *et al.* (2011) *Biomass and Bioenergy*, **35**, 4197–4208.
21. Zhao, X.B., Cheng, K.K., and Liu, D.H. (2009) Organosolv pretreatment of lignocellulosic biomass for enzymatic hydrolysis. *Applied Microbiology and Biotechnology*, **82**, 815.
22. Pan, X.J., Gilkes, N., Kadla, J. *et al.* (2006) Bioconversion of hybrid poplar to ethanol and co-products using an organosolv fractionation process: Optimization of process yields. *Biotechnology and Bioengineering*, **94**, 851.
23. Lloyd, T.A. and Wyman, C.E. (2005) Combined sugar yields for dilute sulfuric acid pretreatment of corn stover followed by enzymatic hydrolysis of the remaining solids. *Bioresource Technology*, **96**, 1967.

24. Schell, D.J., Farmer, J., Newman, M., and McMillan, J.D. (2003) Dilute-sulfuric acid pretreatment of corn stover in pilot-scale reactor – Investigation of yields, kinetics, and enzymatic digestibilities of solids. *Applied Biochemistry and Biotechnology*, **105**, 69.

25. Humbird, D., Davis, R., Tao, L. *et al.* (2011) Process Design and Economics for Biochemical Conversion of Ligno-cellulosic Biomass to Ethanol: Dilute-Acid Pretreatment and Enzymatic Hydrolysis of Corn Stover. NREL Technical Report NREL/TP-5100-47764, National Renewable Energy Laboratory, Golden, CO.

26. Weiss, N.D., Nagle, N.J., Tucker, M.P., and Elander, R.T. (2009) High xylose yields from dilute acid pretreatment of corn stover under process-relevant conditions. *Applied Biochemistry and Biotechnology*, **155**, 418–428.

27. Garlock, R.J., Chundawat, S.P., Balan, V., and Dale, B.E. (2009) Optimizing harvest of corn stover fractions based on overall sugar yields following ammonia fiber expansion pretreatment and enzymatic hydrolysis. *Biotechnology for Biofuels*, **2**, 29.

28. Chundawat, S.P., Vishmeh, R., Sharma, L.N. *et al.* (2010) Multifaceted characterization of cell wall decomposition products formed during ammonia fiber expansion (AFEX) and dilute acid based pretreatments. *Bioresource Technology*, **101**, 8429–8438.

29. Iyer, P.V., Wu, Z., Kim, S.B., and Lee, Y.Y. (1996) Ammonia recycled percolation process for pretreatment of herbaceous biomass. *Applied Biochemistry and Biotechnology*, **57–58**, 121.

30. Kim, T.H. and Lee, Y.Y. (2005) Pretreatment of corn stover by soaking in aqueous ammonia. *Applied Biochemistry and Biotechnology*, **121**, 1119.

31. Bjerre, A.B., Olesen, A.B., and Fernqvist, T. (1996) Pretreatment of wheat straw using combined wet oxidation and alkaline hydrolysis resulting in convertible cellulose and hemicellulose. *Biotechnology and Bioengineering*, **49**, 568.

32. Kaar, W.E. and Holtzapple, M.T. (2000) Using lime pretreatment to facilitate the enzymatic hydrolysis of corn stover. *Biomass and Bioenergy*, **18**, 189.

33. Chang, V.S., Nagwani, M., Kim, C.H., and Holtzapple, M.T. (2001) Oxidative lime pretreatment of high-lignin biomass. *Applied Biochemistry and Biotechnology*, **94**, 1.

34. Cort, J.B., Pschorn, T., and Stromberg, B. (2010) Minimize scale-up risk. *Chemical Engineering Progress*, **106**, 39–49.

35. Linde, M., Galbe, M., and Zacchi, G. (2006) Steam pretreatment of acid-sprayed and acid-soaked barley straw for production of ethanol. *Applied Biochemistry and Biotechnology*, **129–132**, 546–562.

36. Viamajala, S., Selig, M.J., Vinzant, T.B. *et al.* (2005) Catalyst transport in corn stover internodes: Elucidating transport mechanisms using Direct Blue-I. *Applied Biochemistry and Biotechnology*, **129–132**, 509–527.

37. Soderstrom, J., Pilcher, L., Galbe, M., and Zacchi, G. (2003) Combined use of H_2SO_4 and SO_2 impregnation for steam pretreatment of spruce in ethanol production. *Applied Biochemistry and Biotechnology*, **105–108**, 127–140.

38. Tucker, M.P., Kim, K.H., Newman, M.M., and Nguyen, Q.A. (2003) Effects of temperature and moisture on dilute-acid steam explosion pretreatment of corn stover and cellulase enzyme digestibility. *Applied Biochemistry and Biotechnology*, **105–108**, 165–177.

39. Bergeron, P., Benham, C., and Werdene, P. (1989) Dilute sulfuric acid hydrolysis of biomass for ethanol production. *Applied Biochemistry and Biotechnology*, **20–21**, 119–134.

40. Vidal, B.C. Jr., Dien, B.S., Ting, K.C., and Singh, V. (2011) Influence of feedstock particle size on lignocelluloses conversion – a review. *Applied Biochemistry and Biotechnology*, **164**, 1505–1421.

41. Harris Group, Inc. (2001) Acid Hydrolysis Reactors-Batch System, Report 99-10600/18, National Renewable Energy Laboratory Subcontractor Report, Harris Group Inc., Seattle, WA, USA.

42. Hsu, T.A., Himmel, M., Schell, D. *et al.* (1996) Design and initial operation of a high-solids, pilot-scale reactor for dilute-acid pretreatment of lignocellulosic biomass. *Applied Biochemistry and Biotechnology*, **57–58**, 3–18.

43. Stuhler, S.L. and Wyman, C.E. (2003) Estimation of temperature transients for biomass pretreatment in tubular batch reactors and impact on xylan hydrolysis kinetics. *Applied Biochemistry and Biotechnology*, **105–108**, 101–114.

44. Kline, J.E. (1991) *Pulp and Paperboard – Manufacturing and Converting Fundamentals*, 2nd edn, Miller Freeman, San Francisco, CA.

45. Chang, V.S. and Holtzapple, M.T. (2000) Fundamental factors affecting biomass enzymatic reactivity. *Applied Biochemistry and Biotechnology*, **84–86**, 5–37.

46. Zhu, J.Y., Wang, G.S., Pan, X.J., and Gleisner, R. (2009) Specific surface to evaluate the efficiencies of milling and pretreatment of wood for enzymatic saccharification. *Chemical Engineering Science*, **64**, 474–485.

47. Elshafei, A.M., Vega, J.L., Klasson, K.T. *et al.* (1991) The saccharification of corn stover by cellulase from *Penicillium funiculosum*. *Bioresource Technology*, **35**, 73–80.

48. Pordesimo, L.O., Ray, S.J., Buschermohle, M.J. *et al.* (1993) Processing cotton gin trash to enhance in vitro dry matter digestibility in reduced time. *Bioresource Technology*, **96**, 47–53.

49. Kim, K., Tucker, M.P., and Nguyen, Q. (2002) Effects of pressing lignocellulosic biomass on sugar yield in two-stage dilute-acid hydrolysis process. *Biotechnology Progress*, **18**, 489–494.

50. Selig, M.J., Viamajala, S., Decker, S.R. *et al.* (2007) Deposition of lignin droplets produced during dilute acid pretreatment of maize stems retards enzymatic hydrolysis of cellulose. *Biotechnology Progress*, **23**, 1333–1339.

51. Pienkos, P.T. and Zhang, M. (2009) Role of pretreatment and conditioning processes on toxicity of lignocellulosic biomass hydrolysates. *Cellulose*, **16**, 743–762.

52. Overend, R.P. and Chornet, E. (1987) Fractionation of lignocellulosics by steam-aqueous pretreatments. *Philosophical Transactions of the Royal Society of London A*, **A321**, 523–536.

53. Overend, R.P. and Chornet, E. (1989) Steam and aqueous pretreatments: are they prehydrolysis? Proceedings of the 75th Annual Meeting, Technical Section. Canadian Pulp and Paper Association, pp. 327–330.

54. Chum, H.L., Johnson, D.K., and Black, S.K. (1990) Organosolv pretreatment for enzymic hydrolysis of poplars. 2. Catalyst effects and the combined severity parameter. *Industrial and Engineering Chemistry Research*, **29**, 156–162.

55. Nguyen, Q.A., Tucker, M.P., Boynton, B.L. *et al.* (1998) Dilute acid pretreatment of softwoods. *Applied Biochemistry and Biotechnology*, **70–72**, 77–87.

56. Levelton, B.H. *et al.* (1982) Status of Biomass Feeder Technology – Final Report, ENFOR Project C-259. Vancouver, BC, Canada.

57. Rautalin, A. and Wilen, C. (1992) Feeding Biomass into Pressure and Related Safety Engineering. VTT Research Notes 1428. Technical Research Centre of Finland, Espoo, Finland.

58. Metso Paper, Inc. (2011) Plug screw feeder – Data sheet. http://www.metso.com/MP/marketing/Vault2MP.nsf/BYWID/WID-031120-2256C-3F9E3/$File/MPDU_R_2075_091-02.pdf?OpenElement.

59. Heitz, E., Capek-Menard, P., Koeberle, J. *et al.* (1991) Fractionation of *Populus tremuloides* at the pilot plant scale: Optimization of steam pretreatment conditions using the Stake II technology. *Bioresource Technology*, **35**, 23–32.

60. Ropars, M., Marchal, R., Pourquie, J., and Vandecasteele, J.P. (1992) Large scale enzymatic hydrolysis of agricultural biomass, Part 1: Pretreatment procedures. *Bioresource Technology*, **42**, 197–204.

61. Navitel, F., Poruquie, J., Ballerini, D. *et al.* (1992) The biotechnology facilities at Soustons for biomass conversion. *International Journal of Solar Energy*, **11**, 219–229.

62. Christensen, B.H. (27 May, 2010) Methods and devices for continuous transfer of particulate and/or fibrous material between two zones with different temperatures and pressures. World Intellectual Property Organization, International Patent Application WO 2010/058285 A2.

63. Sievers, D.A., Elander, R.T., Kuhn, E.M. *et al.* (2009) Investigating Residence Time Distribution (RTD) and Effects on Performance in Continuous Biomass Pretreatment Reactor Designs. NREL Report No. PO-510-45809. National Renewable Energy Laboratory, Golden, CO, USA. http://www.nrel.gov/docs/fy09osti/45809.pdf.

64. Altomate, R.E. and Ghossi, P. (1986) An analysis of residence time distribution patterns in a twin screw cooking extruder. *Biotechnology Progress*, **2**, 157–163.

65. Yeh, A.I. and Jaw, Y.M. (1998) Modeling residence time distributions for single screw extrusion process. *Journal of Food Engineering*, **35**, 211–232.

66. Lee, S.M., Park, J.C., Lee, S.M. *et al.* (2005) In-line measurement of residence time distribution in a twin-screw extruder using non-destructive ultrasound. *Korea-Australia Rheology Journal*, **17**, 87–95.

67. Katzen, R. and Monceaux, D.A. (1995) Development of bioconversion of cellulosic wastes. *Applied Biochemistry and Biotechnology*, **51–52**, 585–592.

68. Katzen, R. and Schell, D.J. (2006) Lignocellulosic feedstock biorefinery: History and plant development for biomass hydrolysis, in *Biorefineries-Industrial Processes and Products: Status Quo and Future Directions* (eds B. Kamm, P.R. Gruber, and M. Kamm), Wiley-VCH Verlag GmbH, Weinheim, Germany, pp. 129–138.

69. Emert, G.H. and Katzen, R. (1980) Gulf's cellulose to ethanol process. *Chemtech*, **10**, 610–614.

70. Schell, D.J. and Duff, B. (1996) Review of pilot plant programs for bioethanol conversion, Chapter 17, in *Handbook on Bioethanol: Production and Utilization* (ed. C.E. Wyman), Taylor & Francis, Washington, DC, pp. 381–394.

71. Zimbardi, F., Viggiano, D., Nanna, F. *et al.* (1999) Steam explosion of straw in batch and continuous systems. *Applied Biochemistry and Biotechnology*, **77–79**, 117–125.

72. Thomsen, M.H., Thygesen, A., Jorgensen, H. *et al.* (2006) Preliminary results on optimization of pilot scale pretreatment of wheat straw used in coproduction of bioethanol and electricity. *Applied Biochemistry and Biotechnology*, **129–132**, 448–460.

21

Experimental Enzymatic Hydrolysis Systems

Todd Lloyd and Chaogang Liu

Mascoma Corporation, USA

21.1 Introduction

Carbohydrate hydrolyzing enzymes have been known for some time. As early as 1833, Payen and Persoz discovered that malt extract could turn starch to sugar, identifying amylase as the active agent [1]. Beginning in the 1950s at the US Army Natick Laboratory, Reese worked with fungal cellulases to break down cellulose [2,3]. Enzymatic hydrolysis of lignocellulose carbohydrate polymers occurs primarily at the β-1,4 bond that joins sugar molecules, with a review by Walker and Wilson describing mechanisms [4]. Cellulose fibers contain both amorphous and crystalline regions, and crystalline regions are considered to be the more difficult to break down [5–7]. Because cellulose hydrolyzing enzymes are expensive, their economic commercialization requires efficient processes that rapidly convert polymeric sugars to monomers in high yields using minimal enzyme doses, with pretreatment playing a key role in governing the effectiveness of enzymes in releasing sugars from the wide range of cellulosic materials that are potentially attractive for large-scale production of fuels and chemicals.

The focus of this chapter is on experimental systems for application of enzymes to hydrolysis of pretreated biomass to produce sugars from cellulose and hemicellulose. The chapter begins with an overview of sources and key features of cellulose and hemicellulose hydrolyzing enzymes to provide a context from which to view the experimental systems. A summary of empirical and mechanistic kinetic models that can be applied to describe enzyme action on cellulose and hemicellulose follows along with considerations to account for their inhibition, deactivation, and interactions with substrate features. The chapter then describes experimental systems employed to enzymatically hydrolyze pretreated biomass along with summaries of the procedures and references to more detailed protocols. While the information provided here cannot be totally comprehensive, it should give the reader a sense of how cellulolytic enzymes act, quantitative models to describe their action, and key procedures for evaluating the susceptibility of pretreated

Aqueous Pretreatment of Plant Biomass for Biological and Chemical Conversion to Fuels and Chemicals, First Edition.
Edited by Charles E. Wyman.
© 2013 John Wiley & Sons, Ltd. Published 2013 by John Wiley & Sons, Ltd.

biomass to enzymatic hydrolysis, as well as sources for more detailed information than can be covered in a single chapter.

21.2 Cellulases

Bacteria and fungi both produce cellulolytic enzymes, and enzymes that act on macromolecular insoluble substrates, such as cellulose, must be extracellular. Cellulolytic bacteria include aerobic species such as *pseudomonads* and *actinomycetes*, facultative anaerobes including *Bacillus* and *Cellulomonas*, and strict anaerobes such as *Clostridium*, the latter being a potent cellulolytic microorganism that produces large extracellular multienzyme complexes called cellulosomes. In fact, *Clostridium thermocellum* is the best-known member of anaerobic, thermophilic, cellulolytic, ethanol-producing bacteria, and its cellulose system can saccharify both crystalline and amorphous cellulose [8]. Cellulolytic enzyme production among fungi is widespread, including species of *Trichoderma*, *Penicillium*, *Aspergillus*, and *Phanerochaete chrysosporium* (formerly *Sporotrichum pulverulentum*). *Trichoderma* species are the most extensively studied cellulolytic enzyme-producing fungi as a result of successful development of highly productive mutant strains *T. reesei* QM 9414 and *T. reesei* Rut C30 from *Trichoderma reesei* QM 6a (formerly designated *Trichoderma viride* QM 6a) that the US Army Natick Research and Development Command isolated. However, the current emphasis on *Trichoderma* species should not preclude the potential for other cellulolytic enzyme-producing fungi such as various *Penicillium* species to become important sources of cellulolytic enzymes [9].

β-1,4-glycosidic bonds (hence the name, β-1,4-endoglucanase) link together the β-D-glucopyranose units of cellulose. Cellulase enzyme activities have been divided into three main classes (not including cellulosomes) that hydrolyze these bonds to release sugars: (1) endoglucanase, (2) cellobiohydrolase, and (3) β-glucosidase. β-1-4-endoglucanase enzymes specifically cleave the internal bonds of the cellulose chain, and exoglucanases and β-glucosidases are needed to complete the breakdown of cellulose into glucose monomers [10]. These three classes act synergistically, with a hypothesis for their action proposed by Enari and Niku-Paavola [11]. Furthermore, each activity consists of several enzymes. For example, endo-β-glucanase I and II both act on the interior of glucan to create chain-ends and cleave cellodextrins into cellobiose units that comprise two glucose molecules. In turn, cellobiohydrolase I and II attack the exposed cellulose chain-ends to produce soluble cellodextrins and cellobiose. Many endo- and cellobio-hydrolases are motor proteins. Deriving energy from hydrolysis of the glycosidic bonds, motor proteins are able to "walk" along the substrate in a processive manner as hydrolysis proceeds [12]. Finally, β-glucosidase splits cellobiose into the two glucose molecules from which it is made. Unless purified, it is difficult to determine the activity of the individual cellulolytic enzymes due to the synergistic action between them [13].

21.2.1 Endoglucanase

Researchers have shown that endoglucanases cannot break down polysaccharides efficiently without the help of non-catalytic carbohydrate-binding domains (CBD). Endoglucanase is therefore a complex made up of three separate domains: the main catalytic domain, a linker peptide, and a cellulose-binding domain. The main catalytic domain contains a large, globular active site attached to a protein chain. This catalytic domain is attached to a globular CBD by a linker peptide made up of proline, serine, and threonine [14]. The overall shape of the complex looks similar to a tadpole, with the CBD and linker blocks forming the extended tail and the catalytic domain forming the head [15]. The CBD facilitates enzyme action by binding the complex to cellulose, thus maintaining the proximity of the enzyme and the substrate. It can also target areas of the cellulose that are specific to the enzyme complex. In addition, the CBD itself can disrupt the cellulose structure and thus expose the substrate to the active site by "pulling up" the chain and feeding it into the catalytic domain [16]. At this point, the enzyme may release from the substrate and reattach somewhere else, or process down the exposed chain-end in the same way as exoglucanase (described in the following section) [14].

21.2.2 Cellobiohydrolase

Cellobiohydrolase (or exoglucanase or exocellulase) cleaves two to four glucose units from the ends of the exposed cellulose chains produced by endocellulase. There are two major categories of cellobiohydrolase: CBH I that works processively from the reducing end of the cellulose chain and CBH II that works processively from the non-reducing end of the cellulose chain [13]. After the binding domain is secured to the cellulose chain, the chain-end is directed to the tunnel of the catalytic domain where hydrolysis occurs and a cellobiose molecule is liberated. A study using atomic force microscopy (AFM) suggests that CBH attachment is limited to the hydrophobic face of the cellulose chain, limiting opportunities for reaction and hence demonstrating a slower reaction rate [17]. Cellobiohydrolases are produced by both fungi and bacteria. The cellobiohydrolases Cel6A (a CBHII) and Cel7A (a CBHI) are much-studied enzymes of fungal origin and are usually expressed by the ascomycete *Trichoderma reesei* (*Hypocrea jecorina*), which is known to secrete a range of enzymes important for converting biomass to sugars. *Trichoderma* is also widely used for the commercial-scale production of enzymes because of its ability to produce high titers [18]. *CelS*, an extracellular exocellulase and a major enzymatic component of the *Clostridium thermocellum* cellulosome, is an example of a bacterial enzyme with cellobiohydrolase activity [8].

21.2.3 β-glucosidase

β-glucosidases are able to cleave the β-glucosidic linkages in di- and oligo glucosaccharides and several other glycoconjugates. These enzymes are widely distributed and have important roles in many biological processes. In cellulolytic microorganisms, β-glucosidase is involved in cellulose induction and hydrolysis [8]. In plants, the enzyme plays a role in β-glucan synthesis during cell-wall development, pigment metabolism, fruit ripening, and defense mechanisms [19].

β-glucosidase is a glucosidase enzyme that acts upon β1–4 bonds linking two glucose or glucose-substituted molecules (i.e., the disaccharide cellobiose). An exocellulase with specificity for a variety of β-D-glycoside substrates, it catalyzes hydrolysis of terminal non-reducing residues in β-D-glucosides with release of glucose. Cellulose is largely composed of polymers of β-bond linked glucose molecules, and β-glucosidases are required by organisms (some fungi, bacteria, termites) to consume it. Lysozyme, an enzyme secreted in tears to prevent bacterial infection of the eye, is also a β-glucosidase that cleaves β1–4 bonds between N-acetylglucosamine and N-acetylmuramic acid sugars within the peptidoglycan cell walls of gram-negative bacteria.

21.3 Hemicellulases

Hemicelluloses are a group of heterogeneous carbohydrate polymers composed of pentoses (xylose, arabinose), hexoses (mannose, galactose, and glucose), and organic acids (glucuronic, acetic, ferulic, and p-coumaric). The structure of hemicellulose varies among plant species [20]. Hemicellulose is found with lignin, wrapped around glucan fibrils in the secondary cell wall to provide additional structural support. The hemicellulose polymer is highly substituted and cross-linked and, because of this, a large number of enzymes are required for complete hydrolysis [21,22].

Hemicellulases, the enzymes which hydrolyze hemicellulose, are a diverse group that works synergistically to hydrolyze all types of hemicelluloses in nature. Hemicellulases and their accessory enzymes have

Table 21.1   *Classification of typical hemicellulases and accessory enzymes (EC#: Enzyme Commission number).*

Classes	Enzymes	Substrates	EC#
Endo-acting enzymes	Endo-ß-1,4-xylanase	ß-1,4-xylan	3.2.18
	Endo-ß-1,4-mannanase	ß-1,4-mannan	3.2.1.78
	Endo-galactanase	ß-1,4-galactan	3.2.1.89
	Endo-α-1,5-arabinase	α-1,5-arabinan	3.2.1.99
Exo-acting enzymes	Exo-ß-1,4-xylosidase	ß-1,4-xylooligomers	3.2.1.55
	Exo-ß-1,4-mannosidase	ß-1,4-mannooligomers	3.2.1.25
Accessory enzymes	α-L-arabinosidase	α-arabinofuranosyl (1,2) (1,3) xylooligomers α-1,5-arabinan	3.2.1.55
	α-glucuronidase	4-O-methyl-α-glucuronic acid (1,2) xylooligomers	3.2.1.139
	α-galactosidase	α-galactopyranose (1,6) mannooligomers	3.2.1.22
	Acetyl xylan esterase	2- or 3-O-acetyl xylan	3.1.1.72
	Acetyl mannan esterase	2- or 3-O-acetyl mannan	3.1.1.6
	Ferulic and p-cumaric acid esterase	Ferulic and p-cumaric acid-substituted xylooligomers	3.1.1.73

broad industrial applications, including use in the pulp and paper industry where they enhance bleaching and deinking and modify fiber properties. Hemicellulases have also been used to enhance fruit juice production, improve the digestibility of animal feed, and increase recovery in starch mills [23,24].

As summarized in Table 21.1, hemicellulases are generally classified into three categories: (1) endo-acting enzymes that attack polymer chains internally and which have very little activity on short-chain oligomers (DP < 3); (2) exo-acting enzymes that act processively from either the reducing or non-reducing terminal; and (3) accessory enzymes or side-group-cleaving enzymes that help to break down hemicellulose branch-chains [22,23].

Aerobic fungi *Trichoderma* and *Aspergillis* and aerobic bacteria *Bacilli* and *Cellvibrio* are often used commercially to produce hemicellulases [25]. Such fungi can produce a large variety of extracellular hemicellulases that work together to convert complex hemicellulose polymers into both sugar monomers and dimers, while extracellular enzymes of aerobic bacteria only convert hemicellulose to oligomers, completing hydrolysis using intracellular enzymes, [26]. Anaerobic bacteria, such as *Clostridium*, have evolved unique multiple enzyme systems called cellulosomes that combine both cellulolytic and hemicellulolytic activities [27] while some other bacteria, such as *Thermocellum saccharolyticum*, only produce enzymes that degrade hemicelluloses and starch [28–30]. Recently, yeast has been engineered for cellulolytic and hemicellulolytic activity, and a review by van Zyl *et al.* documents progress [31]. Hemicellulase enzymes are expensive to produce, and current research is focused on improving specific activity and efficacy by better understanding enzyme inhibition and enzyme-substrate interactions, and by molecular engineering and development of new fermentation processes [32,33].

21.4   Kinetics of Enzymatic Hydrolysis

Kinetic models are useful to understand and apply data gathered in experimental enzymatic hydrolysis. However, the mechanisms of enzymatic hydrolysis are complicated, and a wide variety of kinetic expressions have been developed to predict performance. These kinetic expressions range from simple to complex, and their value as predictive tools varies from one experimental system to another. Depending on the approach and methodology used, kinetic models can broadly be divided into three classes: empirical, Michaelis–Menten-based, and those that account for adsorption. Literature reviews by Bansal *et al.* [34] and Zhang and Lynd [35] capture the diversity of these models and should be consulted for additional details beyond those outlined here.

21.4.1 Empirical Models

Empirical models are used to quantify the effects of various substrate and enzyme properties in a particular experimental system. Although empirical models are not applicable beyond conditions under which they were developed and do not provide strong insight into mechanistic details of the process, they are helpful in understanding the interactions between enzymatic hydrolysis and substrate properties such as crystallinity, lignin content, and acetyl content [36–38]. Empirical models can also be useful for initial rate estimations, which are important for Lineweaver–Burk plots [39] used in the Michaelis–Menten models. Because the hydrolysis rate decreases continuously over time, an empirical formulation is needed to extrapolate the rate to time zero. This can be illustrated by the empirical expression developed by Ohmine *et al.* [40], where the following equation was found to hold for hydrolysis of Avicel (partially acid-hydrolyzed microcrystalline cellulose) and tissue paper by the cellulase system from *Trichoderma reesei*:

$$P = \left(\frac{S_0}{k}\right)\ln(1 + v_0kt/S_0) \tag{21.1}$$

where P is the product concentration, S_0 is the initial substrate concentration, v_0 is the initial rate, k is the rate constant, and t is time. For enzymatic hydrolysis of cellulose, initial rates are plotted on the y axis versus the reciprocal of the substrate concentration in a Lineweaver–Burk plot to avoid the effects of product inhibition at product concentrations equal to zero.

21.4.2 Michaelis–Menten-based Models

In 1913, Michaelis and Menten proposed a mathematical model to describe enzyme catalyzed reactions [41]. It involves an enzyme E reversibly binding to a substrate S to form an enzyme–substrate complex ES, which in turn is converted by an irreversible reaction into a product P, liberating the enzyme. This may be represented schematically as:

$$E + S \xrightleftharpoons[]{k_{f/r}} ES \xrightarrow{k_{cat}} E + P \tag{21.2}$$

where $k_{f/r}$ are forward (f) and reverse (r) rate constants for the reversible reaction, respectively, and k_{cat} is an irreversible reaction rate constant. Four mass action equations may be written for this expression, and the rate of product P formation can be expressed as

$$\frac{d[P]}{dt} = k_{cat}[ES] \tag{21.3}$$

Making use of the quasi-steady-state assumption of Briggs and Haldane [42] and rearranging the mass action equations, Equation (21.3) becomes

$$v = \frac{d[P]}{dt} = \frac{V_{max}[S]}{K_m + [S]} \tag{21.4}$$

where v is the observed reaction rate, $V_{max} = k_{cat}[E]_0$ and $K_m = (k_r + k_{cat})/k_f$. Because the Michaelis–Menten model is based on homogenous reaction conditions, it cannot be appropriately applied to the heterogeneous reaction conditions of enzymatic hydrolysis of insoluble cellulosic substrates because the excess substrate to enzyme ratio ($[S] \gg [E]$) of the quasi-steady-state assumption is not achieved [43]. Furthermore, the excess substrate condition, even if achieved initially, could not be retained at higher conversions as the substrate is depleted. Monte Carlo simulations have also suggested that the quasi-steady-state assumption is not applicable to heterogeneous systems [44]. However, conversion of cellobiose to glucose by β-glucosidase can be modeled by Michaelis–Menten kinetics since it is a homogeneous reaction.

Although Michaelis–Menten models should not be particularly applicable to modeling cellulose hydrolysis, they nonetheless can provide insight to the mechanisms involved. As an example, Bezerra and Dias tested eight different Michaelis–Menten models against data of Avicel hydrolysis by *T. reesei* Cel7A for 24 different substrate-to-enzyme ratios [45]. Their model that included competitive inhibition by cellobiose was found to fit the data best, and reasons for the observed decreasing hydrolysis rates, such as non-productive cellulase binding, parabolic inhibition (two inhibitor molecules), and enzyme deactivation, were suggested to be insignificant when compared to substrate depletion and competitive inhibition.

21.4.3 Adsorption in Cellulose Hydrolysis Models

To account for the heterogeneity of cellulose hydrolysis, incorporation of adsorbed cellulase concentration into hydrolysis models is usually accounted for by the Langmuir adsorption isotherm [46]. Originally applied to the adsorption of gases on solids, the Langmuir isotherm and its modified forms are widely used to describe heterogeneous hydrolysis reactions. In the Langmuir adsorption isotherm model by Kadam *et al.* [47], the adsorption reaction is given by:

$$E + S_c \xrightleftharpoons{k_1/k_2} ES_c \tag{21.5}$$

and, assuming equilibrium, the bound enzyme concentration can be determined as:

$$E_b = \frac{E_{max} K_{ad} E_f S_c}{1 + K_{ad} E_f} \tag{21.6}$$

where E_b is the bound enzyme concentration, E_f is the free enzyme concentration, K_{ad} is the dissociation constant for adsorption, S is the substrate concentration, and E_{max} is the maximum adsorption capacity in terms of the amount of cellulase per amount of cellulose. By inspection, the similarities between the Michaelis–Menten and Langmuir expressions are evident. However, the enzyme concentration term E_b is a function of surface binding sites whose disappearance during the course of the reaction must be considered.

An example of models based on kinetic equations for the amount of enzyme adsorbed is that by Gan *et al.* built from the following equations [48]:

$$E + S_c \xrightleftharpoons{k_1/k_2} ES_c \xrightarrow{k_p} E + P \tag{21.7}$$

or, in terms of a differential mass balance,

$$\frac{d[ES_c]}{dt} = k_1[E][S_c] - k_2[ES_c] - k_p[ES_c] \tag{21.8}$$

where E is the enzyme concentration, S_c is the active cellulose, ES_c is the enzyme–cellulose complex, k_1 is the adsorption constant on active cellulose, k_2 is the desorption constant on active cellulose, and k_p is the product formation constant.

Some adsorption/kinetic models assume instantaneous substrate–enzyme complex formation (fully productive adsorption), so the amount of adsorbed cellulase is the same as the amount of substrate–enzyme complexes [49–52]. Although the estimated 5–60 min needed to reach adsorption equilibrium is clearly not instantaneous [53–55], the assumption seems reasonable relative to the time required for complete hydrolysis of cellulose (100% conversion), usually 25–100 hours. Other models have accounted for non-instantaneous adsorption by assuming an additional kinetic step on the substrate surface [56–58].

21.4.4 Rate Limitations and Decreasing Rates with Increasing Conversion

Several experimental studies show that the rate of hydrolysis drops by two to three orders of magnitude at high degrees of conversion [58,59], and many reasons for this behavior have been suggested. Some explanations include product inhibition, enzyme deactivation, biphasic composition of cellulose, decrease of substrate reactivity, decrease in substrate accessibility, enzyme jamming, and non-productive binding [60].

Enzyme Inhibition

Inhibition of enzymes by end-product accumulation is a well-known phenomenon, and cellobiose is a particularly strong inhibitor of cellulose hydrolysis by cellulases. Cellobiose inhibition is reduced or eliminated by addition of cellobiase (β-glucosidase), which hydrolyzes the disaccharide to monomers. Other end-products, such as ethanol, glucose, and other sugars, also inhibit enzymatic hydrolysis but to a lesser extent. However, xylooligomers and xylose have recently been shown to be very strong inhibitors of cellulase. End-product inhibition is one reason that most comparative hydrolysis studies are performed with low solids loadings, as low loadings mean low product concentrations.

There are four kinds of reversible enzyme inhibitors. They are classified according to the effect of varying the concentration of the inhibitor and substrate on the reaction [61].

Competitive Inhibition
In competitive inhibition, the substrate and inhibitor cannot bind to the enzyme at the same time and *compete* for access to the enzyme's active site. This type of inhibition can be overcome by sufficiently high concentrations of substrate (V_{max} remains constant), that is, by out-competing the inhibitor. However, the apparent K_m (Equation (21.4)) will increase as it takes a higher concentration of the substrate to reach the K_m point which is equal to half of V_{max}. Competitive inhibitors are often similar in structure to the targeted substrate.

Uncompetitive Inhibition
In uncompetitive inhibition, which should not be confused with non-competitive inhibition, the inhibitor binds only to the enzyme-substrate to be consistent with ES complex. This type of inhibition causes both V_{max} and K_m to decrease as a result of effective elimination of the ES complex, and indicates a higher enzyme-substrate binding affinity.

Mixed Inhibition
In mixed inhibition, the inhibitor can bind to the enzyme at the same time as the enzyme's substrate. However, the binding of the inhibitor affects binding of the substrate and vice versa. This type of inhibition can be reduced but not overcome by increasing concentrations of substrate. Although it is possible for mixed-type inhibitors to bind to the active site, this type of inhibition generally results from

an allosteric effect where the inhibitor binds to a different site on an enzyme. Inhibitor binding to this allosteric site changes the conformation (i.e., tertiary structure or 3D shape) of the enzyme so that the affinity of the substrate for the active site is reduced.

Non-Competitive Inhibition   Non-competitive inhibition is a form of mixed inhibition where the binding of the inhibitor to the enzyme reduces its activity but does not affect the binding of substrate. As a result, the extent of inhibition depends only on the concentration of the inhibitor. V_{max} will decrease due to the inability for the reaction to proceed as efficiently, but K_m will remain the same as the actual binding of the substrate, by definition, will still function properly.

Enzyme Deactivation

As for inhibition, enzyme deactivation can be either reversible or irreversible and leads to decreased hydrolysis. Deactivation is often modeled as a first-order process with respect to the total enzyme concentration, and an example by Converse *et al.* [50] assumes deactivation can be represented by the following reversible reaction for adsorbed enzyme.

$$E_a \xleftrightarrow{k_1/k_2} E_d \tag{21.9}$$

where E_a is the actively adsorbed enzyme, E_d is the inactively adsorbed enzyme, k_1 is the inactivation rate constant, and k_2 is the reactivation rate constant. Others have suggested that irreversible deactivation occurs through mixing shear [48] and some kind of physical hindrance at the surface of the cellulose or by thermal degradation [56,59,62].

Two-Phase Substrate

It has been considered that native cellulose comprises two portions, one more reactive than the other, and models based on dividing cellulose into crystalline and amorphous fractions have been proposed to follow such a pattern. Some have suggested that the amorphous part of cellulose reacts first, increasing crystallinity of the remainder [40,56,63], but constant [60,64] and even decreasing [63] crystallinity have been measured during conversion. Although the reasons for this contradiction are unclear, this observation suggests that the two-phase hypothesis was a simplification of the true physical complexity of cellulose. Although biphasic kinetics may be responsible for some of the observed differences in hydrolysis rates, it therefore seems unlikely to be the only cause for the rate slowdown.

Substrate Reactivity

A change in substrate reactivity with conversion has been included in a number of models to explain the drop in digestibility of both lignocellulosic and pure cellulosic substrates over hydrolysis time. As an example, South *et al.* expressed the reaction rate constant in terms of conversion [52]:

$$k(x) = k(1-x)^n + c \tag{21.10}$$

where k is the reaction rate constant for hydrolysis, x is conversion, $k(x)$ is the reaction rate constant at conversion x, n is the exponent of declining rate constant, and c is a constant. This expression was later used in modeling simultaneous saccharification and fermentation (SSF) with staged reactors and intermediate feeding of enzyme and substrate [61,65]. Although the inclusion of substrate reactivity as a function of conversion may fit the data well, a physical interpretation of the constants in these equations is not possible and the continuous decline in reactivity can be viewed as a continuous multiphasic substrate, similar to the biphasic assumption.

Substrate Accessibility

Due to the insoluble nature of cellulose, large domains are not exposed to cellulases in the reaction mixture during hydrolysis, and cellulases can adsorb only to the accessible portion of the substrate. Consequently, cellulose accessibility can be characterized on the basis of the maximum adsorption capacity of the substrate [43]:

$$F_a = 2\alpha A_{max} M W_a \qquad (21.11)$$

where F_a is the fraction of the β-glycosidic bonds accessible to cellulase, α is the number of cellobiose lattice occupied by the cellulase, A_{max} is the maximum adsorption concentration of cellulase, and MW_a is the molecular weight of anhydro-glucose.

Enzymatic hydrolysis of lignocellulosic biomass is a heterogeneous reaction since it occurs on the substrate surface (large enough to accommodate a large number of enzyme molecules). After adsorption, cellulases have to move on the surface of the substrate to reach the reactive sites (a chain-end in the case of cellobiohydrolases). The inaccessible and non-reactive portions of the substrate can be considered as obstacles, suggesting a fractal character of the hydrolysis reaction, and several recent studies modeled the hydrolysis reaction using fractal kinetics [56,57]. In these models, a fractal parameter is included to account for the dimensional constraints imposed by reactions occurring on surfaces and in channels.

Another barrier to enzymes is the presence of lignin. Not only does lignin limit enzyme access to cellulose but it can also bind enzymes, effectively inhibiting them, both of which result in lower hydrolysis rates [64]. Changes in crystallinity can also be affected by lignin [35], and hence the observation of crystallinity variations along conversion must be interpreted carefully. Furthermore, the extent to which crystallinity limits enzymatic conversion of biomass into sugars can depend on the lignin level and vice versa [66]. Since lignin is not degraded by cellulases, it can act as a barrier preventing access of enzymes to the substrate. In terms of fractal kinetics, lignin and hemicellulose act as obstacles and hence increase the fractal nature of the reaction system. A better understanding of the role of lignin in enzymatic digestion of lignocellulose and its interaction with enzymes is needed, not just to improve pretreatment technologies but also to engineer enzymes that have less affinity for lignin [59].

21.4.5   Summary of Enzyme Reaction Kinetics

Cellulase hydrolysis of cellulose occurs in a heterogeneous medium. Classical homogenous enzyme catalysis is modeled by Michaelis–Menten kinetics, and heterogeneous catalysis on a catalyst support by Langmuir kinetics. Cellulase reactions with insoluble lignocellulosic substrates are a combination of the above two types of reactions and also involve other factors (product inhibition, enzyme deactivation, substrate crystallinity, substrate accessibility changes, substrate reactivity changes, fractal nature of the reaction, changes in enzyme synergism, and lignin inhibition) which can retard rates at higher degrees of conversion. While models in the literature have not pinpointed the exact mechanism of enzyme action on lignocellulosic materials, they have helped to understand the various factors that are at play.

21.5 Experimental Hydrolysis Systems

High enzyme loadings are currently needed to achieve high yields in cellulose hydrolysis due to the resistant character of the substrate, with the results that the cost of enzymes is a major contributor to the overall process cost. The amount of enzyme required to achieve 90% hydrolysis of a 100 g/mL dry weight lignocellulosic biomass slurry within 70–110 hours has been estimated to be at least 10 filter paper units (FPU) per gram dry weight of solids. This loading corresponds to approximately 1 FPU/mL for a 10% solids slurry, and a typical enzyme solution loading in hydrolysis is 5–10% w/w. Thus, a minimum concentration of 10–20 FPU/mL in the cellulose production broth will be required. Techno-economic studies have shown that a productivity of 200 FPU/L/hr is required for enzyme production [10] and, with the growth characteristics of current *Trichoderma* strains, this can only be achieved with an enzyme concentration higher than 20 FPU/mL. A high yield from enzyme production is also necessary for favorable process economics.

Given the importance of enzyme loadings and costs to overall process economics, accurate characterization of enzymes and the susceptibility of pretreated biomass to enzyme action are vital to define pretreatment and associated hydrolysis conditions that result in the lowest possible costs. In this section, methods typically applied to characterize enzymes and pretreated substrates are outlined along with references to more detailed information.

21.5.1 Laboratory Protocols

Quantification of enzymes used in a conversion process is required and Zhang has summarized a number of useful assays [67]. Also, the National Renewable Energy Laboratory (NREL) has issued protocols for determining enzymatic digestibility of lignocellulosic substrates, and the descriptions in the following sections are based on them. Although solids loadings are low and not commercially relevant, these tests provide information on substrate susceptibility to enzymes that is not confused with loss of enzyme activity. Of course, other protocols may be developed and used for individual evaluations, and the methodologies referred to here can serve as a template.

Measurement of Enzyme Activity

To quantify the enzymatic digestibility of lignocellulosic substrates as a function of enzyme dose, standardized enzyme assays are required. However, it is difficult to reliably quantify specific enzymes in enzyme preparations as many different activities are usually present. Enzyme activity of a preparation is therefore used as a proxy for specific enzyme content, where activity is defined as the rate of substrate conversion (or product formation) at a certain temperature, pH, and substrate concentration. Assay values are usually reported as units of activity per mL of aqueous enzyme preparation.

The NREL standardized a test for cellulase activity [68] based on the approach of Ghose [69] and Mandels *et al.* [70], and the resulting protocol is widely used. The standard unit of cellulase activity is defined as the filter paper unit or FPU, which in turn is based on the international unit (IU) defined as 1 μM substrate converted in 1 min. The characterization of cellulase enzymes poses special problems. Kinetic studies are difficult since the natural substrate is both insoluble and structurally variable, and thus relatively undefined with respect to concentration and chemical form. It is a system where many endo- and exoglucanases act synergistically and in a manner that is not well understood. In addition, the variety of end-products that are frequently formed create inhibition through feedback control. The presence of β-glucosidase or other enzymes which are not, strictly speaking, cellulases, further complicates measurements.

At present, the most commonly used assay for comparing cellulolytic enzyme systems is the filter paper assay which measures the hydrolysis of a defined piece of filter paper [71]. However, this assay does not necessarily show the true hydrolyzing capacity of a cellulolytic enzyme system since the activity also depends on the substrate. The filter paper assay for cellulase activity uses a strip of Whatman #1 filter paper ($1.0 \times 6.0 \, cm = 50 \, mg$). Quantities of 1.0 mL of 0.05 M citrate buffer, pH 4.8, and 0.5 mL of enzyme diluted in citrate buffer are added to a test tube (at least two dilutions must be made and the activity is determined by interpolation), along with the filter paper. The tube contents are incubated at 50 °C for one hour, and glucose liberated is determined. 1 FPU is defined as the volume of enzyme solution required to release 2 mg of glucose in 60 min, and the activity of the enzyme solution has reciprocal volume units of mL^{-1}.

Two measurements are commonly used to define cellobiose (β-glucosidase) activity. Analogous to the FPU, the cellobiose unit (CBU) corresponds to the quantity of enzyme that, during a period of 15 min at 40 °C and pH 7, gives 2 mmol of glucose per minute from 4.43 mM cellobiose. In another cellobiase activity assay, 4 mL of MacIlvaine buffer (pH 3.2), 0.5 mL of 0.01 M p-nitrophenyl-,9-o-glucoside (pNPG) substrate, and 0.5 mL of enzyme solution (0.5% by weight) are incubated at 30 °C for 15 min [72]. A 0.5 mL aliquot of this mixture is withdrawn and transferred to a 10 mL volumetric flask containing 1 mL of 1.0 M sodium carbonate solution and diluted to volume with distilled water. The amount of p-nitrophenol liberated is determined by measuring the optical density of the solution at 400 nm, and the cellobiase activity is calculated as for the CBU but reported as pNPGU.

Other hydrolyzing enzyme assays may be determined analogously, with their activities generally reported in terms of the corresponding IU. Clearly, it is important to understand the assay used to determine activity of an enzyme, and the assay must be used consistently if results are to be meaningful. This is particularly true when comparing results from tests performed by different investigators.

Alternatively, enzymes may be dosed on a mass basis instead of an activity basis, such as milligrams of enzyme protein (measured by BCA or other method) per gram substrate [73], and this is left as a matter of preference.

Enzymatic Saccharification of Lignocellulosic Biomass

Developed by the NREL [74], this procedure describes enzymatic saccharification of cellulose from native or pretreated lignocellulosic biomass to glucose in order to determine digestibility. A saturating level of a commercially available or in-house produced cellulase preparation is used for hydrolysis times of up to 1 week. Test specimens unsuitable for analysis by this procedure include acid- and alkaline-pretreated biomass samples that have not been washed, as they will contain free acid or alkali which may change the test solution pH to outside the range of high enzymatic activity. In addition, glucose in the biomass may influence the final result as it can inhibit hydrolysis.

After determining the moisture content for all cellulose-containing samples to be digested (note that all lignocellulosic materials which have undergone some aqueous pretreatment must never be air-dried prior to testing enzyme digestibility, since irreversible pore collapse can occur in the microstructure of the biomass leading to decreased enzymatic release of glucose from the cellulose), a sample equal in mass to the equivalent of 0.1 g of cellulose or 0.15 g total biomass on a 105 °C dry weight basis is weighed out and added to a 20 mL glass scintillation vial. To each vial, 5.0 mL of 0.1 M pH 4.8 sodium citrate buffer is added, followed by 400 µg of tetracycline and 300 µg of cycloheximide to prevent the growth of organisms during digestion. As an alternative, 100 µL of a 2% sodium azide solution may be added instead of the tetracycline/cycloheximide combination (cycloheximide, tetracycline, and sodium azide are hazardous and must be handled with appropriate care; do *not* combine sodium azide with the tetracycline/cycloheximide combination). All solutions and the biomass are assumed to have a specific gravity of 1.000 g/mL. Thus, if 0.200 g of biomass is to be added to the vial, it is assumed to occupy 0.200 mL. Enough water is added at this point so the total volume in each vial will be 10.00 mL after addition of the enzymes in the following step. The appropriate volume of the cellulase enzyme preparation is added to each vial to give an enzyme loading of approximately 60 FPU/g cellulose. Similarly, an appropriate volume of β-glucosidase enzyme is added to give an enzyme loading of 64 pNPGU/g cellulose. The enzymes are always added last since the reaction is initiated by their addition.

Vials are placed in a rack that is transferred to a 50 °C shaking incubator to react for up to a week (or longer). Samples are taken and analyzed at appropriate times to track the rate of sugar release over the course of the digestion. The proportions given here may be altered as long as the relative amounts are the same to give other total volumes as needed for testing purposes.

Simultaneous Saccharification and Fermentation

As the sugar concentration increases during hydrolysis, it inhibits enzymes, and fermenting the sugars to ethanol as they are released can improve hydrolysis rates and yields. The NREL published a standard protocol describing a SSF procedure that is often used for comparative tests to assess conversion yields and ultimate fermentability [75].

Preparation of biomass samples to be evaluated in this protocol is identical to the procedure for enzymatic hydrolysis described in the previous section. However, in addition to substrate preparation, an inoculum of the fermenting organism must be prepared. In the NREL protocol, yeast (*Saccharomyces cerevisiae* D_5A) is specified, but other organisms could be used after appropriately modifying the inoculum protocol to suit the organism chosen.

The first step in this procedure is to prepare a seed culture of the fermentative organism for SSF, with the following example being for inoculum preparation by aerobic growth of yeast on glucose. In this case, yeast extract, peptone, and 5% w/v glucose are added to a sterile inoculum flask (baffled with a Morten cap) to maintain a 1:5 working volume to total flask volume ratio. A thawed stock vial of yeast is added to the flask containing sugar and nutrients, and placed in a rotary incubator shaker operating at the SSF operating temperature for 10–14 hours. The seed culture is ready for use once the glucose concentration falls below 2 g/L. However, a microscope or other measure such as the presence of lactic acid should be employed to check for contamination and ensure the quality of the inoculum. If the inoculum is free from contamination, the washed cell mass can be used for fermentation.

The SSF procedure is almost identical to the saccharification protocol except that a higher initial cellulose content of 6% w/w is applied. Pretreated substrate is washed to remove soluble sugars and inhibitors, with 12 wash volumes generally needed to reduce the glucose concentration to less than 0.1 g/L. The moisture content in the washed solids is measured and used in conjunction with the measured glucan content to calculate the amount of washed solids to add to each sterilized flask to give a 6% w/w cellulose concentration. Then, 1% w/v yeast extract, 2% w/v peptone, and 0.05 M citrate buffer (pH 4.8) are also added to each flask. All flasks are then autoclaved (sterilized) at 121 °C for 30 min, although the times may be increased to accommodate large loads or large volumes of media. It is vital that the vessels can ventilate freely during autoclaving. After the sterilized flasks have cooled, enough cellulase enzyme and seed culture inoculum (starting optical density 0.5, using a spectrophotometer at 630 nm) are added to reach the enzyme and organism loadings desired, respectively. Shake flasks should have a 2:5 culture volume to flask volume ratio and must be equipped with water traps or other devices to release the carbon dioxide produced during fermentation. As for the saccharification with just enzymes, it is important to not dry pretreated biomass prior to SSF as the pores within the biomass may irreversibly collapse, reducing enzyme access and giving deceptive results for hydrolysis and conversion.

Each experiment should include an appropriate control using a reference pretreated substrate and reference cellulase enzyme loaded at a standard level. If a reference pretreated biomass substrate is not available, α-cellulose or another commercially available form of cellulose may be employed. All SSFs should be performed in at least duplicate. It is important to note that media and commercial enzyme preparations often contain fermentable ingredients such as sucrose that yeast can convert to ethanol. A control must therefore be run in parallel with each enzyme preparation to be tested and all medium components except pretreated substrate to determine how much ethanol is made from just the enzyme and medium. Usually the flask is loaded with the highest amount of enzyme used in the tests. The amount of ethanol produced per milliliter of enzyme in the control is then subtracted from the ethanol produced in the SSF flasks to estimate the amount of ethanol produced from hydrolyzed cellulose.

21.5.2 Considerations for Scale-up of Hydrolysis Processes

Rheological characteristics of biomass slurries in large vessels are generally unknown, and empirical studies are required to anticipate their effects on commercial-scale hydrolysis and fermentation results. Although some useful information can be developed at the bench scale, larger (>1000 L) hydrolysis and fermentation tests provide the ability to measure mixing and pumping characteristics that are difficult to differentiate in smaller systems where vessel and pipe wall effects can be considerable.

High-solids Enzymatic Hydrolysis

High-solids biomass hydrolysis or fermentations are generally required to achieve high product concentrations that minimize capital and operating costs for commercial biomass conversion technology. Compared to low-solids hydrolysis protocols, which are usually used to characterize both substrate and enzymes, high-solids hydrolysis/fermentation systems are quite different [33,76]. First, high-solids slurries are very viscous, hindering heat and mass transfer. Also, high-solids substrate concentration means high inhibitor concentrations which reduce yields. Furthermore, a high-solids operation results in high concentrations of end-products, further inhibiting hydrolysis or fermentation. Small-scale high-solids tests are therefore more relevant to understanding how these factors impact commercialization efforts.

Shake flasks and bottles are usually used to characterize enzymatic hydrolysis and fermentation substrates, enzymes, and organisms in the laboratory [74], and 96-well plates have been used as high-throughput devices for parameter screening (pH, temperature, enzyme dose, etc.) and characterization. Commonly used for low-solids hydrolysis, these systems have not been applied to high-solids testing; yields fall as solids content rises for the reasons described previously.

To mitigate the effect of reduced mixing with increasing solids loadings, Roche *et al.* [77] successfully employed roller bottle reactors in the laboratory to provide adequate mixing for enzymatic hydrolysis of high-solids slurries (Figure 21.1a); roller bottle reactors have been shown to be scalable between 125 mL and 2000 mL [77]. Another high-solids lab-scale device consisted of a 500 mL centrifuge bottle adapted for use in a shaking incubator (Figure 21.1b). These modified centrifuge bottles worked well for characterization and hydrolysis and fermentation screening of feedstocks and pretreated substrates at high-solids loadings ($>15\%$). In another configuration, Jørgensen *et al.* used a 250 L horizontal, five-chamber rotating drum to liquefy and ferment up to about 30% pretreated wheat straw and observed a drop in hydrolysis yields for up to about 30% solids [78].

Enzyme Recycle

Because enzyme cost is one of the primary barriers to a commercial lignocellulosic ethanol industry, enzyme recycle and reuse can have a large impact on overall costs. Several recent studies describe the characteristics and efficacy of recycling enzymes [79,80]. Qi *et al.* looked at both adsorption and filtration methods for enzyme recovery and recycle and concluded that, compared to adsorption recycling methods, ultrafiltration recycling retained β-glucosidase, thus further improving the economics [81]. They also found

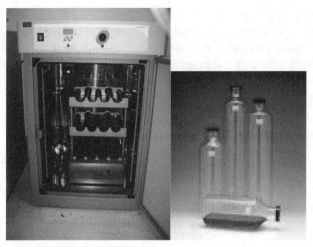

(a) Rotating horizontal reactor system with roller bottles.

(b) Shaking vertical reactor system with modified centrifuge bottles. Photo courtesy of Mascoma Corporation.

Figure 21.1   *Laboratory-scale systems for high-solids enzymatic hydrolysis. (a) Rotating horizontal reactor system with roller bottles. (b) Shaking vertical reactor system with modified centrifuge bottles. (Photo courtesy of Mascoma Corporation). (See figure in color plate section).*

that alkali-treated wheat straw (3.6% lignin) showed better recycling efficiency than acid-treated wheat straw (20.5% lignin), indicating the detrimental effect of lignin on enzyme recycle.

Working with SO_2 pretreated softwood, Lu *et al.* showed that celluloytic activity in the hydrolysis supernatant could be retained for at least three cycles using fresh substrate [82]. Prieto *et al.* worked with a cyclic batch enzyme membrane reactor (CBEMR) operated in three consecutive stages [83]: (1) proteolysis; (2) ultrafiltration; and (3) enzyme recycle and re-use. A proposed mechanistic model comprising zero-order kinetics for hydrolysis and second-order kinetics for enzyme deactivation was experimentally validated, and it was found that the optimum number of enzyme cycles was a function of the productivity required for the CBEMR. The most productive isothermal operation of the CBEMR yielded 4 enzyme cycles at 60 °C, translating into an 18% savings in enzyme costs compared to a traditional discontinuous batch reactor.

Ultrafiltration of hydrolyzates has recovered as much as 50% of the original enzyme activity using a steam-exploded poplar substrate. However, the presence of lignin has been shown to play an important role in limiting the efficiency of enzymatic hydrolysis of lignocellulosic material, and therefore the efficacy of enzyme recycle [84]. It has also been suggested that different pretreatment conditions may affect cellulase adsorption characteristics onto a substrate, reducing the amount of free enzyme available for recycle [85]. However, very little is currently known about these interactions.

High-intensity Mixing

The multiphase nature of biomass hydrolysis necessitates effective mixing for commercially relevant conversion rates and yields. Heat and mass transfer will of course be correlated with mixing intensity but, to the extent that localized high shear can damage enzymes during hydrolysis or organisms during fermentation, mixing optimization is complicated [86].

21.6   Conclusion

Developing processes that efficiently use expensive enzymes (whether produced by the fermenting organism or by a separate organism) is key to an industry that converts lignocellulosic biomass to value-added products using biological systems. This chapter provided a background on key enzyme features and kinetics followed by a summary of current laboratory and small-scale techniques and systems typically used to evaluate the effectiveness of pretreatment for improving enzymatic hydrolysis, as well as developing conversion processes using enzymes and some of their inherent limitations and opportunities.

References

1. Payen, A. and Persoz, J. (1833) Memoire sur la diastase, les principaux produits sses reactions, et leurs applications aux art industriels. *Annales de Chimie et de Physique*, **53**, 73–92.
2. Reese, E.T. (1976) History of the cellulase program at the U.S. Army Natick Development Center. *Biotechnology & Bioengineering Symposium*, (6), 9–20.
3. Reese, E.T. (1956) A microbiological process report; enzymatic hydrolysis of cellulose. *Applied Microbiology*, **4** (1), 39–45.
4. Walker, L.P. and Wilson, D.B. (1991) Enzymatic hydrolysis of cellulose: An overview. *Bioresource Technology*, **36** (1), 3–14.
5. Coughlan, M. (1985) The properties of fungal and bacterial cellulases with comment on their production and application. *Biotechnology & Genetic Engineering Reviews*, **3**, 39–109.
6. Huang, A.A. (1975) Kinetic studies on insoluble cellulose–cellulase system. *Biotechnology and Bioengineering*, **17** (10), 1421–1433.
7. Ladisch, M.R., Lin, K.W., Voloch, M., and Tsao, G.T. (1983) Process considerations in the enzymatic hydrolysis of biomass. *Enzyme and Microbial Technology*, **5** (2), 82–102.
8. Johnson, E.A., Sakajoh, M., Halliwell, G. *et al.* (1982) Saccharification of complex cellulosic substrates by the cellulase system from clostridium thermocellum. *Applied and Environmental Microbiology*, **43** (5), 1125–1132.
9. Persson, I., Tjerneld, F., and Hahn-Hägerdal, B. (1991) Fungal cellulolytic enzyme production: A review. *Process Biochemistry*, **26** (2), 65–74.
10. Kumar, R., Mago, G., Balan, V., and Wyman, C.E. (2009) Physical and chemical characterizations of corn stover and poplar solids resulting from leading pretreatment technologies. *Bioresource Technology*, **100** (17), 3948–3962.
11. Enari, T.-M. and Niku-Paavola, M.-L. (1987) Enzymatic hydrolysis of cellulose: is the current theory of the mechanisms of hydrolysis valid? *Critical Reviews in Biotechnology*, **5** (1), 67–87.

12. Igarashi, K., Koivula, A., Wada, M. *et al.* (2009) High speed atomic force microscopy visualizes processive movement of Trichoderma reesei cellobiohydrolase I on crystalline cellulose. *The Journal of Biological Chemistry*, **284** (52), 36186–36190.

13. Kleman-Leyer, K.M., Siika-Aho, M., Teeri, T.T., and Kirk, T.K. (1996) The cellulases endoglucanase i and cellobiohydrolase ii of trichoderma reesei act synergistically to solubilize native cotton cellulose but not to decrease its molecular size. *Applied and Environmental Microbiology*, **62** (8), 2883–2887.

14. Nimlos, M.R., Matthews, J.F., Crowley, M.F. *et al.* (2007) Molecular modeling suggests induced fit of Family I carbohydrate-binding modules with a broken-chain cellulose surface. *Protein Engineering, Design and Selection*, **20** (4), 179–187.

15. Pilz, I., Schwarz, E., Kilburn, D.G. *et al.* (1990) The tertiary structure of a bacterial cellulase determined by small-angle X-ray-scattering analysis. *The Biochemical Journal*, **271** (1), 277–280.

16. Zhao, X., Rignall, T.R., McCabe, C. *et al.* (2008) Molecular simulation evidence for processive motion of Trichoderma reesei Cel7A during cellulose depolymerization. *Chemical Physics Letters*, **460** (1–3), 284–288.

17. Liu, Y.-S., Baker, J.O., Zeng, Y. *et al.* (2011) Cellobiohydrolase hydrolyzes crystalline cellulose on hydrophobic faces. *The Journal of Biological Chemistry*, **286**(13), 11195–11201.

18. Jeoh, T., Michener, W., Himmel, M.E. *et al.* (2008) Implications of cellobiohydrolase glycosylation for use in biomass conversion. *Biotechnology for Biofuels*, **1** (1), 10.

19. Brzobohatý, B., Moore, I., Kristoffersen, P. *et al.* (1993) Release of active cytokinin by a beta-glucosidase localized to the maize root meristem. *Science*, **262** (5136), 1051–1054.

20. Coughlan, M.P. and Hazlewood, G.P. (1993) *Hemicellulose and Hemicellulases,* Portland Press.

21. Saha, B.C. (2003) Hemicellulose bioconversion. *Journal of Industrial Microbiology & Biotechnology*, **30** (5), 279–291.

22. Shallom, D. and Shoham, Y. (2003) Microbial hemicellulases. *Current Opinion in Microbiology*, **6** (3), 219–228.

23. Brigham, J., Adney, W.S., and Himmel, M.E. (1996) Hemicellulases: Diversity and applications, in *Handbook on Bioethanol: Production and Utilization* (ed. C.E. Wyman), Taylor and Francis.

24. Viikari, L., Tankenan, M., Buchert, J. *et al.* (1993) Hemicellulases for industrial applications, in *Bioconversion of Forest and Agricultural Plant Residues* (ed. J.N. Saddler), CAB International, Wallingford, UK, p. 131–182.

25. Shulami, S., Gat, O., Sonenshein, A.L., and Shoham, Y. (1999) The glucuronic acid utilization gene cluster from Bacillus stearothermophilus T-6. *Journal of Bacteriology*, **181** (12), 3695–3704.

26. Nagy, T., Emami, K., Fontes, C.M.G.A. *et al.* (2002) The membrane-bound alpha-glucuronidase from Pseudomonas cellulosa hydrolyzes 4-O-methyl-D-glucuronoxylooligosaccharides but not 4-O-methyl-D-glucuronoxylan. *Journal of Bacteriology*, **184** (17), 4925–4929.

27. Kosugi, A., Murashima, K., and Doi, R.H. (2002) Xylanase and acetyl xylan esterase activities of XynA, a key subunit of the Clostridium cellulovorans cellulosome for xylan degradation. *Applied and Environmental Microbiology*, **68** (12), 6399–6402.

28. Sizova, M.V., Izquierdo, J.A., Panikov, N.S., and Lynd, L.R. (2011) Cellulose- and xylan-degrading thermophilic anaerobic bacteria from biocompost. *Applied and Environmental Microbiology*, **77** (7), 2282–2291.

29. Shaw, A.J., Hogsett, D.A., and Lynd, L.R. (2010) Natural competence in Thermoanaerobacter and Thermoanaerobacterium species. *Applied and Environmental Microbiology*, **76** (14), 4713–4719.

30. Shaw, A.J., Podkaminer, K.K., Desai, S.G. *et al.* (2008) Metabolic engineering of a thermophilic bacterium to produce ethanol at high yield. *Proceedings of the National Academy of Sciences of the United States of America*, **105** (37), 13769–13774.

31. van Zyl, W.H., Lynd, L.R., den Haan, R., and McBride, J.E. (2007) Consolidated bioprocessing for bioethanol production using Saccharomyces cerevisiae. *Advances in Biochemical Engineering/Biotechnology*, **108**, 205–235.

32. Qing, Q., Yang, B., and Wyman, C.E. (2010) Xylooligomers are strong inhibitors of cellulose hydrolysis by enzymes. *Bioresource Technology*, **101** (24), 9624–9630.

33. Hodge, D.B., Karim, M.N., Schell, D.J., and McMillan, J.D. (2008) Soluble and insoluble solids contributions to high-solids enzymatic hydrolysis of lignocellulose. *Bioresource Technology*, **99** (18), 8940–8948.

34. Bansal, P., Hall, M., Realff, M.J. *et al.* (2009) Modeling cellulase kinetics on lignocellulosic substrates. *Biotechnology Advances*, **27** (6), 833–848.

35. Zhang, Y.-H.P. and Lynd, L.R. (2004) Toward an aggregated understanding of enzymatic hydrolysis of cellulose: noncomplexed cellulase systems. *Biotechnology and Bioengineering*, **88** (7), 797–824.
36. Chang, V.S. and Holtzapple, M.T. (2000) Fundamental factors affecting biomass enzymatic reactivity. *Applied Biochemistry and Biotechnology*, **84–86**, 5–37.
37. Kim, S. and Holtzapple, M.T. (2006) Effect of structural features on enzyme digestibility of corn stover. *Bioresource Technology*, **97** (4), 583–591.
38. O'Dwyer, J.P., Zhu, L., Granda, C.B. *et al.* (2008) Neural network prediction of biomass digestibility based on structural features. *Biotechnology Progress*, **24** (2), 283–292.
39. Lineweaver, H. and Burk, D. (1934) The determination of enzyme dissociation constants. *Journal of the American Chemical Society*, **56** (3), 658–666.
40. Ohmine, K., Ooshima, H., and Harano, Y. (1983) Kinetic study on enzymatic hydrolysis of cellulose by cellulose from Trichoderma viride. *Biotechnology and Bioengineering*, **25** (8), 2041–2053.
41. Michaelis, L. and Menten, M. (1913) Die kinetik der invertinwirkung. *Biochemistry Zeitung*, **49**, 333–369.
42. Briggs, G. and Haldane, J. (1925) A note on the kinetics of enzyme action. *The Biochemical Journal*, **19**, 338–339.
43. Hong, J., Ye, X., and Zhang, Y.-H.P. (2007) Quantitative determination of cellulose accessibility to cellulase based on adsorption of a nonhydrolytic fusion protein containing CBM and GFP with its applications. *Langmuir*, **23** (25), 12535–12540.
44. Berry, H. (2002) Monte Carlo simulations of enzyme reactions in two dimensions: fractal kinetics and spatial segregation. *Biophysical Journal*, **83** (4), 1891–1901.
45. Bezerra, R. and Dias, A. (2004) Discrimination among eight modified Michaelis-Menten kinetics models of cellulose hydrolysis with a large range of substrate/enzyme ratios. *Applied Biochemistry and Biotechnology*, **112** (3), 173–184.
46. Langmuir, I. (1916) The constitution and fundamental properties of solids and liquids. Part I: Solids. *Journal of the American Chemical Society*, **38** (11), 2221–2295.
47. Kadam, K.L., Rydholm, E.C., and McMillan, J.D. (2004) Development and validation of a kinetic model for enzymatic saccharification of lignocellulosic biomass. *Biotechnology Progress*, **20** (3), 698–705.
48. Gan, Q., Allen, S., and Taylor, G. (2003) Kinetic dynamics in heterogeneous enzymatic hydrolysis of cellulose: an overview, an experimental study and mathematical modelling. *Process Biochemistry*, **38** (7), 1003–1018.
49. Brown, R.F. and Holtzapple, M.T. (1990) A comparison of the Michaelis-Menten and HCH-1 models. *Biotechnology and Bioengineering*, **36** (11), 1151–1154.
50. Converse, A.O., Matsuno, R., Tanaka, M., and Taniguchi, M. (1988) A model of enzyme adsorption and hydrolysis of microcrystalline cellulose with slow deactivation of the adsorbed enzyme. *Biotechnology and Bioengineering*, **32** (1), 38–45.
51. Shen, J. and Agblevor, F.A. (2008) Optimization of enzyme loading and hydrolytic time in the hydrolysis of mixtures of cotton gin waste and recycled paper sludge for the maximum profit rate. *Biochemical Engineering Journal*, **41** (3), 241–250.
52. South, C.R., Hogsett, D.A.L., and Lynd, L.R. (1995) Modeling simultaneous saccharification and fermentation of lignocellulose to ethanol in batch and continuous reactors. *Enzyme and Microbial Technology*, **17** (9), 797–803.
53. Bader, J., Bellgardt, K.-H., Singh, A. *et al.* (1992) Modeling and simulation of cellulase adsorption and recycling during enzymatic hydrolysis of cellulosic materials. *Bioprocess and Biosystems Engineering*, **7** (5), 235–240.
54. Ghose, T.K. and Bisaria, V.S. (1979) Studies on the mechanism of enzymatic hydrolysis of cellulosic substances. *Biotechnology and Bioengineering*, **21** (1), 131–146.
55. Medve, J., Ståhlberg, J., and Tjerneld, F. (1994) Adsorption and synergism of cellobiohydrolase I and II of Trichoderma reesei during hydrolysis of microcrystalline cellulose. *Biotechnology and Bioengineering*, **44** (9), 1064–1073.
56. Converse, A.O. and Optekar, J.D. (1993) A synergistic kinetics model for enzymatic cellulose hydrolysis compared to degree-of-synergism experimental results. *Biotechnology and Bioengineering*, **42** (1), 145–148.
57. Ding, Hanshu and Xu, Feng (2004) Productive cellulase adsorption on cellulose. Lignocellulose Biodegradation. *American Chemical Society*, **889**, 154–169.
58. Liao, W., Liu, Y., Wen, Z. *et al.* (2008) Kinetic modeling of enzymatic hydrolysis of cellulose in differently pretreated fibers from dairy manure. *Biotechnology and Bioengineering*, **101** (3), 441–451.

59. Yang, B., Willies, D.M., and Wyman, C.E. (2006) Changes in the enzymatic hydrolysis rate of Avicel cellulose with conversion. *Biotechnology and Bioengineering*, **94** (6), 1122–1128.
60. Mansfield, S.D., Mooney, C., and Saddler, J.N. (1999) Substrate and enzyme characteristics that limit cellulose hydrolysis. *Biotechnology Progress*, **15** (5), 804–816.
61. Eriksson, T., Karlsson, J., and Tjerneld, F. (2002) A model explaining declining rate in hydrolysis of lignocellulose substrates with cellobiohydrolase I (cel7A) and endoglucanase I (cel7B) of Trichoderma reesei. *Applied Biochemistry and Biotechnology*, **101** (1), 41–60.
62. Puls, J. and Wood, T.M. (1991) The degradation pattern of cellulose by extracellular cellulases of aerobic and anaerobic microorganisms. *Bioresource Technology*, **36** (1), 15–19.
63. Shao, X., Lynd, L., and Wyman, C. (2009) Kinetic modeling of cellulosic biomass to ethanol via simultaneous saccharification and fermentation: Part II. Experimental validation using waste paper sludge and anticipation of CFD analysis. *Biotechnology and Bioengineering*, **102** (1), 66–72.
64. Zhu, L., O'Dwyer, J.P., Chang, V.S. *et al.* (2008) Structural features affecting biomass enzymatic digestibility. *Bioresource Technology*, **99** (9), 3817–3828.
65. Berlin, A., Gilkes, N., Kurabi, A. *et al.* (2005) Weak lignin-binding enzymes: a novel approach to improve activity of cellulases for hydrolysis of lignocellulosics. *Applied Biochemistry and Biotechnology*, **121–124**, 163–170.
66. Sison, B.C. and Schubert, W.J. (1958) On the mechanism of enzyme action. LXVIII. The cellobiase component of the cellulolytic enzyme system of Poria vaillantii. *Archives of Biochemistry and Biophysics*, **78** (2), 563–572.
67. Zhang, Y.H.P., Hong, J., and Ye, X. (2009) Cellulase assays, in *Biofuels* (ed. J.R. Mielenz), Humana Press, Totowa, NJ, p. 213–231.
68. Adney, W.S. and Baker, J. (1996) Measurement of Cellulase Activities – Laboratory Analytical Procedure LAP. NREL.
69. Ghose, T.K. (1987) Measurement of cellulase activities. *Pure and Applied Chemistry*, **59** (2), 257–268.
70. Mandels, M., Andreotti, R., and Roche, C. (1976) Measurement of saccharifying cellulase. *Biotechnology & Bioengineering Symposium*, (6), 21–33.
71. Nordmark, T., Bakalinsky, A., and Penner, M. (2007) Measuring cellulase activity. *Applied Biochemistry and Biotechnology*, **137–140** (1), 131–139.
72. Bisaria, V.S. and Mishra, S. (1989) Regulatory aspects of cellulase biosynthesis and secretion. *Critical Reviews in Biotechnology*, **9** (2), 61–103.
73. McMillan, J.D., Jennings, E.W., Mohagheghi, A., and Zuccarello, M. (2011) Comparative performance of precommercial cellulases hydrolyzing pretreated corn stover. *Biotechnology for Biofuels*, **4** (1), 29.
74. Selig, M., Weiss, N., and Ji, Y. (2008) Enzymatic Saccharification of Lignocelluloic Biomass – Laboratory Analytical Protocol (LAP). NREL.
75. Dowe, N. and McMillan, J.D. (2001) SSF Experimental Protocols – Lignocellulosic Biomass Hydrolysis and Fermentation Laboratory Analytical Procedure (LAP). NREL.
76. Zhang, X., Qin, W., Paice, M.G., and Saddler, J.N. (2009) High consistency enzymatic hydrolysis of hardwood substrates. *Bioresource Technology*, **100** (23), 5890–5897.
77. Roche, C.M., Dibble, C.J., and Stickel, J.J. (2009) Laboratory-scale method for enzymatic saccharification of lignocellulosic biomass at high-solids loadings. *Biotechnology for Biofuels*, **2** (1), 28.
78. Jørgensen, H., Vibe-Pedersen, J., Larsen, J., and Felby, C. (2007) Liquefaction of lignocellulose at high-solids concentrations. *Biotechnology and Bioengineering*, **96** (5), 862–870.
79. Gregg, D.J. and Saddler, J.N. (1996) Factors affecting cellulose hydrolysis and the potential of enzyme recycle to enhance the efficiency of an integrated wood to ethanol process. *Biotechnology and Bioengineering*, **51** (4), 375–383.
80. Tu, M., Zhang, X., Paice, M. *et al.* (2009) The potential of enzyme recycling during the hydrolysis of a mixed softwood feedstock. *Bioresource Technology*, **100** (24), 6407–6415.
81. Qi, B., Chen, X., Su, Y., and Wan, Y. (2011) Enzyme adsorption and recycling during hydrolysis of wheat straw lignocellulose. *Bioresource Technology*, **102** (3), 2881–2889.
82. Lu, Y., Yang, B., Gregg, D. *et al.* (2002) Cellulase adsorption and an evaluation of enzyme recycle during hydrolysis of steam-exploded softwood residues. *Applied Biochemistry and Biotechnology*, **98–100**, 641–654.

83. Prieto, C.A., Guadix, E.M., and Guadix, A. (2010) Optimal operation of a protein hydrolysis reactor with enzyme recycle. *Journal of Food Engineering*, **97** (1), 24–30.
84. Knutsen, J. and Davis, R. (2004) Cellulase retention and sugar removal by membrane ultrafiltration during lignocellulosic biomass hydrolysis. *Applied Biochemistry and Biotechnology*, **114** (1), 585–599.
85. Wang, Q.Q., Zhu, J.Y., Hunt, C.G., and Zhan, H.Y. (2012) Kinetics of adsorption, desorption, and re-adsorption of a commercial endoglucanase in lignocellulosic suspensions. *Biotechnology and Bioengineering*, **109** (8), 1965–1975.
86. Samaniuk, J.R., Tim, ScottC., Root, T.W., and Klingenberg, D.J. (2011) The effect of high intensity mixing on the enzymatic hydrolysis of concentrated cellulose fiber suspensions. *Bioresource Technology*, **102** (6), 4489–4494.

22

High-throughput Pretreatment and Hydrolysis Systems for Screening Biomass Species in Aqueous Pretreatment of Plant Biomass

Jaclyn D. DeMartini[1,2,3,*] and Charles E. Wyman[1,2,3]

[1] *Department of Chemical and Environmental Engineering, University of California, Riverside, USA*

[2] *Center for Environmental Research and Technology, University of California, Riverside, USA*

[3] *BioEnergy Science Center, Oak Ridge, USA*

22.1 Introduction: The Need for High-throughput Technologies

The primary barrier to low-cost biological conversion of lignocellulosic biomass to renewable fuels and chemicals is plant recalcitrance, that is to say, resistance of cell walls to deconstruction by enzymes or microbes [1,2]. However, the discovery and use of biomass species with reduced recalcitrance, when combined with optimized pretreatment processes and enzyme mixtures, could potentially improve the commercial viability of fuels and chemicals production from lignocellulosic biomass [3,4]. Unfortunately, the current understanding of biomass recalcitrance is limited, making it difficult to rationally select superior plant species without prior sugar release testing. As a result, there is a need to generate and screen a large variety of plants to identify those that exhibit both superior and sub-par sugar release. To this end, there are two central methodologies in generating and screening plants: (1) generation of mutants to see what effect targeted modifications have and (2) evaluation of natural variants to identify outliers for further characterization, in order to relate observed differences in behavior to structural features and biomass characteristics.

* Present address: DuPont Industrial Biosciences, Palo Alto, USA

Aqueous Pretreatment of Plant Biomass for Biological and Chemical Conversion to Fuels and Chemicals, First Edition.
Edited by Charles E. Wyman.
© 2013 John Wiley & Sons, Ltd. Published 2013 by John Wiley & Sons, Ltd.

Both methodologies require testing of thousands of samples. When considering that each sample should be screened over a range of pretreatment conditions including various times, temperatures, and chemical concentrations, as well as subsequent hydrolysis by a wide range of enzyme sources and formulations, the number of experiments easily reaches tens of thousands.

Due to the time-consuming and laborious nature of biomass analytical techniques, conventional testing of sugar release from pretreatment and enzymatic hydrolysis would be prohibitive. As a result, there has been a recent push to develop high-throughput pretreatment and hydrolysis (HTPH) systems that are capable of providing basic sugar release data in a rapid and automatable manner, while using significantly less biomass than conventional techniques. The benefits of implementing a fast and automatable procedure are clear: the throughput of data can be dramatically increased and the screening of multiple biomass-pretreatment-enzyme formulation combinations can be possible in a greatly reduced timeframe and with lower costs. In addition, high-throughput systems allow more replicate samples to be analyzed, improving estimation of measurement uncertainties.

Equally important benefits can be realized due to the low material requirements of HTPH systems. First, results can be obtained sooner. For genetically modified mutants, it can often take years for plants to adequately mature to produce sufficient amounts of material for analysis; this however is not a concern with downscaled HTPH systems for which only milligram amounts of biomass would be required. Additionally, when screening woody biomass with HTPH systems, entire trees do not have to be sacrificed for analysis; instead, analysis of a small core sample can be adequate without harming the tree. Finally, the reduced material requirement also allows analysis of individual biomass fractions that was previously not possible, in addition to making more biomass material available for other analyses.

22.2 Previous High-throughput Systems and Application to Pretreatment and Enzymatic Hydrolysis

The development and use of high-throughput (HT) technologies is relatively new. Since the 1990s, there has been a push to automate and increase sample throughput in a variety of fields, ranging from pharmaceuticals and drug discovery to genetic sequencing. In fact, efforts have also been directed at downscaling enzymatic hydrolysis of lignocellulosic biomass. Conventionally, enzymatic hydrolysis is typically performed in either 20 mL scintillation vials or 125 mL Erlenmeyer flasks, into which around 0.2 g or 3 g of wet pretreated solids are weighed, respectively [5,6]. In 2005 however, a "rapid microassay to evaluate enzymatic hydrolysis of lignocellulosic substrates" was reported that was based on a 96-well microplate for which each *c.* 300 μL well served as a reactor for enzymatic hydrolysis [7]. In 2007, the use of a standard 96-well plate was similarly reported for cellulase accessibility experiments [8], while more recently, additional reports have been made of microplate-based approaches to evaluate sugar release from the enzymatic hydrolysis of both untreated and pretreated biomass [9,10]. Other downscaled systems, such as 1.5 mL plastic Eppendorf tubes and 2 mL glass high-performance liquid chromatography (HPLC) vials, have also been used [11,12]. Although these systems provided a significant advancement in scaling down and automating biomass analyses, the development of downscaled and high-throughput pretreatment still remained a challenge. Because sugar yields are extremely low without pretreatment, a pretreatment step is required prior to enzymatic hydrolysis in order to adequately screen plant samples for their applicability for conversion to fuels and chemicals [2]. However, extending high-throughput applications to pretreatment processes presents a number of difficulties. Whereas enzymatic hydrolysis is performed at $\leq 50\,°C$ and near-neutral pH, pretreatments typically require high temperatures, high pressures, and the addition of corrosive chemicals, virtually ruling out use of glass HPLC vials and plastic microplates and Eppendorf tubes.

Table 22.1   *Types of batch pretreatment reactors commonly employed in conventional laboratory processes, with pretreatment chemical environment and required reaction mass listed for each.*

Reactor type	Reaction mass (g)	Pretreatment type	Reference
11 mm Pyrex glass tubes	0.4	Dilute acid	[13]
4″ Hastelloy tubing with Swagelok caps	0.4–2	Water only	[14]
300 mL metal Parr stirred tank reactor	2	Dilute acid	[15]
Batch metal tube reactors: 12.5 mm OD, 0.8255 mm wall thickness, 10 cm length	6	Dilute acid	[16]
Batch metal tube reactors: 1.5 in schedule 40 pipe nipples	7.5	Calcium hydroxide	[17]
500 mL metal stirred autoclave	40	Liquid hot water	[18]
1 L Parr metal stirred tank reactor	50	Dilute acid	[19]
2 gal metal Parr stirred tank reactor	500	Dilute acid	[20]

To date, a variety of pretreatment reactors have been utilized at the laboratory bench scale. Metal tubes or stirred tank reactors are often employed for three of the most commonly used pretreatments: hydrothermal (just hot water), dilute acid, and dilute alkali. Such reactors can safely handle acid or base (usually <2% concentrations) at the elevated pressures and temperatures typical for pretreatments (90–220 °C). Table 22.1 summarizes features of a variety of batch pretreatment reactors that have been reported in the literature to demonstrate reactor types, reaction volumes, and biomass material requirements that are often employed for laboratory experiments. The table does not attempt to provide a comprehensive review of all reactor types, but instead demonstrates the range of equipment and conditions that have been used in the past. Such reactor systems are typically heated by steam, fluidized sand baths, or oil baths to achieve fairly uniform heating as well as fast heat-up, although some employ electric heating jackets that suffer from slow heat-up rates. Due to the large vessel size conventionally required for a single pretreatment reaction and the space needed for heating devices, only a limited number of pretreatments can be performed simultaneously.

Although there has been some success in scaling down pretreatments to use less than 1 g [16], a number of time-consuming steps still remained after pretreatment in order to prepare the materials for subsequent enzymatic hydrolysis: (1) separation of pretreated solids and liquids; (2) washing of the solids; and (3) wet chemistry compositional analysis on pretreated washed solids. Enzymatic hydrolysis was then typically performed on the washed pretreated solids, with enzyme addition based on the composition of pretreated biomass [5,6]. The amount of sugars released by both pretreatment and enzymatic hydrolysis was measured, with a post-hydrolysis procedure often applied to determine oligomeric sugar concentrations [21]. The procedures were tedious and time-consuming, required significant amounts of material, and did not easily lend themselves to automation or increased throughput. As a result, unlike the mild reaction conditions of enzymatic hydrolysis that are more readily applied at a small scale, the development of HT pretreatments lagged behind.

22.3   Current HTPH Systems

To date, descriptions of four high-throughput pretreatment and enzymatic hydrolysis systems have been published, all of which were based on the diagram shown in Figure 22.1b [12,22–25]. As opposed to conventional pretreatment and enzymatic hydrolysis (Figure 22.1a) that involved large-scale pretreatments from which pretreated solids were then distributed to multiple hydrolysis experiments, HTPH systems involved downscaled pretreatments that were then used directly for subsequent enzymatic hydrolysis. To accomplish this, three of the HTPH systems employed a process termed "co-hydrolysis" [12,24–26] or "one-tube process" [23] in which both pretreatment and enzymatic hydrolysis were performed in the same

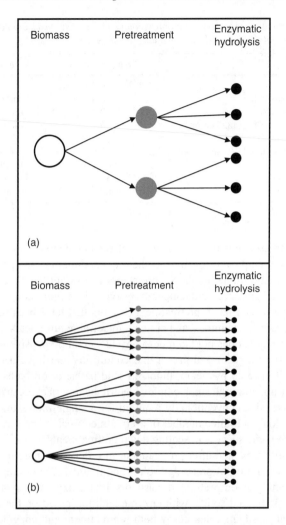

Figure 22.1 *Comparison of (a) conventional pretreatment and subsequent enzymatic hydrolysis laboratory experiments with (b) HTPH approach, in which the same reactor is employed for both pretreatment and enzymatic hydrolysis to avoid processing between the two operations. (Adapted from Studer et al. [26] © BioMed Central Ltd.).*

reactor without typically practiced procedures such as solid/liquid separation and solid washing between the pretreatment and hydrolysis steps.

Figure 22.2 outlines the differences between conventional pretreatment and enzymatic hydrolysis with the new co-hydrolysis or one-tube processes. Work by Santoro *et al.* [23] and Studer *et al.* [25] demonstrated that these processes achieved similar sugar yields to conventional washed solids hydrolysis. More detailed analyses showed that leaving the pretreatment liquid (hydrolyzate) with the pretreated solids could introduce inhibitors to enzymatic hydrolysis that could require lower solids concentrations or higher enzyme loadings for co-hydrolysis results to be more comparable to those from conventional processes [26]. As a result, particular attention must be paid to select conditions that best mimic the sugar yields obtained from conventional pretreatment and enzymatic hydrolysis for HTPH systems that employ co-hydrolysis or one-tube processes. To avoid this concern, one of the HTPH systems [22] took a somewhat different

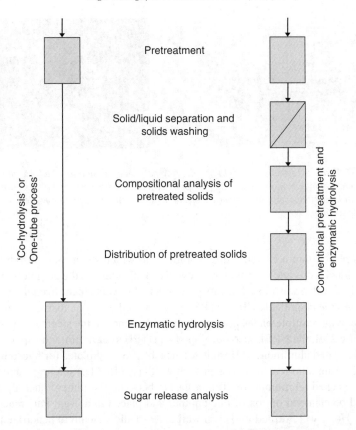

Figure 22.2 *Flow diagrams of conventional pretreatment and enzymatic hydrolysis versus an HTPH approach that employs a co-hydrolysis or one-tube process. (Adapted from Studer et al. [26] © BioMed Central Ltd.).*

approach from the other three: prior to enzymatic hydrolysis, pretreated solids were rinsed several times with a buffered solution to closely mimic the effects of solid/liquid separation and solid washing performed in conventional processes. In this way, the pH of the pretreated material was brought to the same value as that for enzymatic hydrolysis, and the majority of inhibitors generated during pretreatment were removed.

Although all four HTPH systems were based on the same principle (Figure 22.1b), each one varied its reactor configuration and processing conditions. Developed through support of the BioEnergy Science Center (BESC), the HTPH system at the University of California Riverside (UCR) pictured in Figure 22.3 was based on a custom-built 96-well plate design. However, the base plate was constructed of either aluminum or brass, and the reactor wells were made of Hastelloy. The original design [25] included an aluminum base plate into which Hastelloy wells that employed a reaction mass of 250 mg were press fit (Figure 22.3a).

More recently, an updated well plate design was developed with larger Hastelloy wells that held a reaction mass of 450 mg. These wells were also free-standing via a small pin in the well bottom to allow the wells to stand upright on the brass base plate instead of being press fit into the plate. This modification enabled grippers in a robotics platform to pick up and move the individual wells to more accurately tare and add ingredients.

For sealing both designs, the well plate was clamped between a top and bottom stainless steel plate with a flat silicone gasket positioned between the upper plate and the well openings. For pretreatment, the well

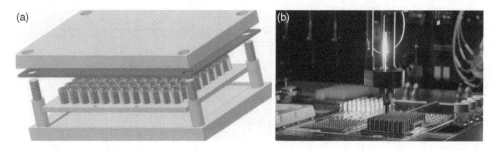

Figure 22.3 *UCR's HTPH reactor system [25] including the original reactor design in which Hastelloy wells with a 250 mg reaction mass were (a) press fit into an aluminum plate clamped between two stainless steel plates during pretreatment, and (b) the updated reactor with larger free-standing Hastelloy wells (450 mg reaction mass) being loaded by a Symyx Core Module. (See figure in color plate section).*

plate assembly was placed into a custom-built chamber into which steam was introduced to penetrate the space between individual wells and distribute heat evenly to all sides. At the completion of the pretreatment reaction, the steam inlet valve was closed, a valve was opened to vent steam from the chamber, and cooling water was flooded into the chamber. UCR's HTPH system was therefore capable of pretreating 96 biomass samples in one plate, with multiple plates potentially heated at once in the steam chamber.

As shown in Figure 22.4, the NREL also developed an HTPH system through support of the BESC based on custom-built gold-plated aluminum or Hastelloy stackable 96-well plates for hydrothermal or dilute acid pretreatments of c. 300 mg total reaction mass in each well [12,24]. After loading, each plate was sealed by placing an adhesive-backed aluminum foil Teflon gasket between the plates. Then up to 20 custom-made 96-well plates could be clamped on top of one another and placed in a 2-gal Parr reactor for pretreatment with indirect steam. Holes were drilled into each well plate to allow steam to penetrate into all of the plates and provide more rapid and uniform heating during pretreatment. After a target pretreatment time was reached, cooling water was forced through the channels in the plate. NREL's HTPH system was capable of pretreating up to 1920 biomass samples simultaneously.

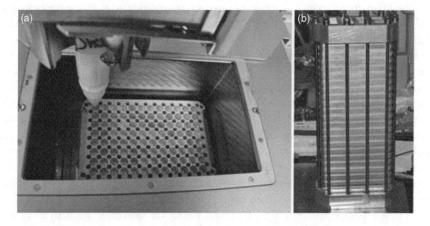

Figure 22.4 *The NREL HTPH reactor system uses the Symyx Powdernium to dispense biomass into (a) the wells of the 96-well reactor plate with a reaction mass of 255 mg per well with (b) 20 reactor plates stacked together in a modified 2-gal Parr reactor for pretreatment. (Reproduced from Decker et al. [12]). (See figure in color plate section).*

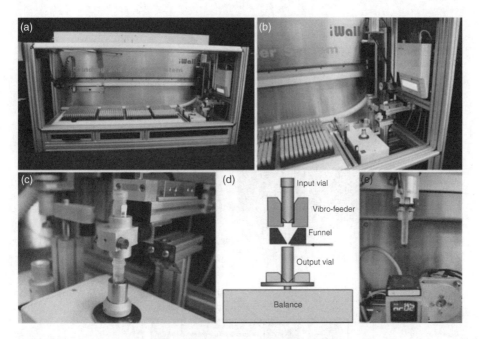

Figure 22.5   *GLBRC's HTPH system, including (a) iWall robotics platform for milling and dispensing. More detailed views of the (b) weighing substation, (c) balance and vibro-feeder dispensing from input (upper) to output (lower) tube, (d) diagram of weighing substation, and (e) bar code scanner substation are also shown. (Reproduced from Santoro et al. [23] with permission from Springer). (See figure in color plate section).*

The Great Lakes Bioenergy Research Center's (GLBRC's) HTPH system shown in Figure 22.5 was based on an off-the-shelf 96-tube Stabo-rack that held 1.4 mL polypropylene microtubes, each of which employed a reaction mass of *c.* 750 mg [23]. After loading, the tube racks were sealed with an elastopolymer seal and placed into a water bath for pretreatment. At the completion of the pretreatment reaction, the tube racks were cooled on ice. GLBRC's HTPH system could pretreat three 96-tube racks simultaneously, that is, 243 biomass samples at a time.

The fourth HTPH system shown in Figure 22.6 was developed by researchers at the University of York and the University of Dundee and was based on a standard off-the-shelf plastic 96-well plate [22]. In this system, each well employed a pretreatment reaction volume of 350 μL. After loading, the plates were sealed with a silicone cover and placed onto a heating block for pretreatment. This system could pretreat 360 biomass samples at a time. However, reaction temperatures were limited to less than 100 °C.

Although it did not measure the sugar release from combined pretreatment and enzymatic hydrolysis, a fifth high-throughput system accomplished downscaled and high-throughput pretreatment coupled with an alternative measure of biomass-pretreatment performance. In this system, a standard off-the-shelf polystyrene 96-well plate was employed (*c.* 200 mg reaction mass per well) in which various ionic liquids could be tested for their ability to dissolve the cellulose portion of biomass samples [27]. For pretreatment, a block containing heating rods through its interspaces was heated by temperature-controlled water. The block itself also acted as a seal to prevent water uptake by the ionic liquids. This HTPH system was capable of pretreating and measuring *in situ* 96 biomass samples at a time.

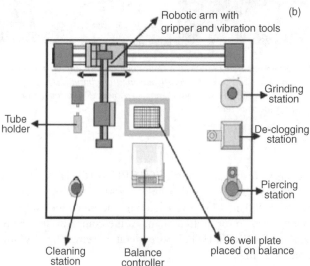

Figure 22.6 *HTPH system described by Gomez et al. including (a) general view of robotics platform for milling and dispensing and (b) schematic of robot's different substations. (Figure obtained from Gomez et al. 2010 [22]). (See color figure in plate section).*

22.4 Key Steps in HTPH Systems

Four basic steps have to be accomplished for all HTPH systems: (1) material preparation; (2) material distribution; (3) pretreatment reaction; and (4) sample preparation and analysis. Each step is described in the following sections, and the techniques applied to accomplish these steps are summarized.

22.4.1 Material Preparation

Preparing biomass materials for high-throughput pretreatment and enzymatic hydrolysis experiments is a crucial step that impacts all downstream processes, as well as the validity of any results. The material has to be homogeneous so that each small sample is representative of the larger mass. The three primary steps involved include sampling, particle size reduction, and any subsequent conditioning to obtain the prepared sample. In NREL's HTPH system, biomass material was knife-milled until it passed through a 1 mm screen [24]. No further conditioning was reported. Similarly, UCR's HTPH system employed knife-milled material; however, milling was typically followed by sieving to obtain a 20–80 mesh fraction (0.18–0.85 mm) [25]. Zavrel *et al.* [27] reported use of wood chips produced by sawing in their HT pretreatment system; the resulting particle sizes were in the range of 1–2 mm in length.

Unlike the above three systems that required time-consuming manual material preparation steps, the HTPH systems reported by Santoro *et al.* [23] and Gomez *et al.* [22] were unique in that they performed size reduction automatically via grinding robotics platforms, namely iWALL (TECAN, Mannedorf, Switzerland) and Labman Automation (Stokesley, North Yorkshire, UK), respectively. With the former, dried plant material (20–40 mg) was manually loaded into 2 mL screw cap microtubes along with three 7/32 inch stainless steel balls. These tubes were then loaded into racks and placed in the robotics system. Pulverization of the biomass was accomplished by ball milling to a fine powder with a particle geometric mean diameter of between 0.034 and 0.055 mm (depending on biomass type); more than 90% of the particles were smaller than 0.35 mm for all plant types tested. The system described by Gomez *et al.* [22] similarly employed automated shaking at *c.* 5000 rpm with three ball bearings within biomass-filled vials, but the particle size range of the resulting material was not reported.

22.4.2   Material Distribution

Following preparation, the materials had to be distributed to the pretreatment reactors. For the small amounts of biomass that were used in HT technologies, accurate solids distribution could be a tedious and time-consuming step. Prior to the development of HTPH systems, some of the downscaled and HT enzymatic hydrolysis set-ups discussed above applied creative methods to accomplish small-scale solids distribution. For example, Berlin *et al.* [7] took advantage of the uniform nature of biomass paper to produce "handsheets" of ethanol organosolv pulped poplar, from which 6 mm disks that weighed on the order of 1–2 mg were produced by a paper punch. Each disk was then manually placed in each individual well for subsequent enzymatic hydrolysis. This method was similar to that developed by Decker *et al.* [28] in which 2.65 mg filter paper disks were produced and placed in the wells of a standard microtiter plate to automate the filter paper assay of cellulase activity. Others [9,10] distributed small amounts of biomass into 96-well plates by suspending the solids in water or buffer (1–5% w/w) and transferring the slurry into individual wells of a 96-deepwell microplate while mixing the slurry to ensure uniformity during transfer.

For HT pretreatment and enzymatic hydrolysis systems, some workers reported that the solids were manually weighed into individual wells of microtiter plates [25,27]. In the case of the initial UCR HTPH design described by Studer *et al.* [25], a small brass weighing cup that held a volume corresponding to the target mass to be dispensed (2.5 mg) facilitated distribution of biomass to the individual wells. However, weighing of milligram quantities of biomass was time-consuming and dramatically reduced the overall throughput of these systems. As a result, all four HTPH systems migrated to solids-dispensing robotics platforms. At NREL, a Symyx MTM Powdernium powder dispensing system (Symyx, Sunnyvale, CA) distributed 5 mg of biomass to each well of their 96-well plate reactor [12,24]. The deck of this robotic platform could accommodate up to 80 plastic 10 mL biomass-dispensing hoppers at a time, each of which could handle small amounts of biomass (50–100 mg). For this system, the entire reactor plate was moved to a modified Sartorius LP330 balance that recorded the final weight of biomass dispensed into each well to 0.1 mg accuracy. After solids distribution, an automated pipetting system (Biomek FX) added liquids for all subsequent steps. The total reaction mass employed in each well was 255 mg, comprising 250 µL of water or dilute acid added to 5 mg of milled biomass.

At UCR, a Standard Configuration 2 Symyx Core Module dispensed 4.5 mg of biomass into each well for the updated well plate design [29]. The robotics platform could accommodate 10 metal 25 mL biomass-dispensing hoppers, each of which could contain 5–5000 mg of biomass. The UCR system differed from NRELs in that the individual wells of the microtiter reactor were moved into a Sartorius WZA65-CW balance one at a time, allowing determination of the amount of each ingredient added to a single well to within 0.01 mg. Subsequent liquid handling steps were accomplished either with the liquid handling set-up on the same Core Module robotics platform or with multichannel pipettes (8 channel pipetter, Eppendorf, Hamburg, Germany). In this system, 445.5 µL of water was typically added to 4.5 mg of milled biomass for hydrothermal pretreatment. For dilute acid and dilute alkali pretreatments, the reaction was performed with 85.5 µL of acid or 40.5 µL of base added to 4.5 mg of biomass.

In GLBRC's HTPH system, the same iWALL robotics platform accomplished both biomass milling and solids dispensing [23]. After automated milling in a 1.4 mL tube, vials were sent to a de-clogging station to break up clumped material, followed by transfer to a piercing station where a 1 mm hole was bored into the base of each vial. Next, 1.5 mg of this milled biomass was dispensed through a funnel by the action of a vibro-feeder into an empty vial located below the original vial and placed on a Mettler Toledo SAG 205 balance. The balance recorded the amount of biomass dispensed into each tube to within 0.01 mg. A PerkinElmer (Waltham, MA) Janus workstation subsequently added 750 µL of pretreatment liquid, typically dilute NaOH, into each tube.

The system reported by Gomez *et al.* [22] was similar to that described immediately above. However, important differences included use of a different robotics platform, namely a Labman Automation platform (Stokesley, North Yorkshire, UK), that dispensed 4.0 mg of material from the original vials filled with milled biomass into the individual wells of a 96-well microplate. The amount of biomass in each plate was monitored to within 0.1 mg by placing the entire microplate on a balance during dispensing. Subsequently, 350 µL of either dilute NaOH or H_2SO_4 were added by the same robotics platform for pretreatment.

22.4.3   Pretreatment and Enzymatic Hydrolysis

After material distribution, the next key steps in any HTPH system were pretreatment and enzymatic hydrolysis. NREL performed either hydrothermal (just hot water) or dilute acid pretreatment in their system at about a 2% solids loading. As an example, they reported hydrothermal pretreatment for 40 min at 180 °C [24] and, more recently, applied dilute acid pretreatment typically with 0.3% H_2SO_4 for 30 min at 180 °C (SR Decker, personal communication, 2011).

UCR's HTPH system is also compatible with hydrothermal [25], dilute acid [30], and dilute alkali pretreatments (H. Li, personal communication, 2011). For hydrothermal pretreatments, a 1% solids loading was used, while for dilute acid and alkali, the solids loadings were 5% and 10%, respectively. Temperatures between 120 and 180 °C and times from 10 to 300 min were applied, with exact conditions determined for the biomass to be tested. For dilute acid and alkali pretreatments, concentrations of 0.5–1% H_2SO_4 and 1% NaOH, respectively, were applied.

Both dilute acid and dilute alkali pretreatments could also be performed in GLBRC's HTPH system. For example, they reported pretreatment at 90 °C in 0.025% NaOH for 3 hours with a solids loading of 0.2% [23]. Similarly, Gomez *et al.* [22] employed both dilute acid and alkali pretreatments but at lower temperatures due to pressure constraints for their system, for example, 1% H_2SO_4 or 0.5N NaOH at 90 °C for 30 min at a solids loading of about 1.1%. Finally, Zavrel *et al.* [27] added a total of 200 µL of various imidazolium-based ionic liquids to between 4 and 12 mg of biomass per well to give a solids loading of 2–6%. In this case, pretreatments were performed at 50 °C for 8–24 hours.

Following pretreatment, enzymatic hydrolysis was performed in the same reactor in a co-hydrolysis or one-tube processing approach for three of the HTPH systems [23–25]. In general, a higher enzyme loading was employed for these methods compared to conventional washed solids hydrolysis to offset the effects of inhibitors possibly present in the pretreated biomass slurry as a result of not removing the pretreated liquids from the solids. For hydrothermal pretreatment with the NREL system, an enzyme loading of 70 mg cellulase per g initial biomass supplemented with 2.5 mg/g β-glucosidase was applied for a 72 hr static incubation at 40 °C [24]. The enzyme addition was made to the entire pretreated biomass slurry along with 1 M sodium citrate buffer to bring the pH of the pretreatment slurry to *c.* 5.0. UCR employed a similarly high enzyme loading, typically 75 mg cellulase per g glucan + xylan in the initial biomass, supplemented with 25 mg/g of xylanase for a 72 hr incubation at 50 °C and with shaking at 150 rpm [25]. Diluted enzyme was added to the pretreated biomass slurry in combination with 1 M citrate buffer and 1 g/L of the biocide sodium azide. For dilute acid and dilute alkali pretreatments, the pretreated slurry was diluted with water 5 or 10 times, respectively, prior to enzyme addition to achieve a total reaction mass of 450 mg. GLBRC's HTPH system employed an enzyme loading of 30 mg cellulase per g glucan in the initial biomass, which was added to the pretreated biomass slurry in each vial as a solution that also contained 30 mM citrate buffer and 0.01% sodium azide. Hydrolysis was performed at 50 °C with end-over-end rotation for a 20 hr incubation time [23].

As mentioned previously, the HTPH system developed at the University of York and University of Dundee [22] more closely mimicked conventional solid/liquid separation and solid washing by applying several rinses with a buffered solution prior to enzymatic hydrolysis. As a result, a lower enzyme loading of 6.3 FPU/g of material (*c.* 14 mg cellulase/g biomass [31]) was typically employed using a mixture of cellulase (Celluclast, Novozymes, Bagsvaerd, Denmark) supplemented with β-glucosidase (Novozyme 188, Novozymes, Bagsvaerd, Denmark) at a 4 : 1 ratio, respectively. Enzymatic hydrolysis was reported to be carried out in 25 mM sodium acetate buffer at 50 °C for 8 hours with constant shaking at 120 rpm.

22.4.4 Sample Analysis

The four HTPH systems reported here all measured the amount of sugar released from combined pretreatment and enzymatic hydrolysis. To accomplish this, UCR employed an HPLC equipped with a refractive index detector [25], using hydrolyzates transferred to HPLC-compatible vials or microplates following the completion of enzymatic hydrolysis. Benefits of this approach included that it was both well established and enabled measurement of sugars in addition to glucose and xylose. However, HPLC was time-consuming and could require between 15 and 30 min per sample, equating to 24–48 hours for analysis of all 96 samples from a single well plate. To avoid this HPLC load, both the NREL and GLBRC systems employed enzyme-based assays [23,24]. For the NREL system, hydrolyzates following enzymatic hydrolysis were diluted and transferred to 96-well flat-bottomed polystyrene plates in which glucose was detected via a modified glucose oxidase/peroxidase (GOPOD) assay and xylose was detected with a xylose dehydrogenase assay (Megazyme International Ireland, Wicklow, Ireland) by measuring their absorbance at 510 and 340 nm, respectively [24]. GLBRC's HTPH system similarly employed GOPOD and xylose dehydrogenase assays following transfer to 384-well microtiter plates [23]. The benefit of sugar analysis by enzyme-based assays was the speed at which they could be completed: it was estimated that a plate of 96 samples could be analyzed in 20 min [12]. However, it has also been reported that, due to the specificity of the assays, certain sugars such as xylooligomers could interfere with xylose measurements [24].

Colorimetric assays, which could easily be applied in high-throughput set-ups, have also been employed for sugar release measurements. In HT enzymatic hydrolysis systems [10,32], a standard dinitrosalicylic acid (DNS) assay was used to measure reducing sugars released during enzymatic hydrolysis. Alternatively, the HTPH system reported by Gomez *et al.* [22] measured glucose equivalents released by enzyme action

Table 22.2 Summary of key characteristics for each of the four HTPH systems including the reaction volumes utilized in pretreatment and enzymatic hydrolysis, solids loading, chemical type and concentration, temperature, and heat source.

Institution		Reaction volume (μL) (pretreatment/EH[a])	Solids loading (%[b]/mg)	Pretreatment type	Temp/acid concentration	Heating medium	Ref
NREL[d]	A	300/300	1.7/5.0	Hydrothermal	180 °C/0%	Steam	[24]
	B	250/250	2.0/5.0	Dilute acid/H_2SO_4	180 °C/0.3%		
UCR[d]	A	450/450	1.0/4.5	Hydrothermal	120–180 °C/0%	Steam	[25]
	B	85.5/450	5.0/4.5	Dilute acid/ H_2SO_4	120–180 °C/0.5–1%		
	C	40.5/450	10.0/4.5	Dilute alkali/ NaOH	120–180 °C/1%		
GLBRC[e]	A	750/750	0.2/1.5	Dilute alkali/ NaOH	90 °C/0.025%	Water bath	[23]
	B			Dilute acid/ H_2SO_4	90 °C/2.0%		
UY[c, e]	A	350/750	1.1/4.0	Dilute acid/ H_2SO_4	90 °C/1.0%	Heating block	[22]
	B			Dilute alkali/ NaOH	90 °C/0.5N		

[a] The reaction volume for enzymatic hydrolysis is the total slurry volume (water, dilute acid, or dilute alkali) prior to enzyme, buffer, biocide addition.

[b] Solids loading is described as % biomass weight per reaction volume.

[c] UY: University of York/University of Dundee at Scottish Crops Research Institute.

[d] Employs custom-built reactor.

[e] Employs off-the-shelf reactor.

using a modified 3-methyl-2-benzothiazolinonehydrozone (MTBH) assay. In this approach, a mixture of hydrolzate, NaOH, MBTH, and dithiothreitol (DTT) with a final volume of 250 μL was incubated at 60 °C for 20 min, after which an oxidizing reagent was added and optical measurements were taken at 620 nm in an optical well plate. As with enzyme-based assays, colorimetric measurements such as the modified MTBH assay employed by Gomez *et al.* [22] had the benefit of fast detection; however, possible interference among sugars and differing response for various sugars could be concerns [12,22].

Although also testing for sugar release from combined pretreatment and enzymatic hydrolysis, Zavrel *et al.* [27] applied a different approach to evaluating pretreatment-biomass combinations by monitoring *in situ* dissolution of cellulose by ionic liquids. In their HT pretreatment system, scattered and transmitted light was measured continuously during pretreatment to follow the size and number of cellulose particles as a function of pretreatment time. Although this system was unable to quantitatively measure sugar release, greater solubilization of crystalline cellulose was associated with enhanced sugar release in subsequent enzymatic hydrolysis, providing an alternative measure of cellulose digestibility.

Table 22.2 summarizes key characteristics for each of the four HTPH approaches.

22.5 HTPH Philosophy, Difficulties, and Limitations

HTPH technologies provided a significant step for screening large numbers of plant samples for their recalcitrance to sugar release. However, it was important to recognize the distinctions between conventional and HT pretreatment and enzymatic hydrolysis testing, particularly that the latter was primarily a screening tool that provided a platform from which sugar release trends, as well as superior and sub-par outliers, could be identified for further analysis. Along these lines, HTPH processes typically measure only total monomeric sugar release from combined pretreatment and enzymatic hydrolysis. On the other hand, conventional methods track sugar release from these two steps individually by separately analyzing the solid and liquid phase from pretreatment and also from enzymatic hydrolysis, thereby revealing more details about the sources and fates of sugars, including those in both monomers and oligomers [33], and facilitating mass balance closure.

However, it is important to note that HTPH could be extended to measuring total monomeric plus oligomeric sugars from pretreatment and enzymatic hydrolysis or easily adapted to measure sugar release from individual stages if such details are needed. It should also be kept in mind that solid/liquid separation is not likely to be desirable commercially due to higher capital and operating costs, as well as introducing additional opportunities for contamination and sugar loss. Although HTPH configurations based on co-hydrolysis or one-tube processes may differ from conventional methods, they may more closely simulate commercial practice.

The lack of solid/liquid separation following pretreatment in HTPH systems also has strong implications on enzyme loadings used in hydrolysis of the pretreated biomass slurry. Since the purpose of HTPH systems was primarily to screen multiple plants for reduced recalcitrance, this required the selection of conditions that best highlighted differences in substrate features and changes in substrate with pretreatment. High enzyme loadings were therefore applied to allow determination of differences in substrate digestibility, as opposed to enzyme inhibition or activity. Another key point in ensuring that differences in substrate digestibility were revealed was selection of proper pretreatment conditions. When the goal was to identify less recalcitrant plants, lower pretreatment severities than those identified for maximum sugar release from baseline substrates were typically applied to facilitate identification of biomass samples that were amiable to high sugar release from enzymatic hydrolysis at conditions that sacrificed less sugar to degradation during pretreatment.

Although the development of HTPH systems is a major step to accelerating screening of large combinations of biomass materials, pretreatment conditions, and enzyme loadings and formulations, important challenges remain. One such limitation, as with any downscaled system, is that errors can be significantly amplified at this small scale. In line with this, the introduction of a single air bubble during pipetting will result in inaccurate sugar concentrations and erroneous sugar release results. Furthermore, glucose released from starch in plants cannot be differentiated from glucose released from cellulose for the HTPH analysis methods discussed previously; as a result, differences in the amount of glucose release observed between samples could be due to varying levels of endogenous sugars in native biomass. Likewise, the fate of extractives and particularly free sugars during pretreatment has not yet been established; as a result, the mass of sugar released per mass of biomass could be influenced.

Unfortunately, the removal of starch and extractives is conventionally performed with 0.1 g and 5–20 g of material, respectively, with equipment and techniques that are not currently designed for large numbers of samples and high-throughput applications [34,35]. A simple way to address possible starch interference in HTPH systems is to test the enzymes used for their ability to hydrolyze starch into glucose. In this regard, both Santoro *et al.* [23] and Studer *et al.* [25] reported non-detectable or minimal levels of starch hydrolysis, respectively. Alternatively, researchers at NREL have developed downscaled and higher-throughput methods to remove starch and extractives prior to HTPH testing to ensure that endogenous sugars do not interfere with sugar release results (SR Decker, personal communication, 2011).

Perhaps the biggest concern with scaling down pretreatment and enzymatic hydrolysis processes is obtaining homogeneous biomass samples, which could be influenced by a range of steps including sampling, biomass inhomogeneity, milling, and sieving. For example, an HTPH system was applied to measure the ring-by-ring composition and sugar release from combined pretreatment and enzymatic hydrolysis across an aspen wood cross-section to show that both varied significantly across the radial direction [36]. Additionally, Garlock *et al.* [37] reported that corn stover composition and performance in ammonia fiber expansion (AFEX) pretreatment varied considerably with anatomical fraction. These findings demonstrate that when only a small portion of a plant was sampled for analysis, such as the case in HTPH testing, the location from which that sample is taken could impact sugar release results.

It has also been noted that although milling and sieving are often necessary to achieve homogeneous samples for biomass analyses, the material preparation process itself could complicate obtaining representative samples [38]. This issue has not been found to be as significant for well-mixed woody samples that are generally quite homogeneous when milled since they contain fewer cell types that segregate by particle size. On the other hand, herbaceous materials are typically composed of diverse cell types that can segregate into distinct particle size fractions. For example, larger particle size fractions of milled corn stover have been reported to be richer in cob and stalk portions that were more recalcitrant to hydrolysis than smaller size fractions, which contained higher amounts of leaves and husk [39]. Furthermore, the fine fractions of herbaceous materials in the size range of <80 mesh or <0.180 mm can contain a large fraction of inorganics. Removal of this fraction could affect composition and sugar release performance and may not provide results that are representative of the entire plant since certain anatomical fractions segregate [38]. Together, these studies all stress the importance of taking great care in sampling and material preparation for HTPH analyses to ensure that the small amounts of materials employed are representative of the entire plant sample being tested. Conversely, due to the downscaled and high-throughput nature of HTPH systems, this limitation can be easily tested by running many replicates to check for biomass inhomogeneity, an approach not taken as easily for previous larger-scale conventional pretreatment and enzymatic hydrolysis methods.

22.6   Examples of Research Enabled by HTPH Systems

To date, HTPH systems have enabled a number of research projects that were previously not possible, including methodologies in both plant development and screening presented in Section 22.1. One of the first reports of a large-scale project to screen sugar release from combined pretreatment and enzymatic hydrolysis of hundreds of plants was by Santoro *et al.* [23] in which 1200 *Arabidopsis* samples were tested. These samples were knock-out and knock-down mutants via T-DNA insertions to genes that were believed or known to play a role in cell-wall metabolism and possibly digestibility. In this case, HTPH successfully identified several *Arabidopsis* lines with significantly higher glucose and xylose release than most others.

Voelker *et al.* [40] also applied an HTPH system [12,24] to screen 4–7 field-grown poplar trees per transgenic event (14 events, 100 samples in total) that had undergone transgenic down-regulation of the Pt4CL1 gene based on previous evidence that this would reduce lignin content in cell walls. However, in contrast to previous studies, they found that trees with reduced lignin contents did not yield substantially higher saccharification potential; instead, very little difference was found in the sugar release from combined hydrothermal pretreatment and enzymatic hydrolysis among all trees tested.

Selig *et al.* [24] similarly employed an HTPH system to screen 755 natural poplar variants from the Pacific Northwest of North America, a feat that would not have been otherwise practical in light of the sheer number of samples. In this study, sugar release from combined hydrothermal pretreatment and enzymatic hydrolysis was independent of total lignin content but strongly related to the lignin syringyl to guaiacyl (S/G) ratio. In an extension of that study, a much smaller subsample of natural poplar variants (47 samples × 3 pretreatment conditions) from the same population was tested under a variety of hydrothermal pretreatment conditions using HTPH technology [41]. In this case, glucose release only had a strong negative correlation to lignin content for trees with low S/G (<2) ratio, while xylose release was dependent on the S/G ratio alone and not lignin content. Furthermore, certain trees featuring average lignin contents and S/G ratios exhibited exceptionally high sugar release, demonstrating that factors beyond lignin content and S/G ratio influenced recalcitrance.

HTPH systems are not only useful for screening large numbers of samples but also for their ability to provide sugar release results from very minimal amounts of biomass material. Along these lines, DeMartini and Wyman [36] applied HTPH technology to test sugar release from combined pretreatment and enzymatic hydrolysis of the individual annual rings of 26- and 8-year-old aspen trees. Although only about 35 samples in total were tested at a single pretreatment condition, downscaled technology was essential to process such small sample amounts (<100 mg). In this case, sugar release (grams of sugar released per gram biomass) varied significantly across the radial direction of the tree, but sugar yields (grams of sugar released per gram available sugar) did not, suggesting that wood maturity impacted composition much more than recalcitrance. A similar study was also undertaken with mixed prairie species in which HTPH technology was utilized to test sugar release from combined pretreatment and enzymatic hydrolysis of grasses and legumes [42]. DeMartini and Wyman [42] found significant differences among these two types of species, as well as among individual anatomical components that influenced the recalcitrance of mixed prairie species.

22.7 Future Applications

The development of HTPH technology has opened the door for reducing sample size and increasing the throughput for a variety of biomass applications. HTPH systems can generate large amounts of sugar release data to better understand factors influencing biomass recalcitrance. However, results from these screening studies typically reported the amount of sugar released per amount of total biomass; at the time that most of the studies were performed, there was no method to accurately determine the carbohydrate content of the large numbers and small amounts of samples available. However, to gain a better sense of a plant's recalcitrance, sugar yields should be determined as the amount of sugar released per amount of sugar available. To meet this need, downscaled and high-throughput compositional analysis approaches have recently been developed [29,43]. Both are based directly on conventional two-stage acid hydrolysis compositional analysis [44] but use significantly less material (between 60 and 100 times less material, equivalent to 3 or 5 mg biomass per test). NREL applied a 96-well plate format similar to their HTPH procedure for downscaled compositional analysis [43], while an array of 48 1.5 mL glass HPLC vials with the same plate-clamping mechanism was used to support the UCR HTPH system [29]. The former method was capable of measuring glucan and xylan contents of much larger sample sets than reported for the latter; however, the latter could also estimate whole ash and Klason lignin contents through measurement of acid insoluble residue (AcIR).

Development of HTPH technologies also enable large-scale pretreatment kinetic studies that require much less time and labor compared to use of conventional reactors. Along these lines, recent work by T. Zhang (personal communication, 2011) employed UCR's HTPH system to study the kinetics of hemicellulose and cellulose conversion to sugar degradation products such as furfural, 5-hydroxymethyl-2-furaldehyde (5-HMF), and levulinic acid, by evaluating a number of different pretreatment times, temperatures, acids, and acid concentrations. Approximately 4000 samples were tested in about 1 month.

In addition to employing HTPH technology to screen sugar release from combined pretreatment and hydrolysis or developing pretreatment degradation kinetics, there are opportunities to expand it to microbial screening. For example, Cianchetta *et al.* [45] reported development of a miniaturized cultivation system based on flat-bottom 24-well plates to quickly and easily identify superior and sub-par cellulase producers on cellulose powder from over 300 *Trichoderma* strains. Extension of this technology to real biomass substrates, including pretreated materials, presents an important opportunity.

22.8 Conclusions and Recommendations

The development of HTPH systems represents a major advancement in biofuels research through providing a platform for rapid screening of large sets of plant samples in order to identify biomass outliers and trends in recalcitrance. Testing sugar release of thousands of samples in combination with various pretreatment and enzymatic hydrolysis conditions is no longer an imposing feat. Downscaling of biomass pretreatment to microplate- or small tube-based formats, in addition to development of co-hydrolysis or one-tube processes, enables automation and increased throughput, as well as greatly reduced material requirements for recalcitrance assays. Although some difficulties arise with scaling down biomass pretreatment and hydrolysis, HTPH systems can be powerful tools if careful attention is paid to biomass sampling and distribution. With the continued application of and improvements in these systems, new insights can be gained into biomass recalcitrance that will aid in identification of superior feedstock candidates, better pretreatment conditions, and improved enzyme formulations.

References

1. Lynd, L.R., Laser, M.S., Brandsby, D. *et al.* (2008) How biotech can transform biofuels. *Nature Biotechnology*, **26**, 169–172.
2. Wyman, C.E. (2007) What is (and is not) vital to advancing cellulosic ethanol. *Trends in Biotechnology*, **25**, 153–157.
3. Lynd, L.R., Cushman, J.H., Nichols, R.J., and Wyman, C.E. (1991) Fuel ethanol from cellulosic biomass. *Science*, **251**, 1318–1323.
4. Lynd, L.R., Wyman, C.E., and Gerngross, T.U. (1999) Biocommodity engineering. *Biotechnology Progress*, **15**, 777–793.
5. Brown, L. and Torget, R. (1996) *Enzymatic Saccharification of Lignocellulosic Biomass Laboratory Analytical Procedure*, National Renewable Energy Laboratory, Golden, Colorado.
6. Selig, M., Weiss, N., and Ji, Y. (2008) *Enzymatic Saccharification of Lignocellulosic Biomass Laboratory Analytical Procedure*, National Renewable Energy Laboratory, Golden, Colorado.
7. Berlin, A., Maximenko, V., Bura, R. *et al.* (2005) A rapid microassay to evaluate enzymatic hydrolysis of lignocellulosic substrates. *Biotechnology and Bioengineering*, **93** (5), 880–886.
8. Jeoh, T., Ishizawa, C.I., Davis, M.F. *et al.* (2007) Cellulase digestibility of pretreated biomass is limited by cellulose accessibility. *Biotechnology and Bioengineering*, **98** (1), 112–122.
9. Chundawat, S.P.S., Balan, V., and Dale, B.E. (2008) High-throughput microplate technique for enzymatic hydrolysis of lignocellulosic biomass. *Biotechnology and Bioengineering*, **99** (6), 1281–1294.
10. Navarro, D., Couturier, M., Damasceno da Silva, G.G. *et al.* (2010) Automated assay for screening the enzymatic release of reducing sugars from micronized biomass. *Microbial Cell Factories*, **9**, 58.
11. Selig, M.J., Vinzant, T.B., Himmel, M.E., and Decker, S.R. (2009) The effect of lignin removal by alkaline peroxide pretreatment on the susceptibility of corn stover to purified cellulolytic and xylanolytic enzymes. *Applied Biochemistry and Biotechnology*, **155**, 397–406.
12. Decker, S.R., Brunecky, R., Tucker, M.P. *et al.* (2009) High-throughput screening techniques for biomass conversion. *Bioengineering Research*, **2**, 179–192.
13. Chen, R.F., Lee, Y.Y., and Torget, R. (1996) Kinetic and modeling investigation on two-stage reverse-flow reactor as applied to dilute-acid pretreatment of agricultural residues. *Applied Biochemistry and Biotechnology*, **57–8**, 133–146.
14. Yang, B., Gray, M.C., Liu, C. *et al.* (2004) Unconventional relationships for hemicellulose hydrolysis and subsequent cellulose digestion, in *Lignocellulose Biodegradation*, ACS Symposium Series (eds B. Saha *et al.*) American Chemical Society, Washington, DC.
15. Foston, M. and Ragauskas, A.J. (2010) Changes in lignocellulosic supramolecular and ultrastructure during dilute acid pretreatment of Populus and swithgrass. *Biomass and Bioenergy*, **34** (12), 1885–1895.
16. Lloyd, T. and Wyman, C.E. (2003) Application of a depolymerization model for predicting thermochemical hydrolysis of hemicellulose. *Applied Biochemistry and Biotechnology*, **105–108**, 53–67.
17. Kaar, W.E. and Holtzapple, M.T. (2000) Using lime pretreatment to facilitate the enzymatic hydrolysis of corn stover. *Biomass and Bioenergy*, **18**, 189–199.
18. Negro, M.J., Manzanares, P., Ballesteros, I. *et al.* (2003) Hydrothermal pretreatment conditions to enhance ethanol production from poplar biomass. *Applied Biochemistry and Biotechnology*, **105–108**, 87–100.
19. Torget, R., Himmel, M., and Grohmann, K. (1992) Dilute-acid pretreatment of two short-rotation herbaceous crops: scientific note. *Applied Biochemistry and Biotechnology*, **34–35**, 115–123.

20. Spindler, D., Wyman, C., and Grohmann, K. (1990) Evaluation of pretreated herbaceious crops for the simultaneous saccharification and fermentation process. *Applied Biochemistry and Biotechnology*, **24–25**, 275–286.

21. Sluiter, A., Hames, B., Ruiz, R. *et al.* (2006) *Determination of Sugars, Byproducts, and Degradation Products in Liquid Fraction Process Samples Laboratory Analytical Procedure*, National Renewable Energy Laboratory, Golden, Colorado.

22. Gomez, L.D., Whitehead, C., Barakate, A. *et al.* (2010) Automated saccharification assay for determination of digestibility in plant materials. *Biotechnology for Biofuels*, **3** (23).

23. Santoro, N., Cantu, S.L., Tornqvist, C.-E. *et al.* (2010) A High-throughput platform for screening milligram quantities of plant biomass for lignocellulose digestibility. *Bioengineering Research*, **3**, 93–102.

24. Selig, M.J., Tucker, M.P., Sykes, R.W. *et al.* (2010) Lignocellulose recalcitrance screening by integrated high-throughput hydrothermal pretreatment and enzymatic saccharification. *Industrial Biotechnology*, **6** (2), 104–111.

25. Studer, M.H., DeMartini, J.D., Brethauer, S. *et al.* (2010) Engineering of a high-throughput screening system to identify cellulosic biomass, pretreatments, and enzyme formulations that enhance sugar release. *Biotechnology and Bioengineering*, **105**, 231–238.

26. Studer, M.H., Brethauer, S., DeMartini, J.D. *et al.* (2011) Co-hydrolysis of dilute pretreated populus slurries to support development of a high throughput pretreatment system. *Biotechnology for Biofuels*, **4**, 19.

27. Zavrel, M., Bross, D., Funke, M. *et al.* (2009) High-throughput screening for ionic liqids dissolving (ligno-)cellulose. *Bioresource Technology*, **100**, 2580–2587.

28. Decker, S.R., Adney, W.S., Jennings, E. *et al.* (2003) Automated filter paper assay for determination of cellulase activity. *Applied Biochemistry and Biotechnology*, **107**, 689–703.

29. DeMartini, J.D., Studer, M.H., and Wyman, C.E. (2011) Small scale and automatable high-throughput compositional analysis of biomass. *Biotechnology and Bioengineering*, **108** (2), 306–312.

30. Gao, X., Kumar, R., DeMartini, J.D., Li, H., and Wyman, C.E. (2012) Application of high throughput pretreatment and co-hydrolysis system to thermochemical pretreatment. Part1: Dilute acid. *Biotechnology and Bioengineering*, accepted for publication.

31. Garcia-Aparicio, M.P., Oliva, J.M., Manzanares, P. *et al.* (2011) Second-generation ethanol production from steam exploded barley straw by *Kluyveromyces marxianus* CECT 10875. *Fuel*, **90**, 1624–1630.

32. Jager, G., Wulfhorst, H., Zeithammel, E.U. *et al.* (2011) Screening of cellulases for biofuel production: Online monitoring of the enzymatic hydrolysis of insoluble cellulose using high-throughput scattered light detection. *Biotechnology Journal*, **6**, 74–85.

33. Wyman, C.E., Dale, B.E., Elander, R.T. *et al.* (2005) Comparative sugar recovery data from laboratory scale application of leading pretreatment technologies to corn stover. *Bioresource Technology*, **96**, 2026–2032.

34. Sluiter, A. and Sluiter, J. (2005) *Determination of Starch in Solid Biomass Samples by HPLC Laboratory Analytical Procedure*, National Renewable Energy Laboratory, Golden, Colorado.

35. Sluiter, A., Ruiz, R., Scarlata, C. *et al.* (2005) *Determination of Extractives in Biomass Laboratory Analytical Procedure*, National Renewable Energy Laboratory, Golden, Colorado.

36. DeMartini, J.D. and Wyman, C.E. (2011) Changes in composition and sugar release across the annual rings of *Populus* wood and implications on recalcitrance. *Bioresource Technology*, **102**, 1352–1358.

37. Garlock, R.J., Chundawat, S.P.S., Balan, V., and Dale, B.E. (2009) Optimization harvest of corn stover fractions based on overall sugar yields following ammonia fiber expansion pretreatment and enzymatic hydrolysis. *Biotechnology for Biofuels*, **2** (29).

38. Sluiter, J.B., Ruiz, R.O., Scarlata, C.J. *et al.* (2010) Compositional analysis of lignocellulosic feedstocks. 1. Review and description of methods. *Journal of Agricultural and Food Chemistry*, **58**, 9043–9053.

39. Chundawat, S.P.S., Venkatesh, B., and Dale, B.E. (2007) Effect of particle size based separation of milled corn stover on AFEX pretreatment and enzymatic digestibility. *Biotechnology and Bioengineering*, **96** (2), 219–231.

40. Voelker, S.L., Lachenbruch, B., Meinzer, F.C. *et al.* (2010) Antisense down-regulation of 4CL expression alters lignification, tree growth, and saccharification potential of field-grown poplar. *Plant Physiology*, **154**, 874–886.

41. Studer, M.H., DeMartini, J.D., Davis, M.F. *et al.* (2011) Lignin content in natural *Populus* variants affects sugar release. *Proceedings of the National Academy of Sciences*, **108** (15), 6300–6305.

42. DeMartini, J.D. and Wyman, C.E. (2011) Composition and hydrothermal pretreatment and enzymatic saccharification performance of grasses and legumes from a mixed species prairie. *Biotechnology for Biofuels*, **4**, 52.

43. Selig, M.J., Tucker, M.P., Law, C. *et al.* (2011) High throughput determination of glucan and xylan fractions in lignocelluloses. *Biotechnology Letters*, **33** (5), 961–967.

44. Sluiter, A., Hames, B., Ruiz, R. *et al.* (2008) *Determination of Structural Carbohydrates and Lignin in Biomass Laboratory Analytical Procedure*, National Renewable Energy Laboratory, Golden, Colorado.

45. Cianchetta, S., Galletti, S., Burzi, P.L., and Cerato, C. (2010) A Novel Microplate-Based Screening Strategy to Assess the Cellulolytic Potential of *Trichoderma* Strains. *Biotechnology and Bioengineering*, **107** (3), 461–468.

23

Laboratory Pretreatment Systems to Understand Biomass Deconstruction

Bin Yang[1] and Melvin Tucker[2]

[1] *Center for Bioproducts and Bioenergy, Department of Biological Systems Engineering, Washington State University, Richland, USA*

[2] *National Bioenergy Center, National Renewable Energy Laboratory, Golden, USA*

23.1 Introduction

The natural recalcitrance of complex biomass polymers to deconstruction into simpler molecules is a major hurdle confronting all processes for the conversion of biomass into useful biofuels and chemicals [1]. This is especially true if monomeric sugars and low-molecular-weight lignin compounds are desired for subsequent processing. The complex biochemical conversion process utilizes the combination of pretreatment and enzymes to deconstruct biomass into simpler compounds that are fermentable into the desired biofuels and bioproducts [2]. Pretreatment refers to the process unit operation that is responsible for disrupting the naturally resistant structure of lignocellulosic biomass to provide reactive intermediates (such as sugars and sugar degradation products) to biochemical or thermochemical processes for production of renewable biofuels and bioproducts [3]. Pretreatment is essential for achieving high yields of desirable products from biochemical processing of naturally resistant cellulosic biomass [4]; it represents about 20% of total costs however [5]. Pretreatment has impacts on the overall conversion process from feedstock size reduction to product recovery and co-product potential [3]. Development of pretreatment technologies offering high yields and low costs is therefore vital to the economic success of cellulosic biomass bioconversion, although recent biotechnological advances in bioengineering energy crops may provide important opportunities to lower costs [6–9].

Various pretreatment technologies, including those with ammonia [10–12], alkali (lime) [13], carbonic acid [14], various dilute acids (e.g., HNO_3) [15–17], peracetic acid [18], maleic acid [19], phosphoric acid [20], sulfuric acid [21–29], organosolv with dilute acid [30], pH-controlled hydrothermolysis [31,32], steaming/steam explosion with or without catalysts (e.g., sulfuric acid) [33], and SO_2 [34–36], have been

Aqueous Pretreatment of Plant Biomass for Biological and Chemical Conversion to Fuels and Chemicals, First Edition.
Edited by Charles E. Wyman.
© 2013 John Wiley & Sons, Ltd. Published 2013 by John Wiley & Sons, Ltd.

developed in conjunction with the production of ethanol and other products. Most pretreatment technologies reported in the literature employ bench-scale pretreatment systems. Although pilot-scale or production-scale reactors are available for some pretreatment technologies, such as steam explosion, extrusion, and percolation, application of large-scale pretreatment reactors in support of cellulosic ethanol production is still very limited.

Bench-scale reactors are widely used for research and development in many scientific and engineering areas for various purposes. Many of the issues regarding particle size, reactor construction materials, heat and mass transfer, and catalyst stability applicable to carrying out laboratory-scale experiments using bench-scale reactors to deconstruct biomass into its desirable components apply across the board for the two biomass conversion platforms: thermochemical conversion (i.e., pyrolysis and gasification) and bio-chemical conversion. However, at the present time the biochemical conversion platform is costly and slow despite the promises of high yields [4,37].

It is easiest to study these various process configurations and unit operations at the laboratory bench scale, especially pretreatment. In general, bench-scale biomass pretreatment reactors are applied to study the mechanisms and kinetics of biomass thermochemical deconstruction reactions in various pretreatment technology options to provide data for the validation of process simulation models. Several hemicellulose hydrolysis models have been reported based on data from bench-scale reactors [38,39]. Within the Biomass Refining Consortium of Applied Fundamentals and Innovation (CAFI) projects, bench-scale data employing promising biomass pretreatment technologies were developed and compared, which led to further process techno-economic evaluations using Aspen plus [40]. The complete cellulosic feedstock conversion to ethanol from the various technology process scenarios were modeled at the commercial scale [41]. Thus, using bench-scale reactors is a cost-effective way to investigate process performance over a range of process variables, resulting in optimization of pretreatment conditions of the various pretreatment technologies without the huge capital and resource expenditures needed for pilot, developmental, or commercial-scale deployment. Prior to beginning construction of commercial-scale pretreatment reactors, bench-scale reactor studies provide important data for improving reactor design [42]. Since bench-scale reactors serve an important function in developing initial datasets that impact development of the whole biomass processing system, the selection of an appropriate bench-scale system is critically important. The type of pretreatment reaction, conversion process requirements, and economics have to be taken into careful consideration when selecting bench-scale reactors.

For biomass pretreatment, the main reactant is the solid biomass such as wood, agricultural residues, or herbaceous substrates, often milled or reduced in size. Other reactants, depending on the pretreatment chemistry, include liquid and/or gaseous catalysts such as water, acid (e.g., H_2SO_4, HNO_3), alkali (e.g., NaOH, lime, NH_3 solutions), and acidic/alkaline gases (e.g., SO_2, CO_2, NH_3), resulting in multiphasic reactions [43]. Most pretreatment technologies require bench-scale reactors to be not only resistant to corrosion by acid or alkali, but also able to operate at elevated temperatures ranging from 100 to 250 °C. Such high temperatures also require reactors to withstand high pressures at temperature without leaking (as opposed to some thermo-chemical reactors such as pyrolysis reactors that operate at high temperatures and near-atmospheric pressures). Accurate temperature control of the reactor contents is vital. Rapid heat-up and cool-down of bench-scale pretreatment reactors is essential in order to precisely control the pretreatment reaction time and obtain meaningful data, especially for kinetic studies. Heat and mass transfer characteristics of the pretreatment systems are key factors that control reactor performance, since solids loadings are usually above 1% [44].

Loading large biomass particles and high solids is difficult because bench-scale reactors are often available in small sizes such as glass vials or small pipe reactors. With small-size reactors, it is important to ensure easy charging of biomass and effective discharging of the pretreated materials from the pretreatment system without significant mass loss. Selection of bench-scale pretreatment systems is further complicated by biomass deconstruction patterns and economic considerations.

Over decades of research and development, substantial progress in biomass pretreatment technologies has been made in terms of understanding the fundamental mechanisms of pretreatment, process development, and integration of pretreatment with downstream processes in the biochemical platform for production of renewable biofuels and chemicals [3,21,22,45–47]. Application of various bench-scale pretreatment reactors has significantly contributed to such progress. Among the various pretreatment reactors reported in the literature, small-diameter reactors made of metal or glass allow rapid heat-up, good temperature control, and accurate closure of material balances even at the high solids loadings favored for a technology at the commercial scale [48]. Larger quantities of biomass can be pretreated in mixed reactors or steam reactors; however, these reactors have their own advantages and limitations. A few laboratory-scale continuous pretreatment systems have been developed that provide valuable insights into biomass deconstruction kinetics that are not possible using batch reactors. In this chapter, we will discuss configurations of commonly used bench-scale reactors and their applications, review biomass deconstruction with different pretreatment systems, offer guidelines for bench-scale reactor selection, and provide references for the reader who seeks more details.

23.2 Laboratory-scale Batch Reactors

Batch reactors are most commonly used in laboratory bench-scale pretreatment research to investigate pretreatment effectiveness of a wide variety of biomass feedstocks, pretreatment chemistries, temperatures, residence times, particle sizes, and impregnation methods. Various batch pretreatment reactors have been employed in different pretreatment systems including sealed glass reactors, metallic tube reactors, mixed reactors, Zipperclave reactors, microwave reactors, steam reactors and, more recently, high-throughput reactors [49,50]. Many pretreatment technologies, such as dilute acid pretreatment, hydrothermal pretreatment, steam explosion, and ammonia fiber expansion (AFEX), have been studied with batch reactors.

Typical features of bench-scale batch reactors and their applications are listed in Table 23.1. These reactors vary from handling greater than 1.5 mg biomass in glass reactors to 2000 g biomass in steam explosion reactors. Some reactors, such as steam explosion reactors, are suitable for pretreatment at high solids (>20% w/w insoluble solids) loadings. For other reactors, such as Parr reactors, the extent of solids loading also depends on the pretreatment mechanism and mixing. While high solids (>40% w/w insoluble solids) loadings of biomass have been employed in Parr reactors in the AFEX process, the mixing abilities of stirred Parr reactors limits solids loadings in pretreatment processes with biomass soaked in liquid, such as dilute acid. These reactors demonstrated consistent performance at elevated temperatures and pressures when the reactors are constructed of various materials that are compatible with the pretreatment technology. As a result, Parr reactors have been utilized for the pretreatment of a wide variety of biomass feedstocks with different pretreatment chemistries. In the following sections, bench-scale pretreatment batch reactors that are commonly used in various pretreatment systems will be described.

23.2.1 Sealed Glass Reactors

Glass is an inexpensive material for use in pretreatment vessels and has advantages such as resistance to most pretreatment chemistries, good sealing via a variety of means from flame sealing to silicone-lined crimp caps, and transparency that allows for observation of the biomass substrate and catalysts before and after pretreatment to gage reactor fill and pretreatment effectiveness [61,62]. However, glass provides poor heat transfer, has limited pressure resistance, and generally results in fragile reactor vessels.

An exception to the rule is for glass vessels used in microwave PT reactors; these 35 ml glass reactor tubes are rated to 300 psig at 200 °C or higher as discussed below. Pressure resistance in glass reactors can be improved by using a reinforcing brass or stainless steel disk between the crimp and the silicone seal [19] or by employing a secondary pressure containment vessel to encase the glass reactors [51].

Table 23.1　Typical features of bench-scale batch reactors.

Type	Size (g dry biomass loaded)	Solids loading (%)	Biomass particle size (mm)	Pretreatment technology	Refs
Sealed glass reactor	0.11 0.05	10% (w/v) —	0.42–1 0.17–0.85	1 M Maleic acid Hot water/dilute H_2SO_4	[19,51]
Tubular reactor	1–2	5–30% (w/w)	0.17–0.85	Dilute acid, water-only	[52,53]
Zipperclave reactor	50–300	Low to high solids (1–50% w/w)	6	Dilute acid-steam pretreatment	[54]
Parr reactor	50–100	5–10% (w/w) High solids (>20% w/w)	0.17–0.85 0.1–2	Dilute acid, water-only AFEX	[55,56] [57]
Steam explosion reactor	400–2000	High solids (>20% w/w)	$4 \times 4 \times 1$ cm Woodchips, 2–10 mm corn stover, wheat straw, etc.	Auto-hydrolysis, SO_2 and H_2SO_4 steam explosion	[35,58–60]

Lu and Mosier [19] used 1.5 mL borosilicate glass high-performance liquid chromatography (HPLC) vials with brass circles held on by modified crimp caps to reinforce the silicone septa seal during pretreatments. Multiple HPLC vials loaded with 1.5 mg of biomass and 1.5 mL of dilute maleic acid were plunged into an air fluidized sand bath at reaction temperature to rapidly heat the HPLC vials and contents to reaction temperature. At the end of pretreatment, the HPLC vials were plunged into an icy water bath to rapidly cool the reaction mixture.

Baugh and McCarty [61] used glass ampoules encased in brass vessels containing water-filled cavities to allow pretreatment conditions to vary between 170 °C and 230 °C at low pH. The brass vessels were lowered into hot oil baths for pretreatment.

Selig *et al.* [51] used 2 mL HPLC vials enclosed in 1 inch Swagelok unions with caps filled with water for dilute acid pretreatment of corn stover. The Swagelok unions containing sealed glass HPLC vials and water were first immersed in an air fluidized sand bath at 220 °C to rapidly raise the temperature in the vessel to near the desired reaction temperature; they were then quickly transferred to another sand bath at the reaction temperature for the desired reaction time. The Swagelok unions were then plunged into an ice-/water bath to cool the reaction quickly. This method allowed the glass ampoules or HPLC vials to come to reaction temperature in 1–2 min and to quickly cool within 15–30 sec.

A similar method was used by Chen *et al.* using Pyrex ampoules immersed in hot oil baths [62]. The first oil bath was set at a higher temperature to rapidly heat the contents, and the second oil bath was set at reaction temperature to control reaction conditions for the target time. Thermocouples were inserted into the center of the secondary containment vessels to allow temperature monitoring and data recording. These small glass reactors are used to obtain kinetic data on pretreatment of a variety of biomass feedstocks with different pretreatment chemistries to obtain kinetic data because of their rapid heat-up and cool-down. However, the biomass should be milled to fit the opening in the vials and the quantities of pretreated biomass slurries obtained for subsequent analyses are limited.

23.2.2　Tubular Reactors

Metallic tube reactors are significantly more robust than glass reactors and will usually withstand higher pressures. However, Hastelloy, zirconium, or tantalum tubing is needed to resist the corrosive effects of acid at higher temperatures. The availability and ease of use of stainless steel and Hastelloy tubing and Swagelok fittings allows customization of the reactors to accommodate available hot sand or hot oil baths [43]. Hastelloy tubing can be used to contain corrosive pretreatment chemicals, and less expensive stainless steel end caps can be used if custom-machined Teflon plugs are inserted in the ends of the tubes and contained by the stainless steel caps, a procedure recommended by Professor Y. Y. Lee from Auburn University in Yang and Wyman [63]. Custom-fabricated Carpenter 20 Cb-3 stainless steel tubing pipe reactors (12.7 mm × 102 mm) were placed in a custom-designed heating block in order to pretreat a number of different biomass samples under the same conditions [53]. The reactors were later gold plated to increase the corrosion resistance of the interior of the tubes for more severe pretreatment conditions. Selig *et al.* used 1 inch Swagelok unions that were gold plated inside to perform dilute-acid pretreatments [51]. Baskerville Reactors and Autoclaves (Manchester, UK) offers a multiple tube reactor system called the Multi-Cell that allows up to 10 pretreatment experiments in one set-up at a time and can be ordered in stainless steel, titanium, Hastelloy, or other materials-of-construction (http://www.baskervilleautoclaves.co.uk). Larger-volume reactors can be fabricated from 1 inch Swagelok unions with caps. The unions and caps can be ordered in Hastelloy, however plating the insides of the unions with gold may be less expensive. Zirconium reactor tubes (66 mL) were used in an Accelerated Solvent Extractor 350 (Dionex Corp) for hot water and dilute acid pretreatment of biomass [64–66].

The relatively small size of tubular reactors offers a low-cost and easy-to-use option for the development of a pretreatment technology, optimization of pretreatment conditions, and the study of reaction kinetics on a wide variety of feedstocks. Research results from tubular reactors have allowed development of several simulation models of hemicellulose hydrolysis [56,67]. Other benefits of tubular reactors include the possibility of rapid heat-up and cool-down of reactor contents and reproducible pretreatment with high solids loadings up to 30% [41]. Although tubular reactors are generally made from stainless steel for water-only and alkali treatments, Hastelloy C276 tubing for dilute acid treatment and glass tubular reactors have been reported for pretreatment with H_2SO_4 at 0.44–1.9%, and temperatures between 120 and 250 °C [68]. Tubular reactors can vary in size depending on the amount of biomass loaded but must be small enough in diameter to ensure that the tubular reactor is evenly heated to reasonably uniform temperatures across the reactor cross-section. For example, modeling of heat transfer within a tubular reactor cross-section showed that tubing diameters of less than 12.7 mm (0.5 inch) are preferred for nearly uniform temperatures throughout the tubular reactor [67,69]. Batch tubular reactors have been employed that use 12.7 mm OD × 0.889 mm wall thickness × 150 mm length (0.5 inch OD × 0.035 inch wall thickness × 6 inch length) with a 14.3 mL total volume, and 25.4 mm OD × 1.65 mm wall thickness × 114 mm length (1.0 inch OD × 0.065 inch wall thickness × 4.5 inch length), giving 45 mL total volume for dilute acid [70] and liquid hot water pretreatments [71]. However, the working volumes of these tubular reactors must be lower than the total geometrical volume of the reactors to allow room for liquid expansion during pretreatment [43].

A schematic of a typical apparatus used for tubular reactor pretreatments is shown in Figure 23.1. The typical set-up consists of one or more temperature-controlled air fluidized sand baths (or hot silicone oil baths) into which the tubular reactors are immersed, and a final ice-water cooling bath into which the reactors are plunged at the end of the reaction time to stop further reactions. The tubing for each reactor is cut to a length, typically 150 mm (6 inches), to be compatible with the sand bath dimensions. A pair of removable Swagelok tube fittings is attached to the ends of each tubular reactor, as described above. These end caps can be made of lower-cost stainless steel for dilute acid pretreatment if Teflon plugs are installed, as described above. To start dilute acid pretreatments, milled biomass is presoaked in dilute acid solution overnight. After excess water is removed to reach the desired solids loading level, the biomass is loaded into the reactor. A thermocouple probe is inserted into the center of each reactor to monitor temperature profiles during pretreatment [48,67]. The tubular reactors are positioned near the thermocouple inside the fluidized sand bath (e.g., TECHNE SBS-4 sand bath, Cole-Parmer, Vernon Hills, IL) to control the

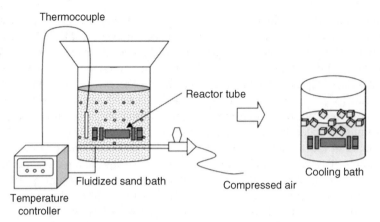

Figure 23.1   *Apparatus for pretreatment using a tubular reactor in a fluidized sand bath. (Reprinted with permission from [71]* © *2005, American Chemical Society).*

temperature. Care must be exercised to ensure the reactor(s) are not positioned too close to the surface of the sand bath, because the reactors may not reach target temperature or temperatures may oscillate.

According to thermal modeling results [67], a three-bath heat-up procedure can be applied in pretreatments over 160 °C in order to minimize the effects of temperature transients [70]. The reactor (0.5 inch OD × 0.035 inch wall thickness × 6 inch length) is preheated in 100 °C water for 2 min and then moved into a sand bath set at 20 °C above the target reaction temperature for 1 min. Finally, the reactor is transferred to a second sand bath at the target reaction temperature. This heat-up approach is particularly important for larger tube diameters of 25.4 mm (1.0 in) or more where the kinetics of the reaction are affected by heat transfer limitations [69]. After the target reaction time is achieved, the tubular reactor is plunged into an ice-water bath for 5 min to quench the reaction. However, it is recommended that the tube should be quenched in a room-temperature water bath before dropping into ice water in order to avoid thermal contraction of the tube and loosening of the end cap fittings due to sudden temperature changes. The ice-water cool-down step can be eliminated if the pretreated biomass slurry will be washed with hot water. After the pretreatment reaction is quenched and the pressure inside the reactor has dropped to ambient levels, the pretreated material is removed from each reactor for subsequent processing and analysis.

A large number of tubular reactors can be fit into an air fluidized bed sand bath for parallel pretreatments with replicates using a porous screened basket with each reactor wired to the screened wall of the basket. However, care must be taken that the number of reactors does not interfere with effective heat transfer with fluidized sand or other heat transfer media. A Multi-Cell reactor (Baskerville Reactor and Autoclaves, Manchester, UK) can be ordered equipped with ten 30 mL volume Hastelloy C-276 tube reactors connected in a ring. The commercially available reactor set is rapidly inserted into a large air fluidized sand bath (Techne Model IFB-121, Techne Inc., Cambridge, UK) set at 230 °C using an overhead hoist until the reactors come to within 10 °C of the desired temperature, and then transferred to a smaller sand bath (Techne SBL-2D) set and maintained at the desired reaction temperature. The temperature of each reactor cell is monitored by Inconel-clad thermocouples inserted into the centers of each reactor. The ring of reactors was removed from each sand bath by an overhead crane following pretreatment and lowered into a bucket of ice water to quench the reactions (N. Nagle, personal communication, 2001). Montane *et al.* used 25.4 mm (1 inch) 100 mL stainless steel pipe reactors for hot water pretreatment of almond shells at a solids loading of 7 wt% at temperatures ranging from 180 °C to 240 °C [72]. These small tube reactors can be used with a variety of biomass feedstocks with different pretreatment chemistries to obtain kinetic data due to their rapid heat-up and cool-down. However, the biomass should be milled to fit the openings in the tube reactors. The small tubular reactors provide limited quantities of pretreated biomass slurries for downstream analyses and fermentation.

23.2.3 Mixed Reactors

Mixed reactors, such as Parr autoclaves (Parr Instruments, Moline, IL) (Figure 23.2), are often used in biomass pretreatments to produce larger quantities of pretreated biomass than possible in the smaller reactor tubes. The volume of available Parr reactors vary from 25 mL up to 5 gallons and can be equipped with magnetic-drive internal stirring to permit long, continuous mixing at pressures up to 5000 psi (345 bar) (http://www.parrinst.com/products/stirred-reactors/). Stirring is needed to maintain uniform mass and temperature conditions but requires that the fluid viscosity be kept low, thereby limiting solids concentrations to less than about 10 wt%. Higher solids concentrations (>10 wt%) are challenging to mix adequately to maintain temperature uniformity [48]. Parr reactors are available in 316 type stainless steel; alloy 20; C1018 low carbon steel; titanium grades 2, 4 and 7; nickel 200; zirconium 702 and 705; and other materials depending upon the physical strength, temperatures, and corrosion resistance required. Typically, 600 mL and under Parr reactors are rated for 3000 psi, and 1 L and larger vessels can handle up to 2000 psi. The maximum operating temperature can be up to 225–500 °C depending on the design of the stirring apparatus, materials of construction, seals, and O-rings.

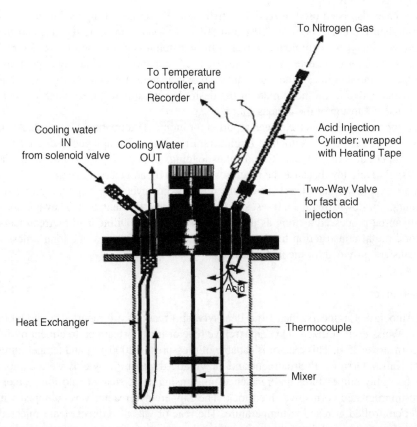

Figure 23.2 *Stainless-steel reactor (Parr model 4563) with acid-injection system, mixer and cooling coil. (Reprinted with permission from [77] © 1997, Elsevier).*

A 1 L Parr reactor constructed of Hastelloy C has been employed for dilute sulfuric acid pretreatment of biomass at 5% solids loading [55,56]. The 920 mL working volume reactor was rated for a maximum working pressure of 2200 psig and was equipped with a 3.5 inch diameter helical impeller on a two-piece shaft driven by a DC motor drive. The reactor was sealed, and agitation was set to start mixing after the presoaked biomass was loaded into the reactor. The reactor system was suspended from a chain hoist and lowered into a 320 °C sand bath to achieve heat-up to reaction temperature within 2 min [74]. The reactor was lifted 1–2 cm out of the sand bath as it approached the target temperature and maintained within ±2 °C of the target temperature by manually adjusting the position of the reactor in the sand bath with the hoist. The reactor temperature was monitored through a type K thermocouple probe inserted through a port in the reactor head plate. The reactor was hoisted into room-temperature water to quench the reaction. The pretreated slurry was removed from the reactor for further analysis and processing.

Esteghlalian *et al.* employed a 600 mL stainless steel Parr reactor equipped with a glass liner and heated by a heat exchanger for the dilute acid pretreatment of switchgrass [39]. A 2 L Parr reactor (304 SS, Model 4843; Moline, IL) was used in the pretreatment of corn fiber using water at 220–260 °C with real-time pH monitoring and adjustment [75]. The reactor was equipped with three turbine impellers and heated by an electrical heating jacket. Cooling water was circulated through a serpentine coil to cool the reactor contents at the end of each run. A bottom port and two inlet ports allowed sampling of the pretreated material and addition of reagents to the reactor.

Parr reactors have also been used in AFEX pretreatment of biomass at high solids loadings [76]. Milled biomass with moisture content between 20% and 80% was loaded into the reactor, and the reactor was slowly charged with anhydrous ammonia delivered from a stainless steel charging vessel through a port in the head plate. The reactor was heated with an electric heating mantle to the target temperature. The heating rate was affected by the amount of moisture and ammonia loaded. Reactor pressure was rapidly relieved through the exhaust valve in the head plate at the desired pretreatment time, which caused an explosive decompression and expansion of the biomass fiber.

Other bench-top mixed reactors have been reported for biomass pretreatment research. A 2 L Hastelloy C stirred reactor (Model EZE-Seal, Autoclave Engineers, Erie, PA) was employed in dilute acid pretreatment of biomass at temperatures of 160–200 °C and solids loadings of 5–10% [77]. However, the heating rate of 2–4 °C/min was relatively low because the reactor was heated by an external electric heating jacket. Mixed reactors are more expensive than glass vials or tube reactors; however, they provide significant quantities of pretreated biomass for downstream analyses, saccharification and fermentation. They are not limited in the feedstock that can be processed as long as the feedstock is milled. Dilute acid pretreatments will require expensive exotic metal construction to resist corrosion. Heat-up times can be long unless special high-temperature baths are provided for the initial heat-up.

23.2.4 Zipperclave

Zipperclave (Autoclave Engineers, Inc., Erie, Pennsylvania) reactors are reported in the biomass pretreatment literature. Weiss *et al.* utilized a 4 L Zipperclave reactor for the steam pretreatment of dilute-acid pre-impregnated corn stover [54]. This reactor is capable of pressures to 2000 psig and temperatures to 230 °C if equipped with Kalrez O-rings. Pretreatment and enzymatic digestibility yields were comparable to pretreatments at the same temperatures and residence times using a 4 L steam explosion reactor on the same dilute-acid pre-impregnated corn stover feedstock. The Zipperclave reactor was equipped with an electric heating blanket controlled at reaction temperature. The reactor was also direct steam injected. As a result of pre-warming the reactor and direct steam injection, the reactor reached the reaction temperature within 5–10 seconds.

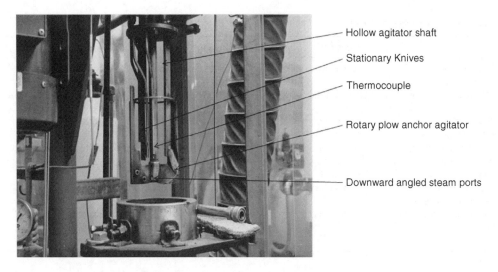

Hollow agitator shaft

Stationary Knives

Thermocouple

Rotary plow anchor agitator

Downward angled steam ports

Figure 23.3   *NREL Zipperclave reactor system. (See figure in color plate section).*

At the end of pretreatment, the reactor was rapidly vented (10–15 s) to heat exchangers to condense the flash vapors for collection and analysis in order to close material balances. The standard anchor impeller was modified to mix high solids feedstocks (>30% w/w) by mimicking a farmer's rotary plow to lift and turn the biomass towards the center of the reactor. Stationary knives help to scrape off biomass sticking to the vertical portions of the agitator. Steam injected through two downward-angled jets near the bottom of the agitator shaft help to mix and evenly heat the biomass charge. A 2.5 L Hastelloy pail containing the biomass charge was inserted into the reactor vessel, isolating the biomass charge from the condensate used to heat the reactor, thus reducing condensate dilution of the pretreated slurry.

The same Zipperclave reactor system was used for dilute ammonia (1–15 wt %) pretreatment of corn stover and cobs at temperatures between 120 °C and 160 °C, resulting in highly digestible slurries [78]. Thermocouples located near the bottom and in the headspace of the reactor monitored temperature during pretreatment (Figure 23.3). The Zipperclave reactors are more expensive than glass vials or tube reactors; however, they provide significant quantities of pretreated biomass for downstream analyses, saccharification and fermentation. They are not limited in the type of feedstock that can be processed as long as the feedstock is milled fine enough [54]. Dilute acid pretreatments will require expensive exotic metal construction to resist corrosion. Heat-up times can be of the order 5–10 s if the reactors are preheated and utilize direct steam injection.

23.2.5   Microwave Reactors

Various types of microwave devices have been used in laboratory research on biomass pretreatment since the mid 1980s (Table 23.2). Many microwave devices reported in the literature were modified from domestic or general purpose laboratory microwave ovens, and often temperature could not be controlled. In addition, the reaction vessels were not sealed. Bench-top microwave reactors are now available.

The CEM Discover single-mode microwave system (CEM Corporation, Matthews, NC) was applied to compare microwave and fluidized sand bath heating for biomass pretreatment using water or dilute sulfuric acid [79]. The microwave system was capable of testing multiple reaction vessels with volumes ranging from 4 to 35 mL (Figure 23.4). Tubular glass vessels were loaded with about 10 mL of presoaked milled

Table 23.2   *Microwave reactors reported in the literature. (Adapted from Shi et al. [80], © 2011, Elsevier).*

Microwave device	Feedstock	Pretreatment techniques	Conditions	Ref.
CEM Discover unit	Corn stover	Dilute sulfuric acid and water-only	5% solids, 140–180 °C, up to 40 min	[79]
MLS-1200 Mega Microwave workstation	Barley husk	Dilute sulfuric acid and water-only	200 °C/5 min, 210 °C/10 min	[81]
WD700 (MG-5062T) type domestic microwave	Rice straw	1% NaOH	uncontrolled, 15 min to 2 h	[82]
Turbora Model TRX-1963 domestic microwave	Rice straw/sugar cane bagasse	Glycerin in water solution	uncontrolled, 10 min	[83,84]
Toshiba Model TMB 3210	Softwoods	Water-only	219–226 °C	[85]
Customized Sharp/R-21 HT domestic microwave	Switchgrass	0.1 g NaOH/g biomass	190 °C, 30 min	[86]
Panasonic Corporation, model NN-S954	Switchgrass	1–3% NaOH	Temperature uncontrolled, 5–20 min	[87]

corn stover slurry before being sealed with Teflon-lined caps. The reaction vessels were automatically placed in the microwave device and irradiated to bring the temperature to a target value for the desired time period while the contents of the reactor were mixed with a multispeed magnetic stirring mechanism. A volume-independent IR temperature sensor was installed to monitor reaction temperature, and power was automatically controlled based on temperature feedback.

Synergy software was employed to monitor reaction conditions, including temperature, irradiation power, time, and pressure. An average microwave power input of 26 W and 75 W was needed to reach target pretreatment temperatures of 140 °C and 180 °C, respectively. Once the biomass pretreatment reactions were completed, the reactors were quickly cooled to 30 °C within 1–2 min by dropping the reactors into a room-temperature water bath. The pretreatment slurry was immediately filtered using glass fiber filters (Whatman GF/F-pore size 0.7 µm, Piscataway, NJ) to separate pretreatment hydrolyzate from pretreated solid residues and then stored at 4 °C for analysis. The pretreated solids were washed with room-temperature de-ionized water for subsequent enzymatic hydrolysis tests.

Custom-built microwave batch reactors equipped with stirrers have recently been reported) [88]. A batch high-pressure stainless steel vessel with a pressure sensor and thermocouple was used for the organosolvolysis of Japanese cedar wood chips (sap wood, 2–6 mm long, 2–4 mm wide, and 1–3 mm thick) [88]. The

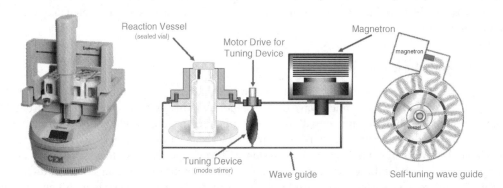

Figure 23.4   *Single-mode microwave system. (Adapted from http://www.cem.com/). (See figure in color plate section).*

vessel was connected with a 2.45 GHz microwave generator, and irradiation power was controlled automatically using feedback from measurements inside the reactor. Microwave reactors are more expensive than glass vials or tube reactors; however, they can be automated. They are not limited in the feedstock that can be processed as long as the feedstock is milled fine enough to be mixed. The biomass slurries that can be mixed are limited to less than 5% (w/v) solids. The glass reactor vessels resist most pretreatment chemistries, but pressure is limited to *c.* 300 psig. Heat-up times can be short (1–2 min) and cool-down can take a couple of minutes.

23.2.6 Steam Reactors

Commercial biomass applications such as Masonite [89–91] and particle board production require reactors capable of handling high solids loadings (>50% w/w) to reduce steam usage, allow rapid heat-up and cooling, and provide uniform temperature distribution throughout the reactor [92,93]. Commercial steam reactors process large amounts of biomass under high solids loadings where the temperature can be well controlled by controlling steam pressure. Direct steam injection rapidly heats biomass to the target temperature (typically 140–260 °C). Steam explosion reactors (Figure 23.5) based on the Masonite patents [89,90] have been utilized by a number of groups to achieve high hemicellulose solubilization yields for pretreatment of wood [94–96], corn stover [54,97], sallow [98] and wheat straw [99].

A 25 L steam explosion reactor was utilized for pretreatment of biomass without added chemicals [100] or pre-impregnated with acidic catalysts (e.g., SO_2, H_2SO_4). In general, steam enters a vertical biomass-filled pipe or chamber that is sealed by two valves. The top valve is used for charging the reactor with biomass. The bottom discharge or blow valve is used to empty the vessel. Depending upon the pressure the discharge can be quite violent, hence the term "steam explosion". The rapid release of pressure causes trapped condensate and steam within the pretreated biomass particles to rapidly expand and escape, causing defibration of the biomass particles [101]. DeLong and Ritchie [102] and DeLong [103] discharged pretreated biomass through a smaller diameter opening in the bottom of their steam explosion reactor to shear particles into smaller fragments following steam pretreatment. Although no increase in pretreatment yields was found, enzymatic digestion was improved substantially by the shearing action in the narrowed discharge valve.

The rapid heat-up possible with steam explosion reactors depends in large part on the piping and control valves leading into the reactor, the qualities and thickness of insulation, the mass of the reactor, and the mass and moisture content of the biomass charge within the reactor. With the NREL 4 L steam explosion reactor, the steam supply is split to enter the reactor from the top and the bottom to take advantage of the energy in the steam to initially fluidize the biomass charge and allow steam to evenly heat the biomass particles. The large automated Bauer rapid cycle digesters have steam entering from the top and the bottom [104].

Rapid cooling of the biomass charge following pretreatment occurs when the bottom blow valve is opened to discharge the pretreated biomass slurry into a flash tank or blow pit where flash steam rapidly removes heat from the charge while dropping the temperature to 100 °C [105]. The very rapid heat-up and cool-down possible using a batch steam explosion reactor allows for very precise control of the residence time within the reactor. The residence time control is much more precise than that available from any other type of pretreatment reactor, except possibly very-small-diameter glass or tubular reactors.

As well as rapid heat-up and cooling, another advantage of steam explosion reactors is that relatively large particle sizes such as wood chips can be directly treated, with the resulting slurry containing small particles of pretreated biomass due to defibration during the steam explosion process. This reduces costs of biomass milling and makes the pretreated slurry more digestible. When acidic catalysts or water are employed, biomass solids are presoaked in dilute acid or water [106] or purged with SO_2 [95] for hours

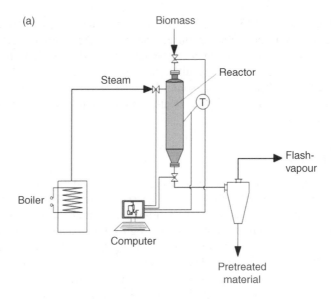

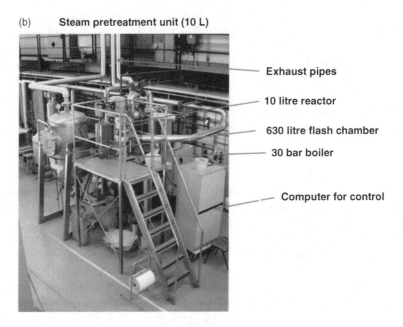

Figure 23.5 *University of Lund 10 L steam explosion reactor. (a) Schematic of the 10 L steam explosion reactor, and (b) photo of 10 L steam explosion reactor, flash tank, boiler and computerized controls. (Used with permission of Professor Guido Zacchi, Lund University, Sweden). (See figure in color plate section).*

prior to loading into the steam explosion reactor. NREL applied dilute-acid steam explosion pretreatments to biomass feedstocks soaked in dilute acid at 40–60 °C for 2–4 h prior to pretreatment [107]. Soaking has been shown to give higher yields in dilute acid pretreatments compared to spraying acid on the biomass particles prior to steam pretreatment [51,106].

Most bench-scale steam explosion reactors are designed by knowledgeable personnel and fabricated by ASME Code shops. Although steam reactors provide many advantages, closing mass balance is often difficult, mainly because of mass loss to the atmosphere in the flash tank during steam explosion. Furfural and acetic acid tend to be lost in the flash steam, unless recovered by condensing the flash vapors. For example, approximately 7.8% of the acetic acid and 61% of the furfural and HMF are recovered in the condensed flash vapor when a steam pretreatment reactor operating at 190 °C and 30% (w/w) solids discharges into a flash tank at 1 atm and 100 °C [108]. However, discharging the steam explosion reactor contents into high-temperature (Hotfill) bags lining the flash tank allows closing mass and component balances to be closed at nearly 100% by capturing and condensing these volatile components [60].

Steam explosion reactors are capable of pretreating a wide variety of biomass feedstocks, using various pretreatment chemistries, in a wide range of particle sizes, at a wide range of temperatures, with well-defined residence times suitable for acquisition of kinetic data. In addition, they can supply enough pretreated slurry for downstream processing, such as enzymatic saccharification and fermentation at the 1 L scale or larger. Because of the dangers associated with working with high-pressure steam and the need for Code-compliant pressure vessel design and construction, steam explosion pretreatment equipment is relatively expensive and requires special knowledge and training to operate.

23.3 Laboratory-scale Continuous Pretreatment Reactors

Only a few bench-top semi-continuous and continuous pretreatment systems have been reported [48]. Batch bench-scale pretreatment systems often meet the needs of academic research on fundamental mechanisms, while development of continuous systems is more expensive to pursue. Some semi-continuous and continuous steam reactors are commercially available [109,110]. Due to the limitations of space, only bench-scale continuous pretreatment systems such as flowthrough systems are discussed in this chapter.

Flowthrough pretreatment systems employ flow of liquid through a packed bed of biomass to constantly remove soluble or low-molecular-weight products from the reaction zone and reactor [48]. Flowthrough pretreatment allows the investigator to determine release patterns of hemicellulose, lignin, cellulose, and other components in biomass during pretreatment, thus providing valuable data on reaction kinetics which other reactor systems, such as batch tubular reactors, Parr reactors, and steam explosion reactors, cannot provide [111]. Flowthrough reactors and other similar flow systems, such as continuous plug-flow reactors [112,113], tubular percolating reactors [114], countercurrent/co-current reactors [115], and countercurrent shrinking bed reactors [116] have been applied in water-only dilute-acid ammonia recycle percolation (ARP) [117], two-stage (water-only followed by ammonia) [118], and other pretreatment processes. Research has shown that flowthrough pretreatment with water resulted in high sugar yields in pretreatment and subsequent enzymatic hydrolysis of the residual cellulose, as well as much higher lignin removal than for batch pretreatment [119]. Most of the xylan solubilized with hot water flowthrough pretreatment was in oligomeric form. Flowthrough systems have proven to be an effective and valuable approach for studying pretreatment mechanisms; however, the high liquid demand hinders commercialization of such systems.

Flowthrough systems have been investigated with various pretreatment techniques. Lee *et al.* utilized 2 inch diameter flowthrough reactors fabricated from titanium with sintered titanium frits to retain biomass [116]. The pretreatment process utilized a flow of hot (170–220 °C), very dilute (0.08 wt% H_2SO_4) acid through the reactor. The acid was heated using a heating coil of 0.25 inch Hastelloy tubing immersed in a sand bath set at the reaction temperature. The flowthrough reactors were equipped with spring-loaded pistons that followed the "shrinking bed" as biomass components were solubilized. The pistons prevented freshly solubilized biomass components from being diluted by the solution that normally would have occupied the void volume created by pretreatment, and are rapidly removed from the reaction zone. This reactor configuration led to very high yields, achieving >92% hemicellulose and cellulose solubilization [116].

Lee's group also utilized a small flowthrough reactor to circulate concentrated solutions of ammonia (10–15 wt%) through packed beds of corn stover biomass within the reactor [120,121]. The reactor was constructed from 25.4 mm OD (22.9 mm ID) heavy-walled 316 stainless steel tubing of 25.4 cm length. The reactor was placed in a convection oven to heat to reaction temperature. Nitrogen was applied to the reactor system (2.5 MPa) to prevent the ammonia from boiling during pretreatment. A 1 L stainless steel cylinder was used as a receiver, and ammonia solution was pumped through the packed biomass bed into the receiver. The reactor was cooled and quenched following reaction.

Small reactors of 3.6 mL (0.5 inch ID × 1.84 inch length) and 14.3 mL (0.5 inch ID × 6 inch length) total volume, made from Hastelloy C-276 for dilute acid pretreatment or 316 stainless steel for hot water, were reported in testing the deconstruction of corn stover [63,122]. In order to deliver liquid through solids, a high-pressure pump was connected to the flowthrough tubular reactor (Figure 23.6). The flow rate could be adjusted from 0 to 40 mL/min. Water with or without catalyst (e.g., very dilute sulfuric acid) was preheated to the target temperature in a preheating coil (316 stainless steel tube, 0.25 inch OD × 0.35 inch wall thickness × 50 inch long) prior to entering the flowthrough tubular reactor. The outlet tubing was a 316 stainless steel tube (0.25 inch OD × 0.35 inch wall thickness × 50 inch long) used to cool down the effluent. The pressure to the reactor was regulated through a back-pressure regulator and monitored by a pressure gage

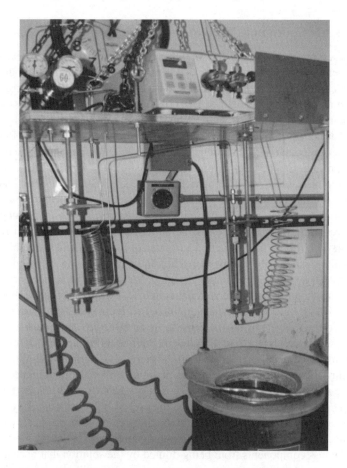

Figure 23.6   *Flowthrough pretreatment reactor system. (Adapted from Yang et al. [111] © 2004, American Chemical Society). (See figure in color plate section).*

(pressure range 0–1500 psi). Temperature was monitored through a 0.25 inch stainless steel thermocouple at the outlet of the reactor. Milled biomass solids were loaded into the reactor. The reactor was covered with two gasket filters (316 stainless steel, average pore size 5 μM) at both ends and connected to the inlet and outlet tubing. Air in the substrate and reactor was then purged out by pumping room-temperature liquid into the reactor for a few minutes until the pressure in the reactor reached the set pressure of 350 to 400 psig. Flow was then stopped to allow soaking of the biomass substrate. After preheating in a 100 °C fluidized sand bath for 2 min, the inlet tubing and the reactor were moved into a second sand bath at the target reaction temperature and liquid started flowing through the reactor at a set flow rate. The reactor and the inlet tubing were soaked in an ice-water bath to quench the reaction at the desired end time. The effluent during pretreatment was collected for further analysis. The pretreated solids were removed from the reactor and washed for compositional analysis and subsequent enzymatic hydrolysis testing (Figure 23.6).

Kim and Lee [118] utilized hot water or hot aqueous ammonia flowing through a packed bed of corn stover biomass in a tubular flowthrough reactor maintained in an oven at temperatures ranging from 170 °C to 200 °C. The flow rate was adjustable. The effluent from the reactor was cooled with a heat exchanger before discharge into one of two receiving vessels. Back pressure was maintained on the system to prevent premature flash cooling by regulating nitrogen pressure on the system. Makishima *et al.* constructed a continuous hydrothermal hot water reactor system that utilized a Moyno pump to push a slurry of corn cobs and water through electrically heated loops of piping that formed the reactor [123]. The heaters maintained temperature between 200 °C and 210 °C. An additional loop of piping was cooled by chilled water to reduce the temperature of the hot pretreated slurry, before discharge from a set of two discharge valves that isolated the high pressure of the reactor piping from the low-pressure product recovery tank.

23.4   Deconstruction of Biomass with Bench-Scale Pretreatment Systems

For bioconversion of cellulosic biomass, deconstruction of biomass through pretreatment often involves hydrolysis of hemicellulose and alteration of lignin, as well as improved conversion of cellulose and residual hemicellulose by enzymes in subsequent enzymatic hydrolysis steps [105,124]. In terms of effective biomass deconstruction, some key areas that should be targeted when developing advanced pretreatment methods include the following: (1) high yields (close to 100%) of fermentable hemicellulose sugars; (2) greater than 90% cellulose-to-glucose conversion from enzymatic saccharification of pretreated biomass in less than 5 days (preferably less than 3 days) using a cellulase enzyme loading of less than 15 FPU/g glucan (c. 20 mg protein/g cellulose based on assays at NREL for Genencor Spezyme CP, Lot No 301-05021-011, [5,125], FPU and protein concentrations of other cellulase enzyme preparations will need to be determined independently); and (3) recovery of lignin and other constituents for conversion to valuable co-products and to simplify downstream processing.

Size reduction of biomass is energy intensive and costly; therefore limited comminution up front should be applied [126]. However, some size reduction is necessary for effective and homogeneous diffusion of steam and catalysts in pretreatment to subsequently and efficiently deconstruct biomass using enzymes [127]. Other factors besides the effectiveness of biomass deconstruction, such as less expensive reactor materials-of-construction, lower chemical costs, and reduction of conditioning chemicals and processing requirements, are important to ensure that data from bench-scale pretreatment systems also lead to cost-effective scale-up.

The effectiveness of biomass deconstruction using enzymes following pretreatment largely depends on the mechanisms of the applied pretreatment technology [3]. In-depth comparative studies of various promising pretreatment technologies have been well documented by the CAFI projects [45]. Table 23.3 lists results from deconstruction of corn stover by some leading pretreatment technologies, including hydrothermal and acidic or alkaline catalysts. All listed pretreatment technologies achieved near 80% or greater

Table 23.3   *Corn stover deconstruction by various bench-scale pretreatment systems.*

Pretreatment technology	Reactor	Conditions	Xylose yield %[a]	Lignin % dw pretreated solids	Cellulose digestibility %[b]	Overall sugar %[c]	Reference
Dilute acid	tubular	0.49% H_2SO4, 160 °C 20 min	85	22.5	91.1	92.4	[52]
Water-only	Flowthrough tube	200 °C, 24 min, water-only	96.3	7.1	95.5	96.6	[128]
AFEX	Parr	90 °C, 15 min, ammonia	—	17.2	96	94.4	[12]
SO_2-steam explosion	Steam gun	190 °C, 5 min, 3% SO_2		27.8	87	78.5	[97]
Controlled pH	Plug flow tube	190 °C, 15 min, water	57.8	25.2	85.2	87.2	[32]
ARP	Flowthrough tube	5%–15% ammonia, 170 °C, 10 min,	47.2	8.7	90.1	89.4	[129]
Lime pretreatment	Packed-bed PVC columns	55 °C, 4 wk, 0.1 g $Ca(OH)_2$/g	24.4	25.2	93	86.8	[130]

[a] Sum of mono- and oligomeric xylan pretreatment yield determined on a basis of original xylan content of raw corn stover.
[b] Cellulose conversion on the basis of glucan content in pretreated solids at 72 hr with 15 FPU/g glucan loading.
[c] Combined xylose and glucose, including mono- and oligomers, yield after pretreatment and enzymatic hydrolysis on basis of original combined xylan and glucan content in the original raw corn stover feedstock.

overall sugar yields (combined glucose and xylose yields after pretreatment and enzymatic hydrolysis, including oligomeric and monomeric sugars). Dilute acid, water-only flowthrough, and AFEX pretreatment technologies achieved 90% or greater yields [131]. Dilute acid and water flowthrough pretreatment technologies both resulted in high xylose yields from pretreatment, but a larger percentage of the xylan was hydrolyzed into oligomeric sugars for hot water flowthrough pretreatments. AFEX pretreatment technology resulted in little solubilization of xylan or lignin during pretreatment; however, the overall enzymatic sugar yield was found to be 94.4%. Flowthrough pretreatment using water and ARP reduced lignin content in the solids from 17.2% of raw corn stover to 7–8% of dry weight in the pretreated solid residues. Cellulose conversion was significantly improved to more than 85% in the pretreatments using a modest enzyme loading of 15 FPU/g glucan for 72 h [120].

On the other hand, the configuration of the pretreatment reactor plays an important role in hemicellulose hydrolysis, alteration of lignin mobility, and subsequent enzymatic hydrolysis of the residual cellulose. Some comparative studies using various pretreatment reactors with similar pretreatment technologies have been reported [132,133]. Dilute sulfuric acid pretreatment in different reactor types has proved to be promising in terms of hemicellulose hydrolysis and enhanced cellulose digestibility in the subsequent enzymatic hydrolysis step [133]. Similarly, high rates of overall xylose and glucose solubilization are achieved using hydrothermal hot water pretreatment. However, the same or higher yields are reached using dilute acid pretreatment with less time and/or lower temperatures. Dilute acid pretreatment has been applied in different pretreatment reactors, including tubular reactors, flowthrough reactors, stirred batch reactors, and steam explosion reactors.

Figure 23.7 compares dilute acid pretreatment under similar pretreatment conditions (dilute acid concentration of 0.5%) for tubular reactors [52] and a steam gun reactor [134] with similar enzyme loadings for

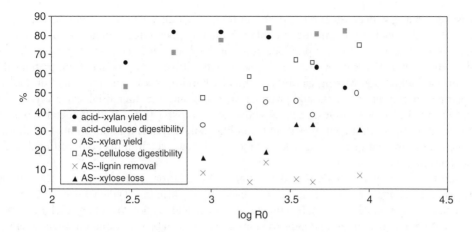

Figure 23.7   *Pretreatment effects on corn stover for dilute acid pretreatment in tubular reactors [52] and acid-steam explosion pretreatment [134] with steam gun at 0.5% H_2SO_4. (AS: acid-steam explosion; xylan yield: sum of monomeric and oligomeric xylan yield based on original xylan in raw corn stover; xylose loss: calculated on basis of original xylan in raw corn stover; lignin removal: based on original component in raw biomass; cellulose digestibility: based on glucan content of pretreated corn stover.)*

subsequent enzymatic hydrolysis (60 FPU/g glucan and 54.2–71.8 FPU/g glucan for acid pretreatment with tubular reactors and acid-steam explosion with steam gun, respectively). The tubular reactor pretreatment temperature (160 °C) was lower than the steam gun pretreatment temperature (190–210 °C). The acid-steam pretreatment time (2–5 min) was much shorter than the time (5–120 min) for the tubular reactor heated by fluidized sand baths. Nevertheless, the pretreatment severities log R_0 (see definition of severity in Section 7.5) for both reactors were similar, in the range 2.5–4. Results show that high xylan removal of more than 75% could be achieved with both reactors although high xylose loss was observed with the steam gun. Such observations are in line with some other studies [97,135] that showed high xylan yields (sum of monomeric and oligomeric xylan) of over 80% for a tubular reactor at around log $R_0 = 3$ and that xylan degradation increased as severity increased. Cellulose digestibility over 70% was achieved for both reactors although lower pretreatment severity was needed for the tubular reactor. Similar to batch tubular acid pretreatment results [119], the low levels of lignin removal by acid-steam explosion suggested possible condensation of lignin.

Other studies also provide comprehensive comparisons of different reactor configurations. Laser *et al.* compared autohydrolysis of sugarcane bagasse with liquid hot water and steam pretreatment in the same direct steam heated reactor [135]. They concluded that both pretreatments resulted in pretreated solids that were highly accessible by enzymes, but liquid hot water treatment led to a much higher sugar yield from hemicellulose [135]. Yang *et al.* reported discernable differences in biomass deconstruction between batch tube and flowthrough tubular systems for autohydrolysis and dilute acid pretreatment [119]. These results suggest that a large portion of the lignin could be removed by liquid flow, while lignin could be recondensed on solids in batch tube pretreatment.

23.5   Heat and Mass Transfer

Heat and mass transfer are critically important for bench-scale reactors. Temperature and pretreatment catalyst concentration gradients lead to lower yields and larger amounts of biomass left under-converted or over-converted to unwanted degradation products [111,136]. Percolation of steam up through the biomass bed into the head space, with constant bleed of the air trapped within the biomass from the reactor, is needed for high solids (>20 wt%) pretreatment to maintain limited temperature gradients within the reactor. For lower solids pretreatments, mixing can provide a more or less homogeneous temperature and chemical catalyst concentration within the reactor. Mixing also helps with diffusion of products from pretreated biomass particles into the bulk liquor. Mixing can reduce product build-up within the pores and on the surface of biomass particles during pretreatment and promotes diffusion into the interstitial space between particles. The use of minimal reactor volumes, such as sealed glass vials [19] or small-diameter tubular reactors [119], allows effective pretreatment at both low and high solids concentrations without mixing or use of steam; however, the amount of pretreated slurry produced is low, which limits compositional analyses and enzymatic digestibility evaluations.

Scaling-up laboratory bench-scale results to pilot, demonstration, and industrial scale is fraught with difficulties as heat and mass transfer characteristics can be quite different. Typical bench-scale pretreatment reactors can take many minutes to reach reaction temperatures and to cool down to quench reactions. On the other hand, heat-up of typical pilot- and industrial-scale reactors can be very rapid for direct steam injection, while cool-down can occur virtually instantaneously as the pretreated slurry is discharged into a flash tank or blow pit. However, the use of bench-scale reactor equipment that is direct steam injected faces challenges such as expensive materials-of-construction, reactor sealing under pressure, requirements for ASME Code-stamped pressure vessels, slurry dilution by steam condensation during pretreatment, bulky flash tanks, the need for a steam boiler, and a lack of personnel experienced in boiler operation.

23.5.1   Mass Transfer

Hemicellulose hydrolysis can be an important pretreatment objective in order to recover high yields of fermentable sugar. Saeman's first-order homogeneous kinetic model of cellulose hydrolysis in a dilute acid batch system [137] was initially adapted to describe hemicellulose hydrolysis. Based on these early models, hemicellulose hydrolysis models were modified to include two different fractions of hemicellulose, one of which is more easily hydrolyzed than the other [138]. These models were further advanced to include other factors, such as oligomeric sugars [139–141]. Most models evolved from the homogeneous two-step first-order reaction kinetic model that assumes these reactions follow a first-order dependence on reactant concentration with an Arrhenius temperature relationship for the rate constant. In fact, biomass pretreatment reactions often involve solids, liquids, and sometimes gases; mass transfer, solubility limitations, and non-homogeneous reactions at the solid-liquid interface could therefore play important roles.

Few models have taken mass transfer effects into account. This may be one of the important reasons why no models are found in the literature that can accurately describe changes in performance observed for different reactor configurations. For example, a model that includes the effects of mass transfer on hemicellulose hydrolysis was developed by Brennan [74]. Table 23.4 shows that the mass transfer parameter k_D determined by this model increased in the following order: batch tubes, stirred batch, and flowthrough reactors. Increasing flow rates also increased the mass transfer parameter.

23.5.2   Direct and Indirect Heating

Direct heating can provide very rapid heat-up to target reaction temperature, which is especially suitable for high solids loadings and treatment of larger quantities of biomass. Direct heating media for bench-scale pretreatment reactors include steam, hot air or gas, hot compressed water, and hot compressed CO_2 [142]. For example, a steam explosion reactor (4 inch ID × 24 inch length) that can process about 1–2 kg of dry biomass per batch can raise the inside temperature of the steam gun to target temperatures of between 140 °C and 260 °C within about 15 s after the jacket is preheated with steam for 2–4 min [48].

Table 23.4 *Mass transfer parameter k_D (cm^2/s) determined for various reactor configurations for corn stover treated at 180°C. (Adapted from Yang et al. [111] © 2004, American Chemical Society).*

Reactor type	Water-only	0.05 wt% H_2SO_4	0.1 wt% H_2SO_4
Batch tube	0.0025	—	0.0036
Stirred batch	0.0058	—	—
Flowthrough: 1 mL/min	0.004	0.0075	0.02
Flowthrough: 10 mL/min	0.011	0.025	0.15

Indirect heating approaches for bench-scale pretreatment reactors reported in the literature include use of electrical heating jackets, steam coils, hot/cold plates, microwaves, hot oil baths, and hot air fluidized sand baths. The heat-up rate by a heating jacket of 3.5–4 °C/min was determined using a 2 L Parr reactor [75]. Heating up a bench-scale reactor in a hot air fluidized sand bath is relatively fast, taking less than 2 min to heat up a 1 L Parr reactor loaded with biomass and water (with or without dilute acid) to more than 160 °C with mixing by a double blade mixer [74].

Unlike convection- or conduction-based heating, microwaves (300 MHz–300 GHz) use an electromagnetic field interacting with molecules containing dipole moments such as water to cause vibrational and rotational motions that release heat to the target object directly through intra-molecular collisions and heat transfer mechanisms [143]. In addition to home usage, microwave technology has been widely applied to various industrial and research areas such as food processing, analytical chemistry, heating and vulcanization of rubber, plasma processing, chemical synthesis and processing, and waste remediation [144]. Thermal and non-thermal effects have been proposed to explain the interaction of microwaves with processed materials [143,145,146]. As well as rapid internal heat-up, a "speed-up effect" was reported whereby the decomposition rate of many difficult-to-dissolve materials was greatly enhanced (by a factor of 10–100) by microwaves [147]. The heating mechanism of microwaves contributes to reducing temperature gradients typically found in conductive methods and overcoming mass transfer limitations, especially with high solids loading.

A comparative study of microwave and air fluidized bed sand bath heating using tubular reactors for hydrothermal and dilute sulfuric acid pretreatment of corn stover under identical conditions (i.e., temperature, acid concentration, time, and solids loading) showed that heat-up and cool-down temperature profiles were similar for both heating options [80]. Similar-sized tubular reactors were made of metal and glass for sand bath heating and microwave heating, respectively. Even though metal usually has better heat transfer than glass, results indicated that it took 2 min and 2.9 min by microwave heating to reach 140 °C and 180 °C, respectively; sand bath heating required 0.1 min less and 0.3 min more to heat up to the same temperatures, respectively. Both heating options have been proven to enable temperature control within ±2 °C. Nevertheless, the effects of biomass deconstruction were quite different: the depolymerization and degradation rates of xylan were faster and glucan conversion by enzymes following pretreatment was higher by microwave heating than by sand bath heating.

The heat-up rate also depends on the energy requirement to heat up the mixture of biomass, water, and/or catalysts as well as the types of indirect heating employed. For example, the amount of low-temperature ammonia loaded in AFEX pretreatment was a key factor that determined the heat-up rate in a Parr reactor [76]. The heat-up time for tube reactors in air fluidized sand baths varies significantly depending on the size of the tube reactors, the solids loading, and the type of cellulosic biomass. Heat-up profiles for a tubular reactor (1.0 inch OD, 0.065 inch wall thickness, 4.5 inch length) containing corn stover and water to a target temperature of 190 °C by a fluidized sand bath showed that the heat-up time varied from *c.* 7 min to 9 min as the solids loading was changed from 20 g/L to 300 g/L [148]. When a tube reactor of 0.5 inch OD was employed, the heat-up time to reach 160 °C was shown to be about 2 min for two sand baths heating in series [44].

As the reactor size increases, a higher indirect heating temperature is most likely required in order to heat up the biomass to the target temperature within a reasonable time. However, too high an indirect heating temperature may result in overheating or even charring the biomass layer next to the reactor walls. This charring can complicate mass transfer issues occurring with high solids loadings in the reactor as well as decrease yields.

23.6   Biomass Handling and Comminuting

The heterogeneity of biomass feedstocks affects pretreatment because different anatomical fractions such as leaves can pretreat differently from other parts such as the rind from the same stalk of the same plant [149,150]. Comminution of biomass at the small scale therefore remains a more or less constant problem due primarily to the heterogeneity of the material and lack of samples of sufficient size to represent the entire crop from the field. A large enough sample of feedstock is needed in order to represent what is found in the fields, yet milling many tens of kilograms of biomass feedstock through a small laboratory mill is time consuming. The large sample of biomass must be well mixed after milling to result in a homogeneous mixture that is then split into smaller containers and the aliquots stored. The aliquots must be mixed just before sampling for use in the pretreatment reactor in order to re-homogenize the mixture due to the segregation and settling of the finer materials that occurs during storage. This can result in mixtures of different proportions and composition from the original sample.

Wet materials are much more difficult to handle with most size reduction mills, and feedstocks must be stored at 4 °C or frozen at −20 °C in order to reduce microbial growth and degradation. To our knowledge, the effects of freezing and thawing of stored biomass on pretreatment has not been evaluated, although freezing and thawing occur for biomass that is stored outside in some climates. Long-term storage of dry (*c.* 20–25% moisture) corn stover in bales can lead to losses of 3.3% for indoor storage and *c.* 18% for outdoor storage [151]. Biomass such as corn stover can be milled at the farm using tub grinders [152], while woody softwood and hardwood logs require chipping to reduce the particle size to manageable pieces [153]. Bale grinders and garden shredders (usually of a hammer mill or swinging bar type of construction) can be utilized to reduce particle size for straws such as wheat and rice [154], although knife mills can also be used [155]. Further size reduction utilizing laboratory knife mills (e.g., Wiley mills) is usually required to reduce the biomass particle size enough to fit into the pretreatment reactor. In general, comminution can increase the enzymatic hydrolysis yields, especially if milling is extensive, but removal of hemicellulose and lignin in pretreatment is more important [126]. Steam explosion reactors can utilize larger particles (10 mm or larger), while dilute acid and alkaline pretreatments generally require particles of size 1–3 mm because of diffusion and other kinetic limits [126,156].

23.7   Construction Materials

23.7.1   Overall Considerations

Operational pH, temperatures, and pressures are essential considerations in selecting reactor materials of construction. The cost of construction materials for the reactor and related accessories can become a large portion of the capital costs, depending upon the pretreatment technology [40,107]. Reducing capital costs associated with the pretreatment reactor is a goal of bench-scale pretreatment; lower severity conditions, temperatures, pressures, and catalyst loadings are targeted in order to develop a cost-effective pretreatment technology. However, tradeoffs are made and various pretreatment technologies compete with each other based on reactor volumes, exotic materials of construction due to highly corrosive chemical environments, operating pressures, recovery and recycling of catalysts, water usage, sugars yields in subsequent enzymatic hydrolysis processes, and fermentability of the pretreated and enzymatically saccharified hydrolyzate liquors and slurries.

23.7.2   Materials of Construction

Glass is relatively inexpensive, has excellent resistance to most corrosive chemicals, and is easily amenable for customized design and repair. However, glass is usually fragile, especially with rapid temperature changes and under high pressure. The heat transfer rate through glass reactor walls is relatively low compared to heat transfer rate in metallic reactors; thus heat-up and cool-down require longer times. Nevertheless, the use of a glass-lined Pfaudler reactor for pretreatment [157] is an excellent alternative to exotic alloys such as Hastelloy, although extreme care is needed to ensure that maximum heating and cooling rates are not exceeded during the heat-up and cool-down cycles of typical glass-lined reactors to prevent cracking of the glass lining. Glass-lined reactors are more suitable for continuous operation than batch operation because the repetitive heating and cooling in batch operation is detrimental to the glass lining.

Type 316 stainless steel (316LSS) is a good material for use with most organic systems. It is resistant to acetic, formic, and many other organic acids. However, a few organic acids and organic halides can hydrolyze under certain conditions to form inorganic halogen acids that rapidly attack all forms of stainless steel, even at low temperatures and in dilute solutions. Although 316LSS is quite resistant to dilute sulfuric, sulfurous, phosphoric, and nitric acids at ambient temperatures, these acids readily attack 316LSS at elevated temperatures and pressures. Type 316SS is therefore not normally the material of choice for inorganic acid systems. Type 316LSS has good resistance to ammonia and to most ammonia compounds. With the exception of halogen salts and especially chlorides, many salt solutions (particularly neutral or alkaline salts) can be routinely handled in stainless vessels. At moderate temperatures and pressures, 316LSS can be used with most commercial gases. Stainless steel has been widely used in reactors and relevant accessories in neutral and alkaline pretreatment systems, such as hot water and AFEX pretreatment.

Nickel-based alloys exhibit higher resistance to corrosion. Hastelloy, a trademark of Haynes Internationals (Kokomo, IN), refers to a range of highly corrosion-resistant metal alloys. As well as the predominant alloy ingredient nickel, other alloy elements including molybdenum, chromium, cobalt, iron, copper, manganese, titanium, zirconium, aluminum, carbon, and tungsten are incorporated to produce a series of corrosion-resistant alloys. Metallic construction materials made of different ingredients exhibit different thermal conductivities, as shown in Table 23.5. Hastelloy alloys are resistant to moderately to severely corrosive and/or erosive, high-temperature, and high-stress environments in which other metallic materials often fail. Hastelloy is therefore widely used for reactors in the chemical industry. Hastelloy series, such as Hastelloy C-276 and C-2000, are reported for pretreatment reactors handling acidic catalysts [48]. Hastelloy C-2000, C-22, and C-276 are highly temperature and corrosion resistant, and the high nickel, molybdenum, and chromium content makes these alloys highly weldable for fabrication purposes.

Table 23.5   *Temperature limits of metallic construction materials (http://parrinst.com).*

Materials	Temperature (°C)	Materials	Temperature (°C)
T316/316L Stainless Steel	600	Nickel 200	316
Alloy 20C	427	Titanium Grade 2	316
Alloy 400	482	Titanium Grade 4	316
Alloy 600	600	Zirconium Grade 702	371
Alloy B-2	427	Zirconium Grade 705	371

23.8   Criteria of Reactor Selection and Applications

23.8.1   Effect of High/Low Solids Concentration on Reactor Choices

The concentration of solids within pretreatment reactors limits available reactors choices because of the limitations on mixing. The maximum solids concentrations that are capable of being mixed with a given pretreatment reactor are feedstock dependent. In general, stringy feedstocks such as straws and corn stover are more difficult to mix than granular feedstocks such as milled corn cobs because stringy materials have a tendency to wrap around agitators and screws and to bridge and clump at discharge chutes and ports. An exception is steam explosion reactors if the steam and discharge systems are designed correctly. Low solids concentrations (<10 wt%) are easier to mix using vertical impeller-type reactor designs. However, higher solids concentrations are more desirable commercially because they should result in decreased reactor size and costs and achieve higher sugar concentrations in the pretreated slurries, leading to increased product concentrations in fermentation and decreased energy costs for distillation [5,107].

23.8.2   Role of Heat-up and Cool-down Rates in Laboratory Reactor Selection

Pretreatments carried out at lower temperatures and/or catalysts concentrations but with longer residence times can be carried out in less-expensive larger reactors or in reactors heated with air fluidized sand baths or hot oil baths suitable for laboratory usage. For rapid heat-up of small pipe and glass reactors tubes, two or more sand baths or hot oil baths are usually required. The first sand or hot oil bath is set at a temperature at least 20 °C higher than the reaction temperature in order to heat the reactors rapidly to almost reaction temperature, followed by moving the vessel to a second hot sand or oil bath that is set at reaction temperature in order to maintain the reaction temperature for the required period of time. Steam pretreatment reactors, either steam explosion, Zipperclave, or modified Parr reactors with direct steam injection [50], are capable of rapid steam heat-up over 10–30 s and rapid cool-down from virtually instantaneous for steam explosion to 30 s if the Zipperclave or Parr reactors are vented carefully to prevent boil-over from clogging the discharge valve. Microwave reactors can heat the contents of their glass vessels to pretreatment temperatures in 1–2 min, with cool-down taking a couple of minutes.

23.8.3   Effect of Mixing and Catalyst Impregnation on Reactor Design

Currently available strategies to improve mass transport in bench-scale pretreatment systems are limited. Small-sized packed bed reactors with narrow diameters that handle relatively small amounts of biomass under low to high solids loadings are commonly employed in pretreatment research so that mass and temperature gradients can be reduced. Mixing with impellers in Parr reactors has been shown to be effective at low solids loadings. With larger-size reactors, steam is applied to heat up the contents and to facilitate heat and mass transport through steam penetration during pretreatment.

When pretreatment involves liquid or gaseous catalysts, such as dilute acid, ammonia, alkali, SO_2, NH_3, or CO_2, proper catalyst impregnation techniques are essential to distribute the pretreatment catalyst(s) at least evenly, if not homogeneously, throughout all of the biomass particles. Selig *et al.* showed that this is especially necessary for pretreatments at higher solids (>10 wt%) loadings where inter-particle diffusion of catalyst is limited due to limited inter-particle surface contact and time for diffusion versus available residence time within the reactor [50]. If the catalyst is not present at every hydrolyzable bond, then the time required for the catalyst to diffuse into the interior of the biomass particles during pretreatment is limited and degradation of the already-hydrolyzed sugars occurs [158], resulting in lower yields. For low solids

($<$10 wt%) loadings, there may be sufficient diffusion of pretreatment catalyst from the bulk solution into the biomass particle even at relatively low pretreatment temperatures, especially for longer residence times. However, heating of the dilute slurry may take a long period of time compared to the desired pretreatment residence time, making application of the data for scale-up difficult. For gaseous catalysts such as CO_2, SO_2, and NH_3, limited inter-particle diffusion may not be a factor as the gas may diffuse rapidly throughout the reactor contents during pretreatment, especially for high solids pretreatment where there is sufficient space between particles.

The use of steam to rapidly heat higher solids ($>$10 wt%) loadings to reaction temperature requires a reactor design that allows the steam to separate the particles and penetrate the pores of the biomass causing rapid heating of the particle from the outside in, as well as the inside out. If the pores of the particles are clogged with catalyst and liquid, then heating of the particles is mainly from the outside in by conduction [60]. If the particles are too dry (at equivalent catalyst loadings), then the effective catalyst concentration is too high, resulting in rapid degradation of the biomass before enough of the penetrating steam has condensed to dilute the catalyst into the proper concentration range and allow the products formed to be diluted and to diffuse from the reaction zone. The combination of high catalyst and product concentrations greatly increases degradation reaction rates and lowers overall pretreatment yields at very high solids concentrations.

Viamajala *et al.* showed by dye penetration studies that soaking overnight under vacuum achieved the highest degree of penetration of dye into the fibers of corn stalks [158]. Soaking of the feedstock prior to pretreatment has been shown to improve yields significantly in steam gun experiments [159]. Linde *et al.*, Soderstrom *et al.* and Sassner *et al.* have shown that dilute acid impregnation by soaking pine chips can increase pretreatment yields by $>$10% over that produced by pretreatment of pine following the usual acid spraying impregnation method [159–161]. Weiss *et al.* showed similar effects for corn stover [54]. NREL's soaking process for acid impregnation utilizes an acid soaking bath at 60 °C for 2–4 h with recirculation of the acid prior to steam explosion pretreatment to achieve high yields [106]. Kim *et al.* have shown that acid diffusion into biomass particles can take several minutes, even at elevated temperatures [162]. Pulp and paper industries utilize a separate impregnation step by steaming at 105 °C, followed by pressurization to force alkali solution deep into pores of wood chips prior to entering pulp digesters [163]. This "squeeze-pumping" action is utilized in US Patent 7,819,976 to increase the extent of impregnation of dilute ammonia into biomass prior to pretreatment [164].

23.8.4   High Temperatures and Short Residence Times Result in High Yields

Dilute-acid pretreatments for poplar and corn stover utilizing high temperatures ($>$170–180 °C) and short residence times (0.5–1 min) lead to high xylan-to-xylose conversion yields [73]. At higher temperatures, xylan conversion to xylose occurs much faster than xylose degradation to furfural and other products, leading to higher yields if reaction times can be kept short enough [73]. High-temperature short-residence-time pretreatments can only be performed in small reactors or steam-gun-type reactors because of the critical rapid heat-up and cool-down that is necessary in order to obtain high yields.

23.8.5   Pretreatment Severity: Tradeoffs of Time and Temperature

The severity factor is derived from the "H" factor used in the pulp and paper industry [163], which trades off residence time in the reactor with temperature of pretreatment to achieve similar yields and paper properties in commercial paper production [165–167]. The kinetics of the reaction appears to be first order and have an Arrhenius-type dependence on reaction temperature. This equation was adapted to steam explosion and hot water pretreatments to explain xylan-to-xylose conversion yields. The severity factor SF is set equal to $\log_{10}$ (R_0), where R_0 is the reaction ordinate:

$$\text{SF} = \log R_0 = \log \left[t \times \exp\left(\frac{T_r - T_b}{\omega}\right) \right] \qquad (23.1)$$

in which t is reaction time (min), T_r is the reaction temperature (°C), and T_b is the reference temperature (°C). The reference temperature T_b is usually set equal to 100 °C, similar to the comparison temperature used in the pulp and paper industries, and ω is usually empirically set equal to 14.75. For pretreatments using dilute acid catalysts, a combined severity factor (CSF) is used to subtract out the effects of pH on pretreatment severity. The combined severity factor (Equation (23.2)) was originally derived for dilute-acid organosolv processes to explain the effects of pH on pretreatment [168], and it has been successfully adapted to dilute-acid steam explosion pretreatment [106]:

$$\text{CSF} = \log \left[t \times \exp\left(\frac{T - 100}{14.75}\right) \right] - \text{pH} \qquad (23.2)$$

The severity factor, or the combined severity factor, is useful for determining tradeoffs among temperature, time, and acid concentration. This approach can be very helpful when dealing with experimental data. It allows selection of new hydrolysis conditions based on limited results. Researchers can use the pretreatment severity factor to select multiple process conditions to achieve target performance. For example, for acid pretreatment, low acid concentration is desired because of the high costs of reactor materials-of-construction. From experimental data under certain acid concentrations, times, and temperatures, an optimum pretreatment severity can be determined based on performance criteria such as high xylose yields or high cellulose digestibility [132]. If the acid concentration is to be lowered, the combination of time and temperature needed to achieve similar pretreatment effectiveness can be estimated from previous severity data, even though the acid concentration is different. According to the pretreatment severity formula, a tradeoff that requires longer time or higher temperature will be needed if acid concentration is lowered. Note that the pretreatment severity only serves as a reference and guide, and the exact pretreatment conditions needed to achieve a given performance have to be confirmed experimentally.

23.8.6 Minimizing Construction and Operating Costs

In general, lower-severity pretreatments require less-expensive materials-of-construction for the pretreatment reactor, saving capital expenditures for the bench-scale reactor as well as for the full-size industrial-scale reactor. However, as illustrated by the severity factor, lower-temperature pretreatments require longer times than higher-temperature pretreatments to achieve approximately the same performance, which increases the necessary reactor volume and thereby the capital expenses required to contain the biomass. Steam explosion reactors, although simple in design, require an auxiliary high-pressure boiler and piping that increase costs significantly. Depending upon the boiler size (>100 boiler hp), in many jurisdictions a stationary engineer may be required to operate the boiler. Alkali pretreatments are less aggressive than dilute acid pretreatments, enabling lower-cost pretreatment reactors [169]. Higher sugar yields and "cleaner" hydrolyzate liquors that are more easily fermentable can result from alkali pretreatments. However, alkali pretreatments require enzymes to take a larger share of the carbohydrate hydrolysis burden, thus likely increasing enzyme costs in the enzymatic hydrolysis step [3,45].

23.9 Summary

Pretreatment is one of the key cost elements in biomass bioconversion and has broad impacts on all upstream and downstream operations [3]. Development of low-cost and effective pretreatment technologies is the key to unlocking low-cost cellulosic ethanol. Fundamental and applied research on various pretreatment technologies, including biological, physical, chemical, and thermal approaches or a combination of these, have been conducted for many purposes such as assessing genetically modified biomass feedstocks, determining biomass deconstruction mechanisms and kinetics, optimizing processes, and simulating scale-up for techno-economic assessments [107].

Batch bench-scale reactors are widely employed in such pretreatment research. Sealed glass reactors and tubular reactors are cost effective and relatively easy to use, even with high solids loadings and other conditions essential to commercial-scale operations. To realize reasonably uniform temperatures in externally heated batch tubular reactors, diameters of less than 0.5 inch ID have been shown to be necessary. Small tubular reactors and larger mixed vessels, such as Parr reactors, enable closing of material balances that are vital for the assessment and commercialization of pretreatment technology. However, solids loadings are limited for the latter due to the strong liquid adsorbing capacity of biomass materials. Tubular reactors and mixed vessels have been heated indirectly with air fluidized sand baths, electrical heating jackets, steam, electrical heating coils, and hot oil baths.

Different indirect heating options offer different heat-up rates. Fluidized sand baths can quickly heat up tubular reactors and mixed Parr reactors although a series of sand baths are recommended in order to reduce temperature transients. Commercially available microwave glass tubular reactors and modified microwave devices have been applied to pretreatment of various woody and herbaceous feedstocks. In a study with tubular reactors, microwave heating gave a similar heat-up profile to that found with fluidized sand baths but led to faster xylan hydrolysis/degradation and higher cellulose conversion by enzymes. Required heating conditions by indirect heating are also affected by the use of chemical catalysts and the solids loading. When chemical catalysts are employed, biomass impregnation prior to loading of biomass into the reactor becomes important, especially for pretreatment at higher solids concentrations. Size reduction is necessary when using tubular reactors and mixed reactors because of their small size and mixing requirements.

Direct heating with steam or hot liquids/gases can provide very rapid heat-up, which is especially suitable for high solids loading and for treatment of large amounts of biomass. However, even though steam explosion reactors are employed commercially, laboratory-scale steam explosion reactors are usually custom built. The fast pressure release in steam explosion is believed to shear biomass into smaller particles, resulting in higher yields in enzymatic saccharification. Zipperclave reactors can be heated with steam; however, specially designed impellers are needed to mix high solids loadings in the vertical reactor configuration. Direct steam injection results in quick heat-up times comparable to that of steam explosion reactors. However, direct steam heating presents challenges to achieving high solids concentrations after pretreatment due to dilution with steam condensate, and achieving accurate material balance closures is more challenging because of loss of volatile components in the flash steam.

Pretreatment in continuous reactors using just water or dilute catalyst (e.g., dilute sulfuric acid or ammonia solutions) that is pumped through a packed bed of biomass at elevated temperatures results in high sugar yields from the hemicellulosic component and high lignin removal, as well as cellulose that is highly accessible by enzymes. Continuous plug-flow reactors, tubular percolating reactors, countercurrent/co-current reactors, and countercurrent shrinking bed reactors have been developed for the various pretreatment chemistries including acid, water, and ammonia. Although flowthrough systems are effective, the copious consumption of water hinders their commercialization. Nevertheless, flowthrough systems offer opportunities for examining aspects of biomass deconstruction kinetics that batch systems are unable to reveal.

Because biomass pretreatment is often conducted under elevated temperatures and involves multiphase reactions, careful attention to mass and heat transfer is vital to success. Mass transfer has proven to be an important factor in biomass deconstruction with bench-top pretreatment systems. Additional research is needed to develop models that take mass transfer effects into account in order to more accurately predict biomass deconstruction during pretreatment.

The basic requirements for all laboratory bench-scale pretreatment technologies include: rapid heat-up and cool-down; accurate temperature control; and accurate closure of material balances, vital to facilitate understanding of reaction kinetics and to ensure consistent kinetic results and interpretation. Reactor design for bench-scale pretreatment reactors should pay careful attention to aspects such as: (1) satisfying the intended application; (2) avoiding inter-and intra-heat and mass transport limitations; (3) minimizing temperature and concentration gradients; (4) maintaining an ideal flow pattern; (5) maximizing the accuracy of concentration and temperature measurements; and (6) minimizing construction and operating costs.

Acknowledgements

We appreciate the assistance of Yanpin Lu and Lishi Yan in collecting literature sources and writing this paper. We also recognize support by the Center for Bioproducts and Bioenergy and Department of Biological Systems Engineering at Washington State University. Dr Tucker is supported by the US Department of Energy, Energy Efficiency and Renewable Energy (EERE), Office of the Biomass Program, Office of Science, Office of Biological and Environmental Research through the BioEnergy Science Center (BESC), a DOE Bioenergy Research Center, and the Center for Direct Catalytic Conversion of Biomass to Biofuels (C3Bio), a DOE Energy Frontiers Research Center.

References

1. Ragauskas, A.J., Williams, C.K., Davison, B.H. *et al.* (2006) The path forward for biofuels and biomaterials. *Science*, **311** (5760), 484–489.
2. Lynd, L.R., Laser, M.S., Bransby, D. *et al.* (2008) How biotech can transform biofuels. *Nature Biotechnology*, **26** (2), 169–172.
3. Yang, B. and Wyman, C.E. (2008) Pretreatment: the key to unlocking low-cost cellulosic ethanol. *Biofuels, Bioproducts, and Biorefining*, **2** (1), 26–40.
4. Wyman, C.E. (2007) What is (and is not) vital to advancing cellulosic ethanol. *Trends in Biotechnology*, **25** (4), 153–157.
5. Humbird, D., Davis, R., Tao, L. *et al.* (2011) Process Design and Economics for Biochemical Conversion of Lignocellulosic Biomass to Ethanol: Dilute-Acid Pretreatment and Enzymatic Hydrolysis of Corn Stover. Technical report May 2011. Report No.: NREL/TP-510-47764.
6. Chen, F. and Dixon, R.A. (2007) Lignin modification improves fermentable sugar yields for biofuel production. *Nature Biotechnology*, **25** (7), 759–761.
7. Huntley, S.K., Ellis, D., Gilbert, M. *et al.* (2003) Significant increases in pulping efficiency in C4H-F5H-transformed poplars: Improved chemical savings and reduced environmental toxins. *Journal of Agricultural and Food Chemistry*, **51** (21), 6178–6183.
8. McCann, M.C. and Carpita, N.C. (2008) Designing the deconstruction of plant cell walls. *Current Opinion in Plant Biology*, **11** (3), 314–320.
9. Fu, C., Mielenz, J.R., Xiao, X. *et al.* (2011) Genetic manipulation of lignin reduces recalcitrance and improves ethanol production from switchgrass. *Proceedings of the National Academy of Sciences of the United States of America*, **108** (9), 3803–3808, S/1-S/6.
10. Dale, B.E. and Bals, B. (2008) Separation of proteins from grasses integrated with ammonia fiber explosion (AFEX) pretreatment and cellulose hydrolysis. Application: WOWO patent 2007-US104102008020901. 20070430.

11. Dale, B.E., Leong, C.K., Pham, T.K. *et al.* (1994) Hydrolysis of lignocellulosics at low enzyme levels: Application of the AFEX process. Liquid Fuels, Lubricants and Additives from Biomass. Proceedings of an Alternative Energy Conference, Kansas City, Mo, June 16–17, 1994, pp. 104–111.

12. Teymouri, F., Laureano-Perez, L., Alizadeh, H., and Dale, B.E. (2005) Optimization of the ammonia fiber explosion (AFEX) treatment parameters for enzymatic hydrolysis of corn stover. *Bioresource Technology*, **96** (18), 2014–2018.

13. Chang, V.S., Burr, B., and Holtzapple, M.T. (1997) Lime pretreatment of switchgrass. *Applied Biochemistry and Biotechnology*, **63–65**, 3–19.

14. van Walsum, G.P. and Shi, H. (2004) Carbonic acid enhancement of hydrolysis in aqueous pretreatment of corn stover. *Bioresource Technology*, **93** (3), 217–226.

15. Brink, D.L. (1993) Method of Treating Biomass Material. US patent 5,221,357.

16. Brink, D.L. (1994) Method of Treating Biomass Material. US patent 5,366,558.

17. Alhasan, A.M., Kuang, D., Mohammad, A.B., and Sharma-Shivappa, R.R. (2010) Combined effect of nitric acid and sodium hydroxide pretreatments on enzymatic saccharification of rubber wood (Heavea brasiliensis). *International Journal of Chemical Technology*, **2** (1), 12–20.

18. Zhao, X.-B., Wang, L., and Liu, D.-H. (2008) Peracetic acid pretreatment of sugarcane bagasse for enzymatic hydrolysis: a continued work. *Journal of Chemical Technology & Biotechnology*, **83** (6), 950–956.

19. Lu, Y. and Mosier, N.S. (2008) Kinetic modeling analysis of maleic acid-catalyzed hemicellulose hydrolysis in corn stover. *Biotechnology and Bioengineering*, **101** (6), 1170–1181.

20. Kim, J.W. and Mazza, G. (2008) Optimization of phosphoric acid catalyzed fractionation and enzymatic digestibility of flax shives. *Industrial Crops and Products*, **28** (3), 346–355.

21. Hsu, T.-A. (1996) Pretreatment of biomass, in *Handbook on Bioethanol, Production and Utilization* (ed. C.E. Wyman), Taylor & Francis, Washington, DC, p. 179–212.

22. McMillan, J.D. (1994) Pretreatment of lignocellulosic biomass, in *Enzymatic Conversion of Biomass for Fuels Production* (eds M.E. Himmel, J.O. Baker, and R.P. Overend), American Chemical Society, Washington, DC, p. 292–324.

23. Grohmann, K., Torget, R., and Himmel, M. (1985) Optimization of dilute acid pretreatment of biomass. *Biotechnology Bioengineering Symposium*, **15**, 59–80.

24. Grethlein, H.E. (1985) The effect of pore size distribution on the rate of enzymatic hydrolysis of cellulosic substrates. *Bio/Technology*, **3**, 155–160.

25. Grohmann, K., Himmel, M., Rivard, C. *et al.* (1984) Chemical-mechanical methods for the enhanced utilization of straw. *Biotechnology Bioengineering Symposium*, **14**, 137–157.

26. Torget, R., Walter, P., Himmel, M., and Grohmann, K. (1991) Dilute-acid pretreatment of corn residues and short-rotation woody crops. *Applied Biochemistry Biotechnology*, **28/29**, 75–86.

27. Torget, R., Himmel, M., and Grohmann, K. (1992) Dilute-acid pretreatment of two short-rotation herbaceous crops. *Applied Biochemistry and Biotechnology*, **34/35**, 115–123.

28. Torget, R., Werdene, P., Himmel, M., and Grohmann, K. (1990) Dilute acid pretreatment of short rotation woody and herbaceous crops. *Applied Biochemistry Biotechnology*, **24–25**, 115–126.

29. Grethlein, H.E. (1978) Acid hydrolysis of cellulosic biomass. Proceedings of Annual Fuels Biomass Symp, 2nd, 1, pp. 461–469.

30. Pan, X., Gilkes, N., Kadla, J. *et al.* (2006) Bioconversion of hybrid poplar to ethanol and Co-products using an organosolv fractionation process: optimization of process yields. *Biotechnology and Bioengineering*, **94** (5), 851–861.

31. Weil, J., Brewer, M., Hendrickson, R. *et al.* (1998) Continuous pH monitoring during pretreatment of yellow poplar wood sawdust by pressure cooking in water. *Applied Biochemistry Biotechnology*, **70–72**, 91–111.

32. Mosier, N., Hendrickson, R., Ho, N. *et al.* (2005) Optimization of pH controlled liquid hot water pretreatment of corn stover. *Bioresource Technology*, **96** (18), 1986–1993.

33. Rosgaard, L., Pedersen, S., and Meyer Anne, S. (2007) Comparison of different pretreatment strategies for enzymatic hydrolysis of wheat and barley straw. *Applied Biochemistry and Biotechnology*, **143** (3), 284–296.

34. Mackie, K.L., Brownell, H.H., West, K.L., and Saddler, J.N. (1985) Effect of sulfur dioxide and sulphuric acid on steam explosion of aspenwood. *Journal Wood Chemistry Technology*, **5** (3), 405–425.

35. Wu, M., Chang, K., Boussaid, A. *et al.* (1998) Optimization of steam explosion to enhance hemicullulose recovery and enzymatic hydrolysis of cellulose in softwoods. *Applied Biochemistry and Biotechnology*, **77–79**, 1–8.
36. Mabee Warren, E., Gregg David, J., Arato, C. *et al.* (2006) Updates on softwood-to-ethanol process development. *Applied Biochemistry and Biotechnology*, **129–132**, 55–70.
37. Wyman, C.E. (2004) Biological processing of cellulosic biomass to fuels and chemicals: progress, challenges, and research opportunities. AIChE Spring National Meeting, Conference Proceedings, New Orleans, LA, United States, Apr 25–29, 2004, pp. 330–335.
38. Jacobsen, S.E. and Wyman, C.E. (2000) Cellulose and hemicellulose hydrolysis models for application to current and novel pretreatment processes. *Applied Biochemistry and Biotechnology*, **84–86**, 81–96.
39. Esteghlalian, A., Hashimoto, A.G., Fenske, J.J., and Penner, M.H. (1997) Modeling and optimization of the dilute sulfuric acid pretreatment of corn stover, poplar and switchgrass. *Bioresource Technology*, **59** (2 & 3), 129–136.
40. Tao, L., Aden, A., Elander, R.T. *et al.* (2011) Process and technoeconomic analysis of leading pretreatment technologies for lignocellulosic ethanol production using switchgrass. *Bioresource Technology*, **102** (24), 11105–11114.
41. Wyman, C.E., Dale, B.E., Elander, R.T. *et al.* (2005) Coordinated development of leading biomass pretreatment technologies. *Bioresource Technology*, **96** (18), 1959–1966.
42. Heitz, M., Capekmenard, E., Koeberle, P.G. *et al.* (1991) Fractionation of populus-tremuloides at the pilot-plant scale – optimization of steam pretreatment conditions using the Stake-Ii technology. *Bioresource Technology*, **35** (1), 23–32.
43. Tanjore, D., Shi, J., and Wyman, C.E. (2011) Dilute acid and hydrothermal pretreatment of cellulosic biomass. RSC Energy and Environment Series. *Chemical and Biochemical Catalysis for Next Generation Biofuels*, **4**, 64–88.
44. Stuhler, L.S. (2002) *Effects of Solids Concentration, Acetylation, and Transient Heat on Uncatalyzed Batch Pretreatment of Corn Stover*, MSc thesis, Dartmouth College, Hanover.
45. Wyman, C.E., Dale, B.E., Elander, R.T. *et al.* (2005) Comparative sugar recovery data from laboratory scale application of leading pretreatment technologies to corn stover. *Bioresource Technology*, **96** (18), 2026–2032.
46. Lynd, L.R., Elander, R.T., and Wyman, C.E. (1996) Likely features and costs of mature biomass ethanol technology. *Applied Biochemistry and Biotechnology*, **57/58**, 741–761.
47. van Walsum, G.P., Allen, S.G., Spencer, M.J. *et al.* (1996) Conversion of lignocellulosics pretreated with liquid hot water to ethanol. *Applied Biochemistry and Biotechnology*, **57/58**, 157–170.
48. Yang, B. and Wyman, C.E. (2009) Dilute acid and autohydrolysis pretreatment. *Methods in Molecular Biology*, **581**, 103–114. (Biofuels).
49. Studer, M.H., De Martini, J.D., Brethauer, S. *et al.* (2009) Engineering of a high-throughput screening system to identify cellulosic biomass, pretreatments, and enzyme formulations that enhance sugar release. *Biotechnology and Bioengineering*, **105** (2), 231–238.
50. Selig, M.J., Tucker, M.P., Law, C. *et al.* (2011) High throughput determination of glucan and xylan fractions in lignocelluloses. *Biotechnology Letters*, **33** (5), 961–967.
51. Selig, M.J., Viamajala, S., Decker, S.R. *et al.* (2007) Deposition of lignin droplets produced during dilute acid pretreatment of maize stems retards enzymatic hydrolysis of cellulose. *Biotechnology Progress*, **23** (6), 1333–1339.
52. Lloyd, T.A. and Wyman, C.E. (2005) Combined sugar yields for dilute sulfuric acid pretreatment of corn stover followed by enzymatic hydrolysis of the remaining solids. *Bioresource Technology*, **96** (18), 1967–1977.
53. Grohmann, K., Torget, R., and Himmel, M. (1986) Dilute acid pretreatment of biomass at high solids concentrations. *Biotechnology and Bioengineering Symposium*, **17**, 137–151.
54. Weiss, N.D., Nagle, N.J., Tucker, M.P., and Elander, R.T. (2009) High xylose yields from dilute acid pretreatment of corn stover under process-relevant conditions. *Applied Biochemistry and Biotechnology*, **155** (1–3), 418–428.
55. Brennan, M.A. and Wyman, C.E. (2004) Initial evaluation of simple mass transfer models to describe hemicellulose hydrolysis in corn stover. *Applied Biochemistry and Biotechnology*, **113–116**, 965–976.
56. Lloyd, T. and Wyman, C.E. (2003) Application of a depolymerization model for predicting thermochemical hydrolysis of hemicellulose. *Applied Biochemistry and Biotechnology*, **105–108**, 53–67.

57. Alizadeh, H., Teymouri, F., Gilbert, T.I., and Dale, B.E. (2005) Pretreatment of switchgrass by ammonia fiber explosion (AFEX). *Applied Biochemistry and Biotechnology*, **121–124**, 1133–1141.
58. Oehgren, K., Galbe, M., and Zacchi, G. (2005) Optimization of steam pretreatment of SO_2-impregnated corn stover for fuel ethanol production. *Applied Biochemistry and Biotechnology*, **121–124**, 1055–1067.
59. Palmqvist, E., Hahn-Hagerdal, H., Galbe, M. *et al.* (1996) Design and operation of a bench-scale process development unit for the production of ethanol from lignocellulosics. *Bioresource Technology*, **58** (2), 171–179.
60. Tucker, M.P., Kim, K.H., Newman, M.M., and Nguyen, Q.A. (2003) Effects of temperature and moisture on dilute-acid steam explosion pretreatment of corn stover and cellulase enzyme digestibility. *Applied Biochemistry and Biotechnology*, **105–108**, 165–177.
61. Baugh, K.D. and McCarty, P.L. (1988) Thermochemical retreatment of lignocellulose to enhance methane fermentation: I. Monosaccharide and furfurals hydrothermal decomposition and product formation rates. *Biotechnology and Bioengineering*, **31**, 50–61.
62. Chen, R.F., Lee, Y.Y., and Torget, R. (1996) Kinetic and modeling investigation on two-stage reverse-flow reactor as applied to dilute-acid pretreatment of agricultural residues. *Applied Biochemistry and Biotechnology*, **57–8**, 133–146.
63. Yang, B. and Wyman Charles, E. (2004) Effect of xylan and lignin removal by batch and flowthrough pretreatment on the enzymatic digestibility of corn stover cellulose. *Biotechnology and Bioengineering*, **86** (1), 88–95.
64. Tunc, M.S., Lawoko, M., and van Heiningen, A. (2010) Understanding the limitations of removal of hemicelluloses during autohydrolysis of a mixture of Southern Hardwoods. *Bioresources*, **5** (1), 356–371.
65. Zhang, B., Shahbazi, A., Wang, L. *et al.* (2011) Hot-water pretreatment of cattails for extraction of cellulose. *Journal of Industrial Microbiology & Biotechnology*, **38**, 819–824.
66. Zhang, B., Wang, L., Shahbazi, A. *et al.* (2011) Dilute-sulfuric acid pretreatment of cattails for cellulose conversion. *Bioresource Technology*, **102** (19), 9308–9312.
67. Stuhler, S.L. and Wyman, C.E. (2003) Estimation of temperature transients for biomass pretreatment in tubular batch reactors and impact on xylan hydrolysis kinetics. *Applied Biochemistry and Biotechnology*, **105–108**, 101–114.
68. Chen, R., Lee, Y.Y., and Torget, R. (1996) Kinetic and modeling investigation on two-stage reverse-flow reactor as applied to dilute-acid pretreatment of agricultural residues. *Applied Biochemistry and Biotechnology*, **57/58**, 133–146.
69. Jacobsen, S.E. and Wyman, C.E. (2001) Heat transfer considerations in design of a batch tube reactor ear biomass hydrolysis. *Applied Biochemistry and Biotechnology*, **91–3**, 377–386.
70. Stuhler, L.S. (2002) *Effects of solids concentration, acetylation, and transient heat on uncatalyzed batch pretreatment of corn stover*. MS thesis, Dartmouth College, Hanover, NH, USA.
71. Kim, Y., Hendrickson, R., Mosier, N., and Ladisch, M.R. (2005) Plug-flow reactor for continuous hydrolysis of glucans and xylans from pretreated corn fiber. *Energy & Fuels*, **19** (5), 2189–2200.
72. Montane, D., Salvado, J., Farriol, X., and Chornet, E. (1993) The fractionation of almond shells by thermomechanical aqueous-phase (Tm-Av) pretreatment. *Biomass & Bioenergy*, **4** (6), 427–437.
73. Esteghlalian, A., Hashimoto, A.G., Fenske, J.J., and Penner, M.H. (1997) Modeling and optimization of the dilute-sulfuric-acid pretreatment of corn stover, poplar and switchgrass. *Bioresource Technology*, **59**, 129–136.
74. Brennan, M.A. (2003) *Predicting Performance of Batch, Flowthrough, and Mixed Batch Hemicellulose Hydrolysis by Coupled Mass Transfer and Reaction Models*, MSc thesis, Dartmouth College, Hanover
75. Weil, J.R., Sarikaya, A., Rau, S.L. *et al.* (1998) Pretreatment of corn fiber by pressure cooking in water. *Applied Biochemistry and Biotechnology*, **73** (1), 1–17.
76. Alizadeh, H., Teymouri, F., Gilbert Thomas, I., and Dale Bruce, E. (2005) Pretreatment of switchgrass by ammonia fiber explosion (AFEX). *Applied Biochemistry and Biotechnology*, **121–124**, 1133–1141.
77. Ballesteros, M., Negro, M.J., Manzanares, P. *et al.* (2007) Fractionation of Cynara cardunculus (cardoon) biomass by dilute-acid pretreatment. *Applied Biochemistry and Biotechnology*, **137–140**, 239–252.
78. Dunson, J.B., Tucker, M., Elander, R., Hennessey, S.M. inventors; (E. I. Du Pont de Nemours and Company, USA). Assignee. (2006) Treatment of biomass to obtain fermentable sugars. Application: WOWO patent 2006-US14146 200611090. 20060412.

79. Shi, J., Yang, B., Pu, Y. *et al.* (2010) Comparison of microwaves to fluidized sand baths for heating tubular reactors for hydrothermal and dilute acid batch pretreatment of corn stover. *Bioresource Technology*, **102** (10), 5952–5962.

80. Shi, J., Yang, B., Pu, Y. *et al.* (2011) Comparison of microwaves to fluidized sand baths for heating tubular reactors for hydrothermal and dilute acid batch pretreatment of corn stover. *Bioresource Technology*, **102** (10), 5952–5962.

81. Palmarola-Adrados, B., Galbe, M., and Zacchi, G. (2005) Pretreatment of barley husk for bioethanol production. *Journal of Chemical Technology & Biotechnology*, **80** (1), 85–91.

82. Zhu, S., Yu, Z., Wu, Y. *et al.* (2005) Enhancing enzymatic hydrolysis of rice straw by microwave pretreatment. *Chemical Engineering Communications*, **192** (10–12), 1559–1566.

83. Ooshima, H., Aso, K., Harano, Y., and Yamamoto, T. (1984) Microwave treatment of cellulosic materials for their enzymatic hydrolysis. *Biotechnology Letters*, **6** (5), 289–294.

84. Kitchaiya, P., Intanakul, P., and Krairiksh, M. (2003) Enhancement of enzymatic hydrolysis of lignocellulosic wastes by microwave pretreatment under atmospheric pressure. *Journal of Wood Chemistry and Technology*, **23** (2), 217–225.

85. Azuma, J., Higashino, J., Isaka, M., and Koshijima, T. (1985) Microwave irradiation of lignocellulosic materials. IV. Enhancement of enzymic susceptibility of microwave-irradiated softwoods. *Wood Research*, **71**, 13–24.

86. Hu, Z. and Wen, Z. (2008) Enhancing enzymatic digestibility of switchgrass by microwave-assisted alkali pretreatment. *Biochemical Engineering Journal*, **38** (3), 369–378.

87. Keshwani Deepak, R. and Cheng Jay, J. (2010) Modeling changes in biomass composition during microwave-based alkali pretreatment of switchgrass. *Biotechnology and Bioengineering*, **105** (1), 88–97.

88. Liu, J., Takada, R., Karita, S. *et al.* (2010) Microwave-assisted pretreatment of recalcitrant softwood in aqueous glycerol. *Bioresource Technology*, **101** (23), 9355–9360.

89. Mason, W.H. (1926) Steam treatment in kilns for preserving resinous lumber and recovering by-products. Application: US patent 1920-4314751577044 19201217.

90. Mason, W.H. (1928) Apparatus for explosion fibration of lignocellulose material. Application: US patent 1655618.

91. Mason, W.H., Boehm, R.M., and Simpson, G.G. (1943) " Double-activation" process of making hard fiber board. Application: US patent 1939-305102 2317394. 19391118.

92. Boehm, R.M. (1930) A note on exploded wood for insulating and structural material. *Journal of Industrial and Engineering Chemistry*, **22**, 493–497.

93. Harris Group I (2001) Acid hydrolysis reactors batch system. Subcontract ACO-9-29067-01, Report 99-10600/18. NREL, Colorado.

94. Yang, B., Boussaid, A., Mansfield, S.D. *et al.* (2002) Fast and efficient alkaline peroxide treatment to enhance the enzymatic digestibility of steam-exploded softwood substrates. *Biotechnology and Bioengineering*, **77** (6), 678–684.

95. Wu, M.M., Chang, K., Gregg, D.J. *et al.* (1999) Optimization of steam explosion to enhance hemicellulose recovery and enzymatic hydrolysis of cellulose in softwoods. *Applied Biochemistry and Biotechnology*, **77–79**, 47–54.

96. Soederstroem, J., Galbe, M., and Zacchi, G. (2004) Effect of washing on yield in one- and two-step steam pretreatment of softwood for production of ethanol. *Biotechnology Progress*, **20** (3), 744–749.

97. Oehgren, K., Bura, R., Saddler, J., and Zacchi, G. (2007) Effect of hemicellulose and lignin removal on enzymatic hydrolysis of steam pretreated corn stover. *Bioresource Technology*, **98** (13), 2503–2510.

98. Galbe, M. and Zacchi, G. (1986) Pretreatment of sallow prior to enzymatic hydrolysis. *Biotechnology and Bioengineering*, **17**, 97–106.

99. Linde, M., Jakobsson, E.-L., Galbe, M., and Zacchi, G. (2008) Steam pretreatment of dilute H_2SO_4-impregnated wheat straw and SSF with low yeast and enzyme loadings for bioethanol production. *Biomass Bioenergy*, **32** (4), 326–332.

100. Ibrahim, M.M., Agblevor, F.A., and El-Zawawy, W.K. (2010) Isolation and characterization of cellulose and lignin from steam-exploded lignocellulosic biomass. *Bioresources*, **5** (1), 397–418.

101. Grous, W.R., Converse, A.O., and Grethlein, H.E. (1986) Effect of steam explosion pretreatment on pore size and enzymic hydrolysis of poplar. *Enzyme and Microbial Technology*, **8** (5), 274–330.

102. DeLong, E.A. and Ritchie, G.S. (1990) Fractionation of lignins from steam-expanded lignocellulosic materials. Application: US patent 1988-246069 4966650 19880919.

103. DeLong, E.A. (1981) Method of rendering lignin separable from cellulose and hemicellulose in lignocellulosic material and the product so produced. Canada patent 1,096,374.

104. Textor, C.K. (ed.) (1957) The Bauer Method of Preparing Furnishes for the Manufacture of Insulation Board and Hardboard. "Fiberboard and Particle board". Proceeding of International Consulation on Insulation board, Hardboard and Particle board, Rome.

105. Grethlein, H.E. (1980) Pretreating cellulosic substrates and producing sugar therefrom. US patent 4237226.

106. Sassner, P., Maartensson, C.-G., Galbe, M., and Zacchi, G. (2007) Steam pretreatment of H_2SO_4-impregnated Salix for the production of bioethanol. *Bioresource Technology*, **99** (1), 137–145.

107. Nguyen, Q.A., Tucker, M.P., Boynton, B.L. *et al.* (1998) Dilute acid pretreatment of softwoods. *Applied Biochemistry and Biotechnology*, **70–72**, 77–87.

108. Aden, A., Ruth, M., Ibsen, K. *et al.* (2002) *Lignocellulosic Biomass to Ethanol Process Design and Economics Utilizing Co-Current Dilute Acid Prehydrolysis and Enzymatic Hydrolysis for Corn Stover*, National Renewable Energy Laboratory, Golden, Colorado.

109. Heitz, M., Capek-Menard, E., Koeberle, P.G. *et al.* (1991) Fractionation of Populus tremuloides in the pilot plant scale: Optimization of steam pretreatment conditions using STAKE II technology. *Bioresource Technology*, **35**, 23–32.

110. Cort, J.B., Pschorn, T., and Stromberg, B. (2010) Minimize scale-up risk. *Chemical Engineering Progress*, **106** (3), 39–49.

111. Yang, B., Gray, M.C., Liu, C. *et al.* (2004) Unconventional relationships for hemicellulose hydrolysis and subsequent cellulose digestion. *ACS Symposium Series*, **889**, 100–125.

112. McParland, J.J., Grethlein, H.E., and Converse, A.O. (1982) Kinetics of acid hydrolysis of corn stover. *Solar Energy*, **28** (1), 55–63.

113. Torget, R., Hatzis, C., Hayward, T.K. *et al.* (1996) Optimization of reverse-flow, two-temperature, dilute-acid pretreatment to enhance biomass conversion to ethanol. *Applied Biochemistry and Biotechnology.*, **57/58**, 85–101.

114. Mok, W.S., Antal, M.J. Jr, and Varhegyi, G. (1992) Productive and parasitic pathways in dilute acid-catalyzed hydrolysis of cellulose. *Industrial & Engineering Chemistry Research*, **31** (1), 94–100.

115. Song, S.K. and Lee, Y.Y. (1982) Countercurrent reactor in acid catalyzed cellulose hydrolysis. *Chemical Engineering Communications*, **17** (1–6), 23–30.

116. Lee, Y.Y., Wu, Z., and Torget, R.W. (1999) Modeling of countercurrent shrinking-bed reactor in dilute-acid total-hydrolysis of lignocellulosic biomass. *Bioresource Technology*, **71** (1), 29–39.

117. Yoon, H.H., Wu, Z., and Lee, Y.Y. (1995) Ammonia-recycled percolation process for pretreatment of biomass feedstock. *Applied Biochemistry Biotechnology*, **51/52**, 5–20.

118. Kim Tae, H. and Lee, Y.Y. (2006) Fractionation of corn stover by hot-water and aqueous ammonia treatment. *Bioresource Technology*, **97** (2), 224–232.

119. Yang, B. and Wyman, C.E. (2004) Effect of xylan and lignin removal by batch and flowthrough pretreatment on the enzymatic digestibility of corn stover cellulose. *Biotechnology and Bioengineering*, **86** (1), 88–95.

120. Yoon, H.H., Wu, Z.W., and Lee, Y.Y. (1995) Ammonia-recycled percolation process for pretreatment of biomass feedstock. *Applied Biochemistry and Biotechnology*, **51/52**, 5–19.

121. Iyer, P.V., Wu, Z.-W., Kim, S.B., and Lee, Y.Y. (1996) Ammonia recycled percolation process for pretreatment of herbaceous biomass. *Applied Biochemistry and Biotechnology*, **57/58**, 121–132.

122. Liu, C. and Wyman, C.E. (2004) Impact of fluid velocity on hot water only pretreatment of corn stover in a flowthrough reactor. *Applied Biochemistry and Biotechnology*, **113–116**, 977–987.

123. Makishima, S., Mizuno, M., Sato, N. *et al.* (2009) Development of continuous flow type hydrothermal reactor for hemicellulose fraction recovery from corncob. *Bioresource Technology*, **100** (11), 2842–2848.

124. Grethlein, H.E. and Converse, A.O. (1991) Continuous acid hydrolysis of lignocelluloses for production of xylose, glucose, and furfural, in *Food, Feed, and Fuel from Biomass* (ed. D.S. Chahal), Oxford & IBH Publishing Company, New Delhi, p. 267–279.

125. Weiss, N.D., Farmer, J.D., and Schell, D.J. (2010) Impact of corn stover composition on hemicellulose conversion during dilute acid pretreatment and enzymatic cellulose digestibility of the pretreated solids. *Bioresource Technology*, **101** (2), 674–678.

126. Himmel, M.E., Tucker, M.P., Baker, J.O., Rivard, C.J., Oh, K.K., and Grohmann, K. (eds) (1986) Comminution of Biomass: Hammer and knife mills. Seventh Symposium for Fuels and Chemicals, Biotechnology and Bioengineering Symposium.

127. Vidal, B.C. Jr, Dien, B.S., Ting, K.C., and Singh, V. (2011) Influence of feedstock particle size on lignocellulose conversion-a review. *Applied Biochemistry and Biotechnology*, **164** (8), 1405–1421.

128. Liu, C. and Wyman, C.E. (2005) Partial flow of compressed-hot water through corn stover to enhance hemicellulose sugar recovery and enzymatic digestibility of cellulose. *Bioresource Technology*, **96** (18), 1978–1985.

129. Kim, T.H. and Lee, Y.Y. (2005) Pretreatment and fractionation of corn stover by ammonia recycle percolation process. *Bioresource Technology*, **96** (18), 2007–2013.

130. Kim, S. and Holtzapple Mark, T. (2005) Lime pretreatment and enzymatic hydrolysis of corn stover. *Bioresource Technology*, **96** (18), 1994–2006.

131. Mosier, N., Wyman, C., Dale, B. *et al.* (2005) Features of promising technologies for pretreatment of lignocellulosic biomass. *Bioresource Technology*, **96** (6), 673–686.

132. Liu, C.G. and Wyman, C.E. (2003) The effect of flow rate of compressed hot water on xylan, lignin, and total mass removal from corn stover. *Industrial & Engineering Chemistry Research*, **42** (21), 5409–5416.

133. Yang, B. and Wyman, C.E. (eds) (2002) The effect of batch and flowthrough reactor pretreatment on the digestibility of corn stover cellulose. Annual Meeting of the American Institute of Chemical Engineers, Indianapolis, IN.

134. Varga, E., Reczey, K., and Zacchi, G. (2004) Optimization of steam pretreatment of corn stover to enhance enzymatic digestibility. *Applied Biochemistry and Biotechnology*, **113–116**, 509–523.

135. Laser, M., Schulman, D., Allen, S.G. *et al.* (2002) A comparison of liquid hot water and steam pretreatments of sugar cane bagasse for bioconversion to ethanol. *Bioresource Technology*, **81** (1), 33–44.

136. Jacobsen, S.E. and Wyman, C.E. (2001) Heat transfer considerations in design of a batch tube reactor for biomass hydrolysis. *Applied Biochemistry and Biotechnology.*, **91–93**, 377–386.

137. Saeman, J.F. (1945) Kinetics of wood saccharification: Hydrolysis of cellulose and decomposition of sugars in dilute acid at high temperature. *Industrial Engineering Chemistry Research*, **37**, 42–52.

138. Kobayashi, T. and Sakai, Y. (1956) Hydrolysis rate of pentosan of hardwood in dilute sulfuric acid. *Bulletin Agricultural Chemical Society Japan*, **20**, 1–7.

139. Conner, A.H. and Lorenz, L.F. (1986) Kinetic modeling of hardwood prehydrolysis. Part III. Water and dilute acetic acid prehydrolysis of southern red oak. *Wood and Fiber Science*, **18** (2), 248–263.

140. Garrote, G., Dominguez, H., and Parajo, J.C. (2001) Study on the deacetylation of hemicelluloses during the hydrothermal processing of Eucalyptus wood. *Holz Roh- Werkst*, **59** (1/2), 53–59.

141. Jacobsen, S.E. (2000) The Effects of Solids Concentration on Sugar Release During Uncatalyzed Pretreatment of Biomass. MSc thesis, Dartmouth College, Hanover, NH.

142. Kim, K.H. and Hong, J. (2001) Supercritical CO_2 pretreatment of lignocellulose enhances enzymatic cellulose hydrolysis. *Bioresource Technology*, **77** (2), 139–144.

143. Newnham, R.E., Jang, S.J., Xu, M., and Jones, F. (1991) Fundamental interaction mechanisms between microwaves and matter. *Ceramic Transactions*, **21**, 51–67.

144. Clark, D.E., Folz, D.C., and West, J.K. (2000) Processing materials with microwave energy. *Materials Science & Engineering, A: Structural Materials*, **A287** (2), 153–158.

145. Lewis, D.A., Summers, J.D., Ward, T.C., and McGrath, J.E. (1992) Accelerated imidization reactions using microwave radiation. *Journal of Polymer Science, Part A: Polymer Chemistry*, **30** (8), 1647–1653.

146. Bond, G., Moyes, R.B., and Whan, D.A. (1993) Recent applications of microwave heating in catalysis. *Catalysis Today*, **17** (3), 427–437.

147. de la Hoz, A., Diaz-Ortiz, A., and Moreno, A. (2005) Microwaves in organic synthesis. Thermal and non-thermal microwave effects. *Chemical Society Reviews*, **34** (2), 164–178.

148. Kim, Y., Hendrickson, R., Mosier, N.S., and Ladisch, M.R. (2009) Liquid hot water pretreatment of cellulosic biomass. *Methods in Molecular Biology*, **581**, 93–102.

149. Zeng, M., Ximenes, E., Ladisch, M.R. *et al.* (2012) Tissue-specific biomass recalcitrance in corn stover pretreated with liquid hot-water: SEM imaging (part 2). *Biotechnology and Bioengineering*, **109** (2), 398–404.

150. Zeng, M., Ximenes, E., Ladisch, M.R. *et al.* (2012) Tissue-specific biomass recalcitrance in corn stover pretreated with liquid hot-water: Enzymatic hydrolysis (part 1). *Biotechnology and Bioengineering*, **109** (2), 390–397.

151. Shinners, K.J., Binversie, B.N., Muck, R.E., and Weimer, P.J. (2007) Comparison of wet and dry corn stover harvest and storage. *Biomass & Bioenergy*, **31** (4), 211–221.

152. Schick, G., Tellefsen, K.A., Johnson, A.J. *et al.* (1991) Hydrogen sources for signal attenuation in submarine optical fiber cables and the effect of cable design. Proceedings of International Wire and Cable Symposium, 40th, pp. 643–652.

153. Tengborg, C., Stenberg, K., Galbe, M. *et al.* (1998) Comparison of SO_2 and H_2SO_4 impregnation of softwood prior to steam pretreatment on ethanol production. *Applied Biochemistry and Biotechnology*, **70–72**, 3–15.

154. Grohmann, K., Himmel, M., Rivard, C. *et al.* (1984) Chemical-mechanical methods for the enhanced utilization of straw. *Biotechnology and Bioengineering Symposium*, **14**, 137–157.

155. Bitra, V.S.P., Womac, A.R., Yang, Y.C.T. *et al.* (2011) Characterization of wheat straw particle size distributions as affected by knife mill operating factors. *Biomass & Bioenergy*, **35** (8), 3674–3686.

156. Hosseini, S.A. and Shah, N. (2009) Multiscale modelling of hydrothermal biomass pretreatment for chip size optimization. *Bioresource Technology*, **100** (9), 2621–2628.

157. Tatsumoto, K., Baker, J.O., Tucker, M.P. *et al.* (1988) Digestion of pretreated aspen substrates. *Applied Biochemistry and Biotechnology*, **17/18**, 159–174.

158. Viamajala, S., Selig Michael, J., Vinzant Todd, B. *et al.* (2006) Catalyst transport in corn stover internodes: elucidating transport mechanisms using Direct Blue-I. *Applied Biochemistry and Biotechnology*, **129–132**, 509–527.

159. Linde, M., Galbe, M., and Zacchi, G. (2006) Steam pretreatment of acid-sprayed and acid-soaked barley straw for production of ethanol. *Applied Biochemistry and Biotechnology*, **129–132**, 546–562.

160. Soderstrom, J., Pilcher, L., Galbe, M., and Zacchi, G. (2003) Combined use of H_2SO_4 and SO_2 impregnation for steam pretreatment of spruce in ethanol production. *Applied Biochemistry and Biotechnology*, **105–108**, 127–140.

161. Sassner, P., Galbe, M., and Zacchi, G. (2005) Steam pretreatment of Salix with and without SO_2 impregnation for production of bioethanol. *Applied Biochemistry and Biotechnology*, **121–124**, 1101–1117.

162. Kim, S.B. and Lee, Y.Y. (2002) Diffusion of sulfuric acid within lignocellulosic biomass particles and its impact on dilute-acid pretreatment. *Bioresource Technology*, **83** (2), 165–171.

163. Rydholm, S.A. (1985) *Pulping Processes*, Robert Krieger Publishing, Malabar.

164. Friend, J., Elander, R.T., Tucker, M.P. III, and Lyons, R.C. (2009) Biomass treatment method, barrel apparatus and subsequent hydrolysis. Application: WOWO patent 2008-US734182009045653. 20080818.

165. Overend, R.P. and Chornet, E. (1987) Fractionation of lignocellulosics by steam-aqueous pretreatments. *Philosophical Transactions of the Royal Society of London, A*, **321** (1561), 523–536.

166. Abatzoglou, N., Chornet, E., Belkacemi, K., and Overend, R.P. (1992) Phenomenological kinetics of complex-systems: the development of a generalized severity parameter and its application to lignocellulosics fractionation. *Chemical Engineering Science*, **47** (5), 1109–1022.

167. Overend, R.P. and Chornet, E. (eds) (1989) Steam and aqueous pretreatments: are they prehydrolysis? Proceedings of 75th Annual Meeting, Technical Section, Canadian Pulp and Paper Assn.

168. Chum, H.L., Johnson, D.K., and Black, S.K. (1990) Organosolv pretreatment for enzymatic hydrolysis of poplars. 2. Catalyst effects and the combined severity parameter. *Industrial & Engineering Chemistry Research*, **29** (2), 156–162.

169. Sousa, LdC, Chundawat, S.P.S., Balan, V., and Dale, B.E. (2009) 'Cradle-to-grave' assessment of existing lignocellulose pretreatment technologies. *Current Opinion in Biotechnology*, **20** (3), 339–347.

Index

Aqueous Pretreatment of Plant Biomass for Biological and Chemical Conversion to Fuels and Chemicals, First Edition.
Edited by Charles E. Wyman.
© 2013 John Wiley & Sons, Ltd. Published 2013 by John Wiley & Sons, Ltd.